BAYESIAN PROBABILITY THEORY

From the basics to the forefront of modern research, this book presents all aspects of probability theory, statistics and data analysis from a Bayesian perspective for physicists and engineers.

The book presents the roots, applications and numerical implementation of probability theory, and covers advanced topics such as maximum entropy distributions, stochastic processes, parameter estimation, model selection, hypothesis testing and experimental design. In addition, it explores state-of-the-art numerical techniques required to solve demanding real-world problems. The book is ideal for students and researchers in physical sciences and engineering.

WOLFGANG VON DER LINDEN is Professor of Theoretical and Computational Physics at the Graz University of Technology. His research area is statistical physics, with a focus on strongly correlated quantum many-body physics based on computational techniques.

VOLKER DOSE is a former Director of the Surface Physics Division of the Max Planck Institute for Plasma Physics. He has contributed to Bayesian methods in physics, astronomy and climate research.

UDO VON TOUSSAINT is a Senior Scientist in the Material Research Division of the Max Planck Institute for Plasma Physics, where he works on Bayesian experimental design, data fusion, molecular dynamics and inverse problems in the field of plasma–wall interactions.

BAYESIAN PROBABILITY THEORY

Applications in the Physical Sciences

WOLFGANG VON DER LINDEN
Graz University of Technology,
Institute for Theoretical and Computational Physics,
Graz, Austria

VOLKER DOSE
Max Planck Institute for Plasma Physics, Garching, Germany

UDO VON TOUSSAINT
Max Planck Institute for Plasma Physics, Garching, Germany

CAMBRIDGE
UNIVERSITY PRESS

CAMBRIDGE
UNIVERSITY PRESS

University Printing House, Cambridge CB2 8BS, United Kingdom

One Liberty Plaza, 20th Floor, New York, NY 10006, USA

477 Williamstown Road, Port Melbourne, VIC 3207, Australia

314-321, 3rd Floor, Plot 3, Splendor Forum, Jasola District Centre, New Delhi - 110025, India

79 Anson Road, #06-04/06, Singapore 079906

Cambridge University Press is part of the University of Cambridge.

It furthers the University's mission by disseminating knowledge in the pursuit of education, learning and research at the highest international levels of excellence.

www.cambridge.org
Information on this title: www.cambridge.org/9781107035904

First published 2014

A catalogue record for this publication is available from the British Library

Library of Congress Cataloging in Publication data
Linden, Wolfgang von der, author.
Bayesian probability theory / Prof. Dr. Wolfgang von der Linden, Graz University of Technology (Austria), Institute for Theoretical and Computational Physics, [and] Prof. Dr. Dr. h.c. Volker Dose Max-Planck-Institute for Plasma Physics (Garching, Germany), [and] Dr. Udo von Toussaint, Max-Planck-Institute for Plasma Physics (Garching, Germany).
pages cm
Includes bibliographical references and index.
ISBN 978-1-107-03590-4 (Hardback)
1. Probabilities. 2. Bayesian statistical decision theory. I. Dose, Volker, author. II. Toussaint, Udo von, author. III. Title.
QC174.85.P76L56 2014
519.5'42–dc23

2013045317

ISBN 978-1-107-03590-4 Hardback

Contents

Preface

The present book is comprehensive and application-oriented, written by physicists with the emphasis on physics-related topics. However, the general concepts, ideas and numerical techniques presented here are not restricted to physics but are equally applicable to all natural sciences, as well as to engineering.

Physics is a fairly expansive discipline in the natural sciences, both financially and intellectually. Considerable efforts and financial means go into the planning, design and operation of modern physics experiments. Disappointingly less attention is usually paid to the analysis of the collected data, which hardly ever goes beyond the 200-year-old method of least squares. A possible reason for this imbalance of efforts lies in the problems which physicists encounter with traditional frequentist statistics. The great statistician G. E. Box hit this point already in 1962: '*I believe, for instance that it would be very difficult to persuade an intelligent physicist that current statistical practise was sensible, but there would be much less difficulty with an approach via likelihood and Bayes' theorem.*' This citation describes fairly precisely the adventure we have experienced with growing enthusiasm during the last 20 years. Bayesian reasoning is nothing but common physicists' logic, however, expressed in a rigorous and consistent mathematical form. Data analysis without a proper background in probability theory and statistics is like performing an experiment without knowing what the electronic devices are good for and how they are used properly. As we will see, guided by numerous examples, the consequent use of probability theory reveals that there is incredibly more information in the data than is usually expected.

More than that: probability theory is at the heart of any science, it represents – in the words of E. T. Jaynes – 'the logic of science' [104]. Besides E. T. Jaynes, M. Tribus, the former director of the 'Centre for Advanced Engineering Study' at MIT, recognized the strength of the Bayesian approach to engineering problems, which he summarized in '*Rational Descriptions, Decisions and Design*' [206]. Strangely enough, most scientists never had a thorough education in probability theory. True enough, probability theory not only plays an – albeit very important – second fiddle, but in theories such as quantum mechanics (QM) and statistical physics it is at the very heart. The importance of Bayesian probability theory, particularly as far as the fundamental interpretation of QM is concerned, has been clearly outlined by L. E. Ballentine in '*Quantum Mechanics: A Modern Development*' [9]. For historical reasons, the traditional approach to statistical physics is based primarily on frequentist statistics. However, a more powerful and systematic derivation, based on Bayesian probability theory, has been presented by W. T. Grandy in '*Foundations of Statistical Mechanics I, II*' [89].

The goal of the present book is to give a comprehensive overview of probability theory and those aspects of frequentist statistics that physicists, both experimentalists and theoreticians, need to know. The book sets a homogeneous framework for all problems occurring in physics and most other sciences that are directly or indirectly amenable to probability theory. Concepts and applications are gathered which are usually fragmented over diverse books and lectures. The first half of this book is based on a course presented to physics students at the Graz University of Technology and represents a comprehensive introduction to probability theory, statistics and data analysis for physicists from a Bayesian perspective.

Probability theory is increasingly important in computational physics or engineering as it forms the basis of various powerful numerical techniques. We will present some of these state-of-the-art techniques which are both interesting from the probabilistic point of view and required to solve challenging data analysis problems. It is worth mentioning that the computational effort involved in the solution of specific data analysis problems may be so high that it would have been prohibitive until a few years ago. Progress in the performance of modern computers renders this problem progressively less important.

The basic concepts and ideas of Bayesian probability theory have already been presented in numerous books for a broad readership. However, in order to stimulate the interest of physicists in the Bayesian choice, it is in our opinion necessary to discuss a wide range of realistic physics applications along with a detailed discussion of the – in some cases elaborate – solution. The second half of this volume is therefore devoted to a wide variety of problems arising in physical data analysis and to Bayesian experimental design. This combination, we hope, will make this volume attractive for advanced students as well as for active researchers.

The table on the following page contains a classification of the content of this book according to the target audience. It is intended to help the reader decide which sections are most appropriate.

Finally, one of us (V. D.) wants to thank Mrs I. Zeising for preparing some parts of the manuscript and S. Gori for his never-ending patience in the preparation of figures. All three of us have benefited from the collaboration with R. Fischer and R. Preuss. The development of Bayesian activity at our institute has also profited very much from the short stays of R. Silver, A. Garrett, T. Loredo, D. Keren and the continuous information exchange with the participants of the workshop series on 'Bayesian Inference and Maximum Entropy Methods in Science and Engineering' and ISBA. We are particularly grateful to J. Skilling for providing valuable insight into nested sampling.

Classification of the content according to the target audience

Target audience	Chapters
Undergraduate students	1–18, without 16
Graduate students	3, 10–25, 30.1–30.4
Readers interested in	
stochastic integration techniques	29–31
Readers interested in concepts	10–13, 14–20
Experimentalists interested in	
simple data analysis problems	3, 7, 10–12, 14, 21
problems beyond the least-squares level	
focus on regression	3, 7, 10–16, 21–26
focus on model comparison	3, 7, 17–20, 27, 29–31
focus on experimental design	3, 7, 10–14, 28, 29–31
Readers interested in the probabilistic foundation	
and instruments of statistical physics	1–12, 29–31

PART I

INTRODUCTION

1
The meaning of 'probability'

Probability theory has a long, eventful, and still not fully settled history. As pointed out in [63]: 'For all human history, people have invented methods for coming to terms with the seemingly unpredictable vicissitudes of existence ... Oracles, amulets, and incantations belonged to the indispensable techniques for interpreting and influencing the fate of communities and individuals alike ... In the place of superstition there was to be calculation – a project aiming at nothing less than the rationalization of fortune. From that moment on, there was no more talk of fortune but instead of this atrophied cousin: chance.'

The only consistent mathematical way to handle chance, or rather probability, is provided by the rules of (Bayesian) probability theory. But what does the notion 'probability' really mean? Although it might appear, at first sight, as obvious, it actually has different connotations and definitions, which will be discussed in the following sections.

For the sake of a smooth introduction to probability theory, we will forego a closer definition of some technical terms, as long as their colloquial meaning suffices for understanding the concepts. A precise definition of these terms will be given in a later section.

1.1 Classical definition of 'probability'

The first quantitative definition of the term 'probability' appears in the work of Blaise Pascal (1623–1662) and Pierre de Fermat (1601–1665). Antoine Gombauld Chevalier de Méré, Sieur de Baussay (1607–1685) pointed out to them that '... mathematics does not apply to real life'. For this nobleman 'real life' at that time meant gambling. He was especially interested in the odds of having at least once the value '6' in four rolls of a die, which was of importance in a common game of chance at the time. The analysis of this problem will be discussed along with equation (4.4) [p. 49]. Pascal and Fermat studied this and related problems and developed the basic concepts underlying classical probability theory that are still used today.

Definition 1.1 (Classical definition of probability) *The probability for the occurrence of a random event is defined as the ratio of the number g of favourable outcomes for the event to the total the number m of possibilities.*

CLASSICAL DEFINITION OF 'PROBABILITY'

$$P = \frac{g}{m}. \tag{1.1}$$

Example 1.1 *The probability for selecting a card of the suit 'spades' from a deck of bridge cards is* $P = \frac{13}{52} = \frac{1}{4}$.

Based on the classical definition we can immediately derive the basic rules of probability theory. Let A and B be arbitrary events and $\vee$ stand for the logical 'or' and $\wedge$ for the logical 'and' (please see Table C.1 for a list of symbols used in the book). Then

$$P(A \vee B) = \frac{n_A + n_B - n_{A\wedge B}}{N} = P(A) + P(B) - P(A \wedge B); \tag{1.2a}$$

$$P(N) = \frac{0}{N} = 0 \qquad N\text{: impossible event;} \tag{1.2b}$$

$$P(E) = \frac{N}{N} = 1 \qquad E\text{: certain event;} \tag{1.2c}$$

$$0 \le P(A) \le 1 \qquad \text{follows from the definition.} \tag{1.2d}$$

We are also prompted to introduce the conditional probability

$$P(A|B) := \frac{n_{A\wedge B}}{n_B} = \frac{P(A \wedge B)}{P(B)}, \tag{1.2e}$$

which is the probability for event A provided B is true. Within the classical definition this can be considered as a kind of pre-selection. Of all possible events only those n_B events are considered which are compatible with the condition implied by B. Of these n_B events only those $n_{A\wedge B}$ are considered favourable which in addition are compatible with the condition implied by A.

Definition 1.2 (Exclusive events) *Events are said to be exclusive if the occurrence of any one of them implies the non-occurrence of any of the remaining events, i.e.* $A \wedge B = N$.

Definition 1.3 (Complementary events) *An event* $\overline{A}$ *is said to be complementary to* A *if*

$$\overline{A} \vee A = E \quad \text{and} \quad \overline{A} \wedge A = N.$$

The general sum rule equation (1.2a) simplifies for exclusive events to

$$P(A \vee B) = P(A) + P(B) \tag{1.3}$$

and thus the relation for complementary events follows:

$$P(\overline{A}) = 1 - P(A). \tag{1.4}$$

These ideas can be generalized to continuous problems. Consider, for example, a square whose edge length is L. Entirely within the square shall be a circle of radius r. Now we generate points inside the square at random; no area element is distinguished from the others. We divide the square into a fine grid of squares and the random points are classified according to which grid point they land in. Now we can apply equation (1.1) again. The total number of possible outcomes is the number of grid points N. The favourable number of outcomes is equal to the number of grid points whose centre lies within the circle. In the limit $N \to \infty$ the sought-for probability is given by the ratio of the corresponding areas. Or in general, we have

$$P = \frac{\text{volume corresponding to favourable events}}{\text{total volume}}. \tag{1.5}$$

For the problem under consideration the probability that a random point lies inside the circle is

$$P = \frac{\pi r^2}{L^2}.$$

Alternatively, we can solve the inverse problem and infer r, if L is given and if we know that of N random points n are inside the circle. As a matter of fact, this is an elementary example of Monte Carlo integration. Both topics, inverse reasoning and Monte Carlo integration, will be discussed in great detail in later chapters.

The classical definition of 'probability' was developed further by Jacob Bernoulli (1654–1705) in his book *Ars Conjectandi* (published posthumously in 1713). This book contains seminal contributions to the theory of probability, among others an extensive discussion on the 'true' meaning of the term probability: *Probability is a measure of certainty.* Bernoulli already distinguished between prior and posterior probabilities. Later, Pierre-Simon Laplace (1749–1827) systematized and extended the field of probability theory. He already applied it to inverse reasoning (e.g. given the street is wet, what is the probability that it has rained). The formula his reasoning was based upon had been derived by Reverend Thomas Bayes (1702–1761). This formula, nowadays known as Bayes' theorem, was published posthumously by Richard Price [10] in 1764. It was quite normal that clergymen at the time were also (amateur) scientists [26]. Remarkable, though, is the result itself. It represents, as we will see, the only consistent solution for inverse problems and it was revealed at a time when inverse conclusions had been drawn based on bizarre logic intertwined with superstition. The original presentation of Bayes' ideas [10], however, was not very revealing and Laplace was the first to restate the theorem in the form known today. In hindsight, it is a very simple application of the product rule.

Laplace applied probability theory to problems in celestial mechanics, games of chance, the needle problem of Buffon, court cases, and many more [123]. He also introduced the principle of indifference to assign prior probabilities.

It was already known to Bernoulli that the definition given in equation (1.1) is not unique, as the total number m of all events and the number g of favourable events are sometimes ambiguous. Consider the following 'two-dice paradox': Two dice are rolled and we are interested in the probability for the sum of the two face values being seven. There are several conceivable approaches:

(a) We consider as possible outcomes the 11 different sums of the face values (2, 3, ..., 12). Of these 11 only one result (7) is favourable. Resulting in $P = 1/11$.
(b) The possible outcomes are the ordered pairs of the face values, i.e. $(1, 1), (1, 2), \dots, (1, 6), (2, 2), (2, 3), \dots, (6, 6)$. Here we do not distinguish between the two dice. Hence there are $6 + 5 + 4 + 3 + 2 + 1 = 21$ possible pairs out of which three are favourable $((1, 6), (2, 5), (3, 4))$. Resulting in $P = 1/7$.
(c) Now we also take into account which die displayed which face value. Then we end up with $N = 36$ and $g = 6$, the favourable events being $(1, 6), (6, 1), (2, 5), (5, 2), (3, 4), (4, 3)$. Resulting in $P = 1/6$.

In this example it appears obvious that the last approach is the correct one. However, as we will see later, there are also situations (especially in quantum physics) where the second approach is to apply. This immediately implies that a refined definition of probability is necessary.

Definition 1.4 (Refined classical definition of 'probability') *The probability for an event is given by the ratio of the number of favourable events to the total number of events, if all events are equally likely.*

This definition is based on circular reasoning and assumes that it is possible to assign 'prior probabilities' to elementary events, those that cannot be decomposed further. Nevertheless, it separates the rules for manipulating probabilities from the assignment of values to the 'prior probabilities' of elementary events. In many cases, it is indeed clear how to assign probabilities to the elementary events and to apply successfully the classical definition of probability. Typical examples are games of chance. The classical definition even forms the basis for statistical physics, as it is presented in most textbooks.

In order to assign probabilities, two principles were suggested early on, the 'Principle of Insufficient Reasoning' by Jacob Bernoulli and the 'Principle of Indifference' by Pierre-Simon Laplace. Both principles state that if there are n possibilities which are in principle indistinguishable, then each possibility should be assigned an equal probability, because a reordering would not alter the situation. Although these principles apply to many situations, there are counterexamples known in statistical physics (e.g. the example given above) and the principles do not generalize easily to continuous problems. A famous example is the Bertrand paradox (1888), which was resolved only in 1968 by E. T. Jaynes (1922–1998).

1.1.1 Bertrand paradox

Suppose straight lines are randomly drawn across a circle. What is the probability that the distance of a line from the centre of the circle is smaller than half of the radius r? Without loss of generality, we set $r = 1$ in suitable length units. To begin with, we have

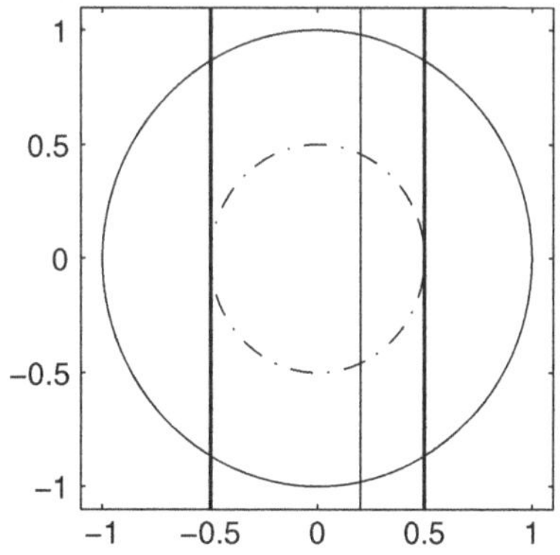

Figure 1.1 Illustration of approach (1) for the Bertrand paradox.

to generalize the classical definition in equation (1.1) to continuous variables. Let the real variable x describe the random event and by construction, all allowed events correspond to $x \in I$ with the size of I being $|I| = L$. The favourable events are given by $x \in I_f$, with $|I_f| = L_f$. Then, according to equation (1.5), the probability that x is in I_f is

$$P = \frac{L_f}{L}. \tag{1.6}$$

Several representations are conceivable. In the following we discuss three possibilities, all apparently valid.

(1) The distance x to the centre of the circle can be assumed to take a value between 0 and 1, i.e. $I = [0, 1]$ and $L = 1$. The interval of favourable events is $I_f = [0, \frac{1}{2}]$ and has $L = \frac{1}{2}$. So, we end up with a probability $P = 1/2$. See Figure 1.1.
(2) A different approach is illustrated in Figure 1.2: Regarding the angle between the straight line and the tangent to the circle as uniformly distributed, the favourable angles are in the range $I_f = [0, \frac{\pi}{3}]$ compared with $I = [0, \pi]$. The resulting probability is therefore $P = 1/3$.
(3) Another possibility is to consider the area A of the concentric disc touching the straight line (see Figure 1.3). If we consider this area to be equally distributed within $I = [0, \pi]$, and taking into account that the favourable area is limited to $I = [0, \frac{\pi}{4}]$, then the probability of interest is $P = 1/4$.

How can a seemingly well-posed problem yield different answers? At the heart of the Bertrand paradox is the problem of describing ignorance about continuous degrees of freedom. Suppose we want to make a statement about a quantity x which may take real values in $I = [0, 1]$. It appears reasonable to use the classical definition of equation (1.6) and assign the probability that x is within $I_f = [x, x + dx]$ as

$$P(x \in (x, x + dx)) = \frac{L_f}{L} = \frac{dx}{1} = p_x(x)dx.$$

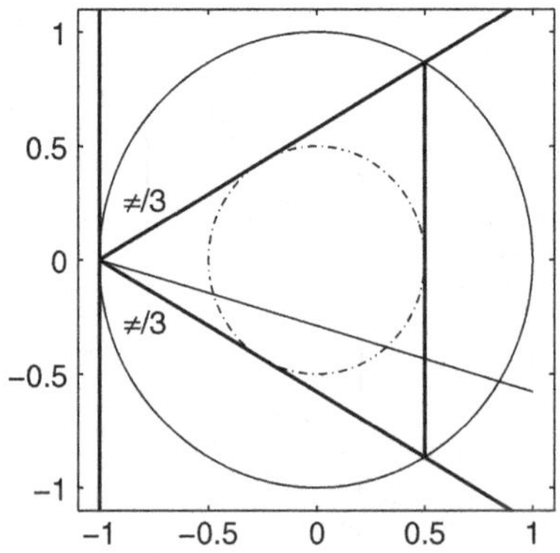

Figure 1.2 Illustration of approach (2) for the Bertrand paradox.

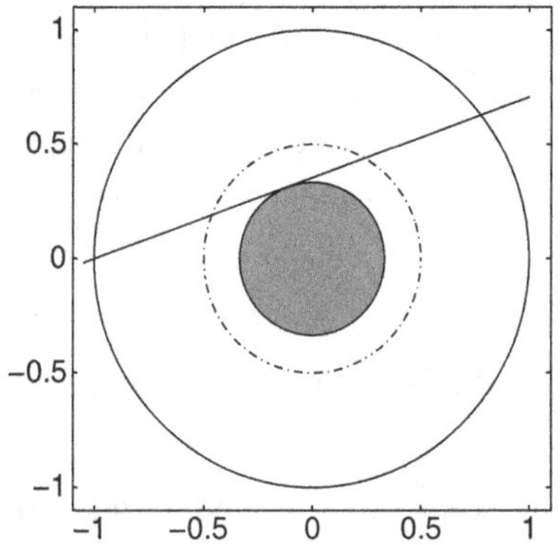

Figure 1.3 Illustration of approach (3) for the Bertrand paradox.

Here, $p_x(x) = 1$ is a 'probability density' which will be treated in more detail in Section 11.1 [p. 178]. Ignorance, or lack of knowledge, results in a uniform assignment. However, now the probability density of x^n is no longer uniform but (as we will see in Chapter 10 [p. 165]) is given by

$$p_z(z = x^n) = p_x(x) \left| \frac{dx}{dz} \right| = \frac{1}{n} z^{\frac{1}{n}-1}.$$

Therefore, the probability density for the variable z has a maximum at $z = 0$. Thus, a uniform probability density for x corresponds to a non-uniform density for the transformed variable z. Quite generally speaking, nonlinear transformations of continuous quantities may cause problems. In which representation are events equally probable (e.g. length, area, volume) and how is complete ignorance best represented? We recognize a similarity to the 'two-dice paradox' but here the situation is much more complex. The problem of 'prior probability assignment' is tackled in Part II [p. 164]. The theory of 'transformation invariance' provides a principled solution to this problem. In the same context, the important concept of 'maximum entropy prior' probabilities is introduced.

Despite all these problems, the classical definition of probability was used for several centuries and is still adequate for many problems, especially in combinatorics. In such problems, there is a close relation between probabilities and relative frequencies of the occurrence of events. Bernoulli was one of the first to analyse this not immediately obvious relation. Let us consider the following example: The probability for rolling an even-face number using a standard die is 1/2. Then, how many even numbers will be obtained in N rolls? We will discuss later in this chapter and in Section 1.3 Bernoulli's 'law of large numbers', which says that the relative frequency approaches the intrinsic probability for $N \to \infty$. However, much more relevant for most applications is the inverse problem: Given a sample of finite size N and n occurrences of the event of interest, what can be inferred about the underlying probability?

1.2 Statistical definition of 'probability'

In order to avoid the need to specify 'prior probabilities,' R. L. Ellis (1772–1842), G. Boole (1815–1864), J. Venn (1834–1923) and R. von Mises (1883–1953) pursued a different direction. Based on Bernoulli's 'law of large numbers', they introduced the following statistical definition.

Definition 1.5 (Statistical definition of probability) *An event A happens at random. The probability for the event is defined by the relative frequency*

$$P(A) = \lim_{N \to \infty} \frac{n}{N} \tag{1.7}$$

that the event occurs n times in N trials, for the limit $N \to \infty$.

In the statistical definition, 'probability' is considered as an intrinsic property of the object under investigation, which is only accessible by an experiment (samples) of infinite size. Since this cannot be realized, the definition is rather hypothetical. Nevertheless, this definition of probability is in widespread use and is the basis of the frequentist statistics. However, avoiding 'prior probabilities' has its price.

- For many problems no frequency distribution or sample is available at all. So it is not possible to define a probability in these cases, for example:
  - Is Mr X guilty?
  - Does the temperature of the fusion plasma $T \in (1.0, 2.0)10^8$ K?
  - Was Julius Caesar left-handed?
- Even if relative frequencies can be determined, only very rarely is $N \gg 1$ achievable. This is typically the case in:
  - large-scale experiments – only a few dedicated experiments are performed due to limited budget;
  - astrophysics – the number of observations is given by nature (e.g. the number of neutrinos from a supernova).

- The limit $N \to \infty$ cannot be accessed in practice and therefore equation (1.7) has to be considered as a hypothesis which defies experimental validation.
- Relative frequencies provide no clear interpretation for specific individual situations:
  - What meaning can be assigned to a statement like: The probability for Mr X being guilty is 0.05? Does it imply that of 100 clones of Mr X, five would be guilty?
  - The probability for drawing the main prize in a lottery is 10^{-8}. Does this mean that buying 10^8 lottery tickets results in a sure win?

The last two examples also reveal how different probabilities are perceived, depending on the context. In the first situation the probability for Mr X's guilt would be considered to be small, whereas in the second example – despite the very small probability for success – a sufficient number of people happily buy lottery tickets. This is because the very high probability for a small loss (the expense of a lottery ticket) seems overcompensated by the potential gain. The idea of assessing probabilities in the light of the outcome leads to 'decision theory'. Within the framework of this theory outcomes are assigned a value (or loss) and the best decision minimizes the expected loss. This topic is addressed in more detail in Chapter 28 [p. 491].

The statistical approach can address only a limited subset of problems directly, those where relative frequencies are appropriate. In many cases only indirect conclusions can be drawn, and that will be discussed later on.

1.3 Bayesian understanding of 'probability'

Bayesian probability theory is the consequent continuation of the Laplacian approach. It is based on propositions, i.e. statements which are either true or false with nothing in between, like

- S_1: It will rain tomorrow.
- S_2: Rolling a fair perfect die N times will yield face value '3' n times.
- S_3: The time between two radioactive decays lies in the interval $[t, t + dt]$.
- S_4: The variance specified by the manufacturer of the device is wrong.
- S_5: Next time the coin will land with heads up.
- S_6: Theory 'T' correctly describes this phenomenon.
- S_7: The coin in my hand is a cent.

In Bayesian reasoning, the probability $P(S_n|\mathcal{I})$ is a measure for the correctness or truth of proposition S_n and Bayesian probability theory provides a consistent calculus for these probabilities. Roughly speaking, it is the generalization of the propositional calculus to partial truth. In Bayesian probability theory all propositions are on par, no matter whether they describe denumerable random events (S_2), continuous random events (S_3), or even situations where the uncertainty originates solely from missing information (S_7).

Above, we have introduced the notation $P(S_n|\mathcal{I})$, which is actually a conditional probability. It stands for the probability for the proposition S_n, given the 'background information' $\mathcal{I}$, which is also called a conditional complex. The background information uniquely specifies the problem under consideration in such a way that even a computer could in principle treat the problem. In many cases the background information is obvious and needs no explicit mention. However, the missing explicit declaration of all assumptions can lead to wrong and misleading conclusions, which are sometimes unperceived for decades (see the paradoxes in [104]). There can be no 'I thought you knew!' In general, probability theory, correctly applied, will very rapidly reveal if there are concealed assumptions or inconsistent data, like erroneous error statistics, incorrectly calibrated experiments, unjustified model assumptions and so on; a list any honest scientist can easily extend. The bottom line is: 'Always state all your assumptions.'

By now it should be clear that

> *There is no such thing as an unconditional probability!*

Even a situation as simple as the coin-tossing experiment is ambiguous. We have to assume that the coin is fair, never lands on its edge, and it is tossed 'randomly'. If this is our background information $\mathcal{I}$, then $P(S_5|\mathcal{I}) = 1/2$. Since probabilities do and have to depend on our background information, like in quantum mechanics, the wavefunction depends on the preparation; E. T. Jaynes called probability a measure of the 'degree of belief' that a proposition is true [104]. A similar connotation had already been given by Laplace [123]. Belief in this context, however, is not to be understood with its religious meaning, but rather as an implication measure, like: 'I believe it will rain', in view of the dark clouds. Therefore, a conditional probability $P(X|Y)$ can be considered as a measure of how strongly proposition Y implies proposition X. This interpretation is also the one which is of relevance in practice. We are always interested in the probability for how strongly the available information (e.g. measured data, prior knowledge) implies that certain parameter values or models are true. We will see that this measure can be defined uniquely and that the calculus for it can be derived based only on the rules of elementary logic [35, 118]. The sum and product rule of probability theory can be derived, using simple consistency requirements of elementary propositional logic. Let A and B be propositions, and $\overline{A}$ stand for the proposition complementary to A.

THE RULES OF PROBABILITY THEORY

$$
\begin{aligned}
P(A \vee B|\mathcal{I}) &= P(A|\mathcal{I}) + P(B|\mathcal{I}) - P(A, B|\mathcal{I}) && \text{(sum rule)},\\
P(A, B|\mathcal{I}) &= P(A|B, \mathcal{I})\, P(B|\mathcal{I}) && \text{(product rule)},\\
1 &= P(A|\mathcal{I}) + P(\overline{A}|\mathcal{I}) && \text{(normalization)},\\
P(A|\mathcal{I}) &= P(A, B|\mathcal{I}) + P(A, \overline{B}|\mathcal{I}) && \text{(marginalization rule)}.
\end{aligned} \tag{1.8}
$$

We use the convention that a comma between propositions stands for the logical 'and'. Derivation and generalizations of the rules of probability theory will be discussed in Chapter 3 [p. 33]. Based on the commutativity of the logical 'and', we obtain $P(A|B,\mathcal{I})\, P(B|\mathcal{I}) = P(B|A,\mathcal{I})\, P(A|\mathcal{I})$ which immediately yields

BAYES' THEOREM

$$P(A|B,\mathcal{I}) = \frac{P(B|A,\mathcal{I})\, P(A|\mathcal{I})}{P(B|\mathcal{I})}. \tag{1.9}$$

1.3.1 Law of succession I

We could not end this section without giving an illustration of the power of the Bayesian approach. To this end we discuss a simplified version of Laplace's 'law of succession'. The original version will be discussed in Section 3.2.4 [p. 40]. Assume there are N regular coins in a bag plus one additional fake coin, with heads on both sides. One of the coins is picked at random. Then it is flipped n times, showing heads each time. What is the probability that the next flip is heads up again. The required propositions are:

- h: Next throw will land heads up.
- N: There are N regular and one fake coin, with heads on both sides.
- n: The coin is flipped n times and always shows heads up.
- f: It's the fake coin.
- $\mathcal{I}$: Background information, i.e. all assumptions and all our prior knowledge.

Using Bayes' theorem we can compute the probability for observing heads in the next flip, based on the data and our prior knowledge. First we invoke the marginalization rule

$$P(h|n,N,\mathcal{I}) = P(h,f|n,N,\mathcal{I}) + P(h,\overline{f}|n,N,\mathcal{I}). \tag{1.10}$$

Then the two terms are computed based on the product rule

$$P(h,x|n,N,\mathcal{I}) = P(h|x,n,N,\mathcal{I})\, P(x|n,N,\mathcal{I}), \tag{1.11}$$

where x stands for f or $\overline{f}$. Now, for the first factor in case of $x = f$ we have $P(h|f,n,N,\mathcal{I}) = 1$, because if the coin is the fake it will always show heads. In the opposite case, if the coin is a regular one, the probability for heads is $P(h|\overline{f},n,N,\mathcal{I}) = 1/2$. In both cases, the additional information (n, N) is irrelevant in this context, as the knowledge of x is an all-determining proposition. Next we need the probability $P(x|n,N,\mathcal{I})$. Now, Bayes' theorem comes into play:

$$P(x|n,N,\mathcal{I}) = \frac{P(n|x,N,\mathcal{I})P(x|N,\mathcal{I})}{P(n|N,\mathcal{I})} = \frac{P(n,x|N,\mathcal{I})}{P(n|N,\mathcal{I})}.$$

In order to satisfy the sum rule

$$P(f|n, N, \mathcal{I}) + P(\overline{f}|n, N, \mathcal{I}) = 1,$$

the denominator is

$$\begin{aligned} P(n|N, \mathcal{I}) &= P(n|f, N, \mathcal{I})P(f|N, \mathcal{I}) + P(n|\overline{f}, N, \mathcal{I})P(\overline{f}|N, \mathcal{I}) \\ &= P(n, f|N, \mathcal{I}) + P(n, \overline{f}|N, \mathcal{I}). \end{aligned}$$

The required elements follow from the classical definition:

$$\begin{aligned} P(n|f, N, \mathcal{I}) &= 1, \\ P(n|\overline{f}, N, \mathcal{I}) &= \frac{1}{2^n}, \\ P(f|N, \mathcal{I}) &= \frac{1}{N+1}, \\ P(\overline{f}|N, \mathcal{I}) &= 1 - P(f|N, \mathcal{I}) = \frac{N}{N+1}. \end{aligned}$$

The first equation is obvious, the fake coin can only show heads. If it is a regular coin, there are in total 2^n possible outcome sequences, all equally likely, however only one is the favourable one with all heads. Consequently, the second equation follows from the classical definition. Similarly for the third equation. All coins are equally likely by symmetry of the problem. Finally, the fourth equation follows from the sum rule. Then

$$\begin{aligned} P(n, f|N, \mathcal{I}) &= 1 \cdot \frac{1}{N+1}, \\ P(n, \overline{f}|N, \mathcal{I}) &= 2^{-n} \cdot \frac{N}{N+1}, \\ P(n|N, \mathcal{I}) &= \frac{1}{N+1} + 2^{-n} \cdot \frac{N}{N+1} = \frac{1}{N+1}\left(1 + N \cdot 2^{-n}\right), \end{aligned}$$

and

$$\begin{aligned} P(f|n, N, \mathcal{I}) &= \frac{P(n, f|N, \mathcal{I})}{P(n|N, \mathcal{I})} = \frac{2^n}{2^n + N}, \\ P(\overline{f}|n, N, \mathcal{I}) &= 1 - P(f|n, N, \mathcal{I}) = \frac{N}{2^n + N}. \end{aligned}$$

Based on equation (1.11) we obtain

$$\begin{aligned} P(h, f|n, N, \mathcal{I}) &= P(h|f, n, N, \mathcal{I})\ P(f|n, N, \mathcal{I}) = 1 \cdot \frac{2^n}{2^n + N}, \\ P(h, \overline{f}|n, N, \mathcal{I}) &= P(h|\overline{f}, n, N, \mathcal{I})\ P(\overline{f}|n, N, \mathcal{I}) = \frac{1}{2} \cdot \frac{N}{2^n + N}. \end{aligned}$$

Eventually, equation (1.10) yields the sought-for result

$$P(h|n, N, \mathcal{I}) = P(h, f|n, N, \mathcal{I}) + P(h, \overline{f}|n, N, \mathcal{I}) = \frac{2^n + N/2}{2^n + N}.$$

For $2^{-n}N \ll 1$, or rather $2^n \gg N$, we find

$$P(h|n, N, \mathcal{I}) \approx 1 - \frac{N}{2}e^{-n}.$$

The probability for heads up in the next flip approaches 1 rapidly as n increases. That is in agreement with common sense, because n times heads only is, for large n, a strong indication that the coin under consideration is the fake one. Less intuitive, though, is the fact that the meaning of 'large' depends on N. Specifically, for the case $N = 2$ it suffices to have a sequence of three times heads up ($n = 3$) in order to have a probability of 0.9 for the next flip being heads up again. For $N = 1000$, however, it takes $n = 12$ repeated heads to obtain the same probability. The origin is the 'prior probability' $P(f|N, \mathcal{I})$. The observation sounds counterintuitive, at first, but we will have ample opportunity in this book to convince ourselves that the 'prior probability' is the key success factor in consistent data analysis.

The present result also shows that for $n \to \infty$ the prior is overruled by the data. But to conclude that it is advisable to perform an experiment long enough for the prior information to be dashed out is an anachronism. If scientists in the past were to have followed this maxim, science would probably not have proceeded beyond the belief that the earth is a disc.

In discussions, some people like to despise the fact that probability theory should contain prior knowledge, which in their opinion makes it subjective and hence unscientific. A brief glance into the history of science should suffice to reveal that the scientific knowledge at a particular point in time is dependent on the global prior knowledge of the scientific community. Only ignorant scientists would ignore this knowledge, but moreover it also depends on global intelligence, mathematical tools, experimental accuracy and even philosophical trends. In any case, a scientist will always use his prior knowledge, however subjective it may be, and it is good scientific conduct to state that knowledge.

2

Basic definitions for frequentist statistics and Bayesian inference

2.1 Definition of mean, moments and marginal distribution

In this chapter we will specify some of the terms encountered so far and we will explore a couple of elementary formulae and concepts, which we will use very often in this book, and which are used both in frequentist statistics and Bayesian inference.

2.1.1 Distribution of discrete random variables

A random variable is a number $x = X(\omega)$ assigned to every outcome ω of an experiment. This number could be the temperature of a random object, the gain in a game of chance, or any other numerical quantity. For the following definitions we assume a countable set $\mathcal{G}$ of elementary events. Each event $\omega \in \mathcal{G}$ occurs with probability P_ω.

Definition 2.1 (Random variable) *A random variable is a functional which assigns to each event $\omega \in \mathcal{G}$ a real number $x = X(\omega)$. x is the realization of X. The set R of all possible realizations x is the range of X. The domain of the random variable $\mathcal{G}$ is mapped to the set R.*

Example:

- In a coin-tossing experiment, we can assign the number 0 to the outcome 'head' and the number 1 to the outcome 'tail'.
- In a die experiment, often the outcome 'face value of n' is assigned to the number n.
- For the experiment of tossing a coin three times, the set of all possible outcomes is given by the following eight possibilities:

$$\begin{aligned}\mathcal{G} =\{&(H, H, H), (H, H, T), (H, T, H), (T, H, H),\\ &(H, T, T), (T, H, T), (T, T, H), (T, T, T)\}.\end{aligned}$$

  A possible assignment is given by the following random variable:

$$3, 2, 2, 2, 1, 1, 1, 0,$$

  i.e. the number of 'heads' in one combined experiment.

Of course, functions $Y = f(X)$ of the random variable X can also be defined and studied. In this case Y is a random variable as well, and to each event ω we now assign the value $f(x)$.

Information about the 'location' of a probability mass function is the mean value.

Definition 2.2 (Discrete random variable) *A random variable is called 'discrete' if it only takes discrete values.*

Throughout this book, we will only discuss such discrete problems that can be described by a countable set of events. Then we can enumerate the events with integers and likewise the random numbers X_n and the corresponding probabilities P_n, with $n \in M$, where M is the set of integers that enumerate the events.

Definition 2.3 (Probability mass function (PMF)) *The PMF is a function that assigns a probability to each value of the discrete random variable.*

For a fair die the PMF assigns $1/6$ to each face value.

Definition 2.4 (Mean of a random variable) *The mean of a discrete random variable is defined as*

MEAN OF A DISCRETE RANDOM VARIABLE

$$\langle X \rangle := \sum_{n \in M} X_n \; P_n. \tag{2.1}$$

Remarks:

(a) The *mean* is also often called an *expectation value*. This term might be misleading, as the result could be a value which we would never expect to occur.

(b) The mean is a fixed, precisely defined value, it is not a random variable. It should not be confused with the sample mean (arithmetic mean) that we will denote by $\overline{x}$, which is a random variable. Its definition will be given below.

(c) The computation of the mean is a linear operation. Let X and Y be random variables for the same set of events, and $\alpha, \beta \in \mathbb{R}$, then

$$\langle \alpha \; X + \beta \; Y \rangle = \alpha \; \langle X \rangle + \beta \; \langle Y \rangle. \tag{2.2}$$

Proof:

$$\begin{aligned} \langle \alpha \; X + \beta \; Y \rangle &= \sum_{n \in M} (\alpha \; X_n + \beta \; Y_n) \; P_n \\ &= \alpha \sum_{n \in M} X_n \; P_n + \beta \sum_{n \in M} Y_n \; P_n \\ &= \alpha \; \langle X \rangle + \beta \; \langle Y \rangle. \end{aligned} \tag{2.3}$$

□

(d) For a function $Y = f(X)$ of a random variable, we get

$$\langle f(X) \rangle = \sum_{n \in M} f(X_n)\, P_n . \tag{2.4}$$

(e) Continuous random variables will be introduced in Chapter 7 [p. 92].

Example 2.1 *Rolling a (fair) die can result in the face values $n \in \{1, 2, 3, 4, 5, 6\}$ with the corresponding probabilities $P_n = \frac{1}{6}$. The random variable X is equal to the face value. This results in a mean $\langle n \rangle = 3.5$. The function $Y = f(X)$ shall define the gain for each outcome. Then the mean $\langle f \rangle$ is the average gain.*

Probability distributions can be characterized in different ways. The most common way is provided by the PMF. Another approach, which is sometimes more convenient, is based on moments. Moments are the means of powers of random variables:

Definition 2.5 (ith moment of a random variable) *The mean of $f(X) = X^i$ is the*

iTH MOMENT OF A RANDOM VARIABLE

$$m_i := \left\langle X^i \right\rangle . \tag{2.5}$$

Remarks: The zeroth moment of a random distribution is equal to one, if the distribution is normalized properly and the first moment m_1 is equivalent to the mean.

Definition 2.6 (ith central moment of a random variable) *The mean of $f(X) = (X - \langle X \rangle)^i$ is the*

iTH CENTRAL MOMENT OF A RANDOM VARIABLE

$$\mu_i := \left\langle (\Delta X)^i \right\rangle = \left\langle (X - \langle X \rangle)^i \right\rangle . \tag{2.6}$$

The second central moment is also known as the variance. Starting from equation (2.6) and making use of equation (2.2), we get

$$\left\langle (X - \langle X \rangle)^2 \right\rangle = \left\langle X^2 - 2\, X\, \langle X \rangle + \langle X \rangle^2 \right\rangle = \left\langle X^2 \right\rangle - \langle X \rangle^2 .$$

Definition 2.7 (Variance) *The second central moment, the variance, is therefore*

VARIANCE OF A RANDOM VARIABLE

$$\operatorname{var}(X) := \sigma^2 := \left\langle (\Delta X)^2 \right\rangle = \left\langle (X - \langle X \rangle)^2 \right\rangle = \left\langle X^2 \right\rangle - \langle X \rangle^2 . \tag{2.7}$$

The variance describes the mean quadratic deviation of the random variable from its mean value. Sometimes it is more convenient to have a linear measure (e.g. to have the same physical units as the mean) of the deviation. This is obtained by introducing the standard deviation.

Definition 2.8 (Standard deviation) *The square root of the variance is defined as*

STANDARD DEVIATION

$$\mathrm{std}(X) := \sigma := \sqrt{\mathrm{var}(X)}. \tag{2.8}$$

A simple approach to estimate the mean of a probability distribution is given by the sample mean. Suppose we want to estimate the mean value of a die. Throwing the die N times yields a sequence (a sample) of face values x_i. If we compute the arithmetic mean (sample mean), we get an estimate of the mean of the probability distribution $\langle x \rangle$:

$$\overline{x} = \frac{1}{N} \sum_{i=1}^{N} x_i .$$

Later we will show the validity of this approach and that the deviation of the sample mean from the true mean is – often but not always – given by

STANDARD ERROR OF A SAMPLE OF SIZE N WITH INDIVIDUAL STANDARD DEVIATION σ

$$\mathrm{SE} = \frac{\sigma}{\sqrt{N}}. \tag{2.9}$$

2.1.2 Multivariate discrete random variables

The following example will be used to guide the extension of the preceding definitions to more than one discrete random variable. In a company, the height and weight of employees have been measured. The results are given in Table 2.1. As a matter of fact, height and weight are actually continuous quantities, but we introduce a discretization, which is frequently very useful. To this end, we have introduced intervals for the possible values. The intervals are centred about the values listed in the table. For example, there are five employees with a weight between 65 kg and 75 kg and a height between 1.65 m and 1.75 m. In the rightmost column of the table, the row totals are given and correspondingly the column totals are listed in the bottom row. In total, the data of 30 employees have been taken. A random variable mass m can be assigned to the rows with range $\{60, 70, 80\}$ kg and another random variable height h with range $\{1.6, 1.7, 1.8\}$ m to the columns. The probability that

Table 2.1 Height and weight of 30 employees of a company

Weight [kg]\height [m]	1.6	1.7	1.8	#
60	3	3	1	7
70	0	5	6	11
80	1	4	7	12
#	4	12	14	30

a randomly selected employee of this company is between 1.65 m and 1.75 m tall and has a weight between 55 kg and 65 kg is, according to Table 2.1, $P = \frac{3}{30}$. Each table entry can be assigned the relative frequency as probability P_{mh}. The values in the rightmost column (margin) allow us to compute marginal probabilities, for example, the probability that an employee's weight lies in the interval $(65, 75]$ kg, irrespective of her/his height, is given by $P = \frac{11}{30}$. That brings us to the general definition of marginal distributions and probabilities, respectively.

Definition 2.9 (Marginal probability) *Suppose $P_{n_1,\ldots,n_N}$ are the probabilities for elementary events, which are uniquely defined by an N-tuple of properties $n_1, \ldots, n_N$. Then a marginal probability of the first j properties is defined by*

MARGINAL PROBABILITY

$$P_{n_1,\ldots,n_j} := \sum_{n_{j+1}} \cdots \sum_{n_N} P_{n_1,\ldots,n_N}. \tag{2.10}$$

Correspondingly, other marginal probabilities can be obtained by different choices of properties to be summed over.

Example 2.2 *In the example in Table 2.1, two marginal distributions are of interest: $P_h = \sum_m P_{mh}$ and $P_m = \sum_h P_{mh}$. For the first distribution we get $P_{h=1.6} = \frac{4}{30}$, $P_{h=1.7} = \frac{12}{30}$ and $P_{h=1.8} = \frac{14}{30}$.*

Similar to equations (2.4)–(2.6), we now define the mean and the moments for functions of several random variables.

Definition 2.10 (Mean value of a function of random variables) *Suppose there are N random variables $X^{(1)}, \ldots, X^{(N)}$, where the ith random variable is enumerated by $n_i \in M_i$. Given a function $Y = f\left(X^{(1)}, \ldots, X^{(N)}\right)$, the mean value of that function is given by*

MEAN VALUE OF A FUNCTION OF SEVERAL RANDOM VARIABLES

$$\left\langle f\left(X^{(1)},\ldots,X^{(N)}\right)\right\rangle := \sum_{n_1\in M_1}\cdots\sum_{n_N\in M_N} f\left(X^{(1)}_{n_1},\ldots,X^{(N)}_{n_N}\right) P_{n_1,\ldots,n_N}. \tag{2.11}$$

Example 2.3 *The body mass index (BMI) is defined as* $BMI = \frac{\textit{mass in kg}}{(\textit{height in m})^2}$. *The average body mass index of the employees is given by the mean of the function* $f(m,h) = \frac{m}{h^2}$. *This results in* $\langle f\rangle = 23.9$.

An important characterization of multivariate stochastic problems is the covariance, which quantifies the correlation between two random variables.

Definition 2.11 (Covariance) *Given N discrete random variables as in Definition 2.10, the covariance of* $X^{(i)}$ *and* $X^{(j)}$ *is defined as*

COVARIANCE OF TWO RANDOM VARIABLES

$$\begin{aligned}
\operatorname{cov}(X^{(i)},X^{(j)}) &:= \left\langle \Delta X^{(i)}\Delta X^{(j)}\right\rangle \\
&= \left\langle X^{(i)}\,X^{(j)}\right\rangle - \left\langle X^{(i)}\right\rangle\left\langle X^{(j)}\right\rangle. \\
\Delta X^{(i)} &:= X^{(i)} - \left\langle X^{(i)}\right\rangle. \\
\left\langle X^{(i)}\,X^{(j)}\right\rangle &= \sum_{n_i\in M_i}\sum_{n_j\in M_j} X^{(i)}_{n_i}\,X^{(j)}_{n_j}\,P_{n_i n_j}.
\end{aligned} \tag{2.12}$$

The computation of the covariance requires the marginal probability $P(n_i,n_j)$, which for $i=j$ is equal to the marginal probability for one variable P_{n_i,n_j}. The covariance describes correlations between fluctuations $\Delta X^{(i)} := X^{(i)} - \langle X^{(i)}\rangle$ of the two random variables under consideration.

If $\Delta X^{(i)}$ and $\Delta X^{(j)}$ always have the same sign, then the covariance is positive. On the contrary, if the fluctuations always have opposite sign, then the random variables are anti-correlated and the covariance is negative.

Example 2.4 *The covariance between mass and height of the employees in Table 2.1 yields* $\operatorname{cov}(m,h) = 0.21$ *kg m. Therefore, on average, an employee with a mass above average is also taller than average.*

Often we will encounter uncorrelated random variables. In this case the joint and marginal probabilities exhibit a simple product structure.

Definition 2.12 (Independent random variables) *If a joint distribution of several random variables can be written as a product of their marginal distributions, then the random variables are independent:*

INDEPENDENT RANDOM VARIABLES

$$P_{n_1,n_2,\ldots,n_N} = \prod_{i=1}^{N} P_{n_i}. \tag{2.13}$$

Consequently, the covariance of independent random variables is zero.

Proof: For $i \neq j$ we have

$$\begin{aligned}
\operatorname{cov}(X^{(i)}X^{(j)}) &= \langle X^{(i)}X^{(j)}\rangle - \langle X^{(i)}\rangle\langle X^{(i)}\rangle \\
&= \sum_{n_1\in M_1}\cdots\sum_{n_N\in M_N} X^{(i)}_{n_i}\, X^{(j)}_{n_j} \prod_{l=1}^{N} P_{n_l} - \langle X^{(i)}\rangle\langle X^{(i)}\rangle \\
&= \underbrace{\Big(\sum_{n_i\in M_i} X^{(i)}_{n_i} P_{n_i}\Big)}_{\langle X^{(i)}\rangle} \underbrace{\Big(\sum_{n_j\in M_j} X^{(j)}_{n_j} P_{n_j}\Big)}_{\langle X^{(i)}\rangle} - \langle X^{(i)}\rangle\langle X^{(j)}\rangle \\
&= \langle X^{(i)}\rangle\langle X^{(j)}\rangle - \langle X^{(i)}\rangle\langle X^{(j)}\rangle = 0.
\end{aligned}$$

□

2.1.3 Generating functions

An important transformation of PMFs is the *generating function*, which is closely related to the characteristic functions used for continuous variables, which will be discussed along with the central limit theorem. The generating function of a PMF P_n for discrete events is defined by

GENERATING FUNCTION

$$\phi(z) := \sum_{n=0}^{\infty} P_n z^n. \tag{2.14}$$

Moments of a probability distribution

The generating function is very powerful, as we will see in later chapters, because it allows for example, the generation of moments. In this context we specify the random variable by the index of the event, i.e. $X_n = n$. Then

$$\phi(z)\Big|_{z=1} = \sum_{n=0}^{\infty} P_n, \tag{2.15a}$$

$$\frac{d}{dz}\phi(z)\Big|_{z=1} = \sum_{n=0}^{\infty} n\ P_n, \tag{2.15b}$$

$$\frac{d^2}{dz^2}\phi(z)\Big|_{z=1} = \sum_{n=0}^{\infty} \left(n^2 - n\right) P_n. \tag{2.15c}$$

So we can determine the low-order moments as follows:

$$\langle n \rangle = \frac{\frac{d}{dz}\phi(z)}{\phi(z)}\Bigg|_{z=1} = \frac{d}{dz}\log[\Phi(z)]\Big|_{z=1}, \tag{2.16a}$$

$$\langle n^2 \rangle = \frac{\frac{d^2}{dz^2}\phi(z)}{\phi(z)}\Bigg|_{z=1} + \langle n \rangle, \tag{2.16b}$$

$$\left\langle (\Delta n)^2 \right\rangle = \frac{d^2}{dz^2}\log[\Phi(z)]\Big|_{z=1} + \langle n \rangle. \tag{2.16c}$$

Convolution of probability distributions

A structure that we will encounter frequently is convolutions. As a concrete example we consider an experiment in which photons stemming from a light source are counted. The measured signal, however, is affected by background radiation. In other words, the measured number of photons n is the sum of photons originating from signal n_s and background n_b:

$$n = n_s + n_b. \tag{2.17}$$

Both counts are Poisson distributed with mean μ_s and μ_b, respectively. The central question is: What is the distribution of the total count n? Owing to equation (2.17), the total count can be obtained from $n+1$ different pairs of exclusive realizations $(n_s, n_b) = (n, 0), (n-1, 1), \dots, (0, n)$. According to the marginalization rule, the probability for the sum being n is

$$\begin{aligned} P(n|\mu_s, \mu_b, \mathcal{I}) &= \sum_{n_s, n_b = 0}^{\infty} \underbrace{P(n|n_s, n_b, \mu_s, \mu_b, \mathcal{I})}_{\delta_{n_s+n_b,n}} P(n_s, n_b|\mu_s, \mu_b, \mathcal{I}) \\ &= \sum_{n_b=0}^{\infty} P(n_s = n - n_b|\mu_s, \mathcal{I}) P(n_b|\mu_b, \mathcal{I}). \end{aligned}$$

We have used the fact that signal and background signals are uncorrelated. The first term actually restricts the sum to $n_b \leq n$. The result is a convolution of the two PMFs. Generally, for two PMFs $P_\alpha(n)$ the convolution $C(n)$ is defined as

Convolution of two probability distributions

$$C(n) := \sum_{m=0}^{\infty} P_1(n-m) P_2(m). \tag{2.18}$$

Next we determine the generating function of the convolution by multiplying both sides by z^n and summing over n:

$$\begin{aligned}
\Phi(z) = \sum_{n=0}^{\infty} C(n)\, z^n &= \sum_{n=0}^{\infty} \sum_{m=0}^{\infty} z^{n-m}\, z^m\, P_1(n-m)\, P_2(m) \\
&= \left(\sum_{l=0}^{\infty} z^l\, P_1(l) \right) \left(\sum_{m=0}^{\infty} z^m\, P_2(m) \right) \\
&= \Phi_1(z)\, \Phi_2(z).
\end{aligned}$$

The generating function of a convolution is therefore generally the product of the individual generating functions. This property will turn out to be very useful in various applications.

Generating function of a convolution of two PMFs $P_\alpha(n)$

$$\Phi(z) = \Phi_1(z) \cdot \Phi_2(z). \tag{2.19a}$$

$$\Phi_\alpha(z) := \sum_{n=0}^{\infty} P_\alpha(n)\, z^n. \tag{2.19b}$$

In case of the photon count composed of Poissonian signal and background, the generating function of the individual Poisson distributions is

$$\Phi_\alpha(z) = \sum_{n=0}^{\infty} z^n\, e^{-\mu_\alpha} \frac{\mu_\alpha^n}{n!} = e^{-\mu_\alpha} \sum_{n=0}^{\infty} \frac{\left(z\mu_\alpha\right)^n}{n!} = e^{-\mu_\alpha(1-z)}.$$

The product of the generating functions of the two Poisson counts, $\Phi_1(z)\Phi_2(z) = e^{-(\mu_1+\mu_2)(1-z)}$, is apparently the generating function of a Poisson distribution with mean $\mu = \mu_1 + \mu_2$. Hence, the sum of Poissonian random numbers is also Poisson distributed with added means.

In the preceding sections we have developed the necessary mathematical tools to tackle typical probabilistic inference problems. This will be demonstrated with a worked example. While addressing this problem we define in a more precise manner some of the recurring terms.

Table 2.2 Three-urn problem. Each urn contains green and red balls with a different percentage

Urn	Percentage of green balls	Percentage of red balls
U_1	$q_1 = 0.2$	0.8
U_2	$q_2 = 0.4$	0.6
U_3	$q_3 = 0.7$	0.3

2.2 Worked example: The three-urn problem

Suppose there are three urns U_1, U_2, U_3, each containing 100 balls. The balls are either red or green and the urns differ in the percentage q_α of green balls. See Table 2.2.

Now, one of the urns, U_α say, is chosen at random. Then 40 balls are drawn with replacement; among them are n_g green balls. In this case the background information $\mathcal{I}$ includes the following information:

- There are three urns with known content and percentage q_α of green balls.
- The urn is selected at random, we don't know which one.
- The urns are identical.
- The urns do not provide a-priori information about the content.
- The balls are drawn at random with replacement.
- Apart from colour, all balls are identical: same size, weight, temperature, etc.
- ...

As pointed out already, only with a sufficiently precise statement of the background information are meaningful answers feasible. Often it is not obvious what information needs to be taken into account. For example, it is of importance that three urns are available. However, the size or colour of the urns can be considered irrelevant – as long as the different properties of the urns do not influence the selection process and we do not see them. Now, several questions may be of interest.

POSSIBLE QUESTIONS

What is the probability that:

- The number of the urn is $\alpha = 1, 2, 3$?
- The next draw will be a green ball?
- An urn with an unknown composition is present?

Definition 2.13 (Experiment) *An experiment or trial is defined as the observation (realization) of an outcome based on the validity of the background information.*

An experiment need not be performed by humans. The observation of an event (e.g. a supernova) can also be considered as an experiment under certain circumstances (e.g. appropriately defined background information $\mathcal{I}$).

Definition 2.14 (Random experiment) *A random experiment is an experiment which can result in different outcomes under repetition, and for which the outcome or event is unknown in advance. Repeating a deterministic experiment, instead, is expected to yield identical outcomes.*

Definition 2.15 (Population) *The set of all possible outcomes of random experiments which are compatible with $\mathcal{I}$ defines the population $\mathcal{G}$, also denoted by the sample space.*

Borel field In order to have a consistent theory, we have to make sure that all valid operations on propositions – such as the complement operation, binary logical operations on two propositions or sequences and series of propositions – are included in the sample space. The precise mathematical structure requires 'Borel sets'. For the examples and applications in this book, these mathematical prerequisites are always fulfilled and are otherwise of minor importance. Henceforward, we will not discuss these aspects further. We refer readers particularly interested in this topic to [118, 120].

The size of the population can be finite or infinite.

Definition 2.16 (Simple events) *Events which are indecomposable are simple events. All other events are compound events.*

For example, the outcome 'red' in roulette may happen in 18 different ways because 18 numbers (2, 4, ..., 34, 36) are labelled in red. It is therefore a compound event. The result '14' (which is also red) is, in this example, a simple event. Events E are subsets of the sample space $\mathcal{G}$, $E \subseteq \mathcal{G}$. There are only simple or compound events. Also, $\mathcal{G}$ and the empty set $\emptyset$ are events, the sure and the impossible event, respectively.

In this book, we consider either discrete sample spaces where the events are countable or continuous sample spaces where the random values are continuous, and can take values in one or more intervals.

The cardinality of the sample space is denoted by $|\mathcal{G}|$. In binary experiments, such as drawing a single ball from the urns, described above, $|\mathcal{G}| = 2$. For simple die experiments we have $|\mathcal{G}| = 6$. For samples of size n the sample space is given by the set of all possible sample results of size n, $\mathcal{G}_n$. For example, in the urn experiment with n draws with replacement, the cardinality of the sample space is given by $|\mathcal{G}_n| = 2^n$.

Definition 2.17 (Bernoulli trials) *Repeated independent binary trials with constant probabilities are called Bernoulli binary trials or Bernoulli experiments.*

The above urn experiment can be considered as a series of Bernoulli trials, with 40 draws with the binary outcome of either red or green in each trial, where the probability for green is the same for all draws.

Definition 2.18 (Samples) *n repeated independent trials (samples) can be bundled together into a sample of size n.*

With this definition the outcome of the urn example provides a sample of size 40.

Definition 2.19 (Events) *The outcome of an experiment is an event. Events happen or they don't. A partial event is undefined.*

A possible event of the urn example is *green ball in the ith draw*. However, the choice of the urn α is also an event. In the urn example we will relate the unknown events to the known events. Besides events, propositions are of crucial importance in Bayesian probability theory. Possible propositions for the urn experiment are:

- The selected urn has label α.
- The next draw will be a green ball.
- Of the 40 balls, there are n_g green balls.
- There is another urn U_4 with unknown q_4.

Definition 2.20 (Hypotheses) *Hypotheses play a crucial role in frequentist statistics. They are propositions concerning unknown facts.*

Frequentist statistics distinguishes between uncertainties due to missing information, 'the chosen urn is number 2', and uncertainties due to randomness, 'the next ball will be red'.

Accordingly, there seem to be two different kinds of probability in the urn example:

- The probability for n_g green balls in a sample of size N, given the urn is U_1,

$$P_1 = P(n_g|U_1, N, \mathcal{I}).$$

- The probability that urn α had been selected, if there are n_g green balls in the sample of size N,

$$P_2 = P(U_1|n_g, N, \mathcal{I}).$$

The first probability is usually referred to as a 'forward probability', because it can easily be computed, and the second as 'backward probability'.

Definition 2.21 (Forward/backward probability) *Let the inference be based on two propositions A and B, with conditional probabilities $P(A|B, \mathcal{I})$ and $P(B|A, \mathcal{I})$, respectively. Usually, one of the two, $P(A|B, \mathcal{I})$ say, is directly accessible based on the background information $\mathcal{I}$. Then $P(A|B, \mathcal{I})$ is referred to as a 'forward probability'. MacKay [136] refers to a forward probability if there is a 'generative model' that defines a process by which the data are generated and which allows us to compute the corresponding PMF by a 'forward calculation'. The 'inverse' probability, here $P(B|A, \mathcal{I})$, is then referred to as a 'backward probability'.*

Quite generally, the distribution of experimental data is a forward probability, which is specified by the error distribution of the experiment. In many cases this is known or otherwise it can be determined experimentally by repeated measurements with known samples.

At any rate, a serious experimentalist would not investigate a sample with a measurement device with unknown systematic or statistical error.

For concreteness, let us compute the forward probability for the urn problem, namely $P_{n_g} := P(n_g|N, U_1, \mathcal{I})$. First of all, the information U_1 is equivalent to specifying the fraction of green balls $q = q_1$. According to the background information, that is the only information content of U_1. So, we can equally well replace U_1 by q_1 in the conditional part of the probability $P_{n_g} := P(n_g|N, q_1, \mathcal{I})$. By construction ($\mathcal{I}$) the balls are drawn one after the other with replacement. Let's assume $N = 4$, the probability for green in a single trial is q, and the colour sequence is e.g. $\boldsymbol{c} = \{g, g, r, g\}$. Since the individual trials are uncorrelated, the joint probability factorizes as

$$\begin{aligned} P(\boldsymbol{c}|N, q, \mathcal{I}) &= P(g|N, q, \mathcal{I})P(g|N, q, \mathcal{I})P(r|N, q, \mathcal{I})P(g|N, q, \mathcal{I}) \\ P(\boldsymbol{c}|N, q, \mathcal{I}) &= q^{n_g}(1-q)^{N-n_g}, \end{aligned} \tag{2.20}$$

with $n_g = 3$ in the sequence $\boldsymbol{c}$. We have used the sum rule to specify the probability for red in a single trial $P(r|N, q, \mathcal{I}) = 1 - q$. One can easily convince oneself that equation (2.20) is valid for arbitrary colour sequences. As a matter of fact, we are not interested in the probability for a specific colour sequence, but rather in the probability that there are n_g green balls. The two probabilities can be linked through the marginalization rule

$$P(n_g|N, q_1, \mathcal{I}) = \sum_{\boldsymbol{c}} P(n_g|\boldsymbol{c}, N, q_1, \mathcal{I})P(\boldsymbol{c}|N, q, \mathcal{I}).$$

The second probability has been given in equation (2.20) already. The first term is also very simple. There is only one value for n_g compatible with the colour sequence, namely the number of green balls in the given sequence, which we denote by $n_g(\boldsymbol{c})$. Then

$$P(n_g|N, q_1, \mathcal{I}) = q^{n_g}(1-q)^{N-n_g} \sum_{\boldsymbol{c}}^{n_g(\boldsymbol{c})=n_g} 1.$$

The remaining task is to determine the number of different colour sequences with fixed n_g. This combinatorial task will be solved in Chapter 4 [p. 47]. The result is the binomial coefficient $\binom{N}{n_g}$ and the sought-for probability is the 'binomial distribution'

$$P(n_g|N, q_1, \mathcal{I}) = \binom{N}{n_g} q^{n_g} (1-q)^{N-n_g}, \tag{2.21}$$

which will be discussed in more detail in equation (4.12a) [p. 54].

The 'backward probability' $P(U_1|n_g, N, \mathcal{I})$ of the three-urn problem involves 'inverse reasoning' or 'inductive reasoning'. The consistent way to handle such problems [35, 118] is provided by Bayes' theorem. For the present problem it reads

$$P(U_1|n_g, N, \mathcal{I}) = \frac{P(n_g, N|U_1, \mathcal{I})P(U_1|N, \mathcal{I})}{P(n_g|N, \mathcal{I})}.$$

In this context, the probabilities $P(U_1|n_g N, \mathcal{I})$ and $P(U_1|N, \mathcal{I})$ are referred to as 'posterior probability' and 'prior probability', as the first is based on the (new) data, so to speak posterior to the experiment, while the second is the probability without the data, so to speak prior to the experiment. But posterior and prior probabilities do not always imply a temporal sequence. The term $P(n_g, N|U_1, \mathcal{I})$ is called the 'likelihood', as we are analysing the proposition U_1. The terminological distinction is reasonable as the likelihood is not a probability as far as proposition U_1 is concerned, because it stands behind the conditional solidus. Eventually, the term in the denominator, $P(n_g|N, \mathcal{I})$, is a normalization factor or the so-called 'data evidence'.

According to E. T. Jaynes [104], there is no well-defined distinction between posterior and prior probability: 'One man's posterior is another man's prior probability.' This is perfectly true, but there is a common situation where the distinction is clear. This is the case in standard data analysis, where a measurement device with known error distribution is used to infer features of a physical/chemical sample of interest.

2.3 Frequentist statistics versus Bayesian inference

There are two controversial views on how to tackle problems requiring inductive logic, frequentist statistics and Bayesian probability theory.

2.3.1 Frequentist statistics

In the frequentist point of view, the forward probability is a genuine probability, as it describes properties of random variables, while the backward probability does not. Therefore, it has been treated differently, by so-called hypothesis tests, which will be discussed in more detail in Chapter 18 [p. 276]. Randomness is considered within the frequentist framework as an intrinsic (physical) property of the experiment. So, conceptually, one has to distinguish between the following situations: (a) a coin is symmetric, (b) the coin is a fake and shows the same image on both sides, but the image (head or tail) is unknown.

In case (a) the uncertainty originates from a random experiment and therefore there exists a probability $P(\text{head}) = 1/2$, while in case (b) the uncertainty is due to missing information, there is nothing random because it will always yield the same result when we actually perform the experiment. In the frequentist definition, the ratios in $\lim_{N\to\infty} n_{head}/N$ are always either 0 or 1. But this will be known only after the experiment is performed. Before, no statements are allowed. Before the selection of the urn, the index α is a random variable with probability $p(\alpha) = 1/3$. However, if the urn has already

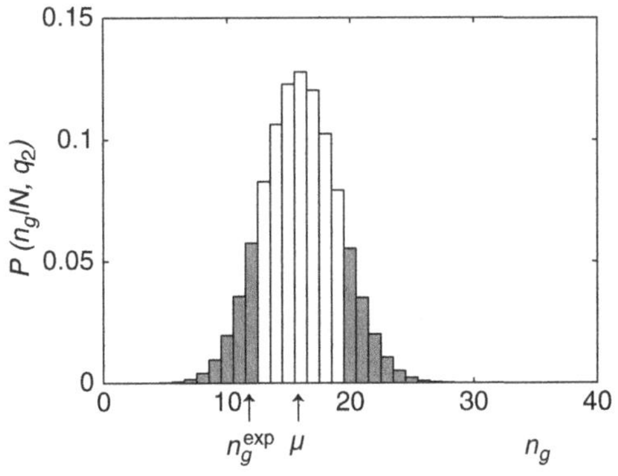

Figure 2.1 Binomial distribution with parameters $q_2 = 0.4$, $N = 40$ and $\mu = 16$. The area of the shaded region yields the P value. Here an experimental value $n_g^{\text{exp}} = 12$ is assumed.

been selected – although we do not know which one – the frequentist statistics does not consider α as a random variable any longer. Within frequentist statistics the probability for hypotheses like 'The urn under consideration has index $\alpha = 2$' can neither be raised nor resolved. A partial remedy was introduced with the concept of significance tests, which will be discussed in more detail in Chapter 18 [p. 276]. In order to assess the validity of a hypothesis H, frequentist statistics uses the following approach. In the present example the hypothesis H specifies the urn, U_2 say. Given the hypothesis is true, one knows the sampling distribution, which in the present example is simply $P(n_g|H, N, \mathcal{I})$, as depicted in Figure 2.1. The hypothesis implies on average $\mu = Nq_2$ green balls. The observed number n_g^{exp} (in Figure 2.1 $n_g^{\text{exp}} = 12$) deviates from this value by $\Delta n^* = \left|\mu - n_g^{\text{exp}}\right|$. Now the probability for a deviation from the mean as large as or even larger than the observed deviation Δn^* is computed:

$$P := \sum_{n_g=0}^{\mu-\Delta n^*} P(n_g|N, q_1) + \sum_{n_g=\mu+\Delta n^*}^{N} P(n_g|N, q_1). \tag{2.22}$$

It corresponds to the shaded area in Figure 2.1. This quantity is called the 'P value'. If the P value is below a given 'significance level' p_S (e.g. 5%) then the data are said to be significant, because the deviations can hardly be caused by chance, and the hypothesis is rejected. The P value for the three-urn example is displayed in Figure 2.2 as a function of the number of green balls n_g^{exp} in the sample. The peaks, denoted by P_α^H, are the P values associated with the hypotheses H_α, saying that U_α is the correct urn. We observe that the P value is one whenever the discrepancy between the hypothesis and the data is zero, and it drops rapidly with increasing discrepancy. Based on the P value and a 5% significance level, we would reject hypothesis H_α according to the values in Table 2.3.

An obvious shortcoming of this approach is the fact that there is a region $n_g^{\text{exp}} \geq 35$ where all hypotheses are to be rejected, although we know that one has to be correct.

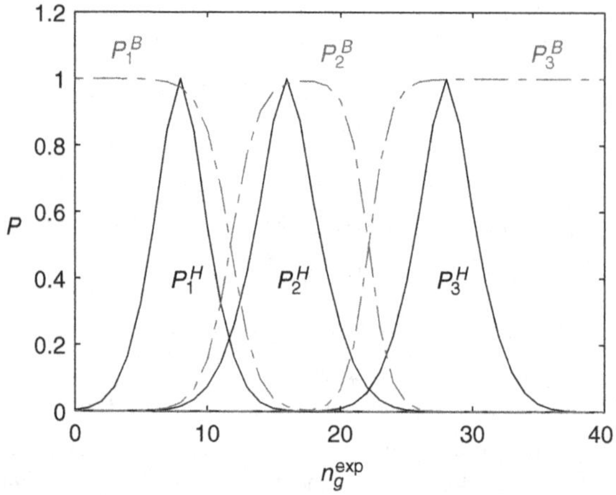

Figure 2.2 Three-urn example with $N = 40$ and q-values provided in the text. Solid lines represent the P values P_α^H under hypothesis U_α. The Bayesian probabilities $P_\alpha^B = P(U_\alpha|N, n_g^{\text{exp}}, \mathcal{I})$ are depicted as dashed lines.

Table 2.3 Rejection regions for the three hypotheses based on a 5% significance level

Urn	Lower tail	Upper tail
U_1		$n_g^{\text{exp}} \geq 8$
U_2	$n_g^{\text{exp}} \leq 8$	$n_g^{\text{exp}} \geq 24$
U_3	$n_g^{\text{exp}} \leq 21$	$n_g^{\text{exp}} \geq 35$

A thorough discussion of this approach will be given in Chapter 18 [p. 276]. Until then we leave it to the reader to reflect on the meaning of this result.

2.3.2 Bayesian inference

Bayesian probability theory differs in two fundamental points from the frequentist statistics: (a) in the interpretation of the term 'probability' and (b) in a different view of the origin of uncertainty. (a) In Bayesian probability theory, probability is considered as a degree of belief in the truth of a proposition. From a different point of view this can be considered as a generalization of propositional calculus to the case of partial truths. (b) Uncertainty is not only due to an intrinsic randomness of the object to be analysed but can also result from incomplete information. Let us consider several examples.

- **Flipping a coin:** Strictly speaking this is not a random event, as the apparatus has a well-defined mechanism to flip the coin. In principle, analytical mechanics would allow us to compute the outcome.

- **Rolling a die:** Same as flipping a coin.
- **(Pseudo-)random numbers:** This is an even more obvious example that randomness is not acting at all. The number sequence is perfectly predictable.
- **Electoral behaviour:** Most voters do not vote at random.
- **Spin:** Of an atom in a Stern–Gerlach (SG) experiment.

In all these cases other than the last one, information is lacking that leads us to believe in chance as the driving mechanism. The quantum-mechanical example is the only one that has no hidden information that would make the outcome predictable. Because uncertainty is mostly not due to intrinsic randomness or another intrinsic property of the subject under consideration, we use the terminology 'probability for E' instead of 'probability of E' as proposed by E. T. Jaynes [104]. For Bayesian inference, however, the origin of the uncertainty does not matter at all. Let us consider a perfectly random coin-tossing experiment, maybe using an apparatus based on quantum-mechanical effects. The probability for heads up in case of a fair coin is then $P(\text{head}|\text{fair}, \mathcal{I}) = 1/2$ due to the random nature of the measurement. In case of a fake coin, which has the same image on both sides, the probability for heads is again $P(\text{head}|\text{biased}, \mathcal{I}) = 1/2$, as we do not know which image is on both sides and both images are equally likely. In this case there is no random element involved at all.

The difference in the two cases is revealed by the Bayesian approach, after the first measurement. Then

$$P(\text{head}|\text{first outcome: tail, fair}, \mathcal{I}) = 1/2,$$

while

$$P(\text{head}|\text{first outcome: tail, biased}, \mathcal{I}) = 0.$$

It has been proved [35, 118] that Bayesian probability theory is the only theory which handles partial truths consistently. It is the content of the next chapter. We want to close this chapter by computing the result of the three-urn problem and comparing it with the frequentist result. First we need to specify all relevant propositions.

- U_α: Urn α has been selected.
- $q = q_\alpha$: The fraction q of green balls is q_α.
- N: The sample has size N.
- n_g: The sample contains n_g green balls.
- $\mathcal{I}$: Background information.

We calculate the probability of interest $P(U_\alpha|N, n_g, \mathcal{I})$ using Bayes' theorem, given in equation (1.9) [p. 12]:

$$p_\alpha := P(U_\alpha|n_g, N, \mathcal{I}) = \frac{P(n_g|U_\alpha, N, \mathcal{I})\; P(U_\alpha|N, \mathcal{I})}{P(n_g|N, \mathcal{I})}. \tag{2.23}$$

The first term in the numerator, $P(n_g|U_\alpha, N, \mathcal{I})$, is the forward probability that a sample of size N from urn U_α contains n_g green balls. This probability is given by the binomial distribution $P(n_g|N, q_\alpha)$ (see equation (2.21)). The second term in the numerator, $P(U_\alpha|N, \mathcal{I})$, is the prior probability that urn U_α had been selected. For this probability the information of the sample size is irrelevant. Hence, $P(U_\alpha|N,\mathcal{I}) = P(U_\alpha|\mathcal{I})$. According to the background information $P(U_\alpha|N,\mathcal{I}) = 1/3$, because one of three urns had been chosen at random. Finally, equation (2.23) is normalized by the denominator which is independent of α. This term ensures that $\sum_{\alpha=1}^{3} p_\alpha = 1$. Finally we obtain

$$P(U_\alpha|n_g, N, \mathcal{I}) := \frac{P(n_g|N, q_\alpha)}{\sum_{\beta=1}^{3} P(n_g|N, q_\beta)}. \tag{2.24}$$

This probability is displayed in Figure 2.2 as a function of $n_g = n_g^{\text{exp}}$. This is a satisfactory outcome. It allows us to decide unambiguously which urn (hypothesis) is preferable given the experimental value n_g^{exp}. In particular, for $n_g^{\text{exp}} > 35$ the result establishes beyond any doubt that urn 3 is the selected one.

3
Bayesian inference

3.1 Propositions

Knuth and Skilling [118] have unified and extended the foundation for probabilistic inference originally outlined by Cox [35] and Kolmogorov [120], which form the basis for Bayesian probability theory. As mentioned earlier, the key elements are propositions. In order to allow a systematic application of Bayesian probability theory we need to introduce a clear classification of propositions.

In many problems it is possible and useful to index propositions A_i ($i = 1, 2, \ldots, N$), where the number of propositions can be either finite ($N < \infty$) or infinite ($N = \infty$). We call such propositions 'discrete'. For the time being, we will only consider discrete propositions. In Section 7.1 [p. 92] we will extend these considerations to the continuous case.

Example: A_i: *The observed face value of a die is i.*

Of particular importance are mutually exclusive propositions:

Mutually exclusive propositions

$$A_i \wedge A_j = \begin{cases} A_i & \text{for } i = j \\ F & \text{otherwise} \end{cases} \qquad \forall\, i, j\ , \tag{3.1}$$

where F stands for the impossible proposition. The propositions given above for the die are mutually exclusive. The face value can be either 1 or 2, etc., but not two values at the same time. If the A_i are in addition complete, then they form a

Partitioning

$$\vee_i\, A_i = T. \tag{3.2}$$

That is, the union of all complete mutually exclusive propositions is the true proposition T. The propositions about the die are complete as the face value will never be anything else but the numbers 1–6. The partitioning is closely related to expansion of the unit operator $\mathbb{1}$ in quantum mechanics in terms of a complete orthonormal basis $|\phi_n\rangle$, as

$$\sum_n |\phi_n\rangle\langle\phi_n| = \mathbb{1}.$$

The product rule simplifies for mutually exclusive propositions to

$$P(A_i, A_j|\mathcal{I}) = \delta_{ij}\; P(A_i|\mathcal{I}),$$

and the sum rule for mutually exclusive propositions is given by

$$1 = P(\vee_i\; A_i|\mathcal{I}) = \sum_i P(A_i|\mathcal{I}). \tag{3.3}$$

Sum rule for discrete propositions

$$\sum_i P(A_i|\mathcal{I}) = 1 \qquad \text{(normalization)}, \tag{3.4a}$$

$$P(B|\mathcal{I}) = \sum_i P(B|A_i, \mathcal{I})\; P(A_i|\mathcal{I}) \qquad \text{(marginalization rule)}. \tag{3.4b}$$

3.2 Selected examples

Before we elaborate the theory further, we will study some selected examples in order to get acquainted with the application of the rules of Bayesian probability theory.

3.2.1 Propagators

The first example is a highly simplified version of propagators, or rather Green's functions, that occur in quantum-mechanical many-body problems. The one-particle Green's function, for instance, describes the propagation of electrons in a crystal, which can be influenced by other electrons, atoms, lattice vibrations and other scattering events. In his book '*A Guide to Feynman Diagrams in the Many-Body Problem*', Mattuck [140] compares the propagation of an electron from point A to B with a party guest, who leaves a party at point A and tries to reach his home at point B. On his way home there are several scattering centres in the form of bars. The situation is sketched in Figure 3.1. At each bar there is a probability P_B that he is tempted and enters. With probability $(1 - P_B)$ he proceeds without entering. There is an additional probability $P_R < 1$ that he will not be able to leave the bar before dawn. For the sake of simplicity, we allow the party guest to enter each bar at most once (no multiple scattering). The sought-for quantity is the probability $P(H|N, \mathcal{I})$ for the proposition H: *He returns home safely the same evening.*

Figure 3.1 Propagation of drunken man. *Source:* R. D. Mattuck, A Guide to Feynman Diagrams in the Many-Body Problem, Dover Publications, Inc., New York, 1992, reprinted with permission.

The propagation home can be partitioned into the following complete mutually exclusive events (propositions) E_n: *He attends n bars*. Using the marginalization rule we find

$$P(H|N,\mathcal{I}) = \sum_{n=0}^{N} P(H|E_n, N, \mathcal{I})\; P(E_n|N, \mathcal{I}).$$

The proposition N indicates that there are N bars and the conditional complex also encodes the information that the decisions to enter the bars shall be independent of each other (uncorrelated). Therefore, the problem can be modelled as a Bernoulli experiment

$$P\left(E_n|N, \mathcal{I}\right) = \binom{N}{n}\; {P_B}^n\; (1 - P_B)^{N-n}\,.$$

The probability that the way home can be continued after a bar has been entered is P_R. In order to reach his apartment the same night, the party guest has to leave all the n visited bars. The probability for that is given by ${P_R}^n$. This results in

$$\begin{aligned} P(H|N,\mathcal{I}) &= \sum_{n=0}^{N}\; P(H|E_n, N, \mathcal{I})\; P(E_n|N, \mathcal{I}) \\ &= \sum_{n=0}^{N} \binom{N}{n}\; (P_B P_R)^n\; (1 - P_B)^{N-n} \\ &= (P_B P_R + 1 - P_B)^N = (1 - P_B(1 - P_R))^N. \end{aligned}$$

The result makes perfect sense: the probability that the guest is 'absorbed' by a bar is $P_B\ (1-P_R)$, therefore the probability that he overcomes a bar is given by $1-P_B\ (1-P_R)$ and this has to happen N times.

The relation to many physical problems is immediately obvious, be it damping of waves, propagation of particles in matter, diffusion, etc. The number N of bars corresponds to the distance travelled x and the probability $P_B(1-P_R)$ of being absorbed by a bar can be related to a decay or damping constant per unit length. If $P_B(1-P_R) \ll 1$ holds, then

$$P(H|N,\mathcal{I}) = e^{N\ \ln(1-P_B(1-P_R))} \approx e^{-P_B(1-P_R)\ N} \triangleq e^{-\alpha x},$$

and we end up with an exponential attenuation law.

3.2.2 The Monty Hall problem

The Monty Hall problem is a probability puzzle loosely based on the American television game show *Let's Make a Deal* and named after the show's original host, Monty Hall. The Monty Hall problem bears some similarity to the much older Bertrand's box paradox. The problem was originally posed in a letter by Steve Selvin to the journal *The American Statistician* in 1975. Suppose you are in a game show, and you are given the choice of three doors. Behind one door is a car as the prize; behind the others is nothing. You pick a door, say No. 1, and the host, who knows what is behind the doors, opens one of the remaining doors but never the door with the prize behind it. He then asks you, 'Do you want to switch to the other closed door?' We face the problem: Is it better to stick to the first choice, is it better to switch, or doesn't it make a difference at all? At first glance, one might argue: Eventually there are two doors, equally likely to hide the prize, so it should not make a difference. To resolve the problem we compute the respective probabilities for winning the prize. The required propositions are

- S: The chosen strategy is to stick to the initial choice.
- G: We win.

Then the probability for winning the prize when sticking to the initial choice is given by $P(G|S,\mathcal{I})$. We know if we win, that the first choice was correct. Therefore, we introduce the additional proposition $\sigma = 1/0$: *The first choice is correct/wrong*. Based on the marginalization rule we then find

$$P(G|S,\mathcal{I}) = \sum_{\sigma=0}^{1} \underbrace{P(G|\sigma,S,\mathcal{I})}_{\delta_{\sigma,1}}\ P(\sigma|S,\mathcal{I}) = P(\sigma=1|S,\mathcal{I}) = 1/3.$$

In the last step, the prior probability that the first choice was correct was required. Since we have not assumed that one of the doors has a higher preference to be chosen for the prize, we consequently have to assign equal prior probability to all three doors. Apparently, retaining the original choice is not the best idea. The probability for success under the strategy 'do not stick to the first door' is 2/3. The outcome becomes perfectly reasonable when we consider it the following way: It is good to stick to the first choice, if it was the correct one from the start. This has a probability $1/3$. It is better to switch if the initial choice was wrong. This is true in two of three cases, i.e. it has a probability $2/3$.

3.2.3 Detection of rare events

What detector performance (i.e. sensitivity, error rate) is necessary to detect rare particles? Questions like this are of importance in the design of detectors. We introduce the following propositions:

- $T/\overline{T}$: A particle is present/no particle is present.
- D: Detector signals a particle.

We are interested in the probability that a particle is indeed present if we receive a detector response. Using Bayes' theorem we can express this probability as

$$P(T|D,\mathcal{I}) = \frac{P(D|T,\mathcal{I})\ P(T|\mathcal{I})}{P(D|T,\mathcal{I})\ P(T|\mathcal{I}) + P(D|\overline{T},\mathcal{I})\ P(\overline{T}|\mathcal{I})}.$$

A numerical example may be

$$\begin{aligned} P(T|\mathcal{I}) &= 10^{-8}, \\ P(D|T,\mathcal{I}) &= 0.99, \\ P(D|\overline{T},\mathcal{I}) &= 10^{-4}. \end{aligned}$$

With these numbers we obtain $P(T|D,\mathcal{I}) = 10^{-4}$. The result may seem disappointing; despite the apparently good performance of the detector a positive response of the detector provides nearly no indication of a successful detection of the rare particles. Nevertheless, prior to the test the probability was 10^{-8} that there is a particle. Now it has increased by four orders of magnitude.

We may ask the question, which detector parameters are required to improve $P(T|D,\mathcal{I})$ to at least $P(T|D,\mathcal{I}) \geq 1/2$? Such a predictive quality needs

$$\begin{aligned} P(D|\overline{T},\mathcal{I})\ P(\overline{T}|\mathcal{I}) &\leq P(D|T,\mathcal{I})\ P(T|\mathcal{I}) \\ P(D|\overline{T},\mathcal{I}) &\leq P(D|T,\mathcal{I})\ \frac{P(T|\mathcal{I})}{1-P(T|\mathcal{I})} \approx P(T|\mathcal{I}). \end{aligned}$$

Therefore the error rate $P(D|\overline{T},\mathcal{I})$ has to be smaller than the prior probability for the particles, namely 10^{-8}!

Example: Clinical trials

Similar considerations can be applied to medical surveys. In this case the meaning of the propositions is

- T: The patient is infected.
- D: The test result is positive.

How large is the probability that a patient is indeed infected if the test has been positive? Realistic parameters for medical surveys are

$$\begin{aligned} P(T|\mathcal{I}) &= 0.001, \\ P(D|T,\mathcal{I}) &= 0.9, \\ P(D|\overline{T},\mathcal{I}) &= 0.01. \end{aligned}$$

This results in

$$\begin{aligned} P(D|T,\mathcal{I})\ P(T|\mathcal{I}) &= 0.9 \times 0.001 = 0.0009, \\ P(D|\overline{T},\mathcal{I})\ P(\overline{T}|\mathcal{I}) &= 0.01 \times (1 - 0.001) = 0.00999, \\ P(T|D,\mathcal{I}) &= \frac{1}{1 + \frac{0.00999}{0.0009}} = \frac{1}{12.1} \approx 0.08. \end{aligned}$$

Therefore, of 100 persons with a positive outcome of the test only eight are infected. However, the probability for being infected has increased by a factor of 80, from initially $P(T|\mathcal{I}) = 0.001$ to $P(T|\mathcal{I}) = 0.08$.

In both of the preceding examples the reason for the small effect for a positive response is the fact that there are many more possibilities for unjustified positive responses $P(D|\overline{T},\mathcal{I})$ $P(\overline{T}|\mathcal{I})$ than for a true detection $P(\overline{T}|\mathcal{I}) \gg P(T|\mathcal{I})$.

One might be bothered that such a 'poor' test may under-diagnose the disease. So let us determine the probability that there is a virus, although the diagnosis indicates the opposite:

$$\begin{aligned} P(T|\overline{D},\mathcal{I}) &= \frac{P(\overline{D}|T,\mathcal{I})P(T|\mathcal{I})}{P(\overline{D}|\overline{T},\mathcal{I})P(\overline{T}|\mathcal{I}) + P(\overline{D}|T,\mathcal{I})P(T|\mathcal{I})} \\ &= \left(1 + \frac{P(\overline{D}|\overline{T},\mathcal{I})\ P(\overline{T}|\mathcal{I})}{P(\overline{D}|T,\mathcal{I})\ P(T|\mathcal{I})}\right)^{-1} \\ &= \left(1 + \frac{0.99}{0.1}\frac{0.999}{0.001}\right)^{-1} \approx 10^{-4}. \end{aligned}$$

Apparently, it is highly unlikely that the test fails to diagnose the presence of the virus.

3.2.4 The fair coin example

Next we want to study the hypothesis H that a coin is fair. Hypothesis tests will be discussed quite generally in Part IV [p. 255]. It is often advantageous to start from the odds ratio

$$o = \frac{P(H|D,\mathcal{I})}{P(\overline{H}|D,\mathcal{I})},$$

where D denotes the available data. The (computational) advantage of the odds ratio is that an often cumbersome computation of normalization constants can be circumvented. Invoking Bayes' theorem:

UPDATE RULE FOR THE ODDS RATIO

$$o = \underbrace{\frac{p(D|H,\mathcal{I})}{p(D|\overline{H},\mathcal{I})}}_{\text{Bayes factor}} \underbrace{\frac{P(H|\mathcal{I})}{P(\overline{H}|\mathcal{I})}}_{\text{prior odds}}, \tag{3.5}$$

we can split the (posterior) odds ratio into two factors, the 'prior odds' (o_P), which represents our assessment of the hypothesis prior to the inclusion of the data D, and the so-called *Bayes factor* (o_{BF}), which quantifies the impact of the data on the inference problem. So we have an update rule for the odds ratio in view of new data D. A simple introductory example is the assessment of a coin. The relevant propositions in this case are

- H: The coin is fair.
- n_K: Heads up occurred n_K times.
- N: The coin has been tossed N times.

We start out with the assumption that we have not the slightest idea whether the coin is fair or not and we therefore assign equal probability to H and $\overline{H}$, or rather

$$o_P = \frac{P(H|\mathcal{I})}{P(\overline{H}|\mathcal{I})} = 1.$$

The marginal likelihood function $P(n_K|N, H, \mathcal{I})$ and $P(n_K|N, \overline{H}, \mathcal{I})$, entering the Bayes factor, can be derived from the likelihood function using the marginalization rule

$$P(n_K|N, A, \mathcal{I}) = \int_0^1 P(n_K|N, q, A, \mathcal{I})\, p(q|N, A, \mathcal{I})\, dq, \tag{3.6}$$

with $A = H$ or $A = \overline{H}$. The likelihood function in this example is the binomial distribution. The remaining task is to clarify what H and in particular $\overline{H}$ really means, given the background information $\mathcal{I}$. This is a crucial step in serious hypothesis testing, as we will discuss in Part IV [p. 255]. In the coin example it corresponds to assigning values to the prior PDF $P(q|N, A, \mathcal{I})$. The present background information is encoded as

$$P(q|A, \mathcal{I}) = \begin{cases} \delta(q - 1/2) & \text{for } A = H \\ \mathbf{1}(0 \le q \le 1) & \text{for } A = \overline{H} \end{cases}.$$

That means for us that a coin is only fair if the probability q is precisely $1/2$. Moreover, as an alternative, $\overline{H}$, everything is conceivable. This might be too extreme in real-world applications, as it implies in some sense that if we collect all coins available on earth, and sort out the fair ones, we are left with a collection of coins with q uniformly spread over the interval $[0, 1]$. Needless to say, it is physically hard to imagine how such coins could be manufactured. We leave the further computational details of this example to Part IV [p. 255].

Laplace's 'law of succession' II

Now we are prepared to consider Laplace's law of succession in its full beauty [123]. Consider an urn with black and white balls. Of n previous draws with replacement, k balls are black. What is the probability that the next draw will be black again? The required propositions are:

- N: N balls have been drawn with replacement.
- n: n of the drawn balls are black.
- B: The next draw will be a black ball.
- E_q: The intrinsic probability for a black ball in a single trial is q.
- $\mathcal{I}$: Background information, i.e. all assumptions and all our prior knowledge.

Invoking the marginalization rule, we can express the probability for B as

$$P(B|n, N, \mathcal{I}) = \int_0^1 dq \; P(B|E_q, n, N, \mathcal{I}) \; P(E_q|n, N, \mathcal{I}).$$

The first term is a sort of tautology; it is the probability that the next ball will be black, given the probability for a black ball in a single trial is q. Then

$$P(B|n, N, \mathcal{I}) = \int_0^1 dq \; q \; P(E_q|n, N, \mathcal{I}) = \langle q \rangle,$$

which is the posterior mean for q. The posterior PDF according to Bayes' theorem is

$$P(E_q|n, N, \mathcal{I}) = \frac{P(n|E_q, N, \mathcal{I}) P(E_q|N, \mathcal{I})}{P(n|N, \mathcal{I})}.$$

The likelihood is the binomial distribution. Next we need to specify the prior probability $P(E_q|N, \mathcal{I})$. Here the knowledge that N draws have been performed with replacement is not of any help. Based on our prior knowledge each value of q is equally likely, which is appropriately described by a uniform prior $p(E_q|\mathcal{I}) = \mathbf{1}(0 \le q \le 1)$. Along with the correct normalization, the resulting posterior is therefore the beta distribution (Section 7.5.2 [p. 100]):

$$P(E_q|n, N, \mathcal{I}) = \mathbf{1}(0 \le q \le 1) \; \frac{q^n (1-q)^{N-n}}{B(n+1, N-n+1)}.$$

Then the posterior mean is, according to equation (7.12c) [p. 100],

$$\langle q \rangle = \frac{n+1}{N+2}$$

which is Laplace's law of succession. Laplace then applied the result to the probability that the sun will rise the next day, if it has risen each day in the past N days. He assumes $N = 182\,623$ days (5000 years) as the most ancient epoch of history. Then, since $n = N$ he concludes that the probability the sun will **not** rise tomorrow is $P = 1/1\,826\,214$. Laplace concludes '... it is a bet of 1 826 214 to one that it will rise again tomorrow. But this number is incomparably greater for him who, recognizing in the totality of phenomena the principal regulator of days and seasons, sees that nothing at the present moment can arrest the course of it' [123].

3.2.5 Estimate of tank production

Suppose a company is producing a product, e.g. tanks. Each tank is labelled with a consecutive serial number. What can be concluded about the total number of manufactured tanks, N, if a sample of size L with serial numbers $\boldsymbol{n}_L := \{n_1, n_2, \dots, n_L\}$ is available? The problem is also known as the 'German tank problem' in the literature because this kind of reasoning was applied in World War II to estimate the production rate of German tanks [139]: The probability $P(N|\boldsymbol{n}_L, L, \mathcal{I})$ can be computed using Bayes' theorem:

$$P(N|\boldsymbol{n}_L, L, \mathcal{I}) = \frac{1}{Z}\, P(\boldsymbol{n}_L|N, L, \mathcal{I})\; P(N|L, \mathcal{I}).$$

The required propositions are:

- N: The number of goods (tanks) is N.
- L: The sample has size L.
- $\boldsymbol{n}_L$: The sample of size L comprises the serial numbers $\{n_1, \dots, n_L\}$.
- n_j: The serial number of sample item j is n_j.
- $\mathcal{I}$: Background information containing the information that the sample is randomly selected.

The probability $P(\boldsymbol{n}_L|N, L, \mathcal{I})$ can be simplified using the product rule

$$P(\boldsymbol{n}_L|N, L, \mathcal{I}) = P(n_L|\boldsymbol{n}_{L-1}, N, L, \mathcal{I})\; P(\boldsymbol{n}_{L-1}|N, L, \mathcal{I}).$$

Repeated application of the product rule to the second factor of these expressions yields

$$P(\boldsymbol{n}_L|N, L, \mathcal{I}) = \prod_{l=1}^{L} P(n_l|\boldsymbol{n}_{l-1}, N, L, \mathcal{I}),$$

with the definition $P(n_1|\boldsymbol{n}_0, N, L, \mathcal{I}) = P(n_1|N, L, \mathcal{I})$. Since the sample is randomly drawn, each number $\{1, 2, \dots, N\}$ is equally likely to occur in the sample. Then

$$P(n_j|\boldsymbol{n}_{j-1}, N, L, \mathcal{I}) = \frac{1}{N - j + 1}\; \mathbb{1}\left(n_j \le N\right),$$

because there are $N - j + 1$ possibilities remaining for serial number n_j, when the serial numbers $n_1, n_2, \dots, n_{j-1}$ have already been selected. Hence, the likelihood function is given by

$$P(\boldsymbol{n}_L|N, L, \mathcal{I}) = \prod_{j=1}^{L} \left(\frac{1}{N + 1 - j}\; \mathbb{1}\left(n_j \le N\right)\right) = \mathbb{1}(N \ge n_{\max})\; \frac{(N - L)!}{N!}.$$

Here $n_{\max}$ is the largest serial number observed in the sample. The prior probability $P(N|L, \mathcal{I})$ takes into account that the sample size is L and that therefore $N \ge L$ has to hold. For the actual computation we will introduce temporarily a maximum number

N_{max}, for example, given by the surface area of the earth divided by the area of a tank. We will see that we can take the limit for $N_{\text{max}} \to \infty$ at the end, if the sample size exceeds a minimal size:

$$P(N|L,\mathcal{I}) = \frac{1}{N_{\text{max}} - L} \; \mathbf{1}(L \leq N < N_{\text{max}})\,.$$

Hence, the probability of interest is

$$P(N|\boldsymbol{n}_L, L, \mathcal{I}) = \frac{1}{Z'} \; \mathbf{1}(n_{\text{max}} \leq N < N_{\text{max}}) \; \frac{(N-L)!}{N!},$$

$$Z' = \sum_{N=n_{\text{max}}}^{N_{\text{max}}} \frac{(N-L)!}{N!}.$$

This PMF and its mean and variance can easily be computed numerically. In order to get a better insight into the result, we proceed analytically based on a quite reasonable assumption, namely $n_{\text{max}} \gg L$ and hence $N \gg L$. Then we can approximate the probability by

$$\frac{(N-L)!}{N!} \approx N^{-L}.$$

With the same justification, the normalization constant Z' can approximate the following integral:

$$Z' \approx \int_{N=n_{\text{max}}}^{N_{\text{max}}} N^{-L}\, dN = \left.\frac{N^{-L+1}}{L-1}\right|_{N_{\text{max}}}^{n_{\text{max}}} \underset{N_{\text{max}} \gg n_{\text{max}}}{\longrightarrow} \frac{n_{\text{max}}^{-L+1}}{L-1}.$$

This final result reads

$$P(N|\boldsymbol{n}_L, L, \mathcal{I}) \approx \mathbf{1}(n_{\text{max}} \leq N < N_{\text{max}}) \; (L-1)\, n_{\text{max}}^{L-1}\, N^{-L}.$$

The probability for N given the sample information is largest for $N = n_{\text{max}}$ and decays for larger values as N^{-L}. Next we compute the mean $\langle N \rangle$, based on the same approximations:

$$\langle N \rangle = \frac{1}{Z'} \sum_{N=n_{\text{max}}}^{N_{\text{max}}} N\, N^{-L} \approx \frac{1}{Z'} \int_{N=n_{\text{max}}}^{N_{\text{max}}} N^{-L+1}\, dN$$

$$\approx n_{\text{max}} \left(1 + \frac{1}{L-2}\right).$$

This result is only sensible for sample sizes $L > 2$. For samples of size 1 and 2 the posterior probability depends on the chosen cutoff value N_{max} and neither the norm nor the first moment exists. Of interest is also the cumulative distribution, i.e. the probability for N being less than a given threshold. With the same approximations as before we obtain

$$
\begin{aligned}
F(N) &:= \sum_{N'=n_{\max}}^{N} P(N'|\boldsymbol{n}_L, L, \mathcal{I}) \\
&\approx \boldsymbol{1}(N \geq n_{\max})\ (L-1)\, n_{\max}^{L-1} \int\limits_{n_{\max}}^{N} N'^{-L} dN' \\
&\approx \boldsymbol{1}(N \geq n_{\max}) \left(1 - \left(\frac{n_{\max}}{N}\right)^{L-1}\right).
\end{aligned}
$$

In order to cover e.g. 90% of the probability mass, $\left(\frac{n_{\max}}{N}\right)^{L-1} = 0.1$ must hold. Therefore 90% of the probability mass is in the interval

$$
I_{90\%} = \left[n_{\max}, n_{\max}\ 10^{\frac{1}{L-1}}\right].
$$

For a sample of size $L = 10$ this implies $I_{90\%} = [1, 1.29]n_{\max}$. The result is quite satisfactory, since in such problems we are usually interested only in the order of magnitude.

3.3 Ockham's razor

Let us now evaluate two theoretical models, M_1 and M_2, in the light of experimental data D. The corresponding odds ratio is given by

$$
o = \frac{P(M^{(1)}|D,\mathcal{I})}{P(M^{(2)}|D,\mathcal{I})} = \underbrace{\frac{P(D|M^{(1)},\mathcal{I})}{P(D|M^{(2)},\mathcal{I})}}_{\text{Bayes factor}} \underbrace{\frac{P(M^{(1)}|\mathcal{I})}{P(M^{(2)}|\mathcal{I})}}_{\text{prior odds}}.
$$

If both models have no adjustable parameters this is the end of the story and we have the competing factors: prior odds (o_{P}) versus Bayes factor (o_{BF}). The situation becomes more realistic and interesting if the models are more *complex* and contain adjustable model parameters. Complexity shall be understood as the variability of the parameters and the richness of the structure of the models. That is, a cosine function $y(x) = A\cos(kx)$ with fixed parameters A and k is understood to be less complex than a constant model $y = a$ with free parameter a. Model $M^{(\alpha)}$ shall have $L^{(\alpha)}$ parameters denoted by $\boldsymbol{a}^{(\alpha)}$, with prior PDF $p(\boldsymbol{a}^{(\alpha)}|M^{(\alpha)},\mathcal{I})$. For the sake of illustration, we assume uniform priors for all parameters, i.e.

$$
\begin{aligned}
p(\boldsymbol{a}^{(\alpha)}|M^{(\alpha)}) &= \frac{1}{V_{\mathrm{pr}}^{(\alpha)}} \prod_{i=1}^{L^{(\alpha)}} \theta\left(a_{i,0}^{(\alpha)} \leq a_i \leq a_{i,1}^{(\alpha)}\right), \\
V_{\mathrm{pr}}^{(\alpha)} &:= \prod_{i=1}^{L^{(\alpha)}} I_i^{(\alpha)}, \\
I_i^{(\alpha)} &:= \left(\alpha_{i,1}^{(\alpha)} - \alpha_{i,0}^{(\alpha)}\right).
\end{aligned}
$$

The marginal likelihood is computed via the marginalization rule, by introducing the prior PDF for the parameters given the respective model. If we assume an informative data set, then the likelihood will be much more restrictive as far as the parameters are concerned than the prior, and the Bayes factor can be approximated reliably by

$$o_{\mathrm{BF}} \approx \frac{\int p(D|\boldsymbol{a}^{(1)}, M^{(1)}, \mathcal{I})\, d\boldsymbol{a}^{(1)}}{\int p(d|\boldsymbol{a}^{(2)}, M^{(2)}, \mathcal{I})\, d\boldsymbol{a}^{(2)}} \frac{V_{\mathrm{pr}}^{(2)}}{V_{\mathrm{pr}}^{(1)}}.$$

The remaining integral should not be confused with the normalization of the likelihood, since the integration variables are the parameters. The integrals can be expressed by the maximum value of the integrand times an effective volume V_{L}^{α} of the likelihood. The maximum value defines the maximum likelihood (ML) value of the parameters. We therefore obtain

$$o_{\mathrm{BF}} \approx \frac{P(D|a_{\mathrm{ML}}^{(1)}, M^{(1)}, \mathcal{I})}{P(D|a_{\mathrm{ML}}^{(2)}, M^{(2)}, \mathcal{I})} \cdot \frac{V_{\mathrm{L}}^{(1)}}{V_{\mathrm{L}}^{(2)}} \cdot \frac{V_{\mathrm{pr}}^{(2)}}{V_{\mathrm{pr}}^{(1)}}.$$

The first ratio compares the goodness of fit of the two models, while the last two factors constitute Ockham's razor, or rather the Ockham factor

$$\mathrm{OF} = \frac{\frac{V_{\mathrm{pr}}^{(2)}}{V_{\mathrm{L}}^{(2)}}}{\frac{V_{\mathrm{pr}}^{(1)}}{V_{\mathrm{L}}^{(1)}}}.$$

The Ockham factor penalizes complexity

We will demonstrate next that Ockham's razor penalizes complexity. The dimensionless quantity $V_{\mathrm{pr}}^{(\alpha)} / V_{\mathrm{L}}^{(\alpha)}$ is the ratio of uncertainties in the parameters before and after the experiment has been taken into account. We can introduce an effective (mean) uncertainty ratio $\kappa^{(\alpha)}$ per parameter in model α and express the ratio by

$$\frac{V_{\mathrm{pr}}^{(\alpha)}}{V_{\mathrm{L}}^{(\alpha)}} = \left(\kappa^{(\alpha)}\right)^{L_\alpha}.$$

The uncertainty ratio κ is certainly significantly greater than 1, because decent data will reduce the uncertainty. The Ockham factor becomes

$$\mathrm{OF} = \frac{\left(\kappa^{(2)}\right)^{L_2}}{\left(\kappa^{(1)}\right)^{L_1}}.$$

We now consider two special cases. Firstly, $\kappa^{(1)} = \kappa^{(2)} := \kappa$, i.e. roughly speaking the prior uncertainty per parameter is the same in both models but the number of parameters differs, $L^{(1)} \neq L^{(2)}$. Secondly, the number of parameters is the same in both models, $L^{(1)} = L^{(2)} := L$, but the prior uncertainty in the parameters shall be different, resulting in $\kappa^{(1)} \neq \kappa^{(2)}$. In the first case, we obtain

$$\text{OF} = \kappa^{(L_2 - L_1)}.$$

Hence, if model II has more parameters (is more complex) than model I, the Ockham factor is significantly greater than 1 and the less complex model I is strongly favoured. In the second case the Ockham factor reads

$$\text{OF} = \left(\frac{\kappa^{(2)}}{\kappa^{(1)}}\right)^{L}.$$

If model II is again the more complex model, i.e. its prior volume is greater $\left(\frac{\kappa^{(2)}}{\kappa^{(1)}} > 1\right)$, then the Ockham factor again favours model I.

The Ockham factor demystified

Somehow it looks like magic that probability theory favours simplicity, a characteristic that has to be put in by hand in frequentist statistics and which strongly influences the result. Where does it really come from? In order to gain a deeper understanding of the origin of Ockham's razor, we simplify the model-selection problem even further and map it onto a box model. Box models are a generalization of classical urn models, introduced by Freedman, Pisani and Purves in [75]. They map all typical statistical problems onto box models, by viewing them as a process of pulling tickets from a box. We suppose there are two types of boxes, type 1 and type 2. A type-1 box contains $L^{(1)}$ different tickets marked by the integers $1, 2, \ldots, L^{(1)}$. Similarly, in a type-2 box, there are tickets labelled by the integers $1, 2, \ldots, L^{(2)}$. Both boxes have the same total number M of tickets, which corresponds to the normalization. M has to be chosen such that $n_1^{(\alpha)} := M/L^{(\alpha)}$ are integers, as they specify how often label 1 occurs in boxes of type α.

Now we consider the following task. A single ticket is selected at random from an unknown box and it carries the integer value 1. Based on this information, we have to infer which type of box it came from. To this end, we identify the types of box with models $M^{(\alpha)}$ and compute the odds ratio

$$o = \frac{P(n|M^{(1)}, \mathcal{I})}{P(n|M^{(2)}, \mathcal{I})} \frac{P(M^{(1)}|\mathcal{I})}{P(M^{(2)}|\mathcal{I})}.$$

Here the Bayes factor is one as both boxes contain label 1. We assume that both boxes are equally likely, which corresponds to prior experience that both types of box are equally often realized in nature. We assume that in nature there are in total $2N$ boxes, N of type 1 and N of type 2. Within the universe of all boxes, tickets with number 1 occur $N_1^{(1)} = Nn_1^{(1)}$ times in boxes of type 1 and $N_1^{(2)} = Nn_2^{(2)}$ times in boxes of type 2. The total number of tickets carrying label 1 is therefore $N_1 = N_1^{(1)} + N_1^{(2)}$. The probability (relative frequency) that the selected ticket comes from a type-α box is therefore, according to the classical definition,

$$P_\alpha = \frac{N_1^{(\alpha)}}{N_1^{(1)} + N_1^{(2)}}.$$

Hence, the odds ratio is

$$o = \frac{N_1^{(1)}}{N_1^{(2)}} = \frac{\frac{NM}{L^{(1)}}}{\frac{NM}{L^{(2)}}} = \frac{L^{(2)}}{L^{(1)}},$$

which is the ratio of the number of different tickets in the two types of box. The greater the variety of tickets ($L^{(\alpha)}$), the more flexible is our model. That is, the more flexible model (of type 2) can explain more experimental data (different integers on the tickets). But if these models are equally often realized in nature, which is claimed by the prior odds, then the more flexible model contributes fewer samples of one species (integers). Since a number 1 ticket can be found in both boxes, both models fit the data equally well and the complexity is penalized, as is the case in the previous example. On the contrary, if the experiment yields an integer not contained in a type-1 box, then the odds ratio is zero as model 1 cannot explain the experiment. So we see that there is a fairly simple explanation for why Ockham's razor is an integral part of probability theory. Ockham's razor depends on the prior probabilities and is therefore missing in approaches like frequentist statistics that disdain any prior knowledge. The data alone can never provide this information and the likelihood will always favour the most complex model. To overcome this shortcoming, frequentist statistics has to invoke an ad hoc element, the so-called Akaike's information criterion [2], in order to enforce the correct behaviour.

4

Combinatorics

Combinatorics is one of the key elements of probability theory, as well as in statistical physics. Although counting the number of elements in a set is a rather broad mathematical topic, many of the problems that arise in applications have a relatively simple combinatorial description. In the following we address the most important ones. A comprehensive treatment can be found in many textbooks, see e.g. [67, 196].

4.1 Preliminaries

4.1.1 Pairs

Definition 4.1 (Pairs) *The number of all possible pairs* (a_i, b_k) *of* m *elements* $a_1, a_2, \ldots, a_m$ *of type a and n elements* $b_1, b_2, \ldots, b_n$ *of type b is given by*

NUMBER OF PAIRS

$$N_P = n * m. \tag{4.1}$$

An example is given in Table 4.1.

Proof: There are m different possibilities to select an element of type a and for each selection there are n different possibilities for an element of type b. □

Example: Playing cards. Each card is defined by a unique combination of card value and suit. As there are 4 suits and 13 different values, a complete deck of cards consists of $4 * 13 = 52$ playing cards.

4.1.2 Multiplets

Definition 4.2 (Multiplets) *The generalization to more than two sets is given by multiplets. The number of multiplets (r-tuple with one element of each of the r different sets)*

Table 4.1 Pairing of the elements $\{a_1, a_2\}$ and $\{b_1, b_2, b_3\}$

1	2	3	4	5	6
(a_1, b_1)	(a_1, b_2)	(a_1, b_3)	(a_2, b_1)	(a_2, b_2)	(a_2, b_3)

Table 4.2 Multiplets of the elements $\{a_1, a_2\}$, $\{b_1, b_2, b_3\}$ and $\{c_1, c_2\}$

1	2	3	4	5	6
(a_1, b_1, c_1)	(a_1, b_1, c_2)	(a_1, b_2, c_1)	(a_1, b_2, c_2)	(a_1, b_3, c_1)	(a_1, b_3, c_2)
7	8	9	10	11	12
(a_2, b_1, c_1)	(a_2, b_1, c_2)	(a_2, b_2, c_1)	(a_2, b_2, c_2)	(a_2, b_3, c_1)	(a_2, b_3, c_2)

$(a_{j_1}, b_{j_2}, \ldots, x_{j_r})$, *provided there are* n_1 *elements* $a_1, a_2, \ldots, a_{n_1}$ *of type* a, n_2 *elements* $b_1, b_2, \ldots, b_{n_2}$ *of type* b, $\ldots$, n_r *elements* $x_1, x_2, \ldots, x_{n_r}$ *of type* x *is given by:*

NUMBER OF MULTIPLETS

$$N_M = n_1 * n_2 * \cdots * n_r = \prod_{i=1}^{r} n_i . \tag{4.2}$$

An example is given in Table 4.2.

Proof: For $r = 2$ the equation holds because it is identical to the previous case for pairs. For $r = 3$ we consider the $n_1 * n_2$ pairs (a_i, b_j) as elements u_l, $l = 1, 2, \ldots, n_1 * n_2$ of a new set. Each triplet (a_i, b_j, c_k) can then be considered as a pair (u_l, c_k). The number of pairs is then given by $n_1 * n_2 * n_3$. Repetition of the argument yields the result for arbitrary r. □

Example 1: We consider the Ising model, which is the simplest model to describe magnetism of localized magnetic moments. The moments are located on a regular lattice with n sites, and each of the moments can only assume one of two values ± 1. The orientation of each moment j corresponds to the two elements of type j, and therefore the total number of configurations is $N_M = 2^n$.

Example 2: We consider the number of different arrangements (multiplets) of r objects in n boxes. For each object there are n possibilities. This results in $n * n * \cdots * n = n^r$ multiplets. We will return to this example in Section 4.3 [p. 59].

Example 3: The same line of reasoning can be used for die problems as well, in particular for the common game of chance that inspired Fermat and Pascal (see Section 1.1 [p. 3]) to invent probability theory. In this game the bank wins if there is at least once face value '6' in four rolls of a die. We solve the problem for the general case that the die is rolled r times. We can identify the sides of the die with the number of boxes $n = 6$ and the number of rolls r with the number of objects. There are in total 6^r different outcomes, out of which 5^r do not contain the face value '6'. If it is a fair die then the probability for having at least once face value '6' in r rolls is given by

$$P_r = 1 - \left(\frac{5}{6}\right)^r . \tag{4.3}$$

Especially for the historic game, the probability is

$$P_4 = \left(\frac{5}{6}\right)^4 = 0.5177 \tag{4.4}$$

and therefore the chance for the bank to win is slightly greater than 50%.

Also interesting is the probability for having at least once face value '6' in six rolls of the die. In this case we obtain $P_6 = 0.6651 < 2/3$. The result might seem counterintuitive at first glance. Doesn't that imply that the other face values occur more frequently? No, because there are sequences with no face value '6', but there are others where it occurs more than once.

Example 4: Another common problem is the following. Find the probability that a selected box stays empty if r particles are randomly distributed among n boxes. The probability is again

$$P(\text{box is empty}) = \left(\frac{n-1}{n}\right)^r = \left(1 - \frac{1}{n}\right)^r$$

and can be approximated by

$$P(\text{box is empty}) = e^{r \ln(1-\frac{1}{n})} = e^{-r/n + O(n^{-2})} \approx e^{-r/n} .$$

4.1.3 Ordered samples

We consider a set of n elements $a_1, a_2, \ldots, a_n$, often also termed a population. From this population a sample of size r is drawn: $a_{j_1}, a_{j_2}, \ldots, a_{j_r}$. Since we are interested in the order in which the elements are drawn, we call the sample an ordered sample.[1] There are essentially two different possibilities:

(a) The elements are copied from the population. This corresponds to sampling with replacement in urn experiments.

(b) The elements are deleted from the population. This corresponds to sampling without replacement in urn experiments. In this case the size of the sample r cannot be larger than the size of the population n.

[1] However, the elements are not sorted!

Table 4.3 Ordered sample of size 2 drawn from the population $\{a_1, a_2, a_3\}$

With replacement								
1	2	3	4	5	6	7	8	9
(a_1, a_1)	(a_1, a_2)	(a_1, a_3)	(a_2, a_1)	(a_2, a_2)	(a_2, a_3)	(a_3, a_1)	(a_3, a_2)	(a_3, a_3)
Without replacement								
1	2	3	4	5	6			
(a_1, a_2)	(a_1, a_3)	(a_2, a_1)	(a_2, a_3)	(a_3, a_1)	(a_3, a_2)			

In Table 4.3, ordered samples are illustrated in terms of a population of three elements from which two are drawn with and without replacement respectively. In the case of sampling with replacement, the number of different samples can be calculated very easily, since for each draw there are n choices, resulting in n^r possibilities.

Without replacement there are n different options for the first draw, then $n-1$ (because one element of the population is already missing) and so on, up to the rth draw for which only $n-r+1$ options remain. The total number of possibilities is

$$n(n-1)(n-2)\cdots(n-r+1) = \frac{n!}{(n-r)!}.$$

In summary, we have

NUMBER OF ORDERED SAMPLES OF SIZE r FROM A POPULATION OF SIZE n

$$N_{\text{op}}^{\text{mz}} = n^r \qquad \text{with replacement,} \qquad (4.5a)$$

$$N_{\text{op}}^{\text{oz}} = \frac{n!}{(n-r)!} \qquad \text{without replacement.} \qquad (4.5b)$$

An example is given in Table 4.3.

A special case of a sample without replacement is given by $r = n$. In this case the sample is also a possible arrangement (permutation) of the elements of the population, implying that

NUMBER OF PERMUTATIONS

$$N_{\text{perm}} = n!. \qquad (4.6)$$

Table 4.4 Permutations of the elements of the population $\{a_1, a_2, a_3\}$

1	2	3	4	5	6
(a_1, a_2, a_3)	(a_2, a_3, a_1)	(a_3, a_1, a_2)	(a_3, a_2, a_1)	(a_2, a_1, a_3)	(a_1, a_3, a_2)

See Table 4.4 .

Examples

Suppose that we take a sample of size r with replacement from a population of size n. What is the probability that no element of the sample has been drawn twice or more? The total number of possible outcomes is given by the number of ordered samples with replacement, equation (4.5a), as n^r. The number of favourable events, where no element occurs twice, is equivalent to the number of ordered samples without replacement, given in equation (4.5b). Then the sought-for probability is

$$P = \frac{n!}{(n-r)!}\left(\frac{1}{n}\right)^r . \tag{4.7}$$

There are various interesting applications of this result.

(a) Fraud detection by the digits Digit distributions are frequently used to detect tax-payers' noncompliance and credit card frauds, etc. [44]. Typically, the first digits are analysed, which are supposed to have a logarithmic distribution (Benford's law) [83]. The last digits, in contrast, are assumed to be uniformly distributed [44]. Let us assume the last digits in a large collection of numbers with many digits are really uniformly distributed, then the preceding result provides a simple (although incomplete) screening procedure. Picking the last 5 digits corresponds to a sample of size $r = 5$ of a population with $n = 10$ elements (the digits $0, 1, \ldots, 9$). For this setting equation (4.7) yields a probability $P = \frac{10!}{5!}10^{-5} = 0.3024$ that no digit occurs twice in a block of 5 digits. We perform repeated experiments based on 100 numbers with 16 digits and count how often the last 5 digits contain no duplicates. The analysis yields the sequence 30, 27, 30, 34, 26, 32, 37, 36, 26, 31, 36, 32. The sample mean is 0.31, in surprising agreement with the preceding result. However, this is just a 'forward' calculation and it would be premature to jump to any conclusions. For this purpose we would need a thorough hypothesis test.

(b) Probability for complete coverage What is the probability for obtaining each element of a population once in a sample that has the size of the population when drawn with replacement? More concretely, what is the probability for getting each face value 1–6 when the die is rolled 6 times? In other words, we are interested in the probability that no duplicates occur in a sample of size n drawn with replacement from a population of size n. The result is given by equation (4.7) with $r = n$, i.e. $P = n!/n^n$. This implies that the probability for obtaining all 6 numbers when rolling a die 6 times is only $P = 0.015$. Therefore, although the probability for obtaining each number is equal, a uniform frequency of the values is highly unlikely.

Table 4.5 Subpopulations of size 2 of the population $\{a_1, a_2, a_3\}$

1	2	3
(a_1, a_2)	(a_1, a_3)	(a_2, a_3)

4.2 Partitions, binomial and multinomial distributions

As before, we consider populations of size n. Two populations are said to be different if there is at least one element in one population which is not present in the other. The number of different subpopulations of size r, with $r \leq n$, of a population of size n is given by:

NUMBER OF SUBPOPULATIONS OF SIZE r OF A POPULATION OF SIZE n

$$N^{\text{nr}} = \binom{n}{r} = \frac{n!}{r!(n-r)!} \qquad \text{without replacement,} \tag{4.8a}$$

$$N^{\text{r}} = \binom{n+r-1}{r} \qquad \text{with replacement.} \tag{4.8b}$$

The proof of equation (4.8b) is postponed to Section 4.3.1 [p. 59].

The proof of equation (4.8a) follows from the number of ordered samples without replacement of size r from populations of size n (equation (4.5b)), which is $N^{\text{oz}}_{\text{op}} = \frac{n!}{(n-r)!}$. However, since we are no longer interested in a specific order, each subpopulation is counted $r!$ times, according to the different permutations that are considered different in an ordered sample. Then we find equation (4.8a) immediately. Expressed differently, there are $\binom{n}{r}$ different subsets of size r of a set of size n. A simple example is given in Table 4.5.

Definition 4.3 (Binomial coefficients) *The quantity*

$$\binom{n}{r} := \frac{n!}{r!(n-r)!} \tag{4.9}$$

is termed a binomial coefficient.

There are as many subpopulations of size r as there are of size $n-r$, because for each subpopulation of size r the remaining elements of the population form a subpopulation of size $n-r$, hence the binomial coefficients have the property

$$\binom{n}{r} = \binom{n}{n-r}.$$

It is expedient to define

$$\binom{n}{0} = 1, \qquad 0! = 1 \qquad \text{and} \qquad \binom{n}{r} = 0 \text{ for } r > n.$$

Binomial coefficients occur in very many applications. One particularly important application is the computation of combinations. C_r^N is also called the 'number of r-combinations' from a population of size N, which is commonly denoted by C_r^N:

NUMBER OF COMBINATIONS OF r ELEMENTS FROM A POPULATION OF SIZE N

$$C_r^N := \binom{N}{r}. \tag{4.10}$$

Suppose there are N basis functions and $n_\uparrow$ electrons with spin up. The number of ways to allocate the up-electrons to the basis functions is

$$N_{\text{many-body basis}} = \binom{N}{n_\uparrow}.$$

This can easily be seen by identifying the distinguishable basis functions as the elements of the population, and the subpopulation of size $n_\uparrow$ contains those functions that are occupied by the electrons. Since electrons are indistinguishable, the order does not matter. If, in addition, there are $n_\downarrow$ electrons with spin down, then the size of the many-body Hilbert space of the particle sector specified by $(n_\uparrow, n_\downarrow)$ is

$$N_{\text{HS}} = \binom{N}{n_\uparrow}\binom{N}{n_\downarrow}.$$

This number increases rapidly with the number of basis functions N. For example, for $N = 16$ and $n_\uparrow = n_\downarrow = 8$, what is called half-filling, the Hilbert space is already $N_{\text{HS}} = 1.7 \times 10^8$, which shows that quantum-mechanical strongly correlated many-body problems are a real challenge. In this book we will have ample opportunity to see further applications of the binomial coefficient in combinatorial problems.

Another occurrence is:

BINOMIAL THEOREM

$$(a+b)^n = \sum_{r=0}^{n} \binom{n}{r} a^r b^{n-r}. \tag{4.11}$$

Proof: Explicit multiplication of the left-hand side of equation (4.11) results in sums of products of the factors a and b in all possible sequences with a total of n factors, e.g.

$$\underbrace{a\ a\ b\ a\ b\ b\ b\ a\ b\ \ldots\ a}_{n \text{ factors}}.$$

For a given number of elements of type a, resulting in $a^r b^{n-r}$, there are $\binom{n}{r}$ different sequences of as and bs. □

Binomial distribution

Another important application of the binomial coefficient is the binomial distribution. Suppose we perform a random experiment with a binary outcome, e.g. red/blue balls, head/tail, defect/working, etc., which is called a Bernoulli trial. Independent repetition of such binary experiments is a Bernoulli process. The probability for the first outcome (red/head/defect) shall be p, then the probability for the alternative outcome is $q = 1 - p$. The probability for observing a specific sequence of r times the first outcome in a sample of size n is given by $p^r q^{n-r}$. If the order is immaterial, then – since there are $\binom{n}{r}$ possible arrangements – we obtain

BINOMIAL DISTRIBUTION

$$P_B(r|n, p) = \binom{n}{r} p^r (1-p)^{n-r}, \tag{4.12a}$$

$$\langle r \rangle = n\ p, \tag{4.12b}$$

$$\mathrm{var}(r) = n\ p\ (1-p). \tag{4.12c}$$

We can verify the correct normalization of the binomial distribution exploiting the binomial theorem in equation (4.11) and $p + q = 1$:

$$\sum_{r=0}^{n} \binom{n}{r} p^r q^{n-r} = (p+q)^n = 1.$$

Proof of the mean:

$$\langle r \rangle = \sum_{r=0}^{n} r \binom{n}{r}\ p^r\ (1-p)^{n-r} = \sum_{r=0}^{n} r\ \frac{n!}{r!\ (n-r)!}\ p^r\ (1-p)^{n-r}$$

$$= \sum_{r=1}^{n} n \frac{(n-1)!}{(r-1)!\,((n-1)-(r-1))!}\, p\, p^{r-1}\,(1-p)^{(n-1)-(r-1)}$$

$$= n\,p \underbrace{\sum_{k=0}^{n-1} \binom{n-1}{k} p^k\,(1-p)^{(n-1)-k}}_{=1} = n\,p.$$

This elementary proof was a bit tedious. For the computation of the variance or even more complex expressions we need a better approach. Let us take a second glance at the computation of the mean:

$$\langle r \rangle = \sum_{r=0}^{n} \binom{n}{r} r\, p^r\,(1-p)^{n-r}. \tag{4.13}$$

Here the term rp^r can also be written as

$$rp^r = p\frac{\partial}{\partial p}\, p^r. \tag{4.14}$$

In order to be able to pull the operator $p\frac{\partial}{\partial p}$ out of the sum, we introduce temporarily a new independent variable $q = (1-p)$. Then

$$\langle r \rangle = p\frac{\partial}{\partial p} \sum_{r=0}^{n} \binom{n}{r} r\, p^r\, q^{n-r} \bigg|_{q=1-p}.$$

It is important to insert $q = 1-p$ only after the derivative has been performed, otherwise we would get an additional term that is not present in equation (4.13). The remaining step is simply the derivative of the binomial theorem:

$$\langle r \rangle = p\frac{\partial}{\partial p}\Big(p+q\Big)^n\bigg|_{q=1-p} = p\,n\Big(p+q\Big)\bigg|_{q=1-p} = p\,n. \qquad \square$$

Proof of the variance: Now we apply this tool to the computation of the variance

$$\left\langle r^2 \right\rangle = \sum_{r=0}^{n} r^2 \binom{n}{r} p^r\,(1-p)^{n-r} = \Big(p\frac{\partial}{\partial p}\Big)\Big(p\frac{\partial}{\partial p}\Big) \sum_{r=0}^{n} \binom{n}{r} p^r\, q^{n-r} \bigg|_{q=1-p}$$

$$= \Big(p\frac{\partial}{\partial p}\Big)\Big(p\frac{\partial}{\partial p}\Big)\Big(p+q\Big)^n\bigg|_{q=1-p} = n\,p\frac{\partial}{\partial p}\, p\Big(p+q\Big)^{n-1}\bigg|_{q=1-p}$$

$$= n\,p\,\left[\Big(p+q\Big)^{n-1} + (n-1)\,p\,\Big(p+q\Big)^{n-2}\right]\bigg|_{q=1-p}$$

$$= n\,p\,\left[1-p+n\,p\right] = n\,p\,(1-p) + \langle r \rangle^2. \qquad \square$$

Multinomial case

In case of more than two alternative outcomes in a single trial, the binomial coefficient can be generalized to the multinomial coefficient. Suppose there are integer numbers $n_1, n_2, \ldots, n_k$ which add up to N. The number of possible partitionings of a population of size N into k subpopulations of size n_i, with $i = 1, \ldots, k$ such that $\sum_{i=1}^{k} n_i = N$, is given by

MULTINOMIAL COEFFICIENTS

$$N(\boldsymbol{n}|n, k) = \binom{n}{\boldsymbol{n}} = \frac{N!}{\prod_{i=1}^{k} n_i!}. \tag{4.15}$$

Proof: We generate all possible subpopulations in the following way. We align the elements of the original population in a row and assign the first n_1 elements to subpopulation 1, the next n_2 elements to subpopulation 2, and so on up to the last n_k elements, which are assigned to subpopulation k. In total there are $n!$ different permutations of the population and hence $n!$ different assignments to the subpopulations. But permutations of the elements within a subgroup are immaterial. Therefore, the $n!$ permutations overcount the contribution of each subpopulation by a factor of $n_\alpha!$ ($\alpha = 1, 2, \ldots, k$). This is accounted for by the denominator in equation (4.15). □

The binomial coefficients are a special case of the multinomial coefficients for $k = 2$, $N(r|n) = N(\{r, n-r\}|n, 2)$.

The generalization of the binomial theorem yields the multinomial theorem:

MULTINOMIAL THEOREM

$$(a_1 + a_2 + \cdots + a_k)^n = \widetilde{\sum_{n_1, n_2, \ldots, n_k = 0}^{n}} \binom{n}{\boldsymbol{n}} a_1^{n_1} a_2^{n_2} \ldots a_k^{n_k}. \tag{4.16}$$

The tilde indicates that the summation runs only over those numbers $\boldsymbol{n}$ which add up to n. The proof is along the same lines as that of the binomial theorem, and we leave it as an exercise for the reader.

Next we generalize the binomial distribution to the multinomial case. The multinomial distribution provides the probability to find n_α objects of n in box α, if p_α is the probability to find an object in box α in a single trial.

Multinomial distribution

$$P(\boldsymbol{n}|n,k) = \binom{n}{\boldsymbol{n}} \prod_{i=1}^{k} p_i^{n_i}, \tag{4.17a}$$

$$\langle n_i \rangle = n\ p_i, \tag{4.17b}$$

$$\text{var}(n_i) = n\ p_i\ (1 - p_i), \tag{4.17c}$$

$$\text{for } i \neq j : \text{cov}(n_i, n_j) = n\ p_i\ (1 - p_j). \tag{4.17d}$$

The normalization of the multinomial distribution follows immediately from equation (4.16). The result for mean and variance of the occupation n_i of box i can be understood very easily by combining all the other boxes to a single one. Then we are back to the binomial situation. The new element is the covariance $\text{cov}(n_i, n_j)$ for $i \neq j$.

Proof for covariance: The computation of $\langle n_i\ n_j \rangle$ can be performed based on equation (4.14), assuming temporarily that the p_j are independent variables. After all derivatives are performed, we use $\sum_j p_j = 1$, then:

$$\begin{aligned}
\langle n_i\ n_j \rangle &= \widetilde{\sum_{n_1,n_2,\ldots,n_k=0}^{n}} n_i\ n_j \binom{n}{\boldsymbol{n}} \prod_{l=1}^{k} p_l^{n_l} \\
&= p_i \frac{\partial}{\partial p_i}\ p_j \frac{\partial}{\partial p_j}\ (p_1 + p_2 + \cdots + p_k)^n \Big|_{\sum_j p_j = 1} \\
&= n(n-1)\ p_i\ p_j = \langle n_i \rangle \langle n_j \rangle - n\ p_i\ p_j.
\end{aligned}$$

□

Examples for binomial and multinomial problems

1-D random walk

The Galton board consists of nails in ordered positions, as visualized in Figure 4.1. Balls are dropped from the top, and bounce left and right with equal probability as they hit the nails. Eventually, they are collected in the trays at the bottom. We are interested in the distribution of the balls in the collecting trays. Initially, the ball is at position $x = 0$. After the first bounce, the ball is equally likely to be at $x = \pm 1$. Later on, at step i, the ball will be at position x_i and will alter its position by $\Delta x = \pm 1$. Therefore, it is identical to a repeated trial where one of the events ($\Delta x = \pm 1$) is happening with the same probability.

Suppose that after n repetitions the event $\Delta x = +1$ has happened k times (which implies that event $\Delta x = -1$ has happened $n - k$ times), then the location of the ball is

$$x_n(k) = (+1) \cdot k + (-1) \cdot (n-k) = 2k - n.$$

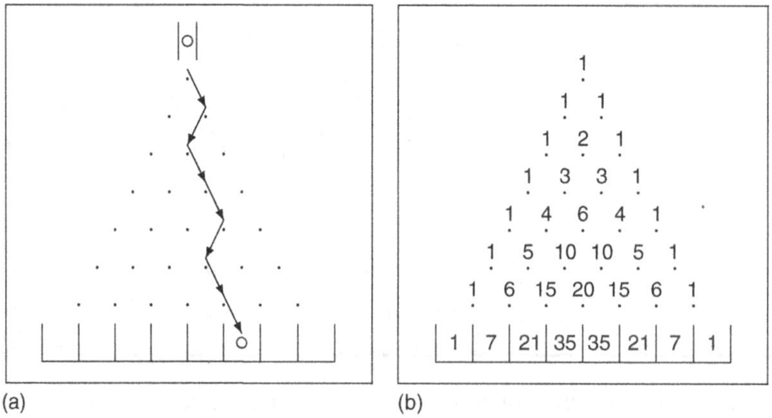

Figure 4.1 (a) Galton's board, (b) Pascal's triangle.

In the following we will assume that probability p for a right move and $q = 1 - p$ for a left move can be different. Therefore, the probability for the ball to be at location $x_n(k)$ is equal to the probability for $k = (x_n(k) + n)/2$ bounces to the right, i.e.

$$P(x_n|n, p) = \binom{n}{k} p^k q^{n-k} \Big|_{k=\frac{x_n+n}{2}} . \tag{4.18}$$

Please note that both n and $x_n(k)$ have to be either even or odd. The binomial coefficient $\binom{n}{k}$ denotes the number of different paths resulting in a displacement of $i = x_n(k)$ in the nth step. To reach position i in the nth step requires the ball to be at position $i - 1$ or $i + 1$ in the previous step. The total number of possibilities, $N_{i,n}$, to be at position i in n steps is therefore given by

$$N_{i,n} = N_{i-1,n-1} + N_{i+1,n-1}.$$

Of course, there is no contribution from outside the edge, i.e.

$$N_{i,n} = 0 \qquad \text{for } |i| > n.$$

So we can successively generate all values $N_{i,n}$ by starting with the uppermost pin ($N_{0,0} = 1$). We obtain the binomial coefficients displayed in Figure 4.1(b), which is also known as Pascal's triangle. The distribution of balls in the nth row is therefore given by the binomial coefficients.

Next we want to modify the process such that we can describe 'Brownian motion'. The ball turns into a particle, which is allowed to rest between decisions. In other words for each iteration step, corresponding to a time step Δt, there are three possibilities: right move with probability p_1, left move with probability p_2, and no move with probability p_3.

The position after N time steps is given by $x = n_1 - n_2$, where n_1 (n_2) is the number of steps taken to the right (left). The mean position is then

$$\langle x \rangle = \langle n_1 \rangle - \langle n_2 \rangle \overset{(4.17b)}{=} N\big(p_1 - p_2\big).$$

Clearly, for small Δt the probability that the particle hops is proportional to Δt. We define $p_1 - p_2 = \Delta t \; v_d$, with v_d being the 'drift velocity'. Then

$$\langle x \rangle = v_d \; N \; \Delta t = v_d \; t,$$

where t is the total time elapsed in N iterations. Next we compute the variance of x:

$$\langle x^2 \rangle = \langle n_1^2 \rangle + \langle n_2^2 \rangle - \langle n_1 n_2 \rangle.$$

We introduce $n_\alpha = \langle n_\alpha \rangle + \Delta n_\alpha$ and find, along with equations (4.17c) and (4.17d),

$$\begin{aligned} \langle x^2 \rangle &= \langle (\Delta n_1)^2 \rangle + \langle (\Delta n_2)^2 \rangle - 2\langle \Delta n_1 \Delta n_2 \rangle + \langle n_1 \rangle^2 + \langle n_2 \rangle^2 - 2\langle n_1 \rangle \langle n_2 \rangle \\ &= N\left(p_1(1-p_1) + p_2(1-p_2) + 2p_1 p_2\right) + \langle x \rangle^2. \end{aligned}$$

That leads to

$$\langle (\Delta x)^2 \rangle = N\left[(p_1 + p_2) - (p_1 - p_2)^2\right].$$

The second contribution in the square bracket vanishes in the limit $\Delta t \to 0$ as it is proportional to Δt^2. The first term contains the probability that the particle hops at all in Δt, which is $p_1 + p_2 := \Delta t D$. The proportionality constant D is the 'diffusion coefficient'. The variance is therefore

$$\langle (\Delta x)^2 \rangle = D \; t.$$

As is usual for diffusion, the width increases as $\sqrt{t}$.

4.3 Occupation number problems

4.3.1 Distribution of objects in boxes

In many applications one has to count the possibilities to distribute objects among boxes under some (problem-specific) constraints. This situation is typical for quantum particles, but it is also common in the statistical analysis of star populations, traffic accidents, and many more. We introduce the k-tuples of occupation numbers $\boldsymbol{n} = \{n_1, n_2, \ldots, n_k\}$, where n_i denotes how many objects are in box i. The total number of particles is fixed:

$$\sum_i n_i = N. \tag{4.19}$$

Two distributions differ only if the k-tuples $\boldsymbol{n}$ of the occupation numbers are different. The number of distinct distributions of N objects in k boxes is given by

$$A_{N,k} = \binom{N+k-1}{N} = \binom{N+k-1}{k-1}. \tag{4.20}$$

Proof: We represent the objects by open circles $\bigcirc$ and the walls between the boxes by vertical bars $\big|$. For example, the distribution of $N = 10$ objects in $k = 5$ boxes with occupation-tuple $= \{3, 0, 2, 1, 4\}$ is represented by

$$\Big[\bigcirc\bigcirc\bigcirc\,|\;\;|\bigcirc\bigcirc|\bigcirc|\bigcirc\bigcirc\bigcirc\bigcirc\Big].$$

The outer walls are depicted by '[' and ']'. We can uniquely generate all possible occupation number distributions (k-tuples) by creating all combinations of the N circles and the $k-1$ inner walls. In total there are $N+k-1$ positions which are occupied by either one of the circles or one of the vertical bars. Consequently, according to equation (4.10) [p. 53], the number of combinations is given by equation (4.20). □

With equation (4.20) we can also prove equation (4.8b) [p. 52] in a straightforward manner.

Proof: The selection of a subpopulation of size r from a population of size n with replacement is equivalent to a distribution of r objects in n boxes, where the occupation number n_j specifies how often the element j of the population occurs in the subpopulation. Using equation (4.20), we immediately get equation (4.8b) [p. 52]. □

4.3.2 *Probabilities for particle distributions*

Next we want to assign probabilities to the occupation number distributions $\boldsymbol{n}$. We start out with classical distinguishable particles. We consider N particles, which shall be distributed among k boxes. If the particles are distinguishable, the situation is that of forming k subpopulations of a population of size N without replacement, as described in equation (4.15). We assume equal prior probability $p_j = 1/k$ for each box. Then the probability for the occupation numbers $\boldsymbol{n}$ is given by the multinomial distribution equation (4.17a), with $p_j = 1/k$. That leads to:

Distribution of identical classical objects (particles)

$$P_B(\boldsymbol{n}|N,k) = \binom{N}{\boldsymbol{n}} k^{-N}. \tag{4.21}$$

Classical statistical physics is based on this distribution. It did come as a big surprise in the early twentieth century that identical particles like electrons or photons did not obey this

Table 4.6 Comparison of different particle statistics (classical, bosonic, fermionic). The symbols a and b stand for distinguishable and x for indistinguishable particles

				Classical particles					
	1	2	3	4	5	6	7	8	9
s_1	a,b	–	–	a	b	a	b	–	–
s_2	–	a,b	–	b	a	–	–	a	b
s_3	–	–	a,b	–	–	b	a	b	a
				Bosons					
	1	2	3	4	5	6			
s_1	x,x	–	–	x	x	–			
s_2	–	x,x	–	x	–	x			
s_3	–	–	x,x	–	x	x			
				Fermions					
	1	2	3						
s_1	x	x	–						
s_2	x	–	x						
s_3	–	x	x						

seemingly 'obvious' distribution. It turned out that quantum particles differ in a fundamental way. Classical particles, even if they are identical in all features like size, mass, charge, etc., are still distinguishable. Identical quantum particles, on the contrary, are strictly indistinguishable. They lose their identity as soon as there is more than one of them in the system. Quantum physics then tells us that there are two types of *particle statistics*, bosonic and fermionic. In the case of bosons, there can still be any number of them in a specific quantum state, while the indistinguishability in the case of fermions entails that there are never two or more of them in the same state, what is also referred to as the 'Pauli principle'. The consequences are illustrated in Table 4.6. There are three states/boxes denoted by s_1, s_2, s_3 and two particles. In the classical case they are distinguishable and hence denoted by different symbols a and b, respectively. In the quantum case identical particles are indistinguishable and they are therefore denoted by the same symbol x. In the classical case there are nine different distributions for the two particles in the three states. The configurations (4) and (5) differ by the individual particles that occupy s_1 and s_2. Such a distinction is not possible for quantum particles. These two configurations correspond to a single configuration (4) in the bosonic case. Similarly for the classical configurations (6), (7) and (8), (9). They are mapped onto (5) and (6) in the bosonic case. Apparently, for bosons all that matters are the occupation numbers. Consequently the number of configurations to distribute N bosons in k states is given by $A_{N,k}$ (equation (4.20)). Each configuration is equally likely, so the probability for one configuration is simply $1/A_{N,k}$.

When it comes to fermions, the bosonic configurations (1)–(3) are forbidden by the 'Pauli principle', which leaves only three possible configurations for fermions. Distributing N fermions in k boxes results in N singly occupied boxes and $N - k$ empty boxes. Consequently, there are $\binom{k}{N}$ different fermionic configurations, which again are all equally likely. The probability for an allowed configuration is therefore $P = 1/\binom{k}{N}$. In summary, the probabilities are

$$
\begin{aligned}
P &= \binom{N}{\boldsymbol{n}} k^{-N} && \text{(classical particles)} \\
P &= \frac{N!(k-1)!}{(N+k-1)!} && \text{(bosons)} \\
P &= \frac{N!(k-N)!}{k!} && \text{(fermions)}
\end{aligned}
\tag{4.22}
$$

with the restriction $n_\alpha \in \{0, 1\}$ in the fermionic case. This shall be illustrated with a worked example. Suppose there are two dice, where the values are interpreted as boxes and the dice as particles, i.e. $N = 2$, $k = 6$. What is the probability P_7 for having a total face value of 7 when both dice are rolled? For classical dice this is the case for the face values $(1, 6)$, $(6, 1)$, $(2, 5)$, $(5, 2)$, $(3, 4)$, $(4, 3)$. The occupation-tuples for e.g. $(1, 6)$ and $(6, 1)$ are given by $\{1, 0, 0, 0, 0, 1\}$. The numerical results are

$$
\begin{aligned}
P_7 &= \frac{2!6^{-2}}{1!0!0!0!0!1!} + \frac{2!6^{-2}}{0!1!0!0!1!0!} + \frac{2!6^{-2}}{0!0!1!1!0!0!} &&= \frac{1}{6} && \text{(classical)} \\
P_7 &= 3\frac{2!5!}{7!} = \frac{3*2}{6*7} &&= \frac{1}{7} && \text{(bosonic)} \\
P_7 &= 3\frac{2!4!}{6!} = \frac{2*3}{5*6} &&= \frac{1}{5} && \text{(fermionic).}
\end{aligned}
$$

This example illustrates nicely the problems underlying the assumption of equal probability for elementary events.

4.4 Geometric and hypergeometric distributions

Suppose, similar to the situation of Bernoulli experiments, that we have a population of size n which is composed of two different subpopulations, e.g. red and black balls of size n_I and $n_{II} = n - n_I$, respectively. Widespread tasks are:

HG1. Objects are drawn sequentially (with or without replacement). Find the probability that the first object of type II occurs in the kth draw.

HG2. A sample of size k is drawn. Find the probability that it has k_I objects of the first type and $k_{II} = k - k_I$ objects of the second type.

4.4.1 First occurrence in sampling without replacement

We begin with HG1 without replacement. Then, the number of events of interest is given by the number of ordered samples without replacement of size $k-1$ from the population of type I and size n_I. According to equation (4.5b) [p. 50], this number is $\frac{n_I!}{(n_I-(k-1))!}$. In addition, the kth draw has to be of type II. This can happen in n_{II} different ways. This has to be related to the total number of all possible events, given by the number of ordered samples without replacement of size k from a population of size n, i.e. by $\frac{n!}{(n-k)!}$. Then we find

$$\frac{n_I!\,(n-k)!\,n_{II}}{n!(n_I-(k-1))!} = \frac{\binom{n_I}{k-1}\,n_{II}}{\binom{n}{k}\,k}. \tag{4.23}$$

4.4.2 First occurrence in sampling with replacement

Next we turn to HG1 with replacement. The calculation is quite similar to the preceding case. Only the expression for the number of ordered samples without replacement (equation (4.5b) [p. 50]) has to be replaced by that for sampling with replacement (equation (4.5a) [p. 50]) and the total number of all possible events is now given by n^k. With $p_I = \frac{n_I}{n}$ and $p_{II} = \frac{n_{II}}{n}$ this results in

$$\frac{{n_I}^{k-1}\,n_{II}}{n^k} = \left(\frac{n_I}{n}\right)^{k-1}\frac{n_{II}}{n} = p_I^{k-1}\,p_{II}. \tag{4.24}$$

An alternative derivation of this result is given by the following observation: The probability to draw $k_I = k-1$ consecutive times objects of type I is given by p_I^{k-1}. The probability for a subsequent draw of an object of type II is $p_{II} = 1 - p_I$. The result is the so-called

GEOMETRIC DISTRIBUTION

$$P(k_I|p_I) = p_I^{k_I}(1-p_I), \tag{4.25a}$$

$$\langle k_I\rangle = \frac{p_I}{1-p_I}, \tag{4.25b}$$

$$\operatorname{var}(k_I) = \frac{p_I}{(1-p_I)^2}. \tag{4.25c}$$

The proper normalization of the geometric distribution can be verified using the geometric series $\sum_{k=0}^{\infty} p^k = 1/(1-p)$. The relation for the mean is derived as follows:

$$\begin{aligned}\langle k_I\rangle &= (1-p_I)\sum_{k=0}^{\infty} k\,p_I^k = (1-p_I)\,p_I\frac{d}{dp_I}\sum_{k=0}^{\infty} p_I^k\\ &= (1-p_I)\,p_I\frac{d}{dp_I}\left(1-p_I\right)^{-1} = \frac{p_I}{1-p_I}.\end{aligned}$$

Similarly for the variance:

$$\left\langle k_I^2 \right\rangle = (1 - p_I)\, p \frac{d}{dp_I} p_I \frac{d}{dp_I} \left(1 - p_I\right)^{-1} = \frac{p_I + p_I^2}{(1 - p_I)^2}\,;$$

$$\Rightarrow \qquad \text{var}(k_I) = \frac{p_I}{(1 - p_I)^2}.$$

4.4.3 Sample probability for sampling without replacement

Now we address the task HG2 in Section 4.4 without replacement. That is, we are interested in the probability for a sample of size k with k_I objects of type I and $k_{II} = k - k_I$ objects of type II, with $k_I \in [0, \min(n_I, k)]$. The number of possibilities of drawing k_α objects of type α is given by $\binom{n_\alpha}{k_\alpha}$. The total number of all possible samples of size k is given by $\binom{n}{k}$. Forming the ratio we find:

HYPERGEOMETRIC DISTRIBUTION
$(k = k_i + k_{II}, n = n_I + n_{II})$

$$P(k_I|k, n_I, n_{II}) = \frac{\binom{n_I}{k_I}\binom{n_{II}}{k_{II}}}{\binom{n_I+n_{II}}{k_I+k_{II}}}, \tag{4.26a}$$

$$\langle k_I \rangle = n_I\, \frac{k}{n}, \tag{4.26b}$$

$$\text{var}(k_I) = \frac{k\,(n-k)\, n_I\, n_{II}}{(n-1)\, n^2}. \tag{4.26c}$$

The normalization is proved in Appendix equation (A.30) [p. 608]. The derivation of the mean and variance is outlined in Appendix equation (A.32) [p. 609] and Appendix equation (A.34) [p. 610], respectively.

Example: Number of fish in a lake

The hypergeometric distribution is of importance in mark-recapture techniques for population estimation, which are used in ecology to estimate population sizes of animals or in epidemiology to track the spread of diseases. For concreteness, let us consider the following example. Assume that the total number of fish in a lake is n. Of these, n_r fishes are caught, tagged with a red dot and released. After a sufficiently long period of time (to allow for mixing), k fishes are caught. What is the probability for having k_r labelled fishes in this sample if the ratio n is known? This probability is given by the hypergeometric distribution $P(k_r|k, n_r, n - n_r)$. But in real life the total number of fish, n, is not known and is actually the key object of this venture. Based on this forward probability, Bayes' theorem allows us to estimate the unknown population size in a straightforward manner.

The probability of interest is therefore $P(n|k_r, k, n_r, \mathcal{I})$, with the associated propositions

- n: The number of fish in the lake is n.
- n_r: The number of fish labelled red is n_r.
- k: The sample size is k.
- k_r: In the sample there are k_r labelled fish.

Using Bayes' theorem we find

$$P(n|k_r, k, n_r, \mathcal{I}) = \frac{1}{Z} P(k_r|n, k, n_r, \mathcal{I})\ P(n|k, n_r, \mathcal{I}).$$

The normalization constant Z is independent of n and can be computed at the very end. The first term $P(k_r|n, k, n_r, \mathcal{I})$ is the hypergeometric distribution (4.26a):

$$P(k_r|n, k, n_r, \mathcal{I}) = \frac{\binom{n_r}{k_r}\binom{n-n_r}{k-k_r}}{\binom{n}{k}}.$$

The prior probability $P(n|k, n_r, \mathcal{I})$ depends on our knowledge about the number of fish in the lake. For sure, the latter is larger than n_r and fish have a certain size so there is an upper limit $n_{\max}$ that fits into the lake. Otherwise we assume perfect ignorance, encoded in a uniform prior

$$P(n|k, n_r, \mathcal{I}) = \frac{1}{n_{\max} - n_r}\ \mathbf{1}(n_r \leq n < n_{\max}).$$

Later we will learn about better methods to derive uninformative prior probability distributions. In the following steps we consider the special case where both catches have the same size, i.e. $k = n_r$. Now we can write the probability of interest by retaining only the n dependent terms (the others are gathered in the normalization constant)

$$P(n|k_r, k, n_r, \mathcal{I}) = \frac{1}{Z'}\ \mathbf{1}(0 < n_{\max})\ \frac{\binom{n-n_r}{n_r-k_r}}{\binom{n}{n_r}}.$$

The first argument of the 'Heaviside function' has been set to zero, because the likelihood contribution vanishes anyway for $n < 2n_r - k_r$. If the sample does not contain a single labelled fish ($k_r = 0$), the probability approaches one for $n \to \infty$. This means that the result depends on the cutoff $n_{\max}$ and is therefore not significant. This can easily be seen because $k_r = 0$ is compatible with $n = \infty$. But already for $k_r = 1$ the probability decays as $1/n$ for $n \to \infty$, although the cutoff ($n_{\max}$) still has some influence. At last, for $k_r > 1$ the cutoff is irrelevant and can be set to infinity. In Figure 4.2 the distribution is plotted for several values of k_r for the specific case of $k = n_r = 40$. The distribution shrinks with increasing values of k_r. The figure also contains the range of probability where the PMF is greater than $1/e$ of its maximum. We observe that the results are very significant. This analysis can also be used to design the experiment, for example, which sample size is necessary to obtain results with a prescribed accuracy (i.e. variance of the probability distribution).

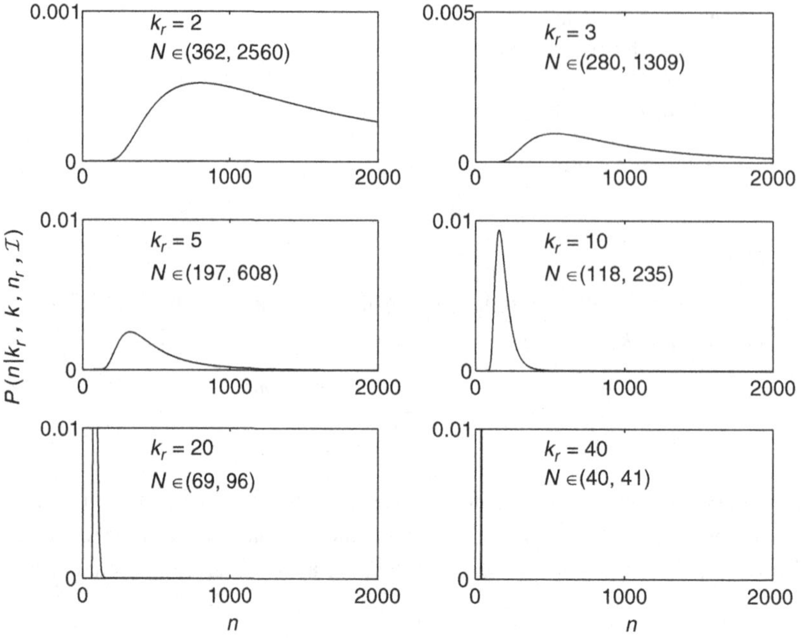

Figure 4.2 Probability $P(n|k_r, k, n_r, \mathcal{I})$ as a function of n for various values of k_r. The parameters are $k = n_r = 40$. In addition, the interval is reported in which the probability is greater than $e^{-1} \cdot P_{\max}$.

4.4.4 Sample probability by sampling with replacement

The final task in Section 4.4 is HG2 with replacement. This topic can be treated in a straightforward manner. The probability to draw an object of type I is given by $p_I = \frac{n_I}{n}$ and similarly for type II. The probability to draw a sample with a specific order of objects of type I and type II is therefore given by

$$p_I{}^{k_I}\, p_{II}{}^{k_{II}} = p_I{}^{k_I}\, (1 - p_I)^{k-k_I}.$$

However, the order is not of relevance and therefore the multiplicity of the partitions (cf. Section 4.2 [p. 52]) has to be taken into account. This results in

$$\binom{k}{k_I} p_I{}^{k_I}\, (1 - p_I)^{k-k_I},$$

which is again the binomial distribution, given in equation (4.12a) [p. 54].

4.5 The negative binomial distribution

Here we address a very useful generalization of the binomial distribution. As a motivation, we start out with the Taylor expansion of the function $(1+x)^{-\alpha}$ with a negative real-valued exponent $\alpha > 0$:

$$
\begin{aligned}
(1+x)^{-\alpha} &= 1 + \frac{-\alpha}{1!}x + \frac{(-\alpha)(-\alpha-1)}{2!}x^2 + \frac{(-\alpha)(-\alpha-1)(-\alpha-2)}{3!}x^3 + \cdots \\
&= \sum_{\nu=0}^{\infty} \frac{(-\alpha)_\nu}{\nu!}x^\nu,
\end{aligned}
$$

where we have introduced a new definition

$$
(\beta)_\nu := \begin{cases} \beta(\beta-1)\ldots(\beta-(\nu-1)) & \text{for } \nu = 1, 2, \ldots \\ 1 & \text{for } \nu = 0. \end{cases}
$$

If β is a positive integer, the coefficients correspond to the binomial coefficients. One is therefore prompted to generalize the binomial coefficients in the following way:

$$
\binom{r}{\nu} := \frac{(r)_\nu}{\nu!}\,; \qquad \text{for } r \in \mathbb{R}.
$$

By virtue of the expansion of $(1+x)^{-\alpha}$, the coefficients are called negative binomial coefficients. The above Taylor expansion is also valid for positive exponents and we therefore end up with the handy generalization

NEWTON'S BINOMIAL THEOREM

$$
(1+x)^\beta = \sum_{\nu=0}^{\infty} \binom{\beta}{\nu} x^\nu \tag{4.27}
$$

for any real number β and all values $|x| < 1$.

We readily see that this formula is also valid for positive integers n, since

$$
(n)_\nu = n(n-1)(n-2)\ldots
$$

will become zero for $\nu > n$, resulting in

$$
\binom{n}{\nu} = 0\,; \qquad \text{for } \nu > n. \tag{4.28}
$$

There exists a wealth of useful formulae for binomial coefficients which can be found in [67]. Here we will summarize some important properties of $\binom{r}{n}$ for integers n.

1. For any real number ν:

$$
\binom{-\nu}{n} = (-1)^n \binom{n+\nu-1}{n} = (-1)^n \binom{n+\nu-1}{\nu-1}. \tag{4.29}
$$

2. For any positive integer r:

$$\binom{r}{n} = 0\,; \qquad \text{for } n > r, \quad \text{or } n < 0. \tag{4.30}$$

3. For any number r and $n > 0$:

$$\binom{r}{n-1} + \binom{r}{n} = \binom{r+1}{n}. \tag{4.31}$$

4. For $n > 0$:

$$\binom{-\frac{1}{2}}{n} = (-2)^{2n}\binom{2n}{n}. \tag{4.32}$$

5. For any number $n > 0$:

$$\binom{\frac{1}{2}}{n} = \frac{1}{2n}\binom{-\frac{1}{2}}{n-1} = 2(n+1)\binom{\frac{1}{2}}{n+1}. \tag{4.33}$$

The proofs are given in Appendix A.5 [p. 603].

4.5.1 Application of negative binomial distribution

A nice application of the negative binomial is the so-called 'stopping time' distribution. Assume an event E (a certain outcome of an experiment) has a probability p and we want to repeat the experiment N times up to the nth occurrence of the event. We ask for the probability for N, i.e. $P(N|n, p, \mathcal{I})$. To be more specific, we introduce the proposition

$$A_N^n : \text{the } N\text{th repetition yields the } n\text{th occurrence of } E.$$

The proposition A_N^n is true if the two following propositions are true:

$$B_{N-1}^{n-1} : \text{There are } n-1 \text{ occurrences of } E \text{ in } N-1 \text{ trials.}$$

$$C : \text{The } N\text{th trial yields the event } E.$$

Hence

$$\begin{aligned} P(N|n, p, \mathcal{I}) &= P(A_N^n|p, \mathcal{I}) = P(B_{N-1}^{n-1}, C|p, \mathcal{I}) = P(B_{N-1}^{n-1}|C, p, \mathcal{I})P(C|p, \mathcal{I}) \\ &= \binom{N-1}{n-1} p^{n-1}(1-p)^{N-n} = \binom{N-1}{n-1} p^n(1-p)^{N-n}. \end{aligned}$$

The result becomes more transparent if we consider the 'failures', i.e. the complementary outcome to E. The probability for a failure is $q = 1 - p$ and the number of failures is $n_f := N - n$. Instead of asking 'how many trials does it take to see the nth event E?', we ask 'how many failures does it take before the nth event E occurs?'. We denote this probability by $P(n_f|n, p, \mathcal{I})$ and it reads

$$P(n_f|n, p, \mathcal{I}) = \binom{n_f + n - 1}{n-1} q^{n_f} p^n = \binom{n_f + n - 1}{n_f} q^{n_f} p^n = P_{\text{nb}}(n_f|n, p).$$

Here we have introduced the *negative binomial* distribution:

NEGATIVE BINOMIAL DISTRIBUTION

$$P_{\mathrm{nb}}(\nu|r,p) = \binom{\nu+r-1}{\nu} q^{\nu} p^{r} \qquad \text{for } r \in \mathbb{R}, \quad \text{and } \nu \in \mathbb{N}_0 \tag{4.34a}$$

$$= \binom{-r}{\nu} (-q)^{\nu} p^{r} \qquad \text{according to (4.29)}, \tag{4.34b}$$

$$\langle \nu \rangle = r\frac{q}{p}, \tag{4.34c}$$

$$\mathrm{var}(\nu) = r\frac{q}{p^2}. \tag{4.34d}$$

The probability for the number of failures n_f is therefore a negative binomial distribution with $\nu = n_f$ and $r = n$. Before we continue the discussion of the stopping-time distribution, we study some of the properties of the negative binomial distribution. We start out with the corresponding generating function

$$\begin{aligned}
\phi_{\mathrm{nb}}(z|r,p) &= \sum_{\nu=0}^{\infty} P_{\mathrm{nb}}(\nu|r,p)\, z^{\nu} \\
&= p^r \sum_{\nu=0}^{\infty} \binom{-r}{\nu}^{\nu} (-qz)^{\nu} \\
&\overset{(4.27)}{=} p^r (1-zq)^{-r}.
\end{aligned}$$

So we readily see that the negative binomial distribution has the correct normalization, which is given by $\phi_{\mathrm{nb}}(z=1|r,p)=1$. Next we prove the formulae given for mean and variance:

$$\begin{aligned}
\langle \nu \rangle &= \frac{d}{dz}\phi_{\mathrm{nb}}(z|r,p)\Big|_{z=1} \\
&= rqp^r(1-qz)^{-(r+1)}\Big|_{z=1} \\
&= rq\frac{p^r}{p^{r+1}} = r\frac{q}{p}, \\
\langle \nu^2 \rangle &= \frac{d^2}{dz^2}\phi_{\mathrm{nb}}(z|r,p)\Big|_{z=1} + \langle \nu \rangle \\
&= r(r+1)q^2 p^r (1-qz)^{-(r+2)}\Big|_{z=1} + r\frac{q}{p} \\
&= \langle \nu \rangle^2 + r\frac{q}{p}\Big(\frac{q}{p}+1\Big),
\end{aligned}$$

$$\mathrm{var}(\nu) = r\frac{q}{p^2}.$$

Now we proceed with the stopping-time problem. The proper normalization to one implies that with probability 1 the goal will be reached, i.e. the nth occurrence of the event will definitely show up. A normalization smaller than 1 would have implied a finite chance that the nth occurrence never shows up, even in an infinite number of repetitions. Average $\langle N \rangle$ and variance $\langle (\Delta N)^2 \rangle$ are related to the corresponding quantities for the number of failures via $N = n + n_f$, resulting in

MEAN AND VARIANCE OF THE STOPPING TIME

$$\langle N \rangle = n + \langle n_f \rangle = n\left(1 + \frac{q}{p}\right) = \frac{n}{p}, \tag{4.35a}$$

$$\langle (\Delta N)^2 \rangle = \langle (\Delta n_f)^2 \rangle = \frac{nq}{p^2}. \tag{4.35b}$$

The negative binomial distribution can also arise from a mixture of a Poisson distribution and a Gamma distribution, i.e. positive integers n are Poisson distributed $P_p(n|\mu)$ with mean μ. The latter itself is distributed according to a Gamma distribution $p_\Gamma(\mu|\alpha, \beta)$ with parameters α and β. Although this problem is an anticipation of continuous distributions, to be discussed in Chapter 7 [p. 92], we will present it here as it fits nicely. The 'mixture model' yields

$$\begin{aligned}
P(n|r, \beta, \mathcal{I}) &= \int_0^\infty P_p(n|\mu)\; p_\Gamma(\mu|r, \beta) d\mu \\
&= \int_0^\infty e^{-\mu} \frac{\mu^n}{n!} \frac{\beta^r}{\Gamma(r)} \mu^{r-1} e^{-\beta\mu}\, d\mu \\
&= \frac{\beta^r}{n!\,\Gamma(r)} \int_0^\infty \mu^{(n+r-1)} e^{-(\beta+1)\mu}\, d\mu \\
&= \frac{\beta^r}{n!\,\Gamma(r)}\, (1+\beta)^{-(n+r)}\; \Gamma(n+r) \\
&= \binom{n+r-1}{n}\; \beta^r\, (1+\beta)^{-(n+r)}.
\end{aligned}$$

With $\beta = \frac{p}{q}$ we obtain, as promised, the negative binomial distribution:

$$P(n|r, \beta = \frac{p}{q}, \mathcal{I}) = \binom{n+r-1}{n}\; p^r\, q^n.$$

5

Random walks

In this chapter, we consider a few random walk problems that are of great importance in statistical physics and in different guises for data analysis. The simplest one has already been discussed before. These problems also allow us to get acquainted with the various combinatorial tools discussed so far, and with the concept of generating functions. A thorough analysis of a wide range of random walk problems, along with useful mathematical tools, can be found in [218].

5.1 First return

As a first example we consider a one-dimensional random walk and determine the probability that a walker returns to his initial position for the first time after n steps. This probability allows us to answer the question of how long a random walk takes to come back to the origin and whether this will happen at all. The number of steps n to return to the origin is apparently even, $n = 2m$. In addition, we consider the probability that the walker returns to his initial position after n steps, no matter if it is the first return. We introduce the corresponding propositions

$$
\begin{aligned}
f_n &: \text{First return after } n \text{ steps.}\\
r_n &: \text{Return after } n \text{ steps.}
\end{aligned}
$$

The sought-for probability is $P(f_n|p,\mathcal{I})$, while the probability for the arbitrary return is $P(r_n|p,\mathcal{I})$. The latter probability is simply a binomial distribution

$$P(r_{2m}|p,\mathcal{I}) = P(N_r = m|N = 2m,\, p,\mathcal{I}) = \binom{2m}{m} p^m q^m .$$

The value of this probability for $m = 0$ is $P(r_0|p,\mathcal{I}) = 1$, which is alright, we only have to stretch the meaning of 'return'. For f_n, however, return really means that the walker has to leave his initial position first, so $P(f_0|p,\mathcal{I}) \overset{!}{=} 0$. The corresponding generating function of $P(r_{2m}|p,\mathcal{I})$ is

$$\phi_r(z) = \sum_{m=0}^{\infty} P(r_{2m}|p,\mathcal{I}) z^{2m} = \sum_{m=0}^{\infty} \binom{2m}{m} (pqz^2)^m .$$

In order to simplify the characteristic function, it is advantageous to express the binomial coefficient with the negative binomial as outlined in Appendix equation (A.23) [p. 605]:

$$\binom{2m}{m} = 2^{2m}(-1)^m \binom{-\frac{1}{2}}{m}.$$

Hence, we have

$$\phi_r(z) = \sum_{m=0}^{\infty} \binom{-\frac{1}{2}}{m} (-4pqz^2)^m = (1-4pqz^2)^{-1/2}.$$

The sum over all propositions yields

$$\sum_{m=0}^{\infty} P(r_{2m}|p, \mathcal{I}) = \phi(z{=}1) = (1-4pqz^2)^{-\frac{1}{2}}\Big|_{z=1} = (1-4pq)^{-\frac{1}{2}}. \tag{5.1}$$

The result can exceed 1, since the propositions are not mutually exclusive. Next, we decompose the proposition r_{2m} such that we are able to compute the probability for f_{2n}. An arbitrary return at time $2m$, for $m \geq 1$, implies that the walker returns for the first time after $2n$ steps and in the remaining time $2(m-n)$ returns again (not necessarily for the first time). A special case is given if $n{=}m$, i.e. the walker returns only once. In mathematical terms, we introduce the mutually exclusive propositions f_{2n} via the marginalization rule, and obtain for $m \geq 1$

$$\begin{aligned} P(r_{2m}|p, \mathcal{I}) &= \sum_{l=1}^{m} P(r_{2m}|f_{2l}, p, \mathcal{I}) P(f_{2l}|p, \mathcal{I}) \\ &= \sum_{l=1}^{m} P(r_{2(m-l)}|p, \mathcal{I}) P(f_{2l}|p, \mathcal{I}). \end{aligned}$$

The last term of the sum, i.e. the one for $l = m$, yields the correct contribution, since $P(r_0|p, \mathcal{I}) = 1$. As we have defined $P(f_0|p, \mathcal{I}){=}0$, we can extend the sum on the right-hand side and include $l{=}0$ as well, resulting in a complete formula for the convolution. For the case $m{=}0$, the equation does not hold. In order to cover also this limiting case we add a Kronecker $\delta_{m,0}$ on the right-hand side and end up with the expression, which is now valid for $m \geq 0$

$$P(r_{2m}|p, \mathcal{I}) = \sum_{l=0}^{m} P(r_{2(m-l)}|p, \mathcal{I}) P(f_{2l}|p, \mathcal{I}) + \delta_{m,0}.$$

The generating function on the left-hand side is $\phi_r(z)$. The generating function on the right-hand side is the sum of the characteristic function of the convolution and that of the Kronecker δ. The latter is 1 and we obtain

$$\phi_r(z) = \phi_r(z) \cdot \phi_f(z) + 1,$$

or rather

$$\phi_f(z) = 1 - \phi_r(z)^{-1}, \tag{5.2a}$$

$$\phi_f(z) \overset{(5.1)}{=} 1 - \sqrt{1 - 4pqz^2}. \tag{5.2b}$$

From the characteristic function we conclude a couple of important results. First of all, the probability that the walker returns at all, either after 2, 4, or any other even number of steps, is equal to the normalization, i.e. $\phi_f(z{=}1) = 1 - \sqrt{1 - 4pq}$, and hence the probability that the walker never returns to his initial position is $\sqrt{1 - 4pq}$. For $p \neq q$ this probability is greater than zero, which is plausible since in that case there is a drift/bias in the random walk. For $p{=}q{=}\frac{1}{2}$, the return probability is 1. Hence, the walker will definitely return to his initial position. An interesting quantity then is the mean waiting time for the first return, or rather the mean number of steps up to the first return:

$$\langle m \rangle = \frac{\partial}{\partial z}\phi_f(z)\bigg|_{z=1} = 2pq(1 - 4pq)^{-1/2}\bigg|_{p=q=\frac{1}{2}} = \infty.$$

Although the walker will definitely return, on average it takes him an eternity. Next we will analyse the result in more detail. To this end we invoke Newton's binomial theorem:

$$\Phi_f(z) = 1 - \sqrt{1 - 4pqz^2} = 1 - \sum_{n=0}^{\infty} \binom{\frac{1}{2}}{n} (-4pqz^2)^n.$$

Comparison with the definition of the generating function

$$\Phi_f(z) = \sum_{m=0}^{\infty} p(f_{2m}|p, \mathcal{I})z^{2m}$$

reveals

$$\text{for } m = 0: \quad P(f_0|p, \mathcal{I}) = 0,$$

$$\begin{aligned}
\text{for } m > 0: \quad P(f_{2m}|p, \mathcal{I}) &= -\binom{\frac{1}{2}}{m}(-4pq)^m \\
&\overset{(A.24a)}{=} \binom{2m}{m}(-4)^{-m}\frac{1}{2m-1}(-4pq)^m \\
&= \frac{1}{2m-1} P_B(m|N{=}2m, p, \mathcal{I}) \\
&\underset{m\gg 1}{=} \frac{1}{2m-1}\frac{(4pq)^m}{\sqrt{4\pi pqm}} \qquad \text{see (6.8).}
\end{aligned}$$

We are merely interested in the case $p{=}q{=}1/2$. In this case we have, for $m \gg 1$:

$$P(f_{2m}|p, \mathcal{I}) = \frac{1}{2m}\frac{1}{\sqrt{\pi m}} \qquad \text{see (6.8).}$$

Apparently, in this case the median is $\check{m}=1$ and the third quartile, which can be obtained by numerical inspection, is at $m \approx 5$. Nonetheless, the mean value $\langle m \rangle$ diverges, since for large m the probability decays only as $m^{-\frac{3}{2}}$.

5.1.1 First return in 2-D and 3-D

Interestingly, equation (5.2a) is independent of the spatial dimension and we will use it for the computation of the first-return probability for two and three dimensions. We will only consider the isotropic case, with equal hopping probability in each direction. The explicit mention of this probability in the conditional part of the probabilities will be omitted. The key quantity that we need to compute is the generating function $\phi_r(z=1)$, which could be determined similarly as in the 1-D case. However, there is actually a more elegant approach, which is particularly preferable for higher dimensions. To begin with, we consider the probability $P_n(\boldsymbol{x})$ that the walker will cover a distance $\boldsymbol{x}$ in n steps, no matter how often the point $\boldsymbol{x}$ occurs throughout this journey. Based on the marginalization rule we can derive a simple recursion relation

$$P_n(\boldsymbol{x}) = \sum_{\Delta \boldsymbol{x}} P_{n-1}(\boldsymbol{x} - \Delta \boldsymbol{x})\; P_1(\Delta \boldsymbol{x}). \tag{5.3}$$

Next we introduce the Fourier transform of these terms, i.e.

$$\tilde{P}_n(\boldsymbol{k}) := \sum_{\boldsymbol{x}} e^{i\boldsymbol{k}\boldsymbol{x}}\; P_n(\boldsymbol{x}), \qquad \varepsilon(\boldsymbol{k}) := \sum_{\boldsymbol{x}} e^{i\boldsymbol{k}\boldsymbol{x}}\; P_1(\boldsymbol{x}), \tag{5.4}$$

where the sum runs over all points of the simple cubic lattice of size $N = L^D$, with D being the spatial dimension of the lattice. Throughout the calculation we assume a finite lattice and at the end we will consider $N \to \infty$. Moreover, we assume periodic boundary conditions, which have no influence on the result for the infinite lattice but simplify the calculations. In the statistical context the Fourier transform is also called the 'characteristic function'. It will be discussed in more detail in Section 7.8 [p. 124], where we will also prove that the characteristic function of a convolution, i.e. an expression like equation (5.3), is simply the product of the characteristic functions of the individual factors. Then

$$\tilde{P}_n(\boldsymbol{k}) = \tilde{P}_{n-1}(\boldsymbol{k})\tilde{P}_1(\boldsymbol{k}) = [\varepsilon(\boldsymbol{k})]^n.$$

The inverse Fourier transform immediately yields the aspired probability

$$P_n(\boldsymbol{x}) = \frac{1}{N} \sum_{\boldsymbol{k}}^{\text{1.b.z.}} e^{-i\boldsymbol{k}\boldsymbol{x}}\; [\varepsilon(\boldsymbol{k})]^n,$$

where the sum covers the first Brillouin zone and the wavevectors are $k_\alpha = \frac{2\pi}{L} n_\alpha$. Since we are interested in an infinite lattice, we can now replace the sum by an integral and we also restrict the further discussion to $\boldsymbol{x} = \boldsymbol{0}$, relevant for the first-return probability

$$P_n(\boldsymbol{x}) = \frac{1}{(2\pi)^D} \int_{-\pi}^{\pi} d^D k \, [\varepsilon(\boldsymbol{k})]^n.$$

We employ this expression for the evaluation of the generating function $\phi_r(z{=}1)$:

$$\begin{aligned}\phi_r(z{=}1) &= \sum_{n=0}^{\infty} P_n(\boldsymbol{0}) = \frac{1}{(2\pi)^D} \int_{-\pi}^{\pi} d^D k \sum_{n=0}^{\infty} [\varepsilon(\boldsymbol{k})]^n \\ &= \frac{1}{(2\pi)^D} \int_{-\pi}^{\pi} d^D k \, \frac{1}{1-\varepsilon(\boldsymbol{k})}. \end{aligned} \tag{5.5}$$

Finally, we compute $\varepsilon(\boldsymbol{k})$:

$$\varepsilon(\boldsymbol{k}) = \sum_{x}^{ikx} P_1(\boldsymbol{x}) = \frac{1}{2D} \sum_{\alpha=1}^{D} \left(e^{ik_\alpha} + e^{-ik_\alpha}\right) = \frac{1}{D} \sum_{\alpha=1}^{D} \cos(k_\alpha).$$

Apart from the prefactor this is the tight-binding dispersion, for which the density of states (dos) can be found in the literature. Now the numerical evaluation of the integral in equation (5.5) is straightforward and not really demanding, but it is worth mentioning that there is an even more elegant way to proceed. As standard in solid-state physics, we introduce the density of states (dos) $\rho(\varepsilon)$, which reduces the integral in equation (5.5) to a one-dimensional integral

$$\phi_r(z{=}1) = \int_{-1}^{1} d\varepsilon \, \frac{\rho(\varepsilon)}{1-\varepsilon} \xi, \qquad \rho(\varepsilon) := \frac{1}{(2\pi)^D} \int_{-\pi}^{\pi} d^D k \, \delta(\varepsilon - \varepsilon(\boldsymbol{k})). \tag{5.6}$$

The dos for the 2-D lattice can be found in [59]. For our purposes, we merely need that $\rho(\varepsilon)$ is finite for $\varepsilon \to D$, resulting in a logarithmic divergence and $\phi_r^{(\text{2-D})}(z=1) = \infty$. According to equation (5.2a), we find $\phi_f^{(\text{2-D})}(z=1) = 1$. In other words, the walker will definitely return to his starting position.

The situation is different in three spatial dimensions, because the dos now vanishes at the band edges $\varepsilon \to \pm 1$, and the integral in equation (5.6) yields a finite value. For the detailed calculation we employed the dos expressed in terms of complete elliptic integrals, as reported in [99]. Straightforward numerical integration of the integral yields $\phi_r(z{=}1) {=} 1.516$, resulting in $\phi_f(z{=}1) {=} 0.341$. Apparently, it is now very likely ($P = 0.659$) that the walker will never return to the origin of his journey. Hence a well-meant piece of advice in [91] reads: 'one should not get drunk in more than two dimensions'.

5.2 First lead

Next we will study a related question, which will become important in the context of hypothesis testing. Given a Bernoulli experiment with two alternative outcomes called success and failure, what is the probability that it takes N trials before the total number of successes exceeds for the first time the total number of the failures. We call this event *first lead*. It is expedient to translate the problem into the random walk (Brownian motion) language. The problem of *first lead* is equivalent to a random walk with up (success) and down (failure) moves, and we seek the probability that the random walk occurs above the abscissa for the first time at step N. For the analysis we introduce the number n_u of up steps and the number n_d of down steps. The total number is $N = n_u + n_d$. Clearly, the 'first lead' can only occur for an odd number $N = 2L + 1$ of trials and, right before the first lead, the random walk has to be balanced, i.e. $n_u = n_d$. This return event need not be the first, but the walker shall never cross the abscissa before. We call this *return from below* in $2L$ steps. Therefore, the probability for the *first lead* in $N = 2L + 1$ steps is equal to the probability (p) for the final step upwards times the probability for a *return from below* in $2L$ steps.

Now we define the relevant propositions:

$$
\begin{aligned}
L_{2L+1} &: \text{First lead at step } 2L + 1. \\
F^b_{2L} &: \text{First return from below at step } 2L.
\end{aligned}
$$

The probability for the 'first return from below' in $2L$ steps is $(pq)^L$ times the number of paths from $(0, 0)$ to $(2L, 0)$ that are never above the abscissa. This number is equivalent to the number $\mathcal{N}(y_0, y_N | y < 0, N = 2L, \mathcal{I})$ of paths from $(0, -1)$ to $(2L, -1)$ that never touch the abscissa. The sought-for probability is therefore

$$
P(L_{2L+1} | p, \mathcal{I}) = p^{L+1} q^L \, \mathcal{N}(y_0, y_N | y < 0, N = 2L, \mathcal{I}). \tag{5.7}
$$

Such restricted random walks can best be analysed based on the 'reflection principle' [170] that we will briefly outline in the next section.

5.2.1 Reflection principle

Here we consider the more general case of a random walk that starts at time $t = 0$ at a negative position $y_0 = -|y_0^*|$ and ends again after N time steps at a negative position $y_N = -|y_N^*|$. We are interested in the number of paths between these points, that never reach $y = 0$. A representative path is depicted in Figure 5.1 as a dashed curve. This number is apparently equal to the number of unrestricted paths minus the number of paths that reach $y = 0$ at least once. The number of unrestricted paths is homogeneous in y, i.e.

$$
\begin{aligned}
&\mathcal{N}(y_0, y_N | N, \mathcal{I}) = \mathcal{N}(0, \Delta y | N, \mathcal{I}), \\
&\Delta y := y_N - y_0.
\end{aligned}
$$

In order to cover a distance of Δy in N steps, the number L of downward steps is given by

$$
\begin{aligned}
\Delta y &= (N - L) - L = N - 2L, \\
L &= \frac{N - \Delta y}{2}.
\end{aligned}
$$

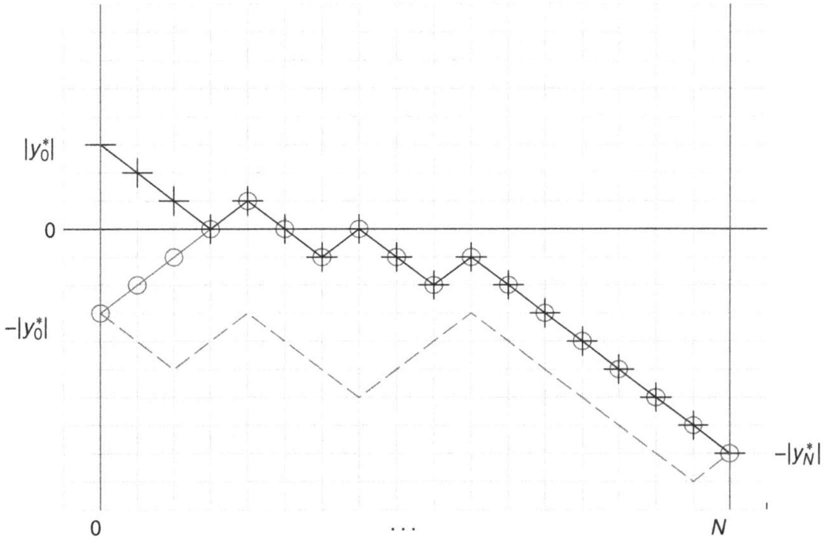

Figure 5.1 Illustration of the reflection principle.

Apparently, we have to take into account that in an even (odd) number N only an even (odd) distance can be covered. In other words, Δy and N need to have the same parity. So we have

NUMBER OF UNRESTRICTED PATHS
(Δy AND N HAVE THE SAME PARITY)

$$\mathcal{N}(0, \Delta y | N, \mathcal{I}) = \binom{N}{L(\Delta y)}, \tag{5.8a}$$

$$L(\Delta y) := \frac{N - \Delta y}{2}. \tag{5.8b}$$

In the present case, we have

$$\Delta y = \Delta y^* := y_N - y_0 = -\Big(\left|y_N^*\right| - \left|y_0^*\right| \Big). \tag{5.9}$$

Next we determine the number of those paths that start at $y_0 = -\left|y_0^*\right|$, end at $y_N = -\left|y_N^*\right|$, and which reach $y = 0$ at least once. A representative path is depicted in Figure 5.1, marked by circles. This number is easily determined based on the reflection principle [170]. To this end, we identify the initial part of the path up to the point where it first reaches the abscissa. Then we reflect this part of the path on the abscissa, as represented in Figure 5.1 with the curve marked with stars. All such reflected paths start at $y_0 = +\left|y_0^*\right|$ and end at $y_N = -\left|y_N^*\right|$. One can easily convince oneself that there is a one-to-one correspondence between the set of restricted paths we are interested in and the set $\mathcal{R}$ of unrestricted paths from $y_0 = |y_0^*|$ to $y_N = -|y_N^*|$. In other words, for each restricted path there is by construction a reflected path. And for each path of the set $\mathcal{R}$ there is a path of the original set of restricted paths that we can construct by simply applying the reflection described before. Therefore, the number N_r of restricted paths, starting at $y_0 = -|y_0^*|$ and reaching at least once $y = 0$, is equal to the number of unrestricted paths starting at $y_0 = |y_0^*|$:

$$N_r = \mathcal{N}(0, -|y_N^*| - |y_0^*|\,|N, \mathcal{I}) = \binom{N}{\frac{N+|y_N^*|+|y_0^*|}{2}} = \binom{N}{L(\Delta y^*) + |y_0^*|}.$$

Finally, the sought-for number of paths starting at $y = -|y_0^*|$ and reaching $y = -|y_N^*|$ in N steps avoiding the abscissa is given by

NUMBER OF ABSCISSA AVOIDING PATHS FROM $-|y_0^*|$ TO $-|y_N^*|$ IN N STEPS

$$\mathcal{N}(-|y_0^*|, -|y_N^*|\,|y < 0, N, \mathcal{I}) = \binom{N}{L(\Delta y^*)} - \binom{N}{L(\Delta y^*) + |y_0^*|}, \tag{5.10}$$

$$L(\Delta y^*) := \frac{N + \left(|y_N^*| - |y_0^*|\right)}{2}.$$

5.2.2 *First lead problem continued*

Now we are prepared to compute the probability for the *first lead*. The key quantity in equation (5.2) is

$$\mathcal{N}(y_0 = -1, y_N = -1|N = 2L, \mathcal{I}).$$

Hence, $|y_N^*| = |y_0^*| = 1$ and we get, according to equation (5.10),

$$\mathcal{N}(y_0 = -1, y_N = -1|N = 2L, \mathcal{I}) = \binom{2L}{L} - \binom{2L}{L+1} = \binom{2L}{L}\frac{1}{L+1}.$$

Eventually we find

PROBABILITY FOR BEING IN LEAD FOR THE FIRST TIME AT STEP $N = 2L + 1$

$$P(L_{2L+1}|p, \mathcal{I}) = \binom{2L}{L}\frac{1}{L+1}p^{L+1}\,q^L. \tag{5.11}$$

Next we will determine the corresponding generating function

$$\Phi_L(z) = \sum_{L=0}^{\infty} \binom{2L}{L} \frac{1}{L+1} p^{L+1} q^L z^L$$
$$= \int_0^p dx \sum_{L=0}^{\infty} \binom{2L}{L} (xqz)^L.$$

We use equation (A.23) [p. 605] to transform the binomial coefficient

$$\Phi_L(z) = \int_0^p dx \sum_{L=0}^{\infty} \binom{-\frac{1}{2}}{L} (-4xqz)^L$$
$$= \int_0^p dx \left(1 - 4xqz\right)^{-1/2},$$

resulting in

GENERATING FUNCTION FOR FIRST LEAD

$$\Phi_L(z) = \frac{1}{2qz}\left(1 - \sqrt{1 - 4pqz}\right). \tag{5.12}$$

The probability $P(L|P, \mathcal{I})$ that a lead occurs at all is, according to equation (5.2), given by

$$P(L|P, \mathcal{I}) = \Phi_L(z = 1) = \frac{1}{2q}\left(1 - \sqrt{1 - 4p(1 - p)}\right)$$
$$= \frac{1}{2q}\left(1 - \sqrt{(1 - 2p)^2}\right)$$
$$= \frac{1}{2q}\left(1 - |1 - 2p|\right).$$

PROBABILITY FOR EVER BEING IN THE LEAD

$$p_L = \begin{cases} 1 & \text{for } p \geq \frac{1}{2} \\ \frac{p}{1-p} & \text{for } p < \frac{1}{2} \end{cases}.$$

The probability is one for $p \geq 1/2$ and it decreases with decreasing $p < 1/2$ and is zero for $p = 0$. It should be noted that the presented problem is closely related to Bertrand's ballot

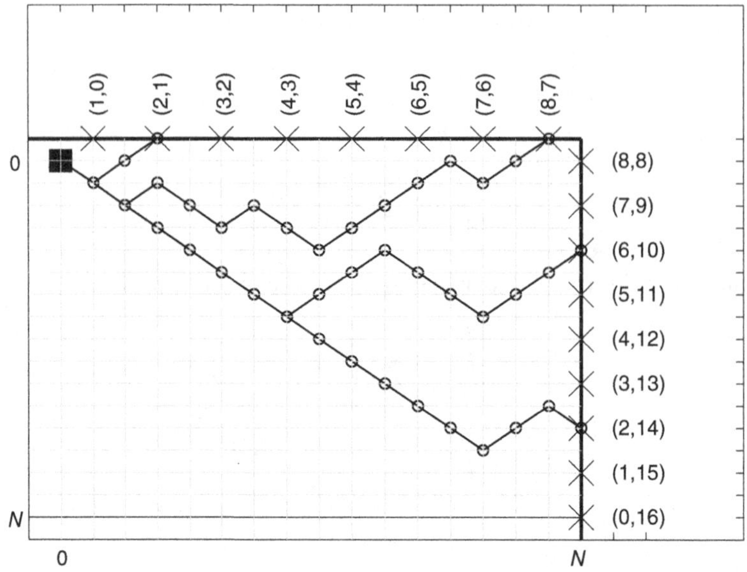

Figure 5.2 Random walk with absorbing wall. The walk starts at the square (0, 0) and ends either at $y = 0$ or $x = N$ (here $N = 16$), whatever happens first. The pairs of numbers at the margins are (u, d), i.e. the number of upward/downward steps required to reach the boundary.

problem [67], which gives the probability that in an election where candidate A receives P and candidate B receives Q votes, with $P > Q$, A will be ahead of B throughout the entire count.

5.3 Random walk with absorbing wall

Next we consider a similar random walk setup, depicted in Figure 5.2. In addition to the absorbing wall[1] at $y = 0$, that we used before, we also introduce an absorbing wall at $x = N^*$. In other words, the random walk introduced in the *first lead* problem will end after N^* steps, at the latest. Here we will denote the number of upward/downward steps by u/d. The pairs (u, d) of steps, required to reach specific points on the boundary, are given in Figure 5.2 as well.

We have to treat the two stopping events separately. The absorption condition at the upper boundary is identical to the stopping in the *first lead* case. In this case, $(u, d) = (L + 1, L)$, with $L \in \{0, 1, \ldots, L^* - 1\}$, where

$$L^* = \left\lceil \frac{N^*}{2} \right\rceil. \tag{5.13}$$

[1] A general discussion about boundaries of random walks is given in [218].

The probability that the walk ends with $(L+1, L)$ at the top boundary is identical to $P(L_{2L+1}|p, \mathcal{I})$ given in equation (5.11):

$$P((L+1,L)|p,\mathcal{I}) = \mathbf{1}(0 \le L < L^*) \binom{2L}{L} \frac{1}{L+1} p^{L+1}\, q^L.$$

If the walk reaches the right boundary, then $(u,d) = (N^* - L, L)$, with $L \in \{L^*, \ldots, N^*\}$. This situation is also covered by equation (5.10), with $N = N^*$ and $L(\Delta y^*) = L$:

$$\begin{aligned} P((N^*-L,L)|p,N^*,\mathcal{I}) &= \mathbf{1}(L^* \le L \le N^*) \left[\binom{N^*}{L} - \binom{N^*}{L+1}\right] p^{N^*-L}\, q^L \\ &= \mathbf{1}(L^* \le L \le N^*) \binom{N^*}{L}\left(1 - \frac{N^*-L}{L+1}\right) p^{N^*-L}\, q^L. \end{aligned}$$

Probability for (u, d)
in random walk with two absorbing walls

$$\begin{aligned} P(u,d|p,N^*,\mathcal{I}) = p^u\, q^d \Bigg\{ &\delta_{d,u-1}\; \mathbf{1}(0 \le d < L^*) \binom{2d}{d} \frac{1}{d+1} p^u\, q^d \qquad (5.14) \\ &+ \delta_{d,N^*-u}\; \mathbf{1}(L^* \le d \le N^*) \binom{N^*}{d}\left(1 - \frac{N^*-d}{d+1}\right)\Bigg\}, \end{aligned}$$

$$L^* := \left\lceil \frac{N^*}{2} \right\rceil.$$

Based on Figure 5.2 we see that the points where the boundary is hit are uniquely specified by the number d of downward moves. The probability to find a walker at such a point is given by

$$\begin{aligned} P(d|p,N^*,\mathcal{I}) = \Bigg\{ &\mathbf{1}(0 \le d < L^*) \binom{2d}{d} \frac{1}{d+1} p^u\, q^d \\ &+ \mathbf{1}(L^* \le d \le 2L^*) \binom{2L^*}{d}\left(1 - \frac{2L^*-d}{d+1}\right)\Bigg\} \times p^u\, q^d. \end{aligned}$$

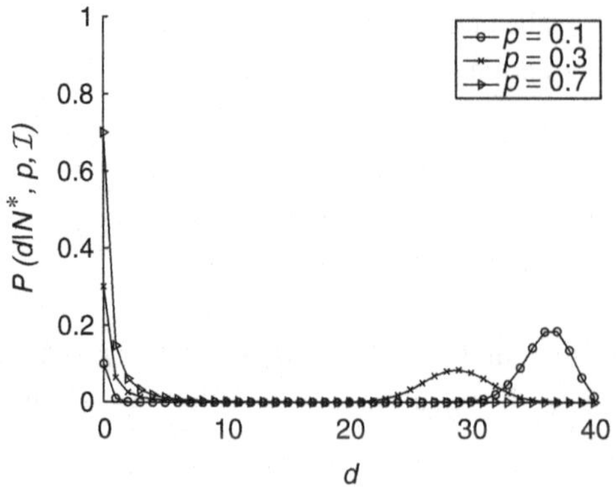

Figure 5.3 Marginal probability $P(d|N^*, p, \mathcal{I})$ for the number d of downward moves for different values of p.

Representative results are depicted in Figure 5.3. Apparently, there is always a peak in the probability distribution for short distances. Another peak crops up for $p \leq 0.5$, i.e. in the regime where there is a finite probability in the *first lead* problem that the walk never stops. We will come back to these results in Section 20.2.3 [p. 329], where they are used to determine the impact of stopping criteria on hypothesis testing.

6

Limit theorems

In the pre-computer era, the handling of binomial coefficients was a nuisance and therefore a couple of approximations were developed. Nowadays, they are less important in numerical evaluations. But they are still important for analytical calculations. In this chapter we will summarize the most important limiting theorems.

6.1 Stirling's formula

It is almost inevitable to encounter the factorial function $n!$ in combinatorial problems, often for large values of n. To handle these problems it is very useful to derive an approximation for $\ln(n!)$ using Stirling's formula, which yields tractable expressions. We will see that there is a close connection between the factorial function and the Gamma function

$$\Gamma(x) = \int_0^\infty t^{x-1}\, e^{-t}\, dt, \tag{6.1}$$

which is given by

$$\Gamma(n+1) = n! \tag{6.2}$$

for integer n. The asymptotic representation of the Gamma function for large values of $|x|$ is given by

$$\Gamma(x) = x^{x-\frac{1}{2}} e^{-x} \sqrt{2\pi} \left\{1 + \frac{1}{12x} + O(x^{-2})\right\} = x^{x-\frac{1}{2}} e^{-x} \sqrt{2\pi} \left\{1 + O(x^{-1})\right\}. \tag{6.3}$$

The derivation in the framework of the 'saddle-point approximation' is given in Section A.2 [p. 598]. Similarly, for large n, the factorial function can be expressed as

Table 6.1 Relative error ε of Stirling's formula

n	$n!$	ε
1	1	0.0779
2	2	0.0405
3	6	0.0273
4	24	0.0206
5	120	0.0165
10	3 628 800	0.0083
20	2 432 902 008 176 640 000	0.0042
30	265 252 859 812 191 058 636 308 480 000 000	0.0028
40	815 915 283 247 897 734 345 611 269 596 115 894 272 000 000 000	0.0021

Stirling's formula

$$n! = n^{\left(n+\frac{1}{2}\right)}\, e^{-n}\, \sqrt{2\pi}\, \left\{1 + O(n^{-1})\right\}, \tag{6.4a}$$

$$\ln(n!) = \left(n + \frac{1}{2}\right) \ln(n) - n + \ln(\sqrt{2\pi}) + O(n^{-1}). \tag{6.4b}$$

We see that the absolute error due to the $O(n^{-1})$ correction term increases rapidly with n, but the relative error decreases as $1/n$. Fortunately, in most applications ratios of factorials are needed, and in this case the absolute error decreases as $1/n$ too. The same is true for $\ln(n!)$. In Table 6.1 the values of the factorial function for different n and the relative errors are shown. Even for $n = 1$ the relative error is already below 8%. And it decreases slowly but steadily.[1]

6.2 de Moivre–Laplace theorem/local limit theorem

The binomial distribution can be well approximated by a Gaussian distribution (normal distribution) with the same mean and variance if the variance is large, i.e. $\sigma^2 = np(1-p) \gg 1$. This is guaranteed by the de Moivre–Laplace theorem [159] or local limit theorem:

[1] Stirling's formula for $n = 1$ yields 0.922.

DE MOIVRE–LAPLACE THEOREM
(LOCAL LIMIT THEOREM)

$$P(k|n,p) = \binom{n}{k} p^k (1-p)^{n-k} \approx g(k|k_0,\sigma) \tag{6.5}$$

with

$$g(k|k_0,\sigma) := \frac{1}{\sqrt{2\pi\sigma^2}} e^{-\frac{(k-k_0)^2}{2\sigma^2}}, \tag{6.6}$$

$$k_0 = np,$$

$$\sigma = \sqrt{np(1-p)}$$

for $\sigma \gg 1$.

The proof can be based on Stirling's formula. For more details, see [159]. The approximation is valid in the neighbourhood of the peak. In the limits $k \to 0$ and $k \to n$, the functional form is different. But these discrepancies are usually immaterial, as the probability is already negligibly small in these regions. For increasing σ^2 the approximation approaches the exact values in the vicinity of the peak, and the size of this vicinity increases at the same time. An overall comparison, depicted in part (a) of Figure 6.1, shows that the approximation already yields reasonable values for moderate n ($n = 100$). A close look at Figure 6.1(b), however, reveals that even close to the peak the deviations are still in the percentage area. In many applications the sample size is small to moderate and this

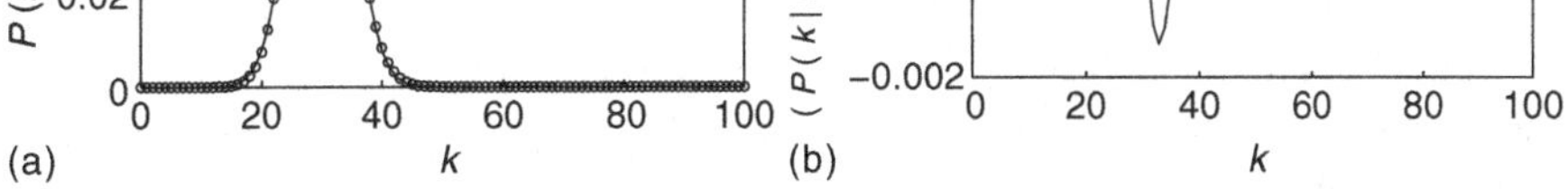

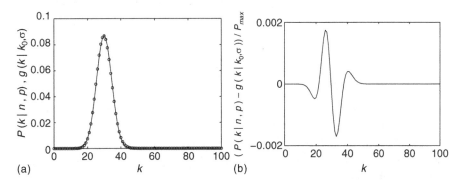

Figure 6.1 (a) Comparison of the binomial distribution (circles) with its Gaussian approximation (solid line) for $n = 100$, $p = 0.3$, resulting in $k_0 = 30$ and $\sigma = 4.58$. (b) Difference between the binomial distribution and the Gaussian approximation, normalized to the maximum of the binomial distribution.

discrepancy is unacceptable. For that reason the local limit theorem has largely lost its relevance for practical purposes, because a numerical evaluation of the binomial distribution is nowadays straightforward and as simple as the computation of the Gaussian distribution. However, for certain analytical derivations it is still of importance, as we shall see below in Section 6.3.

In practical applications it is often only of marginal interest to compute the probability for observing precisely k events. The far more common cases are problems of the following type: *The probability for a product to be deficient is* $p = 0.002$. *How large is the probability that of 5000 elements at most 20 are deficient?* Here, the total probability for the exclusive events $k = 0, 1, 2, \dots, 20$ is of interest. Using the sum rule we can express this probability by

$$P(k \in \{0, 1, 2, \dots, 20\}|n, p) = \sum_{k=0}^{20} P(k|n, p). \tag{6.7}$$

This summation is of course trivial with present-day computers. But in some cases, when n is very large, it may be useful to proceed with an analytic approximation. Application of the de Moivre–Laplace theorem equation (6.5) yields the so-called 'integral theorem of de Moivre', where the binomial is approximated by a Gaussian (equation (6.5)) and the summation is replaced by an integral.

6.3 Bernoulli's law of large numbers

We already know from equation (4.12b) [p. 54] that Bernoulli experiments with n repetitions following a binomial distribution with success probability p exhibit a mean value of $\langle x \rangle = p\ n$. If we now ask for the probability for the occurrence of precisely $k = n\ p$ events, the most likely result (see Figure 6.1), then the local limit theorem yields

$$P(k = n\ p|n, p) = \frac{1}{\sqrt{2\pi np(1-p)}} \underset{n\to\infty}{\longrightarrow} 0. \tag{6.8}$$

Therefore, even the mean value $\langle k \rangle = np$ has in the limit $n \to \infty$ a vanishing probability to be observed. This is compensated by the large number of k-values in the neighbourhood of $k = \langle k \rangle$ of comparable probability. The integral theorem of de Moivre states that the probability to observe a k-value in the range of $\pm$ the standard deviation around the mean (1σ confidence interval)

$$\begin{aligned} I_\sigma &= \left\{k \mid np - \sqrt{np(1-p)} \le k < np + \sqrt{np(1-p)}\right\} \\ &= \left\{k \mid \langle k \rangle - \sigma \le k < \langle k \rangle + \sigma\right\} \end{aligned}$$

is given by

$$P(k \in I_\sigma | n, p) = \frac{1}{\sqrt{2\pi}} \int_{-1}^{1} e^{-\frac{x^2}{2}}\, dx = 0.6827. \tag{6.9}$$

The probability to observe k within the 2σ confidence interval is already close to 1,

$$P(k \in I_{2\sigma} | n, p) = \frac{1}{\sqrt{2\pi}} \int_{-2}^{2} e^{-\frac{x^2}{2}}\, dx = 0.9545.$$

For an arbitrarily small $\varepsilon > 0$, the probability for $|k/n - p| < \varepsilon$ can be computed by

$$P(|k/n - p| < \varepsilon | n, p) = P(pn - \varepsilon n \le k < pn + \varepsilon n) \approx \frac{1}{\sqrt{2\pi}} \int_{a}^{b} e^{-x^2/2}\, dx$$

$$\text{with} \quad \begin{cases} a &= -\frac{\varepsilon n}{\sqrt{np(1-p)}} = -\sqrt{n}\, \frac{\varepsilon}{\sqrt{p(1-p)}} \underset{n\to\infty}{\longrightarrow} -\infty \\ b &= +\sqrt{n}\, \frac{\varepsilon}{\sqrt{p(1-p)}} \underset{n\to\infty}{\longrightarrow} \infty. \end{cases}$$

The approximation by an integral becomes exact in the limit $n \to \infty$ and we end up with

BERNOULLI'S LAW OF LARGE NUMBERS

$$P(|k/n - p| < \varepsilon | n, p) \underset{n\to\infty}{\longrightarrow} 1 \qquad \text{for arbitrary } \varepsilon > 0. \tag{6.10}$$

An important consequence of Bernoulli's law of large numbers (and some generalizations, e.g. the *weak law of large numbers* or the *strong law of large numbers*) is that p is equal to the relative frequency, $p = \frac{k}{n}$, for $n \to \infty$.

It should be noted that the law of large numbers holds under very weak assumptions (essentially for all random variables with distributions of finite variance [159]). However, some relevant distributions do not satisfy these assumptions (e.g. the Cauchy distribution) and here the convergence to the mean in the long run is not guaranteed.

6.4 Poisson's law

The approximation provided by the local limit theorem is reasonable, close to the maximum of the distribution. For very small or very large values of k the error increases considerably. Here Poisson's law provides a much better approximation:

POISSON'S LAW

Given N binary trials, let the probability for event E be p. For fixed mean

$$\mu := N \cdot p = \text{fixed}, \tag{6.11}$$

the limit $N \to \infty$ yields

$$\lim_{N\to\infty} P\left(n|N, p = \frac{\mu}{N}\right) = e^{-\mu}\frac{\mu^n}{n!} := P_P(n|\mu). \tag{6.12}$$

The resulting distribution $P_P(n|\mu)$ is a Poisson distribution. It is interesting to note that Poisson himself never recognized the importance of this distribution and its independent significance. He merely needed it as an auxiliary step in the discussion of the 'law of large numbers' in his book *Sur la probability des judgments* published in 1837. The importance was only recognized some 60 years later, in 1898, by Ladislaus von Bortkiewicz, in a publication entitled *The Law of Small Numbers*.

Proof of Poisson's law: We start out from the binomial distribution

$$B := \frac{N!}{(N-n)!n!}\, p^n(1-p)^{N-n} = \frac{N!}{(N-n)!}\, p^n\, e^{(N-n)\ln(1-p)}\, \frac{1}{n!}.$$

We expand the argument of the exponential in a Taylor series and keep only the first-order term in p as $p \ll 1$, resulting in

$$e^{-Np+np} \approx e^{-\mu}\, e^{np}.$$

The standard deviation of the binomial distribution is $\sigma = \sqrt{\mu(1-p)} \approx \sqrt{\mu}$. Now we define an interval $I_n := (\mu - m\sqrt{\mu}, \mu + m\sqrt{\mu})$ with suitable m such that a desired amount of probability mass lies in this interval. For all $n \in I_n$ we have $np \leq (m+1)\mu p$. Then for $p \ll 1$ we have $e^{np} = 1 + O(p)$ and

$$B = \frac{N!}{(N-n)!}\, p^n\, \frac{1}{n!}\, e^{-\mu}\, \big(1 + O(p)\big) = \underbrace{\frac{N!}{(N-n)!}\, N^{-n}}_{:=C}\, \frac{\mu^n}{n!}\, e^{-\mu}\, \big(1 + O(p)\big).$$

Finally we analyse C:

$$C = \prod_{i=0}^{n-1} \frac{N-i}{N} = \prod_{i=0}^{n-1}\left(1 - \frac{i}{N}\right) = 1 + O(p).$$

We end up with

$$B = \frac{\mu^n}{n!}\, e^{-\mu}\, \big(1 + O(p)\big),$$

which finishes the proof. □

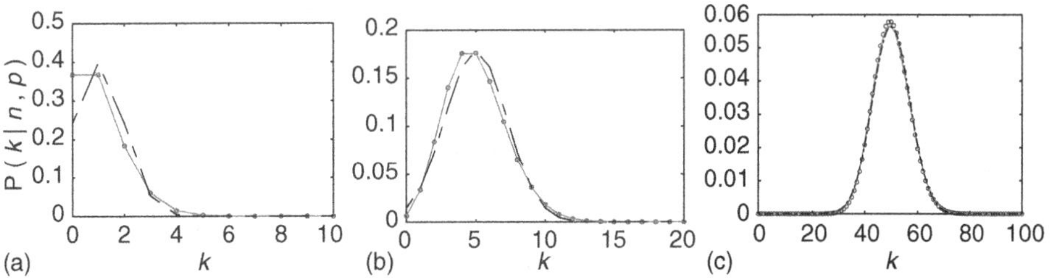

Figure 6.2 Comparison of the binomial distribution (circles) with Poisson (solid line) and Gaussian approximation (dashed lines) displayed for $n = 1000$. (a) $p = 0.001$, $\mu = 1$, $\sigma = 1.00$; (b) $p = 0.005$, $\mu = 5$, $\sigma = 2.23$; (c) $p = 0.05$, $\mu = 50$, $\sigma = 6.89$.

The distribution is normalized to one:

$$\sum_{k=0}^{\infty} e^{-\mu} \frac{\mu^k}{k!} = e^{-\mu} e^{\mu} = 1.$$

As approximation for the binomial distribution, the Poisson distribution is therefore valid for very rare events ($p \ll 1$) as long as the sample size is large ($N \gg 1$). But the Poisson distribution can also be derived independently based on quite simple and general assumptions, as outlined in Chapter 9 [p. 147].

In Figure 6.2 the result of the Poisson approximation is compared to the binomial distribution for different parameter settings. The agreement is in all cases pretty good. In addition, Gaussian distributions are also displayed with corresponding mean and variance. For $\sigma > 1$ the Poisson distribution is reasonably well approximated by a Gaussian.

The mean value of the Poisson distribution is

$$\langle k \rangle = e^{-\mu} \sum_{k=0}^{\infty} k \frac{\mu^k}{k!} = e^{-\mu} \mu \frac{\partial}{\partial \mu} \sum_{k=0}^{\infty} \frac{\mu^k}{k!} = e^{-\mu} \mu \frac{\partial}{\partial \mu} e^{\mu} = \mu.$$

The second moment of the Poisson distribution can be computed similarly, and we obtain

$$\left\langle k^2 \right\rangle = \mu^2 + \mu.$$

Therefore:

POISSON DISTRIBUTION

$$P_P(k|\mu) = e^{-\mu} \frac{\mu^k}{k!}, \tag{6.13a}$$

$$\langle k \rangle = \mu, \tag{6.13b}$$

$$\mathrm{var}(k) = \mu. \tag{6.13c}$$

The expectation value and variance of the Poisson distribution are in agreement with the corresponding values of the binomial distribution (see equation (4.12b) [p. 54] and equation (6.13c)), taking into account that $\mu = p\, n$ and $p \to 0$.

Many experiments are counting experiments, e.g. counting photons or electrons using suitable detectors. Counting experiments obey Poisson statistics. For the sake of convenience, very often the Poisson distribution is approximated by a Gaussian, which is only valid if $\mu \gg 1$. We observe in Figure 6.2 that the Gaussian approximation of the Poisson distribution is not justified for $\mu \leq 5$. For mean values as large as $\mu = 50$, however, there is only a weakly visible discrepancy between the binomial distribution and the Gaussian distribution. Consequently, if the count is sufficiently large ($\mu > 100$), the 'Gaussian approximation' is justified.

Error statistic of counting experiments in the Gaussian approximation

$$P(N|\mu) = \frac{1}{\sqrt{2\pi\sigma^2}}\, e^{-\frac{(N-\mu)^2}{2\sigma^2}}, \qquad \sigma = \sqrt{\mu} \approx \sqrt{N}. \tag{6.14}$$

This implies that the measured number of counts is Gaussian distributed with mean μ and standard deviation $\sqrt{\mu}$. However, the true value of μ is unknown (otherwise the experiment would not be necessary). For that reason equation (6.14) is typically used to estimate the true mean μ from the observed number of counts. Although logically incorrect, the error is in many cases negligible for $N > 10$ and the mean and standard deviation of μ are given as

$$\mu = N \pm \sqrt{N}. \tag{6.15}$$

We have seen in Section 2.1.3 [p. 22] that the sum $k := k_1 + k_2$ of Poissonian random variables k_i is again Poisson distributed $P(k|\mu)$, with $\mu = \mu_1 + \mu_2$. This property is routinely exploited in physics because it allows the summation of different channels in counting experiments without changing the statistics. However, less well known is the fact that the difference of Poisson-distributed random variables is **not** Poisson distributed. This can actually be realized easily by the fact that the difference may become negative. Nevertheless, the subtraction of a measured background from a recorded spectrum, followed by an analysis based on the (wrong) assumption of counting statistics (i.e. a Poisson distribution of the remaining signal) can still be found in many presentations.

The distribution of the count difference r follows instead a 'Skellam distribution'. It can be derived via the marginalization rule, resulting in

$$P(r|\mu_1, \mu_2) = \sum_{k=0}^{\infty} P(k+r|\mu_1)\, P(k|\mu_2)\, \boldsymbol{I}(k+r \geq 0)$$

$$= e^{-(\mu_1+\mu_2)} \sum_{k=0}^{\infty} \frac{\mu_1^{k+r}}{(k+r)!} \frac{\mu_2^k}{k!},$$

where we use the definition $m! = 0$ for negative m. Rearranging the terms leads to

$$P(r|\mu_1, \mu_2) = e^{-(\mu_1+\mu_2)} \left(\frac{\mu_1}{\mu_2}\right)^{r/2} \sum_{k=0}^{\infty} \frac{\sqrt{\mu_1 \mu_2}^{\,2k+r}}{(k+r)!}.$$

The sum on the right-hand side corresponds to the Taylor expansion of the modified Bessel function of the first kind $I_{|r|}$. So we have:

Skellam distribution

$$P(r|\mu_1, \mu_2) = e^{-(\mu_1+\mu_2)} \left(\frac{\mu_1}{\mu_2}\right)^{r/2} I_{|r|}\left(2\sqrt{\mu_1 \mu_2}\right). \tag{6.16}$$

A Bayesian approach to deal with the problem of background subtraction in counting experiments is detailed in [90].

7

Continuous distributions

Up to now we have dealt almost exclusively with discrete problems like die rolls, coin tossing, particles in boxes, etc. However, many problems require a continuous description.

7.1 Continuous propositions

For problems with continuous degrees of freedom (e.g. velocity or temperature) we need to introduce adequate propositions, like

$$A_{<a} : \text{The value of parameter } A \text{ is smaller than } a.$$

The probability for this proposition $P(A_{<a}|\mathcal{I}) = P(A < a|\mathcal{I})$ is a so-called 'cumulative distribution function' (CDF) or in short 'distribution function'. Differentiation of a distribution function results in a 'probability density function' (PDF)

$$p(A_a|\mathcal{I}) = \lim_{da \to 0} \frac{P(A_{<a+da}|\mathcal{I}) - P(A_{<a}|\mathcal{I})}{da}. \tag{7.1}$$

The expression in the numerator can also be understood as the probability $P(dA_a|\mathcal{I})$ for the infinitesimal proposition

$$dA_a : \text{The value of } A \text{ is from the interval } [a, a + da).$$

Therefore, the probability $P(dA_a)$ is the product of the probability density $p(A_a|\mathcal{I})$ and the infinitesimal size of the interval. This general form holds also for dimensions larger than one, $a \in \mathbb{R}^n$.

To distinguish between probability densities and probabilities we will denote probability densities always in lowercase letters. For continuous degrees of freedom the sum and marginalization rule of equation (3.4) [p. 34] turns into

SUM RULE FOR CONTINUOUS DEGREES OF FREEDOM

$$\int p(x|\mathcal{I})\,dx = 1 \qquad \text{(normalization)} \tag{7.2a}$$

$$P(B|\mathcal{I}) = \int P(B|x,\mathcal{I})\,p(x|\mathcal{I})\,dx \qquad \text{(marginalization rule).} \tag{7.2b}$$

7.2 Distribution function and probability density functions

We have just seen that continuous random variables can be introduced with the CDF or cumulative distribution.[1]

Definition 7.1 (Cumulative distribution) *The cumulative distribution function is defined by*

$$F(x) = P(X \leq x|\mathcal{I}),$$

i.e. the probability that a random variable X is smaller than or equal to x.

By construction, the cumulative distribution function is a monotonic increasing function and has the properties

$$F\left(X_{\text{min}}\right) = 0, \qquad F\left(X_{\text{max}}\right) = 1 \tag{7.3}$$

if the range for the random variable is $X_{\text{min}} \leq X \leq X_{\text{max}}$. The PDF is given by the derivative of the cumulative distribution with respect to the random variable.

Definition 7.2 (Probability density function)

$$p(x) = \frac{d}{dx}F(x)$$

with

$$p(x)\,dx = \big(F(x+dx) - F(x)\big) = P\big(x < X \leq x + dx|\mathcal{I}\big).$$

Therefore, the probability density is the probability that X lies in the interval $x < X \leq x + dx$ divided by the infinitesimal interval dx. A generalization of the probability density to higher dimensions is straightforward. In that case it denotes the probability that the random variable X lies within an infinitesimal volume dV_x around x, divided by the volume.

Starting from the probability density function, the cumulative distribution can (in one dimension) be calculated by

[1] In physics a probability density function is often also referred to as a distribution (e.g. temperature distribution, Gaussian distribution), whereas in the statistics literature the term distribution is reserved for the cumulative distribution function. We follow the physics convention and always add 'cumulative' to avoid ambiguities.

$$F(x) = \int_{-\infty}^{x} p(x')\, dx'.$$

For higher dimensions this expression can be adapted correspondingly, e.g. for two dimensions the cumulative distribution is given by

$$F(x_1, x_2) = \int_{-\infty}^{x_1} \int_{-\infty}^{x_2} p(x', x'')\, dx''\, dx'.$$

Cumulative distributions and probability density functions can be defined for discrete problems as well. This allows us to tackle discrete, continuous and mixed problems on the same footing. Assuming that the values of a discrete random variable are given by x_k $(k = 0, 1, \dots, N)$ with probabilities P_k, the cumulative distribution $F(x)$ describes the probability that the value of the random variable x_k is smaller than or equal to x:

$$F(x) = P(x_k \leq x|\mathcal{I}) = \sum_{\substack{k=0 \\ x_k \leq x}}^{N} P_k. \tag{7.4}$$

Based on this CDF, the PDF $p(x)$ is given by

$$p(x) = \sum_{k=0}^{N} P_k\, \delta\,(x - x_k)\,. \tag{7.5}$$

Back-substitution yields

$$F(x) = \int_{-\infty}^{x} p(x)\, dx = \sum_{k=0}^{N} P_k \underbrace{\int_{-\infty}^{x} \delta\left(x' - x_k\right)\, dx'}_{\mathbf{1}(x_k \leq x)} = \sum_{\substack{k=0 \\ x_k \leq x}}^{N} P_k.$$

7.2.1 Example: Dart problem

Let's suppose that darts are randomly and uniformly thrown at a circular target of radius R. For simplicity we assume that there are no darts outside R. We are interested in the probability of hitting within a radius $r \leq R$ (e.g. bull's eye), which corresponds to an area of πr^2. The total area of the target is πR^2. Hence, the probability of interest is given by

$$F(r) = P(X < r|\mathcal{I}) = \begin{cases} 0 & r \leq 0 \\ \frac{r^2}{R^2} & 0 \leq r \leq R \\ 1 & r > R. \end{cases}$$

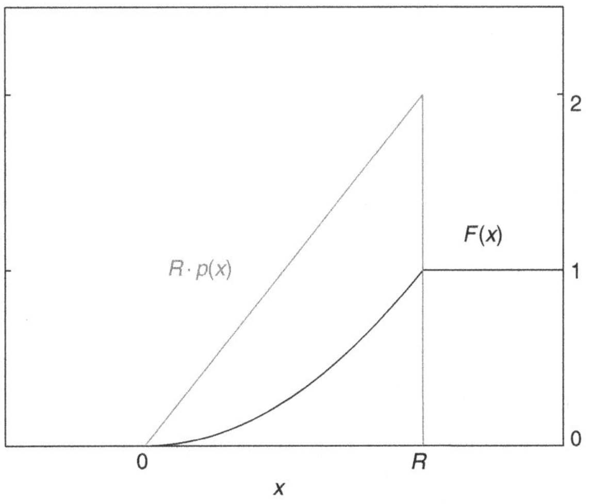

Figure 7.1 Cumulative distribution and probability density of the dart example.

The problem-specific details (i.e. no throws outside the target of radius R, uniform hitting probability) are encoded in the term $\mathcal{I}$. The probability density is provided by the derivative of the CDF:

$$p(r|\mathcal{I}) = \frac{d}{dr}F(r) = \begin{cases} 0 & r \leq 0 \\ \frac{2\,r}{R^2} & 0 < r \leq R \\ 0 & r > R. \end{cases}$$

The probability to hit the infinitesimal annulus of radius $r \in [r, r + dr)$ is zero for all radii with $r < 0$ or $r > R$. For other radii the probability is proportional to the area of the annulus, i.e. it is linearly increasing in r. This is illustrated in Figure 7.1, where the cumulative distribution and probability density of the dart example are displayed.

7.2.2 Example: Coin tossing revisited

We consider a Bernoulli experiment with $n = 6$ repetitions and a probability $p = 1/2$ of the event E. The probability for observing E k times is given by the binomial distribution

$$P(k|n = 6, p = 1/2) = 2^{-6}\binom{6}{k},$$

with the associated cumulative distribution (see equation (7.4))

$$F(x) = P(k \leq x|\mathcal{I}) = 2^{-6} \sum_{\substack{k=0 \\ x_k \leq x}}^{N} \binom{6}{k}.$$

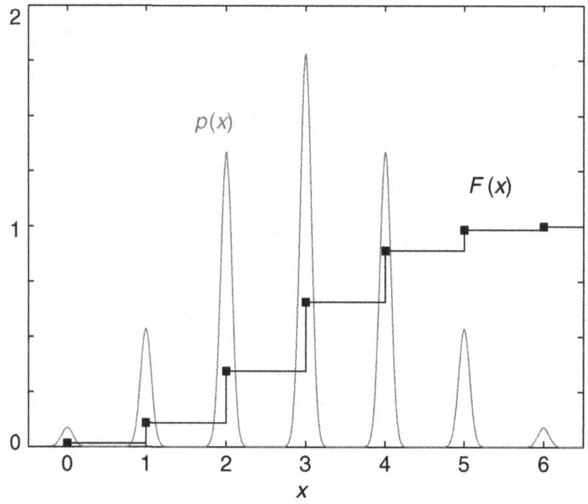

Figure 7.2 Cumulative distribution and probability density function of the Bernoulli distribution for $n = 6$ and $p = 1/2$. For the visualization, δ distributions have been replaced by narrow Gaussians. Solid squares indicate that the intervals are inclusive to the left and exclusive to the right.

The probability density function follows from equation (7.5):

$$p(x|\mathcal{I}) = 2^{-6} \sum_{k=0}^{6} \binom{6}{k} \delta(x-k) .$$

Both the cumulative distribution and the probability density function are displayed in Figure 7.2. Now we are prepared to define another common 'location parameter', the 'median' $\overset{\curlywedge}{x}$ of a random variable. It is constructed such that it splits the probability mass into two equal halves. We will only need it for continuous distributions.

Definition 7.3 (Median of a random variable) *The median of a continuous distribution can best be defined via the cumulative distribution function.*

MEDIAN OF A CONTINUOUS DISTRIBUTION

$$F(\overset{\curlywedge}{x}) := \frac{1}{2}. \tag{7.6}$$

7.3 Application in statistical physics

Classical statistical physics usually deals with high-dimensional problems in phase space. States are mostly represented by vectors $\boldsymbol{x}$ in $\mathbb{R}^N$, where N is the number of particles times twice the spatial dimension. Assume we are interested in some physical quantity $L(\boldsymbol{x})$. The thermodynamic expectation value is defined as

$$\langle L \rangle := \int d^N x \; L(\boldsymbol{x}) \; p(\boldsymbol{x}),$$

where $p(\boldsymbol{x})$ is the PDF in the underlying thermodynamical ensemble. It is often expedient to map this high-dimensional integral to a one-dimensional integral upon introducing the PDF for the L-values:

$$p_L(\lambda) := \int d^N x \; p(\boldsymbol{x}) \; \delta\left(L(\boldsymbol{x}) - \lambda\right).$$

Then we can express the thermodynamic expectation value by

$$\langle L \rangle = \int d\lambda \; \lambda \; p_L(\lambda). \tag{7.7}$$

The proof is as follows:

$$\begin{aligned}\int d\lambda \; p_L(\lambda) \; \lambda &= \int d\lambda \; \lambda \int d^N x \; p(\boldsymbol{x}) \; \delta\left(L(\boldsymbol{x}) - \lambda\right) \\ &= \int d^N x \; L(\boldsymbol{x}) \; p(\boldsymbol{x}) \underbrace{\int d\lambda \; \delta\left(L(\boldsymbol{x}) - \lambda\right)}_{=1}.\end{aligned}$$

Next we introduce the cumulative distribution function

$$F(\lambda) = \int_{-\infty}^{\lambda} p_L(\lambda) \; d\lambda,$$

which can in turn be related to the original integrals in phase space

$$F(\lambda) = \int d^N x \; p(\boldsymbol{x}) \; \boldsymbol{1}\left(L(\boldsymbol{x}) < \lambda\right).$$

It is, so to speak, the probability mass corresponding to $L(\boldsymbol{x}) < \lambda$. Apparently, the CDF is a function in λ that increases monotonically from 0 to 1. Its derivative, the PDF $p_L(\lambda)$, is consequently non-negative. We are sometimes interested in the PDF of the F values, which we obtain due to the monotonicity as follows. For $F \in [0, 1]$

$$p_F(F) = p_L(\lambda) \left| \frac{d\lambda}{dF} \right| = \frac{p_\lambda(\lambda)}{\left|\frac{dF}{d\lambda}\right|} = \frac{p_\lambda(\lambda)}{p_\lambda(\lambda)} = 1.$$

For $F \notin [0, 1]$ clearly $p_F(F) = 0$. Hence the PDF of F-values is:

PDF OF THE CDF VALUES

$$p_F(F) = \mathbf{1}(F \in [0, 1]) \,. \tag{7.8}$$

The same holds true for the complementary CDF, defined as

$$X(\lambda) = \int_{\lambda}^{\infty} p_L(\lambda) \, d\lambda$$
$$= \int d^N x \; p(\boldsymbol{x}) \, \mathbf{1}(L(\boldsymbol{x}) > \lambda) \,.$$

Here we have also:

PDF OF THE COMPLEMENTARY CDF VALUES

$$p_X(X) = \mathbf{1}(X \in [0, 1]) \,. \tag{7.9}$$

7.4 Definitions for continuous distributions

7.4.1 Moments and marginal probabilities

In Section 2.1 [p. 15] we have defined the moments for discrete random variables. Analogously, in the continuous case:

Definition 7.4 (Mean of a continuous random variable) *For a random variable X with probability density $p(X = x) = p(x)$, the mean of X is defined as:*

MEAN OF A CONTINUOUS RANDOM VARIABLE

$$\langle X \rangle := \int_{-\infty}^{\infty} x \; p(x) \, dx \,. \tag{7.10}$$

Again we write $\langle x \rangle$ instead of $\langle X \rangle$ to simplify the notation if there is no risk of confusion. The means of functions of random variables and marginal distributions are defined analogously to equations (2.11) [p. 20] and (2.10) [p. 19], with summations replaced by integrations. The definition of moments can be adapted from Section 2.1 [p. 15].

7.4.2 Definition of a 'sample'

In reality, the density functions of experimental data are almost always unknown. However, realizations (samples, i.e. measured data) of these density functions are experimentally accessible. Samples are therefore of central importance in the estimation and analysis of unknown density functions.

Definition 7.5 (Sample) *Let X be a random variable with cumulative distribution $F(x)$. L independent realizations (e.g. experimental measurements) $\{x_1, x_2, \ldots, x_L\}$ of X represent a sample of size L.*

7.5 Common probability distributions

7.5.1 Uniform distribution

For the representation of the uniform distribution in the interval $[a, b]$ we use the two-sided step function

$$p(x|a, b) = \frac{1}{b-a} \, \mathbf{1}(a \leq x \leq b)\,.$$

The mean is $(b + a)/2$ and the variance yields

$$\left\langle x^2 \right\rangle = \frac{1}{b-a} \int_a^b x^2 \, dx = \frac{(b-a)^2}{12}.$$

Interestingly, the standard deviation is merely $(b - a)/\sqrt{12}$, which is roughly a quarter of the width of the interval $b - a$.

UNIFORM DISTRIBUTION

domain: $x \in [a, b]$

$$p_u(x|a, b) = \frac{1}{b-a}, \tag{7.11a}$$

$$F_u(x|a, b) := \frac{x-a}{b-a}, \tag{7.11b}$$

$$\langle x \rangle = \frac{a+b}{2}, \tag{7.11c}$$

$$\text{mode}(x) = \text{undefined}, \tag{7.11d}$$

$$\text{var}(x) = \frac{(b-a)^2}{12}. \tag{7.11e}$$

The most frequently used uniform distribution is the uniform distribution on the unit interval $p_u(x|0, 1)$ with mean $1/2$ and variance $1/12$. In almost all programming languages a (pseudo-)random number generator for uniform random numbers is provided, typically labelled *rnd* or *rand*. Please be aware that the quality of these internal random number generators may be poor. See, for example, [163] for a detailed discussion of that topic. Random numbers obeying different distributions may be derived using specific transformations, see Section 29.2.2 [p. 517]. A large number of algorithms can also be found in [42].

7.5.2 Beta distribution

In the context of Bernoulli problems the beta distribution (or β distribution) is often of importance.

Beta distribution

domain: $x \in [0, 1]$

$$p_\beta(x|\alpha, \rho) = \frac{1}{B(\alpha, \rho)} \, x^{\alpha-1} \, (1-x)^{\rho-1}, \tag{7.12a}$$

$$F_\beta(x|\alpha, \rho) := \frac{B(x|\alpha, \rho)}{B(\alpha, \rho)}, \tag{7.12b}$$

$$\langle x \rangle = \frac{\alpha}{\alpha + \rho}, \tag{7.12c}$$

$$\text{mode}(x) = \frac{\alpha - 1}{\alpha + \rho - 2} \quad (\alpha, \beta > 1), \tag{7.12d}$$

$$\text{var}(x) = \frac{\langle x \rangle \, (1 - \langle x \rangle)}{\alpha + \rho + 1}. \tag{7.12e}$$

The normalization constant $B(\alpha, \rho)$ is provided by the β integral or β function. If the normalization integral is extended over only the lower part of the domain, the function is referred to as a (lower) 'incomplete β function' $B(x|\alpha, \rho)$.[2] The same wording structure occurs in all distributions. For the X distribution, the normalization integral is the 'X function'. The incomplete normalization is denoted 'incomplete X function'. If the integral runs over the interval $[x_{\text{min}}, x]$ ($[x, x_{\text{max}}]$) it is referred to as a lower (upper) incomplete X function. The CDF of the X distribution is termed a 'regularized incomplete X function' and is the incomplete X function divided by the X function.

[2] Sometimes also called a 'Chebyshev integral'.

BETA FUNCTION AND INCOMPLETE BETA FUNCTION

$$B(\alpha, \rho) := \int_0^1 p^{\alpha-1} (1-p)^{\rho-1} \, dp = \frac{\Gamma(\alpha)\,\Gamma(\rho)}{\Gamma(\alpha+\rho)}, \tag{7.13a}$$

$$B(x|\alpha, \rho) := \int_0^x p^{\alpha-1} (1-p)^{\rho-1} \, dp. \tag{7.13b}$$

Based on equation (7.13a), the mean and variance of the beta distribution can easily be derived:

$$\begin{aligned}\langle x \rangle &= \int_0^1 x \, p_\beta(x|\alpha, \rho) \, dx \\ &= \frac{B(\alpha+1, \rho)}{B(\alpha, \rho)} = \frac{\Gamma(\alpha+1)\,\Gamma(\rho)\,\Gamma(\alpha+\rho)}{\Gamma(\alpha)\,\Gamma(\rho)\,\Gamma(\alpha+\rho+1)} = \frac{\alpha}{\alpha+\rho}, \\ \left\langle x^2 \right\rangle &= \frac{B(\alpha+2, \rho)}{B(\alpha, \rho)} = \frac{\Gamma(\alpha+2)\,\Gamma(\rho)\,\Gamma(\alpha+\rho)}{\Gamma(\alpha)\,\Gamma(\rho)\,\Gamma(\alpha+\rho+2)} = \frac{\alpha\,(\alpha+1)}{(\alpha+\rho)\,(\alpha+\rho+1)}.\end{aligned}$$

The variance follows immediately:

$$\text{var}(x) = \left\langle x^2 \right\rangle - \langle x \rangle^2 = \frac{\langle x \rangle (1 - \langle x \rangle)}{\alpha+\rho+1}.$$

Figure 7.3 displays the CDF and PDF of the β distribution. The parameters α, ρ have been chosen such that $\langle x \rangle = 1/2$ and $\sigma = \text{var}(x) = 0.01, 1/12, 0.2$. The value $\sigma = 1/12$ matches the variance of the uniform distribution on the unit interval. For smaller values of the variance the β density exhibits a maximum at $x = 1/2$, while for $\sigma > 1/12$ the PDF

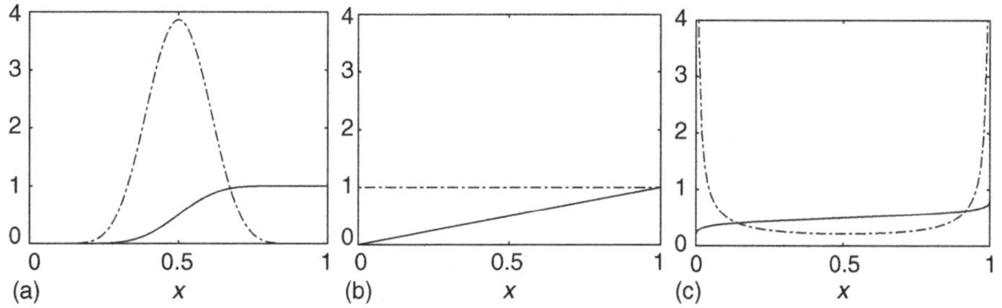

Figure 7.3 Cumulative distribution (solid line) and probability density (dashed line) of the β distribution. The parameters α, ρ have been chosen such that $\langle x \rangle = 1/2$ and $\text{var}(x) = 0.01, 1/12, 0.2$ in (a), (b) and (c).

diverges at the domain boundaries. For $\langle x \rangle = 1/2$ the parameters of the beta distribution are given by

$$\alpha = \rho = \frac{1}{2}\left(\frac{1}{4\sigma} - 1\right).$$

The β functions will be met again in inverse problems based on Bernoulli experiments.

7.5.3 *Gamma and chi-squared distributions*

Another very important distribution, both in physics and probability theory, is the

Γ DISTRIBUTION

domain: $x \in [0, \infty)$

$$p_\Gamma(x|\alpha, \beta) = \frac{\beta^\alpha}{\Gamma(\alpha)}\, x^{\alpha-1}\, e^{-\beta\, x}, \tag{7.14a}$$

$$F_\Gamma(x|\alpha, \beta) := \frac{\beta^\alpha}{\Gamma(\alpha)} \int_0^x t^{\alpha-1}\, e^{-\beta\, t}\, dt, \tag{7.14b}$$

$$\langle x \rangle = \frac{\alpha}{\beta}, \tag{7.14c}$$

$$\text{mode}(x) = \frac{\alpha - 1}{\beta} \qquad (\alpha > 1), \tag{7.14d}$$

$$\text{var}(x) = \frac{\alpha}{\beta^2}. \tag{7.14e}$$

The moments of the Γ distribution can be calculated via

$$\begin{aligned}
\langle x^\nu \rangle = \int_0^\infty x^\nu\, p_\Gamma(x|\alpha, \beta)\, dx &= \frac{\beta^\alpha}{\Gamma(\alpha)} \int_0^\infty x^{(\alpha+\nu)-1}\, e^{-\beta\, x}\, dx \\
&= \frac{1}{\beta^\nu\, \Gamma(\alpha)} \int_0^\infty (\beta x)^{(\alpha+\nu)-1}\, e^{-\beta\, x}\, d(\beta x) \\
&= \frac{1}{\beta^\nu\, \Gamma(\alpha)} \int_0^\infty z^{(\alpha+\nu)-1}\, e^{-z}\, dz \\
&= \frac{\Gamma(\alpha+\nu)}{\beta^\nu\, \Gamma(\alpha)}.
\end{aligned}$$

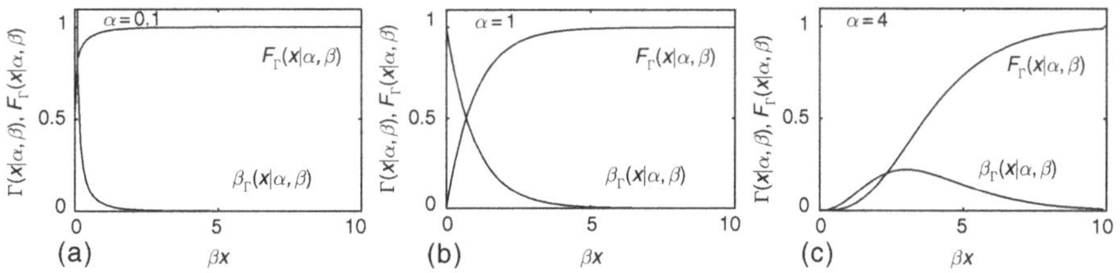

Figure 7.4 CDF and PDF of the Γ distribution for $\alpha = 0.1, 1, 4$.

So we readily confirm the correct normalization and the mean $\langle x \rangle = \alpha/\beta$. Also, the variance is derived easily:

$$\mathrm{var}(x) = \frac{\Gamma(\alpha+2)}{\beta^2\,\Gamma(\alpha)} - \frac{\alpha^2}{\beta^2} = \frac{\alpha(\alpha+1)}{\beta^2} - \frac{\alpha^2}{\beta^2} = \frac{\alpha}{\beta^2}.$$

In Figure 7.4 the cumulative distribution and the probability density of the Γ distribution are displayed for $\alpha = 0.1, 1, 4$. The parameter β affects merely the scaling of the abscissa via βx. For $0 < \alpha < 1$ the density of the Γ distribution diverges at $x = 0$, similarly to Figure 7.4(a). The exponential distribution in Figure 7.4(b) is a special case of the Γ distribution with $\alpha = 1$ (see Section 7.5.4). For parameter values $\alpha > 1$ the Γ distribution exhibits a maximum at $\beta x = \alpha - 1$, similar to Figure 7.4(c). The CDF of the Γ distribution, which is also called the regularized incomplete Γ function, is related to the (lower) 'incomplete Γ function'

$$F_\Gamma(x) = \frac{\Gamma(\beta|x\alpha)}{\Gamma(\alpha)}.$$

Γ FUNCTION AND INCOMPLETE Γ FUNCTION

$$\Gamma(\alpha) := \int_0^\infty t^{\alpha-1}\, e^{-t}\, dt, \tag{7.15a}$$

$$\Gamma(x|\alpha) := \int_0^x t^{\alpha-1}\, e^{-t}\, dt, \tag{7.15b}$$

$$\Gamma(\alpha) = (\alpha-1)\,\Gamma(\alpha-1), \tag{7.15c}$$

$$\Gamma(\alpha) = (n-1)! \quad \text{for } n \in \mathbb{N}, \tag{7.15d}$$

$$\Gamma(1/2) = \sqrt{\pi}. \tag{7.15e}$$

The χ^2 distribution is a special case of the Γ distribution for $(\alpha = \frac{n}{2}, \beta = \frac{1}{2})$.

χ^2 DISTRIBUTION

domain: $x \in [0, \infty)$

$$p_{\chi^2}(x|n) = \frac{2^{-\frac{n}{2}}}{\Gamma(\frac{n}{2})}\, x^{\frac{n}{2}-1}\, e^{-\frac{1}{2}x} \tag{7.16a}$$

$$= p_\Gamma(x|\alpha = \tfrac{n}{2}, \beta = \tfrac{1}{2}),$$

$$F_{\chi^2}(x|n) = \frac{2^{-\frac{n}{2}}}{\Gamma(\frac{n}{2})} \int_0^\infty t^{\frac{n}{2}-1}\, e^{-\frac{1}{2}t}\, dt, \tag{7.16b}$$

$$\langle x \rangle = n, \tag{7.16c}$$

$$\text{mode}(x) = \max(n-2, 0), \tag{7.16d}$$

$$\text{var}(x) = 2n, \tag{7.16e}$$

n : number of degrees of freedom.

7.5.4 Exponential distribution

An important special case is the exponential distribution, given by

EXPONENTIAL DISTRIBUTION

domain: $x \in [0, \infty)$

$$p_e(x|\lambda) = \lambda\, e^{-\lambda x} = p_\Gamma(x|\alpha = 1, \beta = \lambda), \tag{7.17a}$$

$$F_e(x|\lambda) = 1 - e^{-\lambda x}, \tag{7.17b}$$

$$\langle x \rangle = \frac{1}{\lambda}, \quad \text{mode}(x) = 0, \tag{7.17c}$$

$$\text{var}(x) = \frac{1}{\lambda^2}. \tag{7.17d}$$

7.5.5 Inverse Gamma distribution

We will frequently encounter the ‘inverse Gamma distribution’.

INVERSE GAMMA DISTRIBUTION

domain: $x \in [0, \infty)$

$$p_{i\Gamma}(x|\alpha, \beta) = \frac{\beta^\alpha}{\Gamma(\alpha)}\, x^{-\alpha-1}\, e^{-\frac{\beta}{x}}, \tag{7.18a}$$

$$F_{i\Gamma}(x|\alpha, \beta) := \frac{\beta^\alpha}{\Gamma(\alpha)} \int_0^x t^{-\alpha-1}\, e^{-\frac{\beta}{t}}\, dt = \Gamma\left(\frac{\beta}{x}, \alpha\right), \tag{7.18b}$$

$$\langle x \rangle = \frac{\beta}{\alpha - 1}, \quad \text{mode}(x) = \frac{\beta}{\alpha + 1}, \tag{7.18c}$$

$$\text{var}(x) = \frac{\beta^2}{(\alpha - 1)^2(\alpha - 2)}. \tag{7.18d}$$

Moments and CDF can easily be calculated upon substitution $t = \beta/x$.

7.5.6 Gaussian or normal distribution

Definition 7.6 (Gaussian or normal distribution) *By far the most important continuous probability distribution is the Gaussian or normal distribution. A random variable is normally distributed if the probability density takes the form*

GAUSSIAN DISTRIBUTION $\mathcal{N}(x|x_0, \sigma)$

domain : $x \in \mathbb{R}$

$$p(x|x_0, \sigma) = \frac{1}{\sqrt{2\pi\sigma^2}}\, e^{-(x-x_0)^2/2\sigma^2}, \tag{7.19a}$$

$$F(x|x_0, \sigma) = \Phi\left(\frac{x - x_0}{\sigma}\right), \tag{7.19b}$$

$$\langle x \rangle = x_0, \tag{7.19c}$$

$$\text{var}(x) = \sigma^2, \tag{7.19d}$$

$$\left\langle (x - x_0)^n \right\rangle = \begin{cases} \frac{(2\sigma^2)^{\frac{n}{2}}}{\sqrt{\pi}}\, \Gamma\left(\frac{n+1}{2}\right) & \text{for even } n \\ 0 & \text{for odd } n. \end{cases} \tag{7.19e}$$

In this context the function

$$\Phi(x) = \frac{1}{\sqrt{2\pi}} \int_{-\infty}^{x} e^{-t^2/2}\, dt \tag{7.20}$$

is of importance. It is a monotonically increasing function and satisfies the symmetry relation

$$\Phi(-x) = 1 - \Phi(x). \tag{7.21}$$

Special values are $\Phi(-\infty) = 0$, $\Phi(0) = 1/2$ and $\Phi(\infty) = 1$. Closely related to the Φ function is the error function

$$\mathrm{erf}(x) = \frac{2}{\sqrt{\pi}} \int_{0}^{x} e^{-t^2}\, dt = \frac{2}{\sqrt{2\pi}} \int_{0}^{\sqrt{2}x} e^{-t^2/2}\, dt = 2\,\Phi(\sqrt{2}x). \tag{7.22}$$

The error function is implemented in most mathematical software packages and in some programming languages (e.g. Fortran 2008) as a standard function. Algorithms for its computation can be found in most textbooks on mathematical functions or, for example, in [164].

Figure 7.5 displays the probability density and the cumulative distribution function of the Gaussian distribution.

The computation of the moments of the Gaussian distribution is slightly more tedious. The mean is $\langle x \rangle = x_0$ due to the symmetry of the Gaussian function. Using the same symmetry argument it follows that all odd moments of the Gaussian distribution are zero. For the centred even moments we obtain

$$\left\langle (x - x_0)^n \right\rangle = \frac{2}{\sqrt{2\pi\sigma^2}} \int_{0}^{\infty} z^{n+1}\, e^{-z^2/2\sigma^2}\, \frac{dz}{z}.$$

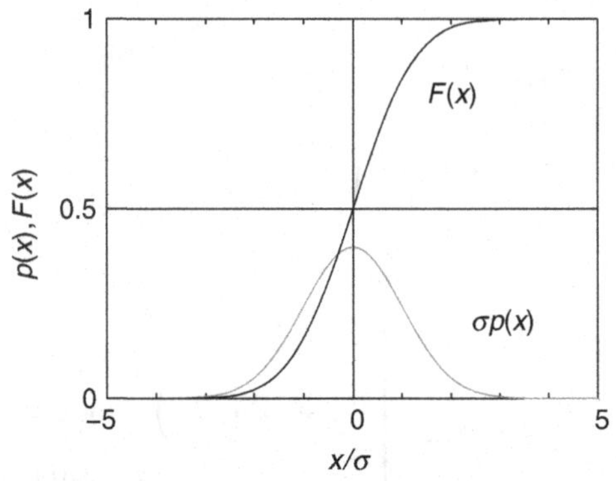

Figure 7.5 PDF and CDF of a Gaussian distribution centred at zero.

Next use the substitution $t = z^2/2\sigma^2$, which will occur repeatedly throughout this book, and find

$$\left\langle (x-x_0)^n \right\rangle = \frac{(2\sigma^2)^{\frac{n+1}{2}}}{\sqrt{2\pi\sigma^2}} \int_0^\infty t^{\frac{n+1}{2}}\, e^{-t}\, \frac{dt}{t} = \frac{1}{\sqrt{\pi}} (2\sigma^2)^{\frac{n}{2}}\, \Gamma\left(\frac{n+1}{2}\right). \tag{7.23}$$

The result for the variance ($n = 2$), $\mathrm{var}(x) = \sigma^2$, follows from equations (7.15c) and (7.15e).

Marginalization of the variance

Very often in data analysis problems there is strong prior information that the data follow a Gaussian probability density function, however, the variance of the density function is unknown. If N data points have been measured to estimate a parameter μ and all data points share a common uncertainty σ, then the likelihood is given by

$$\begin{aligned} p(\boldsymbol{d}|\mu,\sigma,\mathcal{I}) &= \prod_{i=1}^{N} \frac{1}{\sqrt{2\pi\sigma^2}} \exp\left(-\frac{(\mu-d_i)^2}{2\sigma^2}\right) \\ &= (2\pi\sigma^2)^{-\frac{N}{2}} \exp\left\{-\frac{1}{2\sigma^2}\left(\sum_{i=1}^{N}(\mu-d_i)^2\right)\right\}. \end{aligned}$$

We can express the sum over the data

$$\sum_{i=1}^{N}(\mu-d_i)^2 = N\left((\mu-\overline{d})^2 + \overline{(\Delta d)^2}\right)$$

in terms of sample mean $\overline{d}$ and sample variance $\overline{(\Delta d)^2}$, which are defined as

$$\begin{aligned} \overline{d} &:= \frac{1}{N}\sum_{i}^{N} d_i, \\ \overline{(\Delta d)^2} &:= \frac{1}{N}\left(\sum_{i}^{N}(d_i-\overline{d})^2\right). \end{aligned}$$

Then the likelihood reads

$$p(\boldsymbol{d}|\mu,\sigma,\mathcal{I}) = (2\pi\sigma^2)^{-\frac{N}{2}} \exp\left\{-\frac{N}{2\sigma^2}\left((\mu-\overline{d})^2 + \overline{(\Delta d)^2}\right)\right\}. \tag{7.24}$$

Since the variance σ^2 is unknown, we have to compute the marginal likelihood $p(\boldsymbol{d}|\mu,\mathcal{I})$ using the marginalization rule

$$p(\boldsymbol{d}|\mu,\mathcal{I}) = \int_0^\infty p(\boldsymbol{d}|\mu,\sigma,\mathcal{I})\, p(\sigma|\mathcal{I})\, d\sigma. \tag{7.25}$$

The standard deviation σ is a scale variable. Without further information an appropriate (although improper) prior is Jeffreys' prior for scale variables (see Section 10.2 [p. 169]). Then the remaining integral is of the form

$$\mathcal{I}(v, N) := \int_0^\infty \sigma^{-N}\, e^{-\frac{v}{2\sigma^2}}\, \frac{d\sigma}{\sigma}. \tag{7.26}$$

Here we have introduced the abbreviation $v = N\left((\mu - \overline{d})^2 + \overline{(\Delta d)^2}\right)$. Upon substitution with $t = \frac{v}{2\sigma^2}$ the integral can be transformed into a Gamma function and yields

$$\mathcal{I}(v, N) = \int_0^\infty \sigma^{-N}\, e^{-\frac{v}{2\sigma^2}}\, \frac{d\sigma}{\sigma} = \frac{1}{2}\left(\frac{v}{2}\right)^{-\frac{N}{2}} \Gamma\left(\frac{N}{2}\right). \tag{7.27}$$

By now, the marginal likelihood reads

$$p(\boldsymbol{d}|\mu, \mathcal{I}) \propto \left((\mu - \overline{d})^2 + \overline{(\Delta d)^2}\right)^{-\frac{N}{2}} = \frac{1}{Z}\left(1 + \frac{(\mu - \overline{d})^2}{\overline{(\Delta d)^2}}\right)^{-\frac{N}{2}}.$$

Finally we have to normalize the marginal likelihood. The normalization integral has the form

$$Z = \int_{-\infty}^{\infty} dx \left(1 + \frac{x^2}{s^2}\right)^{-\frac{N}{2}} = \sqrt{\pi s^2}\, \frac{\Gamma(\frac{N-1}{2})}{\Gamma(\frac{N}{2})}, \tag{7.28}$$

with $s^2 := \overline{(\Delta d)^2}$. The marginal likelihood is therefore

MARGINAL LIKELIHOOD

$$p(\boldsymbol{d}|\mu, \mathcal{I}) = \frac{\Gamma(\frac{N}{2})}{\sqrt{\pi\, \overline{(\Delta d)^2}}\, \Gamma(\frac{N-1}{2})}\left(1 + \frac{(\mu - \overline{d})^2}{\overline{(\Delta d)^2}}\right)^{-\frac{N}{2}}. \tag{7.29}$$

This is the so-called Student's t-distribution, which we will discuss in the next section. For $N \gg 1$ the distribution converges towards a Gaussian

$$p(\boldsymbol{d}|\mu, \mathcal{I}) \underset{N\gg 1}{\longrightarrow} \frac{1}{\sqrt{2\pi\ \mathrm{SE}^{*2}}} \exp\left\{-\frac{-(\mu - \overline{d})^2}{2\ \mathrm{SE}^2}\right\}, \tag{7.30}$$

$$\text{with}\quad \mathrm{SE}^{*2} := \frac{\overline{(\Delta d)^2}}{N}.$$

7.5.7 Student's t-distribution

Student's t-distribution was introduced by the statistician William Gosset, who was required by his employer Guinness Breweries to publish under the pseudonym 'Student'. As we have seen before, it typically arises when an unknown variance σ^2 is marginalized from a Gaussian distribution $p_{\mathcal{N}}(x|x_0,\sigma)$. The distribution is also of central importance in hypothesis testing if two samples from Gaussian distributions with unknown variance share the same mean value, see Section 19.4 [p. 299].

STUDENT'S t-DISTRIBUTION

---

domain: $t \in \mathbb{R}$

$$p_t(t|\nu) := \frac{1}{\sqrt{\nu}\, B(\frac{1}{2}, \frac{\nu}{2})} \left(1 + \frac{t^2}{\nu}\right)^{-\frac{\nu+1}{2}}, \tag{7.31a}$$

$$F_t(\tau|\nu) = \frac{1}{2}\left(1 + \mathrm{sign}(\tau)\left[1 - \frac{B\left(\left[1+\frac{\tau^2}{\nu}\right]^{-1}|\frac{1}{2}, \frac{\nu}{2}\right)}{2\, B(\frac{1}{2}, \frac{\nu}{2})}\right]\right), \tag{7.31b}$$

$$\langle t \rangle = 0, \tag{7.31c}$$

$$\mathrm{var}(t) = \frac{\nu}{\nu - 2} \quad \text{for } \nu > 2, \tag{7.31d}$$

ν : degrees of freedom.

Proof of CDF: For the computation of the cumulative distribution we consider only $\tau > 0$, since the other case can easily be derived by symmetry considerations. Then

$$F_t(\tau|\nu) = 1 - \left[B\left(\frac{1}{2}, \frac{\nu}{2}\right)\right]^{-1} \underbrace{\int_{\tau}^{\infty} \left(1 + \frac{t^2}{\nu}\right)^{-\frac{\nu+1}{2}} \frac{dt}{\sqrt{\nu}}}_{:=I}.$$

The remaining integral

$$I = \int_{\tau/\sqrt{\nu}}^{\infty} \left(1 + x^2\right)^{-\frac{\nu+1}{2}} dx$$

can be simplified upon substitution with

$$\xi = (1 + x^2)^{-1}\,; \qquad \Rightarrow dx = \frac{1}{2}\,\xi^{-\frac{3}{2}}\,(1-\xi)^{-\frac{1}{2}}\,dx,$$

resulting in

$$I = \frac{1}{2} \int_0^{1/(1+\frac{\tau^2}{\nu})} \xi^{\frac{\nu}{2}-1} (1-\xi)^{\frac{1}{2}-1}\, dx = \frac{1}{2}\, B\left(\left[1+\frac{\tau^2}{\nu}\right]^{-1} \Big| \frac{\nu}{2}, \frac{1}{2}\right).$$

Hence the CDF for $\tau \geq 0$ is

$$F(\tau) = 1 - \frac{B\left(\left[1+\frac{\tau^2}{\nu}\right]^{-1} \big| \frac{1}{2}, \frac{\nu}{2}\right)}{2\, B\left(\frac{1}{2}, \frac{\nu}{2}\right)}.$$

For $-\tau \leq 0$ we can use $F(-\tau) = 1 - F(\tau)$. The two cases can be combined to equation (7.31b). □

7.5.8 Cauchy distribution

The Cauchy distribution is a special case of Student's t-distribution with $\nu = 1$.

CAUCHY DISTRIBUTION

domain: $t \in \mathbb{R}$

$$p_C(x) = \frac{1}{\pi\,(1+x^2)}, \tag{7.32a}$$

$$F_C(x) = \frac{1}{2} + \frac{\arctan(x)}{\pi}, \tag{7.32b}$$

$$\langle x \rangle = \text{undefined}, \tag{7.32c}$$

$$\text{var}(x) = \infty. \tag{7.32d}$$

In spectroscopic applications this probability density is also known as a 'Lorentz function'. Here also the 'Voigt density' arises, which is given by a convolution of the Lorentzian function with a Gaussian function. However, there exists no analytical expression of the Voigt density in closed form.

The PDF and CDF of the Cauchy distribution are displayed in Figure 7.6. The Cauchy distribution decays so slowly in the tails (as $|x|^{-2}$) that moments higher than the zeroth moment, like the mean and variance, do not exist. For this reason the law of large numbers does not apply to Cauchy-distributed random variables. Another distribution where the mean and variance are undefined is, for example, the Landau distribution [122], which plays a role in the description of energy loss of charged particles in matter [141]. At the same time the Landau distribution belongs (like the Gaussian distribution, the Cauchy distribution and the Levy distribution) to the class of stable distributions,

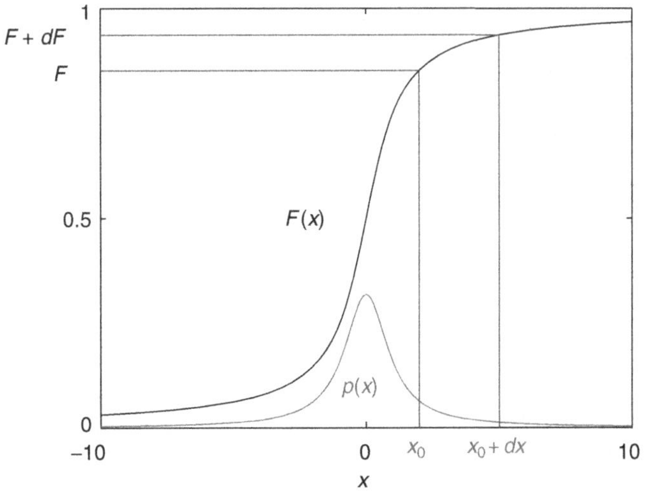

Figure 7.6 Probability density function and cumulative distribution function of the Cauchy distribution.

see Section 8.2 [p. 142]. The probability density function of the Landau distribution is given by

Landau distribution

$$p(x) = \frac{1}{\pi} \int_0^\infty e^{-t \log t - xt} \sin(\pi t)\, dt. \tag{7.33}$$

7.5.9 Dirichlet distribution

Next we consider the generalization of the β distribution to the multivariate case. As motivation, we consider the situation where there are L cells in phase space (boxes) into which N classical particles (balls) are distributed independently at random. The corresponding likelihood for finding n_i particles in cell i with prior probabilities q_i is multinomial:

$$P(\boldsymbol{n}|\boldsymbol{q}, L, N, \mathcal{I}) = \binom{N}{\boldsymbol{n}} \prod_{i=1}^{L} q_i^{n_i}.$$

In many cases, the prior probabilities are not known and their determination is the goal of experiments. Given the experimentally observed occupation numbers $\boldsymbol{n}$ with a total count $\sum_i n_i = N$, we infer the unknown prior probabilities $\boldsymbol{q}$ by virtue of Bayes' theorem:

$$p(\boldsymbol{q}|\boldsymbol{n}, L, N, \mathcal{I}) = \frac{P(\boldsymbol{n}|\boldsymbol{q}, L, N, \mathcal{I})\; p(\boldsymbol{q}|L, \not{N}, \mathcal{I})}{Z}.$$

The most ignorant situation as far as the prior probabilities are concerned is encoded in a uniform prior subject to the normalization constraint $\sum_i q_i = 1$, i.e.

$$p(\boldsymbol{q}|L,\mathcal{I}) = \delta\left(\sum_i q_i - 1\right).$$

The resulting posterior

$$p(\boldsymbol{q}|\boldsymbol{n},L,N,\mathcal{I}) = \frac{1}{Z'}\prod_i q_i^{n_i}\;\delta\left(\sum_i q_i - 1\right)$$

is a Dirichlet distribution, which has the following standard notation:

$$p_D(\boldsymbol{q}|\boldsymbol{\alpha},L,\mathcal{I}) = \frac{1}{Z'}\prod_i q_i^{\alpha_i-1}\;\delta\left(\sum_i q_i - 1\right).$$

The normalization constant is given by

$$Z' = \int_0^1 \cdots \int_0^1 \prod_{i=1}^{L} d^L q\; q_i^{\alpha_i-1}\;\delta\left(\sum_i q_i - 1\right).$$

We will determine it by a recursive approach, that can be reused subsequently to compute marginal probabilities. To this end we introduce a couple of definitions:

$$\begin{aligned}
S_m &:= \sum_{i=m}^{L} q_i,\\
A_m &:= \sum_{i=1}^{m} \alpha_i,\\
Z_1(q_2,\ldots,q_L) &:= \int_0^1 q_1^{\alpha_1-1}\;\delta\left(S_1 - 1\right)\,dq_1,\\
Z_m(q_{m+1},\ldots,q_L) &:= \int_0^1 q_m^{\alpha_m-1} Z_{m-1}(q_m,\ldots,q_L)\,dq_m \qquad \text{for } m = 2,\ldots,L.
\end{aligned}$$

The sought-for normalization is then given by $Z' = Z_L$. Throughout the following steps $L \geq 1$ is arbitrary but fixed. We start out with the computation of Z_1. The allowed values of the variables are $q_i \in [0, 1]$ and hence $S_m \geq 0$. These obvious constraints will be suppressed in the following considerations:

$$\begin{aligned}
Z_1(q_2,\ldots,q_L) &:= \int_0^1 q_1^{\alpha_1-1}\;\delta\left(q_1 - (1 - S_2)\right)\,dq_1\\
&= (1 - S_2)^{\alpha_1-1}\;\boldsymbol{1}(S_2 < 1).
\end{aligned}$$

The Heaviside functions enter since otherwise the peak of the δ function lies outside the domain of integration and the resulting integral vanishes. Now we proceed with Z_2, which will reveal the general structure of the remaining recursion:

$$Z_2(q_3,\dots,q_L) := \int_0^1 q_2^{\alpha_2-1}\left(1-S_2\right)^{\alpha_1-1} \boldsymbol{1}(S_2<1)\; dq_2$$
$$= \int_0^1 q_2^{\alpha_2-1}\left(1-S_3-q_2\right)^{\alpha_1-1} \boldsymbol{1}(q_2<1-S_3)\; dq_2.$$

The integral vanishes for $S_3 \geq 1$. For $S_3 < 1$ we substitute $q_2 = y(1-S_3)$:

$$Z_2 = (1-S_3)^{\alpha_1+\alpha_2-1}\boldsymbol{1}(S_3<1)\int_0^1 y^{\alpha_2-1}\,(1-y)^{\alpha_1-1}\,dy$$
$$= \frac{\Gamma(\alpha_1)\Gamma(\alpha_2)}{\Gamma(A_2)}\,(1-S_3)^{\alpha_1+\alpha_2-1}\boldsymbol{1}(S_3<1)\,.$$

If we are interested in $L = 2$ then we are done, since then $S_3 = \sum_{i=3}^{2} q_i = 0$, and with

$$Z_2 = \frac{\Gamma(\alpha_1)\Gamma(\alpha_2)}{\Gamma(A_2)}$$

the normalized Dirichlet PDF for $L = 2$ is

$$P_D(\boldsymbol{q}|\boldsymbol{\alpha}, L=2,\mathcal{I}) = \Gamma(\alpha_1+\alpha_2)\prod_{i=1}^{2}\frac{q_i^{\alpha_i-1}}{\Gamma(\alpha_i)}\;\delta\left(\sum_i q_i - 1\right).$$

The Dirichlet distribution in the binary case is apparently identical to the β distribution. For $L > 2$ we continue with the integration over q_3, which has the same structure as the integral over q_2, namely

$$Z_3(q_4,\dots,q_L) = \frac{\Gamma(\alpha_1)\Gamma(\alpha_2)}{\Gamma(A_2)}\int_0^1 q_3^{\alpha_3-1}\,(1-S_3)^{A_2-1}\,\boldsymbol{1}(S_3<1)\;dq_3\,\boldsymbol{1}(S_4<1)$$
$$= \frac{\Gamma(\alpha_1)\Gamma(\alpha_2)}{\cancel{\Gamma(A_2)}}\,\frac{\cancel{\Gamma(A_2)}\,\Gamma(\alpha_3)}{\Gamma(A_3)}\,\left(1-S_4\right)^{A_3-1}\,\boldsymbol{1}(S_4<1)$$

and the emerging general form reads

$$Z_m(q_{m+1},\dots,q_L) = \frac{\prod_{i=1}^{m}\Gamma(\alpha_i)}{\Gamma(A_m)}\left(1-S_{m+1}\right)^{A_m-1}\boldsymbol{1}(S_{m+1}<1)\,.$$

For $Z' = Z_L$ we again have $S_L = 0$ and therefore the normalized Dirichlet distribution reads

Dirichlet PDF

$$p(\boldsymbol{q}|\boldsymbol{\alpha}, L, \mathcal{I}) = \Gamma\left(\sum_i \alpha_i\right) \prod_{i=1}^{L} \frac{q_i^{\alpha_i - 1}}{\Gamma(\alpha_i)}\ \delta\left(\sum_i q_i - 1\right). \tag{7.34}$$

A special case of the Dirichlet distribution is the uninformative uniform joint distribution, corresponding to $\alpha_i = 1$.

Marginal Dirichlet probabilities

Based on the recursion formula, we can easily determine the marginal Dirichlet PDF

$$\begin{aligned}
p(q_L|\boldsymbol{\alpha}, L; \mathcal{I}) &= \int_0^1 \dots \int_0^1 p(\boldsymbol{q}|\boldsymbol{\alpha}, \mathcal{I})\, dq_1 \dots dq_{L-1} \\
&= \frac{\Gamma(A_L)}{\prod_{i=1}^{L} \Gamma(\alpha_i)} \int_0^1 \dots \int_0^1 \prod_{i=1}^{L} q_i^{\alpha_i - 1} \delta\left(S_1 - 1\right)\, dq_1 \dots dq_{L-1} \\
&= \frac{\Gamma(A_L)}{\prod_{i=1}^{L} \Gamma(\alpha_i)}\ q_L^{\alpha_L - 1}\ Z_{L-1}(q_L) \\
&= \frac{\Gamma(A_L)}{\prod_{i=1}^{L} \Gamma(\alpha_i)}\ \frac{\prod_{i=1}^{L-1} \Gamma(\alpha_i)}{\Gamma(A_{L-1})}\ q_L^{\alpha_L - 1} \left(1 - S_L\right)^{A_{L-1} - 1} \boldsymbol{1}(S_L < 1) \\
&= \frac{\Gamma(A_L)}{\Gamma(\alpha_L)\Gamma(A_L - \alpha_L)}\ q_L^{\alpha_L - 1} \left(1 - q_L\right)^{A_L - \alpha_L - 1} \boldsymbol{1}(q_L < 1)\,,
\end{aligned}$$

which is again a β distribution. For variables other than q_L, we merely have to relabel the variables and obtain the general formula

Marginal Dirichlet PDF

$$p(q_\nu|\boldsymbol{\alpha}, L; \mathcal{I}) = \frac{\Gamma(A_L)}{\Gamma(\alpha_\nu)\Gamma(A_L - \alpha_\nu)}\ q_L^{\alpha_\nu - 1}\ \left(1 - q_L\right)^{A_L - \alpha_\nu - 1}. \tag{7.35}$$

We have dropped the obvious constraints for $q_\nu \in [0, 1]$ and $\sum_i q_i = 1$. A maybe unexpected result is obtained for the uninformative uniform prior ($\alpha_i = 1$), where the marginal PDF is not uniform any longer but rather given by the distribution $(L - 1)(1 - q_\nu)^{L-2}$, that is peaked at $q_\nu = 0$. On closer inspection the result is very reasonable, as the value $q_\nu = 1$ can only be achieved if all the other variables are zero, i.e. a single point in q-space, whereas the value $q_\nu = 0$ only restricts the sum of the other variables to one, which represents a hyperplane in q-space.

Generating a Dirichlet sample

In many applications one needs samples from a Dirichlet distribution in order to study probabilities or concentrations. A particularly important special case is the uniform prior for L probabilities (concentrations) q_i that add up to one, $\sum_{i=1}^{L} q_i = 1$. For a uniform joint PDF $p(\boldsymbol{q}|L,\mathcal{I}) = \delta\left(\sum_i q_i - 1\right)$ an idea that immediately crosses one's mind is to generate L random numbers x_i from a uniform distribution on the unit interval and to assign

$$q_i = \frac{x_i}{S}, \qquad \text{with } S := \sum_i x_i .$$

This information shall be encoded in the background information $\mathcal{I}_1$. The resulting PDF is determined by introducing the sample $\boldsymbol{x}$ and S via the marginalization rule

$$\begin{aligned}
p(\boldsymbol{q}|L,\mathcal{I}_1) &= \int_0^\infty dS \int_0^1 d^L x \; p(\boldsymbol{q}|\boldsymbol{\alpha},\boldsymbol{x},S,L,\mathcal{I}_1)\; p(S|\boldsymbol{\alpha},\boldsymbol{x},L,\mathcal{I}_1)\; p(\boldsymbol{x}|\boldsymbol{\alpha},L,\mathcal{I}_1) \\
&= \int_0^\infty dS \int_0^1 d^L x \prod_i \delta\Big(q_i - \frac{x_i}{S}\Big)\, \delta\Big(S - \sum_i x_i\Big) \prod_i p(x_i|\alpha_i,\mathcal{I}_1) \\
&= \int_0^\infty dS\; S^{L-1} \int_0^1 d^L x\; \delta\Big(1 - \sum_i \frac{x_i}{S}\Big) \ldots \prod_i \delta\Big(x_i - Sq_i\Big) \prod_i p(x_i|\alpha_i,\mathcal{I}_1) \\
&= \delta\Big(1 - \sum_i q_i\Big) \int_0^\infty dS\; S^{L-1} \prod_i p(x_i = Sq_i|\alpha_i,\mathcal{I}_1).
\end{aligned} \tag{7.36}$$

Now we exploit the individual uniform PDFs for $p(x_i|\alpha_1,\mathcal{I}_1)$:

$$\begin{aligned}
p(\boldsymbol{q}|L,\mathcal{I}_1) &= \delta\Big(1 - \sum_i q_i\Big) \int_0^\infty dS\; S^{L-1} \prod_i \mathbb{1}(0 \le Sq_i \le 1) \\
&= \delta\Big(1 - \sum_i q_i\Big) \int_0^\infty dS\; S^{L-1}\, \mathbb{1}\Big(0 \le S \le \min_i \frac{1}{q_i}\Big) \\
&= \delta\Big(1 - \sum_i q_i\Big) \frac{1}{L} \min_i \frac{1}{q_i^L}.
\end{aligned} \tag{7.37}$$

Apparently, this is not the uniform joint PDF we were aiming at. With hindsight the result is reasonable, since there is more probability mass in x_i-space to generate $\boldsymbol{q} = \frac{1}{\sqrt{L}}(1,\ldots,1)$

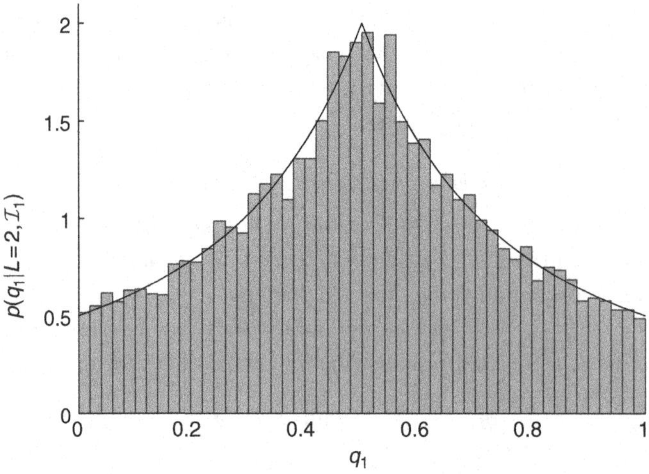

Figure 7.7 Computer simulation (bar diagram) of a naive attempt to generate a 2-D uniform sample of probabilities/concentrations as described in the text. The solid line represents the theoretical prediction of the marginal PDF, according to the naive algorithm.

than there is to generate e.g. $\boldsymbol{q} = (1, 0, \ldots, 0)$. It is instructive to analyse the result for $L = 2$. In this case the δ function forces $q_2 = 1 - q_1$. Then, the PDF is given by

$$p(q_1|L=2,\mathcal{I}_1) = \frac{1}{2}\min_i\left(\frac{1}{q_1^2}, \frac{1}{(1-q_1)^2}\right) = \frac{1}{2}\begin{cases} \frac{1}{(1-q_1)^2} & \text{for } q_1 \le \frac{1}{2} \\ \frac{1}{q_1^2} & \text{for } q_1 \ge \frac{1}{2}. \end{cases} \tag{7.38}$$

In Figure 7.7 the analytic result is compared with that of a computer simulation of the naive approach, in which a sample of size $N = 10\,000$ of pairs (x_1, x_2) of uniform random numbers has been generated and the concentrations were computed by $q_i = x_i/(x_1 + x_2)$. We observe a close agreement between the curve for equation (7.38) and the computer simulation. So, all is well but not what we wanted.

After this digression, we want to reveal a correct approach to sample from the Dirichlet distribution. We will stick to the appealing idea, to generate L independent samples x_i first, and enforce the normalization at the end by the assignment $q_i = x_i / \sum_i x_i$. The problem of the naive approach originates from the shape of the finite domain of the uniform PDF. In order to avoid the geometrically induced non-uniform density, we have to employ a random variable defined on a half-infinite domain. The only distribution with a half-infinite domain that we have encountered so far is the Γ distribution. So, let us give it a try. We encode this information in the background information $\mathcal{I}_2$. The first couple of steps in the computation of the joint probability $p(\boldsymbol{q}|\boldsymbol{\alpha}, L, \mathcal{I}_2)$, up to equation (7.36), are the same as in the case of the uniform distribution. Inserting now the Γ distribution yields

$$
\begin{aligned}
p(\boldsymbol{q}|L,\mathcal{I}_1) &= \delta\Big(1-\sum_i q_i\Big) \int_0^\infty dS\, S^{L-1} \prod_i \left(\frac{(Sq_i)^{\alpha_i-1}}{\Gamma(\alpha_i)} e^{-Sq_i}\right) \\
&= \delta\Big(1-\sum_i q_i\Big) \prod_i \frac{q_i^{\alpha_i-1}}{\Gamma(\alpha_i)} \int_0^\infty dS\, S^{\sum_i \alpha_i - 1} e^{-S} \\
&= \Gamma\Big(\sum_i \alpha_i\Big) \prod_i \frac{q_i^{\alpha_i-1}}{\Gamma(\alpha_i)}\, \delta\Big(1-\sum_i q_i\Big).
\end{aligned}
$$

This approach obviously works and we can generate a sample of the Dirichlet distribution with parameters $(\alpha_1, \ldots, \alpha_L)$ upon drawing a sample $\{x_1, \ldots, x_L\}$ of independent random numbers from a Γ distribution followed by the normalization $q_i = x_i / \sum_i x_i$. In particular, a sample of random numbers which sum up to one with a uniform joint PDF can be generated according to the above considerations using independent and exponentially distributed random numbers ($\alpha_i \overset{!}{=} 1$), followed by normalization. As a check, we perform a computer simulation for the case $L = 2$. Exponential random numbers can easily be produced by computing the logarithm of uniform random numbers (*rand*) on the unit interval, i.e. $r = \log(\textit{rand})$ is an exponential random number.

The result is depicted in Figure 7.8. It illustrates that this approach indeed yields the correct uniform distribution.

7.5.10 Multivariate Gaussian distribution

A particularly interesting distribution is the 'multivariate Gaussian distribution', not only because of its ubiquitous occurrence in applications but also because of the unique mathematical properties. The probability density function of the distribution is given by

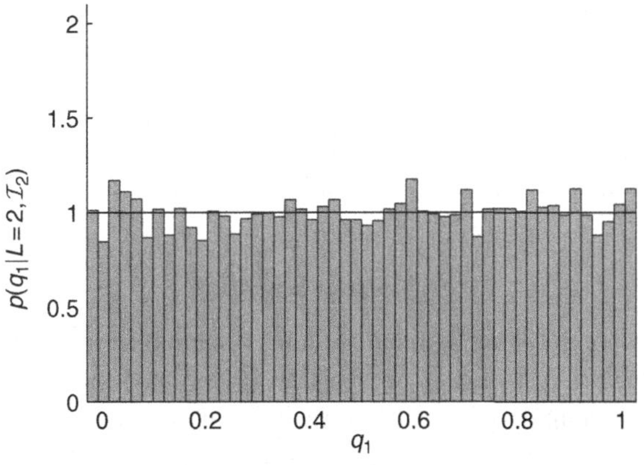

Figure 7.8 Computer simulation (bar diagram) of the correct approach to generate a 2-D uniform sample of probabilities/concentrations by samples of independent exponential random numbers. The solid line represents the sought-for uniform PDF.

Multivariate Gaussian distribution

domain: $x \in \mathbb{R}^n$

$$p_N(\boldsymbol{x}|C, \boldsymbol{x}^0) := \frac{1}{\sqrt{(2\pi)^n\ |\mathbf{C}|}}\ e^{-\frac{1}{2}(\boldsymbol{x}-\boldsymbol{x}^0)^T\ \mathbf{C}^{-1}\ (\boldsymbol{x}-\boldsymbol{x}^0)}, \tag{7.39a}$$

$$\langle x_i \rangle = x_i^0, \tag{7.39b}$$

$$\operatorname{cov}(x_i, x_j) = C_{ij}, \tag{7.39c}$$

$$|\mathbf{C}| = \det(\mathbf{C}). \tag{7.39d}$$

The following formula is often very useful in combination with Gaussian distributions:

Gaussian integrals

$$\int e^{-\frac{1}{2}((\boldsymbol{x}-\boldsymbol{x}^0)^T \mathbf{A}(\boldsymbol{x}-\boldsymbol{x}^0)+2\boldsymbol{x}^T\boldsymbol{b})}\ d^n x = \frac{(2\pi)^{\frac{n}{2}}}{\sqrt{|\mathbf{A}|}}\ e^{\frac{1}{2}\boldsymbol{b}^T\mathbf{A}^{-1}\boldsymbol{b}-\boldsymbol{b}^T\boldsymbol{x}^0}. \tag{7.40}$$

The marginal density of equation (7.39a) is required for the computation of the mean $\langle x_i \rangle$. If the random variables are independent and identically distributed ($p\,(x_1, \ldots, x_n) = p_0\,(x_1) \ldots p_0\,(x_n)$), then the marginal density is simply given by $p_i(x_i) = p_0(x_i)$. In the general case the marginal density is given by (for the derivation see Appendix equation (A.8) [p. 598])

Marginal density of the multivariate Gaussian distribution

$$p_i(x) = \frac{1}{\sqrt{2\ \pi\ C_{ii}}}\ e^{-\frac{(x-x_i^0)^2}{2\,C_{ii}}}. \tag{7.41}$$

7.6 Order statistic

Consider the following problem. Assume we have a sample of size L of independent and identically distributed (i.i.d.) random numbers from a probability density $\rho(x)$ with associated cumulative distribution $F(x)$. What is the PDF of the largest element of the sample? We will derive this density step by step. To begin with, the elements of the sample shall be sorted in increasing order

$$s_1 \leq s_2\ \leq \cdots \leq s_k \leq s_{k+1} \leq \cdots \leq s_L.$$

More generally, we will determine the PDF $P(s_k \in (x, x+dx)|L, \rho, \mathcal{I})$ of the kth element s_k of the sorted sample. It is called the *kth-order statistic*. For this to be the case, the following three propositions must hold simultaneously:

- A: $k-1$ elements are less than or equal to x.
- B: $L-k$ elements are greater than or equal to x.
- C: One element is within the interval $(x, x+dx)$.

The probability for an element to be less than or equal to x is given by the distribution function $p_1 = F(x)$. Correspondingly, the probability for an element to be greater than or equal to x is provided by $p_2 = (1 - F(x))$. Finally, the probability for an element to be in the interval $(x, x+dx)$ is $p_3 = \rho(x)\, dx$. From a different perspective, we have the problem of distributing L balls in three boxes with individual probabilities $p_\alpha, \alpha = 1, \dots, 3$. The probability of $(k-1)$ balls in the first box, $(L-k)$ balls in the second box and one ball in the third box is given by the multinomial distribution (see equation (4.17a) [p. 57]).

kTH-ORDER STATISTIC

$$P(s_k \in (x, x+dx)|L, \rho, \mathcal{I}) = \frac{[F(x)]^{k-1}\,[1-F(x)]^{L-k}}{B(k, L-k+1)} \underbrace{\rho(x)\, dx}_{dF(x)} . \tag{7.42}$$

The resulting PDF is closely related to the beta distribution discussed in Section 7.5.2 [p. 100]. The normalization follows from that of the beta PDF:

$$\begin{aligned}\int P(s_k \in (x, x+dx)|L, \rho, \mathcal{I})\, dx &= \frac{\int [F(x)]^{k-1}\,[1-F(x)]^{L-k}\, dF(x)}{B(k, L-k+1)}\\ &= \frac{B(k, L-k+1)}{B(k, L-k+1)}.\end{aligned}$$

7.6.1 Mean of the order statistic

We assume that the support of $\rho(x)$ is simply connected to ensure that the inverse function F^{-1} is unique. The mean of the order statistic can then be computed as follows:

$$\begin{aligned}\langle x \rangle_k &:= \int x \,\frac{[F(x)]^{k-1}\,[1-F(x)]^{L-k}}{B(k, L-k+1)}\, dF(x)\\ &= \int_0^1 F^{-1}(f)\, \underbrace{\frac{f^{k-1}\,[1-f]^{L-k}}{B(k, L-k+1)}}_{p_\beta(f|k, L-k+1)}\, df\\ &= \int_0^1 F^{-1}(f)\; p_\beta(f|k, L-k+1)\, df.\end{aligned}$$

We summarize the result:

MEAN OF kTH-ORDER STATISTIC FOR A SAMPLE OF SIZE L

$$\langle x \rangle_k = \int_0^1 F^{-1}(f)\; p_\beta(f|k, L-k+1)\, df. \tag{7.43}$$

As an example we consider a uniform PDF $\rho(x) = \mathbf{1}(0 \le x \le 1)$, whose CDF is $F(x) = x\, \mathbf{1}(0 \le x \le 1)$, with inverse $F^{-1}(f) = f$ within the domain $f \in [0, 1]$. The mean is therefore readily computed as

$$\langle x \rangle_k = \int_0^1 f\;\; p_\beta(f|k, L-k+1)\, df \overset{(7.12c)}{=} \frac{k}{L+1}. \tag{7.44}$$

7.6.2 Distribution of the sample maximum and minimum

In practice, the probability distribution of the maximum and minimum value of a sample is of special interest. The sample maximum is the special case of equation (7.42) for $k = L$ and results in

MAXIMUM STATISTIC

$$p(\xi|L, \mathcal{I}) = L\; \rho(\xi)\; F(\xi)^{L-1}. \tag{7.45}$$

The mean of the largest value of uniform random numbers is, according to equation (7.44),

$$\langle \xi \rangle = 1 - \frac{1}{L+1}. \tag{7.46}$$

The PDF of the sample minimum is obtained using equation (7.42) for $k = 1$ and reads

MINIMUM STATISTIC

$$P(\xi|L, \mathcal{I}) = L\; \rho(\xi)\; (1 - F(x))^{L-1}. \tag{7.47}$$

For the uniform distribution in [0, 1] we obtain

$$p(\xi|L, \mathcal{I}) = \frac{1}{L}\; (1-\xi)^{L-1}\; \mathbf{1}(\xi \in [0, 1])\,, \tag{7.48}$$

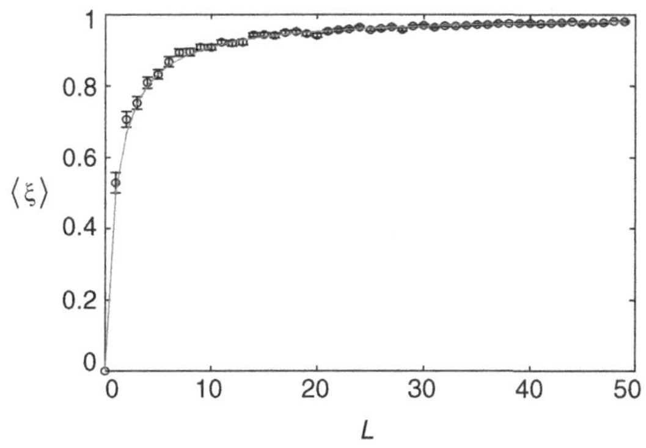

Figure 7.9 Mean of the maxima ξ of L uniform random numbers as a function of L. The solid line represents the analytical solution.

with mean

$$\langle \xi \rangle = \frac{1}{L+1}.$$

This result is related by symmetry to the mean of the sample maximum.

We will verify the result for the sample maximum by computer simulations. To this end we generate L random numbers from the uniform distribution, determine the largest value and repeat the process M times in order to estimate the mean and standard variation of the PDF. The results of such a simulation are displayed in Figure 7.9 as a function of the sample size L. As expected, we find perfect agreement. We also observe how the sample maxima are pushed ever closer to 1 with increasing sample size L.

7.7 Transformation of random variables

Very often, we know the PDF of a random variable x and are interested in the PDF of a random variable y that is related to x through the transformation

$$x \to y = f(x), \qquad x, y \in \mathbb{R}^n.$$

The transformation also affects the infinitesimal volume $dV_x \to dV_y$. If we describe the same infinitesimal event in two different representations, x and y, then

$$p_x(x)\, dV_x = p_y(y)\, dV_y \tag{7.49}$$

should hold, because in both cases the infinitesimal probability 'mass' of the event has to be the same. We find

Transformation of random variables

$$p_y(y) = p_x(x)\left|\frac{\partial x_i}{\partial y_j}\right|. \tag{7.50}$$

The change in volume is given by the 'Jacobian determinant'

$$|J| = \left|\frac{\partial x_i}{\partial y_j}\right| := \left|\begin{pmatrix} \frac{\partial x_1}{\partial y_1} & \frac{\partial x_1}{\partial y_2} & \cdots & \frac{\partial x_1}{\partial y_n} \\ \frac{\partial x_2}{\partial y_1} & \frac{\partial x_2}{\partial y_2} & \cdots & \frac{\partial x_2}{\partial y_n} \\ \vdots & \vdots & & \vdots \\ \frac{\partial x_n}{\partial y_1} & \frac{\partial x_n}{\partial y_2} & \cdots & \frac{\partial x_n}{\partial y_n} \end{pmatrix}\right|. \tag{7.51}$$

7.7.1 Example: Univariate case

Starting from the uniform distribution on the unit interval $x \in (0, 1]$,

$$p_x(x) = \mathbf{1}(0 < x \le 1),$$

let us now consider the new variable

$$y = -\ln(x) \in [0, \infty).$$

The inverse transformation reads $x = e^{-y}$. The probability density of the new variable y is calculated using equation (7.50):

$$p_y(y) = p_x(x)\left|\frac{\partial x}{\partial y}\right| = e^{-y}.$$

As a second univariate example we consider the improper probability density

$$p_x(x) = 1, \qquad x \in (-\infty, \infty).$$

It is again the uniform distribution, but this time defined on $\mathbb{R}$. It is improper because a normalization is impossible on this domain. However, as we will see later, this probability distribution occurs often in practice. To avoid paradoxes it is best to start the calculation with a restricted domain $[-a, a]$, and at the end of the calculation let $a \to \infty$. Here we consider the transformation $\sigma = e^x$, with $\sigma \in (0, \infty)$. The probability density for σ is then given by

$$p_\sigma(\sigma) = p_x(x)\,\frac{dx}{d\sigma} = \frac{1}{\sigma}. \tag{7.52}$$

This implies that a uniform distribution of the logarithm corresponds to a $1/\sigma$ distribution

$$p_\sigma(\sigma)\, d\sigma = \frac{d\sigma}{\sigma}.$$

This probability density is also improper, but it exhibits a very important property. It is scale-invariant, which means that transformations of the form

$$\sigma \to \alpha\, \sigma$$
$$\sigma \to \sigma^n$$

keep the functional form of the PDF invariant. They only modify the prefactor, which is immaterial for an improper prior. We will see later that this probability distribution, 'Jeffreys' scale prior', is suited as a prior distribution for scale-invariant parameters.

7.7.2 Example for the multivariate case

We consider a radioactive point source in a three-dimensional substrate, emitting beta particles. Its location defines the origin of a coordinate system. Let the distance of the source from the surface of the substrate be d, and for simplicity we assume a planar surface. The vector from the origin to the perpendicular foot defines the z-direction. We are interested in the probability density of beta particles on the surface. As beta radiation is isotropic, that is, the probability for beta particles being radiated into the differential solid-angle $d\Omega$ is proportional to $d\Omega$, hence

$$p_\theta(\theta, \phi) = \sin(\theta).$$

Here ϕ is the azimuthal angle about the z-axis and θ stands for the polar angle at the origin. The polar angle is restricted $\phi \in [0, \pi/2]$, as only those particles reach the surface. For the coordinates on the surface we use polar coordinates ϕ (the same as before) and r, the distance of the detector from the perpendicular foot. We use the relations

$$\phi = \text{atan}\begin{pmatrix} r \\ d \end{pmatrix},$$

$$\sin(\theta) = \frac{\tan(\theta)}{\sqrt{1 + \tan(\theta)^2}} = \frac{r}{\sqrt{r^2 + d^2}}.$$

The Jacobian reads

$$|J| = \left| \begin{pmatrix} \frac{\partial \theta}{\partial r} & \frac{\partial \theta}{\partial \phi} \\ \frac{\partial \phi}{\partial r} & \frac{\partial \phi}{\partial \phi} \end{pmatrix} \right| = \left| \begin{pmatrix} \left[d(1 + (r/d)^2)\right]^{-1} & 0 \\ 0 & 1 \end{pmatrix} \right| = \frac{d}{r^2 + d^2}.$$

Then, according to equation (7.50), we have

$$p_r(r, \phi) = \frac{d\, r}{\left(r^2 + d^2\right)^{3/2}}.$$

This could, for instance, be used along with Bayes' theorem to infer the position of the radioactive point source from beta-particle counts on the surface.

7.8 Characteristic function

7.8.1 Alternative description of random variables

In the case of discrete random variables we have introduced the generating function in Section 2.1.3 [p. 21]. We have seen in various applications (e.g. random walk) that this is a very useful concept. A similar transformation is also possible for continuous random variables, which is called the 'characteristic function'.

Definition 7.7 (Characteristic function) *Let $\boldsymbol{x} \in \mathbb{R}^m$ be a vector of random variables. The characteristic function is then defined as*

CHARACTERISTIC FUNCTION

$$\Phi(\boldsymbol{\omega}) = \int_{\mathbb{R}^m} d^m x \; \rho(\boldsymbol{x}) \, e^{i\boldsymbol{\omega}^T \boldsymbol{x}}, \qquad \boldsymbol{x}, \boldsymbol{\omega} \in \mathbb{R}^m. \tag{7.53}$$

The superscript T denotes transposition and $\boldsymbol{\omega}^T \boldsymbol{x}$ is the scalar product of the two column vectors $\boldsymbol{\omega}$ and $\boldsymbol{x}$. The characteristic function can also be considered as the expectation value of the (complex) random variable $e^{i\boldsymbol{\omega}^T \boldsymbol{x}}$:

$$\Phi(\boldsymbol{\omega}) = \langle e^{i\boldsymbol{\omega}^T \boldsymbol{x}} \rangle. \tag{7.54}$$

As a matter of fact, this is nothing else but a Fourier transform. As such, it is a one-to-one mapping of the PDF and has the same information content. The inverse transformation is given by

INVERSE TRANSFORM

$$p(\boldsymbol{x}) = \left(\frac{1}{2\pi}\right)^m \int_{\mathbb{R}^m} d^m \omega \; \Phi(\boldsymbol{\omega}) \, e^{-i\,\boldsymbol{\omega}^T \boldsymbol{x}}. \tag{7.55}$$

7.8.2 Shift theorem

Consider a linear transformation of the random variables

$$\boldsymbol{y} = \mathbf{A}\boldsymbol{x} + \boldsymbol{b},$$

with $\mathbf{A}$ an $(m \times m)$-matrix and $\boldsymbol{y}, \boldsymbol{b} \in \mathbb{R}^m$. The corresponding characteristic function is given by

$$\Phi_y(\boldsymbol{\omega}) = \langle e^{i\boldsymbol{\omega}^T \boldsymbol{y}} \rangle_y = \langle e^{i\boldsymbol{\omega}^T (\mathbf{A}\boldsymbol{x}+\boldsymbol{b})} \rangle_y = e^{i\boldsymbol{\omega}^T \boldsymbol{b}} \, \langle e^{i(\mathbf{A}^T \boldsymbol{\omega})^T \boldsymbol{x}} \rangle_x.$$

So the characteristic function of the linearly transformed random variable $\boldsymbol{y}$ can be expressed by the characteristic function of the original random variable $\boldsymbol{x}$:

$$\Phi_y(\boldsymbol{\omega}) = e^{i\boldsymbol{\omega}^T \boldsymbol{b}}\ \Phi_x(\mathbf{A}^T \boldsymbol{\omega}) \quad \text{with } \boldsymbol{y} = \mathbf{A}\boldsymbol{x} + \boldsymbol{b}. \tag{7.56}$$

The offset (shift) introduces an extra phase factor and a transformed vector $\boldsymbol{\omega}$.

7.8.3 Computation of moments

The characteristic function is particularly useful for the computation of moments. We start out with a single random variable. The nth derivative of the characteristic function with respect to ω yields

$$\Phi^{(n)}(\omega) = i^n \int_{-\infty}^{\infty} dx\ x^n\ \rho(x)\ e^{i\,\omega\,x},$$

which implies

$$\langle x^n \rangle = i^{-n}\ \frac{d^n}{dx^n}\Phi(x)|_{x=0}. \tag{7.57}$$

In particular

$$\int_{-\infty}^{\infty} dx\, \rho(x) = \Phi(0), \tag{7.58a}$$

$$\langle x \rangle = -i\ \Phi^{(1)}(0), \tag{7.58b}$$

$$\left\langle x^2 \right\rangle = -\ \Phi^{(2)}(0), \tag{7.58c}$$

which shows that $\Phi(0)$ provides easy access to the normalization of a PDF and also the lowest moments can easily be computed, once the characteristic function is given.

In the multivariate case, the mean of the individual random variables and the elements of the covariance matrix can be computed as follows:

MEAN AND COVARIANCE IN THE MULTIVARIATE CASE

$$\langle x_i \rangle = -i\ \frac{\partial}{\partial \omega_i}\Phi(\boldsymbol{\omega})\bigg|_{\boldsymbol{\omega}=0}, \tag{7.59}$$

$$\left\langle \Delta x_i \Delta x_j \right\rangle = -\frac{\partial^2}{\partial \omega_i \partial \omega_j}\Phi(\boldsymbol{\omega})\bigg|_{\boldsymbol{\omega}=0} - \langle x_i \rangle \left\langle x_j \right\rangle. \tag{7.60}$$

Further details are given for example, in the books of R. Frieden [76] and A. Papoulis [159].

Example: Γ *distribution*

The characteristic function of the Γ distribution is given by

$$\begin{aligned}
\Phi(\omega) &= \int_{-\infty}^{\infty} dx \; p_{\Gamma}(x|\alpha, \beta)\, e^{i\,\omega\, x} \\
&= \frac{\beta^{\alpha}}{\Gamma(\alpha)} \int_{0}^{\infty} dx\; x^{\alpha-1}\, e^{-(\beta - i\omega)\,x} \\
&= \frac{\beta^{\alpha}}{\Gamma(\alpha)} (\beta - i\omega)^{-(\alpha)} \int_{0}^{(\beta - i\omega)\cdot\infty} dx\; x^{\alpha-1}\, e^{-x}.
\end{aligned}$$

Within the closed contour, defined by $C_1 = (0 \to (\beta - i\omega)\cdot\infty)$, $C_2 = ((\beta - i\omega)\cdot\infty \to \infty)$ and $C_3 = (\infty \to 0)$, there are no poles of the integrand if $\alpha \geq 1$. With Cauchy's integral formula we get

$$\int_{C_1+C_2+C_3} dx\; x^{\alpha-1}\, e^{-x} = 0.$$

Furthermore, if there is no contribution to the integral from the contour C_2, then

$$\int_{0}^{(\beta - i\omega)\cdot\infty} dx\; x^{\alpha-1}\, e^{-x} = \int_{0}^{\infty} dx\; x^{\alpha-1}\, e^{-x} = \Gamma(\alpha),$$

which finally yields the

CHARACTERISTIC FUNCTION OF THE GAMMA DISTRIBUTION

$$\Phi(\omega) = \left(\frac{\beta}{\beta - i\,\omega}\right)^{\alpha} = \left(1 - i\,\frac{\omega}{\beta}\right)^{-\alpha}. \tag{7.61}$$

The characteristic function of an exponential distribution is obtained for $\alpha = 1$. Using the characteristic function it is easy to see that we have a proper normalization $\Phi(0) = 1$. The first two derivatives are given by

$$\begin{aligned}
\Phi^{(1)}(\omega) &= i\,\frac{\alpha}{\beta}\left(1 - i\,\frac{\omega}{\beta}\right)^{-\alpha-1}, \\
\Phi^{(2)}(\omega) &= -\,\frac{\alpha(\alpha+1)}{\beta^2}\left(1 - i\,\frac{\omega}{\beta}\right)^{-\alpha-2}.
\end{aligned}$$

With equations (7.58b) and (7.58c) we find the mean and variance in agreement with the results derived in equations (7.14c) and (7.14e) [p. 102].

Example: Gaussian distribution

The characteristic function of the Gaussian distribution with mean x_0 and standard deviation σ is given by

$$\Phi(\omega) = \frac{1}{\sqrt{2\pi\sigma^2}} \int_{-\infty}^{\infty} dx\; e^{-\frac{1}{2\sigma^2}((x-x_0)^2 - 2\,i\,\sigma^2\omega\,x)}.$$

Completing the square in the exponent results in

$$(x-x_0)^2 - 2\,i\,\sigma^2\omega\,x = (x-(x_0+i\,\sigma^2\,\omega))^2 + (\sigma^2\,\omega)^2 - 2\,i\,\sigma^2\,\omega\,x_0.$$

We introduce a new integration variable $z = x - (x_0 + i\;\sigma^2\;\omega)$, which changes the integration contour to $C = (-\infty - i\,\sigma^2\,\omega \to \infty - i\,\sigma^2\,\omega)$:

$$\Phi(\omega) = \frac{1}{\sqrt{2\pi\sigma^2}}\, e^{-((\sigma^2\,\omega)^2 - 2\,i\,\sigma^2\,\omega)/2\sigma^2} \underbrace{\int_C dx\; e^{-\frac{z^2}{2\sigma^2}}}_{\sqrt{2\pi\sigma^2}}$$

$$= e^{-\frac{1}{2}\sigma^2\,\omega^2 + i\,\omega\,x_0}.$$

The normalization is correct, $\Phi(0) = 1$. Therefore the characteristic function of the Gaussian distribution is given by

CHARACTERISTIC FUNCTION OF THE GAUSSIAN DISTRIBUTION

$$\Phi(\omega) = e^{i\omega x_0}\, e^{-\frac{1}{2}\sigma^2\,\omega^2}. \tag{7.62}$$

Taking advantage of the shift theorem in Section 7.8.2 we can write the characteristic function of the shifted random variable $z = x - x_0$ as

$$\Phi_z(\omega) = \Phi_x(\omega)\, e^{-i\omega\,x_0} = e^{-\frac{1}{2}\sigma^2\,\omega^2}.$$

The moments of the shifted random variables are the central moments:

$$\langle (x-x_0)^n \rangle.$$

The nth derivative of $\Phi_z(\omega)$ at $\omega = 0$ is given by

$$\Phi_z^{(n)}(0) = \begin{cases} (-1)^{\frac{n}{2}}\,\sigma^n\,(n-1)!! & \text{for } n \text{ even} \\ 0 & \text{else.} \end{cases}$$

According to equation (7.57) we get

Central moments of the Gaussian distribution

$$\left\langle (x-x_0)^n \right\rangle = \begin{cases} \sigma^n \; (n-1)!! & n \text{ even} \\ 0 & \text{else.} \end{cases} \tag{7.63}$$

7.8.4 Characteristic function for discrete random variables

The characteristic function can also be defined for discrete distribution random variables. The probability density of a discrete distribution P_n can be written as

$$p(\boldsymbol{x}) = \sum_n P_n \; \delta(\boldsymbol{x} - \boldsymbol{x}_n).$$

The characteristic function then follows from

$$\Phi(\boldsymbol{\omega}) = \int dx \; p(x) \; e^{i\boldsymbol{\omega} \boldsymbol{x}} = \sum_n P_n \left(\int dx^m \; \delta(\boldsymbol{x} - \boldsymbol{x}_n) \; e^{i\boldsymbol{\omega} \boldsymbol{x}} \right)$$
$$= \sum_n P_n \; e^{i\boldsymbol{\omega} \boldsymbol{x}_n}.$$

Characteristic function of discrete distributions

$$\Phi(\boldsymbol{\omega}) = \sum_n P_n \; e^{i\boldsymbol{\omega} \boldsymbol{x}_n}. \tag{7.64}$$

Example: Poisson distribution

The characteristic function of the Poisson distribution is given by

$$\Phi(\omega) = e^{-\mu} \sum_{n=0}^{\infty} \frac{\mu^n}{n!} \; e^{i\omega n} = e^{-\mu} \sum_{n=0}^{\infty} \frac{\left(e^{i\omega}\mu\right)^n}{n!} = e^{-\mu \, (1-e^{i\omega})}.$$

Here we have the rare case of a double exponential exp(exp(.)). The first two derivatives are given by

$$\Phi^{(1)}(\omega) = i\mu \; e^{i\omega} \; e^{-\mu \, (1-e^{i\omega})},$$
$$\Phi^{(2)}(\omega) = -\mu \; \left(\mu e^{i2\omega} + e^{i\omega}\right) \; e^{-\mu \, (1-e^{i\omega})}.$$

Evaluation at $\omega = 0$ yields

$$\Phi^{(1)}(0) = i\mu,$$
$$\Phi^{(2)}(0) = -\mu \; (\mu + 1).$$

The resulting mean and variance are in agreement with the results reported in equations (6.13b) and (6.13c) [p. 89].

Example: Binomial distribution

The characteristic function of the binomial distribution can also be derived straightforwardly, exploiting the properties of the binomial coefficient:

$$\Phi(\omega) = \sum_{n=0}^{N} \binom{N}{n} (p\, e^{i\omega})^n \, q^{(N-n)} = (p\, e^{i\omega} + q)^N .$$

Again, the corresponding mean and variance are correct.

7.8.5 Sum of random variables

The characteristic functions are extremely powerful for the determination of the PDF of sums of independent random variables

$$S = \sum_{n=1}^{N} x_n . \tag{7.65}$$

The PDF of the random variable x_n is $p_n(x)$, which may depend on n. The corresponding characteristic function is denoted by $\Phi_n(\omega)$. The probability density of S follows from

$$\begin{aligned} p(S|N,\mathcal{I}) &= \int d^N x \; p(S|x_1,\ldots,x_N,N,\mathcal{I})\; p(x_1,\ldots,x_N|N,\mathcal{I}) \\ &= \int d^N x \; \delta(S - \sum_n x_n) \prod_n p_n(x_n). \end{aligned}$$

In the last step, we have used the independence of the random variables. The resulting characteristic function becomes

$$\begin{aligned} \Phi(\omega) &= \int dS \; p(S|N,\mathcal{I})\; e^{iS\omega} \\ &= \int d^N x \left(\int dS \; \delta\left(S - \sum_n x_n\right) e^{iS\omega} \right) \prod_n p_n(x_n) \\ &= \int d^N x \; e^{i \sum_n x_n \omega} \prod_n p_n(x_n) = \prod_n \int dx_n \; e^{i\, x_n \omega}\, p_n(x_n) \\ &= \prod_n \Phi_n(\omega). \end{aligned}$$

Characteristic function of a sum of independent random variables

$$S = \sum_{n=1}^{N} x_n,$$
$$\Phi(\omega) = \prod_{n=1}^{N} \Phi_n(\omega). \tag{7.66}$$

The characteristic function of S is the product of the characteristic functions of the individual random summands. The PDF of S is then computed via the inverse Fourier transform. The inverse Fourier transform of a product is equivalent to a convolution.

The characteristic function of N i.i.d. Gamma random variables is, along with equation (7.61),

$$\Phi(\omega) = \left(1 - i\,\frac{\omega}{\beta}\right)^{-N\alpha},$$

and thus we find

PDF of the sum of N i.i.d. random variables from a Gamma distribution

$$P(S|N, \mathcal{I}) = p_\Gamma(S|N\alpha, \beta). \tag{7.67}$$

In particular, for $\alpha = 1$ we obtain

PDF of the sum of N i.i.d. random variables from an exponential distribution

$$P(S|N, \mathcal{I}) = p_\Gamma(S|N, \beta) = \frac{\beta^N}{\Gamma(N)}\, S^{N-1} e^{-\beta S}. \tag{7.68}$$

7.8.6 Moments of a convolution

The sum of two independent random variables $S = x_1 + x_2$, which are distributed according to $p_\alpha(x_\alpha)$, has a PDF that follows from the marginalization rule

$$p(S|, \mathcal{I}) = \int dx_1\, dx_2\; p(S|x_1, x_2, \mathcal{I})\; p_1(x_1)\;\; p_2(x_2)$$

to be a 'convolution'. According to equation (7.66), the characteristic function of S, and hence of a convolution, is the product

$$\Phi(\omega) = \Phi_1(\omega_1)\ \Phi_2(\omega_2)$$

of the characteristic functions associated with the individual PDFs $p_\alpha(x_\alpha)$. This has interesting implications as far as the moments are concerned. The nth moment of S is, according to equation (7.57) [p. 125],

$$\begin{aligned}\langle S^n\rangle &= i^{-n}\frac{d^n}{d\omega^n}\Phi(\omega)\Big|_{\omega=0} i^{-n}\frac{d^n}{d\omega^n}\Phi_1(\omega)\Phi_2(\omega)\Big|_{\omega=0}\\ &= i^{-n}\sum_{k=0}^{n}\binom{n}{k}\ \Phi_1^{(n-k)}(0)\Phi_2^{(k)}(0).\end{aligned}$$

This finally yields the very interesting result

Moments of a convolution

$$\langle S^n\rangle = \sum_{k=0}^{n}\binom{n}{k}\ \langle x_i^{(n-k)}\rangle\langle x_i^{(k)}\rangle. \tag{7.69}$$

With hindsight, the result could have been derived directly as well:

$$\left\langle (x_1+x_2)^n\right\rangle = \sum_{k=0}^{n}\binom{n}{k}\left\langle x_1^{n-k}x_2^k\right\rangle = \sum_{k=0}^{n}\binom{n}{k}\ \langle x_1^{n-k}\rangle\langle x_2^k\rangle.$$

The last step follows from the independence of the two random variables.

7.9 Error propagation

7.9.1 General error propagation

Error propagation is concerned with the question of how uncertainties in a set of parameters are propagated into derived quantities. Very often the transformation functional entering equation (7.49) [p. 121] is not of simple form. As an example, consider the difference $z = x - y$ of two random variables x and y or some more complicated relation $z = f(x, y)$, given that we know the joint probability distribution $p(x, y|\mathcal{I})$. Then, based on the marginalization rule, we can express the PDF for z as

$$p(z|\mathcal{I}) = \int dx \int dy\ p(z|x, y, \mathcal{I})\ p(x, y|\mathcal{I})\,. \tag{7.70}$$

If $z = f(x, y)$ is determined unambiguously by x and y, then $p(z|x, y, \mathcal{I}) = \delta(z - f(x, y))$ holds and we can express $p(z|\mathcal{I})$ by

$$p(z|\mathcal{I}) = \int dx \int dy\, \delta(z - f(x, y))\, p(x, y|\mathcal{I}). \tag{7.71}$$

If the two variables x and y are uncorrelated, $p(x, y|\mathcal{I})$ factorizes into $p(x, y|\mathcal{I}) = p(x|\mathcal{I})\, p(y|\mathcal{I})$. This allows us to simplify equation (7.71) to a convolution of the probability density functions of x and y. For our example $z = f(x, y) = x - y$ we get

$$\begin{aligned} p(z|\mathcal{I}) &= \int dx\ p(x|\mathcal{I}) \int dy\ p(y|\mathcal{I})\, \delta(z - f(x, y)) \\ &= \int dx\ p(x|\mathcal{I}) \int dy\ p(y|\mathcal{I})\, \delta(z - (x - y)) \\ &= \int dx\ p(x|\mathcal{I})\, p(y = x - z|\mathcal{I}). \end{aligned} \tag{7.72}$$

If we further assume that the two parameters x and y follow Gaussian distributions with maxima at x_0 and y_0, and variances σ_x^2 and σ_y^2, respectively, we can evaluate equation (7.72) as

$$\begin{aligned} p(z|\mathcal{I}) &= \int dx\ \frac{1}{\sigma_x\sqrt{2\pi}} \exp\left(-\frac{1}{2}\frac{(x - x_0)^2}{\sigma_x^2}\right) \frac{1}{\sigma_y\sqrt{2\pi}} \exp\left(-\frac{1}{2}\frac{(y - y_0)^2}{\sigma_y^2}\right) \\ &= \frac{1}{\sigma_z\sqrt{2\pi}} \exp\left(-\frac{1}{2}\frac{(z - z_0)^2}{\sigma_z^2}\right), \end{aligned} \tag{7.73}$$

where

$$z_0 = x_0 - y_0 \quad \text{and} \quad \sigma_z^2 = \sigma_x^2 + \sigma_y^2. \tag{7.74}$$

Thus, the difference of two uncorrelated Gaussian-distributed random variables is given by a Gaussian distribution with a maximum at z_0 and with the variance given by the sum of the variances of the random variables x and y. The generalization to more parameters and correlated random variables is straightforward, although the evaluation of the convolution integrals may become complicated. However, the complicated convolution is not always necessary. Often the linear approximation of the error propagation is sufficient and this approximation results in the standard formula for the error propagation known from undergraduate physics, as we will see next.

7.9.2 Linear approximation of error propagation

Consider a linear functional dependence

$$g = a + \sum_n b_n\, x_n \tag{7.75}$$

of individually distributed random variables x_n with means x_n^0 and variances σ_n^2. This situation is quite common when data originating from different experiments are combined: The mean of the sum is readily obtained:

$$\langle g \rangle = a + \sum_n b_n \langle x_n \rangle = a + \sum_n b_n x_n^0.$$

Here we are preliminarily interested in the statistical error of the mean, for which we compute the variance

$$\begin{aligned}
\sigma_g^2 = \left\langle (g - \langle g \rangle)^2 \right\rangle &= \left\langle \left(a + \sum_n b_n x_n - a - \sum_n b_n \langle x_n \rangle \right)^2 \right\rangle \\
&= \left\langle \left(\sum_n b_n \left(x_n - x_n^0 \right) \right)^2 \right\rangle \\
&= \sum_{n,m} b_n b_m C_{nm}, \\
\text{with} \qquad \Delta_n &:= x_n - x_n^0, \\
C_{nm} &:= \langle \Delta_n \Delta_m \rangle.
\end{aligned}$$

For uncorrelated errors the covariance matrix is diagonal, $C_{nm} = \delta_{n,m} \sigma_n^2$, and we obtain

Propagation of uncorrelated errors in a linear model

$$\operatorname{var}(g) = \sum_n b_n^2 \sigma_n^2. \tag{7.76}$$

So, the result is simply the weighted sum of the individual variances.

In case of correlated errors we have, in general,

Propagation of correlated errors in a linear model

$$\sigma_g^2 = \sum_{nm} b_n b_m C_{nm}. \tag{7.77}$$

Only in rare cases is the covariance matrix available. Based on the Cauchy–Schwarz inequality one can at least give an upper bound:

CAUCHY–SCHWARZ INEQUALITY FOR THE COVARIANCE

$$|\langle\Delta_n\ \Delta_m\rangle| \le \sigma_n\ \sigma_m. \tag{7.78}$$

It allows the following estimate:

$$\sigma_g^2 \le \sum_{nm} |b_n|\ |b_m|\ |C_{nm}| \le \sum_{nm} |b_n|\ |b_m|\ \sigma_n\ \sigma_m = \left(\sum_n |b_n|\ \sigma_n\right)^2 .$$

UPPER BOUND OF CORRELATED ERRORS IN A LINEAR MODEL

$$\sigma_g \le \sum_n |b_n|\ \sigma_n. \tag{7.79}$$

This is valid for linear functions of the random variables $\boldsymbol{x} = (x_1, \dots, x_N)$. For general functions $f(\boldsymbol{x})$ in many cases one can still invoke a Taylor expansion about the mean $\boldsymbol{x}^0 = \{x_1^0, x_2^0, \dots, x_N^0\}$, assuming that the errors σ_n are small:

$$f(\boldsymbol{x}) \approx f(\boldsymbol{x}^0) + \sum_n \left.\frac{\partial f(\boldsymbol{x})}{\partial x_n}\right|_{x=x^0} \Delta_n.$$

Thus we again have the linear model of equation (7.75), with

$$a = f(\boldsymbol{x}^0), \qquad b_n = \left.\frac{\partial f(\boldsymbol{x})}{\partial x_n}\right|_{x=x^0},$$

and we can use the previous results.

7.9.3 Limitations of the linear approximation of error propagation

The linear approximation of the error propagation is by far the most common tool for calculating uncertainties of parameters in all branches of science. Furthermore, the resulting equations are very often the only approach which is taught at universities to handle measurement uncertainties in science and engineering. However, this approach may be inadequate in many circumstances. A striking example has been given in [187] in the context of the analysis of powder diffraction data. In essence, the task is to evaluate the structure-factor amplitude $A = \sqrt{I}$ from the intensity

$$I = I_0 \pm \sigma_I, \tag{7.80}$$

where I_0 is the best estimate (i.e. maximum likelihood estimate) of the intensity I based on some data $\boldsymbol{d}$. Using the standard error propagation formula

$$\sigma_A = \sqrt{\left(\frac{\partial A}{\partial I}\right)^2 \sigma_I^2}, \tag{7.81}$$

we can compute the result for the amplitude $A = A_0 \pm \sigma_A$:

$$A = \sqrt{I_0} \pm \frac{\sigma_I}{2\sqrt{I_0}}. \tag{7.82}$$

Unfortunately, this approach fails when I_0 is negative, which can easily happen if the magnitude of the uncertainty σ_I (i.e. due to a high noise level) is comparable to I_0. Such an occurrence is not unusual. But also for values of I_0 where the uncertainty σ_A can be computed, the values obtained for small I_0 can be unphysical, although less obviously so. Why does the standard approach fail? This can be traced to two reasons. The first mistake is the use of the likelihood function of I instead of the posterior distribution $p(I|\boldsymbol{d}, \mathcal{I})$. Equation (7.80) is essentially stating that the likelihood function of I is given by

$$p(\boldsymbol{d}|I, \mathcal{I}) = \frac{1}{\sigma_I\sqrt{2\pi}} e^{-\frac{1}{2}\frac{(I-I_0)^2}{\sigma_I^2}}. \tag{7.83}$$

However, the posterior distribution for I is proportional to the product of the likelihood and prior:

$$p(I|\boldsymbol{d}, \mathcal{I}) \propto p(\boldsymbol{d}|I, \mathcal{I})\, p(I|\mathcal{I}). \tag{7.84}$$

One essential piece of prior information is that the intensity I cannot be negative, which can be incorporated, for example, in the simple prior

$$p(I|\mathcal{I}) = \begin{cases} \text{constant} & \text{for } I \geq 0 \\ 0 & \text{otherwise.} \end{cases} \tag{7.85}$$

This always results in a non-negative estimate for I, even in the case of a noise-influenced negative I_0. The cutoff at zero intensity also invalidates a second assumption underlying the standard error propagation rule: that the expansion around the (centrally located) maximum is providing a reasonable approximation of the underlying probability density distribution.

Within the Bayesian framework, however, the analysis is straightforward even for negative values of I_0. The posterior distribution $p(A|\boldsymbol{d}, \mathcal{I})$ is related to $p(I|\boldsymbol{d}, \mathcal{I})$ via a change of variables

$$p(A|\boldsymbol{d}, \mathcal{I}) = p(I|\boldsymbol{d}, \mathcal{I}) \left|\frac{dI}{dA}\right| = 2|A|\, p(I|\boldsymbol{d}, \mathcal{I}), \tag{7.86}$$

where the Jacobian $\left|\frac{dI}{dA}\right|$ yields $2|A|$. The probability distribution equation (7.86) (which of course contains the complete information on A) can now be used to derive a better summarizing description of the uncertainty of A compared with the standard error propagation rules. For example, a quadratic Taylor expansion of the logarithm of equation (7.86) around its maximum yields a Gaussian of mean A_0 and variance σ_A^2 with

$$A_0 = \frac{1}{2}\sqrt{2I_0 + \sqrt{4I_0^2 + 8\sigma_I^2}} \tag{7.87}$$

and

$$\sigma_A^2 = \frac{1}{\frac{1}{A_0^2} + \frac{2(3A_0^2 - I_0)}{\sigma_I^2}}, \tag{7.88}$$

which reduce to the expressions of the conventional analysis (equation (7.82)) in the limit of $\sigma_I \ll I_0$ but provide a much better description for the case $\sigma_I \approx I_0$ and provide a reasonable approximation even for $I_0 \leq 0$ [187].

7.10 Helmert transformation

The physicist F. R. Helmert investigated several statistical problems occurring in geodesy. In a renowned paper of 1876 he first proved that the sample mean $\overline{x}$ of independent normal random numbers with common mean and variance are independent of any function of the fluctuations $x_i - \overline{x}$, in particular of $X := \sum_i (x_i - \overline{x})^2/\sigma^2$. In this context he introduced a linear transformation, which is now referred to as the *Helmert transformation* [157], in order to show that X is chi-squared distributed. For notational convenience, we combine the independent and normally distributed (i.n.d.) variables $\{x_n\}$ into a column vector $\boldsymbol{x}$. Likewise, we define a column vector $\boldsymbol{\xi}$ of the new variable $\{\xi_n\}$ and write the Helmert transformation as

$$\boldsymbol{\xi}^T = \boldsymbol{x}^T\,\mathbf{M}. \tag{7.89}$$

Here, $\mathbf{M} = (\boldsymbol{m}^{(1)}, \ldots, \boldsymbol{m}^{(N)})$ is an orthogonal matrix whose columns are defined as

HELMERT TRANSFORMATION MATRIX FOR N VARIABLES

$$\text{for } n < N: \qquad \boldsymbol{m}^{(n)} := \frac{1}{\sqrt{n(n+1)}}\begin{pmatrix} 1 \\ \vdots \\ 1 \\ -n \\ 0 \\ \vdots \end{pmatrix} \begin{matrix} \\ \\ \leftarrow n \\ \leftarrow n+1 \\ \\ \\ \end{matrix}, \tag{7.90a}$$

$$\boldsymbol{m}^{(N)} := \frac{1}{\sqrt{N}}\begin{pmatrix} 1 \\ \vdots \\ 1 \end{pmatrix}. \tag{7.90b}$$

All column vectors have zero mean, except the last one, which yields the following important relation instead:

$$\xi_N = \sqrt{N}\,\overline{x}. \tag{7.91}$$

The $\boldsymbol{m}^{(n)}$ form a set of orthonormal vectors, resulting in

$$\mathbf{M}^T\mathbf{M} = \mathbb{1}. \tag{7.92}$$

Since N is finite, $\mathbf{M}$ is orthogonal and has the following properties:

$$|\det \mathbf{M}| = 1, \tag{7.93a}$$

$$\mathbf{M}^{-1} = \mathbf{M}^T, \tag{7.93b}$$

$$\mathbf{M}\mathbf{M}^T = \mathbb{1}. \tag{7.93c}$$

Based on equations (7.91) and (7.93c), we obtain

$$\sum_{n=1}^{N-1} \xi_n^2 = \sum_{n=1}^{N-1} \xi_n^2 - \xi_N^2 = \boldsymbol{x}^T \underbrace{\mathbf{M}^T\mathbf{M}}_{=1} \boldsymbol{x} - N\overline{x}$$
$$= \sum_{n=1}^{N} \left(x_n - \overline{x}\right)^2 = N\overline{(\Delta x)^2}.$$

We summarize these results:

PROPERTIES OF HELMERT TRANSFORM
FOR I.N.D. RANDOM VARIABLES

$$\xi_N = \sqrt{N}\,\overline{x}, \tag{7.94a}$$

$$\sum_n{}' \xi_n^2 := \sum_{n=1}^{N-1} \xi_n^2 = N\overline{(\Delta x)^2}. \tag{7.94b}$$

We are interested in the PDF $p(\boldsymbol{\xi}|x_0, \sigma, N, \mathcal{I})$ of the new variables, given the mean x_0, the variance σ^2 and the number N of i.n.d. variables. We introduce the original set $\boldsymbol{x}$ of i.n.d. variables via the marginalization rule

$$p(\boldsymbol{\xi}|x_0, \sigma, N, \mathcal{I}) = \int d^N x\; p(\boldsymbol{\xi}|\boldsymbol{x}, x_0, \sigma, N, \mathcal{I})\; p(\boldsymbol{x}|x_0, \sigma, N, \mathcal{I}).$$

Inserting the likelihood function for $\boldsymbol{x}$

$$p(\boldsymbol{x}|x_0, \sigma, N, \mathcal{I}) = (2\pi\sigma^2)^{-\frac{N}{2}}\; e^{-\frac{N}{2\sigma^2}\overline{(\Delta x)^2}}\; e^{-\frac{N}{2\sigma^2}(\overline{x}-x_0)^2}$$

along with equations (7.89), (7.93b) and (7.93a), we derive

$$
\begin{aligned}
p(\boldsymbol{\xi}|x_0,\sigma,N,\mathcal{I}) &\propto \int d^N x\, \delta\big(\boldsymbol{\xi}-\mathbf{M}^T\boldsymbol{x}\big) e^{-\frac{N}{2\sigma^2}\overline{(\Delta x)^2}}\; e^{-\frac{1}{2\sigma^2}(\sqrt{N}\overline{x}-\sqrt{N}x_0)^2} \\
&\propto e^{-\frac{1}{2\sigma^2}\sum' \xi_n^2}\; e^{-\frac{1}{2\sigma^2}(\xi_N-\sqrt{N}x_0)^2} \int d^N x\, \delta\big(\boldsymbol{x}-\mathbf{M}\boldsymbol{\xi}\big) \\
&= \mathcal{N}(\boldsymbol{\eta}|0,\sigma)\, \mathcal{N}(\xi_N|\sqrt{N}x_0,\sigma),
\end{aligned}
$$

where $\boldsymbol{\eta}$ stands for the column vector composed of the first $(N-1)$ elements of $\boldsymbol{\xi}$. Apparently, the Helmert-transformed variables are still i.n.d. with the first $(N-1)$ variables having mean zero, while the last has mean $\sqrt{N}x_0$. The variance is the same as that of the original variables.

8
The central limit theorem

8.1 The theorem

In many applications sums of independent and identically distributed random variables occur. Quite generally, we consider random variables x_n, $n = 1, \ldots, N$, all drawn independently from the same probability density. Then the 'central limit theorem' (CLT) claims that the PDF of weighted sums of these variables converges towards a Gaussian for $N \to \infty$.

CENTRAL LIMIT THEOREM

Define the random variable

$$S = \sum_{n=1}^{N} c_n \, x_n$$

with x_n i.i.d. random variables with mean μ_x and variance $\sigma_x^2 < \infty$. The weights c_n fulfil the following conditions:

$$\lim_{N\to\infty} \frac{1}{N} \sum_{n=1}^{N} c_n^{\nu} < \infty, \qquad \nu \in \mathbb{Z}. \tag{8.1}$$

Then

$$\lim_{N\to\infty} p(S|N, \mathcal{I}) = \mathcal{N}(S|\mu_S, \sigma_S)$$

with

$$\mu_S := \langle S \rangle = \mu_x \sum_{n=1}^{N} c_n, \tag{8.2}$$

$$\sigma_S^2 := \operatorname{var}(S) = \sigma_x^2 \sum_{n=1}^{N} c_n^2.$$

A prominent example is the sample mean, where $c_n = 1/N$. In this case we have

CENTRAL LIMIT THEOREM FOR THE SAMPLE MEAN

$$S = \frac{1}{N}\sum_{n=1}^{N} x_n, \tag{8.3a}$$

$$\lim_{N\to\infty} p(S|N,\mathcal{I}) = \mathcal{N}(S|\mu_S,\sigma_S), \tag{8.3b}$$

$$\mu_S := \mu_x, \tag{8.3c}$$

$$\sigma_S^2 := \frac{\sigma_x^2}{N}. \tag{8.3d}$$

As a matter of fact, the relation for mean and variance of S, given in equations (8.3c) and (8.3d), is valid for all values of N (see Section 19.1 [p. 284]), not just for large N, but the PDF in general approaches a Gaussian only for $N \to \infty$. There is an almost obvious exception: the sum of Gaussian random variables always has a Gaussian PDF, as outlined in Section 19.2 [p. 294]. Another interesting special case is the PDF of the sum of N Poissonian random variables. In Section 2.1.3 [p. 22] we have proved that it is Poissonian for $N = 2$. Then, repeatedly adding another Poissonian random variable to the sum reveals that the sum of any number N of Poissonian random variables n_i is always Poisson distributed, with mean $\mu = \sum_{i=1}^{N} \mu_i$, where μ_i is the mean of the individual Poisson random variables. So it is always a Poisson distribution. Is that a contradiction to the central limit theorem? The answer is 'no', as the Poisson distribution approaches a Gaussian one for $\mu \to \infty$. The limiting Gaussian PDF for $N \to \infty$ is not really of great interest for real-world applications, where N is mostly finite. We will therefore study what happens for moderate N guided by two examples, before we take a quick glance at the proof of the CLT.

8.1.1 Example I: Sum of exponential random numbers

First we consider the sample mean S of exponentially distributed random variables. In this case it is quite simple to determine the PDF for any N analytically. The PDF of the sum $\overline{S}$ of exponentially distributed random variables has been determined in equation (7.68) [p. 130]:

$$p(\overline{S}|N,\mathcal{I}) = \frac{\beta^N}{\Gamma(N)}\, \overline{S}^{N-1}\, e^{-\beta\overline{S}}. \tag{8.4}$$

Then the PDF for the sample mean follows from the transformation $\overline{S} = NS$, i.e.

$$p(S|N,\mathcal{I}) = N\,\frac{\beta^N}{\Gamma(N)}\, N^{N-1} S^{N-1}\, e^{-\beta N S}.$$

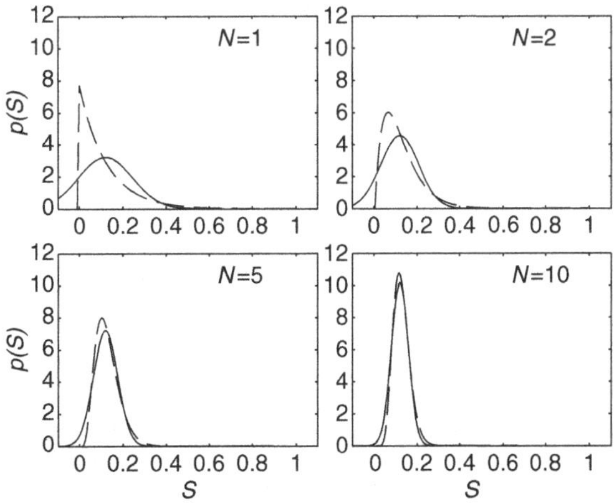

Figure 8.1 CLT in action: The exact distribution of the sample mean of N exponentially distributed random variables (dashed line) is compared with the respective Gaussian distributions (solid line).

In Figure 8.1 the exact PDF is compared with that of the Gaussian of the CLT for different values of N. We observe that the distribution of the sample mean rapidly approaches a Gaussian distribution, already for small values of N.

8.1.2 Example II: Sum of uniform random numbers

A case frequently occurring in practice is the sum $\overline{S}$ of i.i.d. uniform random numbers from the unit interval [0, 1]. Also in this case the PDF for $\overline{S}$ can be computed analytically for arbitrary N. First we determine the characteristic function of the uniform distribution

$$p(x) = \boldsymbol{1}(0 \leq x \leq 1)\,.$$

With equation (7.53) [p. 124] we get

$$\Phi(\omega) = \int_0^1 e^{i\,\omega\,x}\,dx \;=\; \frac{e^{i\,\omega} - 1}{i\,\omega}. \tag{8.5}$$

The characteristic function of the sum $\overline{S} = \sum_{i=1}^{N} x_i$ is the N-power of equation (8.5) (see equation (7.66) [p. 130]). The inverse transformation yields

$$
\begin{aligned}
p(\overline{S}|N,\mathcal{I}) &= \frac{1}{2\pi}\int_{-\infty}^{\infty}\left(\frac{e^{i\omega}-1}{i\omega}\right)^N e^{-i\omega\overline{S}}\,d\omega \\
&= \frac{(-i)^N}{2\pi}\int_{-\infty}^{\infty}\omega^{-N}\left(\sum_{k=0}^{N}\binom{N}{k}e^{ik\omega}\,(-1)^{N-k}\right)e^{-i\omega\overline{S}}\,d\omega \\
&= \frac{i^N}{2\pi}\sum_{k=0}^{N}\binom{N}{k}(-1)^k\underbrace{\int_{-\infty}^{\infty}\omega^{-N}\,e^{i\omega(k-\overline{S})}\,d\omega}_{:=I}.
\end{aligned}
$$

The integral I can be computed using Cauchy's integral theorem and we end up with

$$
I = i^N\,\pi\,\frac{(k-\overline{S})^{N-1}}{(N-1)!}\,\mathrm{sign}(k-\overline{S}).
$$

The PDF of $\overline{S}$ is then

$$
p(\overline{S}|N,\mathcal{I}) = \frac{(-1)^N}{2}\sum_{k=0}^{N}\binom{N}{k}(-1)^k\,\frac{(k-\overline{S})^{N-1}}{(N-1)!}\,\mathrm{sign}(k-\overline{S}).
$$

Therefore, the density $p(\overline{S}|N,\mathcal{I})$ is a piecewise polynomial. The first few ($N = 1, 2, 3$) are given by

$$
p(\overline{S}|N=1,\mathcal{I}) = \begin{cases} 0 & \text{if } \overline{S} \le 0 \\ 1 & \text{if } 0 < \overline{S} \le 1 \\ 0 & \text{if } 1 < \overline{S} \end{cases}
$$

$$
p(\overline{S}|N=2,\mathcal{I}) = \begin{cases} 0 & \text{if } \overline{S} \le 0 \\ \overline{S} & \text{if } 0 < \overline{S} \le 1 \\ -\overline{S}+2 & \text{if } 1 < \overline{S} \le 2 \\ 0 & \text{if } 2 < \overline{S}. \end{cases}
$$

In Figure 8.2 we compare the exact PDF of the sample mean S with the Gaussian approximation given by the CLT for different values of N. We observe that the convergence towards the Gaussian distribution is even faster than in the exponential case.

8.2 Stable distributions

As we have seen, the sum of i.i.d. random variables with finite variance converges towards a Gaussian distribution. But what happens if the random variables violate the finite variance condition? There is an important class of random distributions, the stable distributions, which act as 'attractors' for the sum of i.i.d. random variables. Roughly speaking,

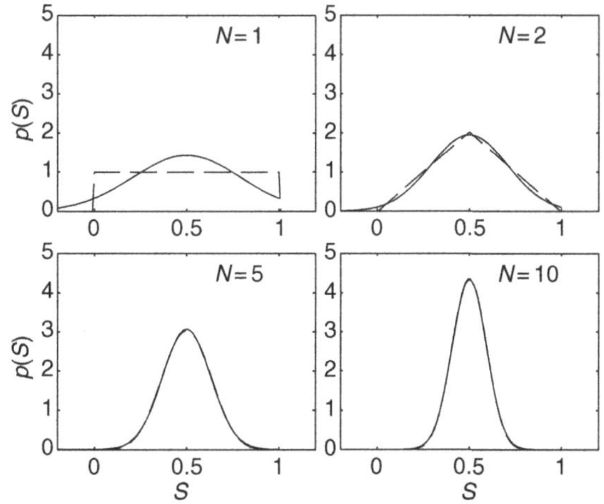

Figure 8.2 For different sample sizes N the exact PDF of the sample mean of N uniformly distributed random numbers (dashed line) is compared with the respective Gaussian distributions (solid line).

a distribution is stable if the (weighted) sum of two random variables (drawn from this distribution) follows the same distribution, only with altered location and scale parameters. Examples for stable distributions are the normal distribution, the Cauchy distribution, the Levy distribution and the Landau distribution. Stable distributions can be considered as a generalization of the central limit theorem to random variables without finite variance. A closed form of the probability density for a general stable distribution does not exist, but the density can be expressed using its specific characteristic function [101]

$$\phi(z) = \exp\left[\left(-c_0 + i\,(z/|z|)\,c_1\right)|z|^{\alpha}\right]. \tag{8.6}$$

The parameter $0 < \alpha \leq 2$ is called the exponent (or index) of the stable distribution. Stable distributions with $\alpha = 1$ are the Cauchy distributions, and stable distributions with $\alpha = 2$ are the Gaussian distributions. The probability density follows from $\phi(z)$ via

$$p(x) = \frac{1}{2\pi} \int_{-\infty}^{\infty} \phi(z)\, e^{-ixz}\, dz. \tag{8.7}$$

Further generalizations to semistable distributions or quasistable distributions are also possible [101].

8.3 Proof of the central limit theorem

Here we will give a rough idea of the proof. A mathematically rigorous treatment can be found in [160]. To understand the principal reason for the CLT it suffices to consider the sample mean with equal weights $c_n = 1/N$. Moreover, we can subtract the mean of the individual random variables and introduce new random variables by

$$y_i = \frac{x_i - \mu_x}{N}.$$

Then the sum $S_y = \sum_i y_i$ of the random variables y_i is related to the original sample mean via

$$S_x = \frac{1}{N} \sum_i x_i = S_y + \mu_x. \tag{8.8}$$

The variables have $\langle y_i \rangle = 0$ and $\sigma_y^2 := \text{var}(y_i) = \sigma_x^2/N^2$. Let the characteristic function of y be $\Phi(\omega)$. According to equation (7.58) [p. 125] we have

$$\Phi(\omega)|_{\omega=0} = 1, \qquad \Phi^{(1)}(\omega)|_{\omega=0} = 0, \qquad \Phi^{(2)}(\omega)|_{\omega=0} = -\frac{\sigma_x^2}{N^2}. \tag{8.9}$$

The characteristic function $\Phi_y(\omega)$ of S_y is, according to equation (7.66) [p. 130], simply

$$\Phi_{S_y}(\omega) = \Phi(\omega)^N = e^{N \ln(\Phi(\omega))}.$$

Next we expand $\ln(\Phi(\omega))$ about $\omega = 0$ up to second order. Owing to equation (8.9) the result is simply

$$\ln(\Phi(\omega)) = -\frac{\sigma_x^2}{2N^2}\, \omega^2.$$

Up to second order in the exponent the characteristic function reads

$$\Phi_{S_y}(\omega) = e^{-\frac{\sigma_x^2 \omega^2}{N}}.$$

The inverse Fourier transform of the Gaussian has been given in equation (7.62) [p. 127] and yields

$$p(S_y|N, \mathcal{I}) = e^{-\frac{S_y^2}{2\sigma_x^2/N}}.$$

Along with equation (8.8) we find the sought-for result

$$p(S_x|N, \mathcal{I}) = \frac{1}{\sqrt{2\pi\sigma_S^2}} e^{-\frac{(S_x - \mu_x)^2}{2\sigma_S^2}},$$

with $\sigma_S^2 = \sigma_x^2/N$, in agreement with equation (8.3d).

8.4 Markov chain Monte Carlo (MCMC)

The central limit theorem is the basis of one of the most important algorithms in the area of computer simulations, the Monte Carlo algorithm. There are a plethora of different problems in physics (statistical physics as well as quantum mechanics), information theory and engineering, which can be tackled with Monte Carlo algorithms. Within classical statistical physics the computation of thermodynamic expectation values in the canonical ensemble is one of the key tasks:

$$\langle O \rangle_T = \int d^m x \; O(\boldsymbol{x}) \; p(\boldsymbol{x}). \tag{8.10}$$

In the case of discrete degrees of freedom, the integration has to be replaced by a summation. The probability density $p(\boldsymbol{x})$ depends only on the energy $E(\boldsymbol{x})$ and is given by the normalized Boltzmann factor

$$p(\boldsymbol{x}) = \frac{1}{Z} \, e^{-E(x)/kT}.$$

The challenge lies in the very high dimensionality of the integrals or sums, which can hardly ever be accessed by analytical means. In order to employ the central limit theorem, we can map the high-dimensional integral to a one-dimensional integral, as outlined in Section 7.3 [p. 96]. To this end we introduce the probability density of O:

$$p(O) = \int d^m x \; \delta(O - O(\boldsymbol{x})) \; p(\boldsymbol{x}).$$

The mean and variance of O can be derived easily:

$$\langle O \rangle = \int dO \; O \; p(O) = \int d^m x \; O(\boldsymbol{x}) \; p(\boldsymbol{x}), \tag{8.11a}$$

$$\text{var}(O) = \int dO \; (O - \langle O \rangle)^2 \; p(O) = \int d^m x \; (O(\boldsymbol{x}) - \langle O \rangle)^2 \; p(\boldsymbol{x}). \tag{8.11b}$$

Therefore the mean of O is the desired integral. Furthermore, as guaranteed by the central limit theorem, the sample mean

$$\mathcal{O} = \frac{1}{N} \sum_{n=1}^{N} O(\boldsymbol{x}_n)$$

of the random numbers $O(\boldsymbol{x}_n)$ is distributed as a Gaussian distribution for sufficiently large N, with

$$\text{var}(\mathcal{O}) = \frac{\text{var}(\mathcal{O})}{\mathcal{N}}.$$

This implies that high-dimensional integrals or sums of type can be estimated by the arithmetic mean of a sample $\{x_1, \ldots, x_N\}$ of random vectors distributed as $p(\boldsymbol{x})$, and

$$\langle O \rangle = \frac{1}{N} \sum_{n=1}^{N} O(x_n) \pm \sqrt{\frac{\text{var}(O)}{N}}. \tag{8.12}$$

This also reveals why it is beneficial to repeat measurements of a quantity distorted by statistical fluctuations. The uncertainty of the arithmetic mean is given by

STANDARD ERROR

$$\mathrm{SE} = \frac{\sigma}{\sqrt{N}}. \tag{8.13}$$

We shall return to MC integration and MCMC in particular in Chapter 30 [p. 537].

8.5 The multivariate case

The CLT can be generalized to the multivariate case. Then it reads, for the sample mean of i.i.d. vectors of random variables,

CENTRAL LIMIT THEOREM IN m DIMENSIONS

Let $\boldsymbol{S} \in \mathbb{R}^m$ be a vector of random variables, defined by

$$\boldsymbol{S} = \frac{1}{N} \sum_{n=1}^{N} \boldsymbol{x}_n$$

with $\boldsymbol{x}_n$ i.i.d. random variables with mean $\boldsymbol{\mu} \in \mathbb{R}^m$ and covariance $\mathbf{C}$. Then

$$\lim_{N\to\infty} p(\boldsymbol{S}|N,\mathcal{I}) = \mathcal{N}(\boldsymbol{S}|\boldsymbol{\mu}, \mathbf{C}/N). \tag{8.14}$$

9

Poisson processes and waiting times

9.1 Stochastic processes

If a classical particle moves in an external force field $\boldsymbol{F}(\boldsymbol{x})$ its motion is deterministic and it could be called a 'deterministic process'. If the particle interacts with randomly distributed obstacles its motion is still deterministic, but the trajectory depends on the random characteristics of the obstacles, and maybe also on the random choice of the initial conditions. Such a motion is an example of a 'stochastic process', or rather a 'random process'.

In a more abstract generalized definition, a stochastic process is a random variable X_ξ that depends on an additional (deterministic) independent variable ξ, which can be discrete or continuous. In most cases it stands for an index $\xi \in \mathbf{N}$ or time t, space x or energy E. An almost trivial ubiquitous stochastic process is given by additive noise $\eta(t)$ on a time-dependent signal $s(t)$, i.e. $X_t = s(t) + \eta(t)$. As a consequence, X_t is no longer continuous. The most apparent applications of stochastic processes are time series of any kind that depend on some random impact. A broad field of applications are time series occurring, for example, in business, finance, engineering, medical applications and of course in physics. Beyond time series analysis, stochastic processes are at the heart of diffusion Langevin dynamics, Feynman's path integrals [43], as well as Klauder's stochastic quantization [62], which represents an unconventional approach to quantum mechanics. Here we will give a concise introduction and present a few pedagogical examples. An important example of stochastic processes, which we have encountered already, are random walks as discussed in Section 4.2 [p. 57] and Chapter 5 [p. 71]. We will have ample opportunity to encounter further examples of stochastic processes in various sections of this book: electron temperature distribution as a function of the laser cord position (Section 23.2), energy dependence of a signal on top of a background (Section 23.3), time dependence of the magnetic flux in a plasma (Section 24.1.2), surface growth as a function of time (Section 24.4), the yearly onset of cherry blossom over time (Section 25.1), and eruptions of Old Faithful geyser as a function of time (Section 25.2.2). There are many more examples of stochastic processes covered in this book, namely all those cases where a function $f(x)$ is measured which is corrupted by (additive) random noise. An important class of stochastic processes are Markov processes. They are introduced in Section 30.3 [p. 544] as the basis for a widespread stochastic integration technique, called 'Markov Chain Monte Carlo'.

For the sake of clarity we consider only a single random variable X_ξ that might, however, depend on an L-component vector $\boldsymbol{\lambda}$ of independent random variables. For fixed ξ, $X_\xi(\lambda)$ is an ordinary random variable, for which the CDF and PDF can be defined as usual:

$$F_\xi(x) := P(X_\xi(\boldsymbol{\lambda}) \le x), \tag{9.1a}$$

$$p_\xi(x) := \frac{d}{dx} F_\xi(x). \tag{9.1b}$$

The really new aspects of stochastic processes concern the dependence on the independent variable ξ, which, as said before, typically represents time, space or energy.

Stochastic processes are referred to as continuous (discrete) stochastic processes if they are based on continuous (discrete) input variables $\boldsymbol{\lambda}$. That does not, however, imply that $X_\xi(\lambda)$ as a function of ξ is necessarily continuous (discrete). As a counterexample consider a (discrete) Bernoulli experiment, whose outcome $o \in \{o_1, o_2\}$ defines the global behaviour

$$X_t(o) := \begin{cases} \cos(t) & \text{for } o = o_1 \\ \exp(-t) & \text{for } o = o_2. \end{cases}$$

This is a case of a discrete stochastic process which is continuous in t. The opposite situation is given by the following process:

$$X_t(\tau) := \boldsymbol{1}(t \le \tau),$$

with τ being a (continuous) exponential random variable. Apparently, X_t is discontinuous if considered as a function of t, but the process is referred to as continuous.

Interesting aspects of stochastic processes are the ξ-dependence of individual realizations (samples)

$$X_\xi^* := X_\xi(\boldsymbol{\lambda}^*)$$

for a fixed $\boldsymbol{\lambda}^*$, or the ξ-dependence of the mean, i.e.

$$\begin{aligned} E(\xi) = \langle X_\xi \rangle &:= \int dx\; x\; p_\xi(x) \\ &= \int d^L\lambda\; p(\boldsymbol{\lambda}) \int dx\; x\; \underbrace{p_\xi(x|\boldsymbol{\lambda})}_{\delta(x - X_\xi(\boldsymbol{\lambda}))} \\ &= \int d^L\lambda\; X_\xi(\boldsymbol{\lambda})\; p(\boldsymbol{\lambda}). \end{aligned}$$

In the second line we have introduced the independent random variable $\boldsymbol{\lambda}$ via the marginalization rule and we have utilized the fact that x is uniquely determined if ξ and $\boldsymbol{\lambda}$ are specified. Another interesting aspect of stochastic processes is the autocorrelation in ξ, i.e.

$$A(\xi,\xi') := \langle \Delta X_{\xi'} \Delta X_{\xi} \rangle = \int d^L\boldsymbol{\lambda}\; X_{\xi'}(\boldsymbol{\lambda})\; X_{\xi}(\boldsymbol{\lambda})\; p(\boldsymbol{\lambda}) \; - \langle X_{\xi'} \rangle \langle X_{\xi} \rangle.$$

The autocorrelation plays a crucial role in MCMC, as outlined in Section 30.3 [p. 544]. As a simple introductory example we consider a one-dimensional random walk on a lattice, with sites $x \in \mathbb{Z}$. In addition, we assume a parabolic potential $v(x) = x^2$, in order to avoid an increasing variance, which we found for the unrestricted random walk in Section 4.2 [p. 57]. Let X_t be the position of the walker at discrete time t and in each time step the walker has only three choices: stay or move one step forward or backward. That is, $X_{t+1} = X_t + \Delta X$. A possible choice for the probability $P(\Delta X|X_t, \mathcal{I})$ that the walker will be at site X_{t+1} in the next time step can be defined via the Metropolis–Hastings algorithm (see Section 30.3 [p. 544]), resulting in

$$P(\Delta X = \pm 1|X_t, \mathcal{I}) = \min\left(\frac{1}{2}\exp\left\{ -\beta \mp 2\beta\; X_t\right\}, 1\right),$$

$$P(\Delta X = 0|X_t, \mathcal{I}) = 1 - P(\Delta X = +1|X_t, \mathcal{I}) - P(\Delta X = -1|X_t, \mathcal{I}).$$

This process is actually a 'Markov process' as the next step depends on the actual position, but otherwise it has no memory. In Figure 9.1 the time evolution of a few individual random walks X_t with $\beta = 0.05$ is depicted. The walks always start at $X_0 = 0$. In addition, the sample mean $\overline{X_t}$ and sample variance $\overline{(\Delta X_t)^2}$, obtained from 1000 independent repetitions of such random walks, are displayed. As expected from symmetry, the sample mean is $\overline{X_t} \approx 0$. The sample variance initially increases linearly with time, according to a diffusion process, but then levels off due to the parabolic potential, which prevents the walkers from deviating too far from the origin. In Figure 9.1(b) the autocorrelation is depicted, which has been computed from the same sample of random walks. Here the autocorrelation of a single random walk is defined as

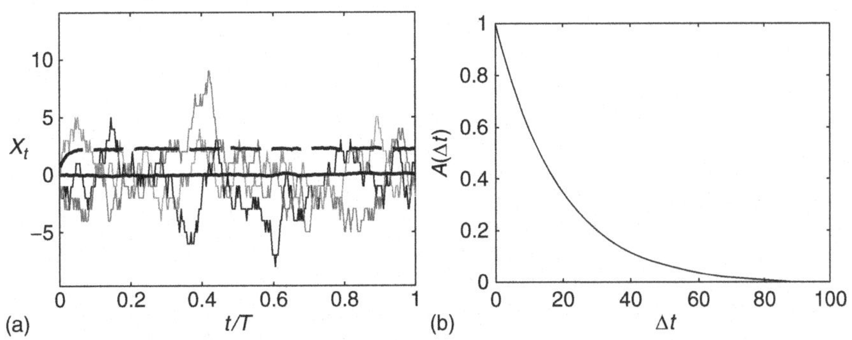

Figure 9.1 (a) Three individual random walks, all starting at $X_0 = 0$, are depicted as jagged lines in different shades of grey. The time is expressed in units of the total time $T = 1000$. In addition, the sample mean (thick solid line) and variance (dashed line) of $N = 1000$ independent random walks are displayed. (b) The corresponding scaled autocorrelation function $A(\Delta t)$ versus lag size Δt.

$$\tilde{A}(\Delta t) := \frac{1}{t_1 - t_0 + 1} \sum_{t=t_0}^{t_1} \Delta X_{t+\Delta t}\, \Delta X_t,$$
$$\Delta X_t := X_t - \overline{X_t}.$$

The autocorrelation is determined in the time window ranging from $t_0 = 100$ to $t_1 = 900$. The first 100 steps have been discarded, in order to enter the region where $\langle X_{t+\Delta t} X_t \rangle$ is independent of t, then the process is said to be 'homogeneous'. There are a couple more subtleties to be accounted for in computing autocorrelations numerically, which are, however, immaterial for the basic understanding of this topic. In Figure 9.1 the scaled autocorrelation function $A(\Delta t) := \tilde{A}(\Delta t)/\tilde{A}(0)$, averaged over the sample, is shown. We see that the autocorrelation decays exponentially, which is a typical behaviour. A fit of $\exp\{\Delta t/\xi\}$ to the data defines the autocorrelation length ξ. We recognize that the random walk has a rather long autocorrelation length $\xi \approx 20$, i.e. there is a long-range correlation between fluctuations. For a small autocorrelation $A(\Delta t) \approx 0$, it is necessary that $\langle X_{t+\Delta t} X_t \rangle$ changes sign frequently as a function of t. Apparently, the autocorrelation length defines a typical time window in which the walk is either above or below the mean value. More details on autocorrelation functions of Markov processes are given in Section 30.3 [p. 544].

In the next sections we focus on Poisson processes and various aspects of Poisson points. As mentioned before, we will encounter further examples of stochastic processes later in this book. For additional details on stochastic processes we refer the reader to [8, 159]. Readers particularly interested in mathematically rigorous definitions and properties of stochastic processes may look at [45].

9.2 Three ways to generate Poisson points

We have introduced the Poisson distribution as a limiting case of the binomial distribution by keeping $\mu = pN$ fixed and considering the limit $N \to \infty$. We use this situation to generate 'Poisson points' as outlined in the next section.

9.2.1 Global generation of Poisson points

We consider a finite interval $\Omega_L = (0, L)$ of length L, in which we generate N independent and uniformly distributed points. They are called 'Poisson points' (PPs). The point density is

$$\rho := \frac{N}{L}. \tag{9.2}$$

We now select an arbitrary interval $I_x \subset \Omega_L$ of length x. The probability for an individual point to appear in I_x is, according to the classic definition,

$$p = \frac{x}{L}.$$

The probability that n PPs land in I_x is therefore binomial. The mean number of points generated on I_x is

$$\mu = p\,N = x\,\rho. \tag{9.3}$$

Next we erase these PPs and start afresh, but this time we double both the length L and the number of points N. We retain the interval I_x. As the point density is the same, the mean number (μ) of points in I_x is also the same. Upon repeating the doubling process we fulfil the requirements of the Poisson theorem ($\mu = pN =$ fixed and $N \to \infty$). Hence, in the limit of $N \to \infty$,[1] we obtain a Poisson distribution

$$P(n|x,\rho,\mathcal{I}) := e^{-\rho x}\,\frac{(\rho x)^n}{n!}. \tag{9.4}$$

Distance law of neighbouring PPs

Next, we are interested in the PDF of the distances between Poisson points $p(\xi|\rho,\mathcal{I})$, where ξ stands for the proposition

ξ : The distance to the next Poisson point is within $[\xi, \xi + d\xi)$.

The proposition requires that there is no Poisson point in the interval $[0,\xi)$, and there must be a Poisson point in the following interval $d\xi$. Both events are logically independent. Both probabilities follow from the Poisson distribution $P(n|x,\rho,\mathcal{I})$ given in equation (9.4) with $(n = 0\,;\, x = \xi)$ and $(n = 1\,;\, x = d\xi)$, respectively. This yields the sought-for probability

$$P(\xi|\rho,\mathcal{I}) = P(n=0|\xi,\rho,\mathcal{I})\;P(n=1|d\xi,\rho,\mathcal{I}) = e^{-\rho\,\xi}\;\rho\;d\xi$$

and the corresponding PDF is exponential.

DISTANCE LAW OF POISSON PROCESS

$$p(\xi|\rho,\mathcal{I}) = \rho\;e^{-\rho\xi}. \tag{9.5}$$

Its mean value is $1/\rho$, in accordance with the mean density of ρ.

Distance law of PPs

The distances between neighbouring PPs are exponentially distributed. Next we want to consider the distribution of the distance d_n between Poisson point m and $m + n$. As the individual intervals are independent, the distances are independent of m and they are a sum of n independent and exponentially distributed random numbers. According to equation (7.68) [p. 130], such a sum is Gamma distributed.

[1] See equation (6.12) [p. 88].

Distance distribution of Poisson points

$$p(p(d_n|\rho,\mathcal{I})) = p_\Gamma(d_n|n,\rho) = \frac{\rho}{\Gamma(n)}\,(\rho\, d_n)^{n-1}\, e^{-\rho\, d_n}. \tag{9.6}$$

9.2.2 Successive generation of Poisson points

The derived exponential distribution suggests an alternative view of the generation process. So far the Poisson points have been generated by distributing the points homogeneously on an interval. In this way it is warranted that the points are independent of each other. Similarly, the distances between the points are i.i.d. For this reason also a sequential generation process can be used. We start at $x = 0$ and determine the next Poisson point x_1 at a distance of ξ_1 according to the exponential distribution equation (9.5). Then the next point x_2 is generated at a distance ξ_2, again derived from $p(\xi|\rho,\mathcal{I})$, and so on. The Poisson points are then located at

$$x_n = \sum_{i=1}^{n} \xi_i .$$

The PDF implies that the probability for the next PP being at a distance $d\xi$ is given by

$$p(d\xi|\rho,\mathcal{I})\, d\xi = (1 - \rho d\xi)\cdot \rho d\xi .$$

This expression can be interpreted as the probability for no event $(1 - \rho d\xi)$ in $d\xi$ times the probability for one event $\rho d\xi$ in the following infinitesimal interval $d\xi$. In other words, the events in infinitesimal intervals are uncorrelated and the probability $P_{d\xi}$ for an event in $d\xi$ is analytic in $d\xi$:

$$P_{d\xi} = \rho\, d\xi .$$

As we will see in the next section, these features entail the Poisson distribution.

9.2.3 Local generation of Poisson processes

The Poisson process can also be defined by the two axioms derived before.

- The probability for an event happening in an infinitesimal interval $d\xi$ is independent of previous PPs.
- The probability is analytic in $d\xi$, i.e. $P(\text{event in } d\xi|\rho,\mathcal{I}) = \rho\, d\xi$.

It would be straightforward to identify the underlying Bernoulli experiment and to derive the Poisson distribution as a limiting case. Here, however, we want to present a different approach, that is over the top for the present problem but turns out to be very useful in more complex situations. For instance, the same approach can be used to derive the 'Fokker–Planck equation' for drift-diffusion processes from random walks. The present problem is nicely suited to illustrate this idea. To begin with, we notice that the axioms imply

- The probability for two events happening in $d\xi$ is negligible $(O(d\xi^2))$.
- The probability for no event in $d\xi$ is therefore $1 - \rho\, d\xi$.

Next, we define the proposition

$$A(n, x) : \text{The number of events in an interval of length } x \text{ is } n.$$

The goal is the probability $P(A(n, x)|\mathcal{I})$ for n Poisson points in the interval of length x. The basic idea of the present approach is to derive a differential equation for the sought-for probability. To this end, we compute the probability for the interval of length $x + dx$ by marginalization over all propositions $A(m, x)$ for the interval of length x:

$$P(A(n, x + dx)|\mathcal{I}) = \sum_{m=0}^{\infty} P(A(n, x + dx)|A(m, x), \mathcal{I})\ P(A(m, x)|\mathcal{I}).$$

We can simplify this expression by taking advantage of the fact that only zero or one events can happen within the infinitesimal interval dx. Hence, only two terms of the summation over m remain: $m = n$ and $m = n - 1$. Contributions of higher order $(O(dx^2))$ can be neglected. Then

$$\begin{aligned} P(A(n, x + dx)|\mathcal{I}) &= P(A(n, x + dx)|A(n, x), \mathcal{I})\ P(A(n, x)|\mathcal{I}) \\ &\quad + P(A(n, x + dx)|A(n - 1, x), \mathcal{I})\ P(A(n - 1, x)|\mathcal{I}) \\ &= (1 - \rho\, dx)\ P(A(n, x)|\mathcal{I}) + \rho\, dx\ P(A(n - 1, x)|\mathcal{I}). \end{aligned}$$

So we obtain the desired differential equation

$$\frac{d}{dx}\ P(A(n, x)|\mathcal{I}) = \rho \left(P(A(n - 1, x)|\mathcal{I}) - P(A(n, x)|\mathcal{I}) \right).$$

Next we introduce a simplified notation $P_n(x) := P(A(n, x)|\mathcal{I})$ and define $P_{-1}(x) = 0$, then the differential equation reads

$$\frac{d}{dx}\ P_n(x) = \rho \left(P_{n-1}(x) - P_n(x) \right). \tag{9.7}$$

The initial condition is given by

$$P_n(0) = \delta_{n,0}, \tag{9.8}$$

because there is no PP in an interval of zero size. This is actually a difference-differential equation, which can be solved by various means. Here we proceed using 'generating functions' because of their importance in many problems of probability theory: We define the *generating function* $\phi(x, z)$ via the following series expansion:

GENERATING FUNCTION

$$\phi(x, z) := \sum_{n=0}^{\infty} P_n(x)\, z^n. \tag{9.9}$$

Apart from the presence of the additional variable x, it is the same object introduced in Section 2.1.3 [p. 21]. For arbitrary but fixed x, $P_n(x)$ represent the coefficients of a Taylor expansion. The representation by $\phi(x, z)$ has the same information as $P_n(x)$ but certain advantages, as we have seen in Section 2.1.3 [p. 21] already. In particular, the probabilities $P_n(x)$ are given by

$$P_\nu(x) = \frac{1}{\nu!} \frac{d^\nu}{dz^\nu} \phi(x, z)\bigg|_{z=0}. \tag{9.10}$$

In order to obtain the differential equation for the generating function, we multiply equation (9.7) by z^n summed over n. The result is

$$\frac{d}{dx} \underbrace{\sum_{n=0}^{\infty} P_n(x)\, z^n}_{\phi(x,z)} = \rho \Bigg(\underbrace{\sum_{n=0}^{\infty} P_{n-1}(x)\, z^n}_{z\,\phi(x,z)} - \underbrace{\sum_{n=0}^{\infty} P_n(x)\, z^n}_{\phi(x,z)} \Bigg),$$

$$\frac{d}{dx}\phi(x, z) = \rho\, (z - 1)\, \phi(x, z). \tag{9.11}$$

Along with the initial condition equation (9.8), which implies $\phi(x, 0) = 1$, the solution of the differential equation (9.11) reads

$$\phi(x, z) = e^{\rho\, (x-1)\, z}.$$

Eventually we obtain the sought-for probabilities $P_n(x)$ via equation (9.10):

$$P_n(t) = \frac{1}{n!} \frac{d^n}{dx^n} e^{\rho\, (x-1)\, t}\bigg|_{x=0} = \frac{1}{n!}\, (\rho\, t)^n\, e^{-\rho\, t}.$$

This is – as expected – the Poisson distribution. We see that the Poisson distribution follows from very general assumptions. For that reason the distribution is ubiquitous in physics.

9.3 Waiting time paradox

If we interpret the distance between neighbouring PPs t in a temporal way we have

$$p(t|\tau) = \frac{1}{\tau} e^{-\frac{t}{\tau}}, \tag{9.12}$$

with τ being the mean 'waiting time'. Let us assume a Poisson-distributed arrival time of buses at a bus stop. What is the mean waiting time for the next bus, when we are at the bus stop at a random instant of time? We could likewise ask: If we switch on a Geiger counter, how long does it take to the next signal, triggered by an alpha particle? Since the mean time between arrivals is τ, it seems reasonable to assume that the mean waiting time should be $\tau/2$. The correct answer, however, is τ, which may appear paradoxical since the remaining time to the next bus should be shorter than the entire interval between successive buses.

In the following, we will see how Bayesian probability theory provides the correct answer. The quantity to be determined is the probability density $p(t|R_I, \tau, \mathcal{I})$ for the waiting time t, given that we arrived at a random instant t^* in time. The time t^* falls into a time interval I between successive buses. This is denoted by the proposition

$$I_{t^*}: \text{The time interval } I \text{ of our arrival is determined by } t^* \in I.$$

We proceed by introducing the length L of the arrival interval I via the marginalization rule

$$p(t|R_I, \tau, \mathcal{I}) = \int_0^\infty \underbrace{p(t|L_I = L, I_{t^*}, \mathcal{I})}_{\frac{1}{L}\, I(0<t\leq L)},\ p(L|I_{t^*}, \tau, \mathcal{I})\, dL,$$

$$p(t|I_{t^*}, \tau, \mathcal{I}) = \int_t^\infty \frac{1}{L}\ p(L|I_{t^*}, \tau, \mathcal{I})\, dL. \tag{9.13}$$

Here we exploited the random arrival time t, resulting in a uniform PDF within the interval I. The quantity $p(L|I_{t^*}, \tau, \mathcal{I})$ is the PDF for the length of the interval selected as specified in I_{t^*}. We proceed with Bayes' theorem:

$$p(L|I_{t^*}, \tau, \mathcal{I}) = \frac{p(I_{t^*}|L, \tau, \mathcal{I})\ p(L|\tau, \mathcal{I})}{p(I_{t^*}|\tau, \mathcal{I})}.$$

The probability to arrive in an interval of length L is, according to the classical definition and in analogy with the dart problem of Section 7.2.1 [p. 94], proportional to its length L

$$p(I_{t^*}|L, \tau, \mathcal{I}) = \kappa\ L,$$

with an unknown parameter κ. The PDF $p(L|\tau, \mathcal{I})$ is the exponential distribution of neighbouring PPs. The expression in the denominator is a constant that can be combined with κ into a constant that is determined by the normalization, resulting in

$$p(L|I_{t^*}, \tau, \mathcal{I}) = \frac{L}{\tau^2}\ e^{-\frac{L}{\tau}}.$$

This is a special case of a Gamma distribution with $\alpha = 2$ and $\beta = 1/\tau$. Hence, the mean interval length L is therefore

$$\langle L \rangle = 2\ \tau.$$

Apparently, intervals selected by I_{t^*} are longer than average.

Now, we proceed with equation (9.13):

$$p(t|I_{t^*}, \tau, \mathcal{I}) = \frac{1}{\tau^2} \int_t^\infty e^{-\frac{L}{\tau}}\, dL = \frac{1}{\tau}\, e^{-\frac{t}{\tau}}.$$

This proves that the waiting time has the same PDF as the interval length between successive buses. As we have seen, this counterintuitive result is due to the larger probability for an arrival in a long interval. This also clarifies why the first interval from an arbitrarily chosen point in time to the first event also has a mean duration of τ.

9.4 Order statistic of Poisson processes

Assume $\{t_n\}$ are Poisson points, obeying equation (9.12), and let $N(t)$ denote the stochastic process which provides the number of Poisson points up to time t:

$$N(t) = \sum_{n=0}^{\infty} \boldsymbol{1}(t \geq t_n)\,.$$

The probability for $N(t) = n$ is of course given by the Poisson distribution. A representative result is depicted in Figure 9.2. Next we want to estimate the PDF for position x of the nth Poisson point. We can use the general considerations on order statistics derived in Section 7.6 [p. 118]. In order to find the nth Poisson point in $[t, t + dt)$, $n - 1$ PPs have to be located in the interval $[0, t)$. The corresponding probability is

$$p(n-1|t, \tau, \mathcal{I}) = e^{-t/\tau}\, \frac{(t/\tau)^{n-1}}{(n-1)!}\,.$$

The probability for the next point being located in $[t, t+dt)$ is given by dt/τ. The product of the two independent probabilities results in a special Gamma distribution, which is also called the 'Erlang distribution' in the present context.

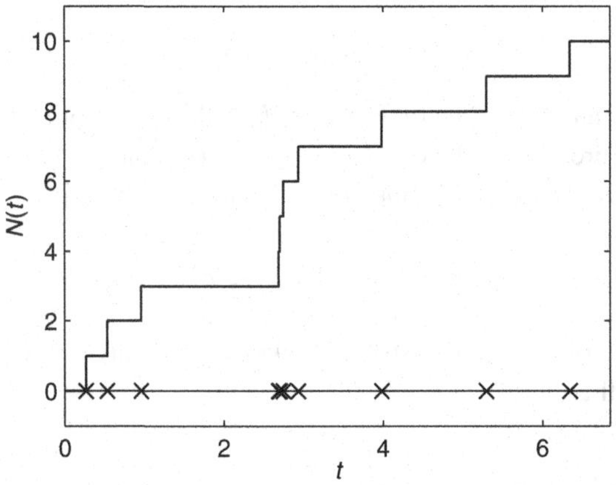

Figure 9.2 Poisson points (crosses) and Poisson process $N(t)$ for $\tau = 1$.

ERLANG DISTRIBUTION

$$p(t_n = t|\tau, \mathcal{I}) = \frac{\tau^{-n}}{\Gamma(n)}\, t^{n-1}\, e^{-t/\tau}. \tag{9.14}$$

The expectation value of this position is

$$\langle t_n \rangle = n\ \tau.$$

This result is in agreement with the expectations, since the mean distance of the points is given by τ. The Erlang distribution has been introduced in studies of the frequency of phone calls and is important for various types of queueing problems.

9.5 Various examples

Next we will give some simple pedagogical examples. Realistic applications will follow in Part V [p. 333].

9.5.1 Shot noise

As a first example we consider shot noise, which is relevant in the measurement of individual photons. Suppose a light source emits photons with a mean rate of λ photons per time interval. The emission of the photons shall be uncorrelated and within an infinitesimal time interval at most one photon is being emitted. Therefore, the probability for the emission of a photon within dt is given by

$$p = \lambda\ dt.$$

These are precisely the conditions for a Poisson process, and the probability to find k photons within a finite interval of length t is therefore

$$P(n|N = t/dt, p = \lambda dt) \underset{dt \to 0}{\longrightarrow} e^{-\lambda t} \frac{(\lambda t)^n}{n!}.$$

Hence, shot noise obeys a Poisson distribution.

9.5.2 The stickiness of bad luck

It is a common impression that all waiting queues proceed faster than our own. Is this just a distorted perception? Let us analyse the problem.

Let t_0 denote *our* waiting time and t_i $(i = 1, \ldots)$ the waiting times in other queues. Let us assume that nobody experiences preferred treatment and that hence all t_i are subject to the same statistical fluctuations and are hence i.i.d.-distributed random variables with probability density $p(t|\mathcal{I})$. We are interested in the probability that only the nth waiting time (t_n) is larger than ours (t_0). For a systematic treatment we define the following proposition.

$$A_n : \text{Only the } n\text{th waiting time } \{t_n\} \text{ is greater than ours.}$$

In order to compute the probability $P(A_n|\mathcal{I})$, we employ the marginalization rule to introduce the value t_0 of our waiting time:

$$P(A_n|\mathcal{I}) = \int P(A_n|t_0, \mathcal{I})\ p(t_0|\mathcal{I})\ dt_0.$$

The expression A_n implies that the waiting times of the queues $1-(n-1)$ are smaller than t_0 and that $t_n > t_0$. Since the waiting times are uncorrelated, we obtain

$$P(A_n|\mathcal{I}) = \int \prod_{i=1}^{n-1} \underbrace{P(t_i < t_0|\mathcal{I})}_{:=q(t_0);\ \text{independent of } i} \quad \underbrace{P(t_n > t_0|\mathcal{I})}_{=(1-q(t_0))}\ p(t_0|\mathcal{I})\ dt_0.$$

Now, $q(t)$ is the CDF corresponding to $p(t|\mathcal{I})$ and therefore $p(t_0|\mathcal{I})dt_0 = dq(t_0)$. So, the integral simplifies significantly as

$$P(A_n|\mathcal{I}) = \int q^{n-1}\ (1-q)\ dq = \frac{\Gamma(n)\Gamma(2)}{\Gamma(n+2)} = \frac{(n-1)!}{(n+1)!} = \frac{1}{n(n+1)}.$$

It is straightforward to verify the correct normalization:

$$\begin{aligned}\sum_{n=1}^{\infty} P(A_n|\mathcal{I}) &= \sum_{n=1}^{\infty} \frac{1}{n(n+1)} = \lim_{L\to\infty} \sum_{n=1}^{L} \left(\frac{1}{n} - \frac{1}{n+1}\right) \\ &= \lim_{L\to\infty} \left(\sum_{n=1}^{L} \frac{1}{n} - \sum_{n=2}^{L+1} \frac{1}{n}\right) = \lim_{L\to\infty} \left(1 - \frac{1}{L+1}\right) = 1.\end{aligned}$$

The surprising aspects of the result are:

- it is independent of the distribution $p(t|\mathcal{I})$;
- the mean value does not exist, $\langle n\rangle = \infty$.

This implies that on average, endless many queues are faster than ours. However, the mean is dominated by the tail of the distribution, because the median is only $n = 1.5$ since $P(A_1|\mathcal{I}) = \frac{1}{2}$.

In conclusion, we want to simulate this queueing problem. As PDF for the waiting times we use uniform random numbers on the unit interval. The algorithm is as follows:

ALGORITHM TO SIMULATE THE QUEUEING PROBLEM

Generate a reference random number t_0.

Generate random numbers t_n until $t_n > t_0$ and record n.

Repeat the experiment L times.

Compute the mean and standard deviation of n. (9.15)

The result of a simulation run up to a maximum length $L = 1000$ is shown in Figure 9.3. Neither the mean value nor the standard deviation converge with increasing simulation length. This is to be expected, since both values diverge for $L \to \infty$. Despite the high probability of $2/3$ that the next or second next element already exceeds the value of the reference element, there is a non-negligible probability for the occurrence of very long waiting times (please note the logarithmic scale of the ordinate in Figure 9.3). In the present example the arithmetic mean at the end of the simulation is 548 and the standard deviation is 17 151, 30 times the mean value!

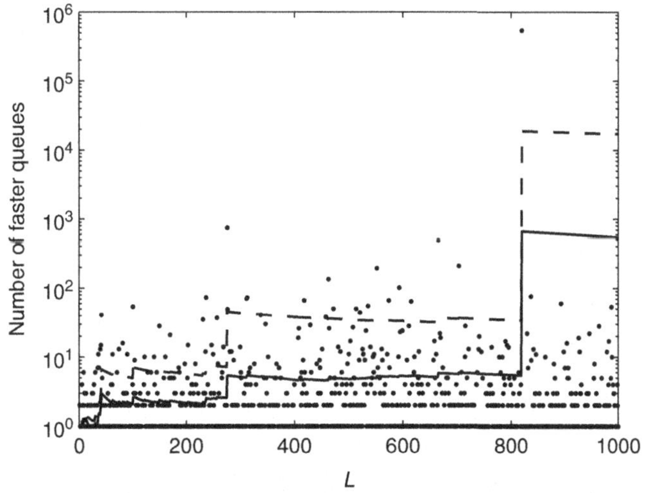

Figure 9.3 Computer simulation of the queueing problem according to algorithm in equation (9.15). The number $n(L)$ of faster queues is marked by dots. The running mean (solid line) and standard deviation (dashed line) are plotted versus the number of repetitions L.

9.5.3 Pedestrian delay problem

In its simplest form the pedestrian delay problem considers pedestrians arriving at random at a street with the goal of crossing it. The traffic is described by a Poisson process, where any structures due to traffic lights or the like are ignored. The central question is how long the pedestrian has to wait on average until the time gap between cars is big enough.

To be more specific, let ΔT be the time the pedestrian needs to traverse the street and τ the mean time gap between the Poisson events, i.e. the PDF for the Poisson events reads

$$p(t|\tau,\mathcal{I}) = \frac{1}{\tau} e^{-\frac{t}{\tau}}.$$

It is expedient to express all times in units of τ. Then the PDF reads

$$p(t|\tau,\mathcal{I}) = e^{-t}.$$

We seek to determine the PDF $p(T|\mathcal{I})$ for T, the waiting time of the pedestrian. To begin with, we introduce the proposition

A_n : The first n intervals are too short and the interval n+1 is long enough.

This proposition implies that the first time intervals $t_1, \ldots, t_n$ are all smaller than ΔT and for the last one we have $t_{n+1} \geq \Delta T$. We will first determine the probability for A_n:

$$\begin{aligned}
P_n := P(A_n|\Delta T,\mathcal{I}) &= \prod_{\nu=1}^{n} P(t_\nu < \Delta T|\mathcal{I})\ P(t_{n+1} \geq \Delta T|\mathcal{I}),\\
P(t < \Delta T|\mathcal{I}) &= \int_0^{\Delta T} e^{-t} dt = 1 - e^{-\Delta T} := F,\\
P(t \geq \Delta T|\mathcal{I}) &= 1 - F,\\
P_n &= F^n\ (1-F).
\end{aligned}$$

This is the geometric distribution. The mean and variance of the number of initially wrong intervals are therefore, according to equations (4.25b) and (4.25c) [p. 63] ,

$$\langle n \rangle = \frac{F}{(1-F)} = e^{\Delta T} - 1, \tag{9.16a}$$

$$\langle (\Delta n)^2 \rangle = \frac{F}{(1-F)^2} = e^{\Delta T}\left(e^{\Delta T} - 1\right). \tag{9.16b}$$

Next we determine the PDF for the waiting time T:

$$p(T|\Delta T,\mathcal{I}) = \sum_{n=0}^{\infty} p(T|\Delta T, A_n,\mathcal{I})\ P_n.$$

We introduce the n random variables $\boldsymbol{t} = \{t_1, \ldots, t_n\}$, corresponding to the interval lengths, by the marginalization rule

$$p(T|A_n, \Delta T, \mathcal{I}) = \int_0^\infty \dots \int_0^\infty dt^n \; p\big(T|\boldsymbol{t}, \cancel{A_n}, \cancel{\Delta T}, \mathcal{I}\big) \;\; p\big(\boldsymbol{t}|A_n, \Delta T, \mathcal{I}\big),$$

$$p\big(T|\boldsymbol{t}, \mathcal{I}\big) = \delta\big(T - \sum_{\nu=1}^{n} t_\nu\big),$$

$$p\big(\boldsymbol{t}|A_n, \Delta T, \mathcal{I}\big) = \prod_{\nu=1}^{n} p(t_\nu|t_\nu < \Delta T).$$

Hence

$$p(T|A_n, \mathcal{I}) = \int_0^{\Delta T} \dots \int_0^{\Delta T} dt^n \; \delta\big(T - \sum_{\nu=1}^{n} t_\nu\big) \prod_{\nu=1}^{n} p\big(t_\nu|t_\nu < \Delta T, \mathcal{I}\big).$$

We will not compute the PDF for T in full, but rather only the mean:

$$\langle T\rangle = \sum_{n=0}^{\infty} \underbrace{\int_0^\infty T \; P(T|A_n, \Delta T, \mathcal{I}) \, dT}_{:=\langle T\rangle_n} \; P_n.$$

For the mean for given n

$$\langle T\rangle_n = \int_0^{\Delta T} \dots \int_0^{\Delta T} dt^n \sum_{\nu=1}^{n} t_\nu \prod_{\nu=1}^{n} p\big(t_\nu|t_\nu < \Delta T, \mathcal{I}\big)$$

we need the integral

$$I := \int_0^{\Delta T} t_\nu \; p\big(t_\nu|t_\nu < \Delta T, \mathcal{I}\big) \, dt_\nu = \int_0^{\Delta T} t \, \frac{e^{-t}}{F} \, dt = \frac{1-F}{F} \left(e^{\Delta T} - 1 - \Delta T\right).$$

Then $\langle T\rangle_n = nI$ and $\langle T\rangle = \langle n\rangle I$. Along with equation (9.16) we obtain the final result, which in original units reads

Mean waiting time

$$\langle T\rangle = \left(e^{\Delta T/\tau} - 1 - \frac{\Delta T}{\tau}\right).$$

This is a rather striking result as the time increases exponentially with the time ΔT needed to cross the road. In particular, it presents an exponential disadvantage for handicapped or elderly people.

PART II

ASSIGNING PROBABILITIES

The rules of probability theory provide the means to transform probabilities. As a direct consequence of the sum rule and the product rule, Bayes' theorem in equation (1.9) [p. 12] has been derived:

$$P(a|d,\mathcal{I}) = \frac{1}{Z}\, P(d|a,\mathcal{I})\; P(a|\mathcal{I}).$$

The theorem allows us to relate the posterior distribution to the product of the likelihood and the prior PDF. The likelihood distribution is usually known and determined by the experimental setup (expressed by $\mathcal{I}$) and the parameters a of the model. Prior probabilities, on the contrary, cannot be assigned solely based on the rules of probability theory, since the priors incorporate the assumptions underlying the inference process. Prior PDFs may also occur in a more indirect way, for example, if not all parameter values entering the likelihood are known. The unknown parameters $\boldsymbol{a}$ can be marginalized:

$$p(\boldsymbol{d}|\boldsymbol{b},\mathcal{I}) = \int p(\boldsymbol{d}|\boldsymbol{a},\boldsymbol{b},\mathcal{I})\; p(\boldsymbol{a}|\boldsymbol{b},\mathcal{I})\; da^m,$$

resulting in a probability distribution on the left-hand side which is independent of $\boldsymbol{a}$. However, the marginalization also requires the specification of a prior PDF $p(\boldsymbol{a}|\boldsymbol{b},\mathcal{I})$ for $\boldsymbol{a}$, which ultimately cannot be provided by the rules of probability theory. A typical example for this situation is provided by Gaussian noise with zero mean and unknown variance. For the *marginal likelihood* the prior PDF for the standard deviation $p(\sigma|\mathcal{I})$ needs to be specified. This is a generic property of Bayesian inference. The rules of probability yield a hierarchy of conditional probabilities, but at the very end there is always a prior probability that has to be provided from outside, based on the available information. There are three generic cases.

- **(a) Ignorant priors:** For parameters or densities without any additional information, except the definition of the problem.
- **(b) Testable information:** In addition to (a) exact information is available, typically in the form of equality constraints, e.g. the moments of the sought-for distribution, $\Phi\{p(\boldsymbol{x}|\mathcal{I})\} = 0$.
- **(c) Priors for positive additive distribution functions.**

Next, we will discuss the first two cases in more detail. The third topic is addressed in Chapter 12 [p. 201] on the 'quantified maximum entropy method' (QME), which has been applied very successfully to form-free image reconstruction problems, for example, in astrophysics and plasma physics.

10

Prior probabilities by transformation invariance

If we have no reasonable expectation about the outcomes $A_1, A_2, \ldots, A_N$ of an experiment, which probabilities should be assigned to the observation A_i? The earliest solution to this problem dates back to Bernoulli (1713). He called it the principle of insufficient reason. It states that if we have an enumerable set of mutually exclusive propositions and there is no reason to believe that any one of these is to be preferred over any other, then we should assign the same probability to all. Since this principle requires an enumerable set of possibilities, its application is limited to discrete problems. Priors $p(A_i) = 1/N$ find frequent application in model-selection problems (see Chapter 27 [p. 470]). The situation is more complicated in the case of continuous degrees of freedom. We have discussed Bertrand's paradox in the introductory section, which reveals that the assignment of PDFs depends on the representation. An interesting solution to this question has been offered by E.T. Jaynes [106]. Jaynes introduced the principle of 'transformation invariance' in order to derive prior probability distributions for continuous degrees of freedom. After considering some elementary cases we will focus on invariance against translations and rotations in order to derive hyperplane priors for later use in Chapter 21 [p. 333] and Chapter 24 [p. 409].

The derivation of Jaynes' invariance prior probabilities starts out from the probability for an infinitesimal event in two different reference frames, as presented in equation (7.49) [p. 121]:

$$p_x(\boldsymbol{x}) = p_y(\boldsymbol{y}) \left| \frac{\partial \boldsymbol{y}}{\partial \boldsymbol{x}} \right|, \tag{10.1}$$

where the last factor is the Jacobian determinant. Now let $\mathbf{T}(\boldsymbol{x})$ be a one-to-one map $\boldsymbol{x} \to \boldsymbol{y} = \mathbf{T}(\boldsymbol{x})$, then

$$p_x(\boldsymbol{x}) = p_y(\mathbf{T}(\boldsymbol{x})) \left| \frac{\partial (\mathbf{T}\boldsymbol{x})}{\partial \boldsymbol{x}} \right|. \tag{10.2}$$

Jaynes' generalization of Laplace's 'principle of insufficient reasoning' claims that if no reference frame is distinguished, then

$$p_x(\boldsymbol{x}) \stackrel{!}{=} p_x(\boldsymbol{x}) := p(\boldsymbol{x}).$$

Along with equation (10.2) we obtain a functional equation for $p(\boldsymbol{x})$:

$$p(\boldsymbol{x}) = p(\mathbf{T}(\boldsymbol{x}))\left|\frac{\partial(\mathbf{T}(\boldsymbol{x}))_i}{\partial \boldsymbol{x}}\right|. \tag{10.3}$$

It suffices to consider infinitesimal transformations $\mathbf{T}_\varepsilon(\boldsymbol{x})$, with

$$\lim_{\varepsilon\to 0} \mathbf{T}_\varepsilon(\boldsymbol{x}) = \boldsymbol{x}. \tag{10.4}$$

We now replace $\mathbf{T}(\boldsymbol{x})$ in equation (10.3) by $\mathbf{T}_\varepsilon(\boldsymbol{x})$ and differentiate w.r.t. ε the derivative of the resulting equation

$$\frac{\partial}{\partial\varepsilon} p(\boldsymbol{x}) = \frac{\partial}{\partial\varepsilon}\left[p(\mathbf{T}_\varepsilon(\boldsymbol{x}))\left|\frac{\partial \mathbf{T}_\varepsilon(\boldsymbol{x})}{\partial \boldsymbol{x}}\right|\right]. \tag{10.5}$$

Since the left-hand side of this equation is independent of ε, the derivative vanishes. Consequently, the right-hand side has to be independent of ε as well, so we can choose $\varepsilon = 0$. Then equation (10.3) has been reduced to a differential equation

$$\frac{\partial}{\partial\varepsilon}\left[p(\mathbf{T}_\varepsilon(\boldsymbol{x}))\left|\frac{\partial \mathbf{T}_\varepsilon(\boldsymbol{x})}{\partial \boldsymbol{x}}\right|\right]_{\varepsilon=0} = 0. \tag{10.6}$$

As a matter of fact, we have to solve this equation simultaneously for all transformations under which the problem is invariant, which results in a set of differential equations. In many respects, it is actually very similar to the Noether theorem in analytical mechanics.

General properties of the solution

Let us denote the Jacobian matrix

$$\mathbf{J}(\varepsilon) := \frac{\partial \mathbf{T}_\varepsilon(\boldsymbol{x})}{\partial \boldsymbol{x}}.$$

It has the property $\mathbf{J}(0) = \mathbb{1}$. By further evaluation:

$$\frac{\partial}{\partial\varepsilon}\Big[p(\mathbf{T}_\varepsilon(\boldsymbol{x}))\Big]_{\varepsilon=0} + p(\boldsymbol{x})\frac{\partial}{\partial\varepsilon}|\mathbf{J}(\varepsilon)|\Big|_{\varepsilon=0} = 0.$$

Moreover, according to equation (A.14) [p. 601], we have

$$\begin{aligned}
\frac{\partial}{\partial \varepsilon}|\mathbf{J}(\varepsilon)|\Big|_{\varepsilon=0} &= |\mathbf{J}(\varepsilon)|\Big|_{\varepsilon=0} \cdot \operatorname{tr}\left\{\mathbf{J}^{-1}(\varepsilon)\mathbf{J}'(\varepsilon)\right\}\Big|_{\varepsilon=0} \\
&= \operatorname{tr}\left\{\mathbf{J}^{-1}(0)\mathbf{J}'(\varepsilon)|_{\varepsilon=0}\right\} \\
&= \operatorname{tr}\left\{\mathbf{J}'(\varepsilon)\right\}\big|_{\varepsilon=0} = \frac{\partial}{\partial \varepsilon}\operatorname{tr}\left\{\mathbf{J}(\varepsilon)\right\}\Big|_{\varepsilon=0}.
\end{aligned}$$

Next we exploit the fact that it suffices to consider the first-order contribution of the transformation, i.e.

$$\boldsymbol{x}' = \boldsymbol{x} + \varepsilon \boldsymbol{\Delta}(\boldsymbol{x}).$$

Consequently, the Jacobian matrix is

$$\begin{aligned}
\mathbf{J} &= \mathbb{1} + \varepsilon \Delta\mathbf{J}, \\
\Delta\mathbf{J} &:= \frac{\partial \boldsymbol{\Delta}}{\partial \boldsymbol{x}}.
\end{aligned}$$

Then we find

JAYNES' TRANSFORMATION INVARIANCE PRINCIPLE

$$\left(\nabla_{\boldsymbol{x}}\, p(\boldsymbol{x})\right)^T \boldsymbol{\Delta}(\boldsymbol{x}) + p(\boldsymbol{x}) \operatorname{tr}\{\Delta\mathbf{J}\} = 0. \tag{10.7}$$

10.1 Bertrand's paradox revisited

As a first example we analyse Bertrand's paradox, which has been outlined in Section 1.1 [p. 6]. There we have found three contradictory solutions, depending on the representation. We shall now show that the unique solution to Bertrand's problem is obtained by employing Jaynes' principle of transformation invariance. E. T. Jaynes published the solution [105] under the title 'The well-posed problem' in 1973. Here we give a somewhat modified derivation. We begin with an unprejudiced representation of the problem. Although it takes a couple more steps, it illustrates more clearly the strength of the 'transformation invariance principle'. The setup is uniquely specified by the coordinates $\boldsymbol{x}$ of the centre of the circle and the coordinates $\boldsymbol{y}$ of the perpendicular foot of $\boldsymbol{x}$ on the straight line. The sought-for probability is $p(\boldsymbol{x}, \boldsymbol{y}|C, \mathcal{I})$, where the proposition C says that the straight line cuts the circle and $\mathcal{I}$ contains further details of the problem. The transformations under which the problem is invariant are homogeneity, isotropy, scale invariance and independence of the generation of the straight lines from the position and size of the circle.

First we exploit the **translational invariance** of the origin of the coordinate system. The transformation reads

$$\begin{aligned} \boldsymbol{x} &\to \boldsymbol{x} + \varepsilon \boldsymbol{n}, \\ \boldsymbol{y} &\to \boldsymbol{y} + \varepsilon \boldsymbol{n}. \end{aligned}$$

The corresponding Jacobian is unity, and hence $\Delta \mathbf{J} = \mathbf{0}$. Then equation (10.7) reads

$$0 = \boldsymbol{n}^T \left(\nabla_{\boldsymbol{x}} p(\boldsymbol{x}, \boldsymbol{y}|C, \mathcal{I}) + \nabla_{\boldsymbol{y}} p(\boldsymbol{x}, \boldsymbol{y}|R, \mathcal{I}) \right).$$

This equation has to hold for all shift vectors $\boldsymbol{n}$ and all positions $\boldsymbol{x}$ and $\boldsymbol{y}$, which implies

$$p(\boldsymbol{x}, \boldsymbol{y}|C, \mathcal{I}) = p(\boldsymbol{x} - \boldsymbol{y}|C, \mathcal{I}).$$

Next we exploit **rotational invariance** and abbreviate $\boldsymbol{x} - \boldsymbol{y}$ by $\boldsymbol{z}$. We introduce the matrix $\mathbf{R}(\varepsilon)$ for rotation by an infinitesimal angle ε:

$$\mathbf{R}(\varepsilon) := \begin{pmatrix} \cos(\varepsilon) & \sin(\varepsilon) \\ -\sin(\varepsilon) & \cos(\varepsilon) \end{pmatrix}, \tag{10.8}$$

$$\mathbf{R}' := \mathbf{R}'(\varepsilon)|_{\varepsilon=0} = \begin{pmatrix} 0 & 1 \\ -1 & 0 \end{pmatrix}.$$

If we apply the derivative matrix $\mathbf{R}'$ to any nonzero vector $\boldsymbol{z}$, the resulting vector $\boldsymbol{z}_\perp$ is perpendicular to $\boldsymbol{z}$. Now the infinitesimal transformation reads

$$\boldsymbol{z} \to \mathbf{R}(\varepsilon)\, \boldsymbol{z} = \boldsymbol{z} + \varepsilon \boldsymbol{z}_\perp.$$

The Jacobian of a rotation is unity. So we have the invariance condition

$$0 = \left(\nabla_{\boldsymbol{z}} p(\boldsymbol{z}|C, \mathcal{I}) \right)^T \boldsymbol{z}_\perp.$$

Obviously, the gradient has to be parallel to $\boldsymbol{z}$ for all $\boldsymbol{z}$. Hence

$$p(\boldsymbol{z}|C, \mathcal{I}) = p(|\boldsymbol{z}|\,|C, \mathcal{I}).$$

Therefore, all that matters is the distance d of the line from the centre of the circle, irrespective of the originally chosen representation.

Now we exploit the fact that the lines are initially drawn independently of the circle. Put differently, the centre of the circle is independent of the position of the line. Translational invariance invoked once more allows us to shift the centre of the circle by an infinitesimal vector $\varepsilon \boldsymbol{n}$. Now it is important to note that $\boldsymbol{z}$ is perpendicular to the line. Its direction does not change by a translation of the centre of the circle, only its length. For $\boldsymbol{z} \neq \mathbf{0}$, the change in length by the infinitesimal shift is given by the projection

$$
\begin{aligned}
& |z| \to |z'| := |z| + \varepsilon \frac{\boldsymbol{n}^T z}{|z|} = |z| \left(1 + \varepsilon \frac{\boldsymbol{n}^T z}{|z|^2}\right) \\
\Rightarrow \quad & z' = z + \varepsilon \kappa \; z, \\
\text{with} \quad & \kappa := \frac{\boldsymbol{n}^T z}{|z|^2}.
\end{aligned}
$$

The corresponding Jacobian matrix is

$$
\begin{aligned}
\mathbf{J} &= \mathbb{1} + \varepsilon\big(\kappa \mathbb{1} + \mathbf{M}\big), \\
M_{ij} &:= z_i \frac{\partial \kappa}{\partial z_j}.
\end{aligned}
$$

So, the invariance condition is

$$
\begin{aligned}
0 &= p'(z|C,\mathcal{I}) \; \kappa \; z + p(z|C,\mathcal{I}) \; \big(n\kappa + \; \mathrm{tr}(\mathbf{M})\big), \\
\mathrm{tr}(M) &= \sum_i z_i \frac{\partial \kappa}{\partial z_i} = 0.
\end{aligned}
$$

Consequently, the PDF is

$$
p(z|C,\mathcal{I}) = \frac{1}{2\pi \; R \; z}.
$$

The correct normalization of the PDF, $p(z|C,\mathcal{I})$ for $\boldsymbol{z} = \boldsymbol{y} - \boldsymbol{x}$ follows from $\int p(z|C,\mathcal{I}) \; d^2z = 1$. Finally, the sought-for probability P that the distance of the line from the centre of the circle is less than half the radius reads

$$
P = \int \boldsymbol{1}\left(|z| \le \frac{R}{2}\right) \; p(z|C,\mathcal{I}) \; d^2z = \frac{2\pi}{2\pi \; R} \int \boldsymbol{1}\left(z \le \frac{R}{2}\right) \; \frac{1}{z} \; z \, dz = \frac{1}{2}.
$$

This corresponds to the first of the three solutions offered in Section 1.1 [p. 7]. In this simple example, most of the steps are obvious and could have been skipped. We have presented them for pedagogical reasons, since in more complex problems the consequences of the transformation invariance are not that obvious.

10.2 Prior for scale variables

In Section 7.5.6 [p. 105] we introduced the normal distribution

$$
p(x|x_0, \sigma) = \frac{1}{\sigma\sqrt{2\pi}} \exp\left\{ -\frac{1}{2} \frac{(x - x_0)^2}{\sigma^2} \right\} \quad . \tag{10.9}
$$

For arbitrary physical variables x, which carry a dimension, σ must have the same dimension in order to avoid a dimensional crime which would render the operation $\exp(\cdot)$ meaningless. Thus, if x is a length then σ is also a length. If σ is chosen in metres then x is also measured in metres. The parameter σ is therefore called a scale variable since it sets the scale for the measurement of x. Imagine that we have no information on the size of σ, for example, we would be equally happy if σ was of the order of Ångströms or parsec. What kind of prior distribution on σ would apply to this situation? Moreover, it is not clear a priori whether we should try to formulate a prior depending on σ or on σ^2. The two possibilities will in general not lead to the same functional dependence. We are therefore led to investigate two transformations, both of which satisfy equation (10.4):

$$x' = T_\varepsilon(x) = (1+\varepsilon)x \; = x + \varepsilon x, \tag{10.10a}$$

$$x' = T_\varepsilon(x) = x^{1+\varepsilon}. \tag{10.10b}$$

The correction of the Jacobian of the first transformation is $\Delta J = 1$ and equation (10.7) becomes

$$p'(x)x + p(x) = 0, \tag{10.11}$$

which leads to the differential equation

$$x\ p'(x) + p(x) = 0 \tag{10.12}$$

with solution

$$p(x) = \frac{c}{x}. \tag{10.13}$$

This prior is also called a 'Jeffreys' prior' for scale variables in the literature. It is not normalizable in $[0, \infty]$, that is, the integration constant c remains undetermined. Such a prior is referred to as an 'improper prior'. The functional form is already uniquely specified due to the transformation in equation (10.10a), and we want to test whether the solution also satisfies equation (10.10b). The corresponding transformation yields

$$p(x') = p(x)\frac{dx'}{dx} = \frac{c}{x^{1+\varepsilon}}\ (1+\varepsilon)\ x^{\varepsilon} = (1+\varepsilon)\ p(x). \tag{10.14}$$

Obviously, the functional form of Jeffreys' prior is indeed invariant under this transformation, and it is therefore irrelevant whether we perform the inference based on σ or σ^2.

Whenever possible, the use of improper priors should be avoided since they can lead to an unnormalizable posterior PDF. In order to keep control on what an improper prior does, it should always be formulated as the limit of a sequence of proper priors. There are many ways to turn equation (10.13) into a proper prior. It is only necessary to stay away from $x = 0$ and $x = \infty$. For some finite lower (upper) cutoff x_l (x_u) we define the normalized version of equation (10.13) as

$$p(x|x_l, x_u) = \frac{1}{\ln(x_u/x_l)}\frac{1}{x}\ \mathbf{1}(x_l \leq x \leq x_u)\,. \tag{10.15}$$

The choice $x_u = B$ and $x_l = 1/B$ simplifies the investigation even further:

$$p(x|B) = \frac{1}{2\ln(B)}\frac{1}{x}\ \mathbf{1}\Big(B^{-1} \leq x \leq B\Big)\,. \tag{10.16}$$

10.3 The prior for a location variable

The Gaussian distribution in equation (10.9) is only fully specified if we fix, apart from the variance σ^2, also its location parameter x_0, which is the position of the peak. The only sensible transformation which may be applied to a location variable is a translation. The transformation is $T_\varepsilon(x_0) = x_0' = x_0 + \varepsilon$, with $\Delta J = 0$. Hence

$$p'(x) = 0, \tag{10.17}$$

with solution $p(x_0) = \text{const}$. Again this prior is improper and should be regarded as the limit of a sequence

$$p(x_0) = \frac{1}{2B}, \quad -B < x_0 < B, \quad B \to \infty. \tag{10.18}$$

10.4 Hyperplane priors

Fitting a straight line in two-dimensional space to a set of experimental data is one of the most important traditional regression problems in physics. The usual procedure employs the least-squares method. In a Bayesian sense this means that the prior for the coefficients of the straight line is taken flat. This choice is often a reasonable approximation, but in a strict sense it violates the requirement of transformation invariance because a flat prior in a given system of coordinates does not in general remain flat after translations and/or rotations of that system. If the prior probability shall be uninformative, then we must insist that it remains invariant under such basic transformations [115]. We shall generalize the example of a straight line in two-dimensional space to a hyperplane in n-dimensional space, which in Hesse normal form reads

$$0 = a_1x_1 + a_2x_2 + \cdots + a_nx_n + 1 \tag{10.19a}$$

or in vector notation

$$0 = \boldsymbol{a}^T\boldsymbol{x} + 1, \tag{10.19b}$$

with $\boldsymbol{a}$ and $\boldsymbol{x}$ being the column vectors containing the elements $\{a_i\}$ and $\{x_i\}$, respectively. In contrast to the examples of priors for scale and location variables, where the transform concerned the parameters directly, we shall now investigate the consequences on $\boldsymbol{a}$ of a transformation applied to $\boldsymbol{x}$. The basic idea is to apply all transformations on $\boldsymbol{x}$ that leave the problem invariant. That is, we consider an infinitesimal transformation $\boldsymbol{x} \to \boldsymbol{x}'$, which implies a transformation $\boldsymbol{a} \to \boldsymbol{a}'$. The normal form shall be form invariant under this transformation, resulting in

$$0 = (\boldsymbol{a}')^T\boldsymbol{x}' + 1. \tag{10.20}$$

We assume that the setup of the problem is rotationally and translationally invariant in $\boldsymbol{x}$.

10.4.1 Simple translation

First we investigate one of the two simplest cases, namely

$$ax_1 + 1 = 0. \tag{10.21}$$

In two-dimensional space this equation can be considered as a straight line parallel to the x_2-axis. The only sensible transformation on x_1 is in this case a translation $x_1' = x_1 + \varepsilon$, or rather $x_1 = x_1' - \varepsilon$. Replacing x_1 by $x_1' - \varepsilon$ in the original normal form yields

$$0 = a\,(x_1' - \varepsilon) + 1 = a\,x_1' + 1 - a\,\varepsilon. \tag{10.22}$$

Division by $1 - a\varepsilon$ yields

$$\frac{a}{1 - a\varepsilon} \cdot x_1' + 1 = 0. \tag{10.23}$$

Comparison with equation (10.20) yields the transformation in a which is implied by the translation of x_1. We merely need the correction in order ε

$$a' = \frac{a}{1 - a\varepsilon} = a + \varepsilon a^2. \tag{10.24}$$

The correction of the Jacobian is $\Delta J = 2a$. Substituting these results into equation (10.7) [p. 167] we arrive at

$$p' a^2 + 2ap = 0. \tag{10.25}$$

Assuming $a \neq 0$, this differential equation yields

$$p(a) = \frac{c}{a^2}. \tag{10.26}$$

This distribution is normalizable, because of the restriction $|a| \geq a_{\min}$. The question of which cutoff $a_{\min}$ to choose will be addressed after treating the general case.

10.4.2 Simple rotation

The second elementary example that we will treat is a straight line passing through the origin. The equation for this case, which is not exactly of the form in equation (10.19), reads

$$ax_1 + x_2 = 0. \tag{10.27}$$

If there is no preferred direction, the Hesse normal form has to be form invariant under rotation. The infinitesimal transformation has been defined in equation (10.8):

$$\boldsymbol{x} = \mathbf{R}(-\varepsilon)\,\boldsymbol{x}' = \begin{pmatrix} x_1' - \varepsilon x_2' \\ \varepsilon x_1' + x_2' \end{pmatrix}. \tag{10.28}$$

Insertion into equation (10.19) yields

$$a(x_1' - \varepsilon x_2') + \varepsilon x_1' + x_2' + 1 = x_1'(a + \varepsilon) + x_2'(1 - a\varepsilon) \stackrel{!}{=} 0 \qquad \Rightarrow \tag{10.29}$$

$$x_1' \underbrace{\frac{a+\varepsilon}{1-a\varepsilon}}_{\stackrel{!}{=}a'} + x_2' \stackrel{!}{=} 0. \tag{10.30}$$

The transformation up to first order in ε is

$$a' = a + \varepsilon(1 + a^2). \tag{10.31}$$

The corresponding Jacobian is $(1 + 2a\varepsilon)$ and $\Delta J = 2a$, yielding the differential equation

$$p'(a^2 + 1) + 2ap = 0, \tag{10.32}$$

which is readily solved to yield

$$p(a) = \frac{1}{\pi}\frac{1}{1+a^2}. \tag{10.33}$$

The factor $1/\pi$ arises of course not from the solution of equation (10.32) but from subsequent normalization. This result can easily be shown to conform with intuition. Noting that the meaning of a is $a = \tan(\psi)$, where ψ is the angle of the straight line with respect to the positive x_1-axis, the corresponding distribution of ψ is obtained from equation (7.49) [p. 121] as $p(\psi) = 1/\pi$. This means that the uninformative prior for a straight line passing through the origin assumes every orientation to be equally probable, in accordance with the rotational invariance in $\boldsymbol{x}$.

10.4.3 The general case

After these two prototypic examples we are well prepared to treat the general case of an $(n - 1)$-dimensional hyperplane in n-dimensional space. The equations for the hyperplane in the primed and unprimed coordinate systems are

$$\boldsymbol{a}^T \boldsymbol{x} + 1 = 0. \tag{10.34}$$

We generalize the infinitesimal matrix $\mathbf{R}(\varepsilon, \boldsymbol{n})$ for the rotation in n-dimensional space about the rotation axis $\boldsymbol{n}$. Then $\boldsymbol{x}' = \mathbf{R}(\varepsilon, \boldsymbol{n})\boldsymbol{x}$. The form invariance requires

$$0 = \boldsymbol{a}^T \boldsymbol{x} + 1 = \underbrace{(\mathbf{R}(\varepsilon, \boldsymbol{n})\boldsymbol{a})}_{=\boldsymbol{a}'}{}^T \boldsymbol{x}' + 1.$$

Consequently,

$$\boldsymbol{a}' = \mathbf{R}(\varepsilon, \boldsymbol{n})\boldsymbol{a}.$$

The Jacobian of a rotation is unity, so the invariance condition reads

$$\Big(\nabla_{\boldsymbol{a}} p(\boldsymbol{a})\Big)^T R'(\varepsilon)\Big|_{\varepsilon=0} \boldsymbol{a} = 0. \tag{10.35}$$

Now $\boldsymbol{r}(\boldsymbol{n}) := R'(\varepsilon)\boldsymbol{a}$ is orthogonal to $\boldsymbol{a}$ and to the rotation axis $\boldsymbol{n}$. The orthogonality on $\boldsymbol{a}$ can easily be seen, because a rotation is norm conserving:

$$|\boldsymbol{a}|^2 = \boldsymbol{a}^T \mathbf{R}^T(\varepsilon)\mathbf{R}(\varepsilon)\boldsymbol{a}.$$

Differentiation with respect to ε at $\varepsilon = 0$ and using $R(0) = \mathbf{1}$ yields

$$0 = 2\,\boldsymbol{a}^T R'(\varepsilon)\Big|_{\varepsilon=0}\,\boldsymbol{a}.$$

Consequently, $\mathbf{R}'\boldsymbol{a}$ is orthogonal to $\boldsymbol{a}$. The orthogonality on the rotation axis $\boldsymbol{n}$ follows from the fact that $\boldsymbol{n}^T\mathbf{R}(\varepsilon, \boldsymbol{n})\boldsymbol{a} = 0$ for arbitrary angle ε. There are n linearly independent rotation axes in n-dimensional space. One of them can be chosen parallel to $\boldsymbol{a}$, then equation (10.35) is trivially fulfilled because $\boldsymbol{r}(\boldsymbol{a}) = \mathbf{0}$. For the other $n-1$ linearly independent rotation axes along with the fact that $\boldsymbol{n}^T\vec{r}(\boldsymbol{n}) = 0$, the vectors $\boldsymbol{r}(\boldsymbol{n})$ span the hyperplane orthogonal to $\boldsymbol{a}$. Then equation (10.35) requires

$$\nabla_{\boldsymbol{a}}\, p(\boldsymbol{a}) \propto \boldsymbol{a} \qquad \Rightarrow \qquad p(\boldsymbol{a}) = p(|\boldsymbol{a}|).$$

The rotational invariance translates into a rotationally invariant PDF for the parameters.

In addition, by assumption the problem is invariant under translation in all directions. This holds in particular for the individual cartesian directions

$$\boldsymbol{x} = \boldsymbol{x}' - \varepsilon \boldsymbol{e}_k, \tag{10.36}$$

where $\boldsymbol{e}_i$ denotes the unit vector in the ith cartesian direction. Then

$$\boldsymbol{a}^T\boldsymbol{x} + 1 = \boldsymbol{a}^T\boldsymbol{x}'1 - \varepsilon a_i = 0, \tag{10.37}$$

$$\left(\frac{\boldsymbol{a}}{(1-\varepsilon a_k)}\right)^T \boldsymbol{x} + 1 = 0, \tag{10.38}$$

and the first-order correction term due to the transformation is

$$\Delta(\boldsymbol{a}) = \boldsymbol{a}\; a_k. \tag{10.39}$$

The corresponding contribution to the Jacobian matrix is

$$\begin{aligned}
(\Delta\mathbf{J})_{ij} &= \delta_{ij}a_k + a_i\delta_{kj},\\
\operatorname{tr}\{\Delta\mathbf{J}\} &= a_k \sum_{i=1}^{n}\delta_{ii} + \sum_{i=1}^{n} a_i\delta_{k,i} = (n+1)\, a_k.
\end{aligned}$$

According to equation (10.7) [p. 167], the unknown function $p(a)$ with $a = |\boldsymbol{a}|$ must therefore obey

$$\begin{aligned}
\left(\nabla_{\boldsymbol{a}} p(a)\right)^T \boldsymbol{a}\, \cancel{a_k} + p(a)\,(n+1)\,\cancel{a_k} &= 0,\\
p'(a)\,\frac{\boldsymbol{a}^T\boldsymbol{a}}{a} + p(a)\,(n+1) &= 0,\\
p'(a)\,a + p(a)\,(n+1) &= 0,
\end{aligned}$$

which is readily integrated to yield

$$p(\boldsymbol{a}) = |\boldsymbol{a}|^{-(n+1)}. \tag{10.40}$$

Note that this general result contains our previous simple case equation (10.26) for $n = 1$ as well as the important cases of priors for the coefficients of a straight line in two dimensions and an ordinary plane in three dimensions.

10.4.4 Normalization

The condition $|a_i| > 0$ in equation (10.40) means of course that the distribution is normalizable. Denote by a_0 the minimum value of a which we allow, based on prior knowledge. Then the normalization integral Z is

$$Z = \int d\Omega_n \int_{a_0}^{\infty} da\, a^{-2} = \frac{2\pi^{n/2}}{\Gamma(n/2)} \cdot \frac{1}{a_0} \tag{10.41}$$

and the normalized distribution becomes, for all $n \geq 1$,

INVARIANT PRIOR FOR HYPERPLANE PARAMETER IN NORMAL FORM

hyperplane: $\boldsymbol{a}^T \boldsymbol{x} = 1$

$$p(\boldsymbol{a}) = \frac{\Gamma(n/2)}{2\pi^{n/2}} \cdot \frac{a_0}{|\boldsymbol{a}|^{n+1}}. \tag{10.42}$$

The remaining question concerns of course the choice of r_0. Let $\boldsymbol{x}^*$ be a point through which the hyperplane is expected to pass, then for this point equation (10.34) holds. We now answer the question: 'What is the minimum value of a^2 given this point?' The Lagrangian optimization problem is to find the minimum of

$$\Phi = |\boldsymbol{a}|^2 + 2\lambda(\boldsymbol{a}^T \boldsymbol{x}^* + 1). \tag{10.43}$$

The derivative with respect to a_k yields

$$\nabla_{\boldsymbol{a}} \Phi = 2(\boldsymbol{a} + \boldsymbol{x}^*) = 0 \quad \rightarrow \quad \boldsymbol{a}^* = -\lambda \boldsymbol{x}^*. \tag{10.44}$$

Insertion into the hyperplane equation yields the condition for the Lagrange parameter λ:

$$-\lambda \left|\boldsymbol{x}^*\right|^2 + 1 = 0 \quad \rightarrow \quad \lambda = \left|\boldsymbol{x}^*\right|^{-2}. \tag{10.45}$$

So, the minimal coefficients for a given data point $\boldsymbol{x}^*$ are

$$\boldsymbol{a}^* = -\frac{\boldsymbol{x}^*}{|\boldsymbol{x}^*|^2},$$

and a_0^2 turns out to be

$$a_0^2 = \left|\boldsymbol{a}^*\right|^2 = \left|\boldsymbol{x}^*\right|^{-2}. \tag{10.46}$$

We can always choose the origin of the coordinate system at will. It is advantageous to always put it at the centre of the data cloud. Then a_0 corresponds to the spread of the data cloud. Based on our prior knowledge we can then give a lower limit for a_0.

Straight line in 2-D

In equation (10.27) we have to consider straight lines in 2-D that pass through the origin. Here we want to consider the general case in its most common representation:

$$x_2 = ax_1 + b.$$

Comparison with equation (10.19) for $n = 2$ yields

$$a_1 = \frac{a}{b},$$
$$a_2 = -\frac{1}{b}.$$

We are interested in $p(a, b)$, which we obtain from $p(a_1, a_2)$ through

$$\begin{aligned} p(a, b) &= p(a_1, a_2) \left| \begin{pmatrix} \frac{\partial a_1}{\partial a} & \frac{\partial a_1}{\partial b} \\ \frac{\partial a_2}{\partial a} & \frac{\partial a_2}{\partial b} \end{pmatrix} \right| \\ &= \frac{a_0}{2\pi \left(\left(\frac{a}{b}\right)^2 + \left(\frac{1}{b}\right)^2 \right)^{3/2}} \left| \begin{pmatrix} \frac{1}{b} & -\frac{a}{b^2} \\ 0 & \frac{1}{b^2} \end{pmatrix} \right|. \end{aligned}$$

INVARIANT PRIOR FOR STRAIGHT-LINE PARAMETER

straight line: $y = ax + b$

$$p(a, b) = \frac{a_0}{2\pi \left(1 + a^2\right)^{3/2}}. \tag{10.47}$$

The invariant prior is uniform in b due to translational invariance. The rotational invariance becomes obvious if we express $a = \tan(\phi)$ and transform to $p(\phi, b)$. This PDF is indeed uniform also in ϕ.

10.5 The invariant Riemann measure (Jeffreys' prior)

The uninformative prior probability for parameters $\boldsymbol{a}$ depends of course on its meaning in the particular problem at hand. The meaning of $\boldsymbol{a}$ is uniquely defined by the likelihood function (the model $\mathcal{B}$) which determines also the invariance properties of $\boldsymbol{a}$. It is therefore meaningless to require a prior probability for $\boldsymbol{a}$ to be uniform without reference to the corresponding likelihood. In contrast, it does make sense to require the prior density to be such that all likelihood functions are equally probable:

$$p(\boldsymbol{d}|\boldsymbol{a},\mathcal{I}) \tag{10.48}$$

in the space of densities $p(\boldsymbol{d})$, which may be considered as Riemannian manifolds. Differential geometric considerations allow us to derive the invariant prior [97, 173], which plays a prominent role in information geometry:

$$\begin{aligned} p(\boldsymbol{a}|\mathcal{I}) &= |g|^{\frac{1}{2}}, \\ g_{ij} &:= \int p(\boldsymbol{d}|\boldsymbol{a},\mathcal{I}) \frac{\partial^2}{\partial a_i \partial a_j} \ln\left(p(\boldsymbol{d}|\boldsymbol{a},\mathcal{I})\right) d^N d \\ &= \left\langle \frac{\partial^2}{\partial a_i \partial a_j} \ln\left(p(\boldsymbol{d}|\boldsymbol{a},\mathcal{I})\right)\right\rangle. \end{aligned} \tag{10.49}$$

This prior was first proposed by Jeffreys [107] and is therefore referred to as 'Jeffreys' prior'. The Riemannian metric g is identical to the Fisher information matrix [77]. The advantage of this approach is that one need not explore the invariance properties of the problem. Invariance information is implicitly provided by the likelihood function and implemented in the Riemannian measure:

$$p(\boldsymbol{a}|\mathcal{I})\, dV_a = |g(\boldsymbol{a})|^{\frac{1}{2}}\, dV_a \tag{10.50}$$

$$\begin{aligned} &= \left|\left\langle \frac{\partial^2}{\partial a_i \partial a_j} \ln\left(p(\boldsymbol{d}|\boldsymbol{a},\mathcal{I})\right)\right\rangle\right|^{\frac{1}{2}} dV_a \\ &= \left|\sum_{mn} \frac{\partial a'_m}{\partial a_i}\frac{\partial a'_n}{\partial a_j}\left\langle \frac{\partial^2}{\partial a'_m \partial a'_n} \ln\left(p(\boldsymbol{d}|\boldsymbol{a}',\mathcal{I})\right)\right\rangle\right|^{\frac{1}{2}} dV_a \\ &= \left|J^{-1}\, g(\boldsymbol{a}')\, J^{-1}\right|^{\frac{1}{2}} dV_a \\ &= \left|g(\boldsymbol{a}')\right|^{\frac{1}{2}} J^{-1}\, dV_a \\ &= \left|g(\boldsymbol{a}')\right|^{\frac{1}{2}} dV_{a'}, \end{aligned}$$

$$p(\boldsymbol{a}|\mathcal{I})\, dV_a = p(\boldsymbol{a}'|\mathcal{I})\, dV_{a'}. \tag{10.51}$$

The preceding equation proves that the density defined by equation (10.49) is indeed invariant under reparametrization.

11

Testable information and maximum entropy

11.1 Discrete case

In this chapter we focus on the derivation of a prior PMF for discrete events and how additional information – *testable information* – can be taken into account. Available prior knowledge is called *testable information* if it is possible to verify whether the derived probability distribution fulfils the constraint(s) set by the information or not.[1]

There are different ways to derive the correct prior distributions: guided by intuition (qualitatively), axiomatic and by consistency requirements. We focus on the first two approaches.

We are looking for the most 'uninformative distribution', or rather the 'least committed distribution', consistent with the available prior knowledge. We will reveal the underlying idea by a simple example. For a die we assume the following testable information:

$$\mu_0 = \sum_i q_i = 1 \qquad \text{normalization,}$$

$$\mu_1 = \sum_i i\, q_i = 3.5 \qquad \text{mean.}$$

These constraints are fulfilled by an infinite number of PMFs. One of them is

$$\{q_i\} = \left(\tfrac{1}{2}, 0, 0, 0, 0, \tfrac{1}{2}\right).$$

This is, however, a highly committing choice. None of the face values 2, 3, 4 or 5 are considered as possible outcomes. The restriction is not forced by the addition (testable) information at all. The choice appears absurd, because also the uniform PMF

$$\{q_i\} = \left(\tfrac{1}{6}, \tfrac{1}{6}, \tfrac{1}{6}, \tfrac{1}{6}, \tfrac{1}{6}, \tfrac{1}{6}\right)$$

is in agreement with the given constraints and obeys Laplace's 'principle of indifference'. The Laplace principle of indifference is equivalent to the least committing, or rather least informative, choice under the present constraints. In the case of more general constraints, the uniform PMF will not be admissible and it appears very sensible to still choose the least committing or rather least informative PMF. So we need a measure for the information

[1] This is in agreement with the requirement put forward by Karl Popper that proper theories should be falsifiable.

content of a probability distribution to be able to rank probability distributions which are compatible with given constraints.

11.1.1 Shannon entropy: Information gain by binary questions

Such a measure is provided by Shannon entropy. As a matter of fact, Shannon's work traces back to the H-theorem, introduced in classical statistical physics by Ludwig Boltzmann in 1872, in order to describe the entropy increase in irreversible processes. Here we will motivate the entropy as an information measure via binary questions [36].

To begin with, how many questions are needed to identify one object of N? For the sake of simplicity, we assume $N = 2^L$. First we enumerate the objects. Let n be the number of the objects to be identified. If all objects are equiprobable, the optimal strategy is given as follows.

- Define the set $\mathcal{M} = \{1, \ldots, N\}$.
  - (a) If $|\mathcal{M}| = 1 \Rightarrow$ then the target has been identified $\Rightarrow$ done!
  - (b) Divide the set $\mathcal{M}$ into two subsets of equal size: $\mathcal{M}_1 = \{1, \ldots, \frac{N}{2}\}$ and $\mathcal{M}_2 = \{\frac{N}{2} + 1, \ldots, N\}$.
  - (c) Ask the binary question: Is the target in the first set?
    If so: $\mathcal{M} = \mathcal{M}_1$;
    else: $\mathcal{M} = \mathcal{M}_2$.
    Go to (a).
- The algorithm terminates after $L = \log_2 N$ questions.

Alternatively, we could use a binary representation of the numbers of the objects

$$n = \sum_{i=0}^{L} 2^i \, n_i, \qquad n_i \in \{0, 1\}.$$

Then we also need L questions. The ith question would be *is* $n_i = 1$? This strategy also succeeds in $L = \log_2 N$ steps. Therefore, $\log_2 N$ can be considered as a measure for uncertainty in the special case of N *equiprobable* events. There are other strategies that are faster on average, but in the worst case there is no better strategy. So, for a pedagogical introduction of entropy this strategy is fine.

As a next step, we need to generalize these ideas to the case where the probabilities differ. As an example we first consider a lottery, where the N lottery tickets are distributed among m boxes such that each box contains $\frac{N}{m}$ lottery tickets (N and m still being a power of two). There shall be one top prize. One possible strategy to find the top prize is to determine the box containing the winning ticket first. This requires $L_1 = \log_2 m$ binary questions. Subsequently the winning ticket is identified with (at most) another $L_2 = \log_2 \frac{N}{m}$ questions. The total number of questions is therefore

$$L_1 + L_2 = \log_2 m + \log_2 \frac{N}{m} = \log_2 N,$$

which is the same number we would have needed without the distribution of the lottery tickets in different boxes. We define the 'number of binary questions' required to solve a given task Q conditioned on the prior information $\mathcal{I}$ as a 'measure of uncertainty'.

$$U(Q|\mathcal{I}) : \text{Uncertainty about } Q \text{ given } \mathcal{I}.$$

Example:

- $\mathcal{I}$: There are N tickets and one top prize.
- Q: Which ticket corresponds to the top prize?

Now we tackle the problem of boxes with different probabilities using the same setting as before except that the boxes may now contain a different number of lottery tickets. Box i shall hold n_i lottery tickets (n_i still being a power of two) and

$$\sum_{i=1}^{m} n_i = N.$$

We introduce the following abbreviations:

- $\mathcal{I}$: Describes the setting.
- B_j: The winning ticket is in box j.
- T: Which ticket is the winner?

It holds that

$$U(T|\mathcal{I}) = \log_2 N,$$
$$U(T|B_i, \mathcal{I}) = \log_2 n_i.$$

The optimal worst-case strategy requires $\log_2 N$ binary questions. We will still stick to the strategy of identifying the box holding the winning lottery ticket first, followed by the determination of the ticket itself. To this end we introduce further abbreviations:

- B: Which box has the winning ticket?
- C: First we have to answer question B.

Owing to the different numbers of lottery tickets in the boxes, the number of binary questions is now

$$U(T|C, B_j, \mathcal{I}) = U(B|\mathcal{I}) + U(T|B_j, \mathcal{I})$$

with B_j indicating which box holds the winning ticket. Let us consider an example. The number of boxes is $m = 8$ and the number of tickets in the boxes is given by

$$\{n_i\} = (2, 2, 2, 2, 2, 2, 4, 16).$$

The total number of lottery tickets is $N = 32$, resulting in

$$U(T|\mathcal{I}) = \log_2 32 = 5.$$

Next we compute $U(T|C, \mathcal{I})$. It should be emphasized that C implies we have to find the correct box first. So we obtain

$$U(T|C, B_j, \mathcal{I}) = U(B|\mathcal{I}) + U(T|B_j, \mathcal{I}) = \log_2 8 + \log_2 n_j = 3 + \log_2 n_j.$$

If the winning lottery ticket is in one of the first six boxes, then

$$U(T|C, B_j, \mathcal{I}) = 3 + \log_2 2 = 4.$$

In the case $j = 7$, we obtain

$$U(T|C, B_j, \mathcal{I}) = 3 + \log_2 4 = 5$$

and in the last case the result is

$$U(T|C, B_j, \mathcal{I}) = 3 + \log_2 16 = 7.$$

So the detour determining the box first results in the worst case in more questions than without it. Here we have introduced B_j as part of the conditional complex, which was a bit schizophrenic, since we pretended in the first step that we didn't know it. The general case is given by $U(T|C, \mathcal{I})$ and the above considerations imply that generally

$$U(T|C, \mathcal{I}) \neq U(B|\mathcal{I}) + U(T|\mathcal{I}).$$

We now define the average uncertainty as

$$\langle U(T|C, \mathcal{I})\rangle = U(B|\mathcal{I}) + \underbrace{\sum_{j=1}^{m} P(B_j|\mathcal{I})\, U(T|B_j, \mathcal{I})}_{:=D},$$

which could also be considered as an application of the marginalization rule. The first term provides the number of binary questions required to determine the correct box and the second term the mean number of questions to determine the winning lottery ticket. The average is with respect to the boxes. Now we transform $U(T|B_j, \mathcal{I})$ as follows:

$$U(T|B_j, \mathcal{I}) = \log_2 n_j = \log_2 N + \log_2 \left(\frac{n_j}{N}\right) = \log_2 N + \log_2 P(B_j|\mathcal{I}).$$

Then

$$D = \log_2 N + \sum_{j=1}^{m} P(B_j|\mathcal{I}) \log_2 P(B_j|\mathcal{I}).$$

We define generally

$$S_2(\{P_j\}) := -\sum_{j=1}^{m} P_j \log_2 P_j,$$

for arbitrary probabilities $P_j \geq 0$ with $\sum_j P_j = 1$. We will see in the next section that S_2 is maximized by $P_j = 1/m$, with

$$S_2 \leq S_{2,\max} := \log_2 m = U(B|\mathcal{I}).$$

Consequently,

$$\begin{aligned}\langle U(T|C,\mathcal{I})\rangle &= U(B|\mathcal{I}) + \log_2 N - S_2(\{P(B_j|\mathcal{I})\}) \\ &\geq U(B|\mathcal{I}) + \log_2 N - \log_2 m = \log_2 N.\end{aligned}$$

We can express the inequality differently:

$$U(B|\mathcal{I}) \geq S_2(\{P(B_j|\mathcal{I})\}) = \frac{1}{\ln(2)} S(\{P(B_j|\mathcal{I})\}),$$

$$\text{with} \qquad S(\{P_i\}) := -\sum_{i=1}^{m} P_i \ln(P_i).$$

The maximization of the uncertainty is therefore equivalent to maximization of the

Shannon entropy

$$S(\{P_i\}) = -\sum_{i=1}^{m} P_i \ln(P_i). \tag{11.1}$$

Obviously, the Shannon entropy S can be considered as a measure of uncertainty or – from a different point of view – a measure of the information content of a probability distribution. Other derivations of the entropy can be found in [104].

11.2 Properties of the Shannon entropy

The entropy has the following properties:

- It is invariant under permutation of the indices i of the events. This is reasonable, since our prior knowledge does not make any distinction between the different events.
- Since $0 \leq P_j \leq 1$, the individual summands $-P_j \ln(P_j) \geq 0$ (see Figure 11.1) are all non-negative. Consequently,

$$S \geq 0.$$

- There is no correlation between any P_i and P_j. The Hesse matrix is diagonal.
- Since the individual summands are non-negative, the minimal value $S = 0$ can only occur if all contributions vanish, which is the case either for $P_j = 0$ or $P_j = 1$. Owing to the normalization,

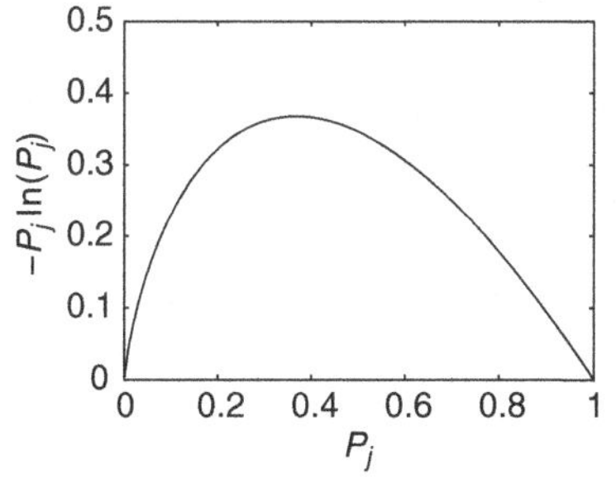

Figure 11.1 Single term of the Shannon entropy.

$S = 0$ happens if one of the probabilities is one and all others are zero. The minimal value of the entropy corresponds to perfect knowledge, or rather no uncertainty, which is in agreement with the definition that S is a measure of uncertainty, or rather a measure of being uncommitted.

> In case of no uncertainty we have $S = 0$.

- S is independent of impossible events $P_j = 0$.
- S is convex, which implies that there is only a global maximum:

$$\frac{\partial^2 S}{\partial P_j \partial P_j} = -\delta_{ij} \, \frac{1}{P_i}.$$

- Without any testable information, the most uncommitted PMF is

$$P_i^{\text{ME}} = \text{constant},$$

in agreement with Laplace's principle of indifference.

Proof: Which set of $\{P_j\}$ has maximal entropy consistent with the normalization constraint $\sum_j p_j = 1$? Including this constraint via a Lagrange multiplier λ results in

$$\frac{\partial}{\partial P_i}\left(S - \lambda \sum_{j=1}^{m} P_j\right) = -\ln(P_i) - 1 - \lambda \stackrel{!}{=} 0 \quad \Rightarrow$$

$$P_i = \frac{1}{m}.$$

As the entropy is convex in all P_j, this is indeed the maximum entropy solution. Therefore, the uniform distribution is the distribution with the maximum entropy under the normalization constraint. The maximum value of the entropy of this 'Maxent' solution is given by

$$S_{\text{Max}} = -\sum_{j=1}^{m} \frac{1}{m} \ln\left(\frac{1}{m}\right) = \ln(m),$$

which corresponds to the Boltzmann entropy in the microcanonical ensemble, introduced in statistical physics. □

11.2.1 Maximum entropy principle

E. T. Jaynes suggested in 1963 [174] that in the presence of testable information the assignment of prior probabilities should be made using the principle of maximum entropy. The prior probability distribution should be chosen which has the maximum entropy S while satisfying all the constraints. Then, in 1984, Shore and Johnson [182] succeeded in deriving the entropy functional as well as the Maxent principle from consistency requirements. In the following we consider testable information of the form

$$\phi_\alpha\{P\} = 0, \quad \alpha = 1, \ldots, L. \tag{11.2}$$

In addition, the probabilities have to fulfil the normalization constraint

$$\textstyle\sum_{j=1}^{N} P_j - 1 = 0. \tag{11.3}$$

The Maxent principle requires the maximization of the entropy under the available constraints. Using the method of Lagrange multipliers we define the functional

$$\mathcal{L} = -\sum_j P_j \ln(P_j) + \lambda_0 \left(\sum_j P_j - 1\right) + \sum_\alpha \lambda_\alpha \phi_\alpha\{P\}. \tag{11.4}$$

The Maxent solution is given by the zero of the derivative with respect to P_j:

$$\begin{aligned}\Psi_i &:= \frac{\partial}{\partial P_i}\mathcal{L} = \frac{\partial}{\partial P_i}\left(-\sum_j P_j \ln(P_j) + \lambda_0\left(\sum_j P_j - 1\right) + \sum_\alpha \lambda_\alpha \phi_\alpha\{P\}\right), \\ \Psi_i &= -\ln(P_i) - 1 + \lambda_0 + \sum_\alpha \lambda_\alpha \frac{\partial}{\partial P_i} \phi_\alpha\{P\} \stackrel{!}{=} 0, \\ \ln(P_i) &= -1 + \lambda_0 + \sum_\alpha \lambda_\alpha \frac{\partial}{\partial P_i}\phi_\alpha\{P\}.\end{aligned} \tag{11.5}$$

The general solution is therefore given by

MAXIMUM ENTROPY SOLUTION

$$P_j^{\mathrm{ME}} = \frac{1}{Z}\, e^{\sum_\alpha \lambda_\alpha \frac{\partial}{\partial P_j} \phi_\alpha\{P\}}, \tag{11.6a}$$

$$Z = \sum_{j=1}^{N} e^{\sum_\alpha \lambda_\alpha \frac{\partial}{\partial P_j} \phi_\alpha\{P\}}. \tag{11.6b}$$

The normalization constraint fixes λ_0 or rather Z. The other Lagrange parameters λ_ν are chosen such that the constraints of equation (11.2) are fulfilled.

Linear constraints

Of particular practical importance are linear constraints

$$\phi_\alpha\{P\} = \sum_{j=1}^{N} K_{\alpha j}\, P_j \;-\; \mu_\alpha. \tag{11.7}$$

In this case the derivatives required in equations (11.6a) and (11.6b) can be computed easily:

$$\frac{\partial}{\partial P_j}\, \phi_\alpha\{P\} = K_{\alpha j} \tag{11.8}$$

and the final result reads

MAXENT SOLUTION FOR LINEAR CONSTRAINTS

constraints:
$$\mu_\alpha = \sum_{j=1}^{N} K_{\alpha j}\, P_j, \tag{11.9a}$$

Maxent solution:
$$P_j^{\mathrm{ME}} = \frac{1}{Z}\, e^{\sum_\alpha \lambda_\alpha K_{\alpha j}}, \tag{11.9b}$$

normalization:
$$Z = \sum_{j=1}^{N} e^{\sum_\alpha \lambda_\alpha K_{\alpha j}}. \tag{11.9c}$$

In this case the constraints can be reformulated as

$$\mu_\alpha = \sum_{j=1}^{N} K_{\alpha j}\, P_j = \frac{1}{Z} \sum_{j=1}^{N} K_{\alpha j}\, e^{\sum_\alpha \lambda_\alpha K_{\alpha j}} = \frac{\partial}{\partial \lambda_\alpha} \ln(Z).$$

The Hessian of the Lagrange function defined in equation (11.4) is solely given by the Hessian of the entropy (see equation (11.5) with equation (11.8)), which is strictly negative. Therefore the Lagrange function is globally convex and exhibits a unique maximum.

As in statistical physics the thermodynamic potential, here $\ln(Z)$, has the property that all other relevant quantities can be determined by differentiation. As a matter of fact, that is no coincidence. There is a formal correspondence to statistical mechanics. The latter can be derived entirely based on the Maxent principle [89]. There are widespread applications of the 'Maxent principle' in many areas of science, for example in pattern recognition, queueing theory, economics, tomography, nonlinear spectral analyses, etc. (see e.g. [112]).

In the following we will discuss a few examples.

11.2.2 PMF of a die

As a first example we consider the PMF of a die, for which additional testable information about the mean face value is provided, i.e.

$$\mu_1 = \sum_{j=1}^{m} j\, p_j.$$

The Maxent solution reads, according to equation (11.9),

$$\begin{aligned} p_j^{\text{ME}} &= \frac{1}{Z}\, e^{\lambda j} := \frac{1}{Z}\, q^j, \\ Z &= \sum_{j=1}^{m} q^j = q\, \frac{1-q^m}{1-q}. \end{aligned}$$

The Lagrange parameter λ, or rather $q = e^{\lambda}$, is determined via the mean (first moment)

$$\begin{aligned} \mu_1 &= \frac{d}{d\lambda} \ln(Z) = q \frac{d}{dq} \ln(Z) \\ &= q \frac{d}{dq} \big(\ln(q) + \ln(1-q^m) - \ln(1-q)\big) \\ &= \left(1 - \frac{m q^m}{1-q^m} + \frac{q}{1-q}\right). \end{aligned}$$

In the limit $\lambda \to 0$ ($q \to 1$), corresponding to effectively no constraint, we find

$$\mu_1 = \frac{m+1}{2} \quad \text{and } Z = m,$$

resulting in a uniform solution

$$\boxed{p_j^{\text{ME}} = \frac{1}{m}.}$$

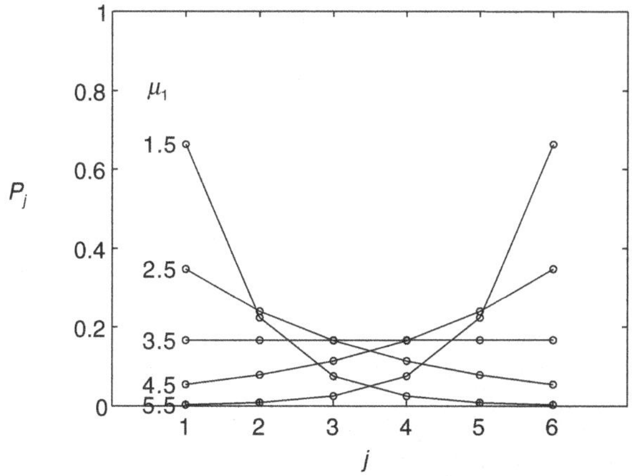

Figure 11.2 Maxent solutions for P_j of the die ($m = 6$) with different values for the constraint μ_1.

For other values of μ_1 the computation of q requires a numerical solution (root-finding) of the constraint equation, for example, based on the Newton algorithm. The result is depicted in Figure 11.2 for various values of the constraint $\mu_1 = \{1.5, 2.5, 3.5, 4.5, 5.5\}$. For $\mu_1 = 3.5$ we find the uniform solution. For $\mu_1 < 3.5$ ($\mu_1 > 3.5$) we observe an exponentially decreasing (increasing) PMF.

11.2.3 Maxwell–Boltzmann distribution

Now we turn to a statistical physics problem. We assume a classical system which can be in different discrete states S_j, with $j = 1, \dots, m$. The corresponding energies are E_j. We still have a single constraint. The testable energy shall be the mean energy. Then

$$\mu_E \overset{!}{=} \langle E \rangle := \sum_{j=1}^{m} P_j \, E_j, \text{ with } E_{\min} \leq E_j \leq E_{\max}. \tag{11.10}$$

Applying equations (11.9b) and (11.9c) results in the Boltzmann distribution for discrete systems:

Maxwell–Boltzmann distribution

$$P_j^{\text{ME}} = \frac{1}{Z} e^{-\beta E_j},$$
$$Z = \sum_j e^{-\beta E_j}.$$

The Lagrange parameter β, which corresponds to the inverse temperature, follows from the energy constraint

$$-\frac{\partial \ln(Z)}{\partial \beta} \stackrel{!}{=} \langle E \rangle .$$

We have shown above that Maxent with linear constraints yields a unique solution, provided the condition $\langle E \rangle = \mu$ can be fulfilled. For obvious reasons, a solution exists if and only if

$$E_{\min} \leq \mu \leq E_{\max} .$$

Other mean values cannot be accommodated within the model. Since Maxent yields by construction the maximum entropy consistent with the constraints, experimental values of the entropy S^{exp} can only be smaller than $S^{\text{ME}} := S(\{P_j^{\text{ME}}\})$. Smaller experimental values $S^{\text{exp}} < S^{\text{ME}}$, then, imply the existence of additional constraints not included in the Maxent solution.

11.2.4 Bose–Einstein distribution

Next we consider a quantum system. As outlined in Section 4.3.2 [p. 60], identical quantum particles are indistinguishable and we can only tell how many particles (n_j) are in a given quantum state S_j but not which particles are involved. Moreover, for the sake of simplicity, we consider the grand-canonical ensemble, in which fluctuations of the particle number are possible. Only the mean number of particles shall be given as testable information in addition to the mean energy of the system. We begin with bosons for which there exists no upper limit in the occupation number of a single quantum state. Let P_{jn} denote the probability to find n particles in the state S_j with energy E_j. The normalization constraints are

$$1 = \sum_{n=0}^{\infty} P_{jn}, \quad \forall j. \tag{11.11}$$

The mean number of particles (*occupation number*) in state S_j is

$$\langle n_j \rangle := \sum_{n=0}^{\infty} n \; P_{jn}, \tag{11.12}$$

which results in the mean total number of particles

$$\mu_N \stackrel{!}{=} \langle N \rangle := \sum_{j=1}^{m} \sum_{n=0}^{\infty} n \; P_{jn}. \tag{11.13}$$

The last constraint is provided by the mean energy

$$\mu_E = \langle E \rangle \stackrel{!}{=} \sum_j E_j \, n_j = \sum_{j=1}^{m} \sum_{n=0}^{\infty} E_j \, n \, P_{jn}. \tag{11.14}$$

We have to adapt the entropy to the situation that events are enumerated by a pair of indices (j, n):

$$S = -\sum_{jn} P_{jn} \, \ln(P_{jn}). \tag{11.15}$$

The functional to be maximized is now

$$\mathcal{L} = S - \sum_j \lambda_{0,j} \left(\sum_m P_{jn} - 1 \right) - \lambda_1 \left(\sum_{jn} n \, P_{jn} - \mu_N \right) - \lambda_2 \left(\sum_{jn} E_j \, n \, P_{jn} - \mu_E \right).$$

The derivative with respect to P_{jn} results in

$$\begin{aligned} \frac{\partial \mathcal{L}}{\partial P_{jn}} &= -\ln P_{jn} - 1 - \lambda_{0,j} - \lambda_1 \, n - \lambda_2 \, n \, E_j \stackrel{!}{=} 0 \Rightarrow \\ P_{jn}^{\mathrm{ME}} &= \frac{1}{Z_j} \, e^{-n(\lambda_1 + \lambda_2 E_j)}. \end{aligned} \tag{11.16}$$

The normalization equation (11.11) is given by

$$Z_j = \sum_{n=0}^{\infty} e^{-n(\lambda_1 + \lambda_2 E_j)} = \sum_{n=0}^{\infty} \left(e^{-(\lambda_1 + \lambda_2 E_j)} \right)^n = \frac{1}{1 - e^{-(\lambda_1 + \lambda_2 E_j)}}.$$

Back-substitution of Z_j into equation (11.16) yields

$$P_{jn}^{\mathrm{ME}} = \left(1 - e^{-(\lambda_1 + \lambda_2 E_j)} \right) \, e^{-n \, (\lambda_1 + \lambda_2 E_j)}. \tag{11.17}$$

The occupation number is therefore given by

$$\begin{aligned} \langle n_j \rangle &= \sum_{n=0}^{\infty} n \, P_{jn} = \left(1 - e^{-\lambda}\right) \sum_{n=0}^{\infty} n \, e^{-n\,\lambda} \Big|_{\lambda = \lambda_1 + \lambda_2 E_j} \\ &= -\left(1 - e^{-\lambda}\right) \frac{\partial}{\partial \lambda} \sum_{n=0}^{\infty} e^{-n\,\lambda} \Big|_{\lambda = \lambda_1 + \lambda_2 E_j} \\ &= \frac{1}{e^{\lambda_1 + \lambda_2 E_j} - 1}. \end{aligned}$$

Substituting the symbols of the Lagrange parameters with those conventionally used in statistical physics we finally arrive at

Bose–Einstein distribution

$$\langle n_j \rangle = \frac{1}{e^{\beta(E_j-\mu)} - 1}. \tag{11.18}$$

The parameters β and μ follow from the constraints on mean energy and mean particle number.

11.2.5 Fermi–Dirac distribution

Finally we turn to fermions. The only difference from the bosonic case is that the occupation of a state is restricted to $n_j \in \{0, 1\}$. We can therefore start with the general solution equation (11.16). However, the normalization now yields

$$\begin{aligned} Z_j &= \sum_{n=0}^{1} e^{-n(\lambda_1+\lambda_2 E_j)} = 1 + e^{-(\lambda_1+\lambda_2 E_j)}, \\ P_{jn}^{\text{ME}} &= \frac{1}{1 + e^{-(\lambda_1+\lambda_2 E_j)}}\, e^{-n\,(\lambda_1+\lambda_2 E_j)}. \end{aligned} \tag{11.19}$$

The mean occupation number is therefore

$$\begin{aligned} \langle n_j \rangle &= \frac{\sum\limits_{n=0}^{1} n\ e^{-n\,(\lambda_1+\lambda_2 E_j)}}{1 + e^{-(\lambda_1+\lambda_2 E_j)}} \\ &= \frac{e^{-\,(\lambda_1+\lambda_2 E_j)}}{1 + e^{-(\lambda_1+\lambda_2 E_j)}} \\ &= \frac{1}{e^{(\lambda_1+\lambda_2 E_j)} + 1}. \end{aligned}$$

Again converting to the standard notation of statistical physics we obtain

Fermi–Dirac distribution

$$\langle n_j \rangle = \frac{1}{e^{\beta(E_j-\mu)} + 1}. \tag{11.20}$$

11.2.6 Random experiments

The Maxent assignment of prior PMFs does not require that the distribution can be associated with frequencies. However, if there is an underlying random experiment, then there should be a close relation to the Maxent solution. To this end we consider a random variable

x that can take the discrete values $\{\xi_j\}$ with probabilities $\{P_j\}$ for $j = 1, \ldots, N$. For convenience, we assume linear constraints (testable information):

$$\mu_\alpha = \sum_j P_j \, K_\alpha(\xi_j), \quad \alpha = 1, \ldots, k.$$

These constraints include the normalization $\sum_j P_j = 1$. A possible PMF fulfilling these constraints is the corresponding Maxent distribution.

Random experiments

Now suppose that a random experiment shall be performed, providing a sample

$$\zeta_1, \ldots, \zeta_M \quad \text{with} \quad \zeta_l \in \{\xi_1, \ldots, \xi_N\}$$

of size M, from which we determine the frequency m_j with which state ξ_j occurs in the sample. The constraints require

$$\mu_\alpha = \sum_{j=1}^{N} K_\alpha(\xi_j) \, \frac{m_j}{M}, \qquad \alpha = 1, \ldots, k. \tag{11.21}$$

What will be the most likely configuration $\boldsymbol{m} := \{m_j\}$ of occupation numbers in these random experiments? To answer this question, we note that the configurations are determined by the samples $\zeta_1, \ldots, \zeta_M$. Each sample is equally likely. Hence, according to the classical definition of probability as outlined in Section 4.3.2 [p. 60], we have

$$P(\boldsymbol{m}) \propto N(\boldsymbol{m}|M) = \frac{M!}{\prod_i m_i}.$$

In addition, we have to take the constraints into account, which prohibit certain configurations $\boldsymbol{m}$. For the allowed configurations, however, the probability is, according to the classical definition, still proportional to $N(\boldsymbol{m}|M)$. Next we assume that the sample size is very large, $M \gg L$ (ideally $M \to \infty$). Then we can employ Stirling's formula:

$$\ln(m_i!) \approx m_i \ln(m_i) - m_i,$$

which yields along with the normalization condition $\sum_j m_j = M$

$$\begin{aligned} \ln\left[N(\boldsymbol{m}|M)\right] &\approx M \ln(M) - \cancel{M} - \sum_{j=1}^{N} \left(m_j \ln(m_j) - \cancel{m_j}\right) \\ &= -M \sum_{j=1}^{N} \frac{m_j}{M} \ln\left(\frac{m_j}{M}\right). \end{aligned}$$

With the definition $\rho_j := m_j/M$ we find

$$N(\boldsymbol{m}|M) \approx e^{M\,S},$$
$$S := -\sum_{j=1}^{N} \rho_j \ \ln(\rho_j).$$

Next we express the constraints in equation (11.21) in vector notation

$$\boldsymbol{\mu} = \mathbf{K}^T \boldsymbol{\rho},$$

where $\boldsymbol{\mu}$ and $\boldsymbol{\rho}$ are column vectors containing the elements $\{\mu_j\}$ and $\{\rho_j\}$, respectively. The matrix $\mathbf{K}$ is defined by the elements $K_{j,\alpha} := K_\alpha(\xi_j)$. In other words, the columns of $\mathbf{K}$ contain a set of k vectors of length N. We assume that these vectors are linearly independent, otherwise we can always reduce the constraints to a set of linearly independent constraints, for example, by employing a **QR**-decomposition for $\mathbf{K}$ resulting in a reduced set of linearly independent constraints

$$\boldsymbol{\mu}' := \mathbf{R}\boldsymbol{\mu} = \big(\mathbf{R}\mathbf{R}^T\big)\mathbf{Q}^T \boldsymbol{\rho}.$$

The linearly independent vectors in $\mathbf{K}$ span a k-dimensional subspace of the N-dimensional vector space for the vectors $\boldsymbol{\rho}$. It is expedient to split $\boldsymbol{\rho}$ accordingly:

$$\boldsymbol{\rho} = \underbrace{\mathbf{K}\boldsymbol{d}}_{:=\boldsymbol{\rho}_{\shortparallel}} + \mathbf{U}\boldsymbol{c}.$$

The columns of matrix $\mathbf{U}$ contain $(N-k)$ orthonormal basis vectors, which in addition are orthogonal to $\mathbf{K}$, i.e.

$$\mathbf{K}^T\mathbf{U} = \mathbf{0}.$$

The vectors $\boldsymbol{d}$ and $\boldsymbol{c}$ are the corresponding expansion coefficients. Then the constraints read

$$\boldsymbol{\mu} = \mathbf{K}^T\mathbf{K}\boldsymbol{d} + \mathbf{K}^T\mathbf{U}\boldsymbol{c} = \mathbf{K}^T\mathbf{K}\boldsymbol{d},$$
$$\boldsymbol{d} = \big(\mathbf{K}^T\mathbf{K}\big)^{-1}\,\boldsymbol{\mu},$$
$$\boldsymbol{\rho}_{\shortparallel} = \mathbf{K}\big(\mathbf{K}^T\mathbf{K}\big)^{-1}\,\boldsymbol{\mu}.$$

In other words, the constraints fix the vector $\boldsymbol{d}$, while $\boldsymbol{c}$ is not affected by the constraints at all. The entropy can be expressed in the new independent variable $\boldsymbol{c}$:

$$S(\boldsymbol{\rho}) = S\big(\boldsymbol{\rho}_{\shortparallel} + \mathbf{U}\boldsymbol{c}\big) := \tilde{S}(\boldsymbol{c}).$$

Then the Maxent solution $\boldsymbol{c}^{\text{ME}}$, based on the constraints in equation (11.21), is obtained through $\frac{\partial}{\partial c_\nu}\tilde{S}(\boldsymbol{c}) = 0$. The corresponding coefficients shall be denoted $\boldsymbol{c}^{\text{ME}}$. Taylor expansion of the entropy up to second order in $\Delta\boldsymbol{c} := \boldsymbol{c} - \boldsymbol{c}^{\text{ME}}$ yields

$$\tilde{S}(\boldsymbol{c}) = S^{\mathrm{ME}} - \frac{1}{2}\Delta\boldsymbol{c}^T\mathbf{H}\Delta\boldsymbol{c},$$
$$H_{\nu,\nu'} := \sum_j \mathbf{U}_{j\nu}\frac{1}{\rho_j^{\mathrm{ME}}}\mathbf{U}_{j\nu'},$$
$$\boldsymbol{\rho}^{\mathrm{ME}} := \boldsymbol{\rho}_{\parallel} + \mathbf{U}\boldsymbol{c}^{\mathrm{ME}}.$$

Finally, the probability for the configuration $\boldsymbol{m}$, or rather $\boldsymbol{\rho}$, is

$$P(\boldsymbol{\rho}|M,\mathcal{I}) \propto \delta\big(\boldsymbol{\rho} - \boldsymbol{\rho}_{\parallel} - \mathbf{U}\boldsymbol{c}\big)\ \exp\left(-\frac{M}{2}\Delta\boldsymbol{c}^T\mathbf{H}\Delta\boldsymbol{c}\right). \tag{11.22}$$

Then

$$\langle\boldsymbol{\rho}\rangle = \boldsymbol{\rho}^{\mathrm{ME}} := \boldsymbol{\rho}_{\parallel} + \mathbf{U}\boldsymbol{c}^{\mathrm{ME}},$$
$$\langle\Delta\boldsymbol{\rho}\Delta\boldsymbol{\rho}^T\rangle = \frac{1}{M}\ \mathbf{U}\mathbf{H}^{-1}\mathbf{U}^T.$$

Obviously, the deviations from the Maxent solution decrease as $1/M$ and for $M \to \infty$ the random experiment will yield the Maxent solution. As a matter of fact, this is the basis for equilibrium statistical physics in the thermodynamic limit ($M \to \infty$).

Based on equation (11.22), it is straightforward to prove the 'entropy concentration theorem' introduced by E. T. Jaynes. To this end we compute

$$\begin{aligned}
p(S|\boldsymbol{\mu}, L, M, k, \mathcal{I}) &= \int d^{L-k}c\ p(S|\boldsymbol{c}, \boldsymbol{\mu}, L, M, k, \mathcal{I})\ p(\boldsymbol{c}|\boldsymbol{\mu}, M, \mathcal{I})\\
&\propto \int d^{L-k}c\ \delta\big[S - \tilde{S}(\boldsymbol{c})\big]\ e^{M\tilde{S}(\boldsymbol{c})}\\
&\propto e^{MS}\int d^{L-k}c\ \delta\left[S - S^{\mathrm{ME}} + \frac{1}{2}\Delta\boldsymbol{c}^T\mathbf{H}\Delta\boldsymbol{c}\right].
\end{aligned}$$

With $\boldsymbol{x} := \mathbf{H}^{1/2}\Delta\boldsymbol{c}$ and $z := 2M(S^{\mathrm{ME}} - S)$, we obtain

$$\begin{aligned}
p(z|\boldsymbol{\mu}, L, M, k, \mathcal{I}) &\propto e^{-M(S^{\mathrm{ME}}-S)}\int d^{L-k}x\ \delta\left[\frac{z}{M} + \frac{1}{2}\boldsymbol{x}\right]\\
&\propto e^{-\frac{1}{2}z}\int dx\ x^{L-k-1}\ \delta\left[\frac{z}{M} - x^2\right]\\
&\propto e^{-\frac{1}{2}z}\int dx\ x^{L-k-2}\ \delta\left[x - \left(\frac{z}{M}\right)^{\frac{1}{2}}\right]\\
&\propto z^{(L-k)/2-1}\ e^{-z/2}.
\end{aligned}$$

So $z = 2M(S^{\mathrm{ME}} - S)$ is chi-squared with $n = L - k$ degrees of freedom (see equation (7.16a) [p. 104]). We define an interval of entropy values $\mathcal{S}(\alpha) := [S^{\mathrm{ME}} - \Delta S(\alpha), S^{\mathrm{ME}}]$ such that $P(S \in \mathcal{S}(\alpha)|\boldsymbol{\mu}, L, M, k, \mathcal{I}) = \alpha$, i.e. the probability mass in $\mathcal{S}(\alpha)$ is α. This corresponds to an interval of z-values $\mathcal{Z}(\alpha) := [0, \Delta z(\alpha)]$ with

$$P(z \in \mathcal{Z}(\alpha)|\boldsymbol{\mu}, L, M, k, \mathcal{I}) = F_{\chi^2}(\Delta z(\alpha)|n) \stackrel{!}{=} \alpha,$$

where F_{χ^2} stands for the CDF of the chi-squared distribution. The inverse yields $\Delta z(\alpha)$ and then

$$\Delta S = \frac{\Delta z(\alpha)}{2M}.$$

So for arbitrary but fixed α the width of the interval $\mathcal{S}(\alpha)$ shrinks to zero as $1/M$, which proves Jaynes' 'entropy concentration theorem'.

11.3 Maximum entropy for continuous distributions

Now we want to generalize the Maxent principle to continuous distributions $p(x)$. To this end we discretize the domain of x-values into intervals Δx_j, centred about values x_j, with $j = 1, \ldots, N$. The only restriction on the intervals is that they shrink to zero as $1/N$. The proportionality is denoted by

$$\Delta x_j = \frac{1}{m(x_j)} \frac{1}{N}.$$

$m(x)$ defines a normalized measure or density in x via

$$m(x_j) = \lim_{N\to\infty} \frac{1/N}{\Delta x_j} = \frac{\text{number of states in } \Delta x_j}{\text{length of interval } \Delta x_j},$$

$$\int m(x)\, dx = \lim_{N\to\infty} \sum_{j=1}^{N} m(x_j)\, \Delta x_j = \lim_{N\to\infty} \sum_{j=1}^{N} \frac{1}{N} = 1.$$

Then we define the probability mass in interval Δx_j as

$$P_j := P(\Delta x_j) := p(x_j)\, \Delta x_j = \frac{p(x_j)}{m(x_j)} \frac{1}{N}.$$

For this PMF we employ the expression for the entropy for discrete distributions

$$S^D = -\sum_{j=1}^{N} P_j \ln(P_j) = -\sum_{j=1}^{N} \Delta x_j\, p(x_j) \ln\left[\frac{p(x_j)}{N\, m(x_j)}\right] \tag{11.23}$$

$$= -\sum_{j=1}^{N} \Delta x_j\, p(x_j) \ln\left[\frac{p(x_j)}{m(x_j)}\right] + \ln(N). \tag{11.24}$$

The constant is irrelevant for Maxent and will be omitted. The entropy for continuous distributions is defined as

$$S^C := \lim_{N\to\infty} \left(S^D - \ln(N)\right) \tag{11.25}$$

and results in

Entropy for continuous distributions

$$S^C = D(p:m) = -\int p(x) \ln\left(\frac{p(x)}{m(x)}\right) dx, \tag{11.26}$$

$$m(x) : \text{invariant measure.} \tag{11.27}$$

We will see in the next section that the Maxent solution $p^{\mathrm{ME}}(x)$ in the absence of testable information (apart from the normalization constraint) is equal to the default model $m(x)$. This situation corresponds to the uninformative prior, discussed in Chapter 10 [p. 165]. For that reason $m(x)$ is also called the *invariant measure*. The entropy in equation (11.26) is also known as relative entropy, cross entropy, Kullback–Leibler entropy or Kullback–Leibler distance. The latter name underlines that

$$D(p:m)$$

provides a measure for the 'distance' between the probability density $p(x)$ and $m(x)$. However, the triangle inequality does not hold, i.e.

$$D(p:m) \neq D(m:p)$$

for general $p(x)$ and $m(x)$. Therefore the correct nomenclature is Kullback–Leibler divergence.

As in the case of discrete distributions, the Maxent solution maximizes the entropy subject to the constraints

$$\phi_\alpha\{p(x)\} = 0, \quad \alpha = 1, \ldots, k.$$

The maximization is similar to the discrete case and yields

$$p(x) = \frac{m(x)}{Z} \exp\left(-\sum_{\alpha=1}^{k} \lambda_\alpha \frac{\delta}{\delta p(x)} \phi_\alpha\{p(x)\}\right).$$

The k Lagrange parameters are fixed by the constraints. In case of linear constraints, i.e.

$$\phi_\alpha\{p(x)\} = \int p(x)\, K_\alpha(x)\, dx - \mu_\alpha, \quad \mu = 1, \ldots, L,$$

the functional derivative yields

$$\frac{\delta}{\delta p(x)} \phi_\alpha\{p(x)\} = K_\alpha(x).$$

The Maxent solution is therefore given by

Maxent solution for continuous distributions for linear constraints

constraints:
$$\mu_\alpha = \int p(x)\, K_\alpha(x)\, dx, \tag{11.28a}$$

Maxent solution:
$$p^{\text{ME}}(x) = \frac{1}{Z}\, m(x)\, e^{-\sum_\mu \lambda_\mu K_\mu(x)}, \tag{11.28b}$$

normalization:
$$Z = \int m(x)\, e^{-\sum_\mu \lambda_\mu K_\mu(x)}\, dx. \tag{11.28c}$$

The determination of the Lagrange multipliers can be achieved as in the discrete case:

$$\mu_\alpha = \int K_\alpha(x)\, p(x) = \frac{1}{Z} \int m(x)\, K_\alpha(x)\, e^{-\sum_\mu \lambda_\mu K_\alpha(x)} = -\frac{\partial}{\partial \lambda_\alpha} \ln(Z).$$

11.3.1 Illustrating examples

We will illustrate the applicability of the Maxent formalism by three examples. For further applications see also [111, 112, 186].

Density reconstruction by the moments

As an example consider the case where the lower moments of a probability density $p(x)$,

$$\mu_n := \langle x^n \rangle = \int x^n p(x) dx,$$

for $n = 0, 1, \ldots, n_{\text{max}}$ are given, either from experiments or from theoretical calculations. In Figure 11.3 the approach is illustrated for a challenging PDF $p(x)$ that consists of two separated rectangular blocks. We can observe how the Maxent solution gradually approaches the exact result. But we also note that 'ringing' occurs at the discontinuities, like the Gibbs oscillations in Fourier series. We will discuss this aspect in the next chapter in more detail.

Barometric formula

Here we want to compute the probability distribution $p(z)$ to find gas particles of mass m in the gravitational field of the earth at an altitude z (with $z \in [0, \infty)$). Besides the normalization constraint for $p(z)$, also the mean potential energy $\langle E_p \rangle$ is given by the underlying physics. With the definition of the inverse temperature $\beta = 1/k_B T$, this constraint reads

$$\frac{\langle E_p \rangle}{m\, g} = \int p(z)\, z\, dz \overset{!}{=} \frac{1}{\beta\, m\, g}.$$

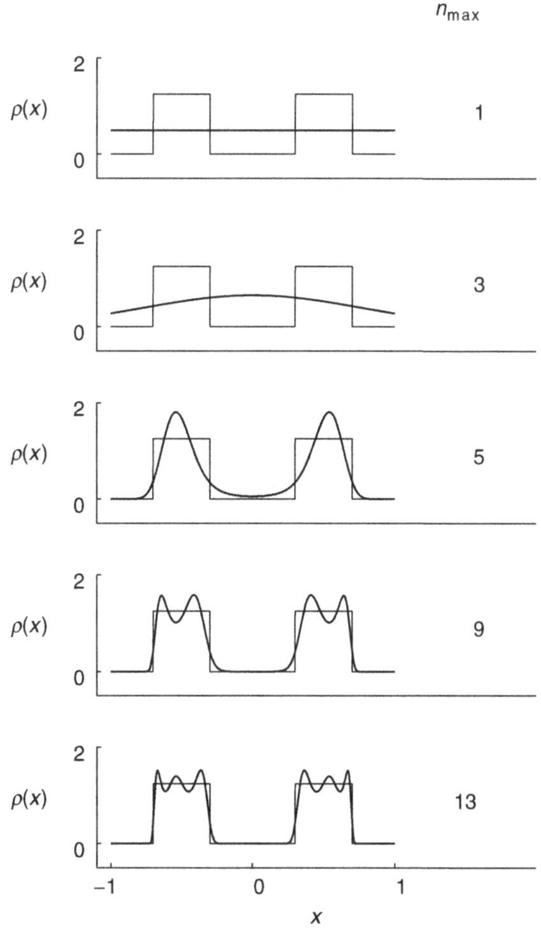

Figure 11.3 Maximum-entropy reconstruction $\rho(x)$ of the density (thick lines) for various moments. The exact test density is depicted as thin lines.

Here the kernel $K_1(z) = z$, and the Maxent solution given in equation (11.28) reads

$$p^{\mathrm{ME}}(z) = \frac{1}{Z} e^{-\lambda z}.$$

The energy constraint results in $\lambda = \beta m g$ and immediately leads to the

Barometric formula

$$p(z) = \beta\, m\, g\, e^{-\beta\, m\, g\, z}, \tag{11.29}$$

which is identical to the distribution derived in statistical mechanics.

Gaussian distribution

Here we want to derive the Maxent solution for a probability density $p(x)$, with $x \in (-\infty, \infty)$ when the first two (central) moments, mean μ and variance σ^2, are given as constraints. Then equation (11.28) yields

$$p(x) = \frac{1}{Z}\, e^{-\lambda_1\, x - \lambda_2\, x^2}.$$

Completing the square in the exponent and taking care of the constraints, we readily obtain

$$p(x) = \frac{1}{\sqrt{2\,\pi\, a}}\, e^{-\frac{(x-\mu)^2}{2\,\sigma^2}}.$$

Therefore the Maxent solution is given by a Gaussian distribution with mean μ and variance σ^2. Hence the least committing density for given mean and variance is the Gaussian. The Maxent property yields a further explanation for the ubiquitous presence of the Gaussian distribution. The Gaussian distribution is not only the limit distribution (under weak technical assumptions) of the sum of random variables. In addition, it is also the least informative probability distribution constrained by the first two moments (mean and variance) – a very generic situation in all natural sciences. Analogously, the Maxwellian velocity distribution can be derived as a Maxent solution. Here the kinetic energy of the particles, i.e. the second moment, is the relevant constraint.

Consider for example the task of signal estimation [19–21] from noisy data $\boldsymbol{d}$. We assume that the data can be separated into a signal part $f\,(x\,|\boldsymbol{\Theta})$ and the random additive noise e_i. The model of the experiment can then be written as

$$\boldsymbol{d} = \boldsymbol{f} + \boldsymbol{e}, \qquad \text{with } f_i = f\,(x_i\,|\boldsymbol{\Theta}), \tag{11.30}$$

where $\boldsymbol{\Theta} = \{\theta_1, \dots, \theta_m\}$ is a set of parameters defining the model. Bayes' theorem tells us that, given the data vector $\boldsymbol{d}$, the probability density for the parameters is

$$p\,(\boldsymbol{\Theta} \mid \boldsymbol{d}, \mathcal{I}) = \frac{p\,(\boldsymbol{d} \mid \boldsymbol{\Theta}, \mathcal{I})\, p\,(\boldsymbol{\Theta} \mid \mathcal{I})}{p\,(\boldsymbol{d} \mid \mathcal{I})}. \tag{11.31}$$

To proceed, we must supply both the problem-specific prior density $p\,(\boldsymbol{\Theta} \mid \mathcal{I})$ and the likelihood distribution $p\,(\boldsymbol{d} \mid \boldsymbol{\Theta}, \mathcal{I})$. We can compute the likelihood of the data if we can assign a prior probability to the noise. Typically, the mean value of additive noise is zero and uncorrelated to the signal – otherwise it would correspond to a systematic effect. Furthermore, the second moment of the noise is finite – otherwise the noise power would be infinite. Then the maximum entropy formalism yields as least informative distribution again a Gaussian distribution as prior probability for the noise distribution:

$$p\,(\boldsymbol{e}|\sigma, \mathcal{I}) = \prod_{i=1}^{N} \frac{1}{\sqrt{2\pi\sigma^2}} \exp\left(-\frac{e_i^2}{2\sigma^2}\right), \tag{11.32}$$

where σ^2 is the noise variance. Any other distribution corresponds to more prior information and would thus result in more precise estimates. With this prior on the noise distribution the likelihood is then given by plugging equation (11.30) into equation (11.32):

$$p\left(\boldsymbol{d} \mid \boldsymbol{\Theta}, \mathcal{I}\right) = \prod_{i=1}^{N} \frac{1}{\sqrt{2\pi\sigma^2}} \exp\left(-\frac{(d_i - f\left(x_i; \boldsymbol{\Theta}\right))^2}{2\sigma^2}\right). \tag{11.33}$$

Multivariate Gaussian distribution

For the multidimensional random variable $x \in \mathbb{R}^N$, the mean vector

$$\boldsymbol{\mu} = \int \boldsymbol{x}\ p(x)\ d^N x, \qquad i = 1, \ldots, N \tag{11.34}$$

and the covariance matrix $\mathbf{C}$

$$C_{ij} = \int \Delta x_i \Delta x_j\ p(\boldsymbol{x})\ d^N x \tag{11.35}$$

with $\Delta x_i = x_i - \mu_i$ shall be given. The 'Lagrange function' is given by

$$\begin{aligned}\mathcal{L} = &-\int p(\boldsymbol{x}) \ln\left(\frac{p(\boldsymbol{x})}{m(\boldsymbol{x})}\right) d^N x - \lambda_0 \int p(\boldsymbol{x})\, d^N x \\ &- \sum_{i=1}^{N} \lambda_i \left(\int x_i\ p(\boldsymbol{x})\ d^N x - \mu_i\right) - \sum_{i,j=1}^{N} \Lambda_{ij} \left(\int \Delta x_i \Delta x_j\ p(\boldsymbol{x})\ d^N x - C_{ij}\right).\end{aligned}$$

The functional derivative with respect to $p(\boldsymbol{x})$ yields

$$\frac{\delta}{\delta p(\boldsymbol{x})} \mathcal{L} = -\ln\left(\frac{p(\boldsymbol{x})}{m(\boldsymbol{x})}\right) - 1 - \lambda_0 - \sum_{i=1}^{N} \lambda_i\ x_i - \sum_{ij} \Lambda_{ij}\ \Delta x_i \Delta x_j.$$

So we obtain

$$p(\boldsymbol{x}) = \frac{m(\boldsymbol{x})}{Z} \exp\left(-\sum_i \lambda_i\ x_i - \sum_{ij} \Lambda_{ij}\ \Delta x_i \Delta x_j\right).$$

Using a uniform invariant measure $m(\boldsymbol{x})$ and rewriting the first term yields

$$p(x) = \frac{1}{Z'} \exp\left(-\sum_i \lambda_i\ \Delta x_i - \sum_{ij} \Lambda_{ij}\ \Delta x_i \Delta x_j\right).$$

The constraint on the mean (equation (11.34)) requires

$$\int \Delta x_i\ p(\boldsymbol{x})\ d^N x = 0,$$

which entails $\lambda_i = 0$. Then the Maxent solution reads

$$p(x) = \frac{1}{Z''}\ e^{-\sum_{ij} \Lambda_{ij}\ \Delta x_i \Delta x_j}. \tag{11.36}$$

This is a multivariate Gaussian distribution given in equation (7.39a) [p. 118]. Finally, matching the covariance constraint equation (11.35), the solution is given by

MAXENT SOLUTION FOR PRESCRIBED MEAN AND COVARIANCE

$$p(x) = (2\pi\ |\mathbf{C}|)^{-\frac{N}{2}}\ e^{-\frac{1}{2}\Delta x^T \mathbf{C}^{-1} \Delta x}. \tag{11.37}$$

12

Quantified maximum entropy

12.1 The entropic prior

The goal of this chapter is the prior PDF for positive and additive distributions (PAD) $P(\rho(\boldsymbol{x})|\mathcal{I})$ $(P(\boldsymbol{\rho}|\mathcal{I}))$ over some D-dimensional continuous (discrete) image field $\boldsymbol{x}$. The following procedure will result in the so-called *quantified maximum entropy* (QME) approach which is due to Steve Gull and John Skilling [27, 94, 95, 189, 190]. Many distributions in natural sciences and engineering are indeed positive and additive. Positivity is self-explanatory and by *additivity* we mean that for non-overlapping elements $\Delta\boldsymbol{x}_1$ and $\Delta\boldsymbol{x}_2$, which may be pixels, voxels or discrete sets, the following relation holds:

$$\rho\big(\boldsymbol{x} \in (\Delta\boldsymbol{x}_1 \cup \Delta\boldsymbol{x}_2)\big) = \rho(\boldsymbol{x} \in \Delta\boldsymbol{x}_1) + \rho(\boldsymbol{x} \in \Delta\boldsymbol{x}_2). \tag{12.1}$$

This could, for example, be the flux of photons hitting two disjoint areas of an image, which is the same as the sum of the individual fluxes. Another example are PDFs. They are additive, positive and in addition normalized to one. In this book we will only consider discrete image fields $\boldsymbol{x}$. We will see shortly that the corresponding prior PDF for $\boldsymbol{\rho}$ is

The entropic prior

$$p(\boldsymbol{\rho}|\alpha, \mathcal{I}) = \frac{1}{Z(\alpha)} \frac{e^{\alpha S}}{\prod_n \sqrt{\rho_n}}, \tag{12.2a}$$

$$S := \sum_n \rho_n - m_n - \rho_n \ln\left(\frac{\rho_n}{m_n}\right). \tag{12.2b}$$

Here m_n is the so-called *default model* and α the 'regularization parameter' that will be determined self-consistently.

12.2 Derivation of the entropic prior

Here we outline a mathematically transparent approach to the entropic prior. A more general derivation is based on measure-theoretical considerations. For our (and most) purposes the present approach is sufficient.

1. Discretization of x

We discretize the underlying field x into N *pixels/voxels* of size ΔV_i and define the corresponding image mass

$$\rho(x) \to \rho_i = \rho(x_i)\, \Delta V_i . \tag{12.3}$$

2. Discretization of ρ_i

Next we quantize the image mass values ρ_n in small units $\Delta\rho$:

$$\rho_i \approx n_i\, \Delta\rho . \tag{12.4}$$

In some situations this quantization is not needed as the images ρ_i are by construction integers.

3. Probabilities for n_i

The remaining task is to determine the PMF for n_i. We assume that there is a total 'intensity' $\sum_i n_i = M$ which is distributed at random among the pixels. The pixels shall have a prior probability q_i. Then the PDF for $\boldsymbol{n}$ is multinomial and, similar to the derivation in Section 11.2.6 [p. 190], we obtain

$$P(\boldsymbol{n} \,|\, \mathcal{I}) = \frac{1}{Z} \frac{e^{\tilde{S}}}{\prod_i \sqrt{n_i}} , \tag{12.5}$$

$$\tilde{S} := \sum_{i=1}^{N} n_i - \mu_i - n_i \log(n_i/\mu_i). \tag{12.6}$$

Here $\mu_i := q_i M$ and $\tilde{S}$ is a generalization of the *Shannon entropy*.

4. PDF for ρ_i

Now we undo the quantization of step 2 by the transformation

$$n_i \to \frac{\rho_i}{\Delta\rho} \quad \Rightarrow \quad \mu_i \to \frac{m_i}{\Delta\rho}, \tag{12.7}$$

where m_i is the so-called *default model*. By inserting equation (12.7) into equation (12.5) we obtain the sought-for prior PDF:

$$
\begin{aligned}
p(\boldsymbol{\rho}|\alpha, \mathcal{I}) &= \frac{1}{Z(\alpha)}\, e^{\alpha S}\, \mu(\boldsymbol{\rho}), \\
S &:= \sum_i \rho_i - m_i - \rho_i \ln\left(\frac{\rho_i}{m_i}\right), \\
\mu(\boldsymbol{\rho}) &:= \prod_i \rho_i^{-1/2}.
\end{aligned}
\tag{12.8}
$$

For notational ease we have suppressed an explicit mention of the default model $\boldsymbol{m}$; it is understood as part of the background information $\mathcal{I}$. We have introduced the parameter $\alpha = \frac{1}{\Delta\rho}$, which is a kind of regularization parameter, *nuisance parameter* or *hyperparameter*, which cannot be fixed a priori, as it results from an artificial quantization for which there is no obvious value. We will see soon how it can be handled self-consistently; meanwhile we add it to the set of conditions. The ρ-dependent factor $\mu(\boldsymbol{\rho})$ is part of the integration measure and $Z(\alpha)$ the normalization that will be determined next.

12.3 Saddle-point approximation for the normalization

The normalization of $P(\boldsymbol{\rho}|\alpha, \mathcal{I})$ can be determined either numerically or analytically. In the latter case one resorts to the *saddle-point approximation*, which in this context is also referred to as the *steepest-descent approximation*. As outlined in Section A.2 [p. 598], we write the integrand as an exponential $\exp(\Phi(\rho))$ and expand the argument about its maximum up to second order. Here we follow the approach commonly used in the Maxent community in which the measure factor $\mu(\boldsymbol{\rho})$ is not included in Φ but rather approximated by the maximum value.[1] The maximum is achieved at the *default model* $\rho_i^* = m_l$. The entropy at the maximum is zero. The second derivative of S at the maximum is

$$
\left.\frac{\partial^2 S}{\partial \rho_i^2}\right|_{\rho_i = m_i} = -\left.\frac{1}{\rho_i}\right|_{\rho_i = m_i} = -\frac{1}{m_i}. \tag{12.9}
$$

Then the Taylor expansion up to second order yields

$$
S = -\frac{1}{2}\sum_i \frac{\Delta\rho_i^2}{m_i}, \qquad \text{with} \quad \Delta\rho_i = \rho_i - m_i. \tag{12.10}
$$

The saddle-point approximation also includes the extension of the integrals over the entire real axis. The resulting Gaussian integrals yield

[1] Inclusion of $\mu(\boldsymbol{\rho})$ consistently in the saddle-point approximation yields generally better results. In particular, when small image values ρ_i occur.

NORMALIZATION OF THE ENTROPIC PRIOR IN SADDLE-POINT APPROXIMATION

$$Z(\alpha) = \prod_{i=1}^{N} \frac{1}{\sqrt{m_i}} \int_{-\infty}^{\infty} e^{-\frac{\alpha}{2}\frac{\Delta\rho_i^2}{m_i}}\, d\rho_i = \left(\frac{2\pi}{\alpha}\right)^{N/2}. \tag{12.11}$$

12.4 Posterior probability density

Now we employ an *entropic prior* in order to determine the posterior probability density $p(\boldsymbol{\rho}|\boldsymbol{d},\mathcal{I})$ for $\boldsymbol{\rho}$ in the light of the data $\boldsymbol{d}$ and prior information $\mathcal{I}$. By application of the marginalization and the product rule we obtain

$$p(\boldsymbol{\rho}|\boldsymbol{d},\mathcal{I}) = \int_{0}^{\infty} d\alpha\; p(\boldsymbol{\rho}|\boldsymbol{d},\alpha,\mathcal{I})\; p(\alpha|\boldsymbol{d},\mathcal{I}). \tag{12.12}$$

As a function of α, the first factor $p(\boldsymbol{\rho}|\alpha,\boldsymbol{d},\mathcal{I})$ is rather flat, whereas $p(\alpha|\boldsymbol{d},\mathcal{I})$ is usually strongly peaked. Therefore, the integral is commonly approximated by

$$p(\boldsymbol{\rho}|\boldsymbol{d},\mathcal{I}) \approx p(\boldsymbol{\rho}|\boldsymbol{d},\alpha^*,\mathcal{I}) \underbrace{\int_{0}^{\infty} d\alpha\; p(\alpha|\boldsymbol{d},\mathcal{I})}_{=1}, \tag{12.13}$$

where α^* denotes the position of the peak of the α-*evidence* $p(\alpha|\boldsymbol{d},\mathcal{I})$. The approximation used in equation (12.13) is called the *evidence approximation*. Along with Bayes' theorem the final result reads

MAXENT POSTERIOR DISTRIBUTION FOR GAUSSIAN NOISE

$$p(\boldsymbol{\rho}|\boldsymbol{d},\mathcal{I}) = \frac{p(\boldsymbol{d}|\boldsymbol{\rho},\alpha^*,\mathcal{I})\; p(\boldsymbol{\rho}|\alpha^*,\mathcal{I})}{p(\boldsymbol{d}|\alpha^*,\mathcal{I})} = \frac{\exp\left\{-\frac{1}{2}\chi^2 + \alpha^* S\right\}}{Z}. \tag{12.14}$$

Here $\mu(\boldsymbol{\rho})$ does not show up as it is part of the integration measure. It will be used when computing posterior expectation values. The maximum of the exponent $\Phi = -\frac{1}{2}\chi^2 + \alpha^* S$ with respect to $\boldsymbol{\rho}$ is the *maximum posterior* (MAP) solution $\boldsymbol{\rho}^*$, introduced earlier.

12.5 Regularization and good data

In order to determine α^* we need the PDF for the regularization parameter, or rather the *α-evidence*

$$p(\alpha|\boldsymbol{d},\mathcal{I}) = \frac{1}{p(\boldsymbol{d}|\mathcal{I})} \int d\rho^N\, p(\boldsymbol{d}|\not\alpha,\boldsymbol{\rho},\mathcal{I})\; p(\boldsymbol{\rho}|\alpha,\mathcal{I})\; p(\alpha|\mathcal{I}). \tag{12.15}$$

The first term under the integral is the likelihood and does not depend on α. As the regularization parameter should be scale invariant, the adequate PDF is a Jeffreys' prior for scale variables (see Section 10.2 [p. 169]), $p(\alpha|\mathcal{I}) = 1/\alpha$. We consider two limiting cases.

$\alpha \to 0$: In this case $p(\boldsymbol{\rho}|\alpha,\mathcal{I}) \approx Z(\alpha)^{-1}$, which implies $p(\alpha|\boldsymbol{d},\mathcal{I}) \propto \alpha^{\frac{N}{2}-1}$.

$\alpha \gg 1$: In this limit the data constraint is overruled by the prior, which can then be replaced by a delta function:

$$p(\boldsymbol{\rho}|\alpha,\mathcal{I}) \to \delta(\boldsymbol{\rho}-\boldsymbol{m}) \qquad \Rightarrow p(\alpha|\boldsymbol{d},\mathcal{I}) \propto p(\alpha\,|\,\mathcal{I}) = \alpha^{-1}. \tag{12.16}$$

The α-evidence is sketched in Figure 12.1. It is typically sharply peaked.

In order to proceed we need the error statistic of the experiment and the underlying model, relating data $\boldsymbol{D}$ and image $\boldsymbol{\rho}$. We will limit the discussion to the most important case, namely that of independent Gaussian noise and a linear model. Then

$$\boldsymbol{d} = \mathbf{M}\boldsymbol{\rho} + \boldsymbol{\eta}, \tag{12.17}$$

where $\boldsymbol{\eta}$ is the unknown noise vector that stems from a Gaussian distribution of zero mean and variance σ^2. The likelihood is therefore

$$p(\boldsymbol{d}|\boldsymbol{\rho},\mathcal{I}) = (2\pi\sigma^2)^{-\frac{N_d}{2}} \exp\left\{-\frac{1}{2}\Delta\boldsymbol{d}^T C^{-1}\Delta\boldsymbol{d}\right\}, \tag{12.18}$$

where N_d denotes the number of data points. The covariance matrix $\mathbf{C}$ is assumed to be known from experiment. In most cases the errors of the individual data points are uncorrelated and the covariance is diagonal, containing the variances. Using the multivariate Gaussian and a linear model in equation (12.15), we obtain

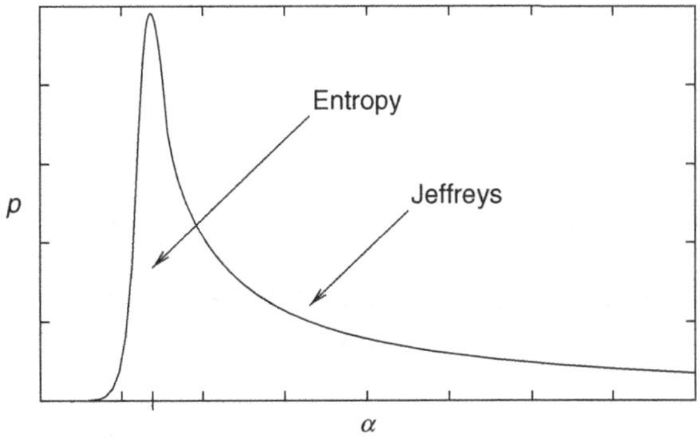

Figure 12.1 Qualitative behaviour of the α-evidence.

$$p(\alpha|\boldsymbol{d},\mathcal{I}) \propto \alpha^{\frac{N}{2}-1} \int\limits_0^\infty \exp\left(-\frac{1}{2}\Delta\boldsymbol{d}^T\mathbf{C}^{-1}\Delta\boldsymbol{d} + \alpha S\right) \mu(\boldsymbol{\rho})\, d\rho^N,$$

with $\mathbf{C} = \text{diag}(\{\sigma_i^2\})$. We invoke the saddle-point approximation again. Firstly, we replace the measure $\mu(\boldsymbol{\rho}) \approx \mu(\rho^*)$, as it has much less variability than the other factors. The *maximum posterior* solution $\boldsymbol{\rho}^*$ is obtained from the maximum of $\Phi(\boldsymbol{\rho})$, which is defined by

$$\Phi(\boldsymbol{\rho}) = -\frac{1}{2}\underbrace{\Delta\boldsymbol{d}^T\mathbf{C}^{-1}\Delta\boldsymbol{d}}_{:=\chi^2} + \alpha\, S.$$

The quantity χ^2 is the 'misfit' that describes the discrepancy between theory and experiment rated by the covariance matrix. This is precisely the same quantity that plays a crucial role in testing hypotheses, outlined in Part IV [p. 255]. For the saddle-point approximation we need the Hessian

$$H_{ij} := \frac{\partial^2\Phi(\boldsymbol{\rho})}{\partial\rho_i\,\partial\rho_j} = -(\mathbf{M}^T\mathbf{C}^{-1}\mathbf{M})_{ij} - \frac{\alpha}{\rho_i}\,\delta_{ij}, \tag{12.19}$$

at the MAP solution, for which the matrix notation reads

$$\begin{aligned}\mathbf{H} &= -\mathbf{M}^T\mathbf{C}^{-1}\mathbf{M} - \alpha\mathbf{D}^{-2},\\ \mathbf{D} &:= \text{diag}\left(\left\{\sqrt{\rho_i^*}\right\}\right).\end{aligned}$$

We will actually need the determinant

$$\begin{aligned}|\mathbf{H}| &= |\mathbf{A} + \alpha\mathbb{1}|\ \mu(\boldsymbol{\rho}^*)^{-2},\\ \mathbf{A} &:= \mathbf{D}\mathbf{M}^T\mathbf{C}^{-1}\mathbf{M}\mathbf{D},\end{aligned} \tag{12.20}$$

where we have used $|D| = \mu(\boldsymbol{\rho}^*)$. By virtue of its structure, the matrix $\mathbf{A}$ can be expressed as

$$\mathbf{A} = \sum_{l=1}^{N_d} \boldsymbol{\xi}_l\boldsymbol{\xi}_l^T,$$

with some suitable vectors $\boldsymbol{\xi}_l$. Consequently, $\text{rank}(\mathbf{A}) \le N_d$. Now, the saddle-point approximation can be carried out, resulting in

PDF FOR THE REGULARIZATION PARAMETER (α-EVIDENCE)

$$p(\alpha|\boldsymbol{d},\mathcal{I}) = (2\pi)^{-\frac{N_d}{2}}\,|\mathbf{C}|^{\frac{1}{2}}\,\alpha^{\frac{N-2}{2}}\,|\mathbf{A} + \alpha\mathbb{1}|^{-\frac{1}{2}}\,\exp\left\{\Phi(\boldsymbol{\rho}^*)\right\}. \tag{12.21}$$

Next we need to maximize the α-evidence to determine α^*. To this end it is better to differentiate the logarithm of the α-evidence:

$$Y := \ln\left[p(\alpha|\boldsymbol{d},\mathcal{I})\right] = \text{const.} - \frac{1}{2}\,\text{tr}\left\{\ln(|\mathbf{A}+\alpha\mathbb{1}|)\right\} + \frac{N-2}{2}\ln(\alpha) + \Phi(\boldsymbol{\rho}^*),$$

$$\frac{dY}{d\alpha} = -\frac{1}{2}\,\text{tr}\left\{\left(\mathbf{A}+\alpha\mathbf{1}\right)^{-1}\right\} + \frac{N-2}{2\alpha} + S(\boldsymbol{\rho}^*) + \left(\frac{d}{d\alpha}\boldsymbol{\rho}^*\right)^T \underbrace{\nabla_\rho \Phi(\boldsymbol{\rho})\Big|_{\boldsymbol{\rho}=\boldsymbol{\rho}^*}}_{=0}$$

$$= -\frac{1}{\alpha} - \frac{1}{2\alpha}\,\text{tr}\left\{\frac{\mathbf{A}}{\mathbf{A}+\alpha\mathbf{1}}\right\} + S(\boldsymbol{\rho}^*) \stackrel{!}{=} 0.$$

Then we find the condition

BAYESIAN CHOICE FOR THE REGULARIZATION PARAMETER

$$2 + 2\alpha\left|S(\boldsymbol{\rho}^*)\right| = \text{tr}\left\{\frac{\mathbf{A}}{\mathbf{A}+\alpha\mathbb{1}}\right\}. \tag{12.22}$$

The choice of regularization parameter has been discussed in great detail in [194, 203]. As outlined in [69] it is, however, not always possible to replace the α-integral in equation (12.12) by a single value. There is actually a lot more to be learnt from the Bayesian analysis [194]. According to Skilling and Gull, the right-hand side of equation (12.22) is what they call the *number of good data*

$$N_d^g := \text{tr}\left\{\frac{\mathbf{A}}{\mathbf{A}+\alpha\mathbb{1}}\right\}, \tag{12.23}$$

which in turn fixes the misfit at the Maxent solution by [194]

$$\chi^2(\boldsymbol{\rho}^*) = N_d - N_d^g. \tag{12.24}$$

We have shown before that $N_d^g \leq N_d$, so the right-hand side is always positive.

12.5.1 Qualitative discussion

Condition (12.24) is different from an erroneous approach frequently encountered, in which the regularization parameter is adjusted such that

$$\left\langle \chi^2 \right\rangle = N_d. \tag{12.25}$$

This choice is based on the fact that the misfit χ^2 is chi-squared distributed, as we will see in Section 19.2.2 [p. 295]. Then, for given image $\boldsymbol{\rho}$ and Gaussian noise, the mean misfit indeed obeys equation (12.25). But the image $\boldsymbol{\rho}$ is not known, it rather has to be inferred from the data. Consequently, the mean chi-squared value is the number of degrees of freedom. For a parametric model, in which the parameters stand for linearly independent directions in data space, the number of degrees of freedom is the number of data reduced by the number of model parameters inferred from the data (see Section 19.6 [p. 305]). In the case of form-free reconstruction of the image and its linear mapping into the data space, given by equation (12.30) later, the situation is rather subtle, because the number N of entries in $\boldsymbol{\rho}$ is not really the number of data that had to be sacrificed to determine the image. The effective number of data lost due to the determination of $\boldsymbol{\rho}^*$ is provided by Bayesian probability theory self-consistently in the form of equation (12.23). It make a lot of sense, because $\alpha|S|$ is a dimensionless measure for the distance of the reconstructed image from the default model. So it is indeed related to the number of independent parameters determined by the data or, equivalently, the number of *good data* as termed by Skilling and Gull. *Good data* pull the image away from the default model and are used to fix $\boldsymbol{\rho}^*$, while poor data are useless for the determination of the image and contribute therefore to the fluctuations entering the misfit.

In order to reveal the meaning of equation (12.23) more clearly, we consider a Gaussian apparatus function (matrix)

$$M_{ij} := \frac{1}{Z} e^{-\frac{(i-j)^2}{2\tau^2}},$$

where Z serves the normalization $\sum_i M_{ij} = 1$. Here, $N = N_d$. We analyse the behaviour in two limiting cases.

12.5.2 $\tau \ll 1$

In this case $M_{ij} = \delta_{ij}$. Each measurement determines an independent part of the image. Then, depending on the signal-to-noise ratio, ρ_i will tend more towards d_i or m_i. The latter image points will not contribute to the entropy. The number of good data is now

$$N_d^g = \sum_{l=1}^{N} \frac{\rho_i}{\rho_i + \alpha\sigma_i^2}$$

and it is roughly given by the number of data points with $\rho_i/\sigma_i > 1$.

12.5.3 $\tau \gg 1$

In this limit, the apparatus function has no focus and each measurement yields

$$d_l = \overline{\rho},$$

i.e. the average image values. Hence $N_d^g = 1$, because one parameter ($\overline{\rho}$) has been determined by the data.

Finally we consider a different representation of equation (12.23), namely

$$N_d^g = \sum_{l=i}^{N} \frac{\lambda_i}{\lambda_i + \alpha},$$

where λ_i are the eigenvalues of **A**. As rank(**A**) $\leq N_d$, the matrix has N − rank(**A**) zero eigenvalues. Moreover, the matrix **A** is positive, i.e. $\lambda_i \geq 0$. So the number of nonzero contributions to the sum is rank(**A**) and each of these terms is $\lambda_i/(\lambda_i + \alpha) \leq 1$. Consequently, $N_d^g \leq N_d$. More precisely, there is a close relationship between the eigenvalues of **A** and the singular values of the apparatus matrix **M**. If **M** corresponds to an ill-posed inversion problem, only very few singular values are significantly different from zero. Suppose this number is N_{nz}, then $N_d^g \approx N_{nz}$.

In summary, one can qualitatively say that good data points are those that have a good signal-to-noise ratio and fix individual parts of the image. Consequently, they do not contribute to the misfit. Poor data points, in contrast, have either a poor signal-to-noise ratio or they stem from a broad apparatus function, which means that several data points measure the same portion of the image. Collectively, these data points only fix a few (n) image points. Then the number of degrees of freedom for the chi-squared statistic is reduced by n.

12.6 A technical trick

Before discussing an example, we will outline one important technical issue. In general, direct maximization of the log-posterior

$$\Phi(\boldsymbol{\rho}, \boldsymbol{d}) = -\frac{1}{2}\chi^2(\boldsymbol{\rho}, \boldsymbol{d}) + \alpha S(\boldsymbol{\rho}) \tag{12.26}$$

poses a numerical challenge, in particular if the data constraints pull some of the reconstructed image values ρ_i^* towards zero. We will show a way to overcome this obstacle. Firstly, we note that $\Phi(\boldsymbol{\rho}, \boldsymbol{d})$ is convex in $\boldsymbol{\rho}$ since the Hessian, given by equation (12.19), is strictly negative. Therefore, there exists a unique maximum position $\boldsymbol{\rho}^*$. It corresponds to data values $\boldsymbol{d}^* = \boldsymbol{d}^{\text{th}}(\boldsymbol{\rho}^*)$. The maximization of $\Phi(\boldsymbol{\rho}, \boldsymbol{d})$ is equivalent to the maximization of the entropy subject to the constraint $\boldsymbol{d}^{\text{th}}(\boldsymbol{\rho}) = \boldsymbol{d}^*$, with yet unknown $\boldsymbol{d}^*$. For the linear model the Maxent solution then reads

$$\rho_i(\boldsymbol{\lambda}) = m_i \exp\left\{\left(\boldsymbol{\lambda}^T \boldsymbol{M}\right)_i\right\}.$$

Now we know already the structure of the result and in particular it is manifestly positive. We merely have to choose $\boldsymbol{\lambda}$ such that it is maximized. To this end we insert the Maxent solution into $\Phi(\boldsymbol{\rho}(\boldsymbol{\lambda}), \boldsymbol{d})$ and differentiate with respect to $\boldsymbol{\lambda}$. Then we find the condition

$$\boldsymbol{\rho}^T(\boldsymbol{\lambda})\mathbf{M}\mathbf{M}^T\mathbf{C}^{-1}\left(\boldsymbol{d} - \mathbf{M}\boldsymbol{\rho}(\boldsymbol{\lambda}) + \alpha\mathbf{C}\boldsymbol{\lambda}\right) \stackrel{!}{=} 0.$$

Since the solution is unique, the bracket has to be zero, resulting in

$$\boldsymbol{\psi}(\boldsymbol{\lambda}) := \boldsymbol{d} - \mathbf{M}\boldsymbol{\rho}(\boldsymbol{\lambda}) + \alpha\mathbf{C}\boldsymbol{\lambda} \stackrel{!}{=} 0. \tag{12.27}$$

This procedure is actually equivalent to a Legendre transform invoked in [194]. The two main advantages of this approach are firstly, there is no constraint any more on the new variables $\boldsymbol{\lambda}$ and secondly, the number of parameters λ_i to be determined is merely N_d instead of originally N, which generally is much greater than N_d. Equation (12.27) can be solved by Newton–Raphson, for which the gradient can even be determined analytically:

$$\frac{\partial \psi_\nu(\boldsymbol{\lambda})}{\partial \lambda_\mu} = -\sum_{i=1}^{N} M_{\nu i}\, M_{\mu i}\, \rho_i(\boldsymbol{\lambda}) - \alpha\, C_{\nu\mu}\,. \tag{12.28}$$

12.7 Application to ill-posed inversion problems

Now we will give an illustrative example for QME. In many applications, the measured data and the sought-for images are related by a linear map. For concreteness we consider a problem occurring in most QMC simulations [207, 212]. In a nutshell, QMC is a simulation technique where the underlying D-dimensional quantum problem is first mapped onto an effective (D+1)-dimensional classical problem, which in turn can be solved by Monte Carlo techniques (see Chapter 30 [p. 537]). The QMC simulation provides the Green's function $g(\tau)$ for imaginary times τ, while the physical quantity of interest is the Green's function $\rho(\omega)$ on the real frequency axis. The relation between the two is a Laplace transform

$$g(\tau) = \int d\omega\, e^{-\tau\omega}\, \rho(\omega). \tag{12.29}$$

In this context, the Laplace transform represents an 'analytic continuation' of imaginary-time data to real frequency response functions. As a matter of fact, the QMC simulation only yields data for a finite set of imaginary times τ_i ($i = 1, \ldots, N_d$) and the reconstruction will therefore also be restricted to discrete frequencies ω_j. Along with the midpoint quadrature formula we obtain

$$g_i = \sum_{j=1}^{N} M_{ij} \rho_j,$$
$$M_{ij} := e^{-\tau_i \omega_j},$$

with the definitions $g_i := g(\tau_i)$ and $\rho_j := \Delta\omega\, \rho(\omega_j)$. In vector notation we have

$$\boldsymbol{g} = \mathbf{M}\, \boldsymbol{\rho}. \tag{12.30}$$

Now the QMC data $\boldsymbol{g}$ suffer from (additive) noise and we actually have the relation

$$\boldsymbol{g} = \mathbf{M}\, \boldsymbol{\rho} + \boldsymbol{\eta}. \tag{12.31}$$

Here $\boldsymbol{\eta}$ stands for the noise vector, which is of course not known. We know, however, that the noise is Gaussian with zero mean and variance σ^2.

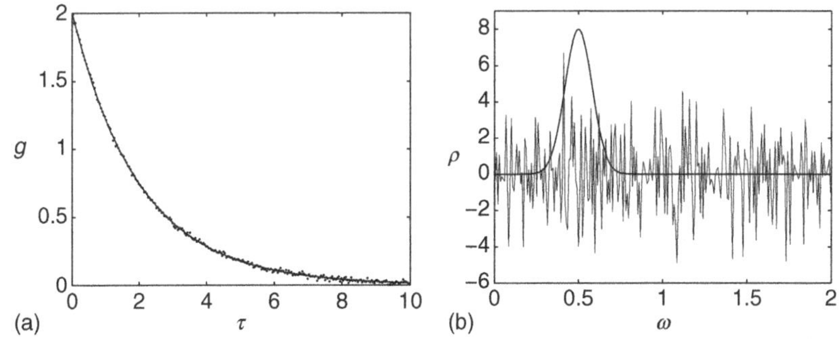

Figure 12.2 (a) The solid line represents the exact data $\boldsymbol{g}$, based on the exact image $\boldsymbol{\rho}$, which is depicted as a (smooth) solid line in (b). Brute-force inversion of the noisy data (dots in (a)) yields the jagged curve in (b). The reconstructed $\boldsymbol{\rho}$ has been multiplied by 10^{-16}! The solid line in (a) is obtained by inserting the inferred image back into equation (12.30).

12.7.1 Brute-force inversion

The brute-force approach would be to ignore the noise, as it is small, choose $N = N_d$ and invert the linear problem. Then

$$\boldsymbol{\rho} = \mathbf{M}^{-1}\boldsymbol{g}.$$

In Figure 12.2 realistic QMC data are presented for a case where the exact result is known. The noisy QMC data $\boldsymbol{g}$ are subjected to brute-force inversion. The result is the jagged curve in the right panel. The structure of the curve has no resemblance to the exact Green's function $\boldsymbol{\rho}$ and – what is worse – it is 16 orders of magnitude too large. The direct approach is completely useless, implying that we are facing an 'ill-posed inversion problem', where the noise is drastically amplified. Nevertheless, in view of the reconstructed $\boldsymbol{g}(\boldsymbol{\rho})$, which is depicted in the left panel, everything seems perfect. This is the deceptive point of ill-posed inversion problems. The origin of the noise amplification is revealed by the spectral representation of $\mathbf{M}$:

$$\mathbf{M} = \sum_{n=1}^{N_d} \varepsilon_n \boldsymbol{\psi}_n \boldsymbol{\psi}_n^T, \tag{12.32}$$

in terms of the eigenvalues ε_n and eigenvectors $\boldsymbol{\psi}_n$. Then the direct inversion of equation (12.31) yields

$$\tilde{\boldsymbol{\rho}} = \mathbf{M}^{-1}(\boldsymbol{g} - \boldsymbol{\eta}) = \underbrace{\sum_n \varepsilon_n^{-1} \boldsymbol{\psi}_n (\boldsymbol{\psi}_n^T \boldsymbol{g})}_{\boldsymbol{\rho}} - \sum_n \varepsilon_n^{-1} \boldsymbol{\psi}_n (\boldsymbol{\psi}_n^T \boldsymbol{\eta})$$

with $\boldsymbol{\rho}$ being the underlying exact solution. The last term covers the impact of the noise. Now the ill-posed nature of the problem comes into play. In Figure 12.3 the log-eigenvalues of $\mathbf{M}$ are depicted in decreasing order. Most of the eigenvalues are below machine accuracy. These eigenvalues explain the drastic amplification of the noise if the corresponding

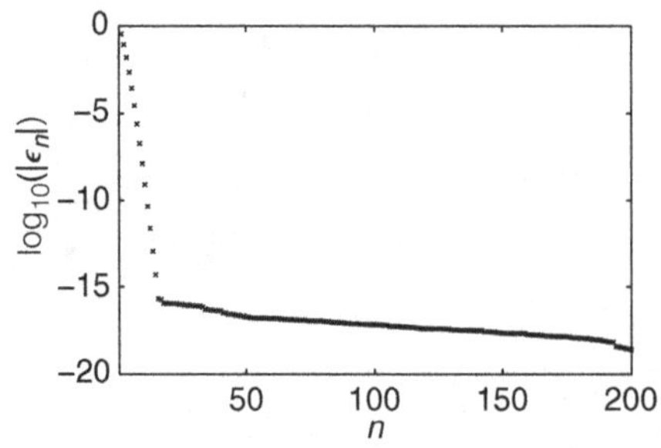

Figure 12.3 Eigenvalues of the kernel matrix **M** for $N_d = 200$ scaled by the largest eigenvalue.

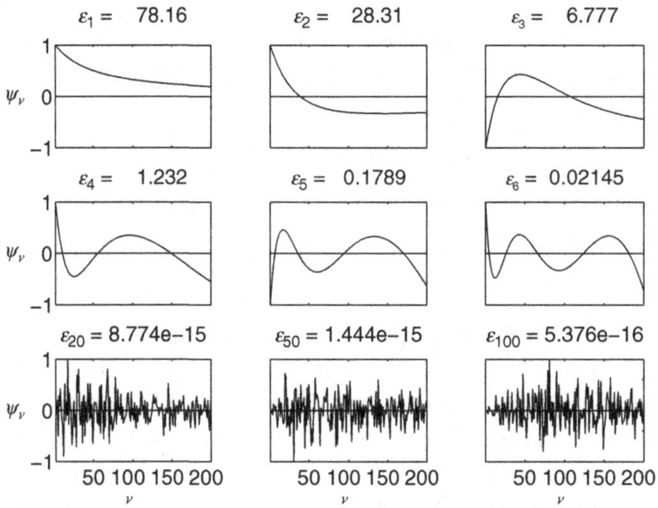

Figure 12.4 Selected eigenvectors of the same kernel matrix as in Figure 12.3.

eigenvectors have a non-negligible overlap with the noise vector. To this end, a few selected eigenvectors are depicted in Figure 12.4. The eigenvectors corresponding to small eigenvalues are characterized by pronounced point-to-point fluctuations that resemble uncorrelated noise, resulting in sizable overlap with the noise vector $\boldsymbol{\eta}$. In more descriptive terms, the exponential decrease of the integral kernel implies that structures in $\rho(\omega)$ at large frequencies have little impact on $g(\tau_i)$. Conversely, based on noisy data $g(\tau_i)$, we can say little about the high-frequency part of $\rho(\omega)$.

12.7.2 The Bayesian choice

Now let us analyse how the Bayesian approach based on the entropic prior performs in such an ill-posed problem. We use the approach outlined before to fix the regularization parameter and determine the MAP solution. For the same Laplacian kernel (that is, for the

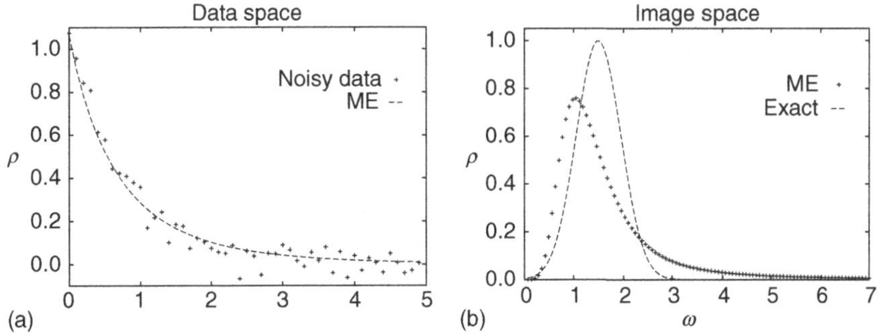

Figure 12.5 QME reconstruction. The crosses in (a) represent the noisy data $\boldsymbol{g}$. The MAP image $\boldsymbol{\rho}$ obtained with QME is marked by + in (b). The dashed line in (a) stands for the data vector calculated with $\boldsymbol{\rho}$ obtained from QME.

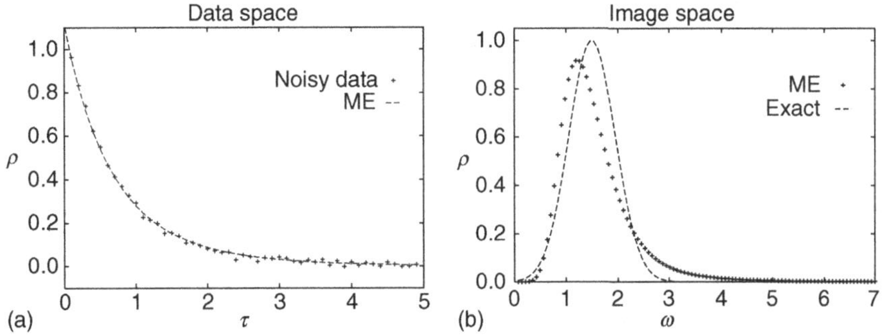

Figure 12.6 Same plots as in the previous figure, only the noise level has been reduced, as apparent from the reduced scatter of data.

same ill-posed inversion problem) we generate mock data $\boldsymbol{g}$ based on an artificial image $\boldsymbol{\rho}$. Then we add Gaussian noise to the data. The noisy data are depicted in Figure 12.5(a). The MAP solution of the QME approach is shown in Figure 12.5(b) along with the underlying true image. It is like a miracle, QME is able to conquer 16 orders of magnitude to achieve an almost precision landing. The QME reconstruction on the image space is depicted by the smooth curve in Figure 12.5(a). It fits nicely into the noisy data. But, as outlined before, that is no quality criterion at all.

In Figure 12.6 we repeat the analysis with a smaller variance in the Gaussian noise. We see clearly that QME converges towards the true image as the noise tends to zero.

Obviously, QME works pretty well for these types of problem. Since Richard Silver introduced QME to the quantum Monte Carlo community in 1990 [184] it has become the standard approach for the 'analytic continuation' of imaginary-time data. Some representative QME applications to real-world QMC data for studies in the field of high-temperature superconductivity can be found in [165, 166]. QME has been applied by us successfully to a wide variety of other problems as well [6, 64, 207–209]. But, there are situations where QME yields results which one finds unsatisfactory. Results that show sharp structures that cannot be. However, this is not a failure of probability theory in general or QME in particular. Unsatisfactory features in the results only mean that some background information has been withheld. In most cases it is prior knowledge that the sought-for image has to be smooth to some degree. In the derivation of the entropic prior, correlation is not an issue. Consequently, the entropy is invariant against the permutations of the indices. The proper way to include local smoothness is discussed in greater detail in the next chapter and, along with various real-world applications, in Part V [p. 333].

13

Global smoothness

We saw in Section 11.2.1 [p. 184] that the principle of maximum entropy is an important tool for formulating prior probabilities given testable information. But we found also that the entropy does not care about smoothness. So we have observed, for example, pronounced 'ringing' at sharp edges. The reason is obvious, the background information did not contain a smoothness requirement. If we are unhappy with the Maxent solution, because it has more structure then expected, we should have included our smoothness expectations in the background information.

More generally, the smoothness Φ of an image $f(x)$ with $x \in \mathbb{R}$ may be quantified by one of the following functionals:

$$\Phi(\boldsymbol{f}) = \int \left[f'(x)\right]^2 dx, \tag{13.1a}$$

$$\Phi(\boldsymbol{f}) = \int \left[f''(x)\right]^2 dx, \tag{13.1b}$$

$$\Phi(\boldsymbol{f}) = \int \left[f'(x)\right]^2 / f(x) dx, \tag{13.1c}$$

or their discrete versions with the integrals replaced by finite sums. Given Φ, the principle of maximum entropy in equation (11.4) [p. 184] assigns to $\boldsymbol{f}$ the distribution

$$p(\boldsymbol{f}) = \exp\left\{-\lambda \Phi(\boldsymbol{f})\right\} / Z(\lambda), \tag{13.2}$$

where $Z(\lambda)$ is the normalization. The functional in equation (13.1b) was employed some time ago for the successful inversion of an autoconvolution integral [50]. All three functionals in equation (13.1) exhibit correlations, meaning that the reordering of the components of $\boldsymbol{f}$ in the respective sums changes the value of Φ. This is in contrast to the Shannon entropy prior in equation (11.1) [p. 182], which is evidently invariant against a reordering of the components of $\boldsymbol{f}$. There are of course problems where this is an advantage. In the recovery of functions from experimental data or of images from blurred data it is, however, a disadvantage. Gull and Skilling [93, 190] have proposed to cure the missing smoothness by representing the image $\boldsymbol{f}$ as a 'pre-blurred' hidden image $\boldsymbol{h}$ through

$$\boldsymbol{f} = \mathbf{B}\,\boldsymbol{h},$$

where **B** is a blur-matrix that could, for example, be the discretized version of a Gaussian convolution. We shall not pursue this approach further, since equation (13.1c) is a functional which ensures correlation and positivity in a natural way.

13.1 A primer on cubic splines

Another way of including smoothness, which is advantageous in certain applications, is by using spline basis functions. Examples will be discussed in Part V [p. 333].

There are various ways of representing a function $f(x)$ on a lattice with pivot points x_j. The simplest approximation is to assume that $f(x)$ is constant between lattice points. An improvement is given by a polygon representation, for example, a sequence of straight-line segments which join continuously the points (x_j, f_j) with $f_j := f(x_j)$. The situation is depicted schematically in Figure 13.1. For a large class of problems one wants not only a continuous model function but one which is once or several times continuously differentiable. Suppose we were given, in addition to the support functional values, also the second derivatives $\{m_j\}$ at positions $\{x_j\}$ and choose to represent $f''(x)$ by a polygon. The first derivative would then be once and the function itself twice continuously differentiable and therefore in colloquial terms quite smooth. The second derivative on the interval $x_j \leq x \leq x_{j+1}$ is given by

$$S_j''(x) = \frac{x_{j+1} - x}{h_{j+1}}\, m_j + \frac{x - x_j}{h_{j+1}}\, m_{j+1}, \tag{13.3}$$

with $h_{j+1} = x_{j+1} - x_j$, and we see immediately that the function must be a third-order polynomial in this interval. The remarkable property of the function called a 'cubic spline' is that its global curvature over the interval $[x_1, x_N]$ is smaller than the curvature of any other twice continuously differentiable function $\varphi(x)$ which passes through the points $\{x_j, f_j\}$ if we require S to satisfy either the boundary condition $S''(x_1) = S''(x_N) = 0$ or prescribe particular values to the first derivative at the endpoints $S'(x_1) = f_1'$ and $S'(x_N) = f_N'$. Integrating equation (13.3) twice yields

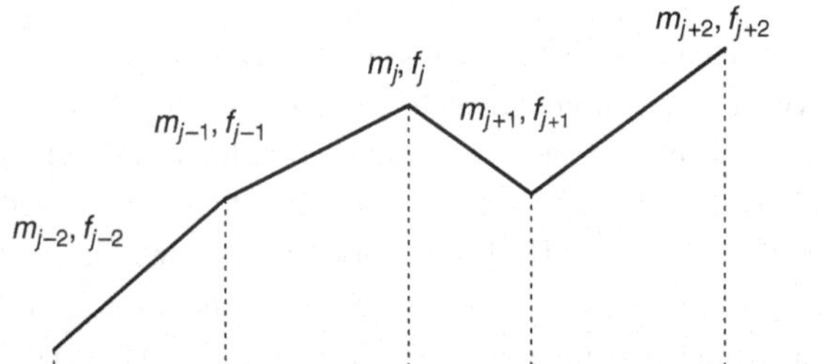

Figure 13.1 The construction of a cubic spline starts with a polygonal representation of its second derivative $m(x)$.

$$S_j(x) = \frac{(x_{j+1} - x)^3}{6h_{j+1}} m_j + \frac{(x - x_j)^3}{6h_{j+1}} m_{j+1} + P_j(x - x_j) + Q_j. \tag{13.4}$$

The integration constants P_j and Q_j result immediately from the requirement that S_j should pass through (x_j, f_j) and (x_{j+1}, f_{j+1}):

$$\begin{aligned} f_j &= \tfrac{1}{6}\, h_{j+1}^2 m_j + Q_j, \\ f_{j+1} &= \tfrac{1}{6}\, h_{j+1}^2 m_{j+1} + P_j h_{j+1} + Q_j, \end{aligned} \tag{13.5}$$

which is readily inverted to yield

$$\begin{aligned} Q_j &= f_j - \tfrac{1}{6}\, h_{j+1}^2 m_j, \\ P_j &= \frac{1}{h_{j+1}}(f_{j+1} - f_j) - \tfrac{1}{6}\, h_{j+1}(m_{j+1} - m_j). \end{aligned} \tag{13.6}$$

The entire function in terms of $\{f_j\}$ and $\{m_j\}$ is then given by

$$\begin{aligned} S_j(x) = {} & \frac{m_{j+1}\, h_{j+1}^2}{6} \left\{ \left(\frac{x - x_j}{h_{j+1}}\right)^3 - \left(\frac{x - x_j}{h_{j+1}}\right) \right\} \\ & + \frac{m_j\, h_{j+1}^2}{6} \left\{ \left(\frac{x_{j+1} - x}{h_{j+1}}\right)^3 + \left(\frac{x - x_j}{h_{j+1}}\right) - 1 \right\} \\ & + f_{j+1} \left(\frac{x - x_j}{h_{j+1}}\right) + f_j \left(\frac{x_{j+1} - x}{h_{j+1}}\right). \end{aligned} \tag{13.7}$$

We have intentionally formulated equation (13.7) such that we recognize clearly the polygonal part in the last line and the correction terms which ensure smoothness in the first two lines. Programming can then be arranged such that a simple switch allows us to choose either the cubical spline option or the polygonal as a model function for nonparametric estimation.

So far, we have assumed that both functional values $\{f_j\}$ and second derivatives $\{m_i\}$ are at our disposal. This is much too strong an assumption since we have so far not used the conditions of continuity of the first derivative of the functions $S_j(x)$, namely

$$S'_{j-1}(x)\Big|_{x_j} = S'_j(x)\Big|_{x_j}, \qquad \text{for } j = 2, \ldots, N-1.$$

This gives us $N - 2$ equations. The missing two equations are obtained from the boundary conditions, and the full system of equations relating the second derivatives $\boldsymbol{m}$ to the functional values $\boldsymbol{f}$ becomes

$$
\begin{aligned}
\tfrac{h_2}{3}\, m_1 + \sigma \tfrac{h_2}{6}\, m_2 &= \sigma \left(\tfrac{f_2 - f_1}{h_2} - f_1' \right), \\
\tfrac{h_j}{6}\, m_{j-1} + \tfrac{h_j + h_{j+1}}{3}\, m_j + \tfrac{h_{j+1}}{6}\, m_{j+1} &= \tfrac{f_{j+1} - f_j}{h_{j+1}} - \tfrac{f_j - f_{j-1}}{h_j}, \\
\sigma\, \tfrac{h_N}{6} m_{N-1} + \tfrac{h_N}{3} m_N &= \sigma \left(f_N' - \tfrac{f_N - f_{N-1}}{h_N} \right).
\end{aligned}
\tag{13.8}
$$

The second equation holds for $j = 2, \ldots, N-1$. A switch variable σ has been introduced. For $\sigma = 0$ the solution of this system of equations corresponds to the boundary conditions $S''(x_1) = S''(x_N) = 0$, referred to as 'natural splines'. For $\sigma = 1$ the solution represents the case of given first derivatives at the endpoints x_1, x_N. Denote the matrix of coefficients on the left-hand side of equation (13.8) by $\mathbf{M}$ and that on the right-hand side by $\mathbf{F}$, then

$$
\mathbf{M}\, \boldsymbol{m} = \mathbf{F}_\sigma \boldsymbol{f} + \sigma \boldsymbol{g}
$$

with $\boldsymbol{g}^T := \left(-f_1', 0, \ldots, 0, f_N'\right)$. The matrix $\mathbf{F}_\sigma$ reads

$$
\mathbf{F}_\sigma = \begin{pmatrix}
-\frac{\sigma}{h_2} & \frac{\sigma}{h_2} & & & & \\
\frac{1}{h_2} & -\frac{h_2 + h_3}{h_2 h_3} & \frac{1}{h_3} & & & \\
& & \ddots & \ddots & \ddots & \\
& & & \frac{1}{h_{N-1}} & -\frac{h_{N-1} + h_N}{h_{N-1} h_N} & \frac{1}{h_N} \\
& & & & \frac{\sigma}{h_N} & -\frac{\sigma}{h_N}
\end{pmatrix}.
\tag{13.9}
$$

This matrix has zero eigenvalues. For $\sigma = 1$ there is one zero eigenvalue corresponding to $f(x) = c$, or rather $f_j = c$. For $\sigma = 0$ there are two zero eigenvalues, one as before for constant $f(x)$ and the other for $f(x) = \alpha x$, or rather $f_j = \alpha x_j = \alpha(x_1 + \sum_{n=2}^{j} h_j)$. The matrix $\mathbf{M}$ has the form

$$
\mathbf{M} = \frac{1}{6} \begin{pmatrix}
2h_2 & \sigma h_2 & & & & \\
h_2 & 2(h_2 + h_3) & h_3 & & & \\
& h_3 & 2(h_3 + h_4) & h_4 & & \\
& & \ddots & \ddots & \ddots & \\
& & & h_{N-2} & 2(h_{N-1} + h_N) & h_N \\
& & & & \sigma h_{N-1} & 2h_N
\end{pmatrix}.
\tag{13.10}
$$

From Gershgorin's circle theorem we see that the eigenvalues are inside the union of the intervals $(h_2, 3h_2)$, $(2h_N - h_{N-1}, 2h_N + h_{N-1})$ and $((h_j + h_{j+1}), 3(h_j + h_{j+1}))$ for $j = 2, \ldots, N-1$. Consequently, all eigenvalues are greater than zero, i.e. the inverse of $\mathbf{M}$ exists. Then we find

$$\boldsymbol{m} = \mathbf{L}\boldsymbol{f} + \sigma\, \boldsymbol{b}, \tag{13.11a}$$

$$\mathbf{L} := \mathbf{M}^{-1}\mathbf{F}, \tag{13.11b}$$

$$\boldsymbol{b} := \mathbf{M}^{-1}\boldsymbol{g}. \tag{13.11c}$$

Note that equation (13.11) simplifies if we choose natural splines ($\sigma = 0$) or $f_1 = f'_N = 0$. This completes the construction of the spline function. The second derivative may now be removed from equation (13.7) if we want a representation of S_j by $\boldsymbol{f}$ only.

13.2 Second derivative prior

A convenient measure of smoothness is the integral over the second derivative of a function squared between the limits of interest. If we invoke this measure as testable information, then the Maxent principle yields the prior

$$p(\boldsymbol{f}|f'', \mu, \mathcal{I}) = \frac{1}{Z(\mu)} \exp\left\{ -\mu \int_{x_1}^{x_N} [f'']^2\, dx \right\}, \tag{13.12}$$

with a 'regularization parameter' μ which has to be treated self-consistently or by marginalization. Let us see what the MAP solution then looks like. We assume a Gaussian experimental noise. The logarithm of the posterior probability, given the data $\boldsymbol{d}$ and variance vector $\boldsymbol{\sigma}$ of the individual measurements, is then

$$\ln[p(f(x)|\boldsymbol{d}, \boldsymbol{\sigma}, \mathcal{I})] = -\mu \int \left[f''(x)\right]^2 dx - \sum_i \frac{1}{2\sigma_i^2}\left[d_i - f(x_i)\right]^2. \tag{13.13}$$

The MAP solution corresponds to minimization of the global curvature of $f(x)$, subject to the data constraints. The right-hand side can be combined in one integral

$$\ln[p(f(x)|\boldsymbol{d}, \boldsymbol{\sigma}, \mathcal{I})] = -\mu \int \mathcal{L}(f(x), f''(x), x)\, dx, \tag{13.14}$$

with

$$\mathcal{L}(f(x), f''(x), x) := \left[f''(x)\right]^2 + \sum_i \omega_i\, \delta(x - x_i)\left[d_i - f(x)\right]^2, \tag{13.15}$$

where the weights ω_i are given by $\omega_i = 1/(2\mu\sigma_i^2)$. Now we can apply the variational calculus in order to find the MAP solution. The corresponding Euler–Lagrange equation reads

$$\frac{\delta\mathcal{L}}{\delta f} - \frac{d}{dx}\frac{\delta\mathcal{L}}{\delta f'} + \frac{d^2}{dx^2}\frac{\delta\mathcal{L}}{\delta f''} = 0, \tag{13.16a}$$

$$2\, f^{(4)}(x) + \sum_i \omega_i\, \delta(x - x_i)\left[f(x_i) - d_i\right] = 0. \tag{13.16b}$$

For x away from the pivot points we obtain the simple differential equation

$$f^{(4)}(x) = 0, \tag{13.17}$$

resulting in a cubic polynomial between the pivot points x_i. Integration of equation (13.16) over an infinitesimal interval $I_{i,\varepsilon} := (x_i - \varepsilon, x_i + \varepsilon)$ about x_i yields

$$f^{(3)}(x_i + \varepsilon) = f^{(3)}(x_i + \varepsilon) + \omega_i \, (d_i - f(x_i)). \tag{13.18}$$

Consequently, the third derivative of $f(x)$ is discontinuous at the pivot points. Repeating the integration over $I_{i,\varepsilon}$ reveals that $f(x)$, $f'(x)$ and $f''(x)$ are continuous. So, the MAP solution for $f(x)$ is cubic splines.

Then we can exploit the spline properties to represent equation (13.12) in terms of the support functional values $\boldsymbol{f}$ only as

$$p(\boldsymbol{f}|\mu, \mathcal{I}) = \frac{1}{Z(\mu)} \exp\left\{ -\frac{\mu}{2} \boldsymbol{f}^T \mathbf{P}_{f''} \boldsymbol{f} \right\}. \tag{13.19}$$

To this end we use equation (13.3) and obtain

$$\left[S_j''(x)\right]^2 = \frac{m_j^2\,(x_{j+1}-x)^2 + 2m_{j+1}m_j(x_{j+1}-x)(x-x_j) + m_{j+1}^2(x-x_j)^2}{h_{j+1}^2}.$$

Straightforward integration between x_j and x_{j+1} yields

$$\int_{x_j}^{x_{j+1}} \left[S_j''(x)\right]^2 dx = \frac{h_{j+1}}{3}\left(m_j^2 + m_j m_{j+1} + m_{j+1}^2\right). \tag{13.20}$$

The entire integral between x_1 and x_N becomes correspondingly

$$\int_{x_1}^{x_N} |S''(x)|^2 dx = \frac{1}{3}\sum_{j=1}^{N-1} h_{j+1}\left(m_j^2 + m_j m_{j+1} + m_{j+1}^2\right). \tag{13.21}$$

The sum on the right-hand side can now be expressed in vector notation as

$$\int_{x_1}^{x_N} \left[S''(x)\right]^2 dx = \boldsymbol{m}^T \mathbf{M}\, \boldsymbol{m}. \tag{13.22}$$

The matrix $\mathbf{M}$ is defined in equation (13.10). Employing finally the explicit equation (13.11) between $\boldsymbol{m}$ and $\boldsymbol{f}$ we find the desired expression of equation (13.19), with

$$\mathbf{P}_{f''} = \mathbf{L}^T\mathbf{M}\mathbf{L} = \mathbf{F}^T\mathbf{M}^{-1}\mathbf{F}.$$

As discussed before, $\mathbf{M}$ is positive definite and $\mathbf{F}$ has one or two zero eigenvalues. For the natural boundary conditions $m_1 = m_N = 0$, matrix $\mathbf{P}_{f''}$ therefore has two eigenvalues equal to zero. Zero eigenvalues of $\mathbf{P}$ imply that equation (13.19) is not normalizable. A simple cure is to replace $\mathbf{P}$ by $\hat{\mathbf{P}} = \mathbf{P} + \varepsilon\mathbb{1}$ where $\mathbb{1}$ is the unit matrix, and at the end of all calculations we consider the limit $\varepsilon \to 0$. The normalization integral is then of the standard multivariate Gaussian type and results in the normalized prior distribution

$$p(\boldsymbol{f}|\mu, \mathcal{I}) = \left(\frac{\mu}{2\pi}\right)^{\frac{N}{2}} |\varepsilon^2 \det{}'(\mathbf{P}_{f''})|^{\frac{1}{2}} \exp\left\{ -\frac{\mu}{2} \boldsymbol{f}^T \hat{\mathbf{P}}_{f''} \boldsymbol{f} \right\}, \tag{13.23}$$

where $\det{}'(\mathbf{P}_{f''})$ is the product over the nonzero eigenvalues of matrix $\mathbf{P}_{f''}$. Instead of choosing natural boundary conditions for the spline function we could as well supply values for the first derivative at the endpoints, in particular $f_1' = f_N' = 0$. This constitutes additional information and reduces the number of zero eigenvalues to one. We anticipate here already that we prefer this case, because the same degree of singularity will show up in the matrices of alternative smoothness priors.

13.3 First derivative prior

The Euler–Lagrange differential equation corresponding to the functional in equation (13.1a) [p. 215] is, in analogy to equations (13.16) and (13.17),

$$\frac{d}{dx} f' = 0 \tag{13.24}$$

for the regions between the pivot points. So the solutions are piecewise linear continuous functions, with discontinuities in the first derivative at the pivot points. We parametrize, in analogy to equation (13.3), in terms of functional values f_k and f_{k+1} in the interval $[x_k, x_{k+1}]$ to obtain

$$F_k(x) = f_k \frac{x_{k+1} - x}{x_{k+1} - x_k} + f_{k+1} \frac{x - x_k}{x_{k+1} - x_k}. \tag{13.25}$$

The functional in equation (13.1a) [p. 215] then becomes

$$\Phi = \sum_{k=1}^{N-1} \left(\frac{f_{k+1} - f_k}{h_{k+1}} \right)^2 h_{k+1} \stackrel{!}{=} \boldsymbol{f}^T \mathbf{P}_{f'} \boldsymbol{f}, \tag{13.26}$$

with $h_{k+1} = x_{k+1} - x_k$ as before. Interestingly, for the matrix $\mathbf{P}_{f'}$ we find $\mathbf{P}_{f'} = -2\mathbf{F}_{\sigma=1}$. So the matrix $\mathbf{P}_{f'}$ has one zero eigenvalue. We again replace $\mathbf{P}_{f'}$ by $\hat{\mathbf{P}}_{f'} = \mathbf{P}_{f'} + \varepsilon \mathbf{1}$, which is positive definite and results in a normalizable prior distribution

$$p(\boldsymbol{f} \mid \mu, \mathcal{I}) = \left(\frac{\mu}{2\pi}\right)^{N/2} |\varepsilon \det{}'\mathbf{P}_{f'}|^{1/2} \exp\left\{ -\frac{\mu}{2} \boldsymbol{f}^T \hat{\mathbf{P}}_{f'} \boldsymbol{f} \right\}. \tag{13.27}$$

13.4 Fisher information prior

A third smoothness prior is the so-called 'Fisher information' which is based on equation (13.3). This prior has been used extensively in [77]. If $f(x)$ were to have a root, then the Fisher information would diverge logarithmically. Hence, $f(x)$ is either strictly positive or negative. Since the sign can be absorbed in μ, we assume that $f(x) > 0$. This also shows that the Fisher information prior is only applicable to problems where the sought-for functions $f(x)$ have a unique sign. The Fisher information prior incorporates positivity, a property shared also by the Shannon entropy. But, contrary to the Shannon entropy, the Fisher information prior also offers smoothness. It is therefore ideally suited for the reconstruction of density functions from possibly noisy experimental or observational data. It is expedient to express $f(x)$ as $f(x) = \Psi(x)^2$. Then the log-posterior for Gaussian noise, along with the Fisher information prior in equation (13.1a) [p. 215], is given by

$$\ln[p(f(x)|\boldsymbol{d}, \boldsymbol{\sigma}, \mathcal{I})] = -\mu \int \left[\Psi'(x)\right]^2 dx - \sum_i \frac{1}{2\sigma_i^2} \left[d_i - \Psi^2(x_i)\right]^2. \tag{13.28}$$

The MAP solution corresponds to the minimal Fisher information, subject to the data constraints. We combine prior and data constraint in one integral, resulting in

$$\ln[p(f(x)|\boldsymbol{d}, \boldsymbol{\sigma}, \mathcal{I})] = -\int \mathcal{L}(\Psi(x), \Psi'(x), x)\, dx \tag{13.29}$$

with

$$\mathcal{L}(\Psi(x), \Psi'(x), x) := \mu \left[\Psi'(x)\right]^2 dx + \sum_i \omega_i\, \delta(x - x_i) \left[d_i - \Psi^2(x)\right]^2. \tag{13.30}$$

Here the weights $\omega_i = 1/(2\sigma_i^2)$ are defined slightly differently. The corresponding Euler–Lagrange equation now reads

$$\frac{\delta \mathcal{L}}{\delta \Psi} - \frac{d}{dx}\frac{\delta \mathcal{L}}{\delta \Psi'} = 0, \tag{13.31a}$$

$$-\mu\, \Psi''(x) + \left(\sum_i \omega_i\, \delta(x - x_i) \left[d_i - \Psi^2(x_i)\right]\right) \Psi(x) = 0. \tag{13.31b}$$

If $f(x)$ has the meaning of a PDF, then we also have the normalization constraint

$$\int \Psi^2(x) = 1,$$

which results in an additional term in the Euler–Lagrange equation:

$$-\mu\, \Psi''(x) + \left(\sum_i \omega_i\, \delta(x - x_i) \left[\Psi^2(x_i) - d_i\right]\right) \Psi(x) = E\, \Psi(x),$$

where E is another Lagrange parameter. The resemblance to quantum-mechanical potential problems is obvious. The regularization parameter μ has the meaning of an inverse mass. The potential consists of delta-peaks and it depends on the local particle density $\Psi^2(x_i)$. For $f(x_i) = \Psi^2(x_i) > d_i$ the potential is repulsive, which results in a reduction of the density at site x_i. The opposite is the case for $f(x_i) < d_i$. In summary, the data constraint tends to pull $f(x_i)$ towards the data value d_i. In contrast, the kinetic energy (first term in (13.31b), resulting from the Fisher information) favours flattening $\Psi(x)$. Further implications of the Fisher information on the foundation of quantum mechanics have been proposed in [77].

Without normalization constraints and between the pivot points x_k and x_{k+1}, equation (13.31b) results in

$$\Psi_k''(x) = 0 \qquad \Rightarrow \Psi_k(x) = a_k(x - x_k) + b_k$$

and it remains to express a_k and b_k in terms of $\{f_j\}$. We require that $\Psi_k^2(x_k) = f_k$ and $\Psi_k^2(x_{k+1}) = f_{k+1}$, yielding the equations

$$f_k = b_k^2, \tag{13.32}$$

$$f_{k+1} = (a_k h_{k+1} + b_k)^2 \tag{13.33}$$

and resulting in

$$b_k = \sqrt{f_k}, \qquad a_k = \left(\sqrt{f_{k+1}} - \sqrt{f_k}\right) / h_{k+1}. \tag{13.34}$$

Comparison with equation (13.26) shows that a_k is the finite difference approximation of the first derivative of $g(x) = \sqrt{f(x)}$, in agreement with equation (13.28). In $\Psi(x)$ the Fisher information prior corresponds to the first derivative prior in equation (13.1a) [p. 215]. As we have seen, there is, however, a price to be paid: The transformation $f(x) = \Psi^2(x)$ results in a non-Gaussian likelihood. Nonetheless, the transformation is advantageous since only few problems in the real world are described by linear functions $f(x)$. For nonlinear $f(x)$ we need to resort to the saddle-point approximation in the likelihood in any case, if the prior is Gaussian. Applications of this procedure will be provided in Chapters 25 [p. 431] and 26 [p. 451].

Common to all smoothness priors, as discussed in this chapter, is the quadrature error. This is, however, of minor importance since our prior knowledge about the smoothness in a particular problem will in general be very vague. This lack of knowledge will transform to corresponding uncertainty in the hyperparameter μ. In fact, we shall sometimes use Jeffreys' prior to marginalize over μ and this prior corresponds to infinite uncertainty (see Chapter 10 [p. 165]).

PART III

PARAMETER ESTIMATION

14

Bayesian parameter estimation

14.1 The estimation problem

Bayesian probability theory provides a straightforward approach to estimate a (vector-valued) parameter $\boldsymbol{a}$ from measured data $\boldsymbol{y}$. The probability density for $\boldsymbol{a}$, given the data $\boldsymbol{y}$, can be expressed using Bayes' theorem as

$$p(a|\boldsymbol{y},\mathcal{I}) = \frac{p(\boldsymbol{y}|a,\mathcal{I})\; p(a|\mathcal{I})}{p(\boldsymbol{y}|\mathcal{I})} \propto p(\boldsymbol{y}|a,\mathcal{I})\; p(a|\mathcal{I}).$$

In many cases one wishes to summarize the full probability distribution of $\boldsymbol{a}$ in just a few characteristic values for $\hat{\boldsymbol{a}}$. Most frequently, the characterization is given by a 'location parameter' and a 'confidence interval'. It depends on the application at hand which characterization is the most appropriate one. In many situations the location parameter is represented by the mode, which is the maximum of the posterior distribution $p(a|\boldsymbol{y},\mathcal{I})$, often denoted as the MAP solution. If the prior influence is negligible, for example in the case of a large number of informative data, then the MAP solution is identical to the 'maximum-likelihood' (ML) estimate. However, besides the mode there are other ways to summarize a probability distribution, which will be discussed in this chapter.

14.2 Loss and risk function

A systematic approach to find the appropriate estimate is provided by a decision-theoretic treatment [12] based on the concept of a 'utility function' or 'cost function'. First we need to define a cost (or loss) function $C(\boldsymbol{a},\hat{\boldsymbol{a}})$, which will depend on the true value of the (vector-valued) parameter $\boldsymbol{a}$ and its estimate $\hat{\boldsymbol{a}}$. The precise form of the cost function depends on the problem and specifies how deviations are penalized. The mean (expected) loss is then given by

$$\langle C\rangle = \int d^{N_p}a\; C(\boldsymbol{a},\hat{\boldsymbol{a}})\; p(a|\boldsymbol{y},\mathcal{I}).$$

Most cost functions C depend only on the difference $C(\boldsymbol{a},\hat{\boldsymbol{a}}) = C(\boldsymbol{a}-\hat{\boldsymbol{a}})$. The most common cost functions are:

- Squared error loss

$$C(\boldsymbol{a}-\hat{\boldsymbol{a}}) = \sum_i (a_i - \hat{a}_i)^2. \tag{14.1}$$

The 'squared error loss' penalizes large deviations more than small ones. The resulting estimate will minimize the mean square deviation, thus being influenced most by the largest deviations.

- Absolute error loss

$$C(\boldsymbol{a}-\hat{\boldsymbol{a}}) = \sum_i \left|a_i - \hat{a}_i\right|. \tag{14.2}$$

In many situations the contribution of a deviation shall be proportional to its magnitude. In that case the absolute error loss is appropriate. It is also less susceptible to the influence of 'outliers' in the data.

- Box error loss (1–0 loss)
If deviations are only acceptable within a certain tolerance region then the 'box error loss' function is appropriate. A typical situation is the production of products. If certain specifications of a product are within tolerance range it is acceptable and can be sold, otherwise it has to be replaced, resulting in a constant loss. The corresponding loss function reads

$$C(\boldsymbol{a}-\hat{\boldsymbol{a}}) = -\boldsymbol{1}\left(\left|\boldsymbol{a}-\hat{\boldsymbol{a}}\right| \leq \varepsilon\right) := -\prod_{i=1}^{N_p} \boldsymbol{1}\left(\left|a_i - \hat{a}_i\right| \leq \varepsilon\right). \tag{14.3}$$

This cost function is also referred to as '1–0 loss'.

The best estimator for a given loss function minimizes the expected loss. Therefore, the derivative of the loss function with respect to $\hat{a}_i$ has to be zero.

14.2.1 Squared error loss

In the case of the squared error loss in equation (14.1), we get

$$\begin{aligned}
\frac{\partial}{\partial \hat{a}_i}\langle C\rangle &= 2\int (a_i - \hat{a}_i)\; p(\boldsymbol{a}\,|\,\boldsymbol{y},\mathcal{I})\; d^{N_p}a \\
&= 2\left(\int a_i\; p(\boldsymbol{a}\,|\,\boldsymbol{y},\mathcal{I})\; d^{N_p}a - \hat{a}_i\right) \stackrel{!}{=} 0 \\
&\Rightarrow \\
\hat{a}_i &= \int a_i\; p(a\,|\,\boldsymbol{y},\mathcal{I})\; d^{N_p}a = \langle a_i\rangle_y \\
\hat{\boldsymbol{a}} &= \hat{\boldsymbol{a}}_{\mathrm{PM}}.
\end{aligned}$$

For the squared error loss the best estimator $\hat{a}$ is given by the 'posterior mean', which is the mean of the posterior distribution $p(\boldsymbol{a}\,|\,\boldsymbol{y},\mathcal{I})$.

14.2.2 Absolute error loss

Now we compute the best estimator for the absolute error loss given in equation (14.2):

$$\begin{aligned}
\frac{\partial}{\partial \hat{a}_i}\langle C\rangle &= \int \sum_j \frac{\partial}{\partial \hat{a}_i} \left|\hat{a}_j - a_j\right| \; p(\boldsymbol{a}\,|\,\boldsymbol{y},\mathcal{I})\; d^{N_p}a \\
&= \int \operatorname{sgn}(\hat{a}_i - a_i)\; p(\boldsymbol{a}\,|\,\boldsymbol{y},\mathcal{I})\; d^{N_p}a \\
&= \int da_i \; \operatorname{sgn}(\hat{a}_i - a_i) \; \underbrace{\int \prod_{j\neq i} da_j \; p(\boldsymbol{a}\,|\,\boldsymbol{y},\mathcal{I})}_{:=p(a_i|\boldsymbol{y},\mathcal{I})} .
\end{aligned}$$

This implies

$$\begin{aligned}
\int_{-\infty}^{\hat{a}_i} p(a_i\,|\,\boldsymbol{y},\mathcal{I})\; da_i &= \int_{\hat{a}_i}^{\infty} p(a_i\,|\,\boldsymbol{y},\mathcal{I})\; da_i \\
\hat{\boldsymbol{a}} &= \hat{\boldsymbol{a}}_{\text{MED}} .
\end{aligned}$$

Consequently, the 'median' is the best estimator under absolute error loss.

14.2.3 Box error loss

Finally let us assume a small tolerance region of size ε in all parameters. Then we derive

$$\begin{aligned}
\frac{\partial}{\partial \hat{a}_i}\langle C\rangle &= \frac{\partial}{\partial \hat{a}_i} \int \left(\prod_i \boldsymbol{1}\left(\left|\hat{a} - a\right| \geq \varepsilon\right) \right) p(\boldsymbol{a}\,|\,\boldsymbol{y},\mathcal{I})\; d^{N_p}a \\
&= -\frac{\partial}{\partial \hat{a}_i} \int_{\hat{a}_i-\varepsilon}^{\hat{a}_i+\varepsilon} da_i \underbrace{\left(\prod_{j\neq i} \int_{\hat{a}_j-\varepsilon}^{\hat{a}_j+\varepsilon} \prod_{j\neq i} da_j \; p(\boldsymbol{a}\,|\,\boldsymbol{y},\mathcal{I}) \right)}_{:=X_i} .
\end{aligned}$$

Let us analyse the integral in the bracket. Since ε is small, we can Taylor expand the PDF about $a_j = \hat{a}_j$ for all $j \neq i$. In order to simplify notation, we define $\hat{\boldsymbol{a}}_i = (\hat{a}_1, \ldots, \hat{a}_{i-1}, a_i, \hat{a}_{i+1}, \ldots, \hat{a}_N)$ and $\Delta a_i = a_i - \hat{a}_i$. Then

$$\begin{aligned}
p(\boldsymbol{a}\,|\,\boldsymbol{y},\mathcal{I}) = p(\hat{\boldsymbol{a}}_i\,|\,\boldsymbol{y},\mathcal{I}) &+ \sum_l^{l\neq i} \frac{\partial}{\partial a_l} p(\hat{a}\,|\,\boldsymbol{y},\mathcal{I})\Big|_{\boldsymbol{a}=\hat{\boldsymbol{a}}_i} \Delta a_l \\
&+ \sum_{l,l'}^{l\neq i} \underbrace{\frac{1}{2}\frac{\partial^2}{\partial a_l \partial a_{l'}} p(\hat{a}\,|\,\boldsymbol{y},\mathcal{I})\Big|_{\boldsymbol{a}=\hat{\boldsymbol{a}}_i}}_{:=c_{ll'}} \Delta a_l \Delta a_{l'} + \cdots
\end{aligned}$$

Since we integrate Δa_l over $(-\varepsilon, \varepsilon)$, linear terms in Δa_l vanish and we obtain

$$X_i = p(\hat{\boldsymbol{a}}_i | \boldsymbol{y}, \mathcal{I}) \, (2\varepsilon)^{N_p - 1} + (2\varepsilon)^{N_p+1} \, \frac{1}{3} \sum_l^{l \neq j} c_{ll}$$
$$= p(\hat{\boldsymbol{a}}_i | \boldsymbol{y}, \mathcal{I}) \, (2\varepsilon)^{N_p - 1} \, \big(1 + O(\varepsilon^2)\big).$$

The remaining integral has the form

$$0 = \frac{\partial}{\partial a_i} \int_{a_i - \varepsilon}^{a_i + \varepsilon} p(a) \, da \quad = \frac{\partial}{\partial a_i} \big(F(a_i + \varepsilon) - F(a_i - \varepsilon)\big) = p(a_i + \varepsilon) - p(a_i - \varepsilon).$$

Consequently, $\frac{\partial}{\partial a_i} p(a) = 0$. For our problem that implies

$$\frac{\partial}{\partial a_i} \, p(\boldsymbol{a} | \boldsymbol{y}, \mathcal{I}) \big|_{\boldsymbol{a} = \hat{\boldsymbol{a}}} = 0.$$

Here the best estimator is given by the maximum posterior solution. It is also referred to as the 'posterior mode'. As we have already seen, the MAP solution and the ML solution coincide for uniform priors. In this case the ML solution can be considered as the best solution under a box error loss.

The loss function, averaged over all data and parameter values with the corresponding joint PDF $p(\boldsymbol{a}, \boldsymbol{y} | \mathcal{I})$

$$R := \int \int K(\boldsymbol{a} - \hat{\boldsymbol{a}}(\boldsymbol{y})) \, p(\boldsymbol{a}, \boldsymbol{y} | \mathcal{I}) \, d^N y \, d^{N_p} a$$

is also termed 'risk'. The product rule yields

$$R = \int p(\boldsymbol{y} | \mathcal{I}) \left(\int K(\boldsymbol{a} - \hat{\boldsymbol{a}}(\boldsymbol{y})) \, p(\boldsymbol{a} | \boldsymbol{y}, \mathcal{I}) \, d^{N_p} a \right) d^N y.$$

The risk is therefore equivalent to the mean loss $\langle C \rangle$ averaged over all data weighted by $p(\boldsymbol{y}|\mathcal{I})$. We are interested in the functional $\hat{a}(\boldsymbol{y})$, which minimizes the risk. The root of the functional derivative is given by

$$0 = \frac{\delta}{\delta \hat{a}(\boldsymbol{y})} R = \int p(\boldsymbol{y} | \mathcal{I}) \left(\frac{\delta}{\delta \hat{a}(\boldsymbol{y})} \int K(a - \hat{a}(\boldsymbol{y})) \, p(a | \boldsymbol{y}, \mathcal{I}) \, d^{N_p} a \right) d^N y.$$

Therefore, the inner bracket has to be zero since $p(\boldsymbol{y}|\mathcal{I}) \geq 0$. This implies that the cost-minimizing estimators minimize the risk also at the same time.

14.3 Confidence intervals

In Bayesian probability theory the confidence interval of a parameter a_i is given by the standard deviation of the posterior probability

$$\langle(\Delta a_i)^2\rangle = \int (\Delta a_i)^2 \; p(a|d, \mathcal{I}) \; d^{N_p}a.$$

Sometimes, to emphasize the different interpretation of confidence intervals in frequentist and Bayesian probability theory, a Bayesian confidence interval is also termed a 'credibility interval'. In order to assess correlated uncertainties it is sometimes useful to consider the posterior covariance

$$C_{ij} := \int \Delta a_i \; \Delta a_j \; p(a|d, \mathcal{I}) \; d^{N_p}a.$$

14.4 Examples

14.4.1 The Bernoulli problem

The binary Bernoulli problem (which we will consider again in Section 15.3.5 [p. 240]) will serve as an example for a typical parameter estimation problem. We consider an urn with red and green balls. Let the relative frequency of green balls be q. Given a sample of size N, n_g green balls are found. The probability density $p(q|N, n_g, \mathcal{I})$ based on Bayes' theorem is:

$$p(q|N, n_g, \mathcal{I}) = \frac{1}{Z} \; q^{n_g} \; (1-q)^{N-n_g} \; p(q|\mathcal{I}).$$

Using a flat (uniform) prior for q, i.e. $p(q|\mathcal{I}) = \mathbf{1}(0 \leq q \leq 1)$, we end up with a β distribution (see equation (7.12a) [p. 100]):

$$p(q|N, n_g, \mathcal{I}) = \frac{(N+1)!}{n_g! \; (N-n_g)!} \; q^{n_g} \; (1-q)^{N-n_g}.$$

Owing to the uniform prior, the MAP solution is identical to the maximum likelihood solution

$$q_{\text{MAP}} = q_{\text{ML}} = \frac{n_g}{N}.$$

The mean of the posterior and the confidence interval are provided by the mean and standard deviation of the β distribution (equations (7.12c) and (7.12e) [p. 100]):

$$q_{\text{PM}} = \langle q \rangle = \frac{n_g + 1}{N + 2},$$

$$\sqrt{\text{var}(q)} = \sqrt{\frac{\langle q \rangle \; (1 - \langle q \rangle)}{N+3}}.$$

In summary, the posterior of a Bernoulli experiment is

POSTERIOR PDF OF A BERNOULLI EXPERIMENT WITH UNIFORM PRIOR

sample: N, n_g

posterior PDF:
$$p(q|N, n_g, \mathcal{I}) = p_\beta(q|\alpha, \rho), \tag{14.4a}$$
$$\alpha = n_g + 1, \quad \rho = N - n_g + 1,$$
$$\langle q \rangle = \frac{n_g + 1}{N + 2}, \tag{14.4b}$$
$$\text{var}(q) = \frac{\langle q \rangle (1 - \langle q \rangle)}{N + 3}. \tag{14.4c}$$

The difference between the MAP solution and the posterior mean is of order $O(1/N)$ (the posterior distribution is almost symmetric). The confidence interval, however, is much larger than the difference between the MAP and the MP solution. It is of order $O(1/\sqrt{N})$. Figure 14.1 displays the probability density for $n_g = 6$ and $N = 20$.

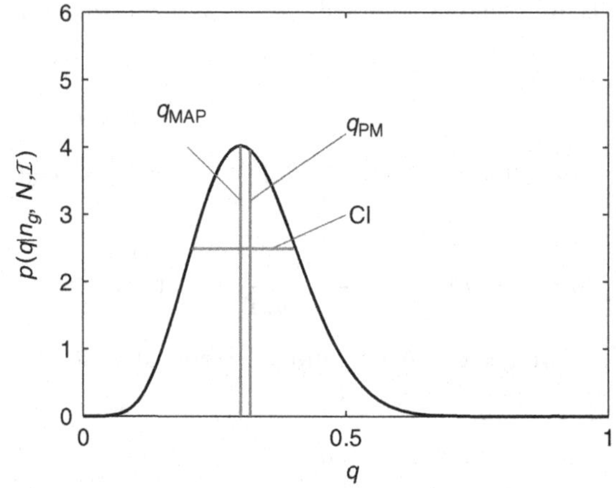

Figure 14.1 Probability density $p(q|N, n_g, \mathcal{I})$ as a function of q for $n_g = 6$ and $N = 20$. Depicted are the posterior mean (PM), maximum posterior estimate (MAP) and confidence interval (CI) given by the one-standard-deviation neighbourhood of MP.

14.4.2 Mean of a Poisson distribution

Given a Poissonian sample of counts $\boldsymbol{l} = \{l_1, \ldots, l_N\}$ that occur in a fixed time window Δt, as the individual counts are uncorrelated the likelihood for $\boldsymbol{l}$ is a product

$$P(\boldsymbol{l}|\mu, \mathcal{I}) = \prod_{i=1}^{N} \ln\left[P(l_i|\mu, \mathcal{I})\right] = \prod_{i=1}^{N} \left\{ e^{-\mu} \frac{\mu^{l_i}}{l_i!} \right\}$$
$$= \frac{1}{Z}\, e^{-N\mu} + \mu^{\sum_i l_i} = \frac{1}{Z} e^{-N\mu}\, \mu^{N\bar{l}},$$

where Z contains μ-independent factors. For the posterior we also need the prior $p(\mu|\mathcal{I})$. Since μ is the mean count in the time window Δt, it is a scale parameter and we use the corresponding Jeffreys' prior $p(\mu|\mathcal{I}) \propto 1/\mu$. Then the posterior reads

$$p(\mu|\boldsymbol{l}, \mathcal{I}) \propto \mu^{N\bar{l}-1}\, e^{-N\mu}.$$

It is a Gamma distribution with $\alpha = \bar{l}$ and $\beta = N$. The posterior mean and variance follow from equation (7.14) [p. 102]. We summarize the result:

Posterior PDF for the mean of a Poisson distribution with Jeffreys' scale prior

sample of counts:	$\boldsymbol{l} = \{l_1, \ldots, l_N\}$,	
posterior PDF:	$p(\mu|\boldsymbol{l}, N, \mathcal{I}) = p_{\Gamma}(\mu|\alpha, \beta)$,	(14.5a)
	$\alpha = N\bar{l}, \quad \beta = N$,	
	$\langle \mu \rangle = \bar{l}$,	(14.5b)
	$\mathrm{var}(\mu) = \frac{\bar{l}}{N}$.	(14.5c)

14.4.3 Estimation of half-lives

The estimation of parameters of systems with exponential behaviour is a common task in many areas of engineering and physics. Here we illustrate the approach with the estimation of the decay parameter of a radioactive source. The data set $= \{t_1, \ldots, t_N\}$ is a sample of recorded times of decay events, e.g. measured by a Geiger–Müller counter. The probability for a decay event is given by

$$p(t|\tau) = \frac{1}{\tau}\, e^{-t/\tau}.$$

We are interested in the probability density $p(\tau|\boldsymbol{t}, \mathcal{I})$ of τ. Bayes' theorem relates this density to the likelihood

$$p(\tau|\boldsymbol{t}, \mathcal{I}) \propto p(\boldsymbol{t}|\tau, \mathcal{I})\; p(\tau|\mathcal{I}).$$

Since all the decay events are uncorrelated, the likelihood is given by the product of the individual PDFs:

$$p(\boldsymbol{t}|\tau,\mathcal{I}) = \prod_{i=1}^{N} \left(\frac{1}{\tau}\, e^{-t_i/\tau}\right) = \tau^{-N}\, e^{-N\bar{t}/\tau}.$$

The symbol $\bar{t}$ stands for the sample mean of the recorded decay times. Next we need to assign the prior for τ. Without further prior knowledge, τ is a scale parameter for which Jeffreys' prior is in most cases appropriate. This finally results in the posterior distribution

$$p(\tau|\boldsymbol{t},\mathcal{I}) = \frac{1}{Z}\tau^{-N-1}\, e^{-N\bar{t}/\tau}.$$

The computation of the normalization constant and the moments of τ requires the integral

$$\begin{aligned} I_\nu &:= \int_0^\infty \tau^{-N+\nu}\, e^{-N\bar{t}/\tau}\, \frac{d\tau}{\tau} \\ &= \int_0^\infty \left(\frac{x}{N\bar{t}}\right)^{N-\nu} e^{-x}\, \frac{dx}{x} \\ &= \left(N\bar{t}\right)^{-N+\nu}\, \Gamma(N-\nu). \end{aligned}$$

Hence, the norm and mean are given by

$$\begin{aligned} Z = I_0 &= \left(N\bar{t}\right)^{-N}\, \Gamma(N), \\ \langle\tau\rangle &= \frac{I_1}{I_0} = \frac{N}{N-1}\,\bar{t}. \end{aligned}$$

The variance can be computed analogously:

$$\begin{aligned} \mathrm{var}(\tau) &= \frac{I_2}{I_0} - \langle\tau\rangle^2 = \frac{N^2}{(N-1)(N-2)}\,\bar{t}^2 - \frac{N^2}{(N-1)(N-1)}\,\bar{t}^2 \\ &= \langle\tau\rangle^2\, \frac{1}{N-2}. \end{aligned}$$

The confidence interval is therefore CI $= \bar{t}/\sqrt{N-2}$. The estimation could be summarized as

$$\tau = \bar{t}\,\frac{N}{N-1}\left(1 \pm \frac{1}{\sqrt{N-2}}\right).$$

Again we find that the difference in the estimators for the location parameter (MAP, PM) is of order $1/N$, while the confidence interval (uncertainty) is of order $1/\sqrt{N}$. The key result is:

POSTERIOR PDF FOR THE HALF-LIFE WITH JEFFREYS' SCALE PRIOR

---

sample of counts: $\boldsymbol{t} = \{t_1, \ldots, t_N\}$,

posterior PDF:
$$p(\tau|\boldsymbol{t}, N, \mathcal{I}) = \frac{(N\bar{t})^N}{\Gamma(N)}\, e^{-N\bar{t}/\tau}, \tag{14.6a}$$

$$\langle\tau\rangle = \bar{t}\, \frac{N}{N-1}, \tag{14.6b}$$

$$\text{var}(\tau) = \bar{t}^2\, \frac{1}{N-2}. \tag{14.6c}$$

15
Frequentist parameter estimation

Now that we have seen how parameters can be estimated in the frame of probability theory, we will outline the frequentist approach. We recall that in the frequentist way of thinking, parameters are not random variables and one cannot assign probabilities to them. Moreover, the parameter estimation shall only be based on the data $\boldsymbol{y}$. The goal is to define a functional $\hat{\boldsymbol{a}}(\boldsymbol{y})$ of the data that estimates the sought-for parameters as well as possible. Two important characteristics of a good estimator are that it is 'unbiased' and 'efficient'. These aspects shall be addressed in the next sections.

15.1 Unbiased estimators

We consider a data analysis problem which depends on the parameters $\boldsymbol{a} \in \mathbb{R}^{N_p}$. The parameters shall be estimated based on a sample of measured data $\boldsymbol{y} = \{y_1, \ldots, y_N\}$. The parameters along with the experimental noise define the probability density of the sample, the likelihood function

$$p(\boldsymbol{y}|\boldsymbol{a}).$$

The most common example of a likelihood function is provided by additive Gaussian errors

$$p(\boldsymbol{y}|\boldsymbol{a}) = (2\pi\sigma^2)^{-\frac{N}{2}}\, e^{-\frac{1}{2}\sum_i \frac{(y_i - f_i(\boldsymbol{a}))^2}{\sigma_i^2}} \,. \tag{15.1}$$

Here $f_i(\boldsymbol{a}) = f(x_i|\boldsymbol{a})$ denotes the theoretical model, which depends on the control variables x_i and on the parameters to be determined, $\boldsymbol{a}$. Let us consider for definiteness a damped oscillation at time t with an elongation

$$f(t|A_0, \omega, \lambda, \phi) = A(t) = A_0 \, \cos(\omega\, t + \phi)\, e^{-\lambda t}.$$

At different times t_i the amplitude $A(t)$ is measured, yielding the value y_i, which differs from $A(t)$ due to noise. In this example the control variables are the measurement times t_i and the unknown parameters are $\boldsymbol{a} = \{A_0, \omega, \lambda, \phi\}$. The measured values are scattered according to a Gaussian distribution with mean zero and standard deviation σ around the model (theoretical) values $y_j = A(t_j)$. Based on the measured data the unknown parameters shall be determined. For simplicity we restrict the discussion first to problems with only one unknown parameter a. Within the frequentist statistics a functional $\hat{a}(\boldsymbol{y})$ has to be derived, which provides an estimator of the parameters for a given set of data. Of special importance are estimators which yield unbiased estimates, so-called *unbiased estimators*. An estimator is unbiased if it yields on average the true parameter value

$$\langle \hat{\boldsymbol{a}}(\boldsymbol{y}) \rangle = \int d^N y \; \hat{\boldsymbol{a}}(\boldsymbol{y}) \; p(\boldsymbol{y}|\boldsymbol{a}) \stackrel{!}{=} \boldsymbol{a}. \tag{15.2}$$

15.2 The maximum likelihood estimator

The most commonly used estimator in frequentist statistics is the so-called 'maximum likelihood' estimator. As implied by the name, it is obtained by maximizing the likelihood

$$\boldsymbol{a}^{\text{ML}} = \arg\max_{\boldsymbol{a}} \; p(\boldsymbol{y}|\boldsymbol{a}).$$

In most cases it is expedient to determine the maximum of the log-likelihood

$$\left. \frac{\partial \ln(p(\boldsymbol{y}|\boldsymbol{a}))}{\partial a_i} \right|_{\boldsymbol{a}=\hat{\boldsymbol{a}}^{\text{ML}}} = 0.$$

Obviously, from the Bayesian perspective it corresponds to the MAP estimator in the case of a uniform prior, or if the peak of the posterior is dominated by the likelihood.

15.3 Examples

In the next sections we will discuss a variety of examples. The generalization to the Bayesian approach is easily obtained by adding the prior PDF.

15.3.1 The univariate Gaussian distribution

As a first example we consider the following simple case of a single parameter a, which is directly accessible by an experiment (a common situation, for example, for length or weight measurements of classical objects). The experimental errors shall be independent Gaussian. The 'log-likelihood' reads

$$\ln(p(\boldsymbol{y}|a)) = C - \frac{1}{2\sigma^2} \sum_i (y_i - a)^2.$$

The root of the derivative with respect to a yields

ML estimate of the centre of a Gaussian

sample of measurements: $\boldsymbol{y} = \{y_1, \ldots, y_N\}$, (15.3a)

$$\hat{a}^{\text{ML}} = \overline{y} = \frac{1}{N} \sum_{i=1}^{N} y_i. \tag{15.3b}$$

In this case the ML estimator coincides with the sample mean. As outlined in Section 19.1 [p. 284], the sample mean yields on average the true value and, therefore, the ML estimator is unbiased in the present case. The ML approach shall be illustrated with a further example.

15.3.2 Mean of a Poisson distribution

Suppose we are given a Poissonian sample of counts $\boldsymbol{l} = \{l_1, \ldots, l_N\}$ that occur in a fixed time window Δt. As the individual counts are uncorrelated, the likelihood for $\boldsymbol{l}$ is a product and the log-likelihood becomes

$$
\begin{aligned}
\ln\left[P(\boldsymbol{l}|\mu, \mathcal{I})\right] &= \sum_{i=1}^{N} \ln\left[P(l_i|\mu, \mathcal{I})\right] = \sum_{i=1}^{N} \ln\left[e^{-\mu}\frac{\mu^{l_i}}{l_i!}\right] \\
&= -N\mu + \sum_i l_i \ln\left[\mu\right] - \sum_i \ln\left[l_i!\right].
\end{aligned}
$$

From the zero of the derivative w.r.t. μ,

$$
\frac{\partial}{\partial\mu}\ln\left[P(\boldsymbol{c}|\mu, \mathcal{I})\right] = -N + \frac{\sum_i l_i}{\mu} \stackrel{!}{=} 0,
$$

we obtain the ML estimate for the parameter μ:

$$
\mu^{\mathrm{ML}} = \frac{\sum\limits_{i=1}^{N} l_i}{N} = \bar{l}.
$$

The ML estimator is again the sample mean and as such is unbiased. In summary, we have

ML ESTIMATE OF THE MEAN OF A POISSON DISTRIBUTION

sample of counts: $\boldsymbol{l} = \{l_1, \ldots, l_N\}$, (15.4a)

$$
\mu^{\mathrm{ML}} = \bar{l} = \frac{1}{N}\sum_{i=1}^{N} l_i. \tag{15.4b}
$$

15.3.3 Half-life distribution of Poisson processes

Assume that a data set of decay times $\{t_1, \ldots, t_L\}$ has been measured. The probability for a decay in $(t, t + dt)$ is provided by the exponential distribution

$$
p_P(t|\tau) = \frac{1}{\tau}\, e^{-t/\tau}.
$$

For uncorrelated decay events the log-likelihood reads

$$\ln(p(t_1, \dots, t_L|\tau, \mathcal{I})) = \sum_{i=1}^{L} \ln\left(\frac{1}{\tau}\, e^{-t_i/\tau}\right) = -L\left(\frac{\bar{t}}{\tau} + \ln(\tau)\right).$$

The derivative vanishes for

ML ESTIMATE OF THE MEAN OF A POISSON DISTRIBUTION

$$\tau^{\mathrm{ML}} = \bar{t}. \tag{15.5}$$

Therefore here also the ML estimator is given by the sample mean. However, the next example will show that this is not always the case.

15.3.4 Cauchy distribution

Let $\boldsymbol{x} = \{x_1, \dots, x_L\}$ be an independent sample of a Cauchy distribution. The probability density of a Cauchy distribution is given by

$$p_C(x|a, b) = \frac{1}{\pi}\, \frac{b}{(x-a)^2 + b^2}.$$

The sought-for parameters are the location parameter a and the width parameter b. The log-likelihood is given by (again exploiting independence of the data)

$$\begin{aligned}\ln(p(\boldsymbol{x}|a, b, \mathcal{I})) &= \sum_{i=1}^{L} \ln\left(\frac{1}{\pi}\, \frac{b}{(x_i-a)^2 + b^2}\right)\\ &= -N\ln(\pi) + L\ln(b) - \sum_{i=1}^{L} \ln\left((x_i-a)^2 + b^2\right).\end{aligned}$$

The derivative with respect to b yields

$$\begin{aligned}\frac{\partial}{\partial b}\ln(p(\boldsymbol{x}|a, b, \mathcal{I})) &= \frac{L}{b} - b\sum_{i=1}^{L} \frac{2}{(x_i-a)^2 + b^2} \stackrel{!}{=} 0\\ \Rightarrow b^2 &= \left(\frac{2}{L}\sum_{i=1}^{L} \frac{1}{(x_i-a)^2 + b^2}\right)^{-1}.\end{aligned}$$

The derivative with respect to a yields

$$\frac{\partial}{\partial a}\ln(p(\boldsymbol{x}|a, b, \mathcal{I})) = 2\sum_{i=1}^{L} \frac{x_i - a}{(x_i-a)^2 + b^2} \stackrel{!}{=} 0.$$

Therefore, the ML solution is given by

ML ESTIMATE OF THE PARAMETERS OF A CAUCHY DISTRIBUTION

$$a^{\mathrm{ML}} = \frac{\frac{1}{L}\sum_{i=1}^{L} \frac{x_i}{(x_i-a)^2+b^2}}{\frac{1}{L}\sum_{i=1}^{L} \frac{1}{(x_i-a)^2+b^2}},$$

$$\left(b^{\mathrm{ML}}\right)^2 = \frac{0.5}{\frac{1}{L}\sum_{i=1}^{L} \frac{1}{(x_i-a)^2+b^2}}.$$

This is a coupled system of two algebraic equations, certainly not of the sample mean type.

15.3.5 Bernoulli problem

We consider again the binary Bernoulli problem with red and green balls. We draw a sample of size N with replacement. Of these N balls are n_g green ones. We want to estimate the relative frequency of the population. The likelihood function is given by

$$P(n_g|q, N, \mathcal{I}) = \binom{N}{n_g} q^{n_g} (1-q)^{N-n_g}.$$

The root of the derivative of the log-likelihood yields

$$\frac{d}{dq}\Big(n_g \ln(q) + (N-n_g)\ln(1-q)\Big) = \frac{n_g}{q} - \frac{N-n_g}{1-q} \overset{!}{=} 0$$

$$\Rightarrow \qquad q_{\mathrm{ML}} = \frac{n_g}{N}. \tag{15.6}$$

ML ESTIMATE OF THE INTRINSIC PROBABILITY OF A BERNOULLI EXPERIMENT

$$q^{\mathrm{ML}} = \frac{n_g}{N}. \tag{15.7}$$

Next we want to address the question of whether this ML estimator is unbiased. In order to answer this question, we need the distribution of the estimator, which we obtain by applying the marginalization rule, i.e. summing over all outcomes:

$$p(q_{\mathrm{ML}}|N,q,\mathcal{I}) = \sum_{k=0}^{N} p(q_{\mathrm{ML}}|k,q,N,\mathcal{I})\; P(k|q,N,\mathcal{I})$$
$$= \sum_{k=0}^{N} \delta\left(q_{\mathrm{ML}} - \tfrac{k}{N}\right)\; P(k|q,N,\mathcal{I}).$$

The moments of this distribution are given by

$$\left\langle q_{\mathrm{ML}}^{\nu}\right\rangle = \int_0^1 q_{\mathrm{ML}}^{\nu}\; p(q_{\mathrm{ML}}|q,N,\mathcal{I})\, dq_{\mathrm{ML}}$$
$$= \int_0^1 \sum_{k=0}^{N} q_{\mathrm{ML}}^{\nu}\; \delta\left(q_{\mathrm{ML}} - \tfrac{k}{N}\right)\; P(k|q,N,\mathcal{I})\, dq_{\mathrm{ML}}$$
$$= \frac{1}{N^{\nu}} \sum_{k=0}^{N} k^{\nu}\; P(k|q,N,\mathcal{I}) = \frac{\langle k^{\nu}\rangle}{N^{\nu}}.$$

In particular we find

$$\langle q_{\mathrm{ML}}\rangle = \frac{\langle k\rangle}{N},$$
$$\mathrm{var}(q_{\mathrm{ML}}) = \frac{\mathrm{var}(k)}{N^2}.$$

The moments are related to the moments of the binomial distribution and we obtain

ML SOLUTION OF THE BERNOULLI PROBLEM

$$\langle q_{\mathrm{ML}}\rangle = q, \tag{15.8a}$$
$$\mathrm{var}(q_{\mathrm{ML}}) = \frac{q\,(1-q)}{N}. \tag{15.8b}$$

The estimator is indeed unbiased and its variance $q(1-q)/N$ decreases with increasing sample size. The variance of the estimator indicates its quality and tells us whether it is an 'efficient estimator', a topic that will be addressed further in Chapter 16 [p. 248].

15.4 Stopping criteria for experiments

Here we want to discuss another important question, which can easily be analysed in terms of the binomial distribution: Can the different stopping criteria of an experiment influence the results of the parameter estimation? Let's assume there are two types of balls, red and green say, and we are interested in the percentage q of green balls. Before performing the experiment it is challenging to determine a suitable sample size N. If N is too small, it may be that we find $n_g = 0$ only, and the ML estimate for q would result in $q^{\mathrm{ML}} = 0$. If N is greater than necessary we might waste precious time. Would it, therefore, be preferable to determine N based on the actual outcome of the experiment? For example, the experiment could be performed until the number of green balls has reached a certain number n_g^*. Does this bias the estimate? A computer simulation can provide a first clue. In this simulation samples of size $N = 10$ were generated, repeatedly. For each sample the estimator $q_{\mathrm{ML}} = \frac{n_g}{N}$ was determined, and the mean value and standard deviation averaged over all repetitions was computed. The results are displayed as the lower curve in Figure 15.1 versus the number of repetitions N_{exp}. It is evident that the sample mean of the ML estimator q^{ML} approaches the true value rapidly and we observe a monotonic decrease of the standard deviation with increasing N_{exp}. The *central limit theorem* ensures the reduction of the standard deviation towards zero as $1/\sqrt{N_{\mathrm{exp}}}$.

Now we repeat the simulation, but this time we stop drawing balls for one sample as soon as there are three green balls in the sample. Again we use $q_{\mathrm{ML}} = \frac{n_g}{N}$ as an estimator of q and compute the average over many repetitions of the simulation. Also here the standard deviation decreases as $1/\sqrt{N_{\mathrm{exp}}}$. However, in this setting the average converges to a different (wrong) result. The experiment exhibits a **bias**!

In the following we will analyse this experiment in more detail. The quantity of interest is the distribution $p(q_{\mathrm{ML}}|q, n_g^*, \mathcal{I})$ of the ML estimator under the stopping rule $n_g = n_g^*$, that is, the experiment is stopped as soon as the number of green balls n_g is equal to n_g^*.

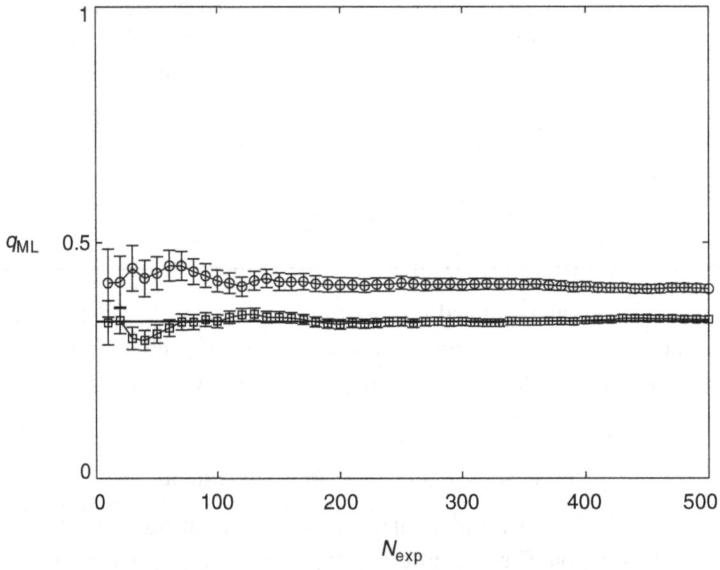

Figure 15.1 ML estimator of q based on (a) sample size $N = 10$ (lower curve) and (b) sample with stopping rule $n_g^* = 3$ (upper curve). The solid line indicates the underlying true parameter value $q = 1/3$.

This information is part of the background information $\mathcal{I}$. Here we want to emphasize again that different background informations typically yields different results:

$$P(A|C,\mathcal{I}) \neq P(A|C,\mathcal{I}').$$

It is therefore important that different background information is adequately reflected in the notation.

For the computation of $p(q_{\mathrm{ML}}|q,n_g^*,\mathcal{I})$ we need the size of the sample N^*, which is the random variable in this case. We introduce this information by applying the marginalization rule

$$p(q_{\mathrm{ML}}|q,n_g^*,\mathcal{I}) = \sum_{N^*=1}^{\infty} p(q_{\mathrm{ML}}|N^*,q,n_g^*,\mathcal{I})\; P(N^*|q,n_g^*,\mathcal{I}). \tag{15.9}$$

The first term is straightforward to derive since the stopping rule is such that the simulation stops if $n_g = n_g^*$ and at the same time we know the size of the sample N^*. Therefore, the ML estimator is $q_{\mathrm{ML}} = \frac{n_g^*}{N^*}$. This yields

$$p(q_{\mathrm{ML}}|N^*,q,n_g^*,\mathcal{I}) = \delta\left(q_{\mathrm{ML}} - \frac{n_g^*}{N^*}\right),$$

and equation (15.9) simplifies to

$$p(q_{\mathrm{ML}}|q,n_g^*,\mathcal{I}) = \sum_{N^*=1}^{\infty} \delta\left(q_{\mathrm{ML}} - \frac{n_g^*}{N^*}\right)\; P(N^*|q,n_g^*,\mathcal{I}). \tag{15.10}$$

The further evaluation of equation (15.10) depends on the probability distribution $P(N^*|q,n_g^*,\mathcal{I})$ for the sample size N^* under the stopping rule $n_g = n_g^*$. The histogram of $P(N^*|q,n_g^*,\mathcal{I})$ based on the performed computer simulations is displayed in Figure 15.2. The shape of the distribution is asymmetric, with a tail towards large values of N^*. The analytical distribution $P(N^*|q,n_g^*,\mathcal{I})$ has already been derived in Section 4.5.1 [p. 68]. In the present notation the PMF and its mean are

$$P(N^*|n_g^*,q,\mathcal{I}) = \binom{N^*-1}{n_g^*-1} q^{n_g^*}\,(1-q)^{N^*-n_g^*}, \tag{15.11a}$$

$$\langle N^* \rangle \overset{(4.35)}{=} \frac{n_g^*}{q}. \tag{15.11b}$$

The analytical PMF is displayed as a solid line in Figure 15.2. The histogram provided by the computer simulation and the theoretical result agree nicely. With the knowledge of the sample size distribution we can compute the distribution of the estimator $q_{\mathrm{ML}} = \frac{n_g^*}{N^*}$ based on equation (15.10). The mean of the estimator is then

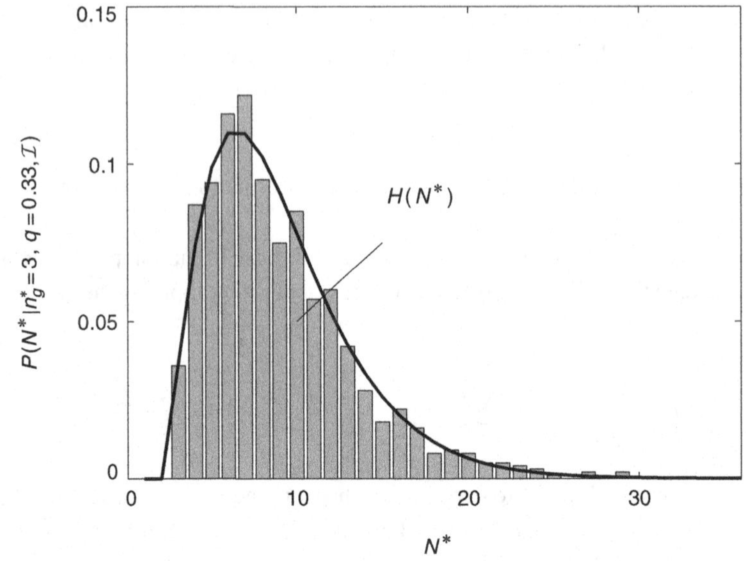

Figure 15.2 Distribution of sample size N^* under the stopping rule $n_g^* = 3$. The percentage of green balls in the simulation is $q = 1/3$. The histogram $H(N^*)$ is derived from computer simulations and the solid line represents the analytic result.

$$
\begin{aligned}
\langle q_{\text{ML}} \rangle &= \int_0^1 q_{\text{ML}}\ p(q_{\text{ML}}|n_g^*, q, \mathcal{I})\ dq_{\text{ML}} \\
&= \sum_{N^*=n_g^*}^{\infty} \frac{n_g^*}{N^*}\ p(N^*|n_g^*, q, \mathcal{I}) \\
&= n_g^* \left(\frac{q}{1-q}\right)^{n_g^*} \sum_{N^*=n_g^*}^{\infty} \binom{N^*-1}{N^*-n_g^*} \frac{1}{N^*} (1-q)^{N^*} \\
&= n_g^* \left(\frac{q}{1-q}\right)^{n_g^*} \sum_{L=0}^{\infty} \binom{L+n_g^*-1}{L} \int_0^{1-q} dx\ x^{L+n_g^*-1} \\
&\overset{(A.22)}{=} n_g^* \left(\frac{q}{1-q}\right)^{n_g^*} \int_0^{1-q} dx\ x^{n_g^*-1} \sum_{L=0}^{\infty} \binom{-n_g^*}{L} (-x)^L \\
&= n_g^* \left(\frac{q}{1-q}\right)^{n_g^*} \int_0^{1-q} dx\ x^{n_g^*-1} (1-x)^{-n_g^*}.
\end{aligned}
$$

The remaining integral is the incomplete beta function, defined in equation (7.13b) [p. 101], and the final result is

$$\langle q_{\text{ML}} \rangle = n_g^* \left(\frac{q}{1-q} \right)^{n_g^*} B(1-q|n_g^*, 1-n_g^*). \tag{15.12}$$

For $n_g^* = 1, 2, 3$ we obtain

$$B(1-q|n_g^*, 1-n_g^*) = \begin{cases} -\ln(q) & \text{for } n_g^* = 1, \\ \frac{1-q}{q} + \ln(q) & \text{for } n_g^* = 2, \\ \frac{(q-1)(3q-1)}{2q^2} - \ln(q) & \text{for } n_g^* = 3. \end{cases} \tag{15.13}$$

Based on equations (15.12) and (15.13), we derive for $q = 1/3$ and $n_g^* = 1$ an average $\langle q_{\text{ML}} \rangle = \frac{1}{2}\ln(3) = 0.549$ which is quite different from the true value $q = \frac{1}{3}$. Again for $q = 1/3$ but with larger stopping value of $n_g^* = 3$ (see Figure 15.1) the result is $\langle q_{\text{ML}} \rangle = \frac{3}{8}\ln(3) = 0.412$, in agreement with the computer simulation in Figure 15.1 but still far away from the correct result. However, the bias shrinks with increasing n_g^*.

15.5 Is unbiasedness desirable at all?

By now, the reader will presumably suspect that the stopping criterion may have a negative impact on the validity of the data. If so, what shall we do with those data for which the stopping criterion is not given by the experimenter? We will postpone this topic to Section 20.2 [p. 325]. Here we will tackle the problem from a different direction, by analysing the deeper aspect of an unbiased estimator. It actually means that we are performing repeated measurements for one and the same quantity, or even one and the same source. In terms of our previous example it means that we are measuring the same or identical boxes with two sorts of objects with the same fraction q, over and over again. More than that, we always use the same stopping criterion: we stop as soon as the number n_g of the selected species is n_g^*. The outcome of these experiments is the sample size N_ν^* for $\nu = 1, 2, \ldots, L$, where L is the number of such identical experiments. For each experiment we use the ratio $q_\nu = n_g^*/N_\nu^*$ as (intermediate) estimator. It would be unbiased if

$$\lim_{L\to\infty} \frac{1}{L} \sum_{\nu=1}^{L} \frac{n_g^*}{N_\nu^*} = q.$$

But why should anyone analyse the data in this way? **If we really have a whole set of data** $\{N_\nu^*\}$, then the correct way in the framework of the maximum likelihood estimate would be to maximize the joint likelihood for all data simultaneously:

$$\begin{aligned} \max_q P(\{N_\nu^*\}|q, n_g^*, \mathcal{I}) &= \max_q \prod_{\nu=1}^{L} \binom{N_\nu^* - 1}{n_g^* - 1} q^{n_g^*}(1-q)^{N_\nu^* - n_g^*} \\ &\propto \max_q q^{Ln_g^*}(1-q)^{\sum_\nu N_\nu^* - Ln_g^*}. \end{aligned}$$

The resulting maximum likelihood estimator is

$$q^{\mathrm{ML}} = \frac{n_g^*}{\overline{N^*}},$$
$$\overline{N^*} := \frac{1}{L} \sum_{\nu=1}^{L} N_\nu^*.$$

Now, the sample mean is an unbiased estimator and therefore

$$\lim_{L\to\infty} \overline{N^*} = \left\langle N^* \right\rangle \overset{(15.11b)}{=} \frac{n_g^*}{q}.$$

Hence this maximum likelihood estimator q^{ML} yields the exact result for $L \to \infty$. If, in contast, **we have only one value** N^*, then it can be misleading to draw conclusions based on hypothetical data sets, as will be discussed in greater detail in Section 30.1.3 [p. 540].

15.6 Least-squares fitting

We now focus on the ubiquitous case of independent additive Gaussian noise, which typically results in the likelihood function given in equation (15.1) [p. 236]. The data are combined in the vector $\boldsymbol{y}$ and the parameters are denoted by $\boldsymbol{a}$. A maximization of the log-likelihood is, in this case, equivalent to the minimization of the weighted mean square deviation

LEAST-SQUARES FIT

$$\boldsymbol{a}^{\mathrm{LS}} = \arg\min_{\boldsymbol{a}} \sum_i \frac{(y_i - f(s_i, \boldsymbol{a}))^2}{\sigma_i^2}. \tag{15.14}$$

So, the ‘least-squares fit’ is appropriate for independent additive Gaussian noise. Equation (15.14) simplifies further if $\sigma_i^2 = \sigma^2$, then the ML estimator follows from

$$\boldsymbol{a}^{\mathrm{LS}} = \min_{\boldsymbol{a}} \sum_i (y_i - f(s_i, \boldsymbol{a}))^2. \tag{15.15}$$

It is straightforward to take correlated noise into account. The log-likelihood of additive multivariate normal/Gaussian noise with covariance matrix $\mathbf{C}$ is

$$\ln[p(\boldsymbol{y}|\boldsymbol{a}, \mathcal{I})] = -\frac{1}{2} (\boldsymbol{y} - \boldsymbol{f}(\boldsymbol{a}))^T \mathbf{C}^{-1}(\boldsymbol{y} - \boldsymbol{f}(\boldsymbol{a})),$$

with $\boldsymbol{f}(\boldsymbol{a}) = (f(s_1, \boldsymbol{a}), \dots, f(s_N, \boldsymbol{a}))$. The likelihood can be transformed into an uncorrelated Gaussian as follows:

$$\ln[p(\boldsymbol{y}|\boldsymbol{a}, \mathcal{I})] = -\frac{1}{2} \left\| \mathbf{C}^{-\frac{1}{2}}[\boldsymbol{y} - \boldsymbol{f}(\boldsymbol{a})] \right\|^2 = -\frac{1}{2} \left\| \tilde{\boldsymbol{y}} - \tilde{\boldsymbol{f}}(\boldsymbol{a}) \right\|^2.$$

Here we have taken advantage of $\mathbf{C} = \mathbf{C}^{\frac{1}{2}}\mathbf{C}^{\frac{1}{2}}$. The transformation by $\mathbf{C}^{-\frac{1}{2}}$ decorrelates the uncertainties and standardizes to variances equal to one. In summary, the result reads

TRANSFORMED MULTIVARIATE NORMAL DISTRIBUTION

$$\begin{aligned} p(\boldsymbol{y}|\mathbf{C},\boldsymbol{a},\mathcal{I}) &= (2\pi)^{-\frac{N}{2}}\ |\mathbf{C}|^{-\frac{1}{2}}\ e^{-\frac{1}{2}\left|\tilde{\boldsymbol{y}}-\tilde{\boldsymbol{f}}(\boldsymbol{a})\right|^2} \\ &\text{with} \\ \tilde{\boldsymbol{y}} &= \mathbf{C}^{-\frac{1}{2}}\ \boldsymbol{y}, \\ \tilde{\boldsymbol{f}}(\boldsymbol{a}) &= \mathbf{C}^{-\frac{1}{2}}\ \boldsymbol{f}(\boldsymbol{a}). \end{aligned} \tag{15.16}$$

Hence, a standard least-squares fit can also be performed, now acting on the transformed variables:

LEAST-SQUARES FIT WITH CORRELATED NOISE

$$\boldsymbol{a}^{\mathrm{LS}} = \min_{\boldsymbol{a}} \sum_i \left(\tilde{y}_i - \tilde{f}_i(\boldsymbol{a})\right)^2. \tag{15.17}$$

16

The Cramer–Rao inequality

16.1 Lower bound on the variance

An unbiased estimator is not unique. Consider, for example, the situation where the sample mean is an unbiased estimator. In this case also each element x_i of the sample provides an unbiased estimator. However, in the latter case the estimator will exhibit a larger scatter around the true value of the parameter than the sample mean. Evidently the best unbiased estimator is the one with the smallest scatter around the true value. This estimator is referred to as the 'efficient estimator' [104]. Immediately two questions spring to mind: Can the mean square error of an unbiased estimator be zero and therefore always yield the true result? And if not, what defines the lower limit? The answer to both questions is provided by the 'Cramer–Rao inequality'. The Cramer–Rao inequality is a fundamental result of 'estimation theory'. Upfront we make the following regularity assumption: The domain of the likelihood is independent of the value of the parameter a and the derivative $\frac{\partial p(\boldsymbol{y}|a,\mathcal{I})}{\partial a}$ exists and is finite. Now assume that the estimator $\hat{a}(\boldsymbol{y})$ of a parameter a, which depends on the data sample $\boldsymbol{y}$ of size N, is unbiased. Then

$$\int p(\boldsymbol{y}|a)\,(\hat{a}(\boldsymbol{y}) - a)\, d^N y = 0. \tag{16.1}$$

Differentiation of equation (16.1) with respect to a and assuming the necessary regularity, we obtain

$$\int \left(\frac{\partial}{\partial a}\, p(\boldsymbol{y}|a)\right)\,(\hat{a}(\boldsymbol{y}) - a)\, d^N y - \underbrace{\int p(\boldsymbol{y}|a)\, d^N y}_{=1} = 0.$$

With $\frac{\partial}{\partial a}\, p = p \frac{\partial}{\partial a}\, \ln(p)$ we can then write

$$1 = \int p(\boldsymbol{y}|a)\,\left(\frac{\partial}{\partial a}\,\ln(p(\boldsymbol{y}|a))\right)\,(\hat{a}(\boldsymbol{y}) - a)\; d^N y. \tag{16.2}$$

This equation holds for all allowed parameter values a and any unbiased estimator. From equation (16.2) and the Schwarz inequality we obtain

$$1 = \left\langle \frac{\partial \ln\left[p(\boldsymbol{y}|a)\right]}{\partial a} \left(\hat{a}(\boldsymbol{y}) - a\right) \right\rangle \tag{16.3a}$$

$$\leq \sqrt{\left\langle \left[\frac{\partial \ln\left[p(\boldsymbol{y}|a)\right]}{\partial a} \right]^2 \right\rangle \left\langle \left[\hat{a}(\boldsymbol{y}) - a\right]^2 \right\rangle}. \tag{16.3b}$$

The second term of the square root is the variance

$$\left\langle \left(\hat{a}(\boldsymbol{y}) - a\right)^2 \right\rangle = \mathrm{var}(\hat{a}(\boldsymbol{y})),$$

since we have assumed that the estimator is unbiased ($\langle \hat{a}(\boldsymbol{y}) \rangle = a$). Then equation (16.3) entails an important lower bound for the uncertainty of an estimator

CRAMER–RAO INEQUALITY

$$\mathrm{var}(\hat{a}(\boldsymbol{y})) \geq \frac{1}{I}, \tag{16.4}$$

$$I = \int p(\boldsymbol{y}|a) \left(\frac{\partial}{\partial a} \ln(p(\boldsymbol{y}|a)) \right)^2 d^N y$$

$$= \int \frac{\left(\frac{\partial}{\partial a} p(\boldsymbol{y}|a)\right)^2}{p(\boldsymbol{y}|a)} d^N y. \tag{16.5}$$

I is called the 'Fisher information'.

Hence, the lower limit depends solely on the form of the likelihood and it is usually straightforward to compute it. We will give two examples in the subsequent sections.

16.2 Examples

16.2.1 Gaussian distribution

For the case of a random sample $\boldsymbol{y}$ of size N of a Gaussian distribution, the log-likelihood is

$$\ln(p(\boldsymbol{y}|a)) = C - \frac{1}{2\sigma^2} \sum_{i=1}^{N} (x_i - a)^2$$

and the derivative yields

$$\frac{d}{da} \ln(p(\boldsymbol{y}|a)) = \frac{N}{\sigma^2} \left(\overline{y} - a\right). \tag{16.6}$$

As before, $\overline{y}$ denotes the sample mean. As proven in Chapter 19 [p. 284], the sample mean is an unbiased estimator of the mean a. Hence the Fisher information is given by

$$I = \frac{N^2}{\sigma^4}\,\mathrm{var}(\overline{y}) = \frac{N^2}{\sigma^4}\frac{\sigma^2}{N} = \frac{N}{\sigma^2}.$$

The variance of any estimator fulfils the inequality

$$\mathrm{var}(\hat{a}(\boldsymbol{y})) \geq \frac{\sigma^2}{N} = \mathrm{var}(\overline{y}).$$

However, the sample mean $\hat{a}(\boldsymbol{y}) = \overline{y}$ achieves the Cramer–Rao lower bound since

$$\mathrm{var}(\hat{a}(\boldsymbol{y})) = \mathrm{var}(\overline{y}) = \frac{\sigma^2}{N}$$

holds. Therefore in this case, the sample mean is an 'efficient estimator'.

16.2.2 Cauchy distribution

As a second example we consider the Cauchy distribution with half-width $\gamma/2 = 1$:

$$p(x|a) = \frac{1}{\pi}\,\frac{1}{(x-a)^2+1}.$$

The log-likelihood reads

$$\ln(p(\boldsymbol{y}|a)) = C - \sum_{i=1}^{L} \ln((y_i - a)^2 + 1)$$

and the associated derivative is given by

$$\frac{d}{da}\ln(p(\boldsymbol{y}|a)) = 2\sum_{i=1}^{N} \frac{y_i - a}{(y_i-a)^2+1}. \tag{16.7}$$

Then the Fisher information can be expressed as the sum of two expectation values:

$$\begin{aligned}
I &= 4\left\langle\left(\sum_{i=1}^{N}\frac{y_i-a}{(y_i-a)^2+1}\right)^2\right\rangle \\
&= 4\sum_{i,j}\left\langle\frac{y_i-a}{(y_i-a)^2+1}\,\frac{y_j-a}{(y_j-a)^2+1}\right\rangle \\
&= 4\sum_{i\neq j}\left\langle\frac{y_i-a}{(y_i-a)^2+1}\right\rangle\left\langle\frac{y_j-a}{(y_j-a)^2+1}\right\rangle + 4\sum_{i}\left\langle\frac{(y_i-a)^2}{((y_i-a)^2+1)^2}\right\rangle \\
&= 4N(N-1)\left(\left\langle\frac{y-a}{(y-a)^2+1}\right\rangle\right)^2 + 4N\left\langle\frac{(y-a)^2}{((y-a)^2+1)^2}\right\rangle.
\end{aligned}$$

Owing to symmetry the first expectation value is identical to zero and the second expectation value can be rewritten as

$$\left\langle \frac{(y-a)^2}{((y-a)^2+1)^2} \right\rangle = \frac{1}{\pi} \int \frac{(y-a)^2}{((y-a)^2+1)^3}\, dy = \frac{1}{8}.$$

Therefore the Fisher information is given by $I = \frac{N}{2}$. The resulting Cramer–Rao bound of the variance reads

$$\mathrm{var}(\hat{a}(\boldsymbol{y})) \geq \frac{2}{N}.$$

In Chapter 19 [p. 284] on sample distributions we will compute properties of the sample median of a Cauchy distribution. The sample median will turn out to be an unbiased estimator. For $N \gg 1$, the variance is given by

$$\mathrm{var}(\hat{a}(\boldsymbol{y})) = \frac{\pi^2}{4N} = \frac{2.47}{N}.$$

The variance is greater than the Cramer–Rao bound and the median is therefore not an 'efficient estimator'. The question arises of whether an unbiased estimator always exists that reaches the Cramer–Rao limit.

16.3 Admissibility of the Cramer–Rao limit

The equal sign in the Schwarz inequality is reached if both vectors entering the equation, $\boldsymbol{a}$ and $\boldsymbol{b}$, are parallel: $\boldsymbol{a} = \alpha\ \boldsymbol{b}$. In the case of the Cramer–Rao bound (equation (16.3)) the random variable is $\boldsymbol{y}$ and wc therefore need

$$\frac{\partial}{\partial a} \ln(p(\boldsymbol{y}|a)) = \alpha\ \big(\hat{a}(\boldsymbol{y}) - a\big) \tag{16.8}$$

as a function of $\boldsymbol{y}$. That is, the constant α may depend on a but not on the sample $\boldsymbol{y}$. In this case the Cramer–Rao equality reads

$$1 = \alpha(a) \left\langle \big(\hat{a}(\boldsymbol{y}) - a\big)^2 \right\rangle \qquad \Rightarrow \alpha(a) \neq 0, \tag{16.9}$$

otherwise the variance of the estimator would have to be infinite, in contradiction to an efficient estimator. If the derivative of the log-likelihood can be cast into the form of the right-hand side of equation (16.8), then an unbiased and efficient estimator does exist. On the contrary, if the derivative cannot be reformulated as the right-hand side of equation (16.8) then no unbiased efficient estimator exists. In this respect, the following result is of practical importance:

If an efficient unbiased estimator exists, then it is the ML estimator.

If an efficient unbiased estimator exists, then equation (16.8) holds for all values of a and for all data vectors $\mathbf{y}$. In particular, for $a = a^{\mathrm{ML}}(\mathbf{y})$, the left-hand side of equation (16.8) is zero by definition of $a^{\mathrm{ML}}(\mathbf{y})$ and we obtain

$$0 = \alpha\left(a^{\mathrm{ML}}(\mathbf{y})\right)\left(\hat{a}(\mathbf{y}) - a^{\mathrm{ML}}(\mathbf{y})\right) \quad \forall \mathbf{y}.$$

According to equation (16.9) ($\alpha(a) \neq 0$; $\forall a$) we immediately find

$$a^{\mathrm{ML}}(\mathbf{y}) = \hat{a}(\mathbf{y}).$$

Therefore the maximum likelihood solution is the most efficient estimator.

PART IV

TESTING HYPOTHESES

17
The Bayesian way

Hypothesis tests are an integral part of conventional statistics and are used in a wide variety of real-world applications. From the Bayesian perspective it is most obvious how to assess an hypothesis H: simply by computing the corresponding probability. From the frequentist way of reasoning it is not that obvious, because an hypothesis is not a random object and thus one cannot assign a probability to it. In this chapter we start out with the Bayesian approach to hypothesis testing.

Let H_α with $\alpha = 1, 2, \ldots$ be some hypotheses to be assessed.[1] In the Bayesian frame an hypothesis is nothing but a proposition and the corresponding probability, in view of some data D and additional background information $\mathcal{I}$, is provided by Bayes' theorem:

$$P(H_\alpha|D,\mathcal{I}) = \frac{p(D|H_\alpha,\mathcal{I})P(H_\alpha|\mathcal{I})}{p(D|\mathcal{I})}.$$

Assuming that the hypotheses are mutually exclusive and complete, i.e. $\vee_\alpha H_\alpha = T$, we can determine the data evidence $p(D|\mathcal{I})$ via normalization:

$$p(D|\mathcal{I}) = \sum_\alpha p(D|H_\alpha,\mathcal{I})P(H_\alpha|\mathcal{I}).$$

In case we consider only one hypothesis and its complement, we discussed already in Section 3.2.4 [p. 38] that it is advantageous to determine the *odds ratio*

$$o := \frac{P(H|D,\mathcal{I})}{P(\overline{H}|D,\mathcal{I})} = \underbrace{\frac{p(D|H,\mathcal{I})}{p(D|\overline{H},\mathcal{I})}}_{o_{\mathrm{BF}}} \underbrace{\frac{P(H|\mathcal{I})}{P(\overline{H}|\mathcal{I})}}_{o_{\mathrm{P}}}, \tag{17.1}$$

as it avoids the computation of the normalization factor, or rather the data evidence. In the above equation o_{BF} stands for the Bayes factor and o_{P} is the prior odds. The probability again follows from the normalization condition.

$$P(H_1|D,\mathcal{I}) = \frac{o}{1+o}. \tag{17.2}$$

The Bayes factor represents how the experimental data advance our state of knowledge. The odds ratio represents the status before the new data are taken into account. In general,

[1] A special case of outstanding importance is when two complementary alternatives, $H_2 = \overline{H}_1$, are tested.

experiments are carried out and analysed by specialists in the field under consideration. That is, there is prior knowledge, maybe due to previous experiments, that has to be taken into account by the prior odds. We see that the Bayesian approach is perfectly suited to treating new and old data information on the same footing. Of course, there may also be situations where no prior knowledge is available, for example, if a new field is entered. Then we have to choose $o_P = 1$, since any other choice implies additional knowledge.

The Bayesian approach has the appealing feature that it is consistent. If we have data sets $D_1, D_2, \ldots, D_N$ of several research groups collected over the years then typically they are analysed sequentially but we could also process them in a single calculation. We introduce a more specific notation for the Bayes factor:

$$o_{BF}(D_1, \ldots, D_N|\mathcal{I}) := \frac{p(D_1, \ldots, D_N|H_1, \mathcal{I}, \mathcal{I})}{p(D_1, \ldots, D_N|H_2, \mathcal{I}, \mathcal{I})}.$$

Owing to the fact that the data sets are independent, the product rule readily yields

$$o_{BF}(D_1, \ldots, D_N|\mathcal{I}) = \prod_{i=1}^{N} o_{BF}(D_i|\mathcal{I}).$$

So the correct procedure is to start a new field with some sort of uninformative prior odds o_P and to analyse the first data set resulting in an odds ratio

$$o(D_1|\mathcal{I}) = o_{BF}(D_1|\mathcal{I}) o_P.$$

If the next data set is analysed we should consistently use the previous odds ratio as prior odds, which would yield

$$o(D_2|\mathcal{I}) = o_{BF}(D_2|\mathcal{I})\, o_{BF}(D_1|\mathcal{I})\, o_P = o_{BF}(D_1, D_2|\mathcal{I})\, o_P.$$

This reveals once more that the Bayesian approach is consistent and it also emphasizes the great importance of the use of prior knowledge. Another strength of the Bayesian approach is that all hypotheses are treated on the same footing.

Next we will give some simple examples in order to illustrate the Bayesian way of testing hypotheses. These examples are closely related to standard frequentist hypothesis tests. In some cases the Bayesian result can be considered as a consistent derivation of the corresponding test statistic. It should be emphasized, however, that this is not always the case. A thorough discussion of this topic can be found in [12].

The examples given in the next sections are quite basic, and should primarily illustrate the Bayesian way of testing hypotheses. Far more realistic examples will be discussed in the ensuing part about real-world problems (Part V).

17.1 Some illustrative examples

17.1.1 Telepathic abilities

Let us suppose Mr T claims to have telepathic abilities. In order to scrutinize Mr T's assertion, we carry out the following experiment. We use a card deck with 32 different cards and shuffle carefully each time before we draw a card. Then we take a look at the selected card and put it back into the deck. Now Mr T has to tell us which card we have seen. This experiment is repeated N times and Mr T's answer appears correct in n cases, say.

The hypothesis under consideration reads

$$H : \text{Mr T has telepathic abilities.}$$

The complement $\overline{H}$ means Mr T has no extraordinary skills and is only guessing. In this case, the probability for a correct answer is $q = q_g = 1/32$. If H is true, the percentage of correct answers has to be greater than $1/32$. The marginalization rule yields

$$P(n|N, H, \mathcal{I}) = \int_0^1 P(n|q, N, \not{H}, \mathcal{I})\; p(q|\not{N}, H, \mathcal{I})\; dq.$$

Hypothesis H says that Mr T has telepathic abilities, but not to what extent. Using the indicator function $\boldsymbol{1}(S)$, which has value 1 if S is true and 0 otherwise, we can write

$$p(q|H, \mathcal{I}) = \frac{32}{31}\; \boldsymbol{1}\left(\tfrac{1}{32} < q \leq 1\right).$$

The likelihood term $P(n|q, N, \mathcal{I}) = P_B(n|q, N)$ is simply binomial:

$$P(n|q, N, H, \mathcal{I}) = \binom{N}{n}\, q^n\, (1-q)^{N-n}.$$

The marginal likelihood is therefore

$$\begin{aligned} P(n|N, H, \mathcal{I}) &= \frac{32}{31}\binom{N}{n} \int_{1/32}^{1} q^n\, (1-q)^{N-n}\, dq \qquad \overset{p=1-q}{\Longrightarrow} \\ &= \frac{32}{31}\binom{N}{n} \int_0^{31/32} p^{N-n}\, (1-p)^n\, dp \\ &= \frac{32}{31}\binom{N}{n}\, B\left(\tfrac{31}{32}; N-n+1, n+1\right), \end{aligned}$$

where the last factor represents the incomplete beta function of equation (7.13b) [p. 101]. In the complementary case, $\overline{H}$, the likelihood is simply

$$P(n|N, \overline{H}, \mathcal{I}) = P_B\left(n|q = \tfrac{1}{32}, N\right) = \binom{N}{n} \left(\tfrac{1}{32}\right)^n \left(\tfrac{31}{32}\right)^{N-n}.$$

So the Bayes factor appears to be

$$o_{\text{BF}} = \frac{32^{N+1}}{31^{N+1-n}}\; B\left(\tfrac{31}{32}; N-n+1, n+1\right).$$

Let us assume a sample size $N = 32$. If the coincidences are random we expect on average $n/32 = 1$ agreements. Let us say, the test yields that n is three times the expectation value of the random coincidence, i.e. $n = 3$. In this case the Bayes factor is $o_{\text{BF}} = 0.51$. If we are completely indifferent as far as telepathic abilities are concerned, we choose a prior odds of one and the probability that Mr T has extraordinary skills is

$$P(H|N, n, \mathcal{I}) = \frac{o_{\text{BF}}}{1 + o_{\text{BF}}} = 0.34.$$

This value would not convince anybody that Mr T's claim is correct. The agreement is very likely due to random fluctuations. In contrast, a probability of 0.34 is not extremely small either, so we decide to repeat the experiment 10 times as long, i.e. $N = 320$. Let us assume that the percentage of agreements is still the same, then $n = 30$. The corresponding Bayes factor is now $o_{\text{BF}} = 3.4610^4$ and the probability, still assuming an indifferent prior odds, is

$$P(H|N, n, \mathcal{I}) = 0.99997.$$

Based on the data and with almost absolute certainty, we can take it for granted that Mr T is right. Nevertheless, hardly anyone would really be convinced of the result and henceforth believe in telepathic abilities. The reason being that we have never met such a skill before. In other words, the prior probability is very small. How can we assess it? If anybody on earth demonstrated reproducibly that he or she possesses telepathic abilities, it would be in the media. On earth there are approximately 10^{10} humans. How many people with such extraordinary skills could remain undetected? Maybe $O(10^3)$. The corresponding prior probability is 10^{-7} and the prior odds reads

$$o_{\text{P}} = \frac{10^{-7}}{1 - 10^{-7}} \approx 10^{-7}.$$

Thus, in the case of $N = 320$ and $n = 30$, the total odds are $o = 3.4610^{-3}$ and the corresponding probability is down to

$$P(n|N, H, \mathcal{I}) = 3.44^{-3}.$$

That makes much more sense. But we also see, as a consequence of the data, that the vanishingly small prior probability of 10^{-7} increases by roughly 4 orders of magnitude and the new value has to be used as prior probability in future inferences. This is just the *learning* quantified by the rules of probability theory. We have seen that the prior odds has an important impact, which is consistent as it correctly includes prior experimental findings.

We should, however, never use $o_{\text{P}} = 0$ or $o_{\text{P}} = \infty$ because that corresponds to absolute certainty about H or $\overline{H}$ and would make the experiment superfluous.

17.1.2 Is the coin fair?

As a second simple example, we consider the hypothesis that a given coin is fair. This example has already been touched upon in Section 3.2.4 [p. 38]. We repeat the relevant propositions:

- H: The coin is fair.
- n_h: The number of times heads shows up is n_h.
- N: The coin has been tossed N times.

We furthermore assume no preference between H and $\overline{H}$, i.e. the prior odds is

$$o_{\mathrm{P}} = \frac{P(H|\mathcal{I})}{P(\overline{H}|\mathcal{I})} = 1.$$

This assumption is of course highly unrealistic, but we are primarily interested here in the Bayes factor containing the data implication. The marginal likelihood functions $P(n_h|N, H, \mathcal{I})$ and $P(n_h|N, \overline{H}, \mathcal{I})$ can be derived from the likelihood function using the marginalization rule

$$P(n_h|N, A, \mathcal{I}) = \int_0^1 P(n_h|N, q, A, \mathcal{I})\; p(q|N, A, \mathcal{I})\, dq, \tag{17.3}$$

with $A = H$ or $A = \overline{H}$. The likelihood function is independent of A since it simply yields the probability of observing heads n_h times in N tosses of the coin if the probability of observing heads in a single toss is q. Under the condition that we know q, proposition A in the conditional complex of the likelihood becomes irrelevant. Therefore, the likelihood function is given by the binomial distribution

$$P(n_h|N, q, \mathcal{I}) = \binom{N}{n_h} q^{n_h}\, (1-q)^{N-n_h}.$$

The dependence on A is relevant in the term $P(q|N, A, \mathcal{I})$. This probability, however, is independent of the number of tosses N. In the case that hypothesis H is true, $q = 1/2$ holds:

$$P(q|H, \mathcal{I}) = \delta(q - 1/2). \tag{17.4}$$

Strictly speaking, $P(n_h|N, q, H, \mathcal{I})$ is only defined for $q = 1/2$ because otherwise the conditional complex is logically inconsistent. However, the delta function in equation (17.4) takes care of this automatically and all contributions with $q \neq 1/2$ are dropped from the integral in equation (17.3). The alternative hypothesis $\overline{H}$ is not that obvious and it depends on the background information. To begin with we assume complete ignorance as far as coin manufacturing is concerned, and choose a uniform prior

$$P(q|\overline{H}, \mathcal{I}) = \mathbf{1}(0 \leq q \leq 1)\,. \tag{17.5}$$

Table 17.1 The probability for the hypothesis 'fair coin' as a function of the number of coin tosses for a coin which yields heads only. The left column contains results for the uniform prior and the right column those for the more informative prior

	$P(H\|n_h = N, N, \mathcal{I})$	
N	prior I	prior II
0	0.50000	0.50000
1	0.50000	0.50000
2	0.42857	0.49751
5	0.15789	0.47586
10	0.01063	0.39765
15	0.00049	0.28341
20	0.00002	0.16650

This results in

$$P(n_h|N, H, \mathcal{I}) = \binom{N}{n_h} \left(\frac{1}{2}\right)^{n_h} \left(1 - \frac{1}{2}\right)^{N-n_h} = \binom{N}{n_h} 2^{-N},$$

$$P(n_h|N, \overline{H}, \mathcal{I}) = \binom{N}{n_h} \int_0^1 q^{n_h} (1-q)^{N-n_h} \, dq = \binom{N}{n_h} \frac{n_h! \, (N-n_h)!}{(N+1)!},$$

and we can compute the odds ratio as

$$o = \frac{2^{-N} \, (N+1)!}{n_h! \, (N-n_h)!}.$$

Now we consider two extreme cases:

- The coin is manipulated and shows heads only.
- The coin is fair and shows heads and tails with equal probability.

In the first case we observe $n_h = N$ and the odds ratio reads

$$o = 2^{-N} \, (N+1).$$

Given the odds ratio we can compute the probability for hypothesis H as

$$P(H|n_h = N, N, \mathcal{I}) = \frac{o}{1+o} = \frac{1}{1+1/o} = \frac{1}{1+2^N/(N+1)}.$$

The results are given in Table 17.1 for different numbers of coin tosses. For $N = 0$ everything is undecided: $P(H|n_h, N, \mathcal{I}) = 1/2$. After the first coin toss the probability is still

Table 17.2 The probability for the hypothesis 'fair coin' as a function of the number of coin tosses for a coin which yields as many heads as tails

	$P(H\|n_h = N/2, N, \mathcal{I})$	
N	prior I	prior II
0	0.500	0.500
2	0.600	0.503
10	0.730	0.512
20	0.787	0.523
40	0.837	0.543
100	0.889	0.587
200	0.919	0.635
2000	0.973	0.822

$P(H|n_h, N, \mathcal{I}) = 1/2$. The probability of the hypothesis 'fair coin' decreases very rapidly and after 10 coin tosses it is almost sure that the coin is biased.

In the case of a fair coin one could expect to observe e.g. $N = 2n_h$. For this result the odds ratio is given by

$$o = 2^{-2\,n_h} \frac{(2\,n_h + 1)!}{n_h!\,n_h!}.$$

The results are given in Table 17.2. Here it is much more challenging to decide if the coin is really fair or if there are any (potentially very small) deviations.

A more restrictive alternative

As pointed out before, the prior PDF $p(q|\overline{H}, \mathcal{I}) = \boldsymbol{1}(0 \leq 1)$ is highly unrealistic, and we want to analyse the impact of a more restrictive alternative. For symmetry reasons it is appropriate to stick to a prior with $\langle q \rangle = 1/2$. The variance should, however, be much smaller, which we encode in a beta prior

$$p(q|\overline{H}, \mathcal{I}) = \frac{1}{B(\alpha, \alpha)}\, q^{\alpha-1}\,(1-q)^{\alpha-1},$$

with a variance given by equation (7.12e) [p. 100]:

$$\sigma^2 := \langle(\Delta q)^2\rangle = \frac{1}{4}\frac{1}{2\alpha+1},$$

or rather

$$\alpha = \frac{1}{8\sigma^2} - \frac{1}{2}.$$

The uniform prior is recovered with $\alpha = 1$ and corresponds to $\sigma = \sqrt{1/12} = 0.29$. For the more restrictive alternative we use $\sigma = 0.05$, or rather $\alpha = 49.5$.

The corresponding marginal likelihood reads

$$P(n_h|N, \overline{H}, \mathcal{I}) = \binom{N}{n_h} \frac{1}{B(\alpha,\alpha)} \int_0^1 q^{n_h} (1-q)^{N-n_h} q^{\alpha-1}(1-q)^{\alpha-1}\, dq$$

$$= \binom{N}{n_h} \frac{1}{B(\alpha,\alpha)} \int_0^1 q^{n_h+\alpha-1} (1-q)^{N-n_h+\alpha-1}\, dq$$

$$= \binom{N}{n_h} \frac{B(n_K+\alpha, N-n_k+\alpha)}{B(\alpha,\alpha)}$$

and we can compute the odds ratio as

$$o = \frac{2^{-N}\, B(\alpha,\alpha)}{B(n_K+\alpha, N-n_k+\alpha)},$$

and subsequently the sought-for probability $P(H|n_k, \mathcal{I})$. We consider the same extreme case as before, i.e. $n_K = N$ or $n_K = N/2$.

17.2 Independent measurements with Gaussian noise

Here we discuss some examples which are particularly suited for a comparison to the frequentist approach, discussed later on. The examples have been introduced and discussed in great detail by L. Bretthorst [24].

17.2.1 Is the mean correct?

Presumably the most straightforward example concerns the analysis of a single feature of a physical object X, e.g. its mass, which can be measured directly.

The hypothesis (H_0) says: *The true value of X is x_0.*

The complementary hypothesis implies that the true value is not x_0, but anywhere in a reasonable interval I. We suppose furthermore that the statistical measurement errors are independent and Gaussian with known variance. All these details are part of the background information $\mathcal{I}$. It also contains all information available on the test object. At least we should know whether we are dealing for example with atomic or extraterrestrial objects, because it implies an order of magnitude estimate of the interval $I = (x_1, x_2)$ in which the value of X is to be expected. The experimental data are collected in a sample $\boldsymbol{x} = \{x_1, \ldots, x_N\}$ of size N. Given the hypothesis H, the true value is x_0 and the likelihood for Gaussian noise, which will be derived in Chapter 19 [p. 284], can be brought into the following form:

$$P(\boldsymbol{x}|H, \mathcal{I}) = C\, e^{-\frac{N}{2\sigma^2}(x_0-\overline{x})^2}.$$

Representative results are shown in Tables 17.1 and 17.2, respectively. The normalization factor C is irrelevant in the present context. For the alternative hypothesis we need the prior for x. If the interval I is all that is known, the prior reads

$$P(x|\overline{H},\mathcal{I}) = \frac{1}{V_p}\ \mathbf{1}(x_1 \leq x \leq x_2)\,,$$

where $V_p = |x_2 - x_1|$ is the accessible volume according to the prior knowledge. The marginal likelihood for the alternative hypothesis is then

$$P(\boldsymbol{x}|\overline{H},\mathcal{I}) = \int P(\boldsymbol{x}|x,\sigma,\mathcal{I})\ P(x|\overline{H},\mathcal{I})\,dx\ = \frac{C}{V_p}\int_{x_1}^{x_2} e^{-\frac{N\,(x-\overline{x})^2}{2\sigma^2}}\ dx.$$

For the sake of computational simplicity, we assume that the distance of the sample mean $\overline{x}$ from the interval boundaries $x_{1/2}$ is much greater than a standard error

$$\mathrm{SE} = \frac{\sigma}{\sqrt{N}}.$$

Anything else would be very surprising, since the interval was obtained by an order of magnitude estimate. This allows us to replace the integration limit in the marginal likelihood by $\pm\infty$:

$$P(\boldsymbol{x}|\overline{H},\mathcal{I}) = \frac{C}{V_p}\int_{-\infty}^{\infty} e^{-\frac{N}{2\sigma^2}\,(x-\overline{x})^2}\ dx = \frac{C\cdot\mathrm{SE}\cdot\sqrt{2\pi}}{V_p},$$

and the Bayes factor is

$$o_{\mathrm{BF}} = \frac{V_p}{\mathrm{SE}}\,\frac{1}{\sqrt{2\pi}}\ e^{-\frac{1}{2}z^2}. \tag{17.6}$$

The final result is closely related to the frequentist *z-test*:

BAYESIAN *z*-TEST

$$o_{\mathrm{BF}} = \frac{V_p}{\mathrm{SE}}\ p_Z(z), \tag{17.7}$$

$$z := \frac{|x_0 - \overline{x}|}{\mathrm{SE}}.$$

The result is extremely satisfying for several reasons:

- Only the probability density at the value of the statistic z enters, that is, only the really measured value matters.
- There is no need for a more or less ad hoc significance level. This part is taken over in some sense by the ratio $\frac{V_p}{\mathrm{SE}}$. This is very satisfying because it depends on the actual subject-specific expert knowledge.
- The decision whether an hypothesis is right or wrong is formally on an equal footing.

Interestingly, the difference between the P-value, which is at the heart of frequentist hypothesis testing, and the value of the probability density of the test statistic at the point z_0 is closely related in case of the z-test. The P-value is defined as the complementary CDF

$$P := \int_{z_0}^{\infty} p_z(z)\, dz.$$

A numerical analysis reveals that the ratio

$$q = \frac{p_z(z_0)}{\int_{z_0}^{\infty} p_z(z)\, dz}$$

ranges from 1 to 4 in the relevant P-value regime, i.e. for $P \in (0.0001, 0.5)$. The real difference in the case of the z-statistic lies in the Ockham factor. More details are discussed in Section 3.3 [p. 43].

17.2.2 Test on the mean with unknown variance

Next we design the problem to be a little more realistic in the sense that the value of the variance is unknown. As before, the hypothesis claims that x_0 is the true value of X. Now the marginal likelihood in the light of H involves an integral over the unknown variance:

$$p(\boldsymbol{x}|H, N, \mathcal{I}) = p(\boldsymbol{x}|x_0, N, \mathcal{I}) = \int p(\boldsymbol{x}|x_0, \sigma, N, \mathcal{I})\; p(\sigma|\cancel{x_0}, \cancel{N}, \mathcal{I})\, d\sigma.$$

In order to proceed, we still need to specify the prior for the variance, in which we have already struck out the logically impossible dependencies. In order to provide a comparison with the frequentist t-test, we employ an entirely ignorant prior. Since the variance is a scale parameter, it corresponds to the improper Jeffreys' scale prior $p(\sigma|\mathcal{I}) \propto 1/\sigma$. We introduce, however, preliminary lower and upper bounds $\sigma_{l/u}$ which will be eliminated in the end. This procedure ensures that the prior can be normalized:

$$\begin{aligned} p_J(\sigma) &= \frac{1}{V_\sigma}\; \mathbf{1}(\sigma_u \le \sigma \le \sigma_o)\; \frac{1}{\sigma}, \\ V_\sigma &= \ln(\sigma_o/\sigma_u). \end{aligned} \tag{17.8}$$

Based on equation (17.6), the marginalization can be carried out:

$$p(\boldsymbol{x}|N, H, \mathcal{I}) \propto \int_{\sigma_u}^{\sigma_o} \sigma^{-N}\, e^{-\frac{N\left((\Delta x)^2 + (x_0 - \overline{x})^2\right)}{2\sigma^2}}\, \frac{d\sigma}{\sigma}.$$

The prefactors have been suppressed as they drop out of the Bayes factor. The integration limits can now be pushed to 0 and ∞, as we have assumed an ignorant prior for σ. The remaining integral yields

$$p(\boldsymbol{x}|N, H, \mathcal{I}) \propto \left(\frac{N \left(\overline{(\Delta x)^2} + (x_0 - \overline{x})^2 \right)}{2} \right)^{-\frac{N}{2}} \Gamma\left(\tfrac{N}{2}\right)$$

$$\propto \left(1 + \frac{(x_0 - \overline{x})^2}{\overline{(\Delta x)^2}} \right)^{-\frac{N}{2}} .$$

For the sake of comparability to the conventional t-test we introduce a new random variable

$$t := \frac{|x_0 - \overline{x}|}{\mathrm{SE}^*}, \tag{17.9}$$

$$\mathrm{SE}^* := \sqrt{\frac{\overline{(\Delta x)^2}}{N-1}},$$

with SE* being the sample estimate of the standard error. So we obtain

$$p(\boldsymbol{x}|N, H, \mathcal{I}) = p(\boldsymbol{x}|N, x_0, \mathcal{I}) = C \left(1 + \frac{t^2}{N-1} \right)^{-\frac{N}{2}} . \tag{17.10}$$

Next we turn to the complementary hypothesis $\overline{H}$. In addition to the case H, we have to integrate over x_0. Along with the same prior for x_0 as in the previous section, we get

$$p(\boldsymbol{x}|N, \overline{H}, \mathcal{I}) = \int p(\boldsymbol{x}|x, \cancel{\overline{H}}, N, \mathcal{I})\ p(x|\cancel{N}, \overline{H}, \mathcal{I})\ dx_0$$

$$= \frac{C}{V_p} \int_{x_1}^{x_2} \left(1 + \frac{t^2}{N-1} \right)^{-\frac{N}{2}} dx$$

$$= \frac{C}{V_p} \int_{x_1}^{x_2} \left(1 + \frac{(x - \overline{x})^2}{\overline{(\Delta x)^2}} \right)^{-\frac{N}{2}} dx.$$

For the sake of convenience we assume that the data constraint is strong enough to confine the integrand to the interval I, so that we can push the integration limits to $\pm\infty$. The remaining integral can be computed analytically:

$$p(\boldsymbol{x}|N, \overline{H}, \mathcal{I}) = C\ \frac{1}{V_p} \int_{-\infty}^{\infty} \left(1 + \frac{y^2}{\overline{(\Delta x)^2}} \right)^{-\frac{N}{2}} dy$$

$$= C\ \frac{\sigma}{V_p}\ \sqrt{\pi}\ \frac{\Gamma\left(\frac{N-1}{2}\right)}{\Gamma\left(\frac{N}{2}\right)}.$$

The Bayes factor finally reads

$$o_{\mathrm{BF}} = \frac{V_p}{\sqrt{\frac{(\Delta x)^2}{N-1}}} \frac{\Gamma\left(\frac{N}{2}\right)}{\Gamma\left(\frac{N-1}{2}\right)} \frac{1}{\sqrt{\pi\,(N-1)}} \left(1 + \frac{t^2}{N-1}\right)^{-\frac{N}{2}}.$$

It can be expressed in terms of the PDF of the Student t-distribution defined with $\nu = N-1$ degrees of freedom in equation (7.31a) [p. 109]:

BAYESIAN t-TEST

$$\begin{aligned} o_{\mathrm{BF}} &= \frac{V_p}{\mathrm{SE}^*}\, p_t(t|N-1), \\ t &= \frac{x_0 - \overline{x}}{\mathrm{SE}^*}, \\ \mathrm{SE}^* &= \sqrt{\frac{(\Delta x)^2}{N-1}}. \end{aligned} \tag{17.11}$$

In spite of the fact that the variance prior is improper, we still get a normalized result and the volume V_σ of the variance prior drops out.

The structure of the result is similar to that of the previous section: the PDF of the test at the corresponding value of the test statistic (here t) multiplied by the ratio of prior and likelihood volume. As before, the straightforward application of the rules of probability theory yields with little effort the relevant statistic.

We will see in the forthcoming chapter that there are two essential differences to the frequentist approach. Firstly, the Bayesian approach contains an Ockham factor, which quantifies naturally and unambiguously the significance level. Secondly, the result depends on the value of the test statistic evaluated for the measured sample, while the frequentist approach relies on hypothetical samples by invoking the complementary CDF. However, also in the present case the ratio of the PDF to the complementary CDF is in the range 0.8–4 for common significance levels and sample sizes $N \geq 10$. So there is still a qualitative similarity between Bayesian and frequentist hypothesis testing. That is, however, not in all cases true.

17.2.3 Do two samples have the same mean?

Consider the following situation. For two different physical objects we want to know whether they have the same mass. To this end appropriate measurements are performed. We obtain two sets of repeated measurements (samples), one for each species. The data are independent and have Gaussian noise with unknown but identical variance, which is summarized in the background information $\mathcal{I}$. Moreover, the goal is to assess the hypothesis that both species have the same mass, or rather

H: The underlying true mean is the same for both species.

We denote the two data sets by $\boldsymbol{d}^{(\alpha)} = \{d_1^{(\alpha)}, d_2^{(\alpha)}, \dots, d_{L^\alpha}^{(\alpha)}\}$, with L^α. Since the hypothesis only specifies the equality of the masses but not their value, we need the marginal likelihood $p(\boldsymbol{d}^{(1)}, \boldsymbol{d}^{(2)}|A, \mathcal{I})$ for $A \in \{H, \overline{H}\}$.

Marginal likelihood of H

We start out with $A = H$. Besides the unknown mass m we also have to marginalize over the unknown standard deviation σ:

$$p(\boldsymbol{d}^{(1)}, \boldsymbol{d}^{(2)}|H, \mathcal{I}) = \int p(\boldsymbol{d}^{(1)}, \boldsymbol{d}^{(2)}|m, \sigma, H, \mathcal{I})\; p(m, \sigma|H, \mathcal{I})\; dm\; d\sigma.$$

As the data sets as well as the value for the mass and the standard deviation are uncorrelated, the integrand factorizes as

$$\begin{aligned} p(\boldsymbol{d}^{(1)}, \boldsymbol{d}^{(2)}|H, \mathcal{I}) &= \int \left[\prod_{\alpha=1}^{2} p(\boldsymbol{d}^{(\alpha)}|m, \sigma, H, \mathcal{I})\right] p(m|H, \mathcal{I}) p(\sigma|H, \mathcal{I})\; dm\; d\sigma \\ &= (2\pi)^{-\frac{L}{2}} \int d\sigma\; \sigma^{-L} \;\; p(\sigma|H, \mathcal{I}) \\ &\quad \times \int dm\;\, p(m|H, \mathcal{I})\; e^{-\frac{1}{2\sigma^2} \sum\limits_{\alpha=1}^{2} \sum\limits_{i=1}^{L^{(\alpha)}} \left(d_i^{(\alpha)} - m\right)^2}, \end{aligned} \tag{17.12}$$

where $L = L^{(1)} + L^{(2)}$. The argument of the exponential can be simplified to

$$\sum_{\alpha=1}^{2} \sum_{i=1}^{L^{(\alpha)}} \left(d_i^{(\alpha)} - m\right)^2 = L\left[\overline{(\Delta d)^2} + \left(m - \overline{d}\right)^2\right].$$

Here $\overline{d}$ stands for the arithmetic mean of all data points, irrespective of which species they belong to. We insert this expression into equation (17.12). As argued in the previous sections, we use a flat prior

$$p(m|\mathcal{I}) = \frac{1}{V_p}\;\; \boldsymbol{1}(m_l \le m \le m_u)\,.$$

We also push the integration limits to $\pm\infty$. As the parts of the integrand sticking out of the interval $I = (m_l, m_u)$ can be neglected in the case of decent experimental data, we obtain

$$\begin{aligned} \int p(m|H, \mathcal{I})\; e^{-\frac{1}{2\sigma^2} \sum\limits_{\alpha=1}^{2} \sum\limits_{i=1}^{L^{(\alpha)}} \left(d_i^{(\alpha)} - m\right)^2} dm &= \frac{e^{-\frac{L}{2\sigma^2}\overline{(\Delta d)^2}}}{V_p} \int_{-\infty}^{\infty} dm\; e^{-\frac{L}{2\sigma^2}(m-\overline{d})^2} \\ &= \frac{e^{-\frac{L}{2\sigma^2}\overline{(\Delta d)^2}}}{V_p} \sqrt{\frac{2\pi\sigma^2}{L}}. \end{aligned} \tag{17.13}$$

Next, we insert this result into equation (17.12) and employ again the normalized Jeffreys' scale prior for σ (equation (17.8)). After pushing the integration limits to 0 and ∞, we obtain

$$p(\boldsymbol{d}^{(1)},\boldsymbol{d}^{(2)}|H,\mathcal{I}) = \frac{(2\pi)^{-\frac{L-1}{2}}}{V_\sigma\ V_p\ \sqrt{L}} \int\limits_0^\infty \frac{d\sigma}{\sigma}\ e^{-\frac{L}{2\sigma^2}\overline{(\Delta d)^2}}\ \sigma^{-(L-1)}$$

$$= \frac{\pi^{-\frac{L-1}{2}}\ L^{-\frac{L}{2}}}{2\ V_\sigma\ V_p}\ \Gamma\left(\tfrac{L-1}{2}\right)\ \left[\overline{(\Delta d)^2}\right]^{-\frac{L-1}{2}}. \tag{17.14}$$

Marginal likelihood of $\overline{H}$

If $\overline{H}$ applies, the underlying masses m_α are different. As outlined at the beginning of this section, the variances of the two data sets are assumed to be equal, albeit unknown. For the sake of simplicity we assume that the prior for m_1 and m_2 is simply

$$p(m_1, m_2|\mathcal{I}) = p(m_1|\mathcal{I})p(m_2|\mathcal{I}).$$

This is certainly not the most realistic prior, but it keeps the calculation transparent and allows us to unravel the key features of the Bayesian hypothesis test. A generalization towards more realistic assumptions is usually straightforward. In analogy to equation (17.12), we now obtain

$$p(\boldsymbol{d}^{(1)},\boldsymbol{d}^{(2)}|\overline{H},\mathcal{I}) = (2\pi)^{-\frac{L}{2}} \int d\sigma\ p(\sigma|\overline{H},\mathcal{I})\ \sigma^{-L}$$

$$\times \prod_{\alpha=1}^{2} \int e^{-\frac{1}{2\sigma^2}\sum\limits_{i=1}^{L^{(\alpha)}}\left(d_i^{(\alpha)}-m_\alpha\right)^2}\ p(m_\alpha|\overline{H},\mathcal{I})\ dm_\alpha. \tag{17.15}$$

The integral over m_α is of the same type as in equation (17.13). We merely have to set $L_{\overline{\alpha}} = 0$, where $\overline{\alpha}$ is the opposite index to α. The integral yields

$$\int e^{-\frac{1}{2\sigma^2}\sum\limits_{i=1}^{L^{(\alpha)}}\left(d_i^{(\alpha)}-m_\alpha\right)^2}\ p(m_\alpha|\overline{H},\mathcal{I})\ dm_\alpha = \frac{\sigma}{V_p}\sqrt{\frac{2\pi}{L^{(\alpha)}}}\ e^{-\frac{L^{(\alpha)}}{2\sigma^2}\overline{(\Delta d^{(\alpha)})^2}},$$

and equation (17.15) turns into

$$p(\boldsymbol{d}^{(1)},\boldsymbol{d}^{(2)}|\overline{H},\mathcal{I}) = \frac{(2\pi)^{-\frac{L-2}{2}}}{V_\sigma\ V_p^2\ \sqrt{L^{(1)}L^{(2)}}} \int\limits_0^\infty \frac{d\sigma}{\sigma}\ \sigma^{-(L-2)}\ e^{-\frac{1}{2\sigma^2}\sum_\alpha L^{(\alpha)}\,\overline{(\Delta d^{(\alpha)})^2}}$$

$$= \frac{\pi^{-\frac{L-2}{2}}}{2\ V_\sigma\ V_p^2\ \sqrt{L^{(1)}L^{(2)}}}\ \Gamma\left(\tfrac{L-2}{2}\right)\ X^{-\frac{L-2}{2}}, \tag{17.16}$$

$$X := \sum_\alpha L^{(\alpha)}\ \overline{(\Delta d^{(\alpha)})^2}. \tag{17.17}$$

The ratio of equation (17.14) to equation (17.16) yields the Bayes factor

$$
\begin{aligned}
o_{\mathrm{BF}} &= \frac{\frac{\pi^{-\frac{L-1}{2}} L^{-\frac{L}{2}}}{2\,V_\sigma\,V_p}\,\Gamma\left(\frac{L-1}{2}\right)\left[\overline{(\Delta d)^2}\right]^{-\frac{L-1}{2}}}{\frac{\pi^{-\frac{L-2}{2}}}{2\,V_\sigma\,V_p^2\,\sqrt{L^{(1)}L^{(2)}}}\,\Gamma\left(\frac{L-2}{2}\right)\,X^{-\frac{L-2}{2}}} \\
&= V_p\,\frac{\Gamma\left(\frac{L-1}{2}\right)}{\Gamma\left(\frac{L-2}{2}\right)}\,\frac{\sqrt{L-2}\,\sqrt{L^{(1)}L^{(2)}}}{\sqrt{\pi(L-2)}\sqrt{L}\sqrt{X}}\left[\frac{L\,\overline{(\Delta d)^2}}{X}\right]^{-\frac{L-1}{2}}.
\end{aligned}
\tag{17.18}
$$

Furthermore, the numerator can be expressed as

$$
L\,\overline{(\Delta d)^2} = X + \frac{L^{(1)}L^{(2)}}{L}\left(\overline{d^{(1)}} - \overline{d^{(2)}}\right)^2.
$$

Proof:

$$
L^2\,\overline{(\Delta d)^2} = L^2\,\overline{d^2} - L^2\,\overline{d}^2.
$$

We compute the two contributions separately:

$$
\begin{aligned}
L^2\,\overline{d^2} &= L\sum_\alpha L^{(\alpha)}\overline{d^{(\alpha)2}} = L\sum_\alpha L^{(\alpha)}\left[\overline{(\Delta d^{(\alpha)})^2} + \overline{d^{(\alpha)}}^2\right] \\
&= L\,X + L\sum_\alpha \overline{d^{(\alpha)}}^2 \\
&= L\,X + \left(L^{(1)}\overline{d^{(1)}}\right)^2 + \left(L^{(2)}\overline{d^{(2)}}\right)^2 + L^{(1)}\,L^{(2)}\left(\overline{d^{(1)}}^2 + \overline{d^{(2)}}^2\right),
\end{aligned}
$$

$$
L^2\,\overline{d}^2 = \left(L^{(1)}\overline{d^{(1)}} + L^{(2)}\overline{d^{(2)}}\right)^2.
$$

Then in total we have

$$
L^2\,\overline{(\Delta d)^2} = L\,X + L^{(1)}L^{(2)}\left(\overline{d^{(1)}} - \overline{d^{(2)}}\right)^2.
$$

□

So, the last factor in equation (17.18) can be reduced to

$$
\left[\frac{L\,\overline{(\Delta d)^2}}{X}\right] = 1 + \frac{\left(\overline{d^{(1)}} - \overline{d^{(2)}}\right)^2}{\frac{L}{L^{(1)}L^{(2)}}X}.
$$

In order to establish ties to the frequentist t-statistic, we introduce

$$
\sigma^{*2} := \frac{X}{L-2} \overset{(17.17)}{=} \frac{1}{L-2}\sum_\alpha L^{(\alpha)}\,\overline{(\Delta d^{(\alpha)})^2}.
\tag{17.19}
$$

This is an unbiased pooled estimator for the common variance of the two samples. As we will see in Section 19.2 [p. 294], the PDF of a sample mean of i.i.d. Gaussian samples of size L and variance σ^2 is also Gaussian, with variance σ^2/L. Applied to the problem of the present section, the two sample means have variance $\frac{\sigma^2}{L^{(\alpha)}}$.

The difference of two Gaussian random numbers with variances σ_1^2 and σ_2^2, respectively, is in general also a Gaussian and its variance is the sum of the individual variances:

$$\sigma_{\text{diff}}^2 = \sigma_1^2 + \sigma_2^2.$$

Therefore, that variance of the difference of the two sample means is

$$\sigma_{\text{diff}}^2 = \frac{\sigma^2}{L^{(1)}} + \frac{\sigma^2}{L^{(2)}} = \sigma^2 \left(\frac{1}{L^{(1)}} + \frac{1}{L^{(2)}} \right) = \sigma^2 \frac{L}{L^{(1)}\, L^{(2)}}.$$

Hence, if we employ σ^{*2} as estimator for the variance, the standard error for the difference in means reads

$$\text{SE}_{\text{diff}}^* := \sqrt{\frac{L}{L_1 L_2}} \sigma^*.$$

Now everything falls neatly into place when we express the Bayes factor of equation (17.18), along with equation (17.19), in terms of $\text{SE}_{\text{diff}}^*$:

$$o_{\text{BF}} = \frac{V_p}{\text{SE}_{\text{diff}}^*} \cdot \frac{\Gamma\left(\frac{L-1}{2}\right)}{\Gamma\left(\frac{L-2}{2}\right)} \frac{1}{\sqrt{\pi(L-2)}} \left(1 + \frac{t^2}{L-2}\right)^{-\frac{L-1}{2}}, \tag{17.20a}$$

$$t := \frac{\overline{d^{(2)}} - \overline{d^{(1)}}}{\text{SE}_{\text{diff}}^*}. \tag{17.20b}$$

The second factor in equation (17.20a) is again the PDF of the Student t-distribution given in equation (7.31a) [p. 109] with $\nu = L - 2$ degrees of freedom. Finally we have

Bayes factor for the hypothesis:
The two samples have the same mean

$$o_{\text{BF}} = \frac{V_p}{\text{SE}_{\text{diff}}^*} \, p_T(t|L-2),$$

$$\left(\text{SE}_{\text{diff}}^*\right)^2 := \frac{L}{L_1 L_2} \frac{1}{L-2} \sum_\alpha L^{(\alpha)} \, \overline{(\Delta d^{(\alpha)})^2}.$$

The result looks similar to that of the previous section, merely the expression for the test statistic and SE has been adjusted.

17.2.4 Do two samples have the same variances?

As before we consider two independent samples $\boldsymbol{d}^{(\alpha)}$ of size L_α, $\alpha = 1, 2$, representing measurement of masses of two different species. The elements within each of the two samples are i.i.n.d. and the corresponding true means are not known. We use the same prior PDF for the two masses.

The hypothesis this time is different. It says

H: The underlying variance of both samples is the same.

The value of the variance is, however, not known. As prior for the common variance we employ again the normalized Jeffreys' scale prior, depicted in equation (17.8) [p. 264]. The marginal likelihood for hypothesis H reads

$$p(\boldsymbol{d}^{(1)}, \boldsymbol{d}^{(2)} | L^{(1)}, L^{(2)}, H, \mathcal{I}).$$

Likelihood for H

Next we employ again the marginalization rule in order to introduce the unknown masses of the two species and the common variance:

$$\begin{aligned} p(\boldsymbol{d}^{(1)}, \boldsymbol{d}^{(2)} | L_1, L_2, H, \mathcal{I}) = \int & p(\boldsymbol{d}^{(1)} | m_1, \sigma, L_1, \mathcal{I})\; p(\boldsymbol{d}^{(2)} | m_2, \sigma, L_2, \mathcal{I}) \\ & \times\; p(m_1|\mathcal{I})\; p(m_2|\mathcal{I})\; p(\sigma|\mathcal{I})\; dm_1\; dm_2\; d\sigma. \end{aligned} \tag{17.21}$$

Now the multivariate Gaussian likelihood can be cast into the following form:

$$\begin{aligned} p(\boldsymbol{d}^{(\alpha)} | m_\alpha, \sigma, L_\alpha, \mathcal{I}) &= (2\pi\sigma^2)^{-\frac{L_\alpha}{2}}\; e^{-\frac{1}{2\sigma^2}\Phi}, \\ \Phi &= \sum_{i=1}^{L^{(\alpha)}} \left(d_i^{(\alpha)} - m_\alpha\right)^2 \\ &= L^{(\alpha)} \left[\overline{(\Delta d^{(\alpha)})^2} + \left(m_\alpha - \overline{d^{(\alpha)}}\right)^2\right]. \end{aligned}$$

For the sake of clarity we introduce the abbreviation

$$v_\alpha = L_\alpha\; \overline{(\Delta d^{(\alpha)})^2}.$$

The integrals over the masses are easily performed if we stretch the integration interval to cover the entire real axis, as argued before. In order to avoid duplications, we introduce different indices for the variance of the two data sets:

$$\begin{aligned} \int & p(\boldsymbol{d}^{(\alpha)} | m_\alpha, \sigma_\alpha, L_\alpha, \mathcal{I})\; p(m_\alpha|\mathcal{I})\; dm_\alpha \\ &= \frac{(2\pi\sigma_\alpha^2)^{-\frac{L_\alpha}{2}}}{V_p}\; e^{-\frac{v_\alpha}{2\sigma_\alpha^2}} \int_{-\infty}^{\infty} e^{-\frac{L_\alpha}{2\sigma_\alpha^2}\,(m_\alpha - \overline{d^{(\alpha)}})^2}\; dm_\alpha \\ &= \frac{(2\pi)^{-\frac{L_\alpha - 1}{2}}}{V_p\,\sqrt{L_\alpha}}\; \sigma_\alpha^{-(L_\alpha - 1)}\; e^{-\frac{v_\alpha}{2\sigma_\alpha^2}} \\ &= C_\alpha\; \sigma_\alpha^{-(L_\alpha - 1)}\; e^{-\frac{v_\alpha}{2\sigma_\alpha^2}}. \end{aligned} \tag{17.22}$$

The constants C_α are immaterial as they vanish in the odds ratio. Now we can tackle the σ integration:

$$p(\boldsymbol{d}^{(1)}, \boldsymbol{d}^{(2)}|L_1, L_2, H, \mathcal{I}) = \frac{C_1 C_2}{V_\sigma} \int_{\sigma_u}^{\sigma_o} \sigma^{-(L-2)}\, e^{-\frac{(v_1+v_2)}{2\sigma^2}}\, \frac{d\sigma}{\sigma}.$$

Based on the reasoning explained before, the integration interval can be extended to $(0, \infty)$. The result of the integral is given in equation (7.27) [p. 108]

$$p(\boldsymbol{d}^{(1)}, \boldsymbol{d}^{(2)}|L_1, L_2, H, \mathcal{I}) = \frac{C_1 C_2}{2V_\sigma}\, 2^{\frac{L-2}{2}}\, (v_1 + v_2)^{-\frac{L-2}{2}}\, \Gamma\left(\tfrac{L-2}{2}\right). \tag{17.23}$$

Likelihood for $\overline{H}$

The complement to hypothesis H assumes different variances for the two data sets. For the marginalization over the two variances we need the joint prior

$$p(\sigma_1, \sigma_2|\mathcal{I}) = p(\sigma_2|\sigma_1, \mathcal{I})\; p(\sigma_1|\mathcal{I}).$$

For $p(\sigma_1|\mathcal{I})$ we will still use the ignorant prior. Now we want to be a bit more realistic as far as the first factor is concerned. So far we have used an entirely ignorant prior for σ, which actually means that the prior volume V_σ eventually becomes infinite. Up to now that was not really a problem, as the data constraints predominated the integrals and the prior volume dropped out. That will not be the case if we compare hypotheses with different numbers of variance integrals. As a matter of fact, in a realistic situation we might be rather ignorant as to the overall scale of the variances, but as soon as we know one variance, σ_1 say, then we know in most cases that the variance σ_2 will not differ too much. We assume here that we know that σ_2 and σ_1 will only differ by a certain percentage. So the prior is actually

$$\begin{aligned} p(\sigma_1, \sigma_2|\kappa, \mathcal{I}) &= p(\sigma_2|\sigma_1, \kappa, \mathcal{I})\; p(\sigma_1|\not\kappa, \mathcal{I}) \\ &= \frac{1}{V_\kappa \sigma_1}\theta\left[\frac{1}{\kappa}\sigma_1 \le \sigma_2 \le \kappa\sigma_1\right]\; p(\sigma_1|\not\kappa, \mathcal{I}) \\ &= \frac{1}{V_\kappa V_\sigma}\theta\left[\frac{1}{\kappa}\sigma_1 \le \sigma_2 \le \kappa\sigma_1\right]\; \frac{1}{\sigma_1\sigma_2}, \\ V_\kappa &:= \kappa - \frac{1}{\kappa}. \end{aligned}$$

The parameter κ is greater than one. For the calculation of the marginal likelihood for $\overline{H}$ we introduce the abbreviation $\nu_\alpha = L^{(\alpha)} - 1$ as it simplifies the resulting expression

$$p(\boldsymbol{d}^{(1)}, \boldsymbol{d}^{(2)}|L_1, L_2, \overline{H}, \mathcal{I}) = \frac{C_1 C_2}{V_\sigma V_\kappa} \int_{\sigma_u}^{\sigma_o} \frac{d\sigma_1}{\sigma_1}\sigma_1^{-n_1}\, e^{-\frac{v_1}{2\sigma_1^2}} \int_{\sigma_1/\kappa}^{\kappa\sigma_1} \frac{d\sigma_2}{\sigma_2}\sigma_2^{-n_2}\, e^{-\frac{v_2}{2\sigma_2^2}}.$$

As before, it is legitimate to spread the integral from 0 to ∞. The integral can then be calculated analytically. Starting from equation (17.22), we derive

$$p(\boldsymbol{d}^{(1)}, \boldsymbol{d}^{(2)}|L_1, L_2, \overline{H}, \mathcal{I}) = \frac{C_1 C_2}{V_\sigma V_\kappa} \int_{-\infty}^{\infty} \frac{d\sigma_1}{\sigma_1} \sigma_1^{-n_1} e^{-\frac{v_1}{2\sigma_1^2}} \int_0^{\infty} \frac{d\sigma_2}{\sigma_2} \sigma_2^{-n_2} e^{-\frac{v_2}{2\sigma_2^2}}.$$

These integrals have been computed in equation (7.27) [p. 108] and yield

$$p(\boldsymbol{d}^{(1)}, \boldsymbol{d}^{(2)}|L_1, L_2, \overline{H}, \mathcal{I}) = \frac{C_1 C_2}{V_\kappa V_\sigma} \frac{1}{4} \left[v_1^{-\frac{n_1}{2}} 2^{\frac{n_1}{2}} \Gamma\left(\tfrac{n_1}{2}\right) \right] \left[v_2^{-\frac{n_2}{2}} 2^{\frac{n_2}{2}} \Gamma\left(\tfrac{n_2}{2}\right) \right].$$

The result in terms of the Bayes factor is gradually assuming shape:

$$o_{\mathrm{BF}} = 4 V_\kappa \, v_1^{\frac{n_1}{2}} \, v_2^{\frac{n_2}{2}} \, (v_1 + v_2)^{-\frac{n_1+n_2}{2}} \frac{\Gamma\left(\frac{L-2}{2}\right)}{\Gamma\left(\frac{n_1}{2}\right) \Gamma\left(\frac{n_2}{2}\right)}.$$

Apparently, the Bayes factor is invariant against exchange of the data sets. This symmetry is to be expected, as both hypotheses make no difference between the two samples. If we introduce the ratio $x := \frac{v_1}{v_2}$, the Bayes factor reads

$$o_{\mathrm{BF}} = 4 V_\kappa \cdot \frac{1}{B(\frac{n_1}{2}, \frac{n_2}{2})} \, x^{\frac{n_1}{2}} \, (1+x)^{-\frac{n_1+n_2}{2}}.$$

The second factor is closely related to the sampling distribution of the ratio x. We consider integrals of the type

$$I(\alpha, \beta) := \int_0^{\infty} x^{\alpha} (1+x)^{-\beta} \, \frac{dx}{x} = \frac{\Gamma(\alpha)\Gamma(\beta-\alpha)}{\Gamma(\beta)}. \tag{17.24}$$

The normalized PDF for x is therefore

$$p_x(x|n_1, n_2) := \frac{1}{B\left(\frac{n_1}{2}, \frac{n_2}{2}\right)} \, x^{\frac{n_1}{2}-1} \, (1+x)^{-\frac{n_1+n_2}{2}}.$$

We will verify this result later on. The first moment of this PDF is easily computed:

$$\begin{aligned}
\left\langle x^n \right\rangle_{n_1,n_2} &= \frac{1}{B\left(\frac{n_1}{2}, \frac{n_2}{2}\right)} \int_0^{\infty} x^{\frac{n_1}{2}+n} \, (1+x)^{-\frac{n_1+n_2}{2}} \, \frac{dx}{x} \\
&\overset{(17.24)}{=} \frac{\cancel{\Gamma\left(\frac{n_1+n_2}{2}\right)}}{\Gamma\left(\frac{n_1}{2}\right)\Gamma\left(\frac{n_2}{2}\right)} \, \frac{\Gamma\left(\frac{n_1}{2}+n\right)\Gamma\left(\frac{n_2}{2}-n\right)}{\cancel{\Gamma\left(\frac{n_1+n_2}{2}\right)}} \\
&= \frac{\left(\frac{n_1}{2}+n-1\right)^n}{\left(\frac{n_2}{2}-1\right)^n}.
\end{aligned}$$

In particular, the lowest two moments are

$$\langle x \rangle = \frac{\left(\frac{n_1}{2}\right)}{\left(\frac{n_2}{2}-1\right)} = \frac{n_1}{n_2-2}, \tag{17.25}$$

$$\left\langle x^2 \right\rangle = \frac{\left(\frac{n_1}{2}+1\right)^2}{\left(\frac{n_2}{2}-1\right)^2} = \frac{(n_1+2)\, n_1}{(n_2-2)\,(n_2-4)},$$

$$\text{var}(x) = \frac{2n_1(n_1+n_2-2)}{(n_2-4)(n_2-2)^2}. \tag{17.26}$$

In order to unravel the relationship to the frequentist F-statistic [157], we introduce the corresponding test statistic

$$f = \frac{\sigma_1^{*2}}{\sigma_2^{*2}} \overset{(19.4)}{=} \frac{n_2}{n_1}\frac{v_1}{v_2}, \tag{17.27}$$

which represents the ratio of the estimators of the two variances. In this case the correctness of H is indicated by the proximity of f to one. Expressed in terms of f, the Bayes factor reads

$$o_{\text{BF}} = 4V_\kappa \cdot f \cdot \frac{n_1^{n_1} n_2^{n_2}}{B\left(\frac{n_1}{2}\,\frac{n_2}{2}\right)}\, f^{\frac{n_1}{2}-1}\,(n_2+n_1 f)^{-\frac{n_1+n_2}{2}}.$$

The second factor is equivalent to the PDF of the F-statistic, or rather for Fisher's variance ratio distribution [157]

PDF OF THE F-STATISTIC
FISHER'S VARIANCE RATIO DISTRIBUTION

$$p_F(f|n_1,n_2) = \frac{n_1^{n_1} n_2^{n_2}}{B\left(\frac{n_1}{2}\,\frac{n_2}{2}\right)}\, f^{\frac{n_1}{2}-1}\,(n_2+n_1 f)^{-\frac{n_1+n_2}{2}}, \tag{17.28a}$$

$$\int_0^\infty p_F(f|n_1,n_2)\, df = 1, \tag{17.28b}$$

$$\langle f \rangle = \frac{n_2}{n_2-2}, \tag{17.28c}$$

$$\text{var}(f) = \frac{2n_2^2(n_1+n_2-2)}{n_1(n_2-2)^2(n_2-4)}. \tag{17.28d}$$

The mean and variance follow directly from the definition of f and equations (17.25) and (17.26). In terms of the PDF of the F-statistic, the Bayes factor finally reads

BAYES FACTOR FOR THE HYPOTHESIS

The two samples have the same variance

$$o_{\mathrm{BF}} = 4V_\kappa \cdot f \cdot p_F(f|n_1, n_2),$$
$$f = \frac{\sigma_1^{*2}}{\sigma_2^{*2}} = \frac{n_2}{n_1}\frac{v_1}{v_2}. \tag{17.29}$$

It is noteworthy that V_κ is dimensionless. Also, this form of the result is invariant against exchange of the samples (i.e. against the transformation $n_1 \Leftrightarrow n_2$ and $f \to f^{-1}$). The extra factor f is essential for this symmetry.

Example: Is an instrument still calibrated?

A measurement device may have a Gaussian noise with zero mean and constant, though unknown, variance. Two series of measurements with $L = 20$ readings are performed for two different targets with fairly long time lapse in between. In addition to the physical question, a check on whether the device is still working with the same variance is important.

From experience with the measuring apparatus we know that once it is misaligned, σ can vary up to one order of magnitude, i.e. $\kappa = 10$ and $V_\kappa = 10 - 1/10 = 9.9$. The hypothesis to be assessed reads:

H: The variance of the experimental noise is the same in both data sets.

We also know from long-standing experience that the failure probability is roughly 1%, i.e. the prior odds is

$$o_{\mathrm{P}} = \frac{P(H|\mathcal{I})}{P(\overline{H}|\mathcal{I})} = \frac{0.99}{0.01} = 99.$$

Let us assume that the measurement provides sample-based estimates for the standard deviation of $\sigma_1^* = 2.5$ and $\sigma_2^* = 4.4$, which implies

$$f = \left(\frac{2.5}{4.4}\right)^2 = 3.1.$$

The numerical value of the odds ratio is

$$\begin{aligned} o &= p_P \cdot 2 \cdot V_\kappa \cdot f \cdot p_F(f|L_1 - 1, L_2 - 1) \\ &= \underbrace{99 \cdot 2 \cdot 9.9 \cdot\ 3.1}_{=6076.62} \cdot \underbrace{p_F(3.1|L_1 - 1, L_2 - 1)}_{=0.015} = 93.40. \end{aligned}$$

The resulting probability is

$$P(H|D, \mathcal{I}) = \frac{o}{1 + o} = 0.99.$$

Based on this result we can safely assume that the measurement apparatus is still working perfectly well.

18

The frequentist approach

As outlined earlier, in the frequentist's reasoning there is no such thing as a probability of or for a hypothesis, as the latter is not a random variable. The basic concept of hypothesis tests in the frequentist theory has been introduced briefly in Section 2.3 [p. 28]. The key idea is fairly simple, though not entirely obvious, and we will start out with a simple and transparent example.

18.1 Introduction

Let us consider a basic example of quality control. A manufacturer sells components of electronic devices. In order to verify that his production line is working properly, he or she periodically takes samples and checks whether all electronic features are correct. If this is the case the component is called intact, otherwise defect. Let the number of elements per test (sample size) be N and the number of defect components be denoted by n. By virtue of an agreement with the clients, the percentage of defective components should not exceed a given threshold q. For the manufacturer, the optimal state of his production line is if the mean number of defective parts in the sample is $\mu = qN$. If there are more defective components, the manufacturer would have to pay a penalty and if there are less, the production line could be modified in one way or another to become more cost-effective. Therefore, the hypothesis relevant for the manufacturer is

H : The production line is working properly

which is equivalent to

H : The mean number of defective parts in a sample of size N is $\mu = qN$.

Naturally, there will be random fluctuations in the number n of defective components. Based on the measured number n^* he wants to assess the status of the production line, or rather test his hypothesis based on frequentist statistics.

Since the probability for H_0 given z^* is a taboo in the frequentist way of reasoning, one considers the analogue of the likelihood, namely the probability density for z^* assuming

H_0 applies. Now one might be prompted to use this very probability density as a measure for the correctness of H_0. But that does not work at all.

A measure for the discrepancy between H_0 and the data is

$$\Delta := \frac{n - \langle n \rangle}{\mathrm{SE}}.$$

Part of the hypothesis is that the defective parts are produced independently and at random with a binomial distribution with failure rate q. Therefore, $\langle n \rangle = qN$ and $\mathrm{SE} = \frac{q(1-q)}{\sqrt{N}}$. For the sake of clarity we assume that qn is an integer. The best situation for H_0 is then represented by $\Delta = 0$, i.e. $n = \mu$, and the corresponding probability is

$$P(\mu = qN|q, N, H) = \binom{N}{\mu} q^{\mu}(1-q)^{N-\mu}.$$

We have seen, in connection with the local limit theorem (equation (6.8) [p. 86]), that for $N \gg 1$ the probability tends to zero as

$$P(\mu|q, N, H) \propto \frac{1}{\sqrt{N}} \quad \underset{N \to \infty}{\longrightarrow} 0.$$

Consequently, the probability for a single value of the test statistic is not suited as a measure for the correctness of H_0. R. A. Fisher proposed a way out by the concept of a *test of significance* [72, 73]. The key quantity in his approach is the so-called P value.

Definition 18.1 (P value) *The P value is the probability for getting a value of the test statistic at least as extreme as that observed solely by chance, provided the null hypothesis H_0 is true.*

In the present example the P value would be

$$P := P(n \geq n^*|q, N, H_0) = \sum_{n=n^*}^{\infty} P(n|q, N, H, \mathcal{I}).$$

Statistical inference in the Fisherian approach is based on the P value and the significance level.

Definition 18.2 (Significance level α) *If the P value is below the significance level α, the null hypothesis is rejected. (R. A. Fisher advocated using 5% as the standard level and 1% in particularly stringent cases.)*

A small P value suggests that the null hypothesis is unlikely to be true. The smaller it is, the more convincing is the rejection of the null hypothesis. Before we can dive into the details of the frequentist approach to testing hypotheses, it is necessary to give some more definitions.

Definition 18.3 (Test statistic) *A test statistic is an appropriate measure, like z or Δ, for the discrepancy between data and the null hypothesis. It is a number that is uniquely specified by the data and H_0.*

Given the correctness of H_0 and a test statistic, t say, one computes the probability for a particular value of $t = t^*$, which is called the *sampling distribution.*

Definition 18.4 (Sampling distribution) *The sampling distribution is the probability (probability density) for a value (t) of a test statistic, given H_0 is true.*

The P value is therefore the probability mass in the tail(s) of the sampling distribution. If the P value is below a predefined threshold α (*the significance level*), the hypothesis is rejected. The data are called statistically *significant* if they are unlikely to have occurred by chance alone.

The reasoning behind this is that if H_0 is correct, the probability for measuring the observed value t^* of the statistic (or an even more extreme value) is too small for the measured data to have occurred merely by chance. This way of reasoning needs getting used to, as we have only measured t^* and nothing more extreme. The idea becomes more digestible if we use the concept of the rejection region.

Definition 18.5 (Critical/rejection region) *The critical or rejection region ($\mathcal{R}_\alpha$) of a hypothesis is the set of all data that correspond to P values smaller than the specific significance level α.*

It should be pointed out that the rejection region depends on the significance level α and on the test statistic. α is also called the *size* of the rejection region.

The procedure of a *significance test* is as follows:

- Formulate the null hypothesis.
- Define the appropriate test statistic, t say, and compute the value t^* corresponding to the measured data.
- Specify the significance level α.
- Determine the rejection region $\mathcal{R}_\alpha$.
- Reject the hypothesis if $t^* \in \mathcal{R}_\alpha$.

Neyman and Pearson [153] showed that in the long run, α is the percentage of cases in which a correct null hypothesis is rejected. In Bayesian reasoning it is the probability for rejecting a true null. This implies the so-called type I error, namely *false rejection*. It should be emphasized that this probability is not to be confused with the probability that the null hypothesis is true, nor whether any specific alternative hypothesis is true. Now, what are the implications if H_0 is not rejected? One option is that it implies no further information. Another tempting but erroneous conclusion would be to accept the hypothesis as being correct. This is a fundamental logical fallacy known as the *argument from ignorance*. An hypothesis is not correct just because it is not contradicted by the data. R. A. Fisher did not tire of emphasizing for decades the fallacy of this idea: 'It would, therefore, add greatly to the clarity with which the tests of significance are regarded if it were generally understood that tests of significance, when used accurately are capable of rejecting or invalidating hypotheses, in so far as they are contradicted by the data: but that they are never capable of establishing them as certainly true ...'.

Neyman and Pearson [153] had severe objections to the 'test of significance' proposed by Fisher. They introduced a different concept which is called **hypothesis testing**. The wording is a bit misleading, as the goal of both approaches is to test hypotheses. From an introductory perspective, there are two main differences: the null hypothesis is always tested against an *alternative hypothesis* H_a, and the null can also be accepted not just rejected. In general the alternative hypothesis is not just the complement of H_0. In the above quality-control example, the alternative hypothesis is

$$H_a : \text{The mean is greater than } qN.$$

In general, the alternative hypothesis depends strongly on the intention of the person invoking the hypothesis testing. For the manufacturer the purpose of the test is not to seek the truth but simply to minimize costs, or rather to avoid a penalty fee.

The Neyman–Pearson approach is tailored to provide a decision rule for accepting or rejecting the null hypothesis. Or alternatively, for accepting or rejecting the alternative hypothesis. This introduces a second, so-called type II error, namely *false acceptance* of the null.

Fisher had a strong opposition to the concept of alternative hypotheses. He wrote: 'The frequency of the 2nd kind (*type II error*) must depend . . . greatly on how close the rival theories resemble the null' [124]. It has been shown, indeed, that the discriminatory power of statistical (frequentist) tests decreases and the test becomes useless if the set of alternatives is too complex. But it is more than that, the discrepancies between the two approaches are particularly apparent in the case of mixture models. There has been a long-lasting controversy about the pros and cons of the two concepts, and this is still not entirely settled. There is a substantial literature about inconsistencies between significance testing and hypothesis testing, and the attempt to unify them [87, 124, 177]. It is, however, not the purpose of this book to dwell more on this – to some extent philosophical – controversy and we will rather concentrate on more practical aspects of these approaches, the way they are commonly used, and we will only consider cases which are undisputed in the frequentist community.

The quality-control example led to a *one-sided* test, which is defined as follows:

Definition 18.6 (One-sided/one-tailed test) *In a one-sided (one-tailed) test the rejection region is located in one tail of the probability (sampling) distribution. That is, the test statistic, t say, is either less or alternatively greater than a critical value t^*, but not both.*

In the quality-control example we have mentioned that it might also be undesirable to have a smaller error rate than demanded by the client, as that might be associated with avoidable production costs. In general, there are circumstances conceivable in which deviations of the test statistic in both directions are to be avoided. In this case one would use a *two-sided test.*

Definition 18.7 (Two-sided/two-tailed test) *In a two-sided (two-tailed) test the rejection region is located in both tails of the probability distribution. In other words, the test statistic is less than a first critical value t_1^* or greater then a second critical value t_2^*.*

If the sample under consideration falls into either of the critical areas, the alternative hypothesis will be accepted instead of the null hypothesis.

In the example studied so far there is only one parameter q and the null hypothesis claims that the parameter has a certain value $q = q_0$. This is also referred to as the *point* hypothesis. In frequentist statistics a distinction is made between *simple* and *compound* hypotheses.

Definition 18.8 (Simple hypothesis) *A simple hypothesis is an hypothesis that completely specifies the probability distribution.*

Examples are:

- Two medical treatments give identical results.
- The sample originates from a binomial distribution and the corresponding parameter is $p = 0.6$.
- The sample is drawn from a normal distribution with mean $\mu = 3.2$ and standard deviation $\sigma = 2.5$.

Definition 18.9 (Compound hypothesis) *A compound hypothesis does not completely specify the distribution.*

This is the case in the following examples:

- The two medical treatments don't give identical results.
- The sample originates from a binomial distribution of unknown parameter p.
- The sample is drawn from a normal distribution with $\sigma = 2.5$ and mean $\mu \in (\mu_1, \mu_2)$.

Finally, the following probabilities are relevant for testing hypotheses.

Various probabilities

Let X be a data sample:

$$\text{probability for type I error} = \alpha = P(X \in \mathcal{R}|H_0), \tag{18.1a}$$

$$\text{probability for type II error} = \beta = P(X \in \overline{\mathcal{R}}|H_1), \tag{18.1b}$$

$$\text{significance} = 1 - P(X \in \mathcal{R}|H_0) = P(X \in \overline{\mathcal{R}}|H_0), \tag{18.1c}$$

$$\text{power} = 1 - P(X \in \overline{\mathcal{R}}|H_1) = P(X \in \mathcal{R}|H_1). \tag{18.1d}$$

The *power* of a statistical hypothesis test is of central importance as it measures the test's ability to make a correct decision, namely to reject the null hypothesis when it is really false. In other words, the *power* is the probability of avoiding a type II error. Ideally, one would like to maximize the power of the test, because it minimizes the probability for a type II error, which is the content of the Neyman–Pearson lemma.

18.2 Neyman–Pearson lemma

Fisher introduced the concept of the P value and applied it in [72] to several test statistics. His approach is based solely on the null hypothesis. A question that Fisher did not address was the origin of these test statistics and the optimal choice of the rejection region. When reflecting upon the meaning of the rejection region, it appears highly ambiguous. The key property of a rejection region of *size* α in frequentist terms is that in the long run the percentage of rejection of a true null is α. A priori there is no reason why this interval should not be somewhere else than in the tails of the distribution.

Neyman and Pearson address precisely this question in [153]. For a simple null hypothesis, and a simple alternative hypothesis, they determined the optimal rejection region which simultaneously yields the underlying test statistic. The optimal NP test follows from

NEYMAN–PEARSON LEMMA [153] OPTIMAL REJECTION REGION

For a data sample X and a predefined significance level α, the optimal rejection region $\mathcal{R}$ which maximizes the power, is given by

$$\mathcal{R} = \left\{ X \,\middle|\, \ln\left[\frac{p(X|H_a,\mathcal{I})}{p(X|H_0,\mathcal{I})}\right] \geq K_\alpha \right\}.$$

The constant K_α follows from the significance constraint:

$$P(X \in \mathcal{R}|H_0) = \alpha.$$

The proof is given in Appendix B.1 [p. 611].

18.2.1 Example of an optimal NP test I

Simple null hypothesis against a simple alternative

We consider a sample $X := \{x_1, x_2, \ldots, x_L\}$ of size L. The simple point null hypothesis claims

$$H_0 : \text{The sample is drawn from } \mathcal{N}(\mu_0, \sigma)$$

while the alternative hypothesis, which is also a simple point hypothesis, favours a different mean:

$$H_a : \text{The sample is drawn from } \mathcal{N}(\mu_a, \sigma).$$

The log-likelihood ratio gives

$$\ln\left[\frac{p(X|H_a,\mathcal{I})}{p(X|H_0,\mathcal{I})}\right] = -\frac{L}{2\sigma^2}\left\{\cancel{\overline{x^2}} + \mu_a^2 - 2\overline{x}\mu_a - \cancel{\overline{x^2}} - \mu_0^2 + 2\overline{x}\mu_0\right\}$$

$$= \frac{L}{\sigma^2}\Big(\mu_a - \mu_0\Big)\left\{\frac{\mu_0+\mu_a}{2} + \overline{x}\right\} \overset{!}{>} K_\alpha.$$

The rejection region is apparently defined by

$$\mathcal{R} = \{X|\text{sign}\ (\mu_a - \mu_0)\ \overline{x} > \tilde{K}_\alpha\}, \tag{18.2}$$

with some constant $\tilde{K}_\alpha$ that is fixed by the significance constraint. The first finding is that the sole information of the sample that matters for the definition of the rejection region is the sample mean. It constitutes the *sufficient statistic* in this case. Moreover, the optimal test rejects the null if $\overline{x}$ is greater, or smaller depending on $\text{sign}(\mu_a - \mu_0)$, than a constant. This is precisely the one-sided z-test. Consequently, the z-test is optimal for testing a point null hypothesis against a point null alternative. According to equation (19.20) [p. 295], the sample mean of a normal sample is also normal with the same mean and the variance given by σ^2/L. That is, the sample distribution is $\mathcal{N}(\mu_0, \frac{\sigma}{\sqrt{L}})$. Consequently, the condition for the constant $\tilde{K}_\alpha$ reads

$$\alpha = P(\overline{x} > \tilde{K}_\alpha|L, \mu_0, \sigma, \mathcal{I}) = \int_{\tilde{K}_\alpha}^{\infty} \mathcal{N}(s|\mu_0, \tfrac{\sigma}{\sqrt{L}})\,ds = \frac{\sqrt{L}}{\sqrt{2\pi\sigma^2}}\int_{\tilde{K}_\alpha}^{\infty} e^{-\frac{(s-\mu_0)^2}{2\sigma^2/L}}\,ds.$$

We substitute $\frac{s-\mu_0}{\sqrt{2/L}\,\sigma} := t$ and obtain

$$\alpha = \frac{1}{\sqrt{\pi}}\int_{\frac{\tilde{K}_\alpha-\mu_0}{\sqrt{2}\sigma}}^{\infty} e^{-t^2}\,dt = \frac{1}{2}\text{erfc}\left(\frac{\tilde{K}_\alpha - \mu_0}{\sqrt{2/L}\,\sigma}\right).$$

The condition for $\tilde{K}_\alpha$ reads

$$\tilde{K}_\alpha = \mu_0 + \sqrt{2/L}\,\sigma\,\text{erfc}^{-1}(\alpha) = \mu_0 + \frac{\sigma}{\sqrt{L}}\underbrace{\sqrt{2}\text{erfc}^{-1}(\alpha)}_{:=C_\alpha}.$$

According to equation (18.2) we would reject the null hypothesis (μ_0) if

$$\overline{x} > \mu_0 + C_\alpha\frac{\sigma}{\sqrt{L}}. \tag{18.3}$$

18.2.2 Example of an optimal NP test II

Simple null hypothesis against a compound alternative

A simple alternative hypothesis is fairly unrealistic. A more realistic situation is, for example, given by the compound alternative

$$\begin{aligned} H_a &: \text{The sample is drawn from } \mathcal{N}(\mu_a, \sigma) \\ \text{with } &: \mu_a > \mu_0. \end{aligned}$$

We just derived in equation (18.3) that the test of H_0 against a simple hypothesis with any μ_a does not depend on the value μ_a at all, as long as $\mu_a > \mu_0$. The z-test with the rejection region defined by equation (18.3) is said to be *uniformly* optimal.

18.2.3 Asymmetry between null and alternative hypotheses

We return to the example of a simple null and alternative hypothesis as outlined in Section 18.2.1. There we have obtained the rejection region for μ_0, which is according to the Neyman–Pearson theory the acceptance region for the alternative, denoted by $\mathcal{A}$:

$$\mathcal{A} = \left\{ X \middle| \overline{x} \geq \mu_0 + \frac{\sigma}{\sqrt{L}}\, C_\alpha \right\}.$$

Now we interchange the meaning of H_0 and H_a and consider μ_a as the null and μ_0 as the alternative hypothesis and repeat the analysis. The rejection condition for μ_a becomes

$$\overline{x} < \mu_a - \frac{\sigma}{\sqrt{L}}\, C_\alpha,$$

and the acceptance region for μ_a is now

$$\mathcal{A}' = \left\{ X \middle| \overline{x} \geq \mu_0 + \frac{\sigma}{\sqrt{L}}\, C_\alpha \right\}.$$

So we find a strange discrepancy, which unravels that the null and the alternative hypothesis are not treated on the same basis. The null is the privileged hypothesis.

A more comprehensive discussion of the fundamental details of hypothesis testing is beyond the scope of this book, which is primarily aimed at scientists who wish to understand the principles of probability theory and statistics in order to effectively apply them in their own research area. Instead, we will discuss the most common test statistics and give some typical applications. Given a particular test statistic, the key quantity is the corresponding sampling distribution. Before discussing the various tests, we will derive the most important sampling distribution of common tests.

19

Sampling distributions and common hypothesis tests

In this chapter we derive the probability distributions of common test statistics of finite samples of random variables and apply them to common hypothesis tests.

19.1 Mean and median of i.i.d. random variables

We start out with the mean and median of a sample $\{x_1, \ldots, x_L\}$ of size L of i.i.d. random numbers drawn from a PDF $\rho(x)$. The mean and variance are as usual

$$\begin{aligned}
\langle x \rangle &= \int x \; \rho(x) \, dx, \\
\text{var}(x) &= \int (x - \langle x \rangle)^2 \; \rho(x) \, dx.
\end{aligned}$$

19.1.1 Distribution of the sample mean

Definition 19.1 (Sample mean) *We define*

SAMPLE MEAN

$$\overline{x} := \frac{1}{L} \sum_{i=1}^{L} x_i. \tag{19.1}$$

The actual value of $\overline{x}$ depends on the sample and is therefore a random variable as well. The corresponding PDF $p(\overline{x}|L, \rho, \mathcal{I})$ can be determined by the marginalization rule

$$p(\overline{x}|L, \rho, \mathcal{I}) = \int \underbrace{p(\overline{x}|x_1, \ldots, x_L, L, \rho, \mathcal{I})}_{\delta\left(\overline{x} - \frac{1}{L}\sum_{i=1}^{L} x_i\right)} \; p(x_1, \ldots, x_L|L, \rho, \mathcal{I}) \, dx_1 \ldots dx_L.$$

As the elements of the sample are i.i.d., we can simplify the expression:

$$p(\overline{x}|L,\rho,\mathcal{I}) = \int \delta\left(\overline{x} - \frac{1}{L}\sum_{i=1}^{L} x_i\right) \rho(x_1)\ldots\rho(x_L)\, dx_1\ldots dx_L.$$

Without specifying the underlying PDF $\rho(x)$, we cannot proceed further with the general expression. We either have to consider the limit $L \to \infty$, in order to use the *central limit theorem*, or we have to specify the PDF.

However, we can derive general results for the lowest moments. The zeroth moment (normalization) is apparently equal to 1, as long as $\rho(x)$ is properly normalized. The first moment, the expectation value of the sample mean, yields

$$\begin{aligned}
\langle\overline{x}\rangle &= \int \overline{x}\; p(\overline{x}|L,\rho,\mathcal{I})\, d\overline{x} \\
&= \int \left(\frac{1}{L}\sum_{i=1}^{L} x_i\right) \rho(x_1)\ldots\rho(x_L)\, dx_1\ldots dx_L \\
&= \frac{1}{L}\sum_{i=1}^{L} \int x_i\; \rho(x_1)\ldots\rho(x_L)\, dx_1\ldots dx_L \\
&= \frac{1}{L}\sum_{i=1}^{L} \Bigg(\underbrace{\int x_i\; \rho(x_i)\, dx_i}_{=\langle x\rangle} \; \prod_{j\neq i} \underbrace{\int \rho(x_j)\, dx_j}_{=1} \Bigg) \\
\langle\overline{x}\rangle &= \langle x\rangle.
\end{aligned}$$

That is to say, the sample mean is an unbiased estimator for the underlying (true) mean of $\rho(x)$.

19.1.2 Variance of the sample mean

Next we will determine the variance of the sample mean. To this end we need the second moment

$$\begin{aligned}
\langle\overline{x}^2\rangle &= \int \overline{x}^2\; p(\overline{x}|L,\rho,\mathcal{I})\, d\overline{x} \\
&= \int \left(\frac{1}{L}\sum_{i=1}^{L} x_i\right)^2 \rho(x_1)\ldots\rho(x_L)\, dx_1\ldots dx_L \\
&= \frac{1}{L^2}\sum_{i,j=1}^{L} \int x_i\, x_j\; \rho(x_1)\ldots\rho(x_L)\, dx_1\ldots dx_L.
\end{aligned}$$

We need to distinguish the cases $i = j$ and $i \neq j$:

$$\begin{aligned}\langle \overline{x}^2 \rangle &= \frac{1}{L^2} \sum_{i=1}^{L} \int x_i^2 \, \rho(x_1) \ldots \rho(x_L) \, dx_1 \ldots dx_L \\ &\quad + \frac{1}{L^2} \sum_{\substack{i,j=1 \\ i \neq j}}^{L} \int x_i \, x_j \, \rho(x_1) \ldots \rho(x_L) \, dx_1 \ldots dx_L \\ &= \frac{1}{L^2} \sum_{i=1}^{L} \langle x^2 \rangle + \frac{1}{L^2} \sum_{\substack{i,j=1 \\ i \neq j}}^{L} \langle x \rangle \, \langle x \rangle \\ &= \frac{1}{L} \langle x^2 \rangle + \frac{L\,(L-1)}{L^2} \langle x \rangle^2 = \frac{1}{L} \left(\langle x^2 \rangle - \langle x \rangle^2 \right) + \langle x \rangle^2 \\ &= \frac{\langle (\Delta x)^2 \rangle}{L} + \langle x \rangle^2 = \frac{\mathrm{var}(x)}{L} + \langle x \rangle^2 .\end{aligned}$$

Because $\langle \overline{x} \rangle = \langle x \rangle$, the variance is

VARIANCE OF THE SAMPLE MEAN

$$\mathrm{var}(\overline{x}) = \frac{\mathrm{var}(x)}{L}. \tag{19.2}$$

In other words, the root mean square deviation of the sample mean from the true mean (statistical error of the mean) is given by

STANDARD ERROR

$$\Delta\overline{x} := \mathrm{SE} := \frac{\sqrt{\mathrm{var}(x)}}{\sqrt{L}}. \tag{19.3}$$

It differs from the *standard deviation* by the factor $1/\sqrt{L}$. The standard deviation is a measure for the scattering of the individual elements of the sample, while the standard error is a measure for the scattering of the sample mean. This is the main advantage of repeated measurements, which suffer from statistical noise. The standard error implies that it is possible to reduce the uncertainty below a desired limit. Another advantage of computing sample means is provided by the central limit theorem. It guarantees that the sample mean for large samples is normally distributed, irrespective of the underlying PDF $\rho(x)$, if the variance is finite, see Section 19.1.3. Needless to say, knowledge of the PDF is important for data analysis and inferences in general.

Up to now, a small deficiency still exists: we use the sample mean in order to estimate the true mean of $\rho(x)$. The uncertainty of the estimator depends on the variance of $\rho(x)$, which we will not know. Where do we get var(x) from? One is prompted to estimate it from the sample, similarly to the sample mean, by

$$v = \frac{1}{L} \sum_{i=1}^{L} \left(x_i - \overline{x}\right)^2.$$

We have replaced the unknown true mean by the sample mean. The estimate for the variance can be rewritten as

$$v = \frac{1}{L} \sum_{i=1}^{L} x_i^2 - 2\,\overline{x} \underbrace{\frac{1}{L} \sum_{i=1}^{L} x_i}_{\overline{x}} + \frac{1}{L} \sum_{i=1}^{L} \overline{x}^2 = \frac{1}{L} \sum_{i=1}^{L} x_i^2 - \overline{x}^2.$$

Also, this estimator is a random variable. Its mean is

$$\langle v \rangle = \underbrace{\frac{1}{L} \sum_{i=1}^{L} \left\langle x_i^2 \right\rangle}_{\langle x^2 \rangle} - \left\langle \overline{x}^2 \right\rangle \overset{(19.2)}{=} \underbrace{\left\langle x^2 \right\rangle - \langle x \rangle^2}_{=\mathrm{var}(x)} - \frac{\mathrm{var}(x)}{L} = \frac{L-1}{L} \,\mathrm{var}(x).$$

Apparently, this estimator exhibits a *bias*, which can easily be eliminated by introducing a slightly different estimator:

UNBIASED ESTIMATOR FOR THE VARIANCE

$$\sigma_{\text{est}}^2 = \frac{1}{L-1} \sum_{i=1}^{L} (x_i - \overline{x})^2. \tag{19.4}$$

The modified prefactor $1/(L-1)$ is understood as follows. The quantities $\Delta_i = x_i - \overline{x}$ are not independent, since $\sum_i \Delta_i = 0$. There are only $L - 1$ independent variables ξ_i that can be formed from the x_i by linear combination, the so-called *Helmert transformation*, as discussed in Section 7.10 [p. 136]. The number of *degrees of freedom* is thereby reduced by one.

19.1.3 Distribution of the sample median

The sample mean has a severe drawback. Its variance is proportional to the variance var(x) of the sample elements. Even though it is reduced by $1/L$, it will be infinite if the underlying PDF has an infinite variance. This is for instance the case if $\rho(x)$ is a Cauchy distribution, which occurs quite often in physics. In this case, the sample mean is meaningless and it is advisable to use the median instead.

In this section we will, therefore, consider the properties of the sample median.

Definition 19.2 (Median) *The median, or central value, of a continuous distribution is defined as*

MEDIAN OF A CONTINUOUS DISTRIBUTION $\rho(x)$ WITH CDF $F(x)$

$$F(\overset{\curlywedge}{x}) = \int_{-\infty}^{\overset{\curlywedge}{x}} \rho(t)\, dt \overset{!}{=} \tfrac{1}{2},$$
$$\overset{\curlywedge}{x} = F^{-1}\left(\tfrac{1}{2}\right). \tag{19.5}$$

Definition 19.3 (Sample median) *Given a sample $\{s_1, \ldots, s_L\}$ of size L drawn from a probability density $\rho(x)$ whose elements are sorted in ascending order $\{s_1, \leq s_2 \ldots \leq s_L\}$, the sample median is defined by*

SAMPLE MEDIAN

$$\breve{x} = \begin{cases} s_{n+1} & \text{for } L = 2n+1, \\ \dfrac{s_n + s_{n+1}}{2} & \text{for } L = 2n. \end{cases} \tag{19.6}$$

We are now interested in the sampling distribution of the sample median, i.e. in $p(\breve{x}\,|\,\rho, L, \mathcal{I})$, of a sample of size L drawn from $\rho(x)$. We confine the discussion to the case of odd sample size $L = 2n + 1$. The sample median is then the element $n + 1$ of the ordered list. The sought-for PDF is nothing but the corresponding $(n + 1)$th-order statistic. According to equation (7.42) [p. 119], the PDF in the present context reads

$$p(\breve{x}\,|\,\rho, L = 2n+1, \mathcal{I}) = \frac{1}{B(n+1, n+1)}\, F(\breve{x})^n\, [1 - F(\breve{x})]^n\, \rho(\breve{x}). \tag{19.7}$$

The average sample median has already been derived in equation (7.43) [p. 120]:

$$\left\langle \breve{x} \right\rangle_n = \int_0^1 F^{-1}(f)\; p_\beta(f\,|\,n+1, n+1)\, df \; = \left\langle F^{-1} \right\rangle_n, \tag{19.8}$$

where the latter mean is that of the associated beta PDF. We restrict the discussion to PDFs with singly connected support, such that F^{-1} is unique. The mean and variance of the beta PDF $p_\beta(q\,|\,n + 1, n + 1)$ are

$$\mu_\beta = \frac{1}{2}, \tag{19.9}$$

$$\sigma_\beta^2 = \frac{1}{4(L+2)}. \tag{19.10}$$

If the density is symmetric, i.e. $\exists x_0 : \rho(x_0 - \Delta x) = \rho(x_0 + \Delta x)$, then the mean and median of $\rho(x)$ are the same: $\langle x \rangle = \overset{\curlywedge}{x} = x_0$. Moreover, the CDF $F(x)$ is antisymmetric w.r.t. point reflection about $(\overset{\curlywedge}{x}, \frac{1}{2})$. Apparently, since it is the same curve merely rotated by 90 degrees, the same holds true for $F^{-1}(x)$. Consequently, in case of a symmetric density, $F^{-1}(x)$ can be decomposed into

$$F^{-1}(x) = \underbrace{F^{-1}\left(\frac{1}{2}\right)}_{=\overset{\curlywedge}{x}} + G(x),$$

where $G(x)$ is an antisymmetric function about $x = \frac{1}{2}$ and the average sample median simplifies to

$$\left\langle \breve{x} \right\rangle_n = \left\langle F^{-1} \right\rangle_n = \overset{\curlywedge}{x} + \left\langle G \right\rangle_n = \overset{\curlywedge}{x},$$

since $p_\beta(f|n+1, n+1)$ entering the definition (equation (19.8)) is symmetric.

MEAN OF THE SAMPLE MEDIAN FOR SYMMETRIC DENSITY

$$\left\langle \breve{x} \right\rangle_n = \overset{\curlywedge}{x}. \tag{19.11}$$

In other words, iff the density is symmetric, the average sample median is equal to both the median and the mean of $\rho(x)$. Consequently, the sample median is an unbiased estimator. **In case of an asymmetric density** we expand $F^{-1}(f)$ about the mean of the beta PDF, $\mu_\beta = \frac{1}{2}$, as it is equivalent to an expansion in $\frac{1}{L}$. Starting from equation (19.8), we obtain

$$\begin{aligned}
\left\langle \breve{x} \right\rangle_n &= \underbrace{\left\langle F^{-1}\left(\tfrac{1}{2}\right)\right\rangle}_{=\overset{\curlywedge}{x}} + \left.\frac{dF^{-1}(f)}{df}\right|_{f=\frac{1}{2}} \underbrace{\left\langle\left(f - \tfrac{1}{2}\right)\right\rangle}_{0} \\
&\quad + \frac{1}{2}\left.\frac{d^2F^{-1}(q)}{dq^2}\right|_{q=\frac{1}{2}} \underbrace{\left\langle\left(q - \tfrac{1}{2}\right)^2\right\rangle}_{\sigma_\beta^2} + O\left(\sigma_\beta^4\right) \\
&= \overset{\curlywedge}{x} + \frac{1}{2}\left.\frac{d^2F^{-1}(q)}{dq^2}\right|_{q=\frac{1}{2}} \sigma_\beta^2 + O\left(\sigma_\beta^4\right).
\end{aligned}$$

According to equation (19.10), $\sigma_\beta^2 = \frac{1}{4(L+2)} = \frac{1}{4L} + O\left(\frac{1}{L^2}\right)$. That is, on average, the sample median provides the true median of the underlying probability density only up to a systematic deviation of order $O(\frac{1}{L})$:

$$\left\langle \breve{x} \right\rangle = \overset{\curlywedge}{x} + \frac{1}{8L} \frac{d^2 F^{-1}(q)}{dq^2}\bigg|_{q=\frac{1}{2}} + O\left(\frac{1}{L^2}\right).$$

For the prefactor we still need the derivatives of $F^{-1}(f)$. By means of the chain rule we obtain

$$\frac{dF^{-1}(f)}{df} = \frac{1}{\rho(F^{-1}(f))},$$

and for the second derivative

$$\frac{d^2 F^{-1}(f)}{df^2} = -\frac{1}{\rho(F^{-1}(f))^3} \frac{d\rho(x)}{dx}\bigg|_{x=F^{-1}(f)}.$$

We merely need these derivatives for $f = \frac{1}{2}$:

$$\frac{dF^{-1}(q)}{dq}\bigg|_{q=\frac{1}{2}} = \frac{1}{\rho(\overset{\curlywedge}{x})}, \tag{19.12a}$$

$$\frac{d^2 F^{-1}(q)}{dq^2}\bigg|_{q=\frac{1}{2}} = -\frac{1}{\rho(\overset{\curlywedge}{x})^3} \frac{d\rho(x)}{dx}\bigg|_{x=\overset{\curlywedge}{x}}. \tag{19.12b}$$

The final result reads

Average sample median in case of asymmetric densities

$$\left\langle \breve{x} \right\rangle = \overset{\curlywedge}{x} - \frac{1}{8\, L\, \rho(\overset{\curlywedge}{x})^3} \frac{d\rho(x)}{dx}\bigg|_{x=\overset{\curlywedge}{x}} + O(L^{-2}). \tag{19.13}$$

19.1.4 Variance of the sample median

Next we determine the variance of the sample median, or rather that of $p(\breve{x}|\rho, L = 2n+1, \mathcal{I})$, given in equation (19.7):

$$\mathrm{var}(\breve{x}) = \left\langle \left[F^{-1}(f) - \left\langle \breve{x} \right\rangle \right]^2 \right\rangle. \tag{19.14}$$

We use the same expansion as before:

$$\begin{aligned}\operatorname{var}(\breve{x}) &= \left\langle \left[\cancel{\hat{x}} + \left.\frac{dF^{-1}(f)}{df}\right|_{f=\frac{1}{2}} \left(f - \tfrac{1}{2}\right) + O(\sigma_\beta^2) - \cancel{\hat{x}} - O(\sigma_\beta^2) \right]^2 \right\rangle \\ &= \left[\left.\frac{dF^{-1}(f)}{df}\right|_{f=\frac{1}{2}} \right]^2 \underbrace{\left\langle \left[f - \tfrac{1}{2} \right]^2 \right\rangle}_{\sigma_\beta^2} + O(\sigma_\beta^4),\end{aligned}$$

and obtain

$$\operatorname{var}(\breve{x}) = \left(\left.\frac{dF^{-1}(f)}{df}\right|_{f=\frac{1}{2}} \right)^2 \frac{1}{4L} + O\left(\frac{1}{L^2}\right).$$

Hence the scattering of the sample median is of order $1/\sqrt{L}$. It therefore predominates the uncertainty of the result for large L, as the bias decreases like $1/L$. Along with equation (19.12), the result is

VARIANCE OF THE SAMPLE MEDIAN
FOR A SAMPLE OF SIZE $L = 2n + 1$ DRAWN FROM $\rho(x)$

$$\operatorname{var}(\breve{x}) = \frac{1}{4L\ \rho(\hat{x})^2} + O(L^{-2}). \tag{19.15}$$

We have seen that in case of a symmetric density, the average sample median agrees with both the true median and the mean. This is corroborated by the above series expansion, as the derivative of $\rho(x)$ vanishes at $\hat{x}$. Interestingly, the variance of the sample median is independent of the variance of the underlying density $\rho(x)$.

Consequently, the sample median is more robust than the sample mean when it comes to experiments with large scattering in the individual values, in other words, when the variance of $\rho(x)$ is large or even infinite.

We want to close the discussion about the sample median with a few examples.

Example: Cauchy distribution

The Cauchy distribution is particularly interesting as it occurs quite frequently in physics and has an infinite variance. Here we use the PDF

$$\rho_C(x) = \frac{1}{\pi\ (1 + x^2)},$$

which has zero mean. It is symmetric and the mean and median are equal. The CDF can be computed analytically:

$$F_C(x) = \frac{1}{2} + \frac{\arctan(x)}{\pi},$$

as well as its inverse:

$$F_C^{-1}(f) = \tan\left[\pi\left(f - \tfrac{1}{2}\right)\right]. \tag{19.16}$$

It is apparently antisymmetric and the median is $\overset{\curlywedge}{x} = F_C^{-1}\left(\frac{1}{2}\right) = 0$. The sample median is unbiased and the leading-order approximation of the variance reads, according to equation (19.15),

$$\operatorname{var}(\breve{x}) = \frac{\pi^2}{4L}. \tag{19.17}$$

The results are depicted in Figure 19.1. The variance diverges for $L = 1$, as it corresponds to that of the original PDF, but it is finite for $L \geq 3$ and approaches rapidly the leading-order approximation of equation (19.17). In addition to the analytic results, Figure 19.1 also contains the data obtained by a straightforward computer simulation. Random Cauchy samples of size L were generated and the corresponding medians determined. The variance of such a list of 10^6 median values is included in the figure. The perfect agreement corroborates the correctness of the analytic result.

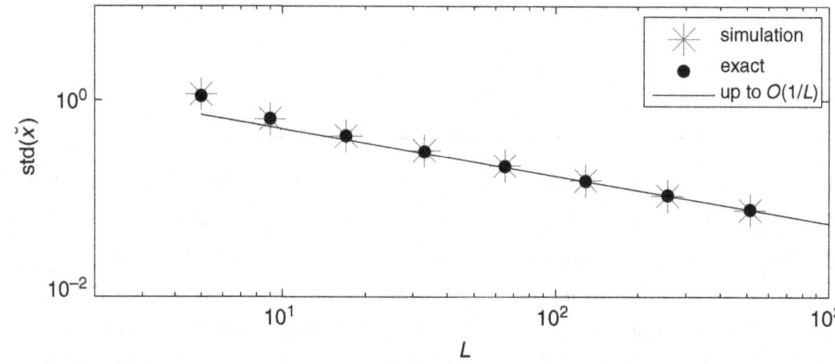

Figure 19.1 Standard deviation of the sample median as a function of sample size L for the Cauchy PDF. The exact result, obtained by numerical integration of equation (19.14) along with equation (19.16), is compared with the leading-order term presented in equation (19.17).

Example: Exponential distribution

As another example we study the sample median of random numbers drawn from an exponential PDF

$$\rho(x) = e^{-x}.$$

In this case, the sample median is biased, as the exponential function is obviously not symmetric. According to equation (7.17b) [p. 104], the CDF of the exponential PDF reads

$$F_e(x|\lambda) = 1 - e^{-x},$$

which has the inverse

$$F_e^{-1}(f|\lambda) = -\ln(1-f).$$

The median is therefore

$$\overset{\curlywedge}{x} = F_e^{-1}\left(\tfrac{1}{2}\right) = \ln(2).$$

At the location of the median the density is $\rho_e(\overset{\curlywedge}{x}) = \frac{1}{2}$ and the derivative reads $\rho_e'(\overset{\curlywedge}{x}) = -\frac{1}{2}$. So, according to equation (19.15), the leading-order approximation of the variance of the sample median is

$$\text{var}(\breve{x}) = \frac{1}{L}.$$

As already mentioned, the sample median is biased and based on equation (19.13) it yields on average

$$\left\langle \breve{x} \right\rangle = \overset{\curlywedge}{x} - \frac{-\frac{1}{2}}{8L\,(\frac{1}{2})^3} = \ln(2) + \frac{1}{2L}.$$

Again, higher-order contributions in $1/L$ have been ignored.

The results for the average sample median and its variance are depicted in Figure 19.2. For samples of finite size there is a systematic deviation from the intrinsic value of the median, which, however, disappears as $1/L$, while the standard error decreases only as $1/\sqrt{L}$.

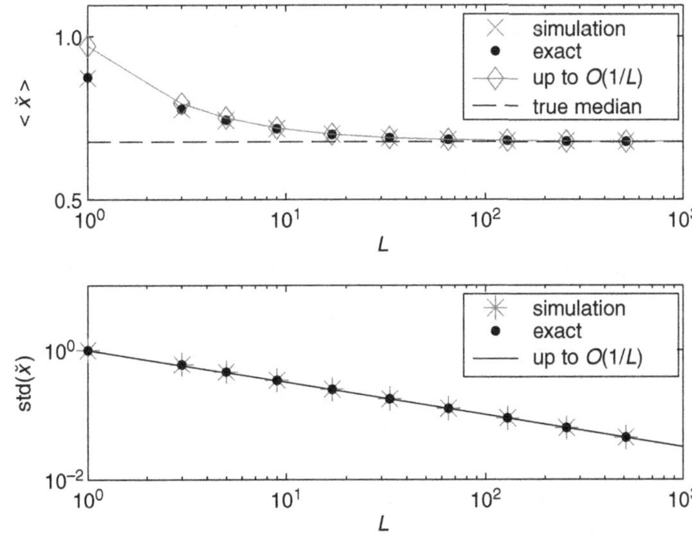

Figure 19.2 Average sample median (upper panel) and its standard deviation (lower panel) of a sample of size L drawn from an exponential PDF.

Also in this case, the analytic result was checked by comparison with results from a computer simulation. The details are the same as in the previous case.

19.2 Mean and variance of Gaussian samples

In the previous sections, we did not specify the PDF from which the independent random numbers are drawn. Owing to the omnipresence of Gaussian random numbers, in particular in conjunction with the central limit theorem, several hypothesis tests are based on normal distributions. The nice thing about this is that the sampling distribution can be given analytically.

Suppose we have samples of size N of i.i.n.d. random variables $\{x_n\}$ with mean x_0 and variance σ^2. For these samples, we determine the sample mean

$$s = \frac{1}{N}\sum_{n=1}^{N} x_n$$

and

$$v = \sum_{n=1}^{N} (x_n - \overline{x})^2,$$

which is proportional to the sample variance. We want to determine the distribution $p(s, v|N, x_0, \sigma, \mathcal{I})$ of s and v. Instead of marginalizing over the samples $\boldsymbol{x}$, we use the Helmert transform (see Section 7.10 [p. 136])

$$p(s, v|x_0, \sigma, N, \mathcal{I}) = \int d^N\xi \; p(s, v|\boldsymbol{\xi}, x_0, \sigma, N, \mathcal{I}) \; p(\boldsymbol{\xi}|x_0, \sigma, N, \mathcal{I}). \tag{19.18}$$

Given the Helmert-transformed sample $\boldsymbol{\xi}$, s and v are uniquely fixed and therefore expressed in terms of Dirac delta functions as

$$p(s, v|\boldsymbol{\xi}, \sigma, \mathcal{I}) = \delta\left(s - \frac{\xi_N}{\sqrt{N}}\right) \delta\left(v - \boldsymbol{\eta}^2\right).$$

Apart from irrelevant prefactors, which we will determine at the end via normalization, we have

$$\begin{aligned} p(s, v|x_0, \sigma, N, \mathcal{I}) &\propto \int d\xi_N \, d\eta^{N-1} \; \mathcal{N}(\boldsymbol{\eta}|0, \sigma) \; \mathcal{N}(\xi_N|\sqrt{N}x_0, \sigma) \ldots \\ &\quad \ldots \delta\left(\xi_N - s\sqrt{N}\right) \delta\left(v - \boldsymbol{\eta}^2\right) \\ &\propto \mathcal{N}\left(\sqrt{N}s|\sqrt{N}x_0, \sigma\right) \int d\eta \; \eta^{N-2} \, e^{-\frac{\eta^2}{\sigma^2}} \; \frac{\delta\left(\eta - \sqrt{v}\right)}{2\eta}. \end{aligned}$$

For the last step we have introduced $(N-1)$-dimensional spherical coordinates. The remaining integral yields

$$p(s, v|x_0, \sigma, N, \mathcal{I}) = \mathcal{N}\left(s|x_0, \frac{\sigma}{\sqrt{N}}\right) \cdot \frac{1}{Z} v^{\frac{N-3}{2}} \, e^{-\frac{v}{\sigma^2}}$$

$$= \mathcal{N}\left(s|x_0, \frac{\sigma}{\sqrt{N}}\right) \cdot p_\Gamma\left(v|\alpha = \frac{N-1}{2}, \beta = \frac{1}{2\sigma^2}\right). \quad (19.19)$$

19.2.1 PDF of the sample mean

The PDF for the sample mean is given by the first factor of equation (19.19).

PDF FOR THE SAMPLE MEAN OF I.I.N.D. RANDOM NUMBERS

---

sample: $\{x_1, \ldots, x_N\}$ drawn from $\mathcal{N}(x|x_0, \sigma)$

parameters: N, x_0, σ

random variable: $s = \frac{1}{N}\sum_{i=1}^{N} x_i$

$$p(s|x_0, \sigma, N, \mathcal{I}) = \frac{1}{\sqrt{2\pi\sigma^2/N}} e^{-\frac{(s-x_0)^2}{2\sigma^2/N}} = \mathcal{N}\left(s|x_0, \frac{\sigma}{\sqrt{N}}\right). \quad (19.20)$$

19.2.2 PDF of the sample variance, chi-squared statistic

The probability density function for the sample variance is obtained from the properly normalized first factor in equation (19.19).

PDF FOR THE SAMPLE VARIANCE FOR I.I.N.D. RANDOM VARIABLES

---

sample: $\{x_1, \ldots, x_N\}$ drawn from $\mathcal{N}(x|x_0, \sigma)$

parameters: N, x_0, σ

random variable: $v = \sum_{n=1}^{N} (x_n - \overline{x})^2$

$$p(v|x_0, \sigma, N, \mathcal{I}) = \frac{(2\sigma^2)^{-\frac{N-1}{2}}}{\Gamma(\frac{N-1}{2})} v^{(\frac{N-1}{2}-1)} \, e^{-\frac{v}{2\sigma^2}}$$

$$= p_\Gamma\left(v|\alpha = \tfrac{N-1}{2}, \beta = \tfrac{1}{2\sigma^2}\right). \quad (19.21)$$

Apparently, it is a Gamma distribution. The mean $\langle v \rangle$ is of particular interest. It follows from equation (7.14c) [p. 102]:

$$\langle v \rangle = \frac{\alpha}{\beta} = (N-1)\sigma^2.$$

The result corroborates that the unbiased estimator for the variance is normalized by $N-1$, as discussed in Section 19.1.2 [p. 285]. In connection with the chi-squared distribution, it is common to use normalized random variables

$$\tilde{x}_n = \frac{x_n}{\sigma}.$$

The sample variance v is transformed into

$$z = \frac{v}{\sigma^2}.$$

The sampling distribution of the variance can be considered as a chi-squared distribution equation (7.16a) [p. 104] with $N-1$ degrees of freedom:

$$p(z|N, x_0, \sigma, \mathcal{I}) = p_{\chi^2}(z|N-1). \tag{19.22}$$

We started out with N degrees of freedom $\{x_n\}$, but lost one as the definition of the sample variance contains the sample mean $\overline{x}$ which had to be determined from the data.

In the above derivation we have assumed i.i.n.d. random numbers, that is, they were all drawn from the same normal PDF. This can easily be generalized, in order to make the result applicable to a wider range of problems. We assume that the random numbers are still independent and normally distributed, but no longer identically. In other words, each random variable is Gaussian with its own mean and variance. Next we define new shifted and scaled random variables

$$\tilde{x}_n = \frac{x_n - x_{0,n}}{\sigma_n},$$

which now all have one and the same normal PDF $\mathcal{N}(\tilde{x}_n|0, 1)$. The previous consideration implies that

$$\chi^2 = \sum_{n=1}^{N} \frac{(x_n - x_{0,n})^2}{\sigma_n^2} \tag{19.23}$$

has a chi-squared distribution. It is therefore called a *chi-squared statistic*.

The sampling distribution of the chi-squared statistic is

SAMPLING DISTRIBUTION OF THE χ^2-STATISTIC

sample: $\{x_1, \ldots, x_N\}$, x_i drawn from $\mathcal{N}(x_i|x_0^i, \sigma_i)$

parameters: $N, x_0^1, \ldots, x_0^N, \sigma_1, \ldots, \sigma_N$

test statistic: $\chi^2 = \sum_{n=1}^{N} \frac{(x_n - x_0^n)^2}{\sigma_n^2}$

$$\begin{aligned} p(\chi^2|N, x_0^1, \ldots, x_0^N, \sigma_1, \ldots, \sigma_N, \mathcal{I}) &= \frac{(2)^{-\frac{N}{2}}}{\Gamma(\frac{N}{2})} (\chi^2)^{(\frac{N}{2}-1)} e^{-\chi^2/2} \\ &= p_{\chi^2}(\chi^2|N). \end{aligned} \tag{19.24}$$

19.3 *z*-Statistic

We consider the situation with two samples X_1 and X_2 of size N_1 and N_2, respectively. The elements of the two samples are drawn independently from two normal distributions with individual variances σ_1^2 and σ_2^2. A null hypothesis in this context might be

H: The two samples have the same mean.

The *z*-statistic in this case is defined as

***z*-STATISTIC**
FOR THE DIFFERENCE IN MEANS

$$\begin{aligned} z &= \frac{\text{sample mean 1} - \text{sample mean 2}}{\text{standard error of difference of sample means}}, \\ z &= \frac{\overline{x}_2 - \overline{x}_1}{\text{SE}}, \\ \text{SE}^2 &= \frac{\sigma_1^2}{N_1} + \frac{\sigma_2^2}{N_2}. \end{aligned} \tag{19.25}$$

This is precisely the sufficient statistic provided by the Bayesian approach to the question 'Are the true means of two normal samples of sizes N_1 and N_2 the same, provided the corresponding variances are σ_1^2 and σ_2^2?' We leave the proof as an exercise for the reader. The relevant quantities are the individual standard errors $\text{SE}_\alpha := \frac{\sigma_\alpha}{\sqrt{N_\alpha}}$ and the resulting standard error of the difference of sample means of i.i.n.d. random variables, i.e. $\text{SE}^2 = \text{SE}_1^2 + \text{SE}_2^2$. For notational ease we use the abbreviations $\mathbf{N} = \{N_1, N_2\}$ and $\boldsymbol{\sigma} = \{\sigma_1, \sigma_2\}$. The PDF for *z* is obtained via the marginalization rule upon introducing the sample means:

$$\begin{aligned}
p(z|\mathbf{N},\boldsymbol{\sigma},\mathcal{I}) &= \int d\overline{x}_1\, d\overline{x}_2\, p(z|\overline{x}_1,\overline{x}_2,\mathbf{N},\boldsymbol{\sigma},\mathcal{I})\, p(\overline{x}_1,\overline{x}_2|\mathbf{N},\boldsymbol{\sigma},\mathcal{I}) \\
&= \int d\overline{x}_1\, d\overline{x}_2\, \delta\left(z - \frac{\overline{x}_2-\overline{x}_1}{\text{SE}}\right) p(\overline{x}_1|N_1,\sigma_1,\mathcal{I})\, p(\overline{x}_2|N_2,\sigma_2,\mathcal{I}) \\
&= \sigma \int d\overline{x}_1\, p(\overline{x}_1|N_1,\sigma_1,\mathcal{I})\, p(\overline{x}_2 = \overline{x}_1 + z\,\text{SE}|N_2,\sigma_2,\mathcal{I}).
\end{aligned}$$

Now we introduce the common mean x_0 of the underlying normal PDFs. Furthermore, we suppress the prefactor and adjust it at the end. Moreover, we use equation (19.20) for the PDF of a sample mean:

$$\begin{aligned}
p(z|\mathbf{N},\boldsymbol{\sigma},\mathcal{I}) &\propto \int dx_0\, p(x_0|\mathcal{I}) \;\ldots \\
&\quad \times \int d\overline{x}_1\, p(\overline{x}_1|N_1,\sigma_1,x_0,\mathcal{I}) p(\overline{x}_2 = \overline{x}_1 + z\,\text{SE}|N_2,\sigma_2,x_0,\mathcal{I}) \\
&\propto \int dx_0\, p(x_0|\mathcal{I}) \int d\overline{x}_1\, e^{-\frac{1}{2}\left(\frac{(\overline{x}_1-x_0)^2}{\sigma_1^2/N_1}+\frac{(\overline{x}_1+z\,\text{SE}-x_0)^2}{\sigma_2^2/N_2}\right)} \\
&\propto \underbrace{\int dx_0\, p(x_0|\mathcal{I})}_{=1} \int ds_1\, e^{-\frac{1}{2}\left(\frac{s_1^2}{\sigma_1^2/N_1}+\frac{(s_1+z\,\text{SE})^2}{\sigma_2^2/N_2}\right)}.
\end{aligned}$$

In the remaining integral the argument of the exponential by completing the square w.r.t. s_1 results in an expression of the form

$$\left(\frac{s_1^2}{\sigma_1^2/N_1} + \frac{(s_1+z\,\text{SE})^2}{\sigma_2^2/N_2}\right) = C\left(s_1 - s_1^*\right)^2 + z^2.$$

The integral over s_1 just modifies the prefactor and we obtain

$$p(z|\mathbf{N},\boldsymbol{\sigma},\mathcal{I}) \propto e^{-\frac{1}{2}z^2},$$

and the sought-for result is

z-STATISTIC

FOR THE DIFFERENCE IN MEANS

sample 1:	$\{x_1^1,\ldots,x_{N_1}^1\}$ drawn from $\mathcal{N}(x|x_0,\sigma_1)$
sample 2:	$\{x_1^2,\ldots,x_{N_2}^2\}$ drawn from $\mathcal{N}(x|x_0,\sigma_2)$
given:	$N_1,\ N_2,\ \sigma_1,\ \sigma_2$
test statistic:	$z = \frac{\overline{x}_2-\overline{x}_1}{\text{SE}}$; $\quad \text{SE}^2 = \frac{\sigma_1^2}{N_1} + \frac{\sigma_2^2}{N_2}$

$$p_z(z) = \frac{1}{\sqrt{2\pi}}\, e^{-z^2/2} = \mathcal{N}(z|\mu=0,\sigma=1). \tag{19.26}$$

It appears that the quantities N_α and σ_α do not enter the PDF of z, they are merely required for the calculation of z. Moreover, the test statistic in equation (19.25) as well as the corresponding PDF are independent of x_0 and we never had to specify the prior PDF for x_0, merely its normalization mattered. It corresponds to a null hypothesis:

H_0 : The means of two normal samples of given variances are the same.

An alternative hypothesis, also leading to a z-statistic, is

H_0 : The true mean μ of a normal sample of given variance is μ_0.

z-STATISTIC
FOR A PREDICTED μ_0

sample:	$\{x_1, \ldots, x_N\}$ drawn from $\mathcal{N}(x\|x_0, \sigma)$
given:	N, σ
test statistic:	$z = \frac{\overline{x}-\mu_0}{\text{SE}}$; $\text{SE}^2 = \frac{\sigma^2}{N}$

$$p_z(z) = \frac{1}{\sqrt{2\pi}} e^{-z^2/2} = \mathcal{N}(z|\mu = 0, \sigma = 1). \tag{19.27}$$

This is a special version of the previous case for $\sigma_2 \to 0$. It implies $\overline{x}_2 = \mu_0$. The rest is obvious and consequently z in this case is also distributed as $\mathcal{N}(z|\mu = 0, \sigma = 1)$.

19.4 Student's t-statistic

Next we generalize the previous discussion towards an unknown variance. The random variables of one sample are still independently and normally distributed with common mean and variance, but the variance is not at our disposal.

19.4.1 Student's t-test for a predicted mean

We start out with the situation where a Gaussian sample is given of unknown variance. This situation has already been examined in the Bayesian framework in Section 17.2.3 [p. 266]. The null hypothesis in this case reads

H_0 : The true mean μ of a normal sample of unknown variance is μ_0.

The suitable test statistic as depicted in equation (17.11) [p. 266] reads

t-Statistic
for a predicted μ_0

$$
\begin{aligned}
t &:= \frac{\overline{x} - \mu_0}{\text{SE}^*}, \\
\text{SE}^{*2} &:= \frac{\overline{(\Delta x)^2}}{(N-1)}.
\end{aligned}
\tag{19.28}
$$

In order to derive the sampling distribution we introduce – via the marginalization rule – the missing values of the sample mean, the sample variance $v := N\ \overline{(\Delta x)^2}$ and the intrinsic standard deviation σ. As several times before, we suppress prefactors that can be determined easily afterwards:

$$
\begin{aligned}
p(t|N, \mu_0, \mathcal{I}) &= \iiint d\sigma\ d\overline{x}\ dv\ p(t|\overline{x}, v, \sigma, N, \mu_0, \mathcal{I})\ p(\overline{x}, v, \sigma|N, \mu_0, \mathcal{I}) \\
&= \iiint d\sigma\ d\overline{x}\ dv\ \delta\left(t - \tfrac{\sqrt{N(N-1)}(\overline{x}-\mu_0)}{\sqrt{v}}\right) \ldots \\
&\quad \ldots \underbrace{p(\overline{x}|\sigma, N, \mu_0, \mathcal{I})}_{=\mathcal{N}\left(\overline{x}|\mu_0, \frac{\sigma}{\sqrt{N}}\right)}\ \underbrace{p(v|\sigma, N, \mu_0, \mathcal{I})}_{=p_\Gamma\left(v|\frac{N-1}{2}, \frac{1}{2\sigma^2}\right)}\ p(\sigma|N, \mu_0, \mathcal{I}).
\end{aligned}
$$

Next it is expedient to introduce new variables $s := (\overline{x} - \mu_0)/\sigma$ and $w := v/\sigma^2$ in which the PDFs are

$$
\mathcal{N}\left(\overline{x}|\mu_0, \frac{\sigma}{\sqrt{N}}\right) \stackrel{(19.20)}{=} \mathcal{N}\left(s|0, \frac{1}{\sqrt{N}}\right), \tag{19.29}
$$

$$
p_\Gamma\left(v|\tfrac{N-1}{2}, \tfrac{1}{2\sigma^2}\right)\ dv \stackrel{(19.21)}{=} p_\Gamma\left(w|\tfrac{N-1}{2}, \tfrac{1}{2}\right)\ dw. \tag{19.30}
$$

We obtain

$$
\begin{aligned}
p(t|N, \mu_0, \mathcal{I}) &= \iint ds\ dw\ \delta\left(t - \tfrac{\sqrt{N(N-1)}\,s}{\sqrt{w}}\right) \ldots \\
&\quad \ldots \mathcal{N}\left(s|0, \tfrac{1}{\sqrt{N}}\right)\ p_\Gamma\left(w|\tfrac{N-1}{2}, \tfrac{1}{2}\right)\ \underbrace{\int d\sigma\ p(\sigma|N, \mu_0, \mathcal{I})}_{=1} \\
&\propto \iint ds\ dw\ \sqrt{w}\ \delta\left(s - \sqrt{w}\tfrac{t}{\sqrt{N(N-1)}}\right) \ldots \\
&\quad \ldots \mathcal{N}\left(s|0, \tfrac{1}{\sqrt{N}}\right)\ p_\Gamma\left(w|\tfrac{N-1}{2}, \tfrac{1}{2}\right) \\
&\propto \int dw\ \sqrt{w}\ \mathcal{N}\left(\sqrt{w}\tfrac{t}{\sqrt{N(N-1)}}|0, \tfrac{1}{\sqrt{N}}\right)\ p_\Gamma\left(w|\tfrac{N-1}{2}, \tfrac{1}{2}\right)
\end{aligned}
$$

$$\propto \int dw\, \sqrt{w}\, e^{-\frac{1}{2}\frac{w\,t^2}{(N-1)}}\, w^{\frac{N-3}{2}} e^{-\frac{w}{2}}$$

$$\propto \int \frac{dw}{w}\, w^{\frac{N}{2}} e^{-\frac{w}{2}(1+\frac{t^2}{N-1})}.$$

Finally, the substitution $w(1+\frac{t^2}{N-1}) = y$ yields the desired result

$$p(t|N,\mu_0,\mathcal{I}) \propto \int \frac{dy}{y}\, y^{\frac{N}{2}} e^{-\frac{y}{2}} \left(1+\frac{t^2}{N-1}\right)^{-\frac{N}{2}} \propto \left(1+\frac{t^2}{N-1}\right)^{-\frac{N}{2}}$$

in the form of the Student t-distribution, already introduced in equation (7.31a) [p. 109].

SAMPLING DISTRIBUTION OF STUDENT'S t-STATISTIC FOR PREDICTED MEAN μ_0

sample:	$\{x_1,\ldots,x_N\}$	drawn from $\mathcal{N}(x|\mu_0,\sigma)$
given:	$N,\ \mu_0$	
test statistic:	$t = \frac{\overline{x}-\mu_0}{\text{SE}^*}$;	$\text{SE}^{*2} := \frac{\overline{(\Delta x)^2}}{(N-1)}$

$$p(t|N,\mu_0,\mathcal{I}) = p_t(t|\nu = N-1),$$

$$p_t(t|\nu) := \frac{\Gamma(\frac{\nu+1}{2})}{\Gamma(\frac{\nu}{2})\sqrt{\pi\,\nu}} \left(1+\frac{t^2}{\nu}\right)^{-\frac{\nu+1}{2}}. \tag{19.31}$$

19.4.2 Student's t-test for the difference in sample means

We generalize the previous analysis to test the hypothesis

H_0: The means of two normal samples of common yet unknown variance are the same.

According to equation (17.20) [p. 270] and equation (17.19) [p. 269], a sensible test statistic is

t-STATISTIC FOR THE DIFFERENCE IN MEANS

$$t := \frac{\overline{x}_2 - \overline{x}_1}{\text{SE}^*_{\text{diff}}},$$

$$\text{SE}^{*\,2}_{\text{diff}} := \frac{N}{(N-2)N_1 N_2} \sum_\alpha N_\alpha\, \overline{(\Delta x_\alpha)^2}. \tag{19.32}$$

The derivation of the sampling distribution is outlined in Appendix B.3 [p. 615]. The result is that t is distributed as $p_{\Gamma}(t|\nu)$, with $\nu = N - 2$. Remarkably, the result is again independent of the prior PDF for the true σ.

The sample size dependence of Student's t-distribution is illustrated in Figure 19.3. It is evident that Student's t-distribution converges rapidly towards the central unit variance norm $N(t|0, 1)$, which is also depicted in the figure.

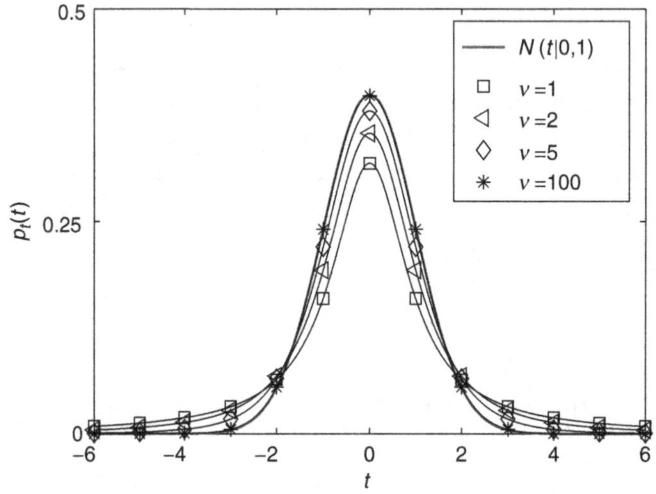

Figure 19.3 Sampling distribution of the t-statistic for various numbers of degrees of freedom (thin lines with markers) compared with the central unit variance norm PDF.

19.5 Fisher–Snedecor F-statistic

The last test statistic we want to investigate is the so-called Fisher–Snedecor F-statistic. As before we start out with two independent samples of size N_1 and N_2, respectively, which are drawn from individual normal distributions. The means of the two distributions are not specified and we wish to analyse whether the variances are the same.

In the Bayesian analysis equation (17.29) [p. 275] we have already encountered the sufficient statistic for this problem, namely

$$f = \frac{\tilde{\sigma_1}^2}{\tilde{\sigma_2}^2} = \frac{n_2}{n_1} \frac{v_1}{v_2} \tag{19.33}$$

$$\text{with:} \qquad n_\alpha := N_\alpha - 1. \tag{19.34}$$

The sought-for sampling distribution again follows from the marginalization rule:

$$
\begin{aligned}
p(f|N_1,N_2,\mathcal{I}) &= \int d\sigma\; dv_1\; dv_2\; p(f|v_1,v_2,N_1,N_2,\mathcal{I})\; p(v_1,v_2,\sigma|N_1,N_2,\mathcal{I}) \\
&= \int d\sigma\; p(\sigma|N_1,N_2,\mathcal{I}) \cdot \\
&\quad \ldots \iint dv_1\; dv_2\; \delta\left(f - \frac{n_2}{n_1}\frac{v_1}{v_2}\right) \underbrace{p(v_1|\sigma,N_1,\mathcal{I})}_{p_\Gamma\left(v_1|\frac{n_1}{2},\frac{1}{2\sigma^2}\right)} \underbrace{p(v_2|\sigma,N_2,\mathcal{I})}_{p_\Gamma\left(v_2|\frac{n_2}{2},\frac{1}{2\sigma^2}\right)}.
\end{aligned}
$$

As before, we introduce the new variables $w_\alpha := \frac{v_\alpha}{\sigma^2}$. Together with equation (19.30) we obtain

$$
\begin{aligned}
p(f|N_1,N_2,\mathcal{I}) &\propto \overbrace{\int d\sigma\; p(\sigma|N_1,N_2,\mathcal{I})}^{=1} \iint dw_1\; dw_2\; w_2\; \delta\big(w_1 - \tfrac{n_1}{n_2}\, f\; w_2\big) \\
&\quad \ldots \cdot p_\Gamma\big(w_1|\tfrac{n_1}{2},\tfrac{1}{2}\big) p_\Gamma\big(w_2|\tfrac{n_2}{2},\tfrac{1}{2}\big) \\
&\propto \int dw_2\; w_2\; p_\Gamma\big(\tfrac{n_1}{n_2}\, f\; w_2|\tfrac{n_1}{2},\tfrac{1}{2}\big)\; p_\Gamma\big(w_2|\tfrac{n_2}{2},\tfrac{1}{2}\big) \\
&\propto \int dw_2\; w_2 \left(\big(\tfrac{n_1}{n_2}\, f\; w_2\big)^{\frac{n_1-2}{2}} e^{-\frac{1}{2}\frac{n_1}{n_2}\, f\, w_2}\right)\left(w_2^{\frac{n_2-2}{2}} e^{-\frac{w_2}{2}}\right) \\
&\propto f^{\frac{n_1-2}{2}} \int \frac{dw_2}{w_2}\; w_2^{\frac{n_1+n_2}{2}}\; e^{-\frac{w_2}{2}(1+\frac{n_1}{n_2}\, f)}.
\end{aligned}
$$

The remaining result is solved by the substitution $w_2(1 + \frac{n_1}{n_2}\, f) := y$:

$$
\begin{aligned}
p(f|N_1,N_2,\mathcal{I}) &\propto f^{\frac{n_1-2}{2}}\; \big(1 + \tfrac{n_1}{n_2}\, f\big)^{-\frac{n_1+n_2}{2}} \left[\int \frac{dy}{y}\; y^{\frac{n_1+n_2}{2}} e^{-\frac{y}{2}}\right] \\
&\propto f^{\frac{n_2-2}{2}}\; \big(n_2 + n_1\, f\big)^{-\frac{n_1+n_2}{2}}.
\end{aligned}
$$

Together with the correct normalization which follows from elementary integration, we obtain

$$
p_F(f|n_1,n_2) = \frac{n_1^{n_1} n_2^{n_2}}{B(\frac{n_1}{2}\,\frac{n_2}{2})}\; f^{\frac{n_1}{2}-1}\, (n_2 + n_1 f)^{-\frac{n_1+n_2}{2}}. \tag{19.35}
$$

SAMPLING DISTRIBUTION OF THE F-STATISTIC

sample 1:	$\{x_1^1, \ldots, x_{N_1}^1\}$ drawn from $\mathcal{N}(x\|x_0^1, \sigma)$
sample 2:	$\{x_1^2, \ldots, x_{N_2}^2\}$ drawn from $\mathcal{N}(x\|x_0^2, \sigma)$
given:	$n_1 = N_1 - 1; \quad n_2 = N_2 - 1$
test statistic:	$f = \frac{n_2}{n_1}\frac{v_1}{v_2}; \qquad v_\alpha = \sum_{i=1}^{N_\alpha} (x_i^\alpha - \overline{x}_\alpha)^2$

$$p(f|N_1, N_2, \mathcal{I}) = p_F(f|n_1, n_2),$$

$$p_F(f|n_1, n_2) = \frac{n_1^{\frac{n_2}{2}} n_2^{\frac{n_1}{2}}}{B(\frac{n_1}{2}, \frac{n_2}{2})} f^{\frac{n_1}{2}-1} (n_2 + n_1 f)^{-\frac{n_1+n_2}{2}}. \tag{19.36}$$

We see again that the prior for σ is not needed for the sampling distribution of the F-statistic. Qualitatively, the curves of the F-distribution are bell-shaped like the chi-squared distribution. The mean and variance have already been discussed in equation (19.35). In particular, the mean is

$$\langle f \rangle = \frac{N_2 - 1}{N_2 - 3} \underset{N_2 \to \infty}{\longrightarrow} 1.$$

It is always greater than 1 and does not depend on N_1, which might appear odd at first glance. On second thoughts, however, it appears quite reasonable, since the sample variances v_α are independently chi-squared distributed:

$$\langle f \rangle = \frac{n_2}{n_1}\left\langle \frac{v_1}{v_2} \right\rangle = \frac{\langle v_1 \rangle}{n_1} \left\langle \frac{n_2}{v_2} \right\rangle.$$

The chi-squared or rather Γ distribution of v_σ is, according to equation (19.21) [p. 295], $p_\Gamma(v|\frac{n_\alpha}{2}, \frac{1}{2\sigma^2})$ and its mean is given by equation (7.14c) [p. 102], i.e.

$$\frac{\langle v_\alpha \rangle}{n_\alpha} = \sigma^2.$$

Thus the mean of v_1/n_1 yields the true variance. However, the nonlinear transformation $1/v_2$ introduces a bias since

$$\left\langle \frac{1}{v_2} \right\rangle \neq \frac{1}{\langle v_2 \rangle},$$

except for $N_2 \to \infty$. The sampling distribution of the F-statistic is depicted in Figure 19.4 for different numbers of degrees of freedom. We can observe that the PDF gets ever narrower with increasing n_2 and will finally end as a delta peak at $f = 1$.

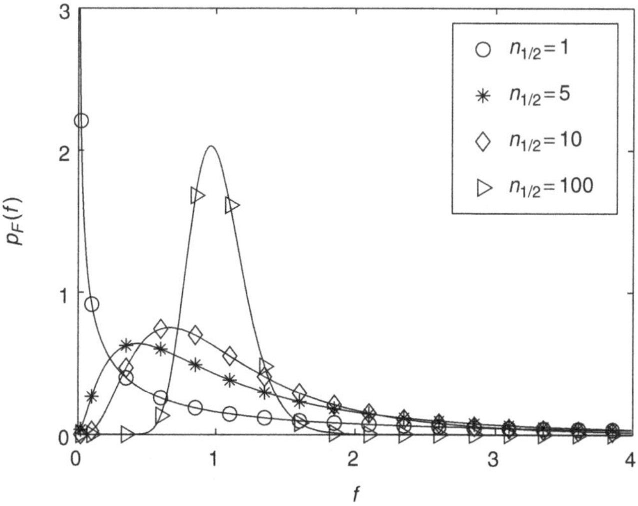

Figure 19.4 Sampling distribution of the F-statistic for various numbers of degrees of freedom.

19.6 Chi-squared in case of missing parameters

The chi-squared statistic has already been touched upon in Section 19.2.2 [p. 295], where all sample elements had a common mean. Here we want to generalize this to prepare the ground for hypothesis testing. To this end we consider a linear model

$$\boldsymbol{g}^{(0)} = \mathbf{M}\,\boldsymbol{a}^{(0)}, \tag{19.37}$$

where $\boldsymbol{a}$ is an N-component column vector comprising the model parameters. The $N_d \times N$ matrix $\mathbf{M}$ represents the model, and $\boldsymbol{g}$ stands for the N_d-component column vector of data values. The superscript (0) in $\boldsymbol{g}^{(0)}$ indicates the underlying true data vector. The basic assumption (null hypothesis) is that the model correctly describes the data. Hence, there is an underlying true set of parameters, denoted by $\boldsymbol{a}^{(0)}$. The actually measured values $\boldsymbol{g}$, however, are distorted by a noise vector $\boldsymbol{\eta}$:

$$\boldsymbol{g} = \boldsymbol{g}^{(0)} + \boldsymbol{\eta}.$$

Needless to say that the noise vector is unknown, but we suppose that it is distributed as a multivariate normal with zero mean:

$$p(\boldsymbol{\eta}|\mathbf{C}) = \frac{1}{Z}\exp\left\{\boldsymbol{\eta}^T\mathbf{C}^{-1}\boldsymbol{\eta}\right\}.$$

The normalization Z is immaterial for the ensuing discussion. Here we even allow for correlations in the measurements, described by the covariance matrix $\mathbf{C}$. The likelihood therefore reads

$$p(\boldsymbol{g}|\boldsymbol{a}, \mathbf{M}, \mathbf{C}) = \frac{e^{\chi^2}}{Z},$$
$$\chi^2 := (\boldsymbol{g} - \mathbf{M}\boldsymbol{a})^T \mathbf{C}^{-1} (\boldsymbol{g} - \mathbf{M}\boldsymbol{a}).$$

The notation simplifies considerably if we introduce the following transformations upfront:

$$\boldsymbol{g} \rightarrow \tilde{\boldsymbol{g}} := \mathbf{C}^{-1/2}\boldsymbol{g},$$
$$\mathbf{M} \rightarrow \tilde{\mathbf{M}} := \mathbf{C}^{-1/2}\mathbf{M},$$
$$\mathbf{C} \rightarrow \mathbb{1}.$$

Henceforward, we suppress the tilde. Then equation (19.37) still holds and the log-likelihood turns into

$$\chi^2 = (\boldsymbol{g} - \mathbf{M}\boldsymbol{a})^T (\boldsymbol{g} - \mathbf{M}\boldsymbol{a}). \tag{19.38}$$

As an example, the model shall be an expansion

$$g_\nu^{(0)} = \sum_{n=1}^{N} a_n \, \Phi_n(x_\nu)$$

in a suitable set of basis functions $\Phi_n(x)$, not necessarily orthonormal. It is not even required that the functions are linearly independent. Then $M_{\nu n} = \Phi_n(x_\nu)$. A simple common example would be a quadratic function

$$g_\nu = a_1 + a_2 x_\nu + a_3 x_\nu^2,$$

and x_ν is the value of a control parameter of the experiment. Usually, some or all model parameters are unknown and have to be inferred from the data. Firstly, we split off that part of the model that contains predetermined parameters and subtract it from the data. This defines a new equation of the form of equation (19.37) in which now all model parameters are unknown. If all parameters are predetermined, the right-hand side is $\mathbf{0}$. Formally, we can use an $N_d \times 1$ model matrix $\mathbf{M} = \mathbf{0}$.

In the frame of frequentist statistics, unknown model parameters are fixed by *maximum likelihood* (see Chapter 15 [p. 236]). The root of the gradient of equation (19.38) yields the ML estimate for the parameters

$$\boldsymbol{a}^{\mathrm{ML}} = \left(\mathbf{M}^T\mathbf{M}\right)^{-1} \mathbf{M}^T \, \boldsymbol{g}. \tag{19.39}$$

As a matter of fact, this is the solution of the *linear regression model* that will be discussed in more detail in Chapter 21 [p. 333]. The data vector corresponding to the ML solution is then

$$\boldsymbol{g}^{\mathrm{ML}} := M\boldsymbol{a}^{\mathrm{ML}} = \underbrace{M\left(\mathbf{M}^T \, \mathbf{M}\right)^{-1} \mathbf{M}^T}_{:=\mathbb{P}} \, \boldsymbol{g} = \mathbb{P}\boldsymbol{g}.$$

We have introduced the matrix $\mathbb{P}$, which is a projection matrix/operator, because $\mathbb{P}^2 = \mathbb{P}$. Applying the projection operator to the true data vector,

$$\mathbb{P}\boldsymbol{g}^{(0)} = \mathbb{P}\mathbf{M}\boldsymbol{a}^{(0)} = M\left(\mathbf{M}^T\mathbf{M}\right)^{-1}\mathbf{M}^T\mathbf{M}\,\boldsymbol{a}^{(0)} = \mathbf{M}\,\boldsymbol{a}^{(0)} = \boldsymbol{g}^{(0)}, \tag{19.40}$$

reveals that $\mathbb{P}$ projects into the subspace

$$V_{\mathbb{P}} = \{\boldsymbol{g}|\boldsymbol{g} = \mathbf{M}\boldsymbol{a};\ \forall \boldsymbol{a}\} \tag{19.41}$$

spanned by the data vectors accessible by the model. So $\boldsymbol{g}^{\mathrm{ML}}$ can be understood as a projection of the measured data into $V_{\mathbb{P}}$. The dimension E of $V_{\mathbb{P}}$ is given by the number of linearly independent vectors in $\mathbf{M}$. Next we compute the difference vector that enters the misfit:

$$\boldsymbol{g} - \mathbf{M}\boldsymbol{a}^{\mathrm{ML}} = (\mathbf{1} - \mathbb{P})\,\boldsymbol{g} := \mathbb{Q}\,\boldsymbol{g}.$$

It depends on the projection operator $\mathbb{Q} = \mathbb{1} - \mathbb{P}$, which projects into the subspace $V_{\mathbb{Q}}$ orthogonal to $V_{\mathbb{P}}$. The misfit

$$\chi^2(\boldsymbol{g}) := \boldsymbol{g}^T\mathbf{Q}^T\mathbf{Q}\boldsymbol{g}$$

is apparently determined by the components of the data vector that cannot be reached by the model. The measured data are distributed as

$$p(\boldsymbol{g}|\boldsymbol{g}^{(0)}, \mathbf{C}, \mathcal{I}) = \frac{1}{Z''}\exp\left\{\left(\boldsymbol{g} - \boldsymbol{g}^{(0)}\right)^T\left(\boldsymbol{g} - \boldsymbol{g}^{(0)}\right)\right\}.$$

Once again, Z'' is an unimportant normalization factor. The noise vector $\boldsymbol{\eta} = \boldsymbol{g} - \boldsymbol{g}^{(0)}$ shall be split into the subspaces $V_{\mathbb{P}}$ and $V_{\mathbb{Q}}$ as follows:

$$\boldsymbol{g} - \boldsymbol{g}^{(0)} = (\mathbb{Q} + \mathbb{P})\left(\boldsymbol{g} - \boldsymbol{g}^{(0)}\right) = \mathbb{Q}\boldsymbol{g} + \mathbb{P}\left(\boldsymbol{g} - \boldsymbol{g}^{(0)}\right).$$

In the last step we have used equation (19.40). Moreover, we have the following orthogonality:

$$\mathbb{Q}^T\,\mathbb{P} = \mathbb{P} - \mathbf{M}\left(\mathbf{M}^T\,\mathbf{M}\right)^{-1}\mathbf{M}^T\mathbf{M}\left(\mathbf{M}^T\mathbf{M}\right)^{-1}\mathbf{M}^T = \mathbb{P} - \mathbb{P} = 0.$$

Together with equation (19.40), it allows us to write

$$\left(\boldsymbol{g} - \boldsymbol{g}^{(0)}\right)^T\left(\boldsymbol{g} - \boldsymbol{g}^{(0)}\right) = (\mathbb{Q}\boldsymbol{g})^T(\mathbb{Q}\boldsymbol{g}) + \left(\mathbb{P}\boldsymbol{g} - \boldsymbol{g}^{(0)}\right)^T\left(\mathbb{P}\boldsymbol{g} - \boldsymbol{g}^{(0)}\right).$$

Now we are prepared to compute the sampling distribution for the misfit through the marginalization rule

$$p(\chi^2|\mathbf{C}, \boldsymbol{g}^{(0)}) \propto \int dg^{N_d}\,\delta(\chi^2 - \chi^2(\boldsymbol{g}))\,e^{-\frac{1}{2}(\mathbb{Q}\boldsymbol{g})^T(\mathbb{Q}\boldsymbol{g})}e^{-\frac{1}{2}\left(\mathbb{P}\boldsymbol{g}-\boldsymbol{g}^{(0)}\right)^T\left(\mathbb{P}\boldsymbol{g}-\boldsymbol{g}^{(0)}\right)}.$$

The integration factors into independent integrals over $V_{\mathbb{P}}$ and $V_{\mathbb{Q}}$, respectively. The former just modifies the prefactor. It also shows that the sampling distribution is independent of the true data vector $\boldsymbol{d}^{(0)}$. The remaining integral over the vectors in $V_{\mathbb{Q}}$ of dimension $n := N_d - E$ has the structure

$$p(\chi^2|\mathbf{C}, \boldsymbol{g}^{(0)}) \propto \int dx^n \delta(\chi^2 - |\boldsymbol{x}|^2)\, e^{-\frac{1}{2}|\boldsymbol{x}|^2} \propto (\chi^2)^{\frac{n}{2}-1} e^{-\frac{1}{2}\chi^2}.$$

We summarize the result:

DISTRIBUTION OF THE CHI-SQUARED STATISTIC
IN CASE OF MISSING PARAMETERS OF A LINEAR MODEL

model:	$\boldsymbol{g} = \mathbf{M}\,\boldsymbol{a}$; $\quad \mathbf{M} : N_d \times N$ matrix
additive noise:	η drawn from $\mathcal{N}(\eta|\mathbf{0}, \mathbf{C})$
statistic:	$\chi^2 := (\boldsymbol{g} - \mathbf{M}\boldsymbol{a})^T \mathbf{C}^{-1} (\boldsymbol{g} - \mathbf{M}\boldsymbol{a})$
E:	number of linearly independent columns in $\mathbf{M}$, associated with unknown parameters

$$p(\chi^2) = p_{\chi^2}\left(\chi^2 | n = N_d - E\right). \tag{19.42}$$

In view of equation (19.39), each parameter corresponds to a particular linear combination of the data. The latter in turn does not contribute to the misfit. That is why each linearly 'independent model parameter' reduces the number of degrees of freedom by one.

19.7 Common hypothesis tests

19.7.1 General procedure

In the introductory part we have outlined the ideas of the two frequentist approaches to hypothesis testing. In this part we limit the discussion to Fisherian significance tests. Up to now we have only considered simple hypotheses in which all parameters are fixed by the null hypothesis. In the case of a compound hypothesis, there are still open parameters. For example, when investigating some crystals with regard to their conductivity, the null hypothesis could be

H_0 : The material has an ohmic resistance.

In other words, the voltage is proportional to the current:

$$U \propto I.$$

The proportionality constant (resistance) is the unknown parameter. The standard approach in dealing with such nuisance parameters in the frequentist approach is to fix the unknown parameters by the maximum likelihood estimator. The number of degrees of freedom, originally the number of data, is reduced by the number of estimated parameters, as these are inferred from the sample.

GENERAL PROCEDURE FOR SIGNIFICANCE TESTS

1. Define the null hypothesis.
2. Choose the test statistic x with sampling distribution $p_x(x)$.
3. Determine the ML estimator for the nuisance parameters $\boldsymbol{\theta}$.
4. Define the significance level (e.g. $\alpha = 0.05$).
5. Determine the rejection region $\mathcal{R}_\alpha$, either one- or two-sided:

$$\int\limits_{\mathcal{R}_\alpha} p_x(x|\boldsymbol{\theta})\, dx.$$

6. Evaluate the test statistic $x = x_0$ for the measured sample.
7. If $x_0 \in \mathcal{R}_\alpha$ reject the null hypothesis.

Next we will discuss the most common significance test in more detail and give some basic examples.

19.7.2 *z*-Tests

Is the mass correctly predicted?

Presumably, one of the simplest types of problem for significance testing is the following: The mass of a particle is measured repeatedly and the data are taken from a sample $\boldsymbol{m} = \{m_1, \ldots, m_N\}$. The statistical errors of the weighing process are additive, uncorrelated and Gaussian with variance σ^2.

The null hypothesis claims *the true value of the mass is* m_0. So we are dealing with a simple null hypothesis and z is again the appropriate test statistic, with

$$z = \frac{\overline{\boldsymbol{m}} - m_0}{\text{SE}},$$

$$\text{SE} = \frac{\sigma}{\sqrt{N}}.$$

The sampling distribution is the centred, unit variance normal distribution, i.e. $\mathcal{N}(z, 0, 1)$. We decide on a two-tailed test as large deviations from $z = 0$ in both directions indicate a discrepancy between null hypothesis and measurement. We choose a significance level α and adjust the rejection region such that the probability mass in both tails is $\alpha/2$, i.e.

$$\int\limits_{z_1^{\text{cr}}}^{\infty} \mathcal{N}(z, 0, 1) = \int\limits_{-\infty}^{z_2^{\text{cr}}} \mathcal{N}(z, 0, 1) = \alpha/2.$$

By symmetry, $z_2^{\mathrm{cr}} = -z_2^{\mathrm{cr}} := -z^{\mathrm{cr}}$ and

$$\int_{-\infty}^{-z^{\mathrm{cr}}} \mathcal{N}(z, 0, 1) = \frac{1}{2}\mathrm{Erfc}\left(\frac{z^{\mathrm{cr}}}{\sqrt{2}}\right) \stackrel{!}{=} \frac{\alpha}{2}.$$

Hence, $z^{\mathrm{cr}} = \sqrt{2}\,\mathrm{InvErfc}(\alpha)$, with 'Erfc' being the complementary error function and 'InvErfc' its inverse. For the two values $\alpha = 0.05$ ($\alpha = 0.01$), recommended by Fisher, we obtain $z^{\mathrm{cr}} = 1.960$ ($z^{\mathrm{cr}} = 2.575$). If the difference $|\overline{m} - m_0|$ is greater than $z^{\mathrm{cr}} \cdot \mathrm{SE}$ the data are called *significant*, since the measurement differs significantly from the null hypothesis by an amount that can hardly be explained by chance alone. And therefore the hypothesis is to be rejected.

Does the control variable have an impact?

We claim that a control parameter S of an experiment does not affect the measurement of quantity X. This assertion is our *null hypothesis* H_0. In order to test the null hypothesis we perform a series of experiments for a specific value of the control variable $S = S_1$. For this data set we determine the sample mean $\overline{x}_1$. Then we repeat the experiment for a second value of the control parameter S_2 and compute the corresponding sample mean $\overline{x}_2$. If the hypothesis is correct, the difference between the two sample means should only be due to statistical fluctuations and they should be small on an appropriate scale. The latter is reasonably defined by the standard error (SE), which for simplicity shall be the same for both samples. According to equation (19.26) [p. 298], the relevant measure for the discrepancy of the two samples is

$$z := \frac{\overline{d}_2 - \overline{d}_1}{\mathrm{SE}}.$$

For simplicity, we assume identical sample sizes $N_\alpha = N$ and variances $\sigma_\alpha = \sigma$. Then

$$\mathrm{SE}^2 = \frac{2\sigma^2}{N}.$$

The rest of the test is the same as in the previous example.

19.7.3 Chi-squared test

The chi-squared test, sometimes also called the *goodness of fit* test, was first introduced by Karl Pearson (1900). It is presumably the most frequently used statistical test. In physics it is commonly utilized for testing whether a theoretical model describes experimental data or whether a frequency distribution fits a predicted probability distribution.

Testing a theoretical model

One class of problems to which the chi-squared test is routinely applied is model testing. Suppose we have a set of values $\boldsymbol{s} = \{s_1, \ldots, s_N\}$ of a control parameter and the corresponding measured values $\boldsymbol{y} = \{y_1, \ldots, y_N\}$ of a physical quantity Y. The data are believed to be describable by a theoretical model

$$y_i = f(s_i|\boldsymbol{\theta}),$$

which may or may not depend on additional known or unknown parameters $\boldsymbol{\theta} \in \mathbb{R}^{N_p}$. The possible null hypothesis could be

$$H_0 : \text{The data obey} \, y_i = f(s_i|\boldsymbol{\theta}), \text{ with unknown parameters } \boldsymbol{\theta}.$$

If the hypothesis and hence the model is correct, the measured data differ from the model only by the noise values

$$y_i = f(s_i|\boldsymbol{\theta}) + \eta_i.$$

The noise is supposed to be additive and i.n.d. with individual known standard deviation σ_i for each data point. The error distribution and its parameter are part of our background knowledge and not to be questioned.[1] If part or all of the model parameters are unknown, they have to be determined by the maximum likelihood estimator (see Section 15.2 [p. 237]). As a result, the number of degrees of freedom (dof) is reduced to $\nu = N - r$, with r being the number of unknown parameters.

The chi-squared test for this problem is thus

CHI-SQUARED TEST FOR MODEL EVALUATION

test statistic:
$$x = \sum_{i=1}^{N} \frac{(y_i - f(s_i|\boldsymbol{\theta}))^2}{\sigma_i^2}, \tag{19.43}$$

PDF:
$$p_{\chi^2}(x|\nu) = \frac{2^{-\frac{\nu}{2}}}{\Gamma(\frac{\nu}{2})} \; x^{\frac{\nu}{2}-1} \, e^{-\frac{x}{2}}, \tag{19.44}$$

number of dof:
$$\nu = N - r. \tag{19.45}$$

r parameters have been fixed by their ML estimate.

In the present context a two-tailed test is appropriate, since deviations of x_0 from the mean $\langle x \rangle_\nu = \nu$ on both sides are equally inconsistent with the null hypothesis. The rejection region is therefore composed of two intervals, with two critical values for the test statistic x_l^{cr} and x_r^{cr}, respectively:

[1] It is an interesting and challenging question whether these assumptions are correct at all. This topic will be discussed further in Part V.

$$\mathcal{R} = \mathcal{R}_{\text{left}} \cup \mathcal{R}_{\text{right}},$$
$$\mathcal{R}_{\text{left}} = \left\{x \middle| x < x_l^{\text{cr}} := F_{\chi^2}^{-1}(\tfrac{\alpha}{2}|N-r)\right\},$$
$$\mathcal{R}_{\text{right}} = \left\{x \middle| x > x_r^{\text{cr}} := F_{\chi^2}^{-1}(1-\tfrac{\alpha}{2}|N-r)\right\}.$$

Table 19.1 contains the critical values x_l and x_r for a two-tailed test for various values of ν and α. The critical values for a right-tailed test are listed in Table 19.2.

Table 19.1 Critical values x_l^{cr} and x_r^{cr} (left and right column for each value of α) for a two-tailed chi-squared test for various values of ν and α

	$\alpha = 0.01$		$\alpha = 0.05$	
ν	x_l^{cr}	x_r^{cr}	x_l^{cr}	x_r^{cr}
1	0.000	7.879	0.001	5.024
6	1.237	14.449	0.676	18.548
7	1.690	16.013	0.989	20.278
10	2.156	25.188	3.247	20.483
20	7.434	39.997	9.591	34.170
30	13.787	53.672	16.791	46.979
40	20.707	66.766	24.433	59.342
50	27.991	79.490	32.357	71.420
100	67.328	140.169	74.222	129.561

Table 19.2 Critical values x^{cr} for a right-tailed chi-squared test dependent on the degrees of freedom ν and the significance level α

ν	$\alpha = 0.01$	$\alpha = 0.05$
1	6.635	3.841
6	12.592	16.812
7	14.067	18.475
10	23.209	18.307
20	31.410	37.566
30	43.773	50.892
40	55.758	63.691
50	67.505	76.154
100	124.342	135.807

By way of an example, we consider the ohmic resistor problem outlined before. Given a set of values for the current (control parameter) $\{I_1, \ldots, I_N\}$ the associated values for the voltage $\{U_1, \ldots, U_N\}$ are determined experimentally. The statistical errors of the voltage measurement are additive, uncorrelated and Gaussian. The variance may depend on I so we have a set of individual standard deviations $\{\sigma_1, \dots, \sigma_N\}$. The theoretical function in this case is $f(U|R) = R \cdot I$. The nuisance parameter R is fixed by its ML estimator. To this end we minimize the misfit

$$x = \sum_{i=1}^{N} \left(\frac{U_i - RI_i}{\sigma_i} \right)^2 = \sum_{i=1}^{N} \left(\frac{U_i}{\sigma_i} \right)^2 - 2R \sum_{i=1}^{N} \left(\frac{U_i I_i}{\sigma_i^2} \right) + R^2 \sum_{i=1}^{N} \left(\frac{I_i}{\sigma_i} \right)^2.$$

In terms of weighted sample means, defined by

$$\langle O \rangle_\sigma := \frac{\sum_{i=1}^{N} O_i \sigma_i^{-2}}{\sum_{i=1}^{N} \sigma_i^{-2}},$$

the misfit reads

$$x = \left(\sum_{i=1}^{N} \sigma_i^{-2} \right) \left(\left\langle U^2 \right\rangle_\sigma - 2R \langle U \cdot I \rangle_\sigma + R^2 \left\langle I^2 \right\rangle_\sigma \right). \tag{19.46}$$

It is minimized by

$$R^{\text{ML}} = \frac{\langle U \cdot I \rangle_\sigma}{\left\langle I^2 \right\rangle_\sigma}.$$

The number of degrees of freedom is reduced by one ($\nu = N - 1$). The value of the test statistic based on the data is obtained by inserting R^{ML} in equation (19.46). In Figure 19.5 mock data are presented for this problem. The sample size is $N = 21$ so that $\nu = 20$. The ML estimate for the resistance yields $R^{\text{ML}} = 25.31$. The corresponding U–I curve is also depicted in the figure and fits the data reasonably well. The mean value of the chi-squared distribution with ν degrees of freedom is generally ν. So in the present example on average we expect $\langle x \rangle = 20$. The numerical example depicted in the figure yields $x_0 = 48.33$, which appears to be in the rightmost part of the chi-squared distribution. From Table 19.1 we read off that x_0 is much greater than x_r^{cr} for both $\alpha = 0.05$ and $\alpha = 0.01$. Consequently, the null hypothesis can safely be rejected. In the mock example we are in the privileged position that we know the true answer, which is usually not the case. The truth is that the data have been generated based on a different theoretical model $U = R \cdot I + 20 \cdot I^2$ that also included an I^2 term. So, rejecting the null hypothesis was the correct decision.

Testing distributions

Frequently one is confronted with the question of whether a measured frequency distribution corresponds to a certain theoretical distribution. For example, are the measured radioactive decays really Poisson distributed? The distribution to be tested can either be discrete or continuous. In the latter case the problem has to be discretized first. This shall

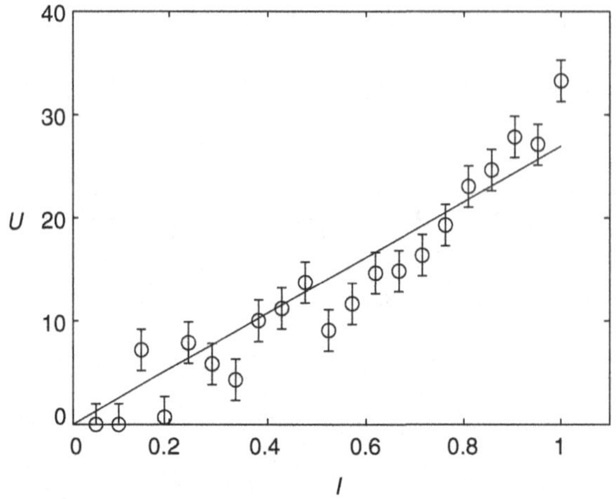

Figure 19.5 Mock data for a U–I curve. Circles and error bars represent the noisy data with the 1-σ interval. Data are in arbitrary units. The solid line represents the ML result. Parameters are $N = 21$, $\sigma/U_0 = 2$, $R = 10$.

be illustrated in terms of a comprehensive example. Let a random variable assume all positive real values $x \in [0, \infty)$. Let the underlying PDF be $\rho(x)$. We discretize the x-axis as

$$0 = x_0 < x_1 < x_2 < \ldots < \infty.$$

The partitioning may, but need not, be equidistant. Next we introduce the probability that x falls into the interval $I_i := [x_i, x_{i+1})$:

$$P_i = P(x \in I_i|\mathcal{I}) = \int_{x_i}^{x_{i+1}} \rho(x)\,dx, \qquad i = 0, 1, \ldots$$

After discretization the continuous problems can be treated like a discrete problem. There are events E_i, for $i = 1, 2, \ldots$; the total number can be finite or infinite. Typical events are: x lies in I_i, and the number of decays in a fixed-energy window is n_i. The total count is $N = \sum_i n_i$. The chi-squared test statistic in this case assumes the form

$$x = \sum_{i=1}^{L} \frac{(n_i - \langle n_i \rangle)^2}{\langle n_i \rangle},$$

$$\langle n_i \rangle := N P_i.$$

Needless to say, impossible events (those with $P_i = 0$) are discarded right from the start. Now the occupation numbers n_i are correlated ($\sum_i n_i = N$) and they obey a multinomial distribution. In other words, the key conditions for the chi-squared test are not fulfilled, namely i.i.n.d. random variables. But we prove in Appendix B.2 [p. 612] that based on the central limit theorem the sampling distribution of x converges for large N towards the chi-squared distribution with $L-1$ degrees of freedom. The reduction by one is due to the fact that the counts n_i sum up to N.

CHI-SQUARED TEST
FOR DISTRIBUTIONS WITH NO FREE PARAMETERS

sample:	$\{n_1, \dots, n_L\}$ (multinomial distribution)
constraint:	$\sum_{i=1}^{L} n_i = N$
given:	$\{P_i\}$
test statistic:	$x = \sum_{i=1}^{L} \frac{(n_i - \langle n_i \rangle)^2}{\langle n_i \rangle} \; ; \quad \langle n_i \rangle = N P_i \; .$

$$p(x|N, L, \{P_i\}, \mathcal{I}) = p_{\chi^2}(x|L-1) \, . \tag{19.47}$$

We will consider two representative examples.

Example: Biased coin

The simplest example for this application of a chi-squared test is the coin experiment. The null hypothesis claims the coin is fair, i.e. $P_{\text{heads}} = P_{\text{tails}} = \frac{1}{2}$. The coin is tossed N times and heads occurs n_1 and tails n_2 times. In total $N = n_1 + n_2$. The chi-squared value is

$$x_0 = \frac{\left(n_1 - \frac{N}{2}\right)^2}{\frac{N}{2}} + \frac{\left(n_2 - \frac{N}{2}\right)^2}{\frac{N}{2}} = \frac{(2n_1 - N)^2}{N} .$$

The number of degrees of freedom is $\nu = 2 - 1 = 1$. In this example a right-tailed test in x_0 is presumably adequate; we would also accept small values of x_0 as it would be difficult to find a sensible alternative hypothesis that could explain small deviations from the theoretical value $n = \frac{N}{2}$ better than the null hypothesis. Hence the data will be rejected if $x_0 > x^{\text{cr}}$, or alternatively if

$$n_1 < \frac{1}{2}\left(N - \sqrt{x^{\text{cr}} N}\right) \quad \text{or} \quad n_1 > \frac{1}{2}\left(N + \sqrt{x^{\text{cr}} N}\right) .$$

According to Table 19.2 the corresponding critical value for e.g. $\alpha = 0.1$ is $x^{\text{cr}}_{1\%} = 6.63$. For a sample of size N that means the hypothesis is rejected if

$$n_1 > 62 \quad \text{or} \quad n_1 < 38.$$

Tests with undetermined parameters

Now we want to generalize the test on distributions if none or not all parameters are specified. As outlined before, the unspecified parameters are replaced by their ML estimates. We use the ML values to determine the theoretical mean occupation

$$\langle n_l \rangle = N p_l(a^{\mathrm{ML}})$$

and compute the chi-squared value as before. The number of degrees of freedom is reduced by the number N_p of ML-estimated parameters

$$\nu = L - 1 - N_p.$$

One degree of freedom is lost by the sum rule $N = \sum_i n_i = \sum_i \langle n_i \rangle$.

Example: Poisson distribution

Let us assume that an experiment is performed in which the number of radioactive decays during a fixed time window is repeatedly determined. The experiment delivers a list of absolute frequencies n_l for the occurrence of l decays, for $l = 0, 1, \ldots$ The data are depicted in Table 19.3. Since they stem from radioactive decays, they should come from a Poisson distribution. In order to test this assumption we formulate the null hypothesis

$$H : \text{The data sample is Poissonian}$$

and perform a chi-squared test. In the first step we determine the missing parameter μ of the Poisson distribution by maximum likelihood. According to equation (15.4) [p. 238] it is given by $\mu^{\mathrm{ML}} = \sum_l l \ f_l$, where f_l are the relative frequencies. The numeric result is $\mu^{\mathrm{ML}} = 2.023$. The histogram of the data is depicted in Figure 19.6 and compared with the maximum likelihood fit to a Poisson distribution (equation (9.4) [p. 151]). The detailed data information is also included in Table 19.3. Both data and ML fit seem reasonably close. Let us check whether the chi-squared test shares this view.

The size of the sample is $N = 8$ and the number of degrees of freedom is therefore $\nu = 8 - 2 = 6$.

The computation of the chi-squared value yields $x_0 = 13.105$. According to Table 19.2, the critical values are $x^{\mathrm{cr}}_{5\%} = 12.592$ and $x^{\mathrm{cr}}_{1\%} = 16.812$. Based on a 5% level we would reject the null hypothesis, saying the data are not Poissonian, whereas on a 1% level we

Table 19.3 The information in the three rows from top to bottom is: number of decays, Poissonian mock data for frequencies, ML estimate for $\langle n_l \rangle^{\mathrm{ML}} = N P_l^{\mathrm{ML}}$

l	0	1	2	3	4	5	6	7
n_l	153	278	288	184	103	33	25	4
$\langle n_l \rangle^{\mathrm{ML}}$	141.2	285.7	289.0	194.9	98.6	39.9	13.5	3.9

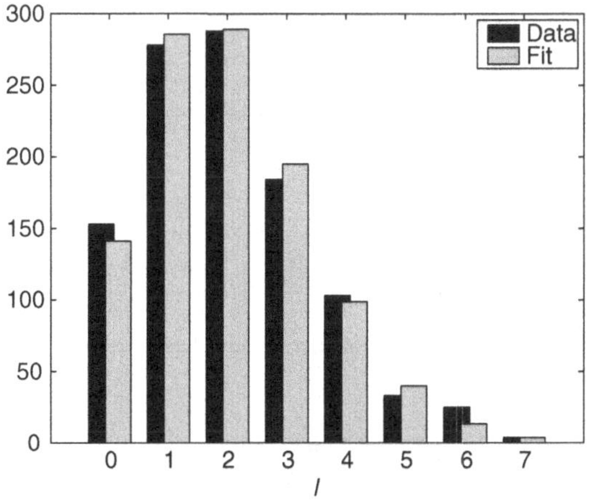

Figure 19.6 Histogram of a counting experiment and ML fit.

would not reject it. So we conclude that the agreement is not as convincing as expected on first glance.

Contingency tables

Another scope of application of the chi-squared test are the so-called contingency tables. The issue covered is as follows. We measure two properties x and y, which, for simplicity, shall only assume discrete values $x_1, x_2, \ldots, x_l$, or rather $y_1, y_2, \ldots, y_k$. The experiment is repeated N times. The number of experiments that yield the data pair (x_i, y_j) is n_{ij}.

We fill the *contingency table* with these numbers, as illustrated in Table 19.4. We have introduced the following abbreviations:

$$n_{i,.} = \sum_j n_{ij},$$
$$n_{.,j} = \sum_i n_{ij},$$
$$N = \sum_{ij} n_{ij}.$$

We denote the probability for the occurrence of the value x_i by p_i and that for the value y_j by q_j. The goal is to test whether there are correlations between the x and y values.

To this end the null hypothesis is

H: The data are uncorrelated.

That means

$$P(x_i, y_j) = p_i\, q_j.$$

Table 19.4 Contingency table

	y_1	y_2	...	y_k	Sum
x_1	n_{11}	n_{12}	...	n_{1k}	$n_{1,.}$
x_2	n_{21}	n_{22}	...	n_{2k}	$n_{2,.}$
$\vdots$	$\vdots$	$\vdots$	...	$\vdots$	$\vdots$
x_l	n_{l1}	n_{l2}	...	n_{lk}	$n_{l,.}$
Sum	$n_{.,1}$	$n_{.,2}$	...	$n_{.,k}$	N

If the hypothesis is correct, the average entry in the contingency table for the pair $(x_i y_j)$ should be $Np_i q_j$. Unfortunately, we don't know the probabilities p_i and q_j and we have to determine them via maximum likelihood. The likelihood function is

$$p(\boldsymbol{n}|\boldsymbol{p},\boldsymbol{q},N) \propto \prod_{ij} (p_i q_j)^{n_{ij}},$$

where the bold symbols stand for the set of corresponding values. We obtain the ML solution by the zeros of the derivative of the log-likelihood subject to the normalization constraints

$$\sum_{i=1}^{l} p_i = 1 \quad \text{and} \quad \sum_{j=1}^{k} q_j = 1.$$

The constraints are incorporated via the Lagrange method and the condition for the ML solution is

$$\frac{\partial}{\partial p_\nu}\left(\sum_{ij} n_{ij}\ \ln(p_i \cdot q_j) - \lambda_1 \sum_i p_i - \lambda_2 \sum_j q_j\right) = \frac{\sum_j n_{\nu j}}{p_\nu} - \lambda_1 \stackrel{!}{=} 0.$$

Repeating the calculation for q_μ and utilizing the normalization condition leads to the desired result

$$p_\nu^{\mathrm{ML}} = \frac{1}{N} n_{\nu,.},$$
$$q_\mu^{\mathrm{ML}} = \frac{1}{N} n_{.,\mu}.$$

The chi-squared statistic therefore reads

$$x = \sum_{ij} \frac{(n_{ij} - Np_i^{\mathrm{ML}} q_j^{\mathrm{ML}})^2}{N(p_i^{\mathrm{ML}} q_j^{\mathrm{ML}})}.$$

The number of degrees of freedom is reduced from $L - 1$, as we have determined the probabilities p_i and q_j. In principle, there are in total $k + l$ unknown parameters, but since both sets of probabilities add up to one, the reduction is only $k + l - 2$. In total we have

$$\nu = l \cdot k - 1 - k - l + 2 = (l - 1)(k - 1).$$

In summary, the chi-squared test for contingency tables is

CHI-SQUARED TEST FOR CONTINGENCY TABLES

sample: $\{n_{ij}\}$ for $i = 1, \ldots, l$, $j = 1, \ldots, k$

no. of dof: $\nu = (l - 1)(k - 1)$

test statistic:

$$x = \sum_{ij} \frac{(n_{ij} - N p_i^{\mathrm{ML}} q_j^{\mathrm{ML}})^2}{N(p_i^{\mathrm{ML}} q_j^{\mathrm{ML}})} \tag{19.48a}$$

$$p_\nu^{\mathrm{ML}} = \frac{1}{N} n_{\nu,.}, \tag{19.48b}$$

$$q_\mu^{\mathrm{ML}} = \frac{1}{N} n_{.,\mu}. \tag{19.48c}$$

Example: Are two histograms based on the same probability distribution?

Suppose we want to test whether two histograms originate from the same unknown distribution, however, without making any assumptions about the underlying distribution. In other words, any set of probabilities $\{P_i\}$ is acceptable. The situation is actually closely related to that described earlier in this section, where data are tested for a particular distribution. There the data are confronted with the ML prediction, as summarized in Table 19.3. As a matter of fact, this is indeed a contingency table, which has the special feature, however, that the row sums are equal. In the case of two arbitrary histograms, the sum of the two rows, N_α say, will be different. The test statistic for contingency tables, given in equation (19.48), simplifies if two histograms are compared:

$$x = \sum_{\alpha=1}^{2} \sum_{j=1}^{k} \frac{\left(n_{\alpha,j} - N_\alpha q_j^{\mathrm{ML}}\right)^2}{N_\alpha q_j^{\mathrm{ML}}} = \sum_{\alpha=1}^{2} N_\alpha \sum_{j=1}^{k} \frac{\left(f_j^{(\alpha)} - q_j^{\mathrm{ML}}\right)^2}{q_j^{\mathrm{ML}}} \tag{19.49a}$$

with

$$q_j^{\mathrm{ML}} = \frac{\sum_{\alpha=1}^{2} n_{\alpha,j}}{N_1 + N_2} \quad \text{and} \quad f_j^{(\alpha)} := \frac{n_{\alpha,j}}{N_\alpha}. \tag{19.49b}$$

As a numeric example we take the data in Table 19.3 in which we replace $\langle n_l \rangle^{\mathrm{ML}}$ by the nearest integer. These values therefore represent frequencies that are actually Poissonian and we compare them with the experimental data in Table 19.3. Hence we are performing a similar test as before, but there is an important difference. In the previous test we assumed the data are Poissonian and, after ML determination of the unknown parameters, the null hypothesis implied that the second row in Table 19.3 contains the correct data and the first row is tested against it. Now, however, we no longer prefer one row over the other and we only assess the mutual proximity. Under these circumstances it is much harder to decide that the underlying true distribution is not the same and, as opposed to the previous test, we expect the data not to be significant. The numerical result for the test statistic given in equation (19.49) is $x_0 = 5.464$. There are $\nu = 7$ degrees of freedom and by means of Table 19.2 [p. 312] the critical values are $x_{5\%}^{\mathrm{cr}} = 14.067$ and $x_{1\%}^{\mathrm{cr}} = 18.475$. Apparently, as expected, the test is much less stringent and the data are not significant (to reject the null hypothesis). So we cannot say that the two histograms are different. For that decision the data have to diverge on a different scale.

19.7.4 Student's t-test

The basic assumptions underlying the t-test are the same as for the z-test, i.e. a sample of i.i.n.d. random numbers. The only difference is that the variance is unknown. Possible applications are the same as for the z-test.

One common application is to test whether a particular μ_0 is the true mean. The Student t-test statistic for this case has been given in equation (19.31) [p. 301]:

$$t_0 = \frac{\overline{x} - \xi}{\mathrm{SE}^*},$$

$$\mathrm{SE}^{*2} = \frac{\sum_{i=1}^{N} (x_i - \overline{x})^2}{N(N-1)}, \qquad \overline{x} = \frac{1}{N} \sum_{i=1}^{N} x_i .$$

The number of degrees of freedom is $\nu = N - 1$. Depending on the application, a one- or two-tailed test will be performed. For a given significance level α and a right-tailed test, the critical value for t is given by

$$t_\alpha^{\mathrm{cr}} = F_t^{-1}(1 - \alpha | \nu),$$

where $F_t^{-1}(t|\nu)$ is the inverse of the cumulative distribution function $F_t(t|\nu)$ of the t-distribution for ν degrees of freedom. The ν dependence of the critical value is depicted in Figure 19.7 for the two standard significance levels recommended by Fisher. In addition, the corresponding critical values for the z-test are also shown. It should be recalled that the sampling distribution of the latter test is independent of the sample size, resulting in horizontal lines in Figure 19.7. The t-test requires the estimation of the variance utilizing the data. Consequently, the test is less stringent than the z-test, which is why in Figure 19.7

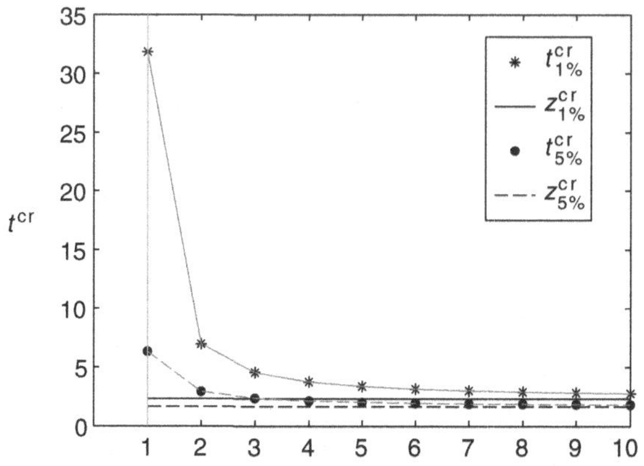

Figure 19.7 Critical value $t_\alpha^{\rm cr}$ of the right-tailed t-test versus the number of degrees of freedom ν for the two standard significance levels $\alpha = 0.01$ and $\alpha = 0.05$. The results are compared with the corresponding critical value of the z-test, which is independent of the sample size.

the critical values of the t-test exceed those of the z-test. The discrepancy is greatest for $N = 2$ and decreases rapidly towards zero with increasing sample size. This is reassuring because the sample estimate of the variance becomes increasingly reliable.

Comparison of sample means

Just like the z-test, the t-test can also be used to test whether the PDF underlying two samples has the same mean. The assumptions are the same as for the z-test, i.e. the data within each sample are i.i.n.d. The variances in both samples are equal. But again, the common variance is replaced by the sample variance. The null hypothesis is:

$$H_0 : \text{The true means of both samples are the same.}$$

The samples are denoted by $\{d_1^\alpha, \ldots, d_{N_\alpha}^\alpha\}$ ($\alpha = 1, 2$) and have the sizes N_1 and N_2. The total number of sample elements is $N = N_1 + N_2$. The sample mean and sample variance are

$$\overline{d^{(\alpha)}} = \frac{1}{N_\alpha} \sum_{i=1}^{N_\alpha} d_i^\alpha,$$

$$\overline{(\Delta d^{(\alpha)})^2} = \frac{1}{N_\alpha} \sum_{i=1}^{N_\alpha} (d_i^\alpha - \overline{d^\alpha})^2.$$

In this case Student's t-statistic is the difference of the sample means scaled by the unbiased sample estimate of the standard error of the difference in means. The number of

Table 19.5 Measured masses of two species in arbitrary units

i	$m_i^{(1)}$	$m_i^{(2)}$
1	11	18
2	14	11
3	8	9
4	12	7
5	9	14
6	—	10
7	—	15
N_α	5	7
$\overline{d^{(\alpha)}}$	10.8	12.0
$\overline{(\Delta d^{(\alpha)})^2}$	4.56	12.57

degrees of freedom is $\nu = N - 2$. The characteristics of the t-test have been given in equation (19.32) [p. 301]:

$$t = \frac{\overline{d^{(2)}} - \overline{d^{(1)}}}{\mathrm{SE}^*_{\text{diff}}},$$

$$\mathrm{SE}^{*\,2}_{\text{diff}} = \frac{N}{(N-2)\, N_1 N_2} \sum_{\alpha=1}^{2} N_\alpha \, \overline{(\Delta d^{(\alpha)})^2}.$$

Example: Comparison of sample means

Table 19.5 contains two samples of normally distributed masses of two species in arbitrary units. We know for sure that the variances are the same but we don't know the value. The null hypothesis claims the masses are actually the same. The numbers in the table yield $\nu = 10$, $\mathrm{SE}^*_{\text{diff}} = 1.95$ and $t_0 = 0.62$. In case of a right-tailed test the critical values can be read off from Figure 19.7. They are $t^{\text{cr}}_{1\%} = 2.82$ and $t^{\text{cr}}_{5\%} = 1.83$. Hence, the data are not significant and the null hypotheses cannot be rejected.

19.7.5 F-Test

As we have seen in Section 19.5 [p. 302], the F-statistic is invoked to test the null hypothesis

$$H_0 : \text{Two samples have the same variance}$$

based on two i.i.n.d. samples $\{d_1^{(\alpha)}, \ldots, d_{N_\alpha}^{(\alpha)}\}$ which have a common variance. The intrinsic variance is not given, nor are the individual means further specified. The test statistic of the F-test is the ratio of the unbiased sample variances

$$f = \frac{\left(\sigma_1^*\right)^2}{\left(\sigma_2^*\right)^2} = \frac{n_2 v_1}{n_1 v_2},$$
$$v_\alpha = \sum_{i=1}^{N_\alpha} \left(d_i^{(\alpha)} - \overline{d}_\alpha\right)^2. \tag{19.50}$$

We have shown in Section 19.5 [p. 302] that f is distributed as $p_F(f|n_1, n_2)$, where $n_\alpha = N_\alpha - 1$ is the number of degrees of freedom in sample α. One degree of freedom is lost in order to fix the unknown mean. Otherwise, the means do not enter the analysis.

Example: Is an instrument still calibrated?

The F-test shall be applied to the question outlined in Section 17.2.4 [p. 275], of whether an instrument is still calibrated, evidenced by a variance ratio of 1. We used two series of measurements of sizes $N_1 = N_2 = 20$. The null hypothesis claimed

H : The variance of the experimental noise is the same in both data sets.

The estimates for the standard deviations were $\sigma_1^* = 2.5$ and $\sigma_2^* = 4.4$, which implied $f_0 = 3.1$. In this context it is reasonable to use a two-tailed test as deviations in both directions from $f = 1$ indicate likewise that the hypothesis is incorrect.

Since the F-distribution is invariant against the interchange of the sample variances, the P value for the two-sided test is

$$P = 2 \cdot \int_{f_0}^{\infty} p_F(f|n_1, n_2)\, df = 0.018.$$

Hence, based on $\alpha = 5\%$ the data are significant and based on $\alpha = 1\%$ the data are not significant. So the result is almost balanced and a decision in either direction is risky. Consequently, in this example the significance of the F-test is marginal.

This is in stark contrast to the Bayesian result, derived in Section 17.2.4 [p. 275], which highly favours the hypothesis ($P(H|D, \mathcal{I}) = 0.99$) leading to the conclusion that the measurement apparatus is still working perfectly well.

20

Comparison of Bayesian and frequentist hypothesis tests

In previous chapters we have applied the Bayesian and the frequentist approach to some basic problems. The key differences of the two approaches are summarized in Table 20.1. The last two rows deserve additional remarks, which will be given in the following sections.

20.1 Prior knowledge is prior data

In scientific problems the present data are certainly not the only information that is known for the problem under consideration. Usually, there exists a wealth of knowledge in the form of previous experimental data and theoretical facts, such as positivity constraints, sum rules, asymptotic behaviour. *A scientist is never in the situation that only the current data count.* The Bayesian approach allows us to exploit all this information consistently. It has been criticized that priors are a subjective element of the theory. This is not really correct, as the Bayesian approach is internally consistent and deterministic. The only part that could be described as subjective is the knowledge that goes into the prior probabilities. *But this degree of subjectivity is actually the foundation of all science. It is the expertise that exists in the respective discipline.* The generation of experimental data is based on the same subjectivity, as it is generally motivated by prior knowledge in the form of previous data or theoretical models.

A great part of the prior knowledge is a summary of a conglomeration of previous measurements, with completely different meanings and sources of statistical errors. The Bayesian approach is proved to be consistent and it is no surprise that it allows a systematic and consistent treatment of these data in the form of prior probabilities. Suppose there is a new data set D and a collection of old data sets $\{D_i\}$ collected over the years. It is just simply a matter of applying the product rule to determine the odds ratio for the null hypothesis:

$$
\begin{aligned}
o(H_0|D, D_1, \ldots, D_N, \mathcal{I}) &= o(D|H_0, \cancel{D_1}, \ldots, \cancel{D_N}, \mathcal{I})\; o(H_0|D_1, \ldots, D_N, \mathcal{I}) \\
&= o(D|H_0, \mathcal{I})\; o(H_0|D_1, \ldots, D_N, \mathcal{I}).
\end{aligned}
$$

The previous data sets are immaterial in the first factor as it is a crucial feature of the null hypothesis that it uniquely fixes the likelihood. The second factor is nothing but

Table 20.1 Comparison of the Bayesian and the frequentist approach

Topic	Bayesian	Frequentist
Test statistic	follows compellingly and uniquely from the rules of probability theory	not unique
Free parameter	fixed by expert knowledge, entering the priors	significance level
Depends on	value of test statistic of measured data	value of test statistic for the entire rejection region
Approximations	not required	sometimes necessary to obtain a standard sampling distribution (e.g. χ^2 for distributions)
Hypotheses	all on the same footing	privileged role of null
Previous data	enter through prior	hard to treat with standard tools
Prior knowledge	consistently included	ignored
Stopping rule	independent	dependent

the prior odds for the current hypothesis test. But it can just as well be viewed as the odds ratio of a hypothesis test performed in the past. The procedure can be repeated for $o(H_0|D_1, \ldots, D_N, \mathcal{I})$ and we get

$$o(H_0|D, D_1, \ldots, D_N, \mathcal{I}) = \prod_{i=\nu}^{N} o(D|H_\nu, \mathcal{I}) \;\cdot\; o_P(H_0).$$

The nice thing about the Bayesian approach is that it puts all prior data on the same footing, even if the various data samples have different error statistics and would have to be treated in frequentist tests with different test statistics. The last factor is the prior odds, which is not based on data information but rather on purely theoretical knowledge. If all data samples are properly accounted for they will predominate the result, unless there is a contradiction (e.g. if the data favour a negative value for a model parameter, while the theory tells us that it has to be positive).

20.2 Dependence on the stopping criterion

We have already addressed the impact of the stopping criterion on the likelihood estimator in Section 15.4 [p. 241]. Here we will deepen the discussion and study the impact of the stopping rule on hypothesis testing. The original idea, outlined here, traces back to L. J. Savage [60] and is discussed in more detail in [13]. We use a similar problem based on Bernoulli experiments. Let H_0 be the hypothesis that the probability for an individual trial is p_0. Based on the data pair (n, N), which is the number of successes n and total trials N, we want to scrutinize the validity of the null hypothesis H_0.

So far it sounds very familiar, but this time the data shall be generated based on different stopping conditions: the experiment was stopped (a) after N trials, or (b) when the nth success was observed.

20.2.1 Bayesian analysis

The Bayesian approach starts in both cases with the odds ratio

$$o = \frac{P(n, N|H_0, \mathcal{I}_\alpha)}{P(n, N|\overline{H}_0, \mathcal{I}_\alpha)} \frac{P(H_0|\mathcal{I}_\alpha)}{P(\overline{H_0}|\mathcal{I}_\alpha)},$$

where the background information $\mathcal{I}_\alpha$ for $\alpha \in \{a, b\}$ encodes the information about the stopping criterion. First of all, the choice of the stopping condition definitely has no influence on our prior knowledge about H_0. In other words, the odds ratio is independent of H_0. For the Bayes factor we investigate the two cases separately.

Predefined sample size

The likelihood conditional on H_0 is

$$P(n, N|H_0, B_a) = P(n|N, H_0, B_a) \underbrace{P(N|H_0, B_a)}_{=1}.$$

The second factor is equal to 1, as for the predefined sample size N it is the true proposition. The first factor is simply the binomial distribution, from which we merely need the p_0 dependence. We therefore write

$$P(n, N|H_0, B_a) = C_a \; p_0^n (1 - p_0)^{N-n}.$$

For the marginal likelihood, conditioned on the complementary hypothesis $\overline{H}_0$, we employ the marginalization rule in order to introduce the unknown success probability p, resulting in

$$P(n, N|\overline{H}_0, B_a) = C_a \underbrace{\int p^n (1-p)^{N-n} \; p(p|B) \, dp}_{:=\mathcal{I}}.$$

The Bayes factor is therefore

$$o_{\mathrm{BF},a} = \frac{p_0^n (1 - p_0)^{N-n}}{\mathcal{I}}, \tag{20.1a}$$

$$\mathcal{I} = \int p^n (1-p)^{N-n} \; p(p|B) \, dp. \tag{20.1b}$$

Predefined number of successes

In this case the likelihood is

$$P(n, N|H_0, B_a) = P(N|n, H_0, B_b) \underbrace{P(n|H_0, B_b)}_{=1}.$$

Here n is the true proposition, as the number of successes is predefined. The first factor is the probability distribution for the *stopping time*, discussed in Section 15.4 [p. 241]. The probability is, according to equation (15.11a) [p. 243],

$$P(N|n, H_0, B_b) = C_b\ p_0^n (1 - p_0)^{N-n}.$$

As a matter of fact, it has the same p_0 dependence. The marginal likelihood required in case of $\overline{H_0}$ therefore reads

$$P(N|n, \overline{H_0}, B_b) = C_b\ \mathcal{I}.$$

So the Bayes factor is the same as in the first case (equation (20.1)):

$$o_{\text{BF},b} = \frac{C_b\ p_0^n (1 - p_0)^{N-n}}{C_b\ \mathcal{I}} = o_{\text{BF},a}.$$

We see that the Bayesian inference does not care about the stopping criterion used. What is the underlying deeper reason why the stopping criterion is irrelevant in the Bayesian approach? It is the fact that the parameter dependence of the likelihood is independent of the choice of the stopping condition. Generally speaking, the likelihood for some data $\boldsymbol{D}$, given all required parameters $\boldsymbol{a}$, has the form

$$p(\boldsymbol{D}|\boldsymbol{a}, \mathcal{I}_\alpha) = f(\boldsymbol{D}, \mathcal{I}_a)\ g(\boldsymbol{a}, \boldsymbol{D}).$$

Consequently, the posterior probability for the parameter(s) is also not affected by the stopping condition, as the prefactor f drops out by the normalization. Similarly, in hypothesis testing, f drops out of the odds ratio, as it occurs both in $p(\boldsymbol{D}|H_0, \mathcal{I}_\alpha)$ and $p(\boldsymbol{D}|\overline{H_0}, \mathcal{I}_\alpha)$.

20.2.2 Frequentist hypothesis testing

Now let's see how the frequentist approach deals with the stopping criterion. Typically, hypothesis testing is based on the P value, which we will determine for the two cases. The point null hypothesis is: *the single-trial probability is* p_0.

P value if the sample size is predefined

If the sample size N is predefined and the null hypothesis specifies p_0, then the expected number of successes and their variance is known. A reasonable measure for the discrepancy between measurement and theory (test statistic) is therefore

$$\begin{aligned}
\xi(n) &:= \frac{|n - \langle n \rangle|}{\sigma_n}, \\
\langle n \rangle &= Np, \\
\sigma_n^2 &= Np_0(1 - p_0).
\end{aligned}$$

We use the modulus since a two-tailed test appears adequate. The PDF for ξ (sampling distribution) is

$$p(\xi|p_0, N, \mathcal{I}) = \sum_{n=0}^{N} p(\xi|n, p_0, N, \mathcal{I}) P(n|p, N, \mathcal{I})$$
$$= \sum_{n=0}^{N} \delta\big[\xi - \xi(n)\big] P(n|p_0, N, \mathcal{I}).$$

Given the experimental count n^*, the value of the test statistic is $\xi^* = \xi(n^*)$ and the P value becomes

$$P = \int_{\xi^*}^{\infty} p(\xi|p, N, \mathcal{I})\, d\xi = \sum_{n=0}^{N} P(n|p, N, \mathcal{I}) \int_{\xi^*}^{\infty} \delta\big(\xi - \xi(n)\big)\, d\xi.$$

The inequality $\xi \geq \xi^*$ corresponds to $|n - \langle n \rangle| \geq |n^* - \langle n \rangle| := \Delta n^*$. The sum over n is therefore restricted to either $n \geq \langle n \rangle + \Delta n^*$ or $n \geq \langle n \rangle - \Delta n^*$, resulting in

$$P = \sum_{n=0}^{\langle n \rangle - \Delta n^*} P(n|p, N, n, \mathcal{I}) + \sum_{n=\langle n \rangle + \Delta n^*}^{N} P(n|p, N, n, \mathcal{I}).$$

As a numerical example we use $N = 100$, $n^* = 5$ and $p_0 = 0.1$. Then $\langle n \rangle = 10$ and $\sigma_n = 3$. The resulting P value is $P = 0.025$. Based on a significance level of 5%, say, we would reject the hypothesis. At first glance, the decision appears reasonable, as the measured value n^* lies clearly outside the 1-σ interval.

P value if the success count is predefined

Now we repeat the analysis for the other stopping criterion in which N is the random variable. It obeys the negative binomial distribution with $\langle N \rangle = n/p_0$ and $\sigma_N^2 = n(1 - p_0)/p_0^2$, according to equation (4.35) [p. 70]. In this case, a reasonable test statistic is

$$\eta(N) := \frac{|N - \langle N \rangle|}{\sigma_N^2}.$$

The corresponding sampling distribution is

$$p(\eta|p, n, \mathcal{I}) = \sum_{N=n}^{\infty} \delta\big(\eta - \eta(n)\big) P(N|p, n, \mathcal{I}).$$

For the experimental count N^* the value of the test statistic is $\eta^* = \eta(N^*)$, and the P value becomes

$$p = \int_{\eta^*}^{\infty} p(\eta|p, n, \mathcal{I})\, d\eta = \sum_{N=n}^{\infty} P(N|p, n, \mathcal{I}) \int_{\eta^*}^{\infty} \delta\big(\eta - \eta(N)\big)\, d\eta.$$

As before, the interval $\eta \geq \eta^*$ corresponds to either $N \geq \langle N \rangle + \Delta N^*$ or $N \leq \langle N \rangle - \Delta N^*$, with $\Delta N^* = |N^* - \langle N \rangle|$. The resulting P value is given by

$$P = \sum_{N=n}^{\langle N \rangle - \Delta N^*} P(N|p,n,\mathcal{I}) + \sum_{n=\langle N \rangle + \Delta N^*}^{\infty} P(N|p,n,\mathcal{I}).$$

For the same set of parameters as before, i.e. $n = 5$, $N^* = 100$ and $p_0 = 0.1$, we obtain $\langle N \rangle = 5/0.1 = 50$ and $\Delta N^* = 50$. Consequently, the first sum does not contribute and the second sum yields $P = 0.13$. Based on the 5% significance level, the hypothesis would clearly not be ruled out. Depending on the stopping criterion, we come to contradictory conclusions about the validity of the null hypothesis H_0, although one and the same set of data is used. The dependence on the stopping condition is rather disturbing, particularly as in most publications the stopping condition is not quoted and presumably the experimenter did not even pay attention to it and might have used different criteria for different experiments.

20.2.3 Yet another stopping criterion

Let us consider an extreme but still legitimate stopping criterion. Suppose somebody is performing quality control and wants to determine the percentage of intact parts in a production, or in case of pharmaceutical research the percentage of successful treatments with a new drug. Generically, we are speaking of a Bernoulli experiment of N repetitions and the outcome consists of n_s successes and n_f failures. The goal is to assess the hypothesis that the failure rate $\frac{n_f}{N}$ is below a certain threshold.

The experimenter could, for whatever reason, decide to stop the experiment when the number of successes exceeds the number of failures (for the first time). However, if that does not happen within a reasonable number of trials, $N_{\max}$ say, he stops in any case.

The reported output is given by $\{n_s, n_f\}$, i.e. the number of successes and failures. In case of the first stopping condition $n_s = n_f + 1$, otherwise $n_s + N_f = N_{\max}$. The likelihood $P(n_s, n_f|A, \mathcal{I})$ ($A = H$ or $A = \overline{H}$) entering the Bayes factor can be determined upon marginalization over the success probability p:

$$\begin{aligned} P(n_s, n_f|A, N^*, \mathcal{I}) &= \int_0^1 P(n_s, n_f|p, \cancel{A}, N^*, \mathcal{I})\; p(p|A, \cancel{N^*}, \mathcal{I})\, dp \\ &= \int_0^1 P(n_s, n_f|p, N^*, \mathcal{I})\; p(p|A, \mathcal{I})\, dp. \end{aligned}$$

The first part of the integrand $P(n_s, n_f|p, N^*, \mathcal{I})$ has already been determined in the context of the first lead problem in Section 5.3 [p. 80], where we studied the Brownian motion in the presence of absorbing walls. Mathematically it is the same setup and we can use the

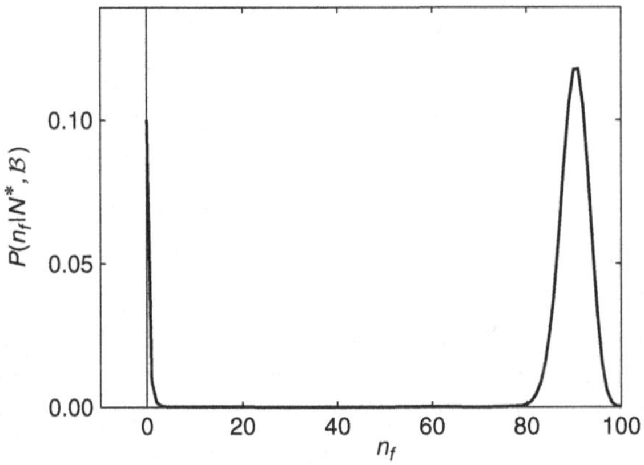

Figure 20.1 Marginal probability $P(n_f|N^*, p_0, \mathcal{I})$ for the number of failures.

result of equation (5.14) [p. 81]. Since the key feature of the result is its q-dependence, we write it as

$$P(n_s, n_f|p, N^*, \mathcal{I}) = p^{n_s}(1-p)^{n_f}\; F(n_s, n_f).$$

Hence, the Bayes factor

$$o_{\text{BF}} = \frac{\int dp\; p^{n_f}(1-p)^{n_s} \cdot p(p|H, \mathcal{I})}{\int dp\; p^{n_f}(1-p)^{n_s} \cdot p(p|\overline{H}, \mathcal{I})} = \frac{p_0^{n_f}(1-p_0)^{n_s}}{\mathcal{I}}$$

is again the same as in equation (20.1) and therefore independent even of this rather extreme stopping criterion.

The situation is entirely different if the frequentist approach is used. In the present case we consider the marginal likelihood for the number of failures as in Figure 5.3 [p. 82] in Section 5.3. Here we again take $p_0 = 0.1$ and $N_{\max} = 100$. The marginal likelihood is plotted in Figure 20.1. First and worst of all, the likelihood is bimodal and the choice of the rejection region requires a more involved discussion. But in any case, we would get yet another P value.

In the Bayesian approach, however, the stopping criterion is irrelevant for the data interpretation [13, 60]. As phrased by Savage: *It is entirely appropriate to collect data until a point has been proved or disproved, or until the data collector runs out of time, money, or patience* [60]. The independence of the stopping criterion is summarized in the so-called

Stopping rule principle

The reason for stopping an experiment should be irrelevant for inferences about underlying parameters or the assessment of hypotheses.

PART V

REAL-WORLD APPLICATIONS

21

Regression

Regression is a technique for describing how a response variable y varies with the values of so-called input variables $\boldsymbol{x}$. There is a distinction between 'simple regression', where we have only one input variable x, and 'multiple regression', with many input variables $\boldsymbol{x}$. Predictions are based on a model function $y = f(\boldsymbol{x}|\boldsymbol{a})$ that depends on the model parameters $\boldsymbol{a}$. At the heart of the regression analysis lies the determination of the parameters $\boldsymbol{a}$, either because they bear a direct (physical) meaning or because they are used along with the model function to make predictions. The reader not familiar with the general ideas of parameter estimation may want to read Part III [p. 227] first. In the literature on frequentist statistics, regression analysis is generally based on the assumption that the measured values of the response variables are independently and normally distributed with equal noise levels. Regression analysis in frequentist statistics boils down to fitting the model parameters such that the sum of the squared deviations between model and data is minimized. A widespread application is the linear regression model, where the function f is linear in the variables $\boldsymbol{x}$ and the parameters $\boldsymbol{a}$.

In the Bayesian framework, there is no need for any restrictions. Here we will deal with the general problem of inferring the parameters $\boldsymbol{a}$ of an arbitrary model function $f(\boldsymbol{x}|\boldsymbol{a})$. In order to cover the bulk of applications we will restrict the following studies to Gaussian errors. It will be a simple exercise for the reader to generalize the approach to other error distributions, if so required. We will also stick to the assumption of independent errors, since only in exceptional cases can the experimental covariance matrix be disposed of. The inclusion of correlated errors in the Bayesian analysis is a simple exercise. We will illustrate the Bayesian approach to regression problems based on a couple of examples, ranging from standard problems to demanding generalizations, usually not encountered in the standard literature on statistics [145].

The general regression problem is greatly simplified if the functional relation between y and $\boldsymbol{x}$ is linear. The simplest member of this class, although it does not really deserve the term 'regression', is the constant model $y_i = \mu + \varepsilon_i$ where ε_i denotes the error on measurement i. We shall treat the estimation of the constant μ for the cases where the error statistic is known or completely unknown. The next simple problem is that of finding the coefficients a and b of a simple linear model $y_i = ax_i + b + \varepsilon_i$. The general linear case is that of a hyperplane in n-dimensional space.

Determination of the parameters of a linear regression problem is in many cases not the end of the story. Though the knowledge of the numerical values of the parameters may in some cases be of importance itself, we face prediction as the true goal of the analysis. Prediction means the estimate of the response variable y for as yet unrealized input variables $\boldsymbol{x}$. We shall demonstrate that the solution to the problem of prediction is not the evaluation of the model function at variable $\boldsymbol{x}$ with the parameters set to their posterior values. This assumption, though wrong, is frequently made in physics. It will be shown that this traditional procedure holds only if the posterior parameter estimation has very small uncertainty.

A variant of the linear regression problem is posed by model functions

$$y = \sum_{\nu} a_{\nu} \phi_{\nu}(\boldsymbol{x}),$$

which are given as expansions in some suitable basis set $\{\phi_{\nu}(\boldsymbol{x})\}$. A familiar member of this class is the polynomial model function, with $\phi_{\nu}(x) = x^{\nu}$. It is linear in the parameters and nonlinear in the variables. The choice of basis functions has no influence on the general procedure of the maximum likelihood analysis. For a Bayesian estimate the basis set influences, however, the choice of the prior probabilities and it indeed makes a difference, whether a linear or a polynomial form is used.

Additional complications of the regression problem arise if the dependence of the model function on the parameters is nonlinear. In particular, the numerical effort may become considerable since the solution of this class of regression problems requires iterative solutions.

So far, we have considered situations where the data y_i were deteriorated by noise while the input variables were given exactly. This is still an idealized case. It is conceivable and happens frequently that the input variables of a regression problem are measurements themselves or, alternatively, that they cannot be realized with the assumed precision. We shall discuss the Bayesian approach to these problems which are slighted in most of the available literature on regression.

21.1 Linear regression

21.1.1 Inferring a constant

We begin with the most simple case, which in the narrower sense is not really a regression problem, but which nonetheless matches this section ideally: the inference of a constant. Consider a series of measurements designed to determine a physical constant μ like the quantum Hall effect, the elementary charge or whatever applies. Each single measurement y_i is corrupted by noise from the experimental apparatus so that the model equation for the process is

$$y_i = \mu + \varepsilon_i . \tag{21.1}$$

Frequently it is assumed that ε_i are samples from a zero mean and σ_i^2 variance Gaussian function. This is a very strong assumption, leading to the likelihood for N data points

$$p(\boldsymbol{y}|\mu, \boldsymbol{\sigma}, \mathcal{I}) = \frac{(2\pi)^{-N/2}}{\prod_i \sigma_i} \exp\left\{-\frac{1}{2}\sum_i \frac{(y_i - \mu)^2}{\sigma_i^2}\right\}. \tag{21.2}$$

From the discussion of maximum entropy-based distributions in Section 11.3 [p. 194], we recall that the same functional form of the likelihood results from the considerably weaker assumption that $\langle\varepsilon_i\rangle = 0$ and $\langle\varepsilon_i^2\rangle = \sigma_i^2$ where σ_i^2 is the true variance of the sampling variable ε_i. The likelihood may be modified in order to emphasize the μ-dependence

$$p(\boldsymbol{y}|\mu, \boldsymbol{\sigma}, \mathcal{I}) = \frac{(2\pi)^{-N/2}}{\prod_i \sigma_i} \exp\left\{-\frac{N}{2\varrho^2}\left(\left(\mu - \overline{Y}\right)^2 + \overline{(\Delta Y)^2}\right)\right\}, \tag{21.3}$$

with weighted averages

$$\overline{Y} = \sum_j^N \omega_j\, y_j,\ \omega_j = \frac{\varrho^2}{N\sigma_j^2},\ \varrho^{-2} = \frac{1}{N}\sum_i \sigma_i^{-2}. \tag{21.4}$$

The quantity ϱ is an effective variance of the sample. The posterior distribution of the parameter μ based on the likelihood derives from Bayes' theorem

$$p(\mu|\boldsymbol{y}, \boldsymbol{\sigma}, \mathcal{I}) = \frac{1}{Z}\, p(\mu|\mathcal{I})\, p(\boldsymbol{y}|\mu, \boldsymbol{\sigma}, \mathcal{I}). \tag{21.5}$$

The normalizing denominator Z does not depend on μ and can therefore be neglected in the following. Consider first the case where we are entirely uncertain as to what values of μ are acceptable. This lack of knowledge is coded in a flat improper prior on μ. The maximum of the posterior density is then equal to the maximum of the likelihood. The maximum likelihood estimate for μ is

$$\mu^{\mathrm{ML}} = \arg\min_{\mu} p(\boldsymbol{y}|\mu, \boldsymbol{\sigma}, \mathcal{I}), \tag{21.6}$$

which due to the Gaussian μ-dependence of equation (21.3) immediately yields

$$\mu^{\mathrm{ML}} = \overline{Y}, \qquad \langle(\Delta\mu)^2\rangle_{\mathrm{ML}} = \varrho^2/N. \tag{21.7}$$

For the special case that $\sigma_i = \varrho\ \forall_i$, this result reduces to $\mu_{\mathrm{ML}} = \overline{y} = \frac{1}{N}\sum y_i$ and $\langle(\Delta\mu)^2\rangle_{\mathrm{ML}} = \varrho^2/N = \mathrm{SE}^2$. This is the standard result, the estimator is the sample mean and its uncertainty the standard error.

We note already here that the above result in equation (21.7) is rather peculiar since the variance of the estimate μ^{ML} depends only on the variances of the individual measurements but not on the scatter of the data y_i. We shall return to this point in detail in Chapter 22 [p. 364].

So far, our posterior estimate was based on a flat improper prior for μ. Of considerable importance is also the situation where we have more specific prior knowledge, perhaps from a previous series of measurements, in the form $\mu = \mu_0 \pm s$. This prior knowledge corresponds to the maximum entropy distribution

$$p(\mu|\mu_0, s, \mathcal{I}) = \frac{1}{s\sqrt{2\pi}} \exp\left\{-\frac{1}{2s^2}(\mu - \mu_0)^2\right\}. \tag{21.8}$$

The posterior probability, as a product of two Gaussians, is again a Gaussian with argument ϕ:

$$\begin{aligned}\phi &= \frac{N}{\varrho^2}\left(\mu^2 - 2\mu\overline{Y} + \overline{Y^2}\right) + \frac{1}{s^2}\left(\mu^2 - 2\mu\mu_0 + \mu_0^2\right)\\ &= \left(\frac{N}{\varrho^2} + \frac{1}{s^2}\right)\left(\mu - \frac{\overline{Y}N/\varrho^2 + \mu_0/s^2}{N/\varrho^2 + 1/s^2}\right)^2 + \text{const.}\end{aligned} \tag{21.9}$$

The posterior mean and variance are therefore

$$\begin{aligned}\mu^p &= \frac{N\overline{Y}/\varrho^2 + \mu_0/s^2}{N/\varrho^2 + 1/s^2} = \overline{Y}\,\frac{1 + \frac{\mu_0}{\overline{Y}}\frac{\varrho^2}{Ns^2}}{1 + \frac{\varrho^2}{Ns^2}},\\ \langle(\Delta\mu)^2\rangle^p &= \left(\frac{N}{\varrho^2} + \frac{1}{s^2}\right)^{-1} = \frac{\varrho^2}{N}\,\frac{1}{1 + \frac{\varrho^2}{Ns^2}}.\end{aligned} \tag{21.10}$$

The prior information is comparable to **one** extra data point, with $\sigma_{N+1} = s$ and $d_{N+1} = \mu_0$. It is interesting to investigate the large-N limit of these expressions:

$$\mu^p \underset{N\to\infty}{\longrightarrow} \overline{Y}, \quad \langle(\Delta\mu)^2\rangle \underset{N\to\infty}{\longrightarrow} \overline{\varrho^2/N}. \tag{21.11}$$

For large N, the posterior mean and variance become independent of the prior knowledge. According to equation (21.10), the prior may become important for moderate to small N. If $\varrho^2/N \gtrsim s^2$, i.e. if the prior constraint (s^2) is more restrictive than the entire data (ϱ^2/N), then $\mu^p \approx \mu_0$ and $\langle(\Delta\mu)^2\rangle \approx s^2$. Hence, the data information is ignored. Another interesting special case concerns the situation where the weighted sample mean $\overline{Y}$ agrees roughly with the prior prediction μ_0. In that case the posterior mean is identical to the sample mean $\mu^p = \overline{Y}$ but the variance is reduced according to equation (21.10). Generally, if $s \ll \varrho$ then a large number of new measurements is required to modify the prior knowledge but, as we have seen before, in the limit $N \gg 1$ the prior is entirely overruled. In general, the posterior variance will be smaller than the sample variance $\langle(\Delta\mu)^2\rangle < \varrho^2/N$, but it is also smaller than the prior width $\langle(\Delta\mu)^2\rangle < s^2$.

The preceding analysis rests on the assumption that the true variances of the measurement errors σ_i^2 are known. This is hardly ever the case. What may be known is a sensible approximation to the true variances. We postpone the mathematical formulation of 'sensible' to Chapter 22 [p. 364] and consider the extreme case that we know very little about σ_i. If we know nothing at all about the experimental errors, the experiment is useless, since we have to marginalize the likelihood for each data point over σ_i with Jeffreys' scale prior. The resulting marginal likelihood

$$p(d_i|\mu, \mathcal{I}) \propto \frac{1}{|d_i - \mu|}$$

is not normalizable, let alone the diverging variance. A less ignorant situation is given if – for some reason – the interrelation of the experimental errors is known, and only the overall scale is undetermined, i.e. $\sigma_k = \varsigma\sigma_k^0$. A special case is the situation where all experimental errors are comparable, i.e. $\sigma_k = \varsigma$, but the common error ς is completely unknown. In the case $\sigma_k = \varsigma\sigma_k^0$, the likelihood is a minor modification of equation (21.3):

$$p(\boldsymbol{y}|\mu, \boldsymbol{\sigma}^0, \mathcal{I}) = \frac{1}{Z}\,\varsigma^{-N}\exp\left\{-\frac{\chi^2}{\varsigma^2}\right\},$$

with

$$\chi^2 = -\frac{N}{2\varrho^2}\left((\mu - \overline{Y})^2 + \overline{(\Delta Y)^2}\right).$$

The averages are defined by equation (21.4) with weights σ_k^0. The normalization constant Z will be determined at the end. Since ς is not even approximately known, we get rid of it by marginalization:

$$p(\boldsymbol{y}|\mu, \boldsymbol{\sigma}^0, \mathcal{I}) = \int p(\varsigma|\mathcal{I})\; p(\boldsymbol{y}|\mu, \boldsymbol{\sigma}^0, \varsigma, \mathcal{I})d\varsigma. \tag{21.12}$$

Since ς is a scale variable the appropriate uninformative prior on ς is Jeffreys' scale prior $p(\varsigma|\mathcal{I}) = 1/\varsigma$:

$$p(\boldsymbol{y}|\mu, \boldsymbol{\sigma}^0, \mathcal{I}) = \frac{1}{Z'}\int_0^\infty \frac{d\sigma}{\sigma}\;\sigma^{-(N-1)}\exp\left\{-\frac{\chi^2}{\sigma^2}\right\}. \tag{21.13}$$

According to equations (7.27) and (7.29) [p. 108], the integral yields

Marginal likelihood for $\sigma_i = \sigma$ with Jeffreys' scale prior for σ

$$p(\boldsymbol{y}|\mu, \boldsymbol{\sigma}^0, \mathcal{I}) = \frac{1}{Z}\left(1 + \frac{(\overline{Y} - \mu)^2}{\overline{(\Delta Y)^2}}\right)^{-\frac{N}{2}}.$$

Assuming a flat improper prior for μ, Bayes' theorem yields with the correct normalization

$$p(\mu|\boldsymbol{y}, \boldsymbol{\sigma}^0, \mathcal{I}) = \frac{\Gamma\left(\frac{N}{2}\right)}{\sqrt{\pi\,\overline{(\Delta d)^2}}\;\Gamma\left(\frac{N-1}{2}\right)}\left(1 + \frac{(\overline{Y} - \mu)^2}{\overline{(\Delta Y)^2}}\right)^{-\frac{N}{2}}. \tag{21.14}$$

This posterior PDF is a 'Student's *t*-distribution'. For large N the posterior probability approaches a Gaussian

$$p(\boldsymbol{y}|\mu, \boldsymbol{\sigma}^0, \mathcal{I}) \underset{N\gg 1}{\longrightarrow} \frac{1}{\sqrt{2\pi}\sigma_\mu} \exp\left\{\frac{-(\overline{Y}-\mu)^2}{2\sigma_\mu^2}\right\},$$

with the standard deviation replaced by the sample standard error

$$\sigma_\mu = \sqrt{\overline{(\Delta Y)^2}/N}.$$

For arbitrary N, the posterior is symmetric in μ around $\overline{Y}$. The mode and mean coincide, therefore

$$\mu^P = \mu^{\text{ML}} = \overline{Y}. \tag{21.15}$$

The local curvature at the maximum is obtained by comparison with a fictitious Gaussian, whose second derivative at maximum is $-\sigma^{-2}$:

$$\frac{1}{\sigma^2} = -\frac{d^2}{d\mu^2} \log p(\mu)|_{\mu^P} = \frac{1}{\sigma_\mu^2}. \tag{21.16}$$

The posterior is therefore conveniently summarized as

$$\mu^P = \overline{y}, \quad \langle(\Delta\mu)^2\rangle_{\text{local}} = \sigma_\mu^2 = \frac{\overline{(\Delta Y)^2}}{N}, \tag{21.17}$$

like in the large-N case. The variance of the posterior in equation (21.14) is readily determined:

$$\langle(\Delta\mu)^2\rangle = \frac{\overline{(\Delta Y)^2}}{N-3}.$$

The problem corresponds to the determination of the sample standard error. The unbiased estimator of the standard error in frequentist statistics is rather given by $\sigma_{\text{SE}}^2 = \frac{\overline{(\Delta Y)^2}}{N-1}$. The difference is due to a different philosophy, as discussed before. Frequentist statistics assume that there is one true parameter μ and seek a sample estimate for μ which yields the correct value, if we average over all conceivable samples. To be honest, in the case of data analysis problems, we will never be in the lucky situation that we can generate all possible samples experimentally. As a matter of fact, there would be a better use of this data wealth. Bayesian probability theory, on the contrary, starts from the realistic situation that there is only one sample at our disposal and we have to make the best of it. Apparently, we don't know the true value of the parameter μ. The Bayesian approach consists of seeking that sample estimate which minimizes the square deviation from the true parameter μ, averaged over all possible values μ can have, based on our prior knowledge. In Section 30.1.3 [p. 540] this topic is studied in more detail.

21.1.2 Parameters of a straight line

A slightly more complicated regression problem is that of fitting a straight line in two dimensions to a series of measurements y_i taken at values of the input variables x_i. We choose to parametrize the straight line in the most popular way as $y = ax + b$. This problem is of eminent importance in all the natural sciences and in econometrics. The likelihood is

$$p(\boldsymbol{y}|\boldsymbol{x}, \boldsymbol{\sigma}, a, b) = \prod_i \left(\frac{1}{\sigma_i\sqrt{2\pi}}\right) \exp\left\{-\frac{1}{2}\sum_i \frac{(y_i - ax_i - b)^2}{\sigma_i^2}\right\}, \tag{21.18}$$

from which we derive the posterior distribution $p(a, b|\boldsymbol{y}, \boldsymbol{x}, \boldsymbol{\sigma})$ by application of Bayes' theorem and the prior for the coefficients of a straight line derived from equation (10.47) [p. 176], which was

$$p(a, b) = \frac{1}{Z_b}\frac{1}{(1 + a^2)^{3/2}}. \tag{21.19}$$

The prior is improper in the parameter b. This is immaterial for parameter estimation. On top of that, there are always upper and lower bounds for real-world parameters; for the length measurements of some classical objects, the lower and upper limits are e.g. [Å, km]. If the reader intuitively winces at these values, then he or she apparently has more precise prior knowlege that can be incorporated into the data analysis. For model comparison we need to restrict the range of b, for which we can take e.g. $B_0 - \Delta b \leq b \leq B_0 + \Delta b$ and obtain the normalization $Z_b = 1/(2\Delta b)$. The log-posterior becomes

$$\ln\left[p(a, b|\boldsymbol{y}, \boldsymbol{x}, \boldsymbol{\sigma}, \mathcal{I})\right] = \text{const.} - \frac{1}{2}\sum_i \frac{(y_i - ax_i - b)^2}{\sigma_i^2} - \frac{3}{2}\ln(1 + a^2). \tag{21.20}$$

We denote this function, without the constant, by ϕ and use the definitions introduced in equation (21.4) of the σ_i^2-weighted averages to obtain

$$\phi = -\frac{N}{2\varrho^2}(\overline{Y^2} + a^2\overline{X^2} + b^2 - 2a\overline{XY} - 2b\overline{Y} + 2ab\overline{X}) - \frac{3}{2}\ln(1 + a^2). \tag{21.21}$$

ϕ is maximized by a, b satisfying the equations

$$\frac{\partial\phi}{\partial b} = 0 \Rightarrow \qquad b = \overline{Y} - a\overline{X}, \tag{21.22}$$

$$\frac{\partial\phi}{\partial a} = 0 \Rightarrow \qquad a\overline{X^2} + b\overline{X} = \overline{XY} - \frac{3\varrho^2}{N}\frac{a}{1 + a^2},$$

$$a = \frac{\overline{\Delta X \Delta Y}}{\overline{(\Delta X)^2}} - \frac{3\varrho^2}{N\,\overline{(\Delta X)^2}}\frac{a}{1 + a^2}. \tag{21.23}$$

The last equation is nonlinear in a due to the prior density in a. We notice that the nonlinear term scales proportionally to the square of the ratio of the sample standard to the width of the input variable, i.e. $\mathrm{SE}/\Delta x$, which in general will be small compared with the slope of the straight line. Hence, for data with small standard error and/or if $\overline{(\Delta X)^2} \gg 1$, this term becomes very small. We shall therefore be content to solve equations (21.22) and (21.23) in perturbation theory, that is we neglect the nonlinear term of lowest order, calculate approximate values for a and b, which will be the maximum likelihood estimates, and finally find the solution of equations (21.22) and (21.23) with a in the nonlinear term replaced by its maximum likelihood value a_{ML}. This is equivalent to solving equation (21.23) iteratively, by starting with $a = 0$. The first iteration yields the ML estimate

$$a_{\mathrm{ML}} = \frac{\overline{\Delta X \Delta Y}}{\overline{(\Delta X)^2}} \tag{21.24}$$

and the second iteration gives

$$a = a_{\mathrm{ML}} + \frac{3\varrho^2}{N\,\overline{(\Delta X)^2}}\,\frac{a_{\mathrm{ML}}}{1 + a_{\mathrm{ML}}^2}. \tag{21.25}$$

In all iterations b is determined by equation (21.22) with a replaced by the respective approximation. In order to determine confidence intervals for the parameter estimates, we need the covariance matrix $C_{ij} = \langle \Delta a_i \Delta a_j \rangle$ for the parameter $a_i \in \{a, b\}$ of the posterior density given by equation (21.20). To this end we approximate the posterior density by a Gaussian, which is legitimate, as only the prior contributions introduce deviations from a Gaussian. The Taylor expansion of the log-posterior ϕ in equation (21.21) about the MAP solution a^P yields

$$\phi(a, b) = \phi(a^P, b^P) - \frac{1}{2}\Big(H_{aa}\,(a - a^P)^2 + H_{bb}\,(b - b^P)^2 + 2H_{ab}\,(a - a^P)\,(b - b^P) \Big)$$

with the Hessian defined by the negative second derivatives of equation (21.21):

$$H_{aa} = \frac{N\overline{X^2}}{\varrho^2} + \frac{3(1 - (a^P)^2)}{((a^P)^2 + 1)^2}, \qquad H_{bb} = \frac{N}{\varrho^2}, \qquad H_{ab} = \frac{N\overline{X}}{\varrho^2}.$$

The elements of the covariance matrix are obtained from $C = H^{-1}$:

$$\begin{aligned} C_{aa} &= \frac{H_{bb}}{H_{aa}H_{bb} - H_{ab}^2} = \frac{\varrho^2}{N}\,\frac{1}{\overline{(\Delta X)^2} + \frac{\varrho^2}{N}\frac{3(1-(a^P)^2)}{((a^P)^2+1)^2}}, \\ C_{bb} &= \frac{H_{aa}}{H_{aa}H_{bb} - H_{ab}^2} = \frac{\varrho^2}{N} + C_{aa}\,\overline{X}^2, \\ C_{ab} &= -\frac{H_{ab}}{H_{aa}H_{bb} - H_{ab}^2} = -C_{aa}\,\overline{X}. \end{aligned} \tag{21.26}$$

If we ignore the prior contribution we obtain the maximum likelihood results C_{ij}^{ML}. Including the prior, the variance C_{aa} shows a peculiar behaviour. For $|a^P| > 1$ we find $C_{aa} > C_{aa}^{\mathrm{ML}}$, while for $|a^P| < 1$, $C_{aa} < C_{aa}^{\mathrm{ML}}$. The variance of C_{bb} and the covariance $|C_{ab}|$ inherit this behaviour. The covariance is negative since a and b are anticorrelated.

21.1.3 Robustness against outliers

As a precursor to a thorough discussion of outliers, we want to study the robustness of the ubiquitous maximum likelihood formula for the slope (21.24) against outliers. In Figure 21.1 the result of a computer simulation is depicted in which one (y_{10}) out of 30 data points has been shifted from its original position by 80 standard deviations. Apparently, but not surprisingly, the maximum likelihood estimate for the slope is highly sensitive to outliers and yields in the present example even a wrong sign for the slope. Let us analyse the behaviour in more detail. We assume that one of the data points, y_i say, is an outlier and its value is modified to $\tilde{y}_i = y_i + \delta y_i$. The deviation leads to the modification $\overline{\Delta X \Delta Y} \longrightarrow \omega_i \Delta x_i \delta y_i$, producing a modified value of the slope $a \longrightarrow a + \omega_i \frac{\Delta x_i \delta y_i}{(\Delta X)^2}$. There are a couple of interesting features. To begin with, the estimate of the slope reacts linearly to the displacement δy_i. The ML estimate does not differentiate between data points which go astray by one standard deviation and those which are off by 80σ, as in the present example. Of course, it is entirely possible that a data point deviates from the true value by 80σ due to statistical fluctuations, without being an outlier, but the probability is extremely small and so is the probability that the ML estimate for the slope is the correct one. We will treat this topic in Chapter 22 [p. 364] with the rules of probability theory. A second likewise disturbing feature concerns the fact that the result is proportional to σ_i^{-2} (via ω_i). As before, a good estimate should exhibit a threshold behaviour. If the deviation of the datum from the fit curve exceeds $O(\sigma_i)$, the influence on the slope has to decline. A third, interesting but decent feature of the result is that outliers have a stronger leverage

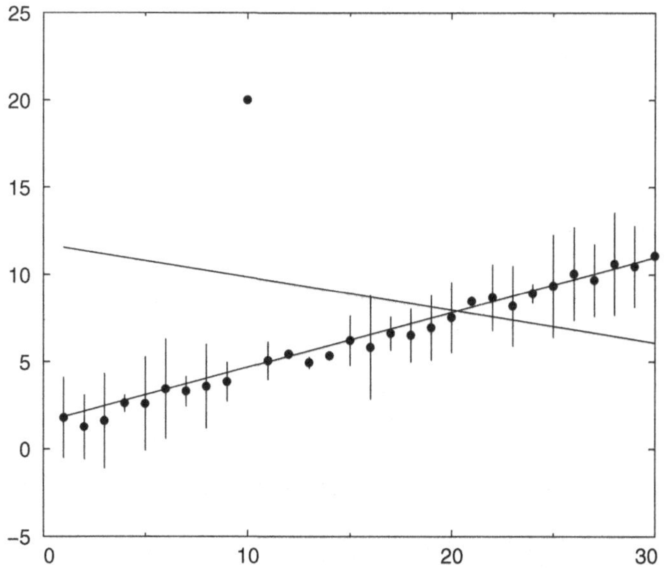

Figure 21.1 Robustness against outliers for 30 data points with Gaussian noise, indicated by dots and error bars. The solid line with positive (negative) slope stands for the fit without (with) outlier.

on the estimate of the slope when they belong to control parameters that are further away from the centre, i.e. the modification of the slope is proportional to $\frac{|\Delta x_i|}{\sqrt{\overline{(\Delta X)^2}}}$.

21.1.4 Prediction by the straight-line model

The determination of the parameters of the regression straight line may be of value and meaning by itself. Very often the interest is, however, to predict response values η for input variables ξ. Here we cover the problem of interpolating and extrapolating predictions. Even for values ξ, contained in the sample $\boldsymbol{x}$, the result is not trivial. For the present task we need the probability density

$$p(\eta|\xi, \boldsymbol{x}, \boldsymbol{y}, \boldsymbol{\sigma}, \mathcal{I}). \tag{21.27}$$

We derive it from the already known distributions by marginalization:

$$\begin{aligned} p(\eta|\xi, \boldsymbol{x}, \boldsymbol{y}, \boldsymbol{\sigma}, \mathcal{I}) &= \iint da\, db\; p(a, b, \eta|\xi, \boldsymbol{x}, \boldsymbol{y}, \boldsymbol{\sigma}, \mathcal{I}) \\ &= \iint da\, db\; p(a, b|\eta, \boldsymbol{x}, \boldsymbol{y}, \boldsymbol{\sigma}, \mathcal{I})\; p(\eta|\xi, a, b, \mathcal{I}). \end{aligned} \tag{21.28}$$

In applying the product rule to the integrand we have suppressed all logically irrelevant conditions. The second term $p(\eta|\xi, a, b, \mathcal{I})$ is of course known. Given the true values of a and b, the distribution of η is a delta function

$$p(\eta|\xi, a, b, \mathcal{I}) = \delta(\eta - a\,\xi - b). \tag{21.29}$$

We use this explicit form to carry out the b-integration and obtain

$$p(\eta|\xi, \boldsymbol{x}, \boldsymbol{y}, \boldsymbol{\sigma}, \mathcal{I}) = \int da\; p(a, b = \eta - a\,\xi \mid \xi, \boldsymbol{x}, \boldsymbol{y}, \boldsymbol{\sigma}, \mathcal{I}). \tag{21.30}$$

In order to keep things manageable we assume that the posterior distribution $p(a, b|\boldsymbol{x}, \boldsymbol{y}, \boldsymbol{\sigma}, \mathcal{I})$ was obtained with a flat prior on a, b, meaning that $p(a, b|\boldsymbol{x}, \boldsymbol{y}, \boldsymbol{\sigma}, \mathcal{I}) \propto p(\boldsymbol{y}|\boldsymbol{x}, \boldsymbol{\sigma}, a, b, \mathcal{I})$. The integrand in equation (21.30) is then a Gaussian with argument ϕ of the exponential:

$$-\phi = \frac{1}{2} \sum_i \frac{(y_i - a x_i - \eta + a\,\xi)^2}{\sigma_i^2}. \tag{21.31}$$

We use our previous definition of the σ_i^2-weighted averages to obtain

$$\begin{aligned} -2\frac{\varrho^2 \phi}{N} &= a^2 \left\{ \overline{X^2} + \xi^2 - 2\overline{X}\,\xi \right\} \\ &\quad - 2a \left\{ \overline{XY} - \xi\overline{Y} - \eta\overline{X} + \xi\,\eta \right\} + \overline{Y^2} + \eta^2 - 2\eta\overline{Y}. \end{aligned} \tag{21.32}$$

The coefficients of the powers of a on the right-hand side can be conveniently expressed as

$$\begin{aligned}
\overline{X^2} + \xi^2 - 2\overline{X}\xi &= \overline{\Delta X^2} + (\Delta\xi)^2, \\
\overline{XY} - \xi\overline{Y} - \eta\overline{X} + \xi\eta &= (\Delta\xi)(\Delta\eta) + \overline{\Delta X \Delta Y}, \\
\overline{Y^2} + \eta^2 - 2\eta\overline{Y} &= (\Delta\eta)^2 + \overline{\Delta Y^2},
\end{aligned} \tag{21.33}$$

where we have introduced $\Delta\xi = \xi - \overline{X}$ and $\Delta\eta = \eta - \overline{Y}$. Next we transform equation (21.32) into a complete square plus a residue:

$$Q(a - a_0)^2 + R = Qa^2 - 2Qaa_0 + Qa_0^2 + R. \tag{21.34}$$

Comparing coefficients with equation (21.32) yields

$$\begin{aligned}
Q &= \overline{\Delta X^2} + \Delta\xi^2, \\
a_0 &= \frac{\Delta\xi\,\Delta\eta + \overline{\Delta X \Delta Y}}{\overline{\Delta X^2} + \Delta\xi^2}, \\
R &= \Delta\eta^2 + \overline{\Delta Y^2} - \frac{(\Delta\xi\,\Delta\eta + \overline{\Delta X \Delta Y})^2}{\overline{\Delta X^2} + \Delta\xi^2}.
\end{aligned} \tag{21.35}$$

The integral in equation (21.30) becomes proportional to $\exp\left[-\frac{N}{2\varrho^2}R\right]$, the unnormalized probability density for η. This facilitates the final step, since all terms in R which do not depend on η can then be neglected and we restore normalization at the end:

$$p(\eta|\xi, \boldsymbol{x}, \boldsymbol{y}, \boldsymbol{\sigma}, \mathcal{I}) \propto \exp\left[-\frac{N}{2\varrho^2}\frac{(\Delta\eta^2\overline{\Delta X^2}) - 2\Delta\xi\,\Delta\eta\overline{\Delta X \Delta Y}}{\overline{\Delta X^2} + \Delta\xi^2}\right]. \tag{21.36}$$

We notice further that $\overline{\Delta X \Delta Y} = a^{\mathrm{ML}}\ \overline{\Delta X^2}$ from equation (21.25) and we obtain, finally and properly normalized,

$$p(\eta|\xi, \boldsymbol{x}, \boldsymbol{y}, \boldsymbol{\sigma}, \mathcal{I}) = \left[\frac{N}{2\pi\varrho^2\left(1 + \frac{\Delta\xi^2}{\overline{\Delta X^2}}\right)}\right]^{\frac{1}{2}} \exp\left[-\frac{N}{2\varrho^2}\frac{(\Delta\eta - a^{\mathrm{ML}}\Delta\xi)^2}{\left(1 + \frac{\Delta\xi^2}{\overline{\Delta X^2}}\right)}\right]. \tag{21.37}$$

The predictive distribution for η peaks at $\Delta\eta = a^{\mathrm{ML}}\Delta\xi$ or, converting back to the original variables,

$$\eta = \overline{Y} + a_{\mathrm{ML}}(\xi - \overline{X}) = a_{\mathrm{ML}}\ \xi + b_{\mathrm{ML}}. \tag{21.38}$$

The most probable predicted response is obtained by evaluating the model function at the posterior values of the parameters (a, b). This is accidental. Generally, as outlined before, this happens only if the posterior distribution of the parameters is of such a high precision that their density $p(a, b|\boldsymbol{x}, \boldsymbol{y}, \boldsymbol{\sigma}, \mathcal{I})$ can be represented by a delta function $\delta(\boldsymbol{a} - \hat{\boldsymbol{a}})$. In this case the integral in equation (21.28) reduces to $p(\eta|\xi, \hat{a}, \hat{b}, \mathcal{I})$. The accidental result in the present case is due to the Gaussian structure of the posterior in both a and b.

The Gaussian structure, however, results from the assumption of flat priors for a and b which we introduced to make the problems analytically solvable.

A more interesting behaviour of equation (21.37) is exhibited by the variance. The most precise prediction is obviously obtained for $\Delta\xi = \xi - \overline{X} = 0$, that is in the weighted centre of the input variables. For ξ sufficiently outside the data cloud such that $\Delta\xi^2/\overline{\Delta X^2} \gg 1$, the standard deviation σ_η behaves like

$$\sigma_\eta = \frac{\varrho}{\sqrt{N}} \frac{\Delta\xi}{\sqrt{\overline{\Delta X^2}}}, \tag{21.39}$$

meaning that in the asymptotic range the uncertainty of the prediction rises linearly with the distance of ξ from the data centre. An example for a predictive function and the associated uncertainty based on a set of measured data is displayed in Figure 21.2.

21.1.5 Multivariate linear regression

The parameter estimation problems of the previous section are the simplest examples of the class of multivariate linear regression problems. Let our model equation for the ith measurement be

$$y_i = x_{i1}\, a_1 + x_{i2}\, a_2 + x_{i3}\, a_3 \, + \cdots + x_{iE}\, a_E + \varepsilon_i . \tag{21.40}$$

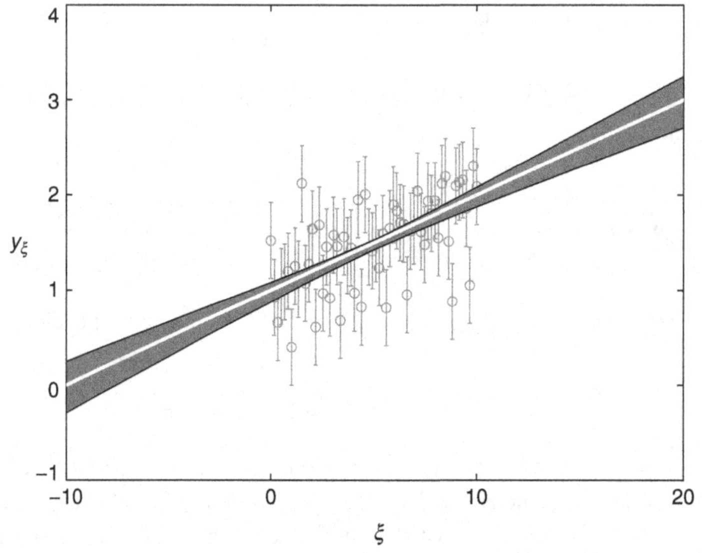

Figure 21.2 Fit of a straight line to a set of mock data generated from a linear function $y = a\xi + b$. The shaded band displays the confidence region for predictions and the white solid line represents the MAP solution.

If we assume that $\langle \varepsilon_i \rangle = 0$ and $\langle \varepsilon_i^2 \rangle = \sigma_i^2$ then the likelihood for N independent measurements becomes

$$\begin{aligned} p(\boldsymbol{y}|\boldsymbol{x}, \boldsymbol{a}, \boldsymbol{\sigma}, E, \mathcal{I}) &= \frac{(2\pi)^{-\frac{N}{2}}}{\prod_i \sigma_i} \exp\left\{ -\frac{1}{2} \sum_{i=1}^{N} \frac{\left(y_i - \sum_{k=1}^{E} x_{ik}\, a_k\right)^2}{\sigma_i^2} \right\} \\ &= \frac{(2\pi)^{-\frac{N}{2}}}{\prod_i \sigma_i} \exp\left\{ -\frac{1}{2}\phi \right\}. \end{aligned} \tag{21.41}$$

The notation simplifies if we introduce the vectors $\boldsymbol{y}$ and $\boldsymbol{a}$ and the matrices $\mathbf{X} = \{x_{ik}\}$ and $\mathbf{S}^{-2} = \text{diag}(\sigma_i^{-2})$ in the argument ϕ of the exponential. Then

$$\phi = (\boldsymbol{y} - \mathbf{X}\boldsymbol{a})^T \mathbf{S}^{-2} (\boldsymbol{y} - \mathbf{X}\boldsymbol{a}).$$

It is convenient to introduce new variables $\boldsymbol{y}' = \mathbf{S}^{-1}\boldsymbol{y}$ and $\mathbf{X}' = \mathbf{S}^{-1}\mathbf{X}$ in the exponent, resulting in

$$\phi = (\boldsymbol{y}' - \mathbf{X}'\boldsymbol{a})^T (\boldsymbol{y}' - \mathbf{X}'\boldsymbol{a}) = \boldsymbol{y}'^{\,2} - 2\boldsymbol{a}^T \mathbf{X}'^T \boldsymbol{y}' + \boldsymbol{a}^T \mathbf{X}'^T \mathbf{X}'\boldsymbol{a}. \tag{21.42}$$

The maximum of the likelihood is achieved for

$$\begin{aligned} & 0 = \nabla\, \phi = -2\mathbf{X}'^T \boldsymbol{y}' + 2\mathbf{X}'^T \mathbf{X}'\boldsymbol{a} \\ \Rightarrow\quad & (\mathbf{X}'^T \mathbf{X}')\, \boldsymbol{a}^{\text{ML}} = \mathbf{X}'^T \boldsymbol{y}'. \end{aligned}$$

In the following, we assume that the inverse of the matrix $\mathbf{X}'^T\mathbf{X}'$ exists. A necessary condition is $N \geq E$, i.e. we have at least as many measurements as there are parameters. If the inverse matrix exists then

$$\boldsymbol{a}^{\text{ML}} = (\mathbf{X}'^T \mathbf{X}')^{-1} \mathbf{X}'^T \boldsymbol{y}'.$$

Since ϕ is quadratic in $\boldsymbol{a}$ we can rewrite ϕ in equation (21.42) as a complete square in $\boldsymbol{a}$ plus a residue

$$\phi = R + (\boldsymbol{a} - \boldsymbol{a}_0)^T \mathbf{Q} (\boldsymbol{a} - \boldsymbol{a}_0). \tag{21.43}$$

This form is achieved via a Taylor expansion about $\boldsymbol{a} = \boldsymbol{a}_{\text{ML}}$. The second derivatives (Hessian) provide

$$\mathbf{Q} = \frac{1}{2}\, \nabla \nabla^T\, \phi(\boldsymbol{a}) \bigg|_{\boldsymbol{a}^{\text{ML}}} = \mathbf{X}'^T \mathbf{X}'. \tag{21.44}$$

The matrix on the right-hand side also shows up in the maximum likelihood solution

$$\boldsymbol{a}^{\text{ML}} = \mathbf{Q}^{-1} \mathbf{X}'^T \boldsymbol{y}'. \tag{21.45}$$

The residue R in equation (21.43) is the constant $R = \phi(\boldsymbol{a}^{\text{ML}})$ of the Taylor expansion, which can be transformed into an interesting expression

$$R = \boldsymbol{y}'^T \Big(\mathbb{1} - \underbrace{\mathbf{X}'\mathbf{Q}^{-1}\mathbf{X}'^T}_{:=\mathbf{P}} \Big) \boldsymbol{y}'. \tag{21.46}$$

Here $\mathbf{P}$ is a projection matrix that filters out those components of $\boldsymbol{y}$ that can be described by the linear model. Consequently, $\mathbf{1} - \mathbf{P}$ is the orthogonal rest of $\boldsymbol{y}$ that cannot be covered by the model. This general result can be cast in an advantageous form by employing singular-value decomposition [162] of the matrix $\mathbf{X}'$. For any $(N \times E)$ matrix $\mathbf{X}'$ there exists a bi-orthogonal expansion of the form

$$\mathbf{X}' = \mathbf{U}\mathbf{D}\mathbf{V}^T. \tag{21.47}$$

The sizes of the matrices $\mathbf{U}$, $\mathbf{D}$ and $\mathbf{V}$ are $(N \times E)$, $(E \times E)$ and $(E \times E)$, respectively. The matrices $\mathbf{U}$ and $\mathbf{V}$ are 'left unitary', by which we mean $\mathbf{U}^T\mathbf{U} = \mathbf{1}$ and $\mathbf{V}^T\mathbf{V} = \mathbf{1}$. In other words, the column vectors of these matrices are orthonormal. In addition, the matrix $\mathbf{V}$ is 'right unitary', i.e. $\mathbf{V}\mathbf{V}^T = \mathbf{1}$, which is not true for $\mathbf{U}$. That is, the column vectors in $\mathbf{V}$ form a complete orthonormal basis. The matrix $\mathbf{D} = \mathrm{diag}(\lambda_i)$ is diagonal and the diagonal elements $\{\lambda_i\}$ are called the singular values. They are equal to the square root of the eigenvalues of the real symmetric matrix $\mathbf{X}'^T\mathbf{X}'$. The transposed matrix $\mathbf{X}'^T$ is simply

$$\mathbf{X}'^T = \mathbf{V}\mathbf{D}\mathbf{U}^T, \tag{21.48}$$

and the product $\mathbf{X}'^T\mathbf{X}'$ becomes, using the unitarity of $\mathbf{U}$,

$$\mathbf{Q} = \mathbf{X}'^T\mathbf{X}' = \mathbf{V}\mathbf{D}\mathbf{U}^T\mathbf{U}\mathbf{D}\mathbf{V}^T = \mathbf{V}\mathbf{D}^2\mathbf{V}^T. \tag{21.49}$$

The last equation is also known as the spectral decomposition of the real symmetric matrix $\mathbf{X}'^T\mathbf{X}'$. We assume that the singular values are strictly positive, in order that the inverse of $\mathbf{Q}$ exists. The virtue of the spectral decomposition is that it yields immediately the inverse $\mathbf{Q}^{-1}$ as

$$\mathbf{Q}^{-1} = \mathbf{V}\mathbf{D}^{-2}\mathbf{V}^T. \tag{21.50}$$

It is easily verified that the matrix product of equations (21.49) and (21.50) yields the identity matrix $\mathbf{1}$, which is a consequence of the left unitarity of $\mathbf{U}$. As a final operation we determine the maximum likelihood estimate $\boldsymbol{a}_{\mathrm{ML}}$:

$$\boldsymbol{a}^{\mathrm{ML}} = \mathbf{Q}^{-1}\mathbf{X}'^T\boldsymbol{y}' = \mathbf{V}\mathbf{D}^{-2}\mathbf{V}^T\mathbf{V}\mathbf{D}\mathbf{U}^T\boldsymbol{y}' = \mathbf{V}\mathbf{D}^{-1}\mathbf{U}^T\boldsymbol{y}' = \sum_i \left(\frac{\boldsymbol{u}_i^T\boldsymbol{y}'}{\lambda_i}\right)\boldsymbol{v}_i, \tag{21.51}$$

where $\boldsymbol{u}_i$ $(\boldsymbol{v}_i)$ are the column vectors of $\mathbf{U}$ $(\mathbf{V})$. The maximum likelihood estimate $\boldsymbol{a}^{\mathrm{ML}}$ is thereby expanded in the basis $\{\boldsymbol{v}_i\}$ with expansion coefficients $(\boldsymbol{u}_i^T\boldsymbol{y}'/\lambda_i)$. Whenever a singular value λ_i becomes very small, the maximum likelihood solution $\boldsymbol{a}^{\mathrm{ML}}$ becomes very sensitive to even small amounts of noise in the data. Singular-value decomposition provides a very valuable diagnostic on the stability of the maximum likelihood estimate $\boldsymbol{a}^{\mathrm{ML}}$.

It appears that the ratio of the largest in magnitude to the smallest singular value is a measure of how badly the result is corrupted by noise. The ideal case is apparently the condition number one, which is given if all eigenvalues have magnitude one. This is the case if $\mathbf{X}^T\mathbf{X} = \mathbf{1}$, corresponding to the expansion in an orthonormal basis.

We now turn to the Bayesian estimation of $\boldsymbol{a}$. The full information on the parameters $\boldsymbol{a}$ is contained in the posterior distribution

$$p(\boldsymbol{a}|\boldsymbol{y}, \mathbf{X}, \boldsymbol{\sigma}, E, \mathcal{I}) = \frac{1}{Z}\, p(\boldsymbol{a}|E, \mathcal{I})\; p(\boldsymbol{y}|\mathbf{X}, \boldsymbol{a}, \boldsymbol{\sigma}, E, \mathcal{I}), \tag{21.52}$$

from which we can determine, for example, the maximum posterior solution via

$$p(\boldsymbol{a}|E, \mathcal{I})\; \nabla_{\boldsymbol{a}} p(\boldsymbol{y}|\mathbf{X}, \boldsymbol{a}, \boldsymbol{\sigma}, E, \mathcal{I}) + p(\boldsymbol{y}|\mathbf{X}, \boldsymbol{a}, \boldsymbol{\sigma}, E, \mathcal{I}) \nabla_{\boldsymbol{a}} p(\boldsymbol{a}|E, \mathcal{I}) = 0.$$

For a sufficient number of well-determined data the likelihood is strongly peaked around its maximum $\boldsymbol{a}^{\mathrm{ML}}$, while the prior will be comparatively flat, in particular if it is chosen uninformative. A reliable approximate solution will then be obtained from

$$p(\boldsymbol{a}|E, \mathcal{I})\; \nabla p(\boldsymbol{y}|\mathbf{X}, \boldsymbol{a}, \boldsymbol{\sigma}, E, \mathcal{I}) = 0, \tag{21.53}$$

which is the maximum likelihood estimate. Similar arguments hold for posterior expectation values of any function $f(\boldsymbol{a})$:

$$\begin{aligned} \langle f(\boldsymbol{a})\rangle &= \frac{1}{Z} \int d^E a\; f(\boldsymbol{a})\; p(\boldsymbol{a}|E, \mathcal{I})\; p(\boldsymbol{y}|\mathbf{X}, \boldsymbol{a}, \boldsymbol{\sigma}, E, \mathcal{I}), \\ Z &= \int d^E a\; p(\boldsymbol{a}|E, \mathcal{I})\; p(\boldsymbol{y}|\mathbf{X}, \boldsymbol{a}, \boldsymbol{\sigma}, E, \mathcal{I}). \end{aligned} \tag{21.54}$$

For general $p(\boldsymbol{a}|E, \mathcal{I})$ the integrals can only be performed numerically. We shall discuss now an interesting approximation which exploits the fact that the likelihood is generally precisely localized compared with the diffuse prior. This suggests that we replace the prior $p(\boldsymbol{a}|E, \mathcal{I})$ by $p(\boldsymbol{a}^{\mathrm{ML}}|E, \mathcal{I})$ and take it out of the integrals:

$$\langle f(\boldsymbol{a})\rangle \approx \frac{p(\boldsymbol{a}^{\mathrm{ML}}|E, \mathcal{I})}{Z} \int d^E a\; f(\boldsymbol{a})\; p(\boldsymbol{y}|\mathbf{X}, \boldsymbol{a}, \boldsymbol{\sigma}, E, \mathcal{I}), \tag{21.55}$$

$$Z \approx p(\boldsymbol{a}^{\mathrm{ML}}|E, \mathcal{I}) \int d^E a\; p(\boldsymbol{y}|\mathbf{X}, \boldsymbol{a}, \boldsymbol{\sigma}, E, \mathcal{I}). \tag{21.56}$$

Now, $p(\boldsymbol{y}|\mathbf{X}, \boldsymbol{a}, \boldsymbol{\sigma}, E, \mathcal{I})$ is a multivariate Gaussian in $\boldsymbol{a}$ according to equations (21.41) and (21.43). The expectation value $\big(f(\boldsymbol{a}) = a_i\big)$ and the covariance $\big(f(\boldsymbol{a}) = \operatorname{cov}(a_i, a_j)\big)$ can easily be determined:

$$\langle \boldsymbol{a}\rangle = \boldsymbol{a}_{\mathrm{ML}}, \tag{21.57}$$

$$\operatorname{cov}(a_i, a_j) = \mathbf{Q}^{-1}_{ji}. \tag{21.58}$$

The result is independent of the choice of the prior, since we have ignored its parameter dependence. The posterior mean and posterior mode are coincident, and the question remains of what has been gained by considering the expectation value. Firstly, we have access to the confidence interval given by the covariance matrix. Secondly, we have derived a reasonable approximation for the normalization Z, which has played a minor role so far. Z is also called the 'prior predictive value', the 'evidence', or the 'global likelihood'. The latter expression is particularly illuminating, since Z is the likelihood of the data conditional on a hyperplane model of order E regardless of what the particular values of the coefficients $\boldsymbol{a}$ are. In other words, Z represents the probability for the data, given the assumed model. It is therefore a measure for how representative a sample is for the model. The global likelihood will therefore become of paramount importance in model comparison and complexity analysis problems. We shall deal with these topics in more detail in Chapter 27 [p. 470]. A typical question in the present problem is of course whether the data really require the expansion order E or are also satisfactorily explained by some lower order $E' < E$. For these problems the full expression for Z is required, including the prior factor. The remaining Gaussian integral in equation (21.56) can be performed easily along with equations (21.41), (21.43) and (21.46), resulting in

$$Z \approx p(\boldsymbol{a}^{\mathrm{ML}}|E,\mathcal{I})\,\frac{(2\pi)^{\frac{E-N}{2}}}{\prod_i^N \sigma_i}|\mathbf{Q}|^{-1/2}\exp\left\{-\frac{1}{2}\boldsymbol{y}'^T\left(\mathbb{1}-\mathbf{X}'\mathbf{Q}^{-1}\mathbf{X}'^T\right)\boldsymbol{y}'\right\}.$$

The singular-value decomposition of the argument of the exponential reveals an interesting aspect of the result:

$$\mathbb{1}-\mathbf{X}'\mathbf{Q}^{-1}\mathbf{X}'^T = \mathbb{1}-\mathbf{UDV}^T\mathbf{VD}^{-2}\mathbf{V}^T\mathbf{VDU}^T = \mathbb{1}-\mathbf{UU}^T = \mathbb{1}-\sum_{i=1}^{E}\boldsymbol{U}_i\boldsymbol{U}_i^T.$$

Since the unit matrix may be expressed as $I = \sum_i^N \boldsymbol{U}_i\boldsymbol{U}_i^T$, the difference $I - \sum_i^E \boldsymbol{U}_i\boldsymbol{U}_i^T$ turns out to be a projection matrix which removes from the data vector $\boldsymbol{y}$ that part which is explained by the model and the global likelihood then reads

$$p(\boldsymbol{y}|\mathbf{X},\boldsymbol{\sigma},E,\mathcal{I}) \approx p(\boldsymbol{a}^{\mathrm{ML}}|E,\mathcal{I})\,\frac{(2\pi)^{\frac{E-N}{2}}}{\prod_k^E \lambda_k \prod_i^N \sigma_i}\exp\left\{-\frac{1}{2}\boldsymbol{y}'^T\left(\mathbb{1}-\sum_{i=1}^{E}\boldsymbol{U}_i\boldsymbol{U}_i^T\right)\boldsymbol{y}'\right\}. \tag{21.59}$$

The exponent in equation (21.59) represents that part of the data that is due to noise or a different model.

An appropriate uninformative prior for the coefficients of an E-dimensional hyperplane

$$a_1x_1 + a_2x_2 + \cdots + a_Ex_E + 1 = 0 \tag{21.60}$$

has been derived in Chapter 10 [p. 165]. Our present model function in equation (21.40) can be cast into this form by the change of variables $\boldsymbol{x} \to \boldsymbol{x}' = -\boldsymbol{x}/y$. The formalism developed so far holds for all models into which the unknown coefficients enter linearly.

A particularly important class of applications is the expansion of a model function $f(x)$ in appropriate basis functions $\{\varphi_i(x)\}$:

$$y(x) = \sum_{i=1}^{E} \varphi_i(x)\,a_i,$$

including of course the system of ordinary polynomials $\varphi_i(x) = t^{i-1}$. For discrete values x_ν of x we have the assignment

$$y_\nu \triangleq y(x_\nu), \quad x_{\nu i} \triangleq \varphi_i(x_\nu), \quad \boldsymbol{y} = \mathbf{X}\boldsymbol{a}.$$

The only thing which changes is the applicable prior density. An uninformative prior density on the expansion coefficients $\boldsymbol{a}$ can be derived from the fact that the L^2-norm of $y(x)$ is finite [22, 23]. Then

$$\mathcal{P}(a) := \int y^2(x)dx = \sum_{ij=1}^{E} a_i a_j \int \varphi_i(x)\varphi_j(x)\,dx = \sum_{ij=1}^{E} a_i a_j S_{ij} = \mu^2, \tag{21.61}$$

where we have introduced the overlap matrix $S_{ij} = \int \varphi_i(x)\varphi_j(x)\,dx$. The maximum entropy distribution of $\boldsymbol{a}$, $p(\boldsymbol{a})$, derives from the variation of the functional (see Section 11.1 [p. 178])

$$\delta\left[-\int p(\boldsymbol{a})\ln p(\boldsymbol{a})d^E a + \lambda\left(\mu^2 - \int \left(\boldsymbol{a}^T\mathbf{S}\boldsymbol{a}\right)p(\boldsymbol{a})d^E a\right)\right] = 0, \tag{21.62}$$

which yields

$$p(\boldsymbol{a}|\lambda, \mathcal{I}) = \frac{1}{Z(\lambda)}\exp\left[-\lambda\boldsymbol{a}^T\mathbf{S}\boldsymbol{a}\right] \tag{21.63}$$

with

$$Z(\lambda) = \int d^E a \exp\left[-\lambda\boldsymbol{a}^T\mathbf{S}\boldsymbol{a}\right] = \left(\frac{\pi}{\lambda}\right)^{E/2}|\mathbf{S}|^{-1/2}. \tag{21.64}$$

The expectation value μ^2 of $\boldsymbol{a}^T\mathbf{S}\boldsymbol{a}$ is therefore given by

$$\mu^2 = \frac{1}{Z(\lambda)}\int d^E a\ \boldsymbol{a}^T\mathbf{S}\boldsymbol{a}\exp\left[-\lambda\boldsymbol{a}^T\mathbf{S}\boldsymbol{a}\right], \tag{21.65}$$

which can be calculated using

$$\mu^2 = -\frac{1}{Z(\lambda)}\frac{d}{d\lambda}Z(\lambda) = \frac{E}{2\lambda} \rightarrow \lambda = \frac{E}{2\mu^2}. \tag{21.66}$$

The probability distribution $p(\boldsymbol{a}|\mu, \mathcal{I})$ is then

$$p(\boldsymbol{a}|\mu, \mathcal{I}) = \left(\frac{E}{2\pi\mu^2}\right)^{E/2}|\mathbf{S}|^{1/2}\exp\left[-\frac{E}{2\mu^2}\boldsymbol{a}^T\mathbf{S}\boldsymbol{a}\right]. \tag{21.67}$$

In the frequent case that μ^2 is not known we remove it from equation (21.67) by marginalization with Jeffreys' scale prior on μ to obtain

$$p(\boldsymbol{a}|\mathcal{I}) = \int p(\boldsymbol{a}|\mu, \mathcal{I})\frac{d\mu}{\mu} = \frac{1}{2}\frac{\Gamma(E/2)\,|\mathbf{S}|^{1/2}}{\pi^{E/2}\left(\boldsymbol{a}^T\mathbf{S}\boldsymbol{a}\right)^{E/2}}. \tag{21.68}$$

This prior $p(\boldsymbol{a}|\mathcal{I})$ is non-integrable singular at $\boldsymbol{a} = 0$. There is nearly always a price to be paid when using the completely ignorant Jeffreys' scale prior. Though the singularity is immaterial in our approximate evaluation of the evidence in equation (21.56), it would be harmful for the exact integral in equation (21.54). The cure for the problem is simple: according to equation (21.65), $\lambda \to \infty$ corresponds to a vanishing signal. However, any real apparatus always has a finite threshold for signal detection: $\mu_0^2 = E/2\lambda_0$. This brakes the scale invariance of the prior knowledge and the prior encoding this information reads $\exp\{-\lambda/\lambda_0\}/\lambda_0$. The prior for the hyperparameter λ is well behaved and modifies the prior on $\boldsymbol{a}$ to

$$p(\boldsymbol{a}|I, \lambda_0) = \frac{|S|^{1/2}}{\lambda_0\, \pi^{E/2}} \frac{\Gamma(E/2+1)}{\{1/\lambda_0 + \boldsymbol{a}^T \mathbf{S} \boldsymbol{a}\}^{\frac{E}{2}+1}}, \tag{21.69}$$

which is well behaved for all $\boldsymbol{a}$ and normalizable. This completes the topic of multivariate linear regression.

21.2 Models with nonlinear parameter dependence

We have seen in Section 21.1.4 [p. 342] that model functions which are linear in the parameters allow a general *closed-form* maximum likelihood estimate of the parameters. This simplicity is generally already distorted for Bayesian estimates since normally the prior density of the parameters will destroy the linear structure in the posterior density. One can save the day in such a case if the prior can be treated in an approximate way based on the assumption that the likelihood is much more informative than the prior. However, this means only deferring a salient problem which we meet in the wide field of problems that contain the parameters already in nonlinear form in the likelihood.

21.2.1 Marginal posterior probability

In many cases one is interested in individual parameters one at a time, for example, if one seeks the posterior mean $\langle a_k \rangle$ or the respective variance $\langle (\Delta a_k)^2 \rangle$. A different situation is encountered if we are interested in expectation values $\langle g(\boldsymbol{a}) \rangle$ of some nonlinear function $g(\boldsymbol{a})$ of the parameters, for example, predictions for the underlying model $f(\boldsymbol{\xi}|\boldsymbol{a})$ for a particular input vector $\boldsymbol{\xi}$. The latter case will be deferred to the next section. For the first class of problems, the proper procedure would be to calculate marginal distributions

$$p(a_k|\mathbf{X}, \boldsymbol{d}, N, \mathcal{I}) = \int D^{\not k} a \; p(\boldsymbol{a}|\mathbf{X}, \boldsymbol{d}, N, \mathcal{I}). \tag{21.70}$$

The symbol $D^{\not k} a$ means integration over all a_i except a_k. These marginal distributions contain the full information which we can derive from the posterior distribution about one parameter, irrespective of the values of all the others. Whether these one-dimensional distributions can meaningfully be summarized, for example, by a mean $\langle a_k \rangle$ and a variance $\langle \Delta a_k^2 \rangle$ is a different question. If so, we have to perform another integral

$$\langle a_k^n \rangle = \int da_k \; a_k^n \; p(a_k|\mathbf{X}, \boldsymbol{d}, N, \mathcal{I}). \tag{21.71}$$

The integration in equations (21.70) and (21.71) is the crucial technical step in the analysis of problems with nonlinear parameter dependencies in the model functions. It causes considerable difficulties if the dimension of the parameter space is high. To understand this let us consider a one-dimensional parameter space. In this case there is no need for marginalization and we can plot the full posterior in order to make our inferences. For two dimensions we can still explore the posterior in the form of a contour plot of $p(\boldsymbol{a}|\mathbf{X}, \boldsymbol{d}, N, \mathcal{I})$. For dimensions three and higher no sensible means of visualization without marginalization exist. Only in outstanding cases can the necessary integrals be performed analytically. A common approximation is given by the 'saddle-point approximation' described in Appendix A.2 [p. 598], which works well as long as the posterior distribution can be approximated by a multivariate Gaussian. But, of course, numerical integration would do as well. Numerical integration by traditional methods requires the evaluation of the integrand on some not necessarily regular grid with the number of function evaluations rising as m^E with the dimension E of the parameter, where m is the average number of function evaluations along one coordinate a_k of $\boldsymbol{a}$. It is obvious that the computational effort rapidly becomes prohibitive for high-dimensional parameter spaces due to its exponential dependence on dimension E. The way out of this problem is provided by 'stochastic integration techniques', or rather Monte Carlo integration, to be discussed in Chapter 29 [p. 509]. Unlike the exponential dependence on the dimension encountered in conventional quadrature methods, stochastic techniques exhibit typically only a power-law dependence of the computational effort on the dimension of the parameter space.

21.2.2 Predictions by nonlinear models

Starting from the posterior probability for the vector $\boldsymbol{a}$ of L parameters, given data and noise, we aim to predict the value η of the model function $\eta = f(\xi|\boldsymbol{a})$ for an input value ξ. To this end we require the multidimensional integral

$$p(\eta|\xi, \boldsymbol{x}, \boldsymbol{y}, \boldsymbol{\sigma}, \mathcal{I}) = \int d^L a \; p(\eta|\boldsymbol{a}, \xi, \boldsymbol{x}, \boldsymbol{y}, \boldsymbol{\sigma}, \mathcal{I}) \; p(\boldsymbol{a}|\xi, \boldsymbol{x}, \boldsymbol{y}, \boldsymbol{\sigma}, \mathcal{I}). \tag{21.72}$$

As outlined before, it is mostly not possible to evaluate the posterior probability $p(\boldsymbol{a}|\boldsymbol{x}, \boldsymbol{y}, \boldsymbol{\sigma}, \mathcal{I})$ analytically and one has to resort to numerical techniques. By far the most powerful approach is given by Markov chain Monte Carlo (see Chapter 30 [p. 537] and [74, 84, 96]) and 'nested sampling', which is less known at present but very powerful for really high-dimensional problems. Details are given in Chapter 31 [p. 572]. If the parameters $\boldsymbol{a}$ and the input ξ are explicitly given, the value of the output is uniquely fixed as $\eta = f(\xi|\boldsymbol{a})$ and the first probability in equation (21.72) reads

$$p(\eta|\boldsymbol{a}, \xi, \boldsymbol{x}, \boldsymbol{y}, \boldsymbol{\sigma}, \mathcal{I}) = \delta\big(\eta - f(\xi|\boldsymbol{a})\big),$$

resulting in

$$p(\eta|\xi, \boldsymbol{x}, \boldsymbol{y}, \boldsymbol{\sigma}, \mathcal{I}) = \int d^L a \; \delta\big(\eta - f(\xi|\boldsymbol{a})\big) \; p(\boldsymbol{a}|\xi, \boldsymbol{x}, \boldsymbol{y}, \boldsymbol{\sigma}, \mathcal{I}). \tag{21.73}$$

If one is really interested in the distribution of η one can numerically determine the corresponding histogram by Monte Carlo simulations based on the sampling distribution $p(\boldsymbol{a}|\xi, \boldsymbol{x}, \boldsymbol{y}, \boldsymbol{\sigma}, \mathcal{I})$. In general, however, the prime interest will be in posterior moments $\langle\eta^{\nu}\rangle$, which follow from equation (21.73):

$$\langle\eta^{\nu}\rangle = \int d^L a \; f^{\nu}(\xi|\boldsymbol{a}) \; p(\boldsymbol{a}|\xi, \boldsymbol{x}, \boldsymbol{y}, \boldsymbol{\sigma}, \mathcal{I}). \tag{21.74}$$

For the evaluation of this multidimensional integral one is, in general, still forced to rely on Monte Carlo simulations. If the posterior probability $p(\boldsymbol{a}|\xi, \boldsymbol{x}, \boldsymbol{y}, \boldsymbol{\sigma}, \mathcal{I})$ for the parameters $\boldsymbol{a}$ is unimodal and essentially confined to a region centred about the posterior mean $\boldsymbol{a}^p := \langle\boldsymbol{a}\rangle$, in which $f^{\nu}(\boldsymbol{a})$ is weakly varying, we can utilize the Taylor expansion

$$f^{\nu}(\boldsymbol{a}) = f^{\nu}(\boldsymbol{a}^p) + \sum_i \Delta a_i \left.\frac{\partial f^{\nu}(\boldsymbol{a})}{\partial a_i}\right|_{\boldsymbol{a}=\boldsymbol{a}^p} + \frac{1}{2}\sum_{ij} \Delta a_i \; \Delta a_j \left.\frac{\partial^2 f^{\nu}(\boldsymbol{a})}{\partial a_i \partial a_i}\right|_{\boldsymbol{a}=\boldsymbol{a}^p} + \cdots, \tag{21.75}$$

with the definition $\Delta a_i = a_i - a_i^p$. In the integral of equation (21.74) the linear term vanishes on account of the definition of the posterior mean $\boldsymbol{a}^p$ and we obtain

$$\langle\eta^{\nu}\rangle = f^{\nu}(\boldsymbol{a}^p) + \frac{1}{2}\sum_{ij} \left.\frac{\partial^2 f^{\nu}(\boldsymbol{a})}{\partial a_i \partial a_i}\right|_{\boldsymbol{a}=\boldsymbol{a}^p} \int d^L a \; \Delta a_i \; \Delta a_j \;\; p(\boldsymbol{a}|\xi, \boldsymbol{x}, \boldsymbol{y}, \boldsymbol{\sigma}, \mathcal{I}).$$

The remaining integral is the covariance matrix C_{ij} of the parameters $\boldsymbol{a}$, which can be determined by MCMC, and the result simplifies further to

$$\langle\eta^{\nu}\rangle = f^{\nu}(\boldsymbol{a}^p) + \frac{1}{2}\sum_{ij} \left.\frac{\partial^2 f^{\nu}(\boldsymbol{a})}{\partial a_i \partial a_i}\right|_{\boldsymbol{a}=\boldsymbol{a}^p} C_{ij}. \tag{21.76}$$

21.2.3 Saddle-point approximation

During the last decade, MCMC has become an increasingly powerful and popular numerical technique for the approximation-free determination of high-dimensional integrals. Depending on the structure of the posterior and the dimension of the parameter space, the saddle-point approximation can be very useful. As outlined in Appendix A.2 [p. 598], it can be used to perform high-dimensional integrals analytically. It is based on the assumption that the integrand has no sign changes, is unimodal, and its logarithm can be approximated by a Gaussian. In many cases, this assumption is increasingly satisfied with increasing dimension. For example, in statistical physics, the saddle-point approximation forms the basis for the derivation of the Boltzmann factor in the canonical ensemble. For low dimensions it can still be used in many cases as a quick and dirty approximation in order to gain a qualitative result. In the context of data analysis, the saddle-point approximation is also referred to as steepest descent approximation (SDA). In SDA the posterior probability density for the parameters $\boldsymbol{a}$ is rewritten in exponential form as

$$p(\boldsymbol{a}|\mathbf{X}, \boldsymbol{d}, N, \mathcal{I}) = \exp\{-\phi(\boldsymbol{a})\}, \qquad \boldsymbol{a} \in V_a. \tag{21.77}$$

Here N is the number of measurements, carried out at choices $\boldsymbol{x}_i$ of the input variables, and $\boldsymbol{d}$ is the data vector. Such a reformulation of the posterior density is always possible since the posterior density is a positive (semi-)definite function. SDA is now obtained upon replacing $\phi(\boldsymbol{a})$ by its Taylor expansion around the maximum posterior position[1] $\boldsymbol{a}_{\text{MAP}}$ up to the quadratic term:

$$\begin{aligned}
\phi(\boldsymbol{a}) &\approx \phi(\boldsymbol{a}_{\text{MAP}}) + \frac{1}{2}\Delta\boldsymbol{a}^T \mathbf{H} \Delta\boldsymbol{a}, \\
\Delta\boldsymbol{a} &= \boldsymbol{a} - \boldsymbol{a}_{\text{MAP}}, \\
H_{ij} &= \left. \frac{\partial^2 \phi(\boldsymbol{a})}{\partial a_i \partial a_j} \right|_{\boldsymbol{a}=\boldsymbol{a}_{\text{MAP}}}.
\end{aligned} \tag{21.78}$$

The matrix $\mathbf{H}$ is called the Hesse matrix. In some applications it may happen that all or some of the parameters a_i have a finite support V_a. Reasons may be the definition of the parameter, prior constraints, or the finite support of some functions, entering prior or likelihood. If the posterior is sufficiently localized, such that it has decayed on the boundary ∂V_a to negligible values, then the support can safely be extended to $V_a \to \mathbb{R}^L$, since the Gaussian approximation is well defined in $\mathbb{R}^L$ and the added volume does not contribute to any sensible posterior expectation value. In this case we obtain as approximate posterior density

$$p(\boldsymbol{a}|\mathbf{X}, \boldsymbol{d}, N, \mathcal{I}) = (2\pi)^{-N/2} |\mathbf{H}|^{1/2} \exp\left\{ -\frac{1}{2}\Delta\boldsymbol{a}^T \mathbf{H} \Delta\boldsymbol{a} \right\}, \qquad \boldsymbol{a} \in \mathbb{R}^L, \tag{21.79}$$

a multivariate normal distribution. The expectation values of the parameters are therefore identical to the MAP values $\langle \boldsymbol{a} \rangle = \boldsymbol{a}_{\text{MAP}}$ in this approximation. Further, we know from Section 11.3 [p. 194] that the Hesse matrix $\mathbf{H}$ is the inverse of the covariance matrix $\mathbf{C}$. This provides the covariances of the parameters

$$\text{cov}(a_i, a_j) = (\mathbf{H}^{-1})_{ij}. \tag{21.80}$$

It seems that we have solved the problem of nonlinear parameter dependencies by reduction to the linear case. However, remember that equation (21.79) is an approximation which makes the strong assumption that the hypersurfaces $\phi(\boldsymbol{a}) = \text{const.}$ are close to ellipsoids in multidimensional parameter space. This may happen sometimes but is not at all guaranteed. The Gaussian approximation equation (21.79) is by no means a panacea and it will be shown in Section 22.1.1 [p. 364] that it can be considerably in error even in a one-dimensional problem.

21.3 Errors in all variables

In the preceding sections we have investigated the estimation of parameters from linear and nonlinear models assuming that a measurement error is only associated with the

[1] Note that it is expedient in the present context to expand about the MAP solution, instead of the mean, in order to get rid of the linear term.

'dependent' (response) variable. This assumption will hold to good approximation in many cases. However, there are also cases where the setting of the 'independent' (input) variables is only possible within an appreciable error margin.

21.3.1 Straight lines and hyperplanes

In order to acquaint ourselves with erroneous input variables and their respective consequences, we start by considering the estimation of the coefficients of a straight line $y_i = ax_i + b$. It is usually implicitly assumed that the errors in $\{x_i\}$ are so much smaller than the errors in $\{y_i\}$ that they can be neglected. The other extreme lends itself also to a simple treatment: if the errors in $\{x_i\}$ exceed the errors in $\{y_i\}$ then the role of input and response is exchanged and we investigate $x' = a'y' + b'$ instead of $y = ax + b$. Between the two extremes a large number of cases is conceivable and does occur. It is this most general case which we shall now focus on. We assume that together with a data point $\boldsymbol{x}_i = \{x_i, y_i\}$, estimates for the covariance matrices $\mathbf{C}_i$ of the data point are provided. In most cases, the covariance will be diagonal: $\mathbf{C}_i = \text{diag}(\sigma_{x,i}, \sigma_{y,i})$. Given the 'true' coordinates $\boldsymbol{x}_i^{\text{e}}$ of the data points, the likelihood reads correspondingly

$$p(\{\boldsymbol{x}_i\}|\{\boldsymbol{x}_i^{\text{e}}\}, \{\mathbf{C}_i\}, \mathcal{I}) \propto \prod_{i=1}^{N} \exp\left\{-\frac{1}{2}\,(\boldsymbol{x}_i - \boldsymbol{x}_i^{\text{e}})^T \mathbf{C}_i^{-1} (\boldsymbol{x}_i - \boldsymbol{x}_i^{\text{e}})\right\}. \tag{21.81}$$

We have suppressed the normalization as it is immaterial for the following considerations. Needless to say, the true data coordinates are not at our disposal and parameter estimation will be based on the marginal likelihood $p(\{\boldsymbol{x}_i\}|a, b, \mathbf{C}, \mathcal{I})$ instead. The marginal likelihood can be derived via the marginalization over the true $\boldsymbol{x}_i^{\text{e}}$ coordinate. Unfortunately, the prior distribution $p(x^{\text{e}}|a, b, \boldsymbol{\sigma}, \mathcal{I})$ poses severe problems, as discussed by Gull [92]. There is, however, a different approach that poses no problems at all. The true data point, which lies on the straight line, can be expressed as

$$\boldsymbol{x}_i^{\text{e}} = \boldsymbol{x}_0 + s_i \boldsymbol{n} \quad \text{with} \quad \boldsymbol{x}_0 = (0, b)^T \quad \text{and} \quad \boldsymbol{n} = \frac{1}{\sqrt{1+a^2}}\,(1, a)^T. \tag{21.82}$$

Then the marginalization reads

$$p(\{\boldsymbol{x}_i\}|a, b, \{\mathbf{C}_i\}, \mathcal{I}) = \int ds^N \; p(\{\boldsymbol{x}_i\}|\{s_i\}, a, b, \{\mathbf{C}_i\}, \mathcal{I})\; p(\{s_i\}|a, b, \{\mathbf{C}_i\}, \mathcal{I}). \tag{21.83}$$

Now the data points are supposed to be uncorrelated, that is

$$p(\{s_i\}|a, b, \{\mathbf{C}_i\}, \mathcal{I}) = \prod_{i=1}^{N} p(s_i|a, b, \mathbf{C}_i, \mathcal{I}).$$

Based on 'Jaynes' transformation principle', we infer that the probability for a data point lying in a specific line segment ds is invariant against shift and rotation of the straight line.

Moreover, the probability that the true data point is in $(s, s+ds)$ is logically independent of the covariance of the measured data point. Therefore,

$$p(s_i|a, b, \mathbf{C}_i, \mathcal{I}) = p(s_i|\mathcal{I}).$$

Finally, no position on the line shall be preferred. Consequently, $p(s_i|\mathcal{I})$ is a uniform prior. Usually we would have to introduce finite bounds for s_i, e.g. $s_i \in (-S, S)$, and consider $S \to \infty$ at the end of the calculation. However, since the normalization of the marginal likelihood poses no problem, we can even use an improper prior $p(s_i|\mathcal{I}) = \text{const}$. Then the marginal likelihood in equation (21.83) is

$$p(\{\boldsymbol{x}_i\}|a, b, \{\mathbf{C}_i\}, \mathcal{I}) \propto \prod_{i=1}^{N} \underbrace{\int ds_i \; p(\boldsymbol{x}_i|s_i, a, b, \mathbf{C}_i, \mathcal{I})}_{:=I_i}. \tag{21.84}$$

For the sake of notational ease, we will omit the subscript i in the following calculations, as long as we are considering a single data point. The integral for a single data point has the form

$$I_i = \int ds \; e^{-\frac{1}{2}\Phi(s)}, \tag{21.85a}$$

$$\Phi(s) := (\boldsymbol{x} - \boldsymbol{x}_0 - s\,\boldsymbol{n})^T \mathbf{C}^{-1} (\boldsymbol{x} - \boldsymbol{x}_0 - s\,\boldsymbol{n}). \tag{21.85b}$$

Now $\Phi(s)$ is a quadratic form in s. Its minimum is obtained for

$$\frac{d}{ds}\Phi(s) = 2\,s\,\boldsymbol{n}^T\mathbf{C}^{-1}\boldsymbol{n} - 2\,\boldsymbol{n}^T\mathbf{C}^{-1}(\boldsymbol{x} - \boldsymbol{x}_0) \stackrel{!}{=} 0$$

$$\Rightarrow \quad s^* = \frac{\boldsymbol{n}^T\mathbf{C}^{-1}(\boldsymbol{x} - \boldsymbol{x}_0)}{\boldsymbol{n}^T\mathbf{C}^{-1}\boldsymbol{n}}.$$

The corresponding point on the line is

$$\boldsymbol{x}^* = \boldsymbol{x}_0 + \boldsymbol{n}\,\frac{\boldsymbol{n}^T\mathbf{C}^{-1}(\boldsymbol{x} - \boldsymbol{x}_0)}{\boldsymbol{n}^T\mathbf{C}^{-1}\boldsymbol{n}}.$$

This point has an interesting feature:

$$\left(\boldsymbol{x} - \boldsymbol{x}^*\right)^T \mathbf{C}^{-1}\boldsymbol{n} = 0.$$

That is, $\boldsymbol{x}_*$ is the foot point of the measured data point $\boldsymbol{x}$ on the straight line, under the metric defined by $\mathbf{C}^{-1}$. Put differently, it is the point $\boldsymbol{x}^*$ that yields the minimal misfit

$$\Phi(\boldsymbol{x}^*) := \left(\boldsymbol{x} - \boldsymbol{x}^*\right)^T \mathbf{C}^{-1}\left(\boldsymbol{x} - \boldsymbol{x}^*\right).$$

Now we proceed with $\Phi(s)$ in equation (21.85):

$$\Phi(s) = \Phi(\boldsymbol{x}^*) + \frac{1}{2}\boldsymbol{n}^T\mathbf{C}^{-1}\boldsymbol{n}\;(s - s^*)^2.$$

The remaining integral over s in equation (21.85) merely modifies the proportionality constant and the result for the sought-for marginal likelihood in equation (21.84) reads

$$
\begin{aligned}
p(\{\boldsymbol{x}_i\}|a, b, \{\mathbf{C}_i\}, \mathcal{I}) &= \frac{1}{Z} \prod_{i=1}^{N} \exp\left\{ -\frac{1}{2} \Phi_i^* \right\}, \\
\Phi_i^* &= \min_{\boldsymbol{x}^* \in \text{line}} \left(\boldsymbol{x}_i - \boldsymbol{x}^*\right)^T \mathbf{C}^{-1} \left(\boldsymbol{x}_i - \boldsymbol{x}^*\right).
\end{aligned}
\tag{21.86}
$$

It is straightforward to generalize these ideas to hyperplanes. In this case the vectors $\boldsymbol{x}$ stand for $\boldsymbol{x} = (x_1, \ldots, x_n, y)$. The model equation for the hyperplane can be expressed as $\Psi(\boldsymbol{x}) = 0$, with

$$
0 = \boldsymbol{a}^T \boldsymbol{x} + b, \tag{21.87a}
$$

or alternatively

$$
\boldsymbol{x} = \boldsymbol{x}_0 + \sum_{i=1}^{n} s_i\, \boldsymbol{n}_i. \tag{21.87b}
$$

The model parameters are b and the vector $\boldsymbol{a} := \{a_1, \ldots, a_n\}$, which also defines the direction normal to the hyperplane. The vector $\boldsymbol{n}_i$ is any set of n normalized vectors orthogonal to $\boldsymbol{a}$ and $\boldsymbol{x}_0$ is an arbitrary point in the plane. The derivation of the marginal likelihood is analogous to the straight-line case, and we obtain

MARGINAL LIKELIHOOD FOR THE HYPERPLANE MODEL WITH ERRORS IN ALL VARIATES

$$
0 = \boldsymbol{a}^T \boldsymbol{x} + b \qquad \text{(hyperplane)}
$$

$$
\begin{aligned}
p(\{\boldsymbol{x}_i\}|\boldsymbol{a}, b, \{\mathbf{C}_i\}, \mathcal{I}) &= \frac{1}{Z} \prod_{i=1}^{N} \exp\left\{ -\frac{1}{2} \Phi_i^* \right\}, \\
\Phi_i^* &= \min_{\boldsymbol{x}^* \in \text{plane}} \left(\boldsymbol{x}_i - \boldsymbol{x}^*\right)^T \mathbf{C}^{-1} \left(\boldsymbol{x}_i - \boldsymbol{x}^*\right).
\end{aligned}
\tag{21.88}
$$

Now we will determine the minimum misfit Φ^* for one data point (the index i will be suppressed again):

$$
\Phi(\boldsymbol{x}^*) = (\boldsymbol{x} - \boldsymbol{x}^*)^T \mathbf{C}^{-1} (\boldsymbol{x} - \boldsymbol{x}^*), \tag{21.89}
$$

for straight lines and hyperplanes in one go. The common model constraint reads

$$
0 = \boldsymbol{a}^T \boldsymbol{x}^* + b. \tag{21.90}
$$

The minimal misfit, subject to the model constraint, can easily be obtained by the vanishing gradient of the functional

$$\mathcal{L} = \Phi(\boldsymbol{x}^*) + 2\lambda\Big(\boldsymbol{a}^T\boldsymbol{x}^* + b\Big),$$

with Lagrange multiplier λ. The gradient $\nabla_{\boldsymbol{x}^*}\mathcal{L} = \mathbf{C}^{-1}\big(\boldsymbol{x}^* - \boldsymbol{x}\big) + \lambda\boldsymbol{a}$ vanishes for

$$\big(\boldsymbol{x}^* - \boldsymbol{x}\big) = -\lambda\mathbf{C}\boldsymbol{a}. \tag{21.91}$$

Upon inserting this $\boldsymbol{x}^*$ into the hyperplane equation (21.90), we obtain the Lagrange multiplier

$$\lambda = \frac{b + \boldsymbol{a}^T\boldsymbol{x}}{\boldsymbol{a}^T\mathbf{C}\boldsymbol{a}}.$$

Along with equations (21.89) and (21.91), the minimal misfit is simply

$$\Phi^* = \frac{\big(b + \boldsymbol{a}^T\boldsymbol{x}\big)^2}{\boldsymbol{a}^T\mathbf{C}\boldsymbol{a}}. \tag{21.92}$$

For the simplest case of a straight line in two dimensions, with $\boldsymbol{a} = (a, -1)^T$, and diagonal covariance matrix, we obtain

$$\Phi^* = \frac{(b + ax - y)^2}{\sigma_y^2 + a^2\sigma_x^2}. \tag{21.93}$$

We insert equation (21.93) into the expression for the marginal likelihood given in equation (21.88) to obtain

MARGINAL LIKELIHOOD FOR STRAIGHT LINE WITH ERRORS IN x AND y

$$p(x, y|a, b, \boldsymbol{\sigma}_x, \boldsymbol{\sigma}_y, \mathcal{I}) = \frac{1}{Z}\prod_{i=1}^{N} \exp\left\{-\frac{1}{2}\frac{\big(y_i - ax_i - b\big)^2}{\sigma_{y,i}^2 + \sigma_{x,i}^2 a^2}\right\}. \tag{21.94}$$

The result is formally identical to that of the case with errors only in the y-component. Here we merely have an effective error $\sigma_{\text{eff}}^2 = \sigma_y^2 + a^2\sigma_x^2$, which is very plausible. The uncertainty in the x-position of the data point adds an uncertainty $a\sigma_x$ in the y-position. Since the errors are uncorrelated, they add quadratically. Another difference from the previous case in equation (21.18) [p. 339], where only the experimental y-coordinates were assumed to be erroneous, is that the maximum likelihood equations for determination of (a, b) now become nonlinear, which complicates the solution.

It should also be pointed out that the likelihood is not normalizable in the parameter a, as the argument of the exponential approaches a constant for $a^2 \to \infty$. There is however no need to worry. The behaviour is actually very reasonable, since an infinite slope ($a = \infty$)

does not imply an infinite distance from the data point, and, therefore, the likelihood still has a finite value. This observation confirms the distinction between *likelihood* and *probability*. A probability is called a likelihood if we consider its functional dependence on variables behind conditional solidus. There is no rule that probabilities are normalized or normalizable in such variables. The normalization is only guaranteed after applying Bayes' theorem, that is, in combination with a proper prior. Equation (21.94) is the basis for an estimation of the straight-line parameters (a, b). A possible uninformative prior to arrive at the posterior density of (a, b) has already been given in equation (21.19) [p. 339]. The result is

POSTERIOR PDF OF STRAIGHT-LINE PARAMETERS WITH ERRORS IN x AND y

$$p\left(a, b|\boldsymbol{x}, \boldsymbol{y}, \boldsymbol{\sigma}_x, \boldsymbol{\sigma}_y, \mathcal{I}\right) = \frac{\mathbf{1}(|b| < B)}{Z\,\left(1 + a^2\right)^{3/2}} \exp\left[-\frac{1}{2}\sum_i \frac{(y_i - a x_i - b)^2}{\sigma_{y,i}^2 + a^2 \sigma_{x,i}^2}\right]. \tag{21.95}$$

In Figure 21.3 a representative data set is depicted for which the Bayesian analysis is performed. Point estimates and standard deviations of (a, b) are now calculated from equation (21.95) by standard numerical integration (Chapter 29 [p. 509]) and collected in Table 21.1. The prior width B has been chosen large enough in order to ensure that $\langle b \rangle$ and $\langle \Delta b^2 \rangle$ become independent of B. The numerical results of the analysis of the data in Figure 21.3 are summarized in Table 21.1. The prominent message from Table 21.1 is the enlarged uncertainty of the parameters when both errors in $\boldsymbol{x}$ and $\boldsymbol{y}$ are included in the calculation. In the estimate of a we notice a 6% reduction upon introducing the prior in (a, b). A big influence of the prior on (a, b) is seen in the estimate of b. This signals a sizable correlation of a and b. Figure 21.3 shows the straight line obtained from a calculation similar to that in Section 21.1.4 [p. 342]. Errors in x and y are considered. The shaded band indicates the confidence region of the prediction. Note the widening of the

Table 21.1 Results for straight-line parameters (a, b). (1) Standard least-squares analysis with errors only in y, (2) errors in x and y plus uniform prior, (3) errors in x and y plus correct prior

	a	b	Method
(1)	1.996 ± 0.156	-2.772 ± 2.438	errors in y
(2)	2.082 ± 0.326	-2.770 ± 3.965	errors in x, y
(3)	1.966 ± 0.303	-1.504 ± 3.833	errors in x, y and prior for (a, b)

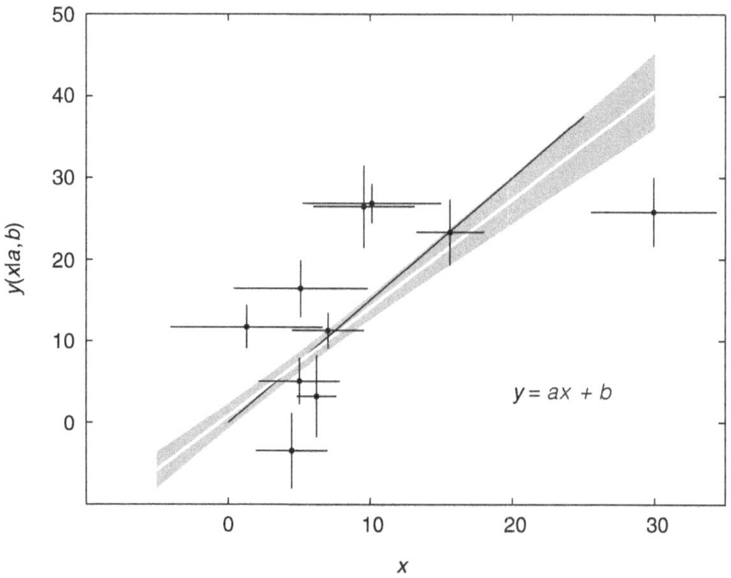

Figure 21.3 Straight-line fit to data (dots) with uncertainties in x and y (error bars). The solid line is the posterior prediction of the linear function obtained from a calculation similar to Section 21.1.4. The shaded band defines the confidence region of the prediction, computed as for Figure 21.2.

band at the ends of the data range. A comparative calculation with errors in y only results in nearly the same straight line but in a sizable shrinking of the confidence band by 35% at the lower end and by 54% at the upper end of the data range.

The general result of equation (21.92) applies of course also to multivariate linear problems. A representative case is met in the problem of energy confinement in toroidal magnetic confinement fusion devices [200]. In the absence of a fundamental theory of energy confinement it has become common practice in fusion research to represent the energy content W of the device as

$$W = a^{\alpha_a} B^{\alpha_B} n^{\alpha_n} p^{\alpha_p} e^{\alpha_c}, \tag{21.96}$$

where a is the minor toroidal radius, B the confining magnetic field, n the plasma particle number density, p the injected power and e^{α_c} a constant. It is common practice to transform this highly nonlinear model equation to a linear one by taking the logarithm on both sides of equation (21.96):

$$\ln W = \alpha_a \ln a + \alpha_B \ln B + \alpha_n \ln n + \alpha_p \ln p + \alpha_c. \tag{21.97}$$

This nonlinear transformation has been shown to lead to unbiased estimates of $\boldsymbol{\alpha}$ except for α_c [51]. Equation (21.97) now meets the model equation (21.90) with $b = 0$. The likelihood in the variables $y = \ln W, x_1 = \ln a, x_2 = \ln B, x_3 = \ln n, x_4 = \ln p, x_5 = 1$ reads, apart from an uninteresting constant,

$$p\left(\boldsymbol{x}_i, y_i | \boldsymbol{\alpha}, \boldsymbol{\sigma}_i, \mathcal{I}\right) \propto \exp\left[-\frac{1}{2}\sum_i \frac{\left(y_i - \boldsymbol{\alpha}^T \boldsymbol{x}_i\right)^2}{\sigma_{\mathrm{eff},i}^2}\right] \tag{21.98}$$

with

$$\sigma_{\mathrm{eff},i}^2 = \sigma_{y,i}^2 + \sum_k \alpha_k^2 \sigma_{k,i}^2. \tag{21.99}$$

An uninformative prior for the multivariate linear case has been presented in equation (10.42) [p. 175]:

$$p\left(\boldsymbol{\alpha} | \mathcal{I}\right) = \left(\boldsymbol{\alpha}^T \boldsymbol{\alpha}\right)^{-\frac{N+1}{2}}, \tag{21.100}$$

with $N = 5$ in the present case. An estimate of $\boldsymbol{\alpha}$ and the corresponding uncertainties can be obtained from minimizing the log-posterior

$$\ln P = -\frac{1}{2}\left(\boldsymbol{y} - \tilde{\mathbf{X}}\boldsymbol{\alpha}\right)^T \mathbf{C}_{\mathrm{eff}}^{-1}\left(\boldsymbol{y} - \tilde{\mathbf{X}}\boldsymbol{\alpha}\right) - \frac{6}{2}\ln\left(\boldsymbol{\alpha}^T\boldsymbol{\alpha}\right), \tag{21.101}$$

where $\tilde{\mathbf{X}}$ is a matrix made up from the rows $\boldsymbol{x}_i$ and $\mathbf{C}_{\mathrm{eff}} = \mathrm{diag}(\sigma_{i,\mathrm{eff}}^2)$. We define $\boldsymbol{d} = \mathbf{C}_{\mathrm{eff}}^{-1/2}\boldsymbol{y}$ and

$$\mathbf{X} = \mathbf{C}_{\mathrm{eff}}^{-1/2}\tilde{\mathbf{X}}. \tag{21.102}$$

Please note that $\mathbf{X}$ actually depends on $\boldsymbol{\alpha}$ via $\sigma_{\mathrm{eff},i}$. The minimum of equation (21.101) is then found from

$$\frac{d}{d\alpha^T}\left[-\frac{1}{2}\left(\boldsymbol{d} - \mathbf{X}\boldsymbol{\alpha}\right)^T\left(\boldsymbol{d} - \mathbf{X}\boldsymbol{\alpha}\right) - \frac{6}{2}\ln\left(\boldsymbol{\alpha}^T\boldsymbol{\alpha}\right)\right] = 0. \tag{21.103}$$

This yields

$$\mathbf{X}^T\mathbf{X}\boldsymbol{\alpha} = \mathbf{X}^T\boldsymbol{d} - 6\frac{\boldsymbol{\alpha}}{\boldsymbol{\alpha}^T\boldsymbol{\alpha}}. \tag{21.104}$$

This nonlinear equation may be solved by iteration:

$$\boldsymbol{\alpha}_{k+1} = \left(\mathbf{X}_k^T\mathbf{X}_k\right)^{-1}\mathbf{X}_k^T\boldsymbol{d} - \frac{6}{\boldsymbol{\alpha}_k^T\boldsymbol{\alpha}_k}\left(\mathbf{X}_k^T\mathbf{X}_k\right)^{-1}\boldsymbol{\alpha}_k. \tag{21.105}$$

The dependence of $\mathbf{X}_k$ on iteration step k arises from its dependence on Σ_{eff} in equation (21.102). The numerical part of the calculations is greatly simplified by employing singular-value decomposition of $\mathbf{X}$ (equation (21.47) [p. 346]). The data used for this calculation are taken from [167]. The results are collected in Table 21.2. Since the input variables a, B, n, p in SI units in equation (21.96) span a range of 20 orders of magnitude, they have been scaled to the unit mean prior in numerical work. Inclusion of uncertainties in all variables increases again the uncertainties of the parameters compared with the traditional least-squares case. The mean values are again much less affected, as in the previous univariate case, and the prior has nearly no influence on the result. This is due to the 153 data sets in the multivariate linear case compared with 10 in Figure 21.3.

Table 21.2 Posterior estimates of the parameters in equation (21.97) in various approximations: (1) least squares, (2) errors in all variables, (3) errors in all variables including a hyperplane prior

	α_a	α_B	α_n	α_p	α_c
(1)	2.230 ± 0.047	0.585 ± 0.070	0.422 ± 0.012	0.470 ± 0.014	8.072 ± 0.055
(2)	2.257 ± 0.057	0.575 ± 0.077	0.424 ± 0.014	0.474 ± 0.017	8.069 ± 0.062
(3)	2.256 ± 0.057	0.577 ± 0.077	0.424 ± 0.014	0.474 ± 0.017	8.066 ± 0.062

21.3.2 Nonlinear functions

In the last section we dealt with model functions which were linear in the variables as well as in the parameters. We have shown already how to deal with the case of nonlinearity in the parameters in Section 21.2.2 [p. 351]. Nonlinearity in the variables is a new case which requires attention if we consider errors in all variables. For example, a polynomial is a model function which is linear in the parameters but nonlinear in the input variable x. The input variables are $x_i,\ i = 1, \dots, n$, and the response variable is y. As before, we combine all variables in a vector $\boldsymbol{x}^T = (x_1, \dots, x_n, y)$. Let the model equation, which actually defines a hypersurface, read

$$\Psi(\boldsymbol{\xi}) := f(\xi_1, \dots, \xi_n|\boldsymbol{a}) - \xi_{n+1} \stackrel{!}{=} 0, \tag{21.106}$$

with the sought-for model parameters $\boldsymbol{a}$. The marginalization integral, analogous to equation (21.83) [p. 354], reveals that for each data point $\boldsymbol{x}_i$ only a small patch of the hypersurface in the vicinity of the corresponding foot point $\boldsymbol{x}_i^*$ contributes to the marginalization, provided the signal-to-noise ratio is much greater than one. Then the hypersurface can safely be replaced by the tangent plane at the point $\boldsymbol{x}_i^*$. All remaining steps are the same as before, and we obtain

Marginal likelihood for a nonlinear model with errors in all variates

$$p(\boldsymbol{x}|\boldsymbol{a}, \boldsymbol{\sigma}, \mathcal{I}) = \frac{1}{Z} \exp\left\{-\frac{1}{2}\Phi^{*2}\right\},$$
$$\Phi^* = \min_{\boldsymbol{x}^*:\ \Psi(\boldsymbol{x}^*)=0} \left(\boldsymbol{x} - \boldsymbol{x}^*\right)^T C^{-1}\left(\boldsymbol{x} - \boldsymbol{x}^*\right). \tag{21.107}$$

Application of equation (21.107) requires then determining the coordinates for the foot points $\boldsymbol{x}_i^*$ to the data points $\boldsymbol{x}_i$. For definiteness we consider here the case of a nonlinear function $y = f(x|u, v) = u \ln(x + v)$. Moreover, we assume diagonal covariance matrices $\mathbf{C}_i = \text{diag}(\sigma_{x,i}^2, \sigma_{y,i}^2)$. Let x_i, y_i be the coordinates of the data point under consideration. The foot points are determined by brute-force numerical minimization of the misfit $\Phi(\boldsymbol{x}^*)$ in equation (21.107). Given the coordinates of a foot point (x_i^*, y_i^*), the equation for the tangent at that point is

$$y = y_i^* + f'\left(x_i^*|u, v\right)\ (x - x_i^*), \tag{21.108}$$

which has the form $y = a_i x + b_i$ with $a_i = f'\left(x_i^*|u, v\right)$ and $b_i = y_i^* - x_i^* f'\left(x_i^*|u, v\right)$. The single likelihood term is then identical to equation (21.94) and the full likelihood becomes

$$p(\boldsymbol{x}, \boldsymbol{y}|u, v, \boldsymbol{\sigma}_x, \boldsymbol{\sigma}_y, \mathcal{I}) = \frac{1}{Z} \exp\left[-\frac{1}{2}\sum_i \frac{(y_i - a_i x_i - b_i)^2}{\sigma_{y,i}^2 + a_i^2 \sigma_{x,i}^2}\right]. \tag{21.109}$$

Inferring the values of u, v, which enter the definition of $f(x|u, v)$, requires the specification of a prior distribution for u, v. In the absence of special knowledge we choose flat improper priors. Then

$$p(u, v|\boldsymbol{x}, \boldsymbol{y}, \boldsymbol{\sigma}_x, \boldsymbol{\sigma}_y, \mathcal{I}) = \frac{1}{Z'} \exp\left[-\frac{1}{2}\sum_i \frac{(y_i - a_i x_i - b_i)^2}{\sigma_{y,i}^2 + a_i^2 \sigma_{x,i}^2}\right]. \tag{21.110}$$

Figure 21.4 shows representative mock data for this data analysis problem. The predicted function is depicted as a white solid line which agrees fairly well with the underlying 'true'

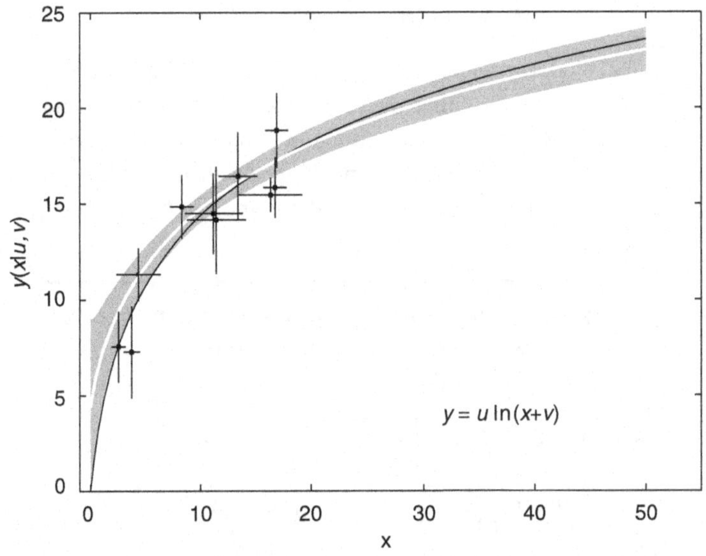

Figure 21.4 Mock data for the nonlinear function $f(x|u, v)$ (dots) along with individual errors $\sigma_{x,i}$ and $\sigma_{y,i}$ (error bars). The underlying 'true' function is depicted as a black solid line. The predicted function is shown as white solid line. The shaded band is the confidence region of the prediction based on equation (21.110).

Table 21.3 Numerical values of the parameters (u, v) in $y = u \ln(x + v)$ with two approximations: (1) only errors in y, (2) results obtained based on equation (21.110) for the case of errors in x and y

	u	v	
(1)	5.546 ± 0.032	1.869 ± 0.203	errors in y
(2)	5.883 ± 0.075	1.042 ± 0.448	errors in x, y

curve, represented by the black solid line. The shaded band indicates the confidence region of the prediction. As in the linear model, we find that the confidence band increases outside the data cloud.

The numerical results of this application are summarized in Table 21.3. The true parameters are $u = 6$ and $v = 1$. We find a fairly good agreement between the Bayesian estimate and the true values. A comparative calculation where the errors $\sigma_{x,i}$ are ignored is also given, termed 'error in y'. The deviations are clearly worse and the errors underestimated. The treatment of the multivariate nonlinear case is a simple extension of the above univariate case. The central problem is the calculation of the foot point of the normal passing through a given data point. Having accomplished this, the problem is reduced to a single factor in equation (21.98). It is obvious that the calculation becomes demanding for problems with more independent variables and possibly a greater number of parameters. Also, with a complicated multivariate surface the number of data points must increase for a reliable parameter estimation.

Finally we recall that our treatment of the nonlinear case is an approximation which will hold only in cases with low curvature. The exact case has been treated in [217]. It is considerably more complicated than the tangential approximation.

22
Consistent inference on inconsistent data

It is common experience of all experimental scientists that repeated measurements of supposedly one and the same quantity result occasionally in data which are in striking disagreement with all others. Such data are usually called *outliers*. There are numerous conceivable reasons for such outliers. Formally we may consider the sequence of measurements d_j with confidence limits $d_j \pm \sigma_j$. If the distance $|d_j - d_k|$ of any two data points d_j and d_k becomes larger than the sum $(\sigma_k^2 + \sigma_j^2)^{1/2}$ of the associated errors, then the data start to become inconsistent and finally at least one of them will become an outlier. There are no strict meanings to the terms 'inconsistent' and 'outliers' and it is exactly this lack of uniqueness in the definition which enforces a treatment by probability theory. We shall consider two different cases. Inconsistency of the data may result from a wrong estimate of the measurement error σ_k. Already Carl Friedrich Gauss was concerned about measurement uncertainties and stated that 'the variances of the measurements are practically never known exactly' [81]. Inconsistency may also arise from measurements distorted by signals from some unrecognized spurious source, leading to 'strong' deviations of d_k from the mainstream. We shall begin with the first case.

22.1 Erroneously measured uncertainties

In the preceding sections we have boldly assumed that the standard deviation σ_i of the error distribution for the measured quantity d_i is known exactly. This assumption almost never applies. What we may know at least are apparent – measured – errors s_i. The connection between the unknown standard deviation σ_i and the measured errors s_i is coded in a distribution $p(\boldsymbol{\sigma} \,|\, \boldsymbol{s}, \mathcal{I})$ which expresses the confidence in the precision of the s_i. In the following discussion we assume Gaussian errors for the data. In the following sections we analyse four different ways to encode prior knowledge in the form of $p(\boldsymbol{\sigma} \,|\, \boldsymbol{s}, \mathcal{I})$, representing common situations as far as the reliability of experimental error estimate s_i is concerned. The various error corrections are summarized in Table 22.1.

22.1.1 Errors uncertain by a common factor

We start with the simple inconsistency assumption that all measured errors s_i are uncertain by the same unknown factor α [47]:

Table 22.1 Various approaches to include uncertainties in the error assignment. The priors are: Jeffreys' scale prior (JP), conjugate prior (CP) and inverse Gamma prior (iΓ), which has 45% of its prior mass between the half-maximum positions

Classification	Characterization	Prior
(1) global	$\sigma_i = \sigma$	JP
(2) global correction	$\sigma_i = \alpha s_i$	JP
(3) individual correction	$\langle \sigma_i \rangle = s_i$, $\mathrm{var}(\sigma_i) = \infty$	CP
(4) individual correction	$P(0.24 s_i \le \sigma_i \le 1.31 s_i) = 0.45$	iΓ

GLOBAL ERROR CORRECTION

$$p(\boldsymbol{\sigma} | s, \alpha, \mathcal{I}) = \prod_i \delta(\sigma_i - \alpha \cdot s_i). \tag{22.1}$$

The global hyperparameter α will be marginalized based on the uninformative Jeffreys' scale prior $p(\alpha|\mathcal{I})$. It is expedient to perform the marginalization at the end of all transformations, because an upfront marginalization results in an unenjoyable expression. The choice of a global hyperparameter implies that all experimental errors suffer from the same uncertainty. Consequently, all experimental errors are corrected by the same correction factor. In other words, we rule out that some of the errors are correct and others are wrong.

Here we will investigate the effect of this assumption on the determination of the sample mean. The likelihood for the constant model $d_i = \mu + \varepsilon_i$ is, assuming $\langle \varepsilon_i \rangle = 0$ and $\langle \varepsilon_i^2 \rangle = \sigma_i^2$,

$$p(\boldsymbol{d}|\mu, \boldsymbol{\sigma}, \mathcal{I}) = \left(\prod_i \frac{1}{\sigma_i \sqrt{2\pi}} \right) \exp\left\{ -\frac{1}{2} \sum_i \frac{(d_i - \mu)^2}{\sigma_i^2} \right\}. \tag{22.2}$$

We have treated the problem of estimating μ in Section 21.1.1 [p. 334] assuming that $\boldsymbol{\sigma}$ is known. The result was unsatisfactory insofar as the variance of the posterior for μ depended only on the standard deviation σ_i of the contributing data and not on their scatter. This is strange, to say the least, and will be shown to be a consequence of assuming the exact σ_i to be known precisely.

Marginalization helps to reformulate the likelihood in equation (22.2) in terms of the experimental error estimates s_i with the help of equation (22.1):

$$p(\boldsymbol{d}|\mu, s, \alpha, \mathcal{I}) = \int d^N\sigma \left\{ \prod_i \delta(\sigma_i - \alpha s_i) \right\} p(\boldsymbol{d}|\mu, \boldsymbol{\sigma}, \mathcal{I}). \tag{22.3}$$

The integration is trivial and results in the replacement of σ_i by αs_i in equation (22.2). Employing the previously defined σ-weighted averages of equation (21.4) [p. 335], the marginal likelihood becomes

$$p(\boldsymbol{d}|\mu, \boldsymbol{s}, \alpha, \mathcal{I}) = \frac{1}{\alpha^N}\left(\prod_i \frac{1}{s_i\sqrt{2\pi}}\right) e^{-\frac{N(\mu-\overline{d})^2}{2\alpha^2\varrho^2}}\, e^{-\frac{N\overline{(\Delta d)^2}}{2\alpha^2\varrho^2}}. \tag{22.4}$$

The errors entering the σ-weighted averages in equation (21.4) [p. 335] are the standard deviations of the error distribution of the measured data, which seems to pose problems since we do not know these in advance. However, these standard deviations of the error distribution deviate from the measured ones merely by a global factor α, which cancels in the expressions for $\overline{d}$ and $\overline{(\Delta d)^2}$. We can therefore compute the σ-weighted averages also with the measured errors. One aspect of the problem is to use equation (22.4) to estimate α as a measure of the degree of consistency of the data. Bayes' theorem

$$p(\alpha|\boldsymbol{d}, \boldsymbol{s}, \mathcal{I}) = \frac{p(\alpha|\mathcal{I}) \cdot p(\boldsymbol{d}|\alpha, \boldsymbol{s}, \mathcal{I})}{p(\boldsymbol{d}|\boldsymbol{s}, \mathcal{I})} \tag{22.5}$$

requires the marginal likelihood $p(\boldsymbol{d}|\alpha, \boldsymbol{s}, \mathcal{I})$, which we obtain from equation (22.4) by marginalizing over μ. Since μ is a location parameter we choose, if no prior information on μ is available, the uniform prior

$$p(\mu) = \frac{1}{W}\, \mathbf{1}(\mu_{\min} \leq \mu \leq \mu_{\max}), \qquad W = \mu_{\max} - \mu_{\min}. \tag{22.6}$$

The marginal likelihood $p(\boldsymbol{d}|\alpha, \boldsymbol{s}, \mathcal{I})$ becomes

$$p(\boldsymbol{d}|\alpha, \boldsymbol{s}, \mathcal{I}) = \frac{1}{W}\int_{\mu_{\min}}^{\mu_{\max}} d\mu\; p(\boldsymbol{d}|\mu, \alpha, \boldsymbol{s}, \mathcal{I}). \tag{22.7}$$

The likelihood under the integral is more or less strongly peaked as a function of μ. The amount of localization depends of course on the number and quality of the data. Making this assumption, the integral in equation (22.7) will change insignificantly if we extend the limits of integration to $[-\infty, \infty]$. The resulting integral is standard Gaussian and yields, as far as its α-dependence is concerned,

$$p(\boldsymbol{d}|\alpha, \boldsymbol{s}, \mathcal{I}) = C\, \alpha^{-N+1} e^{-\frac{N\phi^2}{2\alpha^2}} \tag{22.8}$$

with

$$\phi^2 = \frac{1}{N}\sum_i \frac{(d_i - \overline{d})^2}{s_i^2}. \tag{22.9}$$

The next probability we need for equation (22.5) is the prior $p(\alpha|\mathcal{I})$. Since α is a parameter which sets the scale for the data dispersion $\overline{(\Delta d)^2}$, we choose Jeffreys' scale prior as the appropriate uninformative distribution. Jeffreys' scale prior is of course the most ignorant choice. It will hardly ever be sensible to assume a data dispersion which may extend from zero to infinity with equal weight on a logarithmic scale. However, we stay with Jeffreys' scale prior because the further integrals remain simple and the result shows the main effect of introducing uncertainty in the errors. More generally, the impact of the prior is comparable to one additional data point; that is, if there are enough data points, the precise form of the prior in irrelevant. The standard deviation of $\langle \mu \rangle$ will become dependent on the data scatter. We discussed the choice of informative priors in Section 21.1.2 [p. 339]. Here we continue with Jeffreys' scale prior, which yields the posterior probability of equation (22.5) for α:

$$p(\alpha|\boldsymbol{d}, \boldsymbol{s}, \mathcal{I}) = C\, \alpha^{-N}\, e^{-\frac{N\phi^2}{2\alpha^2}}. \tag{22.10}$$

From the posterior we want to derive estimates for $\langle \alpha \rangle$ and $\langle \Delta\alpha^2 \rangle$. The computation of the moments requires integrals which have been introduced in equation (7.27) [p. 108]:

$$\int_0^\infty \alpha^{-(N-\nu)}\, e^{-\frac{[N\phi^2/2]}{\alpha^2}}\, d\alpha = \frac{1}{2}\left[\frac{N\phi^2}{2}\right]^{-\frac{N-\nu-1}{2}} \Gamma\left(\frac{N-\nu-1}{2}\right). \tag{22.11}$$

The moments are therefore

$$\langle \alpha \rangle = \phi\, \frac{\Gamma\left(\frac{N-2}{2}\right)}{\Gamma\left(\frac{N-1}{2}\right)}\sqrt{\frac{N}{2}} \qquad \xrightarrow[N\to\infty]{} \phi, \tag{22.12a}$$

$$\langle (\Delta\alpha)^2 \rangle = \phi^2 \left(\frac{2}{N-3} - \left[\frac{\Gamma\left(\frac{N-2}{2}\right)}{\Gamma\left(\frac{N-1}{2}\right)}\right]^2\right)\frac{N}{2} \qquad \xrightarrow[N\to\infty]{} \frac{\phi^2}{2N}, \tag{22.12b}$$

$$\phi = \frac{1}{N}\sum_i \frac{(d_i - \overline{d})^2}{s_i^2} \qquad \xrightarrow[N\to\infty]{} \left(\frac{1}{N}\sum_i \frac{\sigma_i^2}{s_i^2}\right)^{1/2}. \tag{22.12c}$$

The result is very reasonable. According to equation (22.1), the parameter α is the ratio of the true noise level to the measured one. In the limit $N \to \infty$ the mean of α tends to the root mean square ratio, given by equation (22.12c). At the same time, the variance of α decreases proportionally to $1/N$. The posterior for α is not a Gaussian. There is no need in the present case for the Gaussian approximation, since the exact result can be derived easily. However, in many real-world applications the exact evaluation is not possible and one has to resort to Gaussian-type approximations. Therefore, we will scrutinize this ubiquitous approximation in the present example. First we determine the mode

$$\frac{d}{d\alpha}\ln p(\alpha|\boldsymbol{d}, \boldsymbol{s}, \mathcal{I}) = \left(\frac{N}{\alpha^3}\phi^2 - \frac{N}{\alpha}\right)\Bigg|_{\alpha=\alpha_{\max}} = 0, \tag{22.13}$$

with solution $\alpha_{\max} = \phi$. Then the variance is

Table 22.2 Comparison of exact parameter estimates with the Gaussian approximation

	Gauss		Exact	
N	α_G/Φ	$\Delta\alpha_G^2/\Phi^2$	$\langle\alpha\rangle/\Phi$	$\langle\Delta\alpha^2\rangle/\Phi^2$
4	1	0.125000	1.59576	1.45352
8	1	0.062500	1.20360	0.15133
16	1	0.031250	1.08829	0.04639
32	1	0.015625	1.04143	0.01885

$$\begin{aligned}\langle(\Delta\alpha_G)^2\rangle &= -\left(\frac{d^2 \ln p(\alpha|\boldsymbol{d}, s, \mathcal{I})}{d\alpha^2}\right)^{-1}\Bigg|_{\alpha=\alpha_{\max}} \\ &= -\left(-\frac{3N\phi^2}{\alpha^4} + \frac{N}{\alpha^2}\right)^{-1}\Bigg|_{\alpha=\alpha_{\max}} = \frac{\phi^2}{2N}. \end{aligned} \tag{22.14}$$

In Table 22.2 we present numerical values for the exact moments in units of ϕ and ϕ^2, respectively, compared with the Gaussian approximation as a function of the number of data N. For $N = 32$ we find good agreement between the exact solution and the Gaussian approximation. This becomes rapidly worse as we reduce the number of data. The reason is very simple. The posterior distribution in α becomes progressively more strongly asymmetric as the number of data is reduced. It is not surprising that the Gaussian approximation to an asymmetric function does not perform too well. In particular, for an asymmetric function the mode and the mean are usually numerically different while they coincide in the Gaussian approximation.

The second aspect of our problem is the estimation of $\langle\mu\rangle$ and its variance $\langle\Delta\mu^2\rangle$. The posterior distribution for μ is given by Bayes' theorem:

$$p(\mu|\boldsymbol{d}, s, \mathcal{I}) = \frac{p(\mu|\mathcal{I}) \cdot p(\boldsymbol{d}|\mu, s, \mathcal{I})}{p(\boldsymbol{d}|s, \mathcal{I})}, \tag{22.15}$$

in terms of the marginal likelihood $p(\boldsymbol{d}|\mu, s, \mathcal{I})$ and the prior $p(\mu|\mathcal{I})$ for which we shall again use a uniform density in equation (22.6). Before evaluating the moments of μ, we will determine the marginal likelihood $p(\boldsymbol{d}|\mu, s, \mathcal{I})$. The α as marginalization, along with Jeffreys' scale prior, yields a Student's t-distribution:

Marginal likelihood with global error correction ($\sigma_i = \alpha s_i$) and Jeffreys' scale prior for α

$$p(\boldsymbol{d}|\mu, s, \mathcal{I}) = C\left(1 + \frac{(\mu - \overline{d})^2}{(\Delta d)^2}\right)^{-N/2}, \tag{22.16}$$

with a constant C that is independent of μ. The moments of μ require integrals of the form

$$\int_{-\infty}^{\infty} \mu^{\nu}\left(1+\frac{\mu^2}{\overline{(\Delta d)^2}}\right)^{-N/2} d\mu = \left(\overline{(\Delta d)^2}\right)^{\nu+1} \frac{\Gamma\left(\frac{\nu+1}{2}\right)\Gamma\left(\frac{N-\nu-1}{2}\right)}{\Gamma\left(\frac{N}{2}\right)} \delta_{\nu,\text{even}},$$

which results in

$$\langle\mu\rangle = \overline{d}, \quad \langle(\Delta\mu)^2\rangle = \frac{\overline{(\Delta D)^2}}{N-3}. \tag{22.17}$$

Recall the definitions of $\overline{D^2}$ and $\overline{d}$ from equation (21.4) [p. 335] to realize that the variance $\langle(\Delta\mu)^2\rangle$ depends now on the errors of the individual measurements as well as on the scatter of the data, since the error scale is determined from the data. There are two instructive special cases, namely the experimental errors are much smaller than the data scatter or the opposite limit, the scatter in the data is much smaller than expected from the experimental errors. We consider the instructive limiting cases: (a) all data have the same experimental error $s_i = s_0 = \rho$ and (b) all data values are the same $d_i = d_0$, i.e. the scatter in the data is much smaller than pretended by the experimental errors s_i. In case (a) $\overline{d}$ corresponds to the unweighted sample mean and $\overline{(\Delta d)^2}$ to the unweighted sample variance. Both expressions are independent of the assigned experimental error s_0. By contrast, the standard approach, discussed in equation (21.7) [p. 335], yields $\langle(\Delta\mu)^2\rangle = \rho^2/N$, which is independent of the scatter in the data. In case (b) the present analysis yields $\overline{d} = d_0$ and $\langle(\Delta\mu)^2\rangle = 0$, while the standard approach still yields the same finite value for $\langle(\Delta\mu)^2\rangle$, unimpressed by the fact that all data values are the same. If all data values are the same, the underlying intrinsic value of μ has to be identical to the data value. It is entirely impossible to always measure the same value if the likelihood has a finite width. The compelling conclusion is that the experimental errors are wrong. Obviously, there is more information in the data, which sampling mean and variance cannot reveal.

We shall now discuss an application of the developed formulae. The system of physical constants and conversion factors arises from a critical joint evaluation of a series of measurements like the quantum Hall effect (QHE), calibration of the ampère (A), gyromagnetic ratio of the proton γ_p and many more. The algorithms for deducing a system of constants on the basis of these data are essentially maximum likelihood. This means that the errors quoted for the individual measurements determine the weight of the respective quantity in the overall evaluation. Out of the many possibilities we choose the data on the QHE and the calibration factor for the ampère (A) to investigate the consistency of data and quoted errors. Table 22.3 is reproduced from [33]. As a result of our analysis we find that the error estimate for the QHE data is too optimistic and needs correction by a factor of $\langle\alpha\rangle = 1.2$. The standard deviation on this is fairly large, namely $\sqrt{\langle(\Delta\alpha)^2\rangle} = 0.51$. This is not unexpected for Table 22.3 since we are dealing with a very small number of data. The ampère data require a correction of 0.77 ± 0.33. This means that in a least-squares

Table 22.3 Measurements of the quantum Hall effect (QHE) and a quantity related to the definition of the ampère (A), after [33]

QHE			A		
Value	Error		Value	Error	
25812.8469	0.0048		0.9999974	0.0000084	
25812.8495	0.0031		0.9999982	0.0000059	
25812.8432	0.0040		0.9999986	0.0000061	
25812.8427	0.0034		1.0000027	0.0000097	
25812.8397	0.0057		1.0000032	0.0000079	
25812.8502	0.0039		1.0000062	0.0000041	
mean	error		mean	error	
25812.8461			1.0000021		
	0.0016	[(21.7)]		0.0000025	[(21.7)]
	0.0021	[(22.17)]		0.0000021	[(22.17)]

adjustment the weight of the QHE data is too large by a factor of 1.44 while the A data are underweighted by a factor of 0.57.

In both applications, the case $\alpha = 1$ is still within the error bars and one is prompted to compute the odds ratio for the hypothesis H that the experimental error bars are correct, i.e. $\alpha = 1$. The corresponding odds ratio is

$$o = \frac{P(H|\boldsymbol{d}, s, \mathcal{I})}{P(\overline{H}|\boldsymbol{d}, s, \mathcal{I})} = \frac{p(\boldsymbol{d}|H, s, \mathcal{I})}{p(\boldsymbol{d}|\overline{H}, s, \mathcal{I})} \frac{P(H|s, \mathcal{I})}{P(\overline{H}|s, \mathcal{I})}. \tag{22.18}$$

Without any prejudice we assume that the prior odds ratio is 1. The Bayes factor depends upon the marginal likelihood in the case H is true or false. In the first case

$$\begin{aligned} p(\boldsymbol{d}|H, s, \mathcal{I}) &= p(\boldsymbol{d}|\alpha = 1, s, \mathcal{I}) \\ &= \int p(\boldsymbol{d}|\alpha = 1, s, \mathcal{I})\; p(\mu|\alpha = 1, s, \mathcal{I}) d\mu. \end{aligned}$$

In order to recycle the result for the case $\overline{H}$, we will keep α variable and plug in $\alpha = 1$ at the end. As before, we assume a flat prior for μ within an interval $\mu \in (\mu_0, \mu_1)$ and obtain

$$p(\boldsymbol{d}|H, s, \mathcal{I}) = \frac{\alpha^{-N}}{V_\mu V_d} \int_{\mu_0}^{\mu_1} e^{-\frac{N(\mu-\overline{d})^2}{2\alpha^2 \varrho^2}}\, e^{-\frac{N\overline{(\Delta d)^2}}{2\alpha^2 \varrho^2}}\, d\mu, \tag{22.19}$$

where $V_\mu = (\mu_1 - \mu_0)$ and $V_d = \prod_i \sqrt{2\pi s_i^2}$ is the normalization of the likelihood, or rather the respective effective integration volume. Both factors cancel in the odds ratio and need no further consideration. We again assume that the contributions to the integral outside the limits of integration are negligible and that we can extend the integral over the entire real axis, yielding

$$p(\boldsymbol{d}|H, s, \mathcal{I}) = \alpha^{-N+1} \frac{\sqrt{2\pi\rho^2/N}}{V_\mu V_d} e^{-\frac{N\overline{(\Delta d)^2}}{2\alpha^2\varrho^2}}\bigg|_{\alpha=1} \tag{22.20a}$$

$$= \frac{\sqrt{2\pi\rho^2/N}}{V_\mu V_d} e^{-\frac{N\overline{(\Delta d)^2}}{2\varrho^2}} \tag{22.20b}$$

$$= \frac{\sqrt{2\pi\rho^2/N}}{V_\mu V_d} e^{-\frac{N\Phi^2}{2}}. \tag{22.20c}$$

Now we turn to the complementary hypothesis $\overline{H}$, which differs in that α is not fixed to $\alpha = 1$:

$$p(\boldsymbol{d}|\overline{H}, s, \mathcal{I}) = \int p(\boldsymbol{d}|\alpha, s, \mathcal{I})\ p(\alpha|\overline{H}, s, \mathcal{I})d\alpha. \tag{22.21}$$

Along with equation (22.20), we have

$$p(\boldsymbol{d}|\overline{H}, s, \mathcal{I}) = \frac{\sqrt{2\pi\rho^2/N}}{V_\mu V_d} \int \alpha^{-N+1}\ e^{-\frac{N\overline{(\Delta d)^2}}{2\alpha^2\varrho^2}}\ p(\alpha|\overline{H}, s, \mathcal{I})d\alpha. \tag{22.22}$$

As far as the prior for α is concerned, we have to employ a proper prior, which poses no problems in the present application as we can certainly assume that the experimental errors should not be off by more than a factor of 10, i.e. $\alpha \in (0.1, 10)$. Hence, Jeffreys' scale prior with upper (lower) limit α_0 (α_1) becomes

$$p_J(\alpha) = \frac{1}{V_\alpha}\ \boldsymbol{1}(\alpha_0 \le \alpha \le \alpha_1)\ \frac{1}{\alpha},$$

$$V_\alpha = \frac{1}{\ln(\alpha_1/\alpha_0)},$$

with normalization constant V_α. Another choice of prior in the case of ignorance, apart from boundaries, is a flat prior

$$p_f(\alpha) = \frac{1}{V_\alpha}\ \boldsymbol{1}(\alpha_0 \le \alpha \le \alpha_1),$$

$$V_\alpha = (\alpha_1 - \alpha_0).$$

Based on this prior, the marginal likelihood is

$$p(\boldsymbol{d}|\overline{H}, s, \mathcal{I}) = \frac{1}{V_\alpha} \frac{\sqrt{2\pi\rho^2/N}}{V_\mu V_d} \int_{\alpha_0}^{\alpha_1} \alpha^{-N+\nu}\ e^{-\frac{N\overline{(\Delta d)^2}}{2\alpha^2\varrho^2}}\ d\alpha$$

$$\overset{(7.27)}{=} \frac{1}{V_\alpha} \frac{\sqrt{2\pi\rho^2/N}}{V_\mu V_d}\ \Gamma\left(\frac{N-1-\nu}{2}\right) \left[\frac{N\phi^2}{2}\right]^{-\frac{N-1-\nu}{2}}.$$

Here the value of ν depends on the choice of the prior for α. We have $\nu = 0$ for Jeffreys' scale prior and $\nu = 1$ for the flat prior. Apparently, the prior volume also depends on this prior. Finally, the odds ratio becomes

$$o = \frac{\left(\frac{N-1-\nu}{2}\right)}{V_\alpha} \frac{1}{\Gamma\left(\frac{N+1-\nu}{2}\right)} \left(\frac{N\Phi^2}{2}\right)^{\frac{N+1-\nu}{2}-1} e^{-\frac{N\Phi^2}{2}} \tag{22.23}$$

$$= \frac{\left(\frac{N-1-\nu}{2}\right)}{V_\alpha} p_{\chi^2}\left(N\Phi^2 | n = N+1-\nu\right). \tag{22.24}$$

Interestingly but not surprisingly, the odds ratio is proportional to the χ^2 density with $n = N+1-\nu$ degrees of freedom.

22.1.2 Errors with individual uncertainty

The assumption on the errors of the previous section is naïve and rather artificial. In particular, if the data stem from different sources, it is not sensible to assume a uniform uncertainty for all experimental errors. Under these or similar circumstances, it is much more reasonable to introduce individual error densities $p(\sigma_i|s_i, \mathcal{I})$. In this case, however, the assumption of a completely uncertain individual scaling constant α would result in Jeffreys' scale prior for σ_i, which would ignore the experimental errors s_i altogether and since each data point is treated separately, the marginal likelihood $p(d|\boldsymbol{\sigma}, \mathcal{I})$ would be improper and useless.

Approach I

It is therefore much more realistic to assume that the distribution $p(\boldsymbol{s}|\boldsymbol{\sigma}, \mathcal{I})$ of the experimental errors s_i is at least unbiased, i.e. $\langle s_k \rangle = \sigma_k$. The uncertainty about the precision of the error estimates $\boldsymbol{s}$ can then be expressed by a parameter describing the width of the posterior distribution $p(\boldsymbol{\sigma}|\boldsymbol{s}, \mathcal{I})$. As long as there is no additional knowledge about the width, we will use a prior probability density with infinite variance to encode our ignorance in this respect. It is convenient to invoke a two-parameter 'conjugate prior' to accommodate this knowledge, namely

INDIVIDUAL ERROR CORRECTION I

$$p(\sigma_j|s_j, \alpha, \beta, \mathcal{I}) = 2\frac{\beta^\alpha}{\Gamma(\alpha)\, s_j} \cdot \left(\frac{s_j}{\sigma_j}\right)^{2\alpha+1} \exp\left\{-\beta\frac{s_j^2}{\sigma_j^2}\right\}, \tag{22.25}$$

which is related to a Γ density via the substitution $s^2/\sigma^2 \to t$. Note that this distribution contains Jeffreys' scale prior as the limiting case for $\alpha, \beta \to 0$. The expectation value and variance of σ_j are given by

$$\frac{\langle\sigma_j\rangle}{s_j} = \sqrt{\beta}\cdot\frac{\Gamma(\alpha-1/2)}{\Gamma(\alpha)}, \qquad \frac{\langle\Delta\sigma_j^2\rangle}{s_j^2} = \frac{\beta}{\alpha-1} - \frac{\langle\sigma_j\rangle^2}{s_j^2}. \tag{22.26}$$

The likelihood obtained from equation (22.2) [p. 365] then reads

MARGINAL LIKELIHOOD
WITH INDIVIDUAL ERROR CORRECTION I

$$p(\boldsymbol{d}|\mu, \boldsymbol{s}, \alpha, \beta, \mathcal{I}) = \frac{1}{Z}\prod_{i=1}^{N}\left\{1 + \frac{(d_i-\mu)^2}{2\beta s_i^2}\right\}^{-\left(\alpha+\frac{1}{2}\right)}. \tag{22.27}$$

The result is a product of Student's t-densities, which is quite reasonable, since we marginalized individually the variances of Gaussian densities. Based on the choice of a flat prior for μ, the posterior PDF for μ reads

$$p(\mu|\boldsymbol{d}, \boldsymbol{s}, \alpha, \beta, \mathcal{I}) = \frac{1}{Z'}\prod_{i=1}^{N}\left\{1 + \frac{(d_i-\mu)^2}{2\beta s_i^2}\right\}^{-\left(\alpha+\frac{1}{2}\right)}, \tag{22.28}$$

with the normalization constant Z' that is yet to be determined. It is important to note that contrary to the original likelihood in equation (22.2) [p. 365], the posterior probability equation (22.28) is in general multimodal in the variable μ, which means that already a plot of the (marginal) likelihood as a function of μ would reveal whether the scatter of the data $\{d_i\}$ is compatible with their quoted errors or not. Data points which are consistent with their corresponding error s_i will add up to a major peak, while inconsistent data points will lead to minor secondary structures.

Next we compute the posterior moments of σ_k:

$$\begin{aligned}\langle\sigma_k^\nu\rangle &= \int_0^\infty \sigma_k^\nu\ p(\boldsymbol{\sigma}|\boldsymbol{d}, \boldsymbol{s}, \mathcal{I})\ d\sigma^N \\ &= \int d\mu\ p(\mu|\boldsymbol{d}, \boldsymbol{\sigma}, \mathcal{I})\ \underbrace{\int_0^\infty \sigma_k^\nu\ p(\boldsymbol{\sigma}|\boldsymbol{d}, \boldsymbol{s}, \mu, \mathcal{I})\ d\sigma^N}_{=:\langle\sigma_k^\nu\rangle_\mu}.\end{aligned} \tag{22.29}$$

In order to keep the notation as transparent as possible, we hide the propositions concerning the values of the hyperparameters α and β in the background information $\mathcal{I}$. The posterior probability for $\boldsymbol{\sigma}$ is again obtained from Bayes' theorem:

$$p(\boldsymbol{\sigma}|\boldsymbol{d}, \boldsymbol{s}, \mu, \mathcal{I}) \propto p(\boldsymbol{\sigma}|\boldsymbol{s}, \mathcal{I})\cdot p(\boldsymbol{d}|\boldsymbol{\sigma}, \mu, \mathcal{I}).$$

Both the likelihood $p(\boldsymbol{d}|\boldsymbol{\sigma}, \mu, \mathcal{I})$ of equation (22.27) and the prior $p(\boldsymbol{\sigma}|\boldsymbol{s}, \mathcal{I})$ of equation (22.25) factorize:

$$p(\boldsymbol{\sigma}|\boldsymbol{d}, \boldsymbol{s}, \mu, \mathcal{I}) = \prod_{i=1}^{N} p(\sigma_i|d_i, s_i, \mu, \mathcal{I}), \tag{22.30}$$

leading to

$$\langle \sigma_k^\nu \rangle_\mu = \int_0^\infty d\sigma_k\, \sigma_k^\nu\, p(\sigma_k|d_k, s_k, \mu, \mathcal{I}). \tag{22.31}$$

According to equations (22.30), (22.25) and (22.2) [p. 365], the posterior probability $p(\sigma_k|d_k, s_k, \mu, \mathcal{I})$ is related to a Γ-density and the moments $\langle \sigma_k^\nu \rangle$ can be computed easily:

$$\langle \sigma_k^\nu \rangle_\mu = \left(s_k \sqrt{\beta}\right)^\nu \frac{\Gamma\left(\alpha - \frac{\nu-1}{2}\right)}{\Gamma\left(\alpha + \frac{1}{2}\right)} \left\{1 + \frac{(d_k - \mu)^2}{2\beta s_k^2}\right\}^{\frac{\nu}{2}}. \tag{22.32}$$

Combining equations (22.32) and (22.31) yields, for the posterior mean,

$$\langle \sigma_k^\nu \rangle = \left(s_k \sqrt{\beta}\right)^\nu \frac{\Gamma\left(\alpha - \frac{\nu-1}{2}\right)}{\Gamma\left(\alpha + \frac{1}{2}\right)} \cdot \int d\mu\, p(\mu|\boldsymbol{d}, \boldsymbol{\sigma}, \mathcal{I}) \left\{1 + \frac{(d_k - \mu)^2}{2\beta s_k^2}\right\}^{\frac{\nu}{2}}, \tag{22.33}$$

with $p(\mu|\boldsymbol{d}, \boldsymbol{\sigma}, \mathcal{I})$ being defined in equation (22.28). We can introduce an interesting approximation if μ is well determined by the data, that is, if $p(\mu|\boldsymbol{d}, \boldsymbol{s}, \alpha, \beta)$ is a function with a narrow peak around $\mu = \langle \mu \rangle$. The expression in curly brackets in equation (22.33) is comparatively slowly varying. Hence, we can replace μ in this expression by $\langle \mu \rangle$. The remaining integral is one and the posterior moments are approximately

$$\langle \sigma_k^\nu \rangle \approx (s_k \sqrt{\beta})^\nu \frac{\Gamma\left(\alpha - \frac{\nu-1}{2}\right)}{\Gamma\left(\alpha + \frac{1}{2}\right)} \left\{1 + \frac{(d_k - \langle \mu \rangle)^2}{2\beta s_k^2}\right\}^{\frac{\nu}{2}}. \tag{22.34}$$

At first glance it seems as if the data points are analysed separately. This is not entirely true, though, since the mean $\langle \mu \rangle$ depends on all data points. The posterior mean and variance are now

$$\langle \sigma_k \rangle \approx s_k \sqrt{\beta}\, \frac{\Gamma(\alpha)}{\Gamma(\alpha + \frac{1}{2})} \left\{1 + \frac{(d_k - \langle \mu \rangle)^2}{2\beta s_k^2}\right\}^{\frac{1}{2}},$$

$$\langle (\Delta\sigma_k)^2 \rangle \approx s_k^2\, \beta \left\{1 + \frac{(d - \langle \mu \rangle)^2}{2\beta s_k^2}\right\} \left\{\frac{\Gamma(\alpha - 1/2)}{\Gamma(\alpha + 1/2)} - \left(\frac{\Gamma(\alpha)}{\Gamma(\alpha + 1/2)}\right)^2\right\}.$$

In order that this approach may be applied to a particular data set, a choice of parameters α, β must be made. One equation arises from the previously required consistency argument $\langle \sigma_i \rangle = s_i$. The second equation is in general much harder to formulate, since it characterizes the width of the distribution in equation (22.25). Let us assume that we are dealing with a data set which suggests a large scatter. A coding for this would be an infinite variance, i.e. $\alpha = 1$, which in turn specifies immediately $\beta = 1/\pi$. We summarize the results for this case:

POSTERIOR VARIANCES WITH INDIVIDUAL ERROR CORRECTION I

$$\langle \sigma_k \rangle \approx s_k \, \frac{2}{\pi} \left\{ 1 + \frac{(d_k - \langle \mu \rangle)^2}{2 s_k^2/\pi} \right\}^{1/2}, \tag{22.35}$$

$$\langle (\Delta\sigma_k)^2 \rangle \approx \langle \sigma_k \rangle^2 \left(\frac{\pi}{2} - 1 \right). \tag{22.36}$$

For data points close to $\langle \mu \rangle$, that is $(d_k - \langle \mu \rangle)^2 \ll s_k^2$, we have

$$\langle \sigma_k \rangle \approx s_k \, \frac{2}{\pi}, \tag{22.37}$$

which deviates merely by a factor $2/\pi$ from the experimental value s_k. In the opposite limit, for data points far from the mainstream such that $(d_k - \langle \mu \rangle)^2 \gg s_k^2$, we have

$$\langle \sigma_k \rangle \approx \sqrt{\frac{2}{\pi}} \, \big| d_k - \langle \mu \rangle \big|. \tag{22.38}$$

In this case the error estimate is independent of the experimentally assigned error and depends only on the difference $\big| d_k - \langle \mu \rangle \big|$. From equation (22.36) we see that the posterior uncertainty in σ_k is reduced from an infinite value for the prior to a finite value, comparable with the error estimate itself. Finally, we want to present the posterior PDF for μ for this set of parameters (α, β). From equation (22.28) we obtain

POSTERIOR PDF FOR μ WITH INDIVIDUAL ERROR CORRECTION I

$$p(\mu | \boldsymbol{d}, \boldsymbol{s}, \alpha, \beta, \mathcal{I}) = \frac{1}{Z'} \prod_{i=1}^{N} \left\{ 1 + \frac{(d_i - \mu)^2}{2 s_i^2/\pi} \right\}^{-\frac{3}{2}}. \tag{22.39}$$

Analytic expressions for the mean and variance are not possible and we have to rely on a numerical evaluation. As mentioned earlier, the PDF will be multimodal if there are discrepancies in the data (outlier) or in the quoted variances (s_i^2). In the case of a pronounced multimodality the approximation in equation (22.34) is not justified. Otherwise, for sufficiently many data points the assumptions are justified, which the approximation rests upon.

Approach II

So far we have replaced the assumption of exactly known measurement uncertainties by a distribution of the unknown true uncertainties, which reproduces the experimentally determined uncertainties with minimum precision, that is with infinite variance. This is still a fairly strong belief in the correctness of the experimental value s_i. An even more uninformative prior that only vaguely relies on s_i is the following inverse Gamma distribution, defined in equation (7.18) [p. 105]:

INDIVIDUAL ERROR CORRECTION II

$$p(\sigma_i|s_i) = s_i\, \sigma_i^{-2} \exp\left\{-\frac{s_i}{\sigma_i}\right\} = p_{i\Gamma}(\sigma_i|\alpha = 1, \beta = s_i). \tag{22.40}$$

Different from Jeffreys' scale prior it is properly normalized, but all other moments are infinite as well. In this respect it is a highly uninformative prior. Nonetheless, it is constructed such that 45% of its probability mass is localized in the interval $[0.24 s_i, 1.31 s_i]$, corresponding to the points at which the prior has decreased to half maximum.

Let us first analyse the posterior information about $\boldsymbol{\sigma}$. We again use the traditional Gaussian for $p\,(d_i|\mu, \sigma_i)$, given in equation (22.2) [p. 365]. In the frame of the 'evidence approximation', that is replacing μ by its MAP estimate μ^*, the posterior PDF for the error variance $\boldsymbol{\sigma}$ is

$$p(\boldsymbol{\sigma}|\boldsymbol{d}, \boldsymbol{s}, \mu^*, \mathcal{I}) \propto p(\boldsymbol{d}|\boldsymbol{\sigma}, \not{\boldsymbol{s}}, \mu^*, \mathcal{I}) p(\boldsymbol{\sigma}|\boldsymbol{s}, \not{\mu^*}, \mathcal{I}) \propto \prod_i \sigma_i^{-3} \exp\left\{-\frac{x_i^2 s_i^2}{\sigma_i^2} - \frac{s_i}{\sigma_i}\right\},$$

with $x_i = (d_i - \mu^*)\,/s_i\sqrt{2}$. The present prior information results in

$$\langle\sigma_i\rangle = \begin{cases} s_i & \text{for } \frac{|d_i - \mu^*|}{s_i} \ll 1 \\ \sqrt{\frac{\pi}{2}}\,|d_i - \mu^*| & \text{for } \frac{|d_i - \mu|}{s_i} \gg 1 \end{cases}.$$

This is a very pleasing result. It resembles that found in the previous section. Contrary to the previous case, however, the posterior variance of the individual σ_i is infinite. This is more reasonable than before, since a single data point cannot contain information about the variance, nor does the point estimate s_i. In view of the bad data information, any finite value for $\mathrm{var}(\sigma_i)$ would come from the prior, but we are assuming no such prior knowledge.

Next we turn to the marginal likelihood, which follows from the marginalization over σ_i:

$$p\,(\boldsymbol{d}|\mu, \boldsymbol{s}) = \prod_{i=1}^{N} \int d\sigma_i\; p(\sigma_i|s_i, \mathcal{I})\; p\,(d_i|\mu, \sigma_i)\,. \tag{22.41}$$

Utilizing the substitution $\omega_i := s_i/\sigma_i$, we find

$$p(\boldsymbol{d}|\mu, \boldsymbol{s}) = \prod_{i=1}^{N} \frac{1}{s_i\sqrt{2\pi}} \int_0^\infty d\omega_i \; \omega_i \exp\left\{-\omega_i - \omega_i^2 x_i^2\right\}. \tag{22.42}$$

The ω-integration can be performed analytically and we end up with

Likelihood
with individual error correction II

$$p(\boldsymbol{d}|\mu, \boldsymbol{s}) = \prod_i \frac{x_i^{-2}}{2s_i\sqrt{2\pi}} \left(1 - \frac{\sqrt{\pi}}{2|x_i|} \exp\left\{\frac{1}{4x_i^2}\right\} \operatorname{erfc}\left\{\frac{1}{2|x_i|}\right\}\right), \tag{22.43}$$

where $\operatorname{erfc}(z) = 1 - \operatorname{erf}(z)$ is the complementary error function. For large $|x|$ the leading-order terms [1] read

$$p(d_i|\mu, s_i, \mathcal{I}) \propto \frac{1}{|d_i - \mu|^2} + O(x_i^{-3}).$$

For a uniform prior on μ and for μ far away from the data cloud, the posterior PDF has the asymptotic behaviour

$$p(\mu|\boldsymbol{d}, \boldsymbol{s}, \mathcal{I}) \propto \prod_{i=1}^{N} \frac{1}{|d_i - \mu|^2} \rightarrow \mu^{-2N}.$$

For a single data point ($N = 1$) the resulting PDF is normalizable but the higher moments are infinite. In other words, we cannot infer μ from a single data point. But already two data points suffice to obtain a finite value for the posterior mean $\langle\mu\rangle$ and variance $\operatorname{var}(\mu)$.

For more details we refer to numerical integration. Numerical evaluation of equation (22.43) requires special attention for large $z = 1/2x$. In this range a useful asymptotic expansion is [1]

$$\sqrt{\pi} z e^{z^2} \operatorname{erfc}(z) \sim 1 - \frac{1}{2z^2} + \frac{1 \cdot 3}{(2z^2)^2} - \frac{1 \cdot 3 \cdot 5}{(2z^2)^3} + \cdots \tag{22.44}$$

Single factors from equations (22.2) [p. 365], (22.39) and (22.43) are shown in Figure 22.1. Both equations (22.39) and (22.43) show appreciably heavier tails than the plain Gaussian of equation (22.2) [p. 365]. It is interesting to compare the performance of the three choices in the evaluation of the Newtonian constant of gravitation G from the data set given in [143] and displayed in the upper panel of Figure 22.3 later. Figure 22.2 shows posterior distributions obtained with flat improper priors. Interesting is the close similarity of the posteriors from equations (22.39) and (22.43). Both distributions show again much heavier tails than the Gaussian and are – unlike the Gaussian – strongly asymmetric. While the

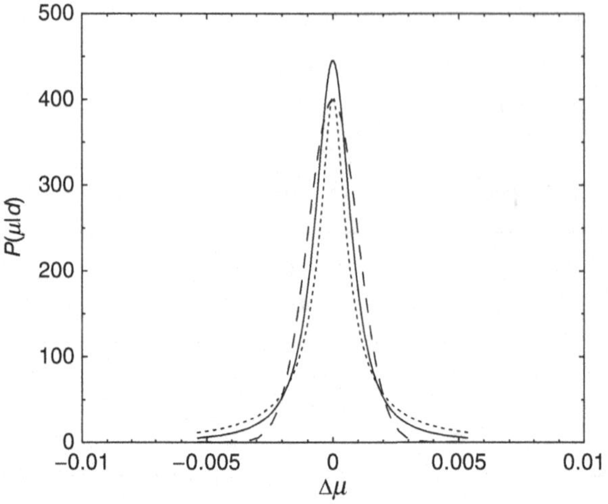

Figure 22.1 Single factors from equation (22.2) (dashed), equation (22.39) (continuous) and equation (22.43) (dotted). Note the progressively heavier tails of the distributions.

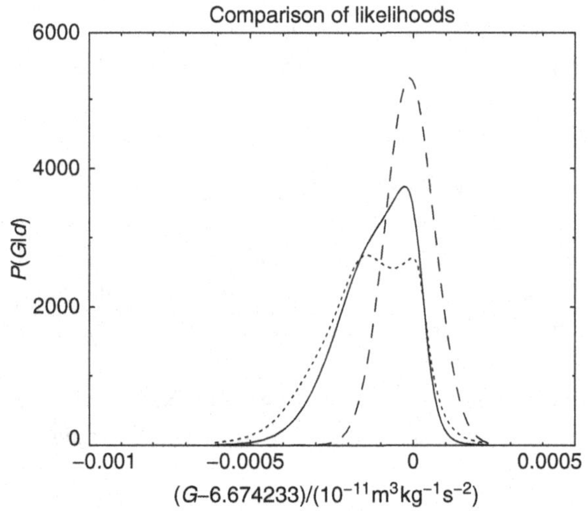

Figure 22.2 Posterior distribution according to equation (22.2) (dashed), equation (22.39) (continuous) and equation (22.43) (dotted). Unlike the Gaussian, the latter two distributions are strongly asymmetric.

mean and standard deviation are, in many cases of symmetric Gaussian-like functions, satisfactory summaries of the full posterior, they give a wrong impression for asymmetric posterior distributions. A better choice in this situation is the median instead of the mean. The standard deviation is replaced by the right and left arguments of tail areas of 15.9%. This definition coincides with the 1σ uncertainty when applied to a plain Gaussian. For asymmetric posteriors the uncertainty is then specified by the plus and minus limits of

Table 22.4 Newtonian constant of gravitation in units of $10^{11}\ G/\mathrm{m}^3\ \mathrm{kg}^{-1}\ \mathrm{s}^{-2}$

	Least squares	Equation (22.39)	Equation (22.43)
Median	6.67422	6.67411	6.67414
Uncertainty	+0.00001	+0.00013	+0.00010
	−0.00001	−0.00015	−0.00013

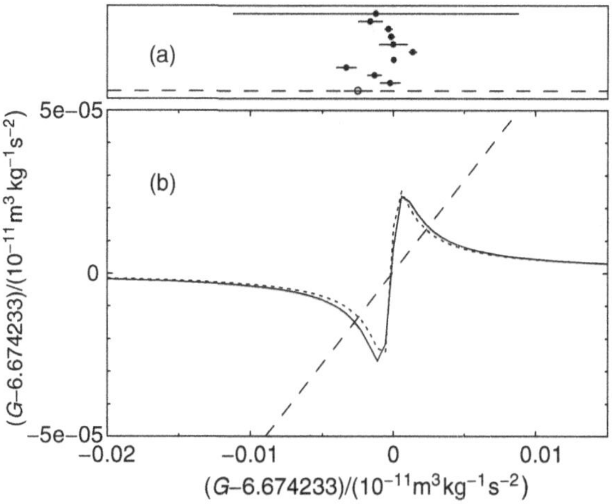

Figure 22.3 G data set (full dots with error bars in (a)) plus an additional fictitious data point (circle with error bar in (a)). In panel (b) the dependence of the estimate for G (ordinate) on the position of the additional fictitious data point (abscissa) is plotted. Compared are the estimates of the bare Gaussian in equation (22.2) (dashed (b)), equation (22.39) (continuous (b)) and equation (22.43) (dotted (b)).

the confidence interval. Results on point estimates and uncertainties for the Newtonian constant of gravitation are collected in Table 22.4. Note the ridiculously low uncertainty of the least-squares estimate. It is a consequence of equation (21.7) [p. 335], which shows that the uncertainty in the least-squares evaluation does not depend on the data scatter but is given uniquely by the data uncertainties. This difference between the least-squares result and the treatment by equations (22.39) and (22.43) is therefore a measure of the inconsistency of the G data set. These two results are quite similar, the smaller uncertainty range being offered by equation (22.43).

22.1.3 Sensitivity analysis

We turn now to the question of robustness of the three approaches. The upper panel of Figure 22.3 shows the G data set as full dots with error bars plus an additional fictitious data point (open circle) with error bar. This fictitious point is now moved across the real data cloud. Evaluation of G from the real data plus the fictitious point as a function of the numerical value of the latter is shown in panel (b). While the least-squares approach shows a strong linear response to the fictitious point, the two other approaches exhibit a dispersion-like behaviour rendering the influence of the outlier less important with increasing distance from the real data. Gauss's hope that the method of least squares would render data selection or tailoring obsolete [88] has not materialized. Instead the method is very outlier sensitive, unlike equations (22.39) and (22.43) which provide excellent stability. Let us analyse what we will get on average, when we study many samples. For convenience we will focus on $\langle \sigma_k^2 \rangle$. The mean over all possible samples yields

$$E\Big(\langle \sigma_k^2 \rangle\Big) = \frac{2}{\pi} s_k^2 + \sigma_k^2. \tag{22.45}$$

If the experimentally assigned variance s_k^2 is much smaller than the true variance σ_k^2, then the estimated variance will on average be identical to the true variance σ_k^2. In the opposite extreme $\sigma_k \ll s_k$, with the true variance much smaller than the experimental one, the estimate is $\frac{2}{\pi} s_k^2$. The value is reduced from the experimental assignment but only slightly. The lesson we learn is that underestimated experimental errors can be corrected better than overestimated errors. An application of the foregoing analysis to the chemical erosion rates of carbon materials in fusion devices has been given in [52].

22.2 Combining incompatible measurements

The idea we have explored so far assumes that the errors on all data are more or less ill specified. This assumption is too rigorous for many applications in the physical sciences, and we shall relax it now. We consider the situation where experimental errors are either correctly specified or suffer from misspecification.

22.2.1 Mixture model

A mixture model that describes this situation was first proposed by Press [161] for a robust estimation of the Hubble constant. Press assumes

$$p(\boldsymbol{\sigma}|\boldsymbol{s}, \beta, \boldsymbol{\omega}, \mathcal{I}) = \prod_j \big\{\beta \; p(\sigma_j|s_j, \omega_j, \mathcal{I}) + (1-\beta)\; \delta(\sigma_j - s_j)\big\}, \tag{22.46}$$

where β is the probability that an error is ill specified and $(1-\beta)$ the probability that it is correctly specified. In the case that the error is ill specified, the probability $p(\sigma_j|s_j, \omega_j, \mathcal{I})$ can for example be $p(\sigma_j|s_j, \mathcal{I}) = \delta(\sigma_j - \omega_j s_j)$ as in equation (22.1) [p. 365]. Here the factor ω_j characterizes by what amount misspecification is assumed. We can also use the probability defined in equation (22.25) [p. 372] for $p(\sigma_j|s_j, \omega_j, \mathcal{I})$, or any other density encoding the information we have about the error distribution of the outliers. If we marginalize the likelihood in equation (22.2) [p. 365] over $\boldsymbol{\sigma}$, we obtain

$$p(\boldsymbol{d}|\mu, s, \beta, \boldsymbol{\omega}, \mathcal{I}) = \prod_{j}^{N} \left\{\beta A_j + (1-\beta)B_j\right\}, \tag{22.47}$$

where $B_j = \mathcal{N}(d_j|\mu, s_j)$ is a Gaussian density, while A_j depends on the choice of $p(\sigma_j|s_j, \omega_j, \mathcal{I})$. For the case where the assigned error is wrong by a given factor ω_k (equation (22.1) [p. 365]), $A_j = \mathcal{N}(d_j|\mu, \omega_j s_j)$ is also a Gaussian.

22.2.2 What happens if we ignore the outliers?

Before we dive into the outlier-tolerant approach, it is instructive to study what happens if we ignore the outliers and assume that all variances are correctly assigned, i.e. $\sigma_i^2 = s_i^2$. According to equations (21.3) and (21.5) [p. 335], the posterior $p(\mu|d, \mathcal{I})$ is a Gaussian, peaked at the sample mean $\overline{d}$, and the width of the peak is given by the standard error

$$\text{SE} = \rho/\sqrt{N}.$$

Both quantities are given in equation (21.4) [p. 335]:

$$\overline{d} = \sum_{j}^{N} \omega_j \, y_j, \quad \omega_j = \frac{\varrho^2}{N} \sigma_j^{-2}, \quad \varrho = \left(\frac{1}{N} \sum_{i} \sigma_i^{-2}\right)^{-\frac{1}{2}}. \tag{22.48}$$

Hence we conclude

$$\mu = \overline{d} \pm \rho/\sqrt{N}.$$

Needless to say, a new sample will yield a new sample mean. The average of $\overline{d}$ over all conceivable samples is, according to equation (21.4) [p. 335], given by

$$\langle \overline{d} \rangle = \sum_{j} \omega_j \langle d_j \rangle = \mu,$$

corroborating our previous finding that the sample mean is an unbiased estimator. The scatter of the individual values $\overline{d}$ is described by

$$\langle (\Delta \overline{d})^2 \rangle = \sum_{j} c_j^2 \langle (\Delta d_j)^2 \rangle,$$

where we have used the fact that the data points are uncorrelated. If, contrary to our assumption, the data points indeed suffer from outliers as described by the likelihood in equation (22.47), then

$$\langle (\Delta d_j)^2 \rangle = \Big(\beta \omega_j^2 + (1-\beta)\Big) s_j^2.$$

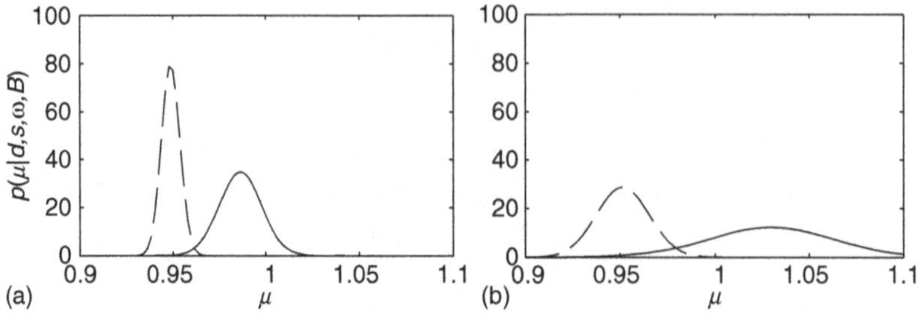

Figure 22.4 Posterior probability for μ, based on data generated by the mixture model likelihood with parameters $\beta = 0.1, \mu = 1, \omega = 10, \sigma = 0.1$ and $N_{\text{data}} = 100$ (a) or rather $N_{\text{data}} = 13$ (b). The result of the naïve approach (dashed line) is compared with that of the outlier-tolerant procedure (solid line).

If we assume for simplicity that the correction factor $\omega_j = \omega$ is independent of j, then we obtain

$$\langle(\Delta\overline{d})^2\rangle = \left(\beta\omega^2 + (1-\beta)\right) \sum_j \omega_j s_i^2.$$

The scatter of the sample means is therefore a factor $C = \sqrt{\beta\omega^2 + (1-\beta)}$ bigger than the standard error, which would be used as error bar in the naïve estimate. In Figure 22.4, results of a computer simulation are depicted in which samples of size $N_{\text{data}} = 100$ are generated according to the mixture model likelihood with $\beta = 0.1, \mu = 1, \sigma = 0.1$ and $\omega = 10$. Then the posterior PDF is computed in two different ways, firstly by ignoring the presence of outliers and using a bar Gaussian and secondly based on equation (22.47). The simulation illustrates impressively that ignoring outliers leads to misleading conclusions: the PDF is peaked with unjustified confidence at the wrong position.

22.2.3 Inference based on the mixture model

Inferences shall now be drawn from the likelihood equation (22.47). We consider in turn μ and the prior probability β that a data point is ill specified. To this end we need $p(\mu, \beta | \boldsymbol{d}, \boldsymbol{s}, \boldsymbol{\omega}, \mathcal{I})$. From Bayes' theorem and the product rule we get

$$p(\mu, \beta|\boldsymbol{d}, \boldsymbol{s}, \boldsymbol{\omega}, \mathcal{I}) \propto p(\boldsymbol{d}|\mu, \beta, \boldsymbol{s}, \boldsymbol{\omega}, \mathcal{I})\ p(\mu|\beta, \boldsymbol{s}, \boldsymbol{\omega}, \mathcal{I})\ p(\beta|\boldsymbol{s}, \boldsymbol{\omega}, \mathcal{I}). \tag{22.49}$$

The parameters β and μ are logically uncorrelated and they are, in addition, not influenced by prior knowledge about the errors $\boldsymbol{s}$ and $\boldsymbol{\omega}$. Moreover, we consider the situation where the prior knowledge encoded in $\mathcal{I}$ does not contain further information about the parameters and we have to employ uninformative uniform priors for both, resulting in

$$p(\mu, \beta|\boldsymbol{d}, \boldsymbol{s}, \boldsymbol{\omega}, \mathcal{I}) = \frac{1}{Z}\ p(\boldsymbol{d}|\mu, \beta, \boldsymbol{s}, \boldsymbol{\omega}, \mathcal{I}) = \frac{1}{Z} \prod_j \left\{\beta A_j + (1-\beta) B_j\right\}. \tag{22.50}$$

The marginal posterior probabilities $p(\mu|\boldsymbol{d}, \boldsymbol{s}, \boldsymbol{\omega}, \mathcal{I})$ and $p(\beta|\boldsymbol{d}, \boldsymbol{s}, \boldsymbol{\omega}, \mathcal{I})$, respectively, follow from integrating equation (22.50) over β and μ, respectively. Generally these integrals have to be calculated numerically. In the present case, however, an analytic computation of the β-marginalization is possible. This is based on the observation that the likelihood in equation (22.47) is an Nth-order polynomial P_N in β, which when expanding the product in equation (22.50) may be written as

$$P_N(\beta) := p(\boldsymbol{d}|\mu, \boldsymbol{s}, \beta, \boldsymbol{\omega}, \mathcal{I}) = \sum_{\nu=0}^{N} D_\nu^{(N)}\ \beta^\nu (1-\beta)^{N-\nu}. \tag{22.51}$$

The β-dependent terms are proportional to the β-distribution equation (7.12) [p. 100] and we can, therefore, alternatively write

$$P_N(\beta) = \sum_{\nu=0}^{N} C_\nu^{(N)}\ p_\beta(\beta|\nu+1, N-\nu+1). \tag{22.52}$$

The coefficients $C_j^{(N)}$ are derived from two alternative equivalent representations of the polynomial $P_N(\beta)$:

$$\sum_{j=0}^{N} C_j^{(N)} \binom{N}{j} \beta^j (1-\beta)^{N-j} = P_N(\beta) = P_{N-1}(\beta)\big(\beta A_N + (1-\beta)B_N\big). \tag{22.53}$$

Comparing coefficients of equal powers of β on both sides of equation (22.53) yields the recurrence relation

$$C_\nu^{(n)} = \frac{\nu}{n+1} A_N C_{\nu-1}^{(n-1)} + \frac{n-\nu}{n+1} B_N C_\nu{}^{(n-1)}, \qquad \text{for} \quad \nu = 1, \ldots, n \tag{22.54}$$

with $n = 2, \ldots, N$ and with the definitions $C_{-1}^{(n-1)} = 0$ and $C_n^{(n-1)} = 0$. Initial values are obtained from

$$P_1(\beta) = \beta\ A_1 + (1-\beta)\ B_1 = C_0^{(1)} p_\beta(\beta|1, 2) + C_1^{(1)} p_\beta(\beta|2, 1). \tag{22.55}$$

That is, along with the normalization factor entering the beta PDF ($B(1, 2) = B(2, 1) = 2$), the initial values are $C_0^{(1)} = B_1/2$ and $C_1^{(1)} = A_1/2$.

On account of the representation in beta PDFs, given in equation (22.51), the posterior PDF for μ and β can be expressed as

$$p(\mu, \beta|\boldsymbol{d}, \boldsymbol{s}, \boldsymbol{\omega}, \mathcal{I}) = \frac{\sum_{\nu=0}^{N} C_\nu^{(N)}(\mu)\ p_\beta(\beta|\nu+1, N-\nu+1)}{\sum_{\nu=0}^{N} \overline{c}_\nu^{(N)}}. \tag{22.56}$$

Here we have explicitly stated the μ-dependence of the coefficients $C_\nu^{(N)}(\mu)$. Since the beta PDF is already normalized to one, the denominator, along with the definition

$$\overline{c}_\nu^{(N)} = \int C_\nu^{(N)}(\mu)\ d\mu, \tag{22.57}$$

ensures the correct normalization of the posterior PDF.

Posterior PDF for μ

The posterior PDF for μ is readily derived from equation (22.56) by the marginalization over β:

$$p(\mu|\boldsymbol{d}, \boldsymbol{s}, \boldsymbol{\omega}, \mathcal{I}) = \frac{\sum_{\nu=0}^{N} C_\nu^{(N)}(\mu)}{\sum_{\nu=0}^{N} \overline{c}_\nu^{(N)}}. \tag{22.58}$$

In Figure 22.4 the posterior PDF of the present outlier-tolerant approach has already been presented and compared with that of the naïve procedure in which the experimental error bars are taken at face value. The superiority of the consistent outlier-tolerant approach is overwhelming.

From the posterior probability it is straightforward to calculate numerically $\langle\mu\rangle$ and $\langle\mu^2\rangle$ through

$$\langle\mu^n\rangle = \frac{1}{Z_\mu} \sum_{j=0}^{N} \int \mu^n \, C_j^{(N)}(\mu) \, d\mu. \tag{22.59}$$

Posterior PDF for β

Next we will determine the probability for β, that is for the prior probability that an error is ill specified. It follows from μ-marginalization of equation (22.56) that

$$p(\beta|\boldsymbol{d}, \boldsymbol{s}, \boldsymbol{\omega}, \mathcal{I}) = \sum_{\nu=0}^{N} P_\nu \, p_\beta(\beta|\nu+1, N-\nu+1). \tag{22.60}$$

In this pleasant mixture model expression we have introduced the probability P_ν that there are ν ill-specified data points in the sample defined by

$$P_\nu := \overline{c}_\nu^{(N)} \Big(\sum_j \overline{c}_j^{(N)} \Big)^{-1}. \tag{22.61}$$

The posterior PDF for β is depicted in Figure 22.5. The Bayesian outlier-tolerant analysis is capable of detecting correctly the percentage of outliers. Now the moments $\langle\beta^n\rangle$ can be derived quite easily from the corresponding moments of the beta PDF:

$$\int \beta^n \, p_\beta(\beta|\nu+1, N-\nu+1) \, d\beta = \frac{B(\nu+1+n, N-\nu+1)}{B(\nu+1, N-\nu+1)},$$

leading to

$$\langle\beta^n\rangle = \sum_{\nu=0}^{N} P_\nu \, \frac{(\nu+n)!(N+1)!}{\nu!(N+n+1)!}. \tag{22.62}$$

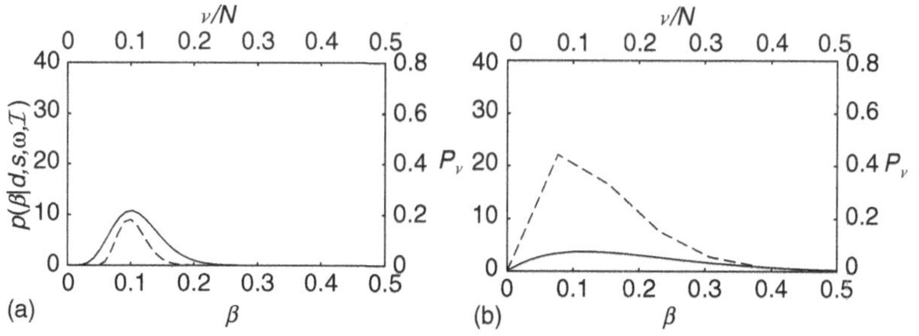

Figure 22.5 Posterior PDF (solid lines) for β based on the same data as in Figure 22.4, with parameters $\beta = 0.1$, $\mu = 1$, $\omega = 10$, $\sigma = 0.1$ and $N_{\text{data}} = 100$ (a) or rather $N_{\text{data}} = 13$ (b). Dashed lines represent the probability P_ν (right y-axis) versus ν/N (top x-axis), which is given in equation (22.61).

We have specifically

$$\langle \beta \rangle = \frac{1}{N+2} \sum_{\nu=0}^{N} P_\nu \, (\nu + 1) = \frac{\langle \nu \rangle + 1}{N+2}, \tag{22.63}$$

$$\langle \beta^2 \rangle = \frac{1}{(N+2)(N+3)} \sum_{\nu=0}^{N} P_\nu \, (\nu+1)(\nu+2) = \frac{\langle \nu^2 \rangle + 3\langle \nu \rangle + 2}{(N+2)(N+3)}. \tag{22.64}$$

Here the moments $\langle \nu^\eta \rangle$ of the number of outliers are defined by $\sum_\nu \nu^\eta p_\nu$.

Posterior outlier probability for a data point

Finally, we consider the probability that a particular data point is an outlier. Let Q_k be the proposition that data point d_k is an outlier. We are interested in $P(Q_k|\boldsymbol{d}, \mathcal{I})$. In order to simplify the notation, we include the error vectors $\boldsymbol{s}$ and $\boldsymbol{\omega}$ in the background information $\mathcal{I}$, since these values are fixed throughout the entire calculation. The sought-for probability may be expressed as a marginalization integral

$$P(Q_k|\boldsymbol{d}, \mathcal{I}) = \int_{-\infty}^{\infty} d\mu \int_0^1 d\beta \; P(Q_k|\boldsymbol{d}, \mu, \beta, \mathcal{I}) \; p(\mu, \beta|\boldsymbol{d}, \mathcal{I}), \tag{22.65}$$

$$P(Q_k|\boldsymbol{d}, \mu, \beta, \mathcal{I}) = \frac{P(\boldsymbol{d}|Q_k, \mu, \beta, \mathcal{I}) \; \overbrace{P(Q_k|\mu, \beta, \mathcal{I})}^{\beta}}{P(\boldsymbol{d}|\mu, \beta, \mathcal{I})}. \tag{22.66}$$

Again the likelihood terms factorize and the numerator and denominator only differ in the factors for data point d_k, i.e.

$$P(Q_k|\boldsymbol{d}, \mu, \beta, \mathcal{I}) = \frac{\beta A_k(\mu)}{\beta A_k(\mu) + (1-\beta) B_k(\mu)}. \tag{22.67}$$

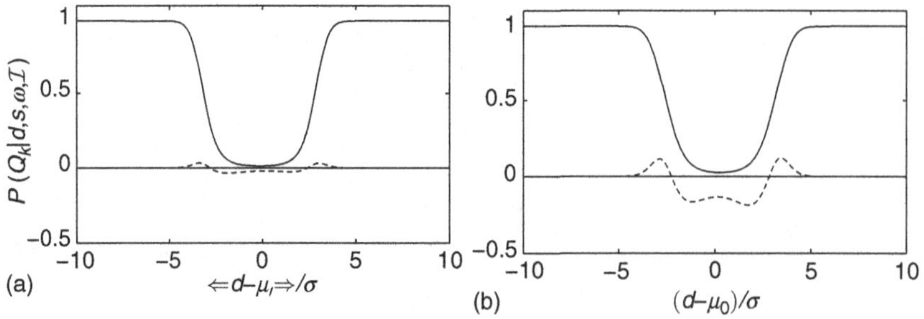

Figure 22.6 Exact posterior outlier probability according to equation (22.67) for the same data as in Figure 22.4 (solid line). The dashed line represents the relative error of the result of the approximation given in equation (22.68). Panel (a) is for $N_{\mathrm{d}} = 100$ and panel (b) for $N_{\mathrm{d}} = 13$.

The integral over μ and β in equation (22.65) has to be computed numerically. We may, however, exploit the fact that the posterior $p(\mu, \beta|\boldsymbol{d}, \mathcal{I})$ is generally sharply peaked at the mean values $\langle\beta\rangle$ and $\hat{\mu}$, respectively. We can then approximate the outlier probability by

$$P(Q_k|\boldsymbol{d}, \mathcal{I}) = \frac{\langle\beta\rangle A_k(\hat{\mu})}{\langle\beta\rangle A_k(\hat{\mu}) + (1 - \langle\beta\rangle) B_k(\hat{\mu})}. \tag{22.68}$$

In this approximation the odds ratio reads

Odds ratio for d_k being an outlier

$$o := \frac{P(Q_k|\boldsymbol{d}, \mathcal{I})}{P(\overline{Q}_k|\boldsymbol{d}, \mathcal{I})} = \frac{\langle\beta\rangle}{1 - \langle\beta\rangle} \frac{A_k(\hat{\mu})}{B_k(\hat{\mu})}. \tag{22.69}$$

The result is fairly plausible. The quantity $\langle\beta\rangle$ is the mean outlier probability. It can be considered as the prior probability that a data point is an outlier. In this respect, $\langle\beta\rangle / (1 - \langle\beta\rangle)$ represents the prior odds. Likewise, the ratio $A_k(\hat{\mu})/B_k(\hat{\mu})$ corresponds to the Bayes factor for the data point under consideration. We see nicely how Ockham's factor appears automatically. If, for example, the sample mean $\hat{\mu}$ lies in the peaks of both likelihoods, then the Bayes factor corresponds to the ratio of the likelihood volumes, i.e. $A_k(\hat{\mu})/B_k(\hat{\mu}) \approx 1/\omega_k$, and Ockham's razor favours the simpler model, without outlier. Only if a data point differs strongly from the sample mean is it considered an outlier. In Figure 22.6 the outlier probability is depicted for the data set described along with Figure 22.4, where one of the data values d is scanned over the interval shown. For the given parameter set of Figure 22.4 and for $N_{\mathrm{d}} = 100$ (left panel) the outlier probability is one if $|d - \mu_0| > 5\sigma$, while it is close to zero for $|d - \mu_0| < 2\sigma$. The dashed line in Figure 22.6 represents the relative error of the approximate formula equation (22.68). In the present case with many data points ($N_{\mathrm{d}} = 100$), the approximation works fairly well. A reduction of the sample size to $N_{\mathrm{d}} = 13$ has strikingly little impact on the exact values, while it clearly deteriorates the approximation, which is represented by the dashed line. Next we will analyse the outlier selectivity in more detail. The odds ratio (given in equation (22.69)) for data point d_k being an outlier depends only through the sample mean on the other data points in the sample. Once we have the outlier probability, how do we decide whether a point is characterized as an outlier or not? A simple decision criterion could be: The data point d_k is an outlier if the odds ratio is greater than one, which leads to

$$|d_k - \hat{\mu}|^2 > \frac{2s^2\omega_k^2}{(\omega_k^2 - 1)} \ln\left(\frac{(1-\beta)\omega_k}{\beta}\right). \tag{22.70}$$

For the parameter set that we used before, $\beta = 0.1$, $\omega_k = 10$, the outlier criterion reads $|d - \mu| > 3\sigma$. It corroborates the qualitative observation in Figure 22.6 and appears fairly reasonable for the present parameter set. The criterion in equation (22.70) can, however, be highly misleading. The lower bound of the outlier odds ratio is $o_{\text{lb}} := \langle\beta\rangle/(\omega_k(1 - \langle\beta\rangle))$. For obvious reasons, the outlier odds ratio is the smallest for data points which are in perfect agreement with the sample mean, i.e. for $d_k = \hat{\mu}$. It may happen that $o_{\text{lb}} > 1$, that is the outlier probability is always greater than 1/2. This happens if $\beta > \omega_k/(1 + \omega_k)$, e.g. for $\omega_k = 2$ and $\beta > 2/3$. In this case the above-mentioned criterion would classify all data points as outliers, although on average only a fraction β really are. The criterion is obviously too restrictive. A better approach is to generate a uniform random number r. The data point is considered an outlier if r is smaller than the outlier probability. The procedure guarantees that on average, outliers are classified with the corresponding probability.

Performance of the mixture model approach

Now we will determine the average rate at which the mixture model procedure properly or wrongly classifies a data point. To keep things simple and transparent, we assume that both μ and β are given. The probability that a data point is classified as an outlier is then given by $P(Q_k|\boldsymbol{d}, \mu, \beta, \mathcal{I})$ in equation (22.67), which actually depends merely on d_k and not on the entire sample. Let us now generate regular test data according to the PDF $B_k(\mu)$, and test whether they are erroneously identified as outliers. In the following calculation we will explicitly need the dependence on μ, σ and ω_k and it is therefore favourable to introduce the centred normal PDF

$$\tilde{N}(x, \sigma) := \frac{1}{\sigma\sqrt{2\pi}} e^{-\frac{x^2}{2\sigma^2}},$$

for which we have $A_k(\mu) = \tilde{N}((d_k - \mu), \sigma_k\omega_k)$ and $B_k(\mu) = \tilde{N}((d_k - \mu), \sigma_k)$. Now we compute the average rate r_{eo} for erroneous outliers (that is, regular data points erroneously classified as outliers):

$$r_{\text{eo}} = \int P(Q_k|d_k = y, \mu, \beta, \mathcal{I})\ \tilde{N}((y - \mu), \sigma_k)\ dy.$$

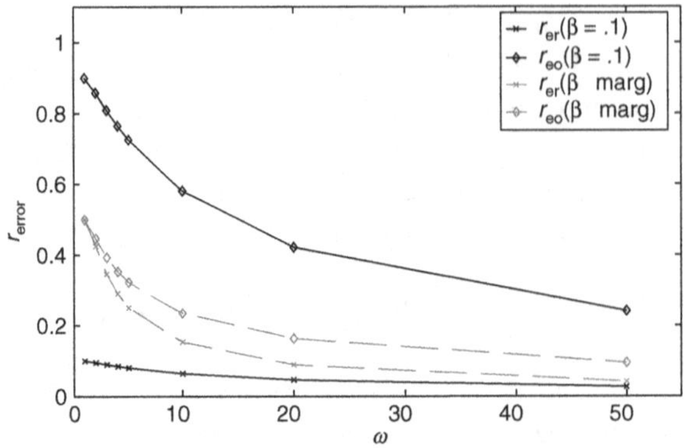

Figure 22.7 Study of the outlier selectivity as a function of ω for $\beta = 0.1, \mu = 1$ and $\sigma = 0.1$. Solid lines stand for results if β is known, while dashed lines represent the outcome of the β-marginalization.

Since the data index k is irrelevant, we will suppress it from now on. Along with the substitution $x = (y - \mu)/\sigma$ we have

$$r_{\text{eo}} = \beta \int \frac{\tilde{N}(x,\omega)\tilde{N}(x,1)}{\beta\tilde{N}(x,\omega) + (1-\beta)\tilde{N}(x,1)}\, dx. \tag{22.71}$$

The probability for the complementary event (that is, a regular data point r_{er} identified correctly) is apparently $r_{\text{er}} = 1 - r_{\text{eo}}$. The limiting case ($\omega = 1$) can readily be computed from equation (22.71), yielding $r_{\text{eo}} = \beta$. For $\omega = 1$ there is no difference in the data, regular or outlier, and we will classify regular data points according to the prior probability β. An analogous derivation yields the rate r_{er} for erroneous regular data points (that is, an actual outlier classified as regular):

$$r_{\text{er}} = (1-\beta) \int \frac{\tilde{N}(x,1)\tilde{N}(x,\omega)}{\beta\tilde{N}(x,\omega) + (1-\beta)\tilde{N}(x,1)}\, dx. \tag{22.72}$$

Again the sum rule yields the rate for correct identification of outliers $r_{\text{eo}} = 1 - r_{\text{er}}$. According to equations (22.71) and (22.72), there is a simple relation between erroneous identification of outliers and regular data points:

$$\beta\, r_{\text{er}} = (1-\beta)\, r_{\text{eo}}.$$

So if β, the prior probability for outliers, is smaller than 0.5, which should be the case for thorough measurements, then $r_{\text{eo}} < r_{\text{er}}$, that is, outliers are more often classified as regular data points than the other way round.

Figure 22.7 shows the result for mock data with the same parameters as in Figure 22.4 [p. 382], i.e. $\beta = 0.1, \mu = 1$ and $\sigma = 0.1$, as a function of ω. We see that there is a factor $(1-\beta)/\beta = 9$ between r_{er} and r_{eo}, irrespective of ω. The error rate drops rapidly. At the same rate the selectivity of the procedure increases, as ω increases.

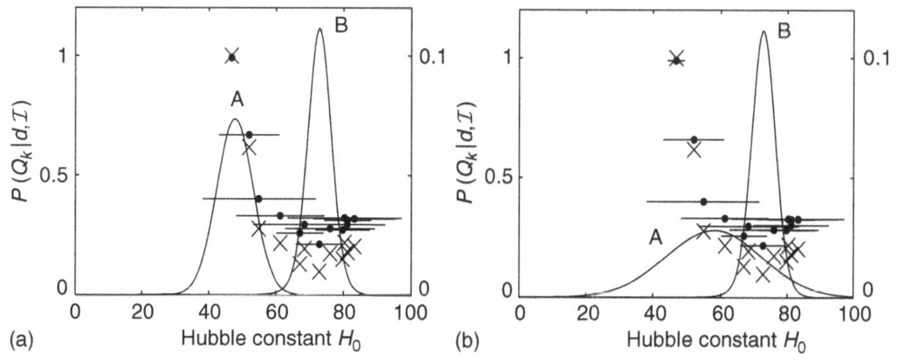

Figure 22.8 Measurements (full circles) of the Hubble constant H_0 with error bars (horizontal scale). The vertical position (left axis) represents the probability that the specified error is an outlier. Crosses indicate the result of the large-N approximation (equation (22.68)). Continuous lines stand for the posterior PDF for H_0 (right axis). The result of the present approach (B) is compared with that of the standard procedure (A), according to equation (21.5) [p. 335]. (a) Analysis of the published data. (b) The 'experimental error' of the leftmost data point has been increased from 0.6 to 2.5. The analysis has been performed with $S = 30$.

This behaviour is qualitatively reasonable, as regular data points and outliers are better separated with increasing ω.

In most applications we don't really know the prior outlier probability β and we have to marginalize the results derived so far with respect to β. The result of the numerical integration for a flat prior PDF $p(\beta|\mathcal{I})$ is also depicted in Figure 22.7. The result is both striking and reasonable. For $\omega = 1$ there is no distinction between outliers and regular data points. We therefore expect the same error rate in both cases and on top of that also the correct identification and the erroneous one should be the same due to symmetry; we have $r_{\text{er}} = 0.5$, as corroborated by the numerical result. For $\omega > 1$ the error rates differ, which is due to the fact that the marginalized error rates are related through

$$\int r_{\text{er}}(\beta)d\beta = \int \frac{\beta}{1-\beta}\, r_{\text{eo}}(\beta)d\beta.$$

Press's estimate of the Hubble constant

Next we will apply this analysis to a real-world problem. Figure 22.8 shows results of the present analysis for 13 'reputable' measurements of the Hubble constant H_0 taken from [161]. Press assumed in his analysis of the Hubble constant for the as yet unspecified constants ω_j that $\omega_j s_j = S = 30$ independent of j, which yields a posterior expectation value $\langle H_0 \rangle = (72.4 \pm 3.2)$ km/(s Mpc), while the naïve approach would yield $\langle H_0 \rangle = (47.8 \pm 5.4)$. The mean outlier probability is $\langle \beta \rangle = 0.34 \pm 0.2$. In Figure 22.8 we give the results of the Bayesian analysis for the Hubble constant. We see in the left panel that there is a tremendous difference as far as the probability density for H_0 is concerned between the naïve and the outlier-tolerant approach. There is one data point that stands out against the others. It is the leftmost point, which has by far the smallest experimentally assigned error bar. At the same time the data point is far away from the bulk of the sample. The probability density of the naïve approach is obviously dominated by this data point. The outlier-tolerant approach, on the contrary, considers this point an outlier and discards it from the evaluation of H_0. In the right panel of Figure 22.8 we depict the results we get if we increase the error bar of the data point under consideration from $\sigma = 0.6$ to $\sigma = 2.5$. We see that it has an enormous effect in the naïve approach, $\langle H_0 \rangle = 58.09 \pm 14.20$, while it has no influence in the outlier-tolerant approach.

Though Press points out that the intent of his paper was to discuss a statistical method rather than the details of its application to the problem of the Hubble constant, it seems noteworthy that a more recent measurement [161] yielded (71 ± 7) km /(s Mpc). He points out that the results are not particularly sensitive to the choice $S = 30$. This is certainly true for the point estimate $\langle H_0 \rangle$ and its standard deviation. For example, for $S = 50$ we obtain $\langle H_0 \rangle = (72.2 \pm 3.1)$ km/(s Mpc). The choice of S has, however, significant influence on β, which for $S = 50$ becomes $\langle \beta \rangle = 0.24 \pm 0.1$. It does likewise influence the outlier probability, as can be seen from equation (22.69) for the odds ratio. Consider a data point coincident with the posterior expectation $\langle H_0 \rangle$. For such a data point the ratio A_k / B_k (remember both are Gaussian functions) is σ_k / S. Apparently, $p(Q_k)$ is sensitive to the choice of S.

There is another strong influence on S, namely the quality of the large-N approximation given in equation (22.68). In Figure 22.8 we observe that there is a considerable difference between the exact and the approximate values for the outlier probabilities. This discrepancy is significantly reduced if we use $S = 50$. The relative errors are roughly halved. The reason is that the peak of the probability density for β is much sharper for $S = 50$.

The assumption encoded in the prior PDF in equation (22.46) [p. 380] is that for good data the errors are Gaussian distributed with variance ${s_j}^2$, while for bad data the variance of the Gaussian is $\omega_j^2 s_j^2$. We will hardly ever have such detailed knowledge as to specify the factor ω_j individually for each point. In fact, this inability led Press to the choice $\omega_j s_j = S$ for all j, meaning that the 'bad' data stem from one and the same normal distribution with variance S^2. This assumption is, however, rather special and may be abandoned. Returning to the error model for the mixture model of equation (22.46) [p. 380], we investigate the problem of introducing a distribution for ω for each error. Thus, for one data point we have (data index i is suppressed)

$$p(\sigma \,|s, \beta, \mathcal{I}) = \int p(\sigma, \omega|s, \beta, \mathcal{I}) d\omega = \int p(\sigma \,|s, \omega, \beta, \mathcal{I})\; p(\omega|\mathcal{I}) d\omega \qquad (22.73)$$

where in the last equality we have assumed that the scaling factor ω is independent of β and s, respectively. $p(\omega|\mathcal{I})$ is of course a normalized distribution. Inserting $p(\sigma \,|s, \beta, \omega, \mathcal{I})$ from equation (22.46) [p. 380], we obtain

$$p(\sigma \,|s, \beta, \mathcal{I}) = \int p(\omega|\mathcal{I}) \left\{\beta\delta(\sigma - \omega s) + (1 - \beta)\delta(\sigma - s)\right\} d\omega. \qquad (22.74)$$

Integration of the second term yields $(1-\beta)\delta(\sigma-s)$ since $p(\omega|\mathcal{I})$ is normalized. The first integral can easily be performed by the substitution $\omega s = x$. Then the joint PDF for all data points in equation (22.46) [p. 380] yields

$$p(\boldsymbol{\sigma}|\boldsymbol{s},\beta,\mathcal{I}) = \prod_i \left\{\beta \frac{1}{s}\, p\left(\frac{\sigma_i}{s_i}\right) + (1-\beta)\,\delta(\sigma_i - s_i)\right\}. \tag{22.75}$$

Calculation of the marginal likelihood $p(\boldsymbol{d}|\mu,\boldsymbol{s},\beta,\mathcal{I})$ will now result in the same distribution $B_i(\mu)$ as defined along with equation (22.47) [p. 381]. The function $A_i(\mu)$, however, becomes

$$A_i(\mu) = \frac{1}{s_i}\int p\left(\frac{\sigma_i}{s_i}\right)\frac{1}{\sigma_i\sqrt{2\pi}}\exp\left\{-\frac{1}{2\sigma_i^2}(d_i-\mu)^2\right\}d\sigma_i \tag{22.76}$$

and $p(\sigma_i/s_i)$ specifies our knowledge about the distribution of misspecified errors. An example for a distribution of the family $p(\sigma/s)$ was already given in equation (22.25) [p. 372]. In that case the parameters were chosen such that the distribution represented regular as well as irregular data. The case is different for a mixture model where we represent the regular data by an explicit term $B_i(\mu)$. The choice of $p(\sigma/s)$ should then only represent the outlying data. The choice of $p(\sigma/s)$ depends of course on the knowledge available about the outliers. A fairly uninformative distribution which codes just one piece of information is the Pareto distribution:

$$p(\sigma|s,\mathcal{I}) = \frac{1}{s}p\left(\frac{\sigma}{s}\right) = (\beta-1)\,\frac{s^{\beta-1}}{\sigma^\beta}, \qquad \infty > \sigma \geq s. \tag{22.77}$$

Normalizability is ensured for this distribution for all $\beta > 1$. Convergence of the first moment would require even $\beta > 2$. More interesting is the median $\hat{\sigma}$ of $p(\sigma|s,\mathcal{I})$:

$$\hat{\sigma}/s = 2^{1/(\beta-1)}. \tag{22.78}$$

Assuming $p(\sigma|s)$ of equation (22.77), we may calculate the marginal likelihood

$$\begin{aligned} A_i(\mu) &= \frac{\beta-1}{s_i\sqrt{2\pi}}\int_{s_i}^{\infty}\frac{d\sigma_i}{\sigma_i}\frac{s_i^\beta}{\sigma_i^\beta}\exp\left\{-\frac{1}{2\sigma_i^2}(d_i-\mu)^2\right\} \\ &= \frac{\beta-1}{2s_i\sqrt{2\pi}}\left(\frac{d_i-\mu}{\sqrt{2}s_i}\right)^{-\beta}\Gamma\left(\frac{(d_i-\mu)^2}{2s_i^2};\frac{\beta}{2}\right) \end{aligned} \tag{22.79}$$

with the incomplete Gamma function discussed in equation (7.15b) [p. 103].

An interesting special case arises for $\beta = 2$, a distribution characterizing the case where half of the data with irregular errors have a true error larger than twice the specified error. In this case we obtain

$$A_i(\mu) = \frac{1}{s_i\sqrt{2\pi}}\frac{s_i^2}{(d_i-\mu)^2}\left(1-\exp\left\{-\frac{(d_i-\mu)^2}{2s_i^2}\right\}\right), \tag{22.80}$$

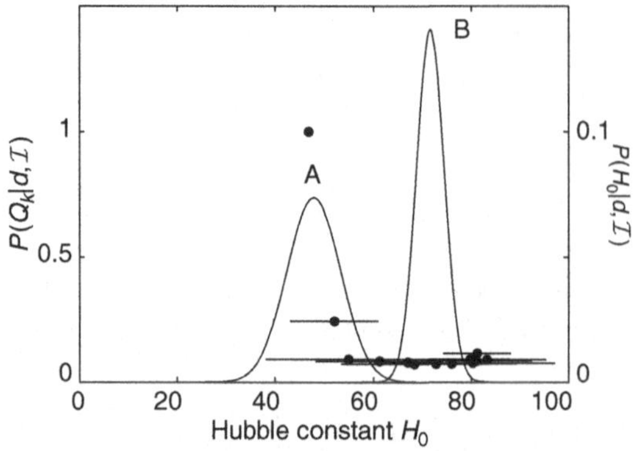

Figure 22.9 Meaning of symbols and curves same as left panel of Figure 22.8. Here misspecified errors are marginalized, as explained in the text.

a distribution with quite a heavy tail proportional to μ^{-2}. For the problem of outlier classification the 50% probability level (unit odds ratio) is obtained for that value $\overline{\mu}$ for which $A_i(\overline{\mu}) = B_i(\overline{\mu})$ according to equation (22.69) [p. 386]. This equality is reached approximately for $|\mu - d_i| = \sqrt{2}\ s_i$, which seems quite reasonable in view of the frequently considered 1σ region of confidence. The result of the present approach for the Hubble data is depicted in Figure 22.9. Owing to the more reluctant outlier assignment of the present approach, the individual outlier probabilities $P(Q_k|\boldsymbol{d},\mathcal{I})$ and the mean prior outlier probability ($\langle\beta\rangle_{\hat{\mu}} = 0.13$) are strongly reduced.

The importance of robust estimation

Before we discuss two further applications of the above-developed formalism we return to the question of why outlier-tolerant analysis methods are important. An important field of application of an algorithm for robust estimation of a generalized sample mean would be the evaluation of physical constants and conversion factors periodically performed by CODATA. The CODATA committee draws conclusions (which become compulsory for the whole physics community) from selected sets of raw data. We illustrate by citation how CODATA proceeds: 'After a thorough analysis using a number of least squares algorithms, the initial group of 38 items of stochastic input data was reduced to 22 by deleting those that were either highly inconsistent with the remaining data or had assigned uncertainties so large that they carried negligible weight.' The latter reason may have been of (doubtful) importance before the advent of electronic computation facilities. The selection procedure, however, is more serious. One might feel confident in rare cases that a single measurement sticks out of a whole lot which seems consistent. Deleting this measurement may turn out to be dangerous. This has also been experienced by the CODATA committee. We quote again from the CODATA report: 'The large change in K_v (a constant relating the SI unit of volt to a calibration standard) and hence in many other quantities between 1973 and 1986 would have been avoided if two determinations of F (the Faraday number)

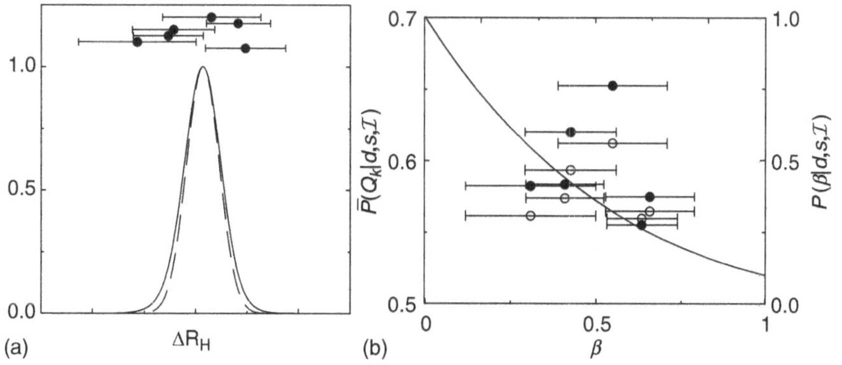

Figure 22.10 Least squares (dashed in (a)) and mixture model (solid curve in (a)) estimation of the quantum Hall constant. The vertical position of the experimental data has no specific meaning. The outlier probability of the data determined by equation (22.68) as well as $p(\beta|\boldsymbol{d}, \boldsymbol{s}, \mathcal{I})$ evaluated according to equation (22.60) are depicted in (b).

which seemed to be discrepant with the remaining data had not been deleted in the 1973 adjustment.'

It is of course remarkable that, while deleting seemingly discrepant data in 1973 had turned out to be misleading, the committee chose to proceed in exactly the same way in 1986. The outstanding property of the above-developed outlier-tolerant estimation is that data pre-selection is avoided and all available knowledge about measurement errors may be included in the calculation.

The quantum Hall effect

The first data set which we investigate for illustration is the stochastic input data for the quantum Hall effect selected by CODATA for the 1986 least-squares adjustment. The left panel of Figure 22.10 shows the original data and the posterior distribution of their mean value calculated by least-squares analysis (dashed curve) and by the mixture model in equations (22.46) [p. 380], (22.47) [p. 381] and (22.58) [p. 384], represented by the continuous curve. The two distributions are quite similar in this case. Their different widths signal that the specified errors in the primary data are slightly smaller than expected on the grounds of data dispersion. In fact, the least-squares-derived error is too small by about 16%. It is instructive to realize that the mean values derived from the two distributions differ only in insignificant digits. This suggests that the data set can also be analysed with the assumption that all errors are misspecified by a common factor. This case was considered previously and processing the data according to Section 22.1.1 [p. 364] would have yielded a value of 1.2 for such a common factor.

The mean value of the QHE constant evaluated with the mixture model is $(25\,812.8462 \pm 0.0018)\ \Omega$. This result should be compared with results in Table 22.3 [p. 370]. The right-hand panel of Figure 22.10 shows as a continuous curve the distribution $p(\beta|\boldsymbol{d}, \boldsymbol{s}, \mathcal{I})$. For the case of six data points it is a polynomial of degree six. The function peaks at $\beta = 0$,

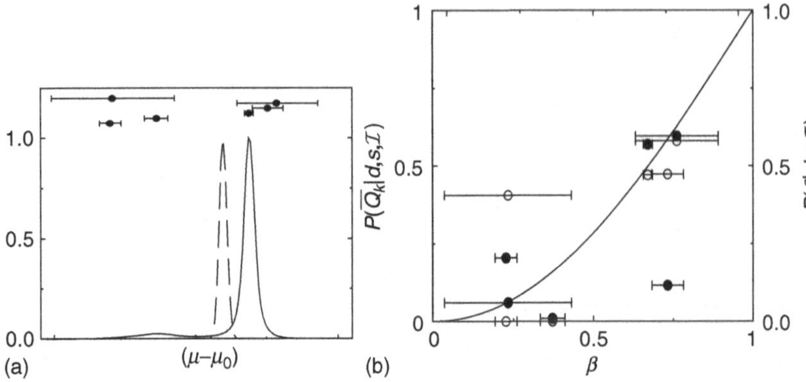

Figure 22.11 Least squares (dashed) and mixture model (solid curve) estimation of the low-field proton gyromagnetic ratio.

indicating that the data are highly compatible with each other. Also shown in this figure are again the input data with their respective outlier probability. Full dots refer to the numerically exact calculation based on equation (22.65) and open circles to the large-N approximation in equation (22.68) [p. 386]. The agreement is quite good and signals that the data set contains a sufficient number of regular, consistent data.

The gyromagnetic ratio of the proton

As a second example we consider estimation of the low-field proton gyromagnetic ratio $\gamma_p{}'$ from experimental data. The data are shown in the left-hand panel of Figure 22.11 together with posterior distributions for $\gamma_p{}'$ either from least-squares analysis (dashed curve) or from the more demanding mixture model. The interesting feature is that the least-squares approach predicts a narrow distribution at a value of $\gamma_p{}'$ not covered by any of the primary experimental data and their error bars; not a very plausible result, to say the least. The mixture model, however, peaks at the data point with by far the lowest error margin. The distribution is bimodal due to the two relatively precise data points on the left, and signals clearly that the data are basically inconsistent. This is a case where it makes little sense to summarize a distribution in terms of mean value and standard deviation. The CODATA committee has also realized this inconsistency and decided to resolve the case in a not very convincing way: 'Of all the $\gamma_p{}'$ data, the most glaring discord comes from the NPL (UK) low field value (the leftmost data point in Figure 22.11). The measurements of the proton resonance frequency were completed in December 1975 after which the coil dimensions were measured, but no verification was made, that the measurement process did not affect the coils. Because this result is so discrepant, and because the measurements were forced to terminate prematurely, we consider it to be an incomplete effort which should not be included in the final adjustment.' Though it seems perfectly acceptable to reject a measurement for reasons of an erroneous or incomplete procedure, the reason that it is discrepant is unjustified. We note without presenting details that the measurements remain discrepant even after deleting the NPL value. A least-squares treatment of the truncated data set remains unsatisfactory. The right-hand panel of Figure 22.11 shows $p(\beta|\boldsymbol{d}, s, \mathcal{I})$. The distribution peaks at $\beta = 1$ and tends to zero at $\beta = 0$, meaning that the data are highly inconsistent. This is also reflected in the large difference in the outlier probability, obtained from either equation (22.65) or from the approximation equation (22.68) [p. 386]. This data set is small and inconsistent, hence the failure of the approximation does not come as a surprise.

The discussion up to now was limited to the estimation of a sample mean. Such a limitation may easily be relaxed. Let, for example, $d_i = f(\boldsymbol{x}_i, \boldsymbol{\lambda}) + \varepsilon_i$ where x_i is an input variable which may be a vector in general, $\boldsymbol{\lambda}$ is a set of say q parameters and f a functional relationship mapping $\boldsymbol{x}_i, \boldsymbol{\lambda}$ onto the response variable d_i. All our previous analysis remains valid by replacing μ with $f(\boldsymbol{x}_i, \boldsymbol{\lambda})$ and $d\mu$ by $d^q\lambda$. Our generalization has thus led to the problem of robust parameter estimation.

23

Unrecognized signal contributions

Our discussion of inconsistent data has so far focused on the assumption that the inconsistency between the scatter of the data and the quoted errors is due to an inappropriate error assignment. A different assumption may apply, however. It is conceivable that the error determination is experimentally well under control and the data scatter results from accidental signal contributions from an unexpected and unrecognized source. There are many possibilities for such effects. Examples are ill-specified background, dead time in counting experiments, sample impurities, calibration errors and many more. We are still interested in estimating a single physical quantity μ. Instead of a model $\mu = d_i + \sigma_i$ we must then invoke a model which accounts for the additional signal in the form $\mu = d_i + \varepsilon_i + \sigma_i$.

23.1 The nuclear fission cross-section ^{239}Pu (n, f)

For reasons of simplicity we assume that the Gaussian-distributed true errors σ are known and we define slip variables ε_i which represent the unrecognized signal contributions. We assume further that the unrecognized signal contributions $\boldsymbol{\varepsilon} = \{\varepsilon_1, \ldots, \varepsilon_N\}$ obey the distribution

$$p(\boldsymbol{\varepsilon}|\boldsymbol{\tau}, \mathcal{I}) = \left(\prod_i \frac{1}{\tau_i\sqrt{2\pi}}\right) \exp\left\{-\frac{1}{2}\sum_i \frac{\varepsilon_i^2}{\tau_i^2}\right\}. \tag{23.1}$$

To begin with, we assume that τ_i shall be at our disposal. The likelihood corresponding to our model is

$$p(\boldsymbol{d}|\mu, \boldsymbol{\sigma}, \boldsymbol{\varepsilon}, \mathcal{I}) = \left(\prod_i \frac{1}{\sigma_i\sqrt{2\pi}}\right) \exp\left\{-\frac{1}{2}\sum_i \frac{(d_i - \mu - \varepsilon_i)^2}{\sigma_i^2}\right\}. \tag{23.2}$$

The goal of our analysis is of course again the determination of $\langle\mu\rangle$ and $\langle\Delta\mu^2\rangle$. These can be calculated from the distribution

$$p(\mu|\boldsymbol{d}, \boldsymbol{\sigma}, \boldsymbol{\tau}, \mathcal{I}) = \frac{p(\mu|\not\boldsymbol{\sigma}, \not\boldsymbol{\tau}, \mathcal{I}) \cdot p(\boldsymbol{d}|\mu, \boldsymbol{\sigma}, \boldsymbol{\tau}, \mathcal{I})}{p(\boldsymbol{d}|\boldsymbol{\sigma}, \boldsymbol{\tau}, \mathcal{I})}. \tag{23.3}$$

We have struck out immaterial dependencies from the prior $p(\mu|\mathcal{I})$. Next we need the marginal likelihood $p(\boldsymbol{d}|\mu, \boldsymbol{\sigma}, \boldsymbol{\tau}, \mathcal{I})$. This is obtained by marginalization over $\boldsymbol{\varepsilon}$:

$$p(\boldsymbol{d}|\mu, \boldsymbol{\sigma}, \boldsymbol{\tau}, \mathcal{I}) = \int d^N\varepsilon \; p(\boldsymbol{d}|\boldsymbol{\varepsilon}, \mu, \boldsymbol{\sigma}, \boldsymbol{\tau}, \mathcal{I}) \; p(\boldsymbol{\varepsilon}|\not{\mu}, \not{\boldsymbol{\sigma}}, \boldsymbol{\tau}, \mathcal{I}). \tag{23.4}$$

The prior for $\boldsymbol{\varepsilon}$ is logically independent of μ and $\boldsymbol{\sigma}$. The integral factorizes into one-dimensional integrals

$$p(\boldsymbol{d}|\mu, \boldsymbol{\sigma}, \boldsymbol{\tau}, \mathcal{I}) = \prod_i \frac{1}{2\pi \, \sigma_i \tau_i} \int d\varepsilon_i \; e^{-\frac{1}{2}\Phi_i} \tag{23.5a}$$

with

$$\begin{aligned} \Phi_i &:= \frac{(d_i - \mu - \varepsilon_i)^2}{\sigma_i^2} - \frac{\varepsilon_i^2}{\tau_i^2} \\ &= \frac{\sigma_i^2 + \tau_i^2}{\sigma_i^2 \tau_i^2} \left(\varepsilon_i - \frac{d_i - \mu}{1 + \sigma_i^2/\tau_i^2} \right)^2 + \frac{(d_i - \mu)^2}{\sigma_i^2 + \tau_i^2}, \end{aligned} \tag{23.5b}$$

and the marginal likelihood becomes

Marginal likelihood in the presence of unrecognized signals

$$p(\boldsymbol{d}|\mu, \boldsymbol{\sigma}, \boldsymbol{\tau}, \mathcal{I}) = \left(\prod_i \frac{1}{\sqrt{2\pi} \cdot \sqrt{\sigma_i^2 + \tau_i^2}} \right) \exp\left\{ -\frac{1}{2} \sum_i \frac{(d_i - \mu)^2}{\sigma_i^2 + \tau_i^2} \right\}. \tag{23.6}$$

The estimate of $\langle \mu \rangle$ and $\langle \Delta\mu^2 \rangle$ remains simple if we assume a uniform prior for μ in equation (22.6) [p. 366] since the posterior is Gaussian in μ and has the same structure as equation (21.2) [p. 335] with σ^2 replaced by $\sigma^2 + \tau^2$. We can therefore also use the results of equation (21.7) [p. 335]:

$$\varrho^{-2} = \frac{1}{N} \sum_i \frac{1}{\sigma_i^2 + \tau_i^2}, \quad \text{var}(\mu) = \frac{\varrho^2}{N}, \quad \langle \mu \rangle = \frac{\varrho^2}{N} \sum_i \frac{d_i}{\sigma_i^2 + \tau_i^2}. \tag{23.7}$$

The result in equation (23.7) justifies a comment on systematic errors in physical measurements. There is some confusion as far as terminology is concerned. The difference in performance between a real and an ideal instrument is a correction which must be calculated and included in the calibration. The remaining error, which may for example be the zero point precision of a balance, the precision of baseline restoration in electronics or others, is systematic with zero mean and finite variance. Our calculation has shown that this apparatus-based variance has to be added to the measurement variance to obtain the total. Systematic errors are treated in the same manner as measurement errors.

We now proceed to an estimation of the 'unrecognized' errors $\boldsymbol{\varepsilon}$. They are obtained from the marginal distribution $p(\boldsymbol{\varepsilon}|\boldsymbol{d},\boldsymbol{\sigma},\boldsymbol{\tau},\mathcal{I})$, which we obtain from marginalization over μ:

$$p(\boldsymbol{\varepsilon}|\boldsymbol{d},\boldsymbol{\sigma},\boldsymbol{\tau},\mathcal{I}) = \int d\mu \; p(\boldsymbol{\varepsilon}|\mu,\boldsymbol{d},\boldsymbol{\sigma},\boldsymbol{\tau},\mathcal{I}) \; p(\mu|\boldsymbol{d},\boldsymbol{\sigma},\boldsymbol{\tau},\mathcal{I}). \tag{23.8}$$

We have

$$p(\boldsymbol{\varepsilon}|\mu,\boldsymbol{d},\boldsymbol{\sigma},\boldsymbol{\tau},\mathcal{I}) = \frac{1}{Z}\, p(\boldsymbol{\varepsilon}|\not{\mu},\boldsymbol{\tau},\mathcal{I}) \cdot p(\boldsymbol{d}|\mu,\boldsymbol{\varepsilon},\boldsymbol{\sigma},\not{\boldsymbol{\tau}},\mathcal{I}) \tag{23.9}$$

which, according to equations (23.1) and (22.4) [p. 366], factorizes into

$$p(\boldsymbol{\varepsilon}|\mu,\boldsymbol{d},\boldsymbol{\sigma},\boldsymbol{\tau},\mathcal{I}) = \prod_i p(\varepsilon_i|\mu,d_i,\sigma_i,\tau_i,\mathcal{I}), \tag{23.10a}$$

$$p(\varepsilon_i|\mu,d_i,\sigma_i,\tau_i,\mathcal{I}) := \frac{1}{Z_i}\exp\left\{-\frac{1}{2}\left(\frac{(d_i-\mu-\varepsilon_i)^2}{\sigma_i^2}+\frac{\varepsilon_i^2}{\tau_i^2}\right)\right\} \tag{23.10b}$$

$$= \frac{1}{\sqrt{2\pi\rho_i^2}}\exp\left\{-\frac{\left(\varepsilon_i-\varepsilon_i^{(0)}\right)^2}{2\rho_i^2}\right\}, \tag{23.10c}$$

$$\varepsilon_i^{(0)} := \frac{\tau_i^2}{\tau_i^2+\sigma_i^2}\Big(d_i-\mu\Big), \tag{23.10d}$$

$$\rho_i^2 := \frac{\sigma_i^2\tau_i^2}{\sigma_i^2+\tau_i^2}. \tag{23.10e}$$

Then the expectation value $\langle\varepsilon_k\rangle$ is, according to equations (23.8) and (23.10),

$$\langle\varepsilon_i\rangle = \int d\mu \; p(\mu|\boldsymbol{d},\boldsymbol{\sigma},\boldsymbol{\tau},\mathcal{I})\; \varepsilon_k^{(0)} = \frac{\tau_i^2}{\tau_i^2+\sigma_i^2}\Big(d_i-\langle\mu\rangle\Big). \tag{23.11}$$

The posterior estimate for ε_k is just the deviation of the data point d_k from the posterior estimate of the weighted sample mean multiplied by a compression factor which describes the relative contribution τ_k^2 of the variance of the unrecognized errors to the total variance $(\sigma_k^2+\tau_k^2)$.

At this stage we must remember that our analysis rests on the assumption that we know the variance τ_j^2 of the errors ε_j. In case we face a systematic error this may be a justified assumption which simplifies in many cases to $\tau_j \equiv \tau\forall j$. If, in contrast, the origin of the unrecognized signal is unknown it is also close to impossible to estimate its variance.

An example for the former case is the optimum estimate of the nuclear fission cross-section for ^{239}Pu(n,f) [78]. Table 23.1 shows six measurements of this cross-section at a neutron energy of 14.7 MeV with associated measurement errors. If we apply our previous procedure of scaling all errors to make the data consistent, we find a factor $\langle\alpha\rangle = 2.14$ and a best estimate of the cross-section with equation (22.17) [p. 369]: $Q = (2.416 \pm 0.046)$ barn. Such a scaling is, however, inappropriate. Cross-section measurements are typically affected by systematic errors like uncertainty in the projectile current or the number

Table 23.1 Measurement of the cross-section for $^{239}\mathrm{Pu}(n, f)$ after [78]

Measurement	Cross-section	$\langle\varepsilon\rangle$
1	2.37 ± 0.09	-0.019
2	2.29 ± 0.05	-0.086
3	2.51 ± 0.05	$+0.060$
4	2.53 ± 0.05	$+0.060$
5	2.44 ± 0.09	$+0.007$
6	2.39 ± 0.03	-0.025

density of the target. These errors persist even if the statistical error is reduced by prolonged measurements. It is therefore a typical case for application of the developed formulae for unrecognized errors. Let $\tau = 0.07$ barn, which is typical for the kind of measurement we discuss. Then equation (23.7) yields a cross-section of $Q = (2.423 \pm 0.037)$ barn. So, both procedures lead to very similar results as far as expectation value and standard deviation of the cross-section is concerned. Conceptually they are, however, very different. While in the former case the expectation value is independent of the scaling factor α, in the latter case the effective variances of the measurements are changed differently from a simple scaling and consequently also the weighted average changes.

23.2 Electron temperature in a tokamak edge plasma

As a special example we discuss the measurement of the electron temperature at the edge of a tokamak plasma [53]. We start the discussion by inspection of Figure 23.1. Circles with error bars represent the measurements of the electron temperature in the plasma edge employing Thompson scattering. The Thompson scattering system in this experiment was under very good control and particular care had been devoted to the determination of errors. The data, taken as a function of distance from the plasma edge, do show some mainstream trend. At the same time the measured data show clear evidence for outliers. Remember that the term 'outlier' has to be understood as the distance of a data point from the mainstream in terms of the measurement error. A data point near the mainstream with a very small quoted error may be more of an outlier than a less precise data point farther away. An elementary theory of heat transport in the edge plasma of a divertor tokamak predicts, for the electron temperature T as a function of distance from the edge x,

$$T(x) = c\lambda^{-2/7} \cdot \left\{1 - \frac{8}{3}\frac{x - x_0}{\lambda}\right\}^{1/2}, \quad x \leq x_0, \tag{23.12}$$

$$T(x) = c\lambda^{-2/7} \cdot \left\{1 + \frac{x - x_0}{\lambda}\right\}^{-4/3}, \quad x \geq x_0. \tag{23.13}$$

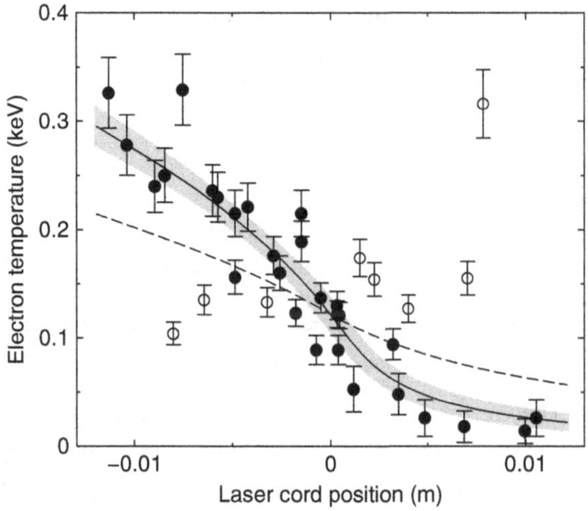

Figure 23.1 Data represent electron temperature measurements from the edge of a tokamak plasma. The continuous curve is the outlier-tolerant fit while the dashed curve is the model function evaluated at the least-squares values of the model parameters. Open circles are outliers according to equation (22.67).

Here x_0 denotes the true separatrix position, λ a decay length and c an amplitude constant. The model equation for interpreting the measured data becomes, if all data are regular,

$$d_i = T(x_i, c, \lambda, x_0) + \varepsilon_i. \tag{23.14}$$

Assuming that $\langle \varepsilon_i \rangle = 0$ and $\langle \varepsilon_i^2 \rangle = \rho_i^2$, this yields the likelihood

$$p(d_i | x_i, c, x_0, \lambda, \rho_i, \mathcal{I}) = \frac{1}{\rho_i \sqrt{2\pi}} \exp\left\{ -\frac{(d_i - T_i)^2}{2\rho_i^2} \right\}, \tag{23.15}$$

with the obvious notation $T_i = T(x_i, c, \lambda, x_0)$. Equation (23.15) contains the true errors ρ_i, which are never known. We have previously considered probability distributions for the error in terms of the experimentally specified error (equation (22.46) [p. 380]) in order to accommodate the likelihood function to the problem of outliers. The present case is different from the previous problems because of the extreme care which has been taken to obtain a reliable error estimate. We express this care by assuming $p(\rho|\sigma) = \delta(\rho - \sigma)$. This does not leave room to account for outliers on the grounds of erroneously specified confidence limits. Instead, in the present case we are forced to assume that the system under investigation, the tokamak edge plasma, is not stable, but fluctuates about a mainstream behaviour irregularly in time. It is in fact known that the plasma edge of a tokamak undergoes turbulent flows under special conditions. We now account for this irregular behaviour in the nonstationary phases by introducing a new variable S which accounts for the additional signal due to the irregular state. Our model equation and likelihood must be modified accordingly for the irregular data points:

$$d_i = \underbrace{T(x_i, c, \lambda, x_0)}_{:=T_i} \pm S_i + \varepsilon_i, \tag{23.16}$$

$$p(d_i|x_i, c, x_0, \lambda, \sigma_i, S_i, \mathcal{I}) = \frac{1}{\rho_i\sqrt{2\pi}} \exp\left\{-\frac{(d_i - T_i \mp S_i)^2}{2\rho_i^2}\right\}. \tag{23.17}$$

Equation (23.17) contains one parameter which we do not know yet, namely S_i. It has to be removed by marginalization after specifying an appropriate distribution. We assume that we know a typical signal size, which we denote by ξ. The principle of maximum entropy yields, on the basis of this knowledge, the distribution

$$p^{\pm}(S_i|\xi, \mathcal{I}) = \frac{1}{Z_i} \exp\left\{\mp \frac{S_i}{\xi}\right\}. \tag{23.18}$$

The plus sign in equation (23.18) applies for $d > T$ in the whole range $0 \leq d - T < \infty$. The minus sign applies correspondingly to $d < T$. In this case the support of the function in equation (23.18) is finite and limited to $0 \leq T - d \leq T$. This limit follows from the fact that the Thompson scattering system does not produce negative data d_i, regardless of which source (quiet or turbulent plasma edge) they came from. The normalization Z_i in equation (23.18) respecting these limits follows immediately as

$$Z_i = \xi\left[2 - \exp\left\{-\frac{T_i}{\xi}\right\}\right]. \tag{23.19}$$

Marginalization of S_i in equation (23.17) using the distribution of equation (23.18) requires distinguishing the cases of plus and minus sign in equation (23.18), resulting in

$$p^{-}(d_i|x_i, c, \lambda, x_0, \sigma_i, \xi, \mathcal{I}) = \exp\left\{\frac{d_i - T_i}{\xi} + \frac{\sigma_i^2}{2\xi^2}\right\} \times \frac{1 + \operatorname{erf}\left[\frac{(d_i - T_i - \sigma_i^2/\xi)}{\sigma_i\sqrt{2}}\right]}{2\xi\left(2 - \exp\left(-\frac{T_i}{\xi}\right)\right)}, \tag{23.20a}$$

$$p^{+}(d_i|x_i, c, \lambda, x_0, \sigma_i, \xi, \mathcal{I}) = \exp\left\{-\frac{d_i - T_i}{\xi} + \frac{\sigma_i^2}{2\xi^2}\right\} \times \frac{\operatorname{erf}\left[\frac{d_i + \sigma_i^2/\xi}{\sigma_i\sqrt{2}}\right] - \operatorname{erf}\left[\frac{d_i - T_i + \sigma_i^2/\xi}{\sigma_i\sqrt{2}}\right]}{2\xi\left[2 - \exp\left(-\frac{T_i}{\xi}\right)\right]}. \tag{23.20b}$$

Now the stage is set to identify the likelihood for the regular data in equation (23.15) with our previously defined function B_i:

$$B_i(c, \lambda, x_0, \mathcal{I}) = p(d_i|x_i, c, \lambda, x_0, \sigma_i, \mathcal{I}) \tag{23.21}$$

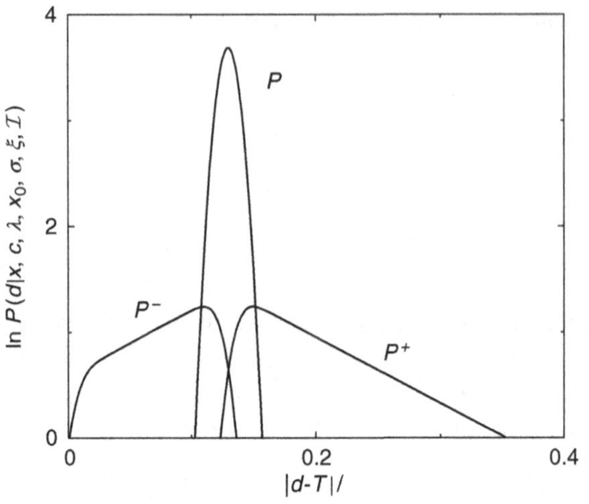

Figure 23.2 Schematic graph of the functions $p^{\pm}$ (equation (23.22)) and p (equation (23.21)).

and the corresponding function in equation (23.20) representing the likelihood for the irregular data with the previously defined A_i:

$$A_i(c, \lambda, x_0) = p^{\pm}(d_i|x_i, c, \lambda, x_0, \sigma_i, \xi, \mathcal{I}) \qquad (23.22)$$

and estimate parameters c, λ, x_0 on the basis of the previously introduced mixture model in equation (22.47) [p. 381]. This requires an estimate of the scale parameter ξ, which turns out not to be very critical. For the data in Figure 23.2, ξ was chosen to be the average signal size $\overline{d}$. We shall not discuss the choice of priors on the other parameters c, λ, x_0 since that would require an in-depth discussion of the experiment.

Another topic is, however, worth pointing out. It is of course not only interesting to infer the values of the parameters c, λ, x_0 from the mixture model based on equations (23.22) and (23.21) but also of course interesting to estimate, on the basis of the available data, electron temperatures $T(z)$ at arbitrary positions z, in particular in the space between experimental data. This requires calculation of the distribution $p(T|\boldsymbol{d}, \boldsymbol{x}, \boldsymbol{\sigma}, z, \mathcal{I})$, which can be constructed using marginalization:

$$p(T|\boldsymbol{d}, \boldsymbol{x}, \boldsymbol{\sigma}, z, \mathcal{I}) = \int p(c, \lambda, x_0|\boldsymbol{d}, \boldsymbol{x}, \boldsymbol{\sigma}, z, \mathcal{I}) \times p(T|c, \lambda, x_0, \boldsymbol{d}, \boldsymbol{x}, \boldsymbol{\sigma}, z, \mathcal{I})\, dc\, d\lambda\, dx_0. \qquad (23.23)$$

The first factor in the integrand is the posterior distribution of the parameters c, λ, x_0. This can be obtained by Bayes' theorem from the mixture likelihood by introducing suitable priors on c, λ, x_0. The second factor may be simplified. Since the electron temperature at the position z is uniquely specified by equation (23.12) or equation (23.13), if c, λ, x_0 are given, $p(T|c, \lambda, x_0, z)$ is logically independent of $\boldsymbol{d}, \boldsymbol{x}, \boldsymbol{\sigma}$ and given by

$$p(T|c, \lambda, x_0, z, \mathcal{I}) = \delta(T - T(z|c, \lambda, x_0)). \tag{23.24}$$

One realizes immediately that back-substitution of equation (23.24) into equation (23.23) yields an expression containing a δ function. Therefore, the moments $\langle T^\nu \rangle$ can easily be computed:

$$\langle T^\nu \rangle = \int p(c, \lambda, x_0|\boldsymbol{d}, \boldsymbol{x}, \boldsymbol{\sigma}, \mathcal{I})\big(T(z|c, \lambda, x_0)\big)^\nu \, dc \, d\lambda \, dx_0. \tag{23.25}$$

The special cases $\nu = 1, 2$ allow us then to evaluate the expectation value $\langle T \rangle$ and variance $\langle \Delta T^2 \rangle = \langle T^2 \rangle - \langle T \rangle^2$ for the posterior temperature. These results are shown in Figure 23.1 as a continuous curve and a shaded band, respectively. Our derivation of the posterior temperature estimates seems to be unduly complicated. Why should we not simply use the model function evaluated with the parameters set to their posterior values? Equation (23.25) becomes exactly of this form if

$$p(c, \lambda, x_0|\boldsymbol{d}, \boldsymbol{x}, \boldsymbol{\sigma}, \mathcal{I}) = \delta(c - \hat{c})\delta(\lambda - \hat{\lambda})\delta(x_0 - \hat{x}_0). \tag{23.26}$$

In other words, evaluating the model function with the parameters set to their posterior expectation values $\hat{c}, \hat{\lambda}, \hat{x}_0$ is only equivalent to equation (23.25) if the precision of the posterior estimates of the parameters is very high, as expressed by the δ functions in equation (23.26). It may comfort the reader to learn that the two ways of estimating the posterior temperature cannot be distinguished within the drawing line width for the case displayed in Figure 23.1. Note, finally, that the dashed curve in Figure 23.1 represents the model function evaluated with the parameters set to their least-squares values. It is satisfactory to see that the Bayesian solution is well in accord with intuition, while no experienced physicist would be prepared to accept the dashed curve as a fit of a model function to the data.

23.3 Signal–background separation

In the last application of the mixture model, we considered the case of a constant model μ corrupted by an unrecognized signal. The logic and procedure of that example lend themselves readily to the solution of a ubiquitous problem of data analysis in experimental physics: the proper separation of a signal from an underlying background. We shall discuss this problem with reference to Figure 23.3. The figure shows a particle-induced X-ray spectrum from an ivory sample. 'Particles' in this context means beams of protons or helium ions with energies of several MeV. The advantage of heavy particle excitation of the X-rays is that the primary Bremsstrahlung contribution to the spectrum is relatively small. Appreciable Bremsstrahlung contributions to the slowly varying background stem from secondary electrons produced by the primary beam. Superimposed on the Bremsstrahlung background we see narrow lines due to characteristic X-ray emission from the sample. They constitute the sought-for analytical information on the elemental composition of the

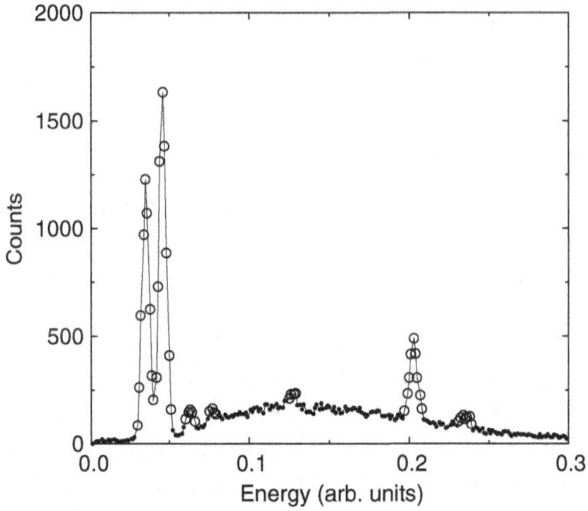

Figure 23.3 PIXE spectrum from an ivory sample. Open circles represent data points which are identified to carry a signal contribution [211].

sample. Intuitively it seems quite clear which data points are to be regarded as background and which regions contain signal. The background is of course the slowly varying contribution to the data while the signal varies comparatively rapidly. The strategy to model this perception in a Bayesian framework will be to represent the background by smooth, slowly varying functions and consider the signal as outlying data points from the mainstream background. We represent the background b_i at energy x_i by an expansion

$$b_i = \sum_{\nu=1}^{E} \phi_\nu(x_i) c_\nu, \tag{23.27}$$

where $\{\phi_\nu\}$ is some appropriate smooth basis function, $\{c_\nu\}$ are the expansion coefficients and E is the expansion order. The basis function $\{\phi_\nu\}$ can, for example, be polynomials. Alternatively, one could consider cubic splines. We shall leave the question open at this point and return to the spline alternative in Section 26.2 [p. 456].

We further introduce the proposition $\overline{Q}_i$ that the data point d_i is background and correspondingly the proposition Q_i that the data point d_i contains signal. The probability distribution $p(d_i|b_i, Q_i, \mathcal{I})$ is then given by the Poisson distribution

$$p(d_i|b_i, \overline{Q}_i, \mathcal{I}) = \frac{b_i^{d_i}}{\Gamma(d_i+1)}\, e^{-b_i}. \tag{23.28}$$

The choice of the Poisson distribution derives from the nature of the present data, which come from a counting experiment. It is of course known that the Gaussian distribution $\exp\left\{-(d_i - b_i)^2/2\sigma_i^2\right\}$ is the large-count approximation to the Poisson distribution. However, in the present case many channels do not fulfil the large-count assumption. The alternative likelihood for a data point d_i containing both signal and background is

$$p(d_i|b_i, Q_i, S_i, \mathcal{I}) = \frac{(b_i + S_i)^{d_i}}{\Gamma(d_i + 1)} \, e^{-(b_i+S_i)}. \tag{23.29}$$

Remember that the sum of Poissonian random numbers is again Poissonian. For more details see Section 6.4 [p. 87]. The newly introduced variable S_i has again to be removed by marginalization. As before we introduce a common scale ξ for the signal contributions $\{S_i\}$. The distribution corresponding to the choice of scale ξ is

$$p(S_i|\xi, \mathcal{I}) = \frac{1}{\xi} \, e^{-\frac{S_i}{\xi}}. \tag{23.30}$$

This prior is simpler than that in equation (23.18), since in the present case sensible values of S_i are restricted to $S_i \geq 0$ for all i. The marginal likelihood from equations (23.29) and (23.30) becomes

$$p(d_i|b_i, Q_i, \xi, \mathcal{I}) = \int p(S_i|\xi, \mathcal{I}) \, p(d_i|b_i, Q_i, S_i, \mathcal{I}) \, dS_i, \tag{23.31}$$

and after performing the S-integration

$$p(d_i|b_i, Q_i, \xi, \mathcal{I}) = \frac{e^{\frac{b_i}{\xi}}}{\xi} \, \frac{\Gamma\big(d_i + 1|b_i[1 + 1/\xi]\big)}{(1 + 1/\xi)^{d_i+1} \, \Gamma(d_i + 1)}, \tag{23.32}$$

which depends on the incomplete Gamma function defined in equation (7.15b) [p. 103]. Identifying equation (23.32) with the earlier defined function $A_i(d_i, b_i, \xi)$ and equation (23.28) with the corresponding $B_i(d_i, b_i)$, the total likelihood becomes a mixture model

$$p(\boldsymbol{d}|\boldsymbol{b}_i, \xi, \beta, \mathcal{I}) = \prod_i (\beta A_i(d_i, b_i, \xi) + (1 - \beta) B_i(d_i, b_i)) \tag{23.33}$$

where we have introduced as before the probability β that a particular data point is described by the function A_i. The functions A_i and B_i which enter equation (23.33) are shown in Figure 23.4. For the chosen parameters B_i is very much concentrated and very similar to a Gaussian, while A_i decays only exponentially. For data points which fall close to b_i the likelihood is mainly represented by B_i while for the peaks in the spectrum in Figure 23.3 the likelihood takes on values according to A_i, meaning that the signal contributions are largely neglected in the estimation of the background in equation (23.31). The likelihood in equation (23.33) is the basis for further inferences. In particular, we want to classify the measured data as belonging either to the signal or to the background. The analysis for this problem was already presented earlier (equation (22.67) [p. 385]) and, since we have many data, the asymptotic formula equation (22.68) [p. 386] may be employed. The result is contained in Figure 23.3. The full dots in this figure represent the data points which contain only background, the open circles carry also signal information. It is again highly satisfactory that the results of an involved Bayesian analysis comply well with intuition.

The second result to be drawn from equation (23.33) is an estimate of the parameters describing the background in equation (23.27). The desired probability distribution $p(\boldsymbol{b}|\boldsymbol{d}, \xi, \mathcal{I})$ is obtained from Bayes' theorem:

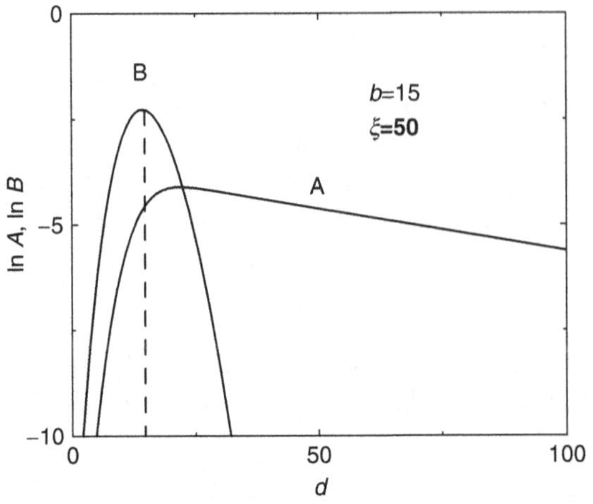

Figure 23.4 The functions $A_i(d_i, b_i, \xi)$ and $B_i(d_i, b_i)$ in equation (23.33). The scale ξ was chosen to be $\xi = \overline{d}$.

$$p(\boldsymbol{b}|\boldsymbol{d}, \xi, \mathcal{I}) = \frac{p(\boldsymbol{b}|\mu, \mathcal{I}) \cdot p(\boldsymbol{d}|\boldsymbol{b}, \xi, \mathcal{I})}{p(\boldsymbol{d}|\xi, \mathcal{I})}, \tag{23.34}$$

where we have introduced the prior on the background $\boldsymbol{b}$ given a hyperparameter μ that will be specified below. The likelihood showing up in equation (23.34) is obtained from equation (23.33) by marginalization over β. The required calculations have been described earlier and the results are summarized in equations (22.47) [p. 381] to (22.58). The initial assumption of the signal–background separation problem was that the background is a smooth function of energy. Hence we want the prior on $\boldsymbol{b}$ to express this smoothness property. Suitable functionals which represent smoothness have been discussed in Chapter 13 [p. 215]. $p(\boldsymbol{b}|\mu, \mathcal{I})$ may therefore be chosen as

$$p(\boldsymbol{b}|\mu, \mathcal{I}) = \frac{1}{Z} \exp\left\{-\frac{\mu}{2} \sum_i (b''(x_i))^2\right\}. \tag{23.35}$$

Since the background $\boldsymbol{b}$ is alternatively fully specified in terms of the expansion coefficients $\boldsymbol{c}$ in equation (23.27), once the expansion basis has been chosen the likelihood in equation (23.34) and the prior in equation (23.35) may be rewritten in terms of $\boldsymbol{c}$:

$$p(\boldsymbol{c}|\boldsymbol{d}, \xi, \mathcal{I}) = \frac{p(\boldsymbol{c}|\mu, \mathcal{I}) \cdot p(\boldsymbol{d}|\boldsymbol{c}, \xi, \mathcal{I})}{p(\boldsymbol{d}|\xi, \mathcal{I})}, \tag{23.36a}$$

$$p(\boldsymbol{c}|\mu, \mathcal{I}) = \frac{1}{Z} \exp\left\{-\frac{\mu}{2}\boldsymbol{c}^T \mathbf{D}\boldsymbol{c}\right\}, \tag{23.36b}$$

where $\mathbf{D}$ is the overlap matrix of the second derivatives of the basis functions $\{\phi_\nu\}$ given by

$$D_{l1,l2} = \sum_i^N \phi''_{l1}(x_i)\phi''_{l2}(x_i). \tag{23.37}$$

This completes the identification of the background in the data. Our main interest is, however, in the signal. It is tempting to define the signal as the difference between the data and the background. Generations of physicists have made this choice. The objection against such a procedure is that the result may be unphysical since it is by no means guaranteed that the difference between data and background is positive. Negative signals from a counting experiment such as that shown in Figure 23.3 are of course unphysical. Bayesian probability theory resolves the difficulty. We return to the probabilities that the data point contains only background in equation (23.28) and signal and background in equation (23.29), respectively. The corresponding mixture likelihood is

$$p(\boldsymbol{d}|\boldsymbol{b}, \boldsymbol{S}, \beta, \mathcal{I}) = e^{-\sum_i b_i} \prod_i \frac{\beta(b_i + S_i)^{d_i} e^{-S_i} + (1-\beta) b_i^{d_i}}{\Gamma(d_i + 1)} \tag{23.38}$$

where β is, as before, the prior probability that the data point contains both signal and background contributions. In order to infer the signal $\{S_i\} = \boldsymbol{S}$ we need the distribution $p(\boldsymbol{S}|\boldsymbol{d}, \mathcal{I})$, which is given by Bayes' theorem as

$$p(\boldsymbol{S}|\boldsymbol{d}, \boldsymbol{x}, \mathcal{I}) = \frac{p(\boldsymbol{S}|\xi, \boldsymbol{x}, \mathcal{I}) \cdot p(\boldsymbol{d}|\boldsymbol{S}, \boldsymbol{x}, \mathcal{I})}{p(\boldsymbol{d}|\xi, \boldsymbol{x}, \mathcal{I})}. \tag{23.39}$$

The symbol $\boldsymbol{x}$ stands for the pivot points. Moments of a particular signal value S_k result from

$$\langle S_k^\nu \rangle = \int d^N S \; S_k^\nu \; \frac{p(\boldsymbol{S}|\xi, \boldsymbol{x}, \mathcal{I}) \cdot p(\boldsymbol{d}|\boldsymbol{S}, \boldsymbol{x}, \mathcal{I})}{p(\boldsymbol{d}|\xi, \boldsymbol{x}, \mathcal{I})}. \tag{23.40}$$

The prior PDF $p(\boldsymbol{S}|\xi, \mathcal{I})$ is derived from equation (23.40):

$$p(\boldsymbol{S}|\xi, \not{\boldsymbol{x}}, \mathcal{I}) = \frac{1}{\xi^N} e^{-\frac{1}{\xi}\sum_i S_i}. \tag{23.41}$$

More complicated is the calculation of the marginal likelihood $p(\boldsymbol{d}|\boldsymbol{x}, \boldsymbol{S}, \mathcal{I})$:

$$p(\boldsymbol{d}|\boldsymbol{x}, \boldsymbol{S}, \mathcal{I}) = \int d^E c \; p(\boldsymbol{c}|\boldsymbol{x}, \not{\boldsymbol{S}}, \mathcal{I}) \int d\beta \; p(\beta|\not{\boldsymbol{x}}, \not{\boldsymbol{S}}, \mathcal{I}) \; P(\boldsymbol{d}|\boldsymbol{x}, \boldsymbol{S}, \boldsymbol{c}, \beta, \mathcal{I}). \tag{23.42}$$

We have chosen here to represent the background instead of the numerical values b_i at pivotal points $\{x_i\}$ by the pivot points $\boldsymbol{x}$ and the expansion coefficients $\boldsymbol{c}$. Logically irrelevant conditionals have been struck out. The inner β integral has been performed earlier (see equations (22.47) [p. 381] to (22.58)) for a uniform prior $p(\beta)$. It yields the marginal likelihood $p(\boldsymbol{d}|\boldsymbol{x}, \boldsymbol{S}, \boldsymbol{c}, \mathcal{I})$. For the moments of S_k we then have

$$\langle S_k^\nu \rangle = \int d^N S \, d^E c \;\; S_k^\nu \; \rho(\boldsymbol{S}, \boldsymbol{c}, \boldsymbol{d}) \tag{23.43}$$

with

$$\rho(\boldsymbol{S}, \boldsymbol{c}, \boldsymbol{d}) = \frac{p(\boldsymbol{S}|\xi, \mathcal{I}) \cdot p(\boldsymbol{c}|\boldsymbol{x}, \mathcal{I}) \cdot p(\boldsymbol{d}|\boldsymbol{x}, \boldsymbol{S}, \boldsymbol{c}, \mathcal{I})}{p(\boldsymbol{d}|\xi, \mathcal{I})},$$

and $p(\boldsymbol{c}|\boldsymbol{x}, \mathcal{I})$ given by equation (23.36b). It is clear that $\rho(\boldsymbol{S}, \boldsymbol{c}, \boldsymbol{d})$ is a positive definite function and the meaningful range of integration over S is $S_i \geq 0$. As a consequence, the expectation values of the signal $\langle S_k \rangle$ are all positive semidefinite. Our Bayesian analysis has therefore overcome the difficulty of negative signals which arise in the conventional procedure. At the same time, signal and background no longer add up exactly to the measured data but only within experimental error margins.

This completes the specification of both signal and background in a given data set $\boldsymbol{d}$. We have only given an outline here. We shall return to the problem of determining the background function in more detail in Section 27.2 [p. 474].

24
Change point problems

24.1 The Bayesian change point problem

The general problem of detecting and estimating the position of change points arises in many physics problems. The change point problem may occur as the change of a piecewise constant signal level with possibly associated piecewise constant noise levels. However, more complicated scenarios are possible such as for example the piecewise linear function model. This can be mapped to the previous class if we consider the first derivative. A piecewise linear model is equivalent to a piecewise constant first derivative. In the same sense a cubic spline interpolation problem reduces to the piecewise constant third derivative model. We shall investigate these examples in the following sections, and start out with the prototype case of a piecewise constant signal which has exactly one change point of unknown position.

24.1.1 Change in mean or variance

Let μ_1 and μ_2 denote the signal strengths before and after the change point. We shall assume that the signals are deteriorated by Gaussian noise with variances σ_1^2 and σ_2^2, respectively. Sample data for such a scenario are shown in the left panel of Figure 24.1. The likelihood function describing these data is

$$p(\boldsymbol{d} \mid \boldsymbol{\mu}, \boldsymbol{\sigma}, m, N) = \prod_{\nu=1}^{2} \left(\frac{1}{\sigma_\nu \sqrt{2\pi}} \right)^{N_\nu} \exp\left\{ -\frac{\sum_{i \in I_\nu} (d_i - \mu_\nu)^2}{2\sigma_\nu^2} \right\}, \tag{24.1}$$

with $I_1 = \{1, \dots, m\}$ and $I_2 = \{m+1, \dots, N\}$ being the indices of points belonging to the two sections separated by the change point with index m. The number of points in the first section is $N_1 = m$ and N is the total number of points. We can rewrite equation (24.1):

$$p(\boldsymbol{d} \mid \boldsymbol{\mu}, \boldsymbol{\sigma}, m, N) = \prod_{\nu=1}^{2} \left(\frac{1}{\sigma_\nu \sqrt{2\pi}} \right)^{N_\nu} \exp\left\{ -\frac{\overline{(\Delta d_\nu)^2} + (\mu_\nu - \overline{d_\nu})^2}{2\sigma_\nu^2 / N_\nu} \right\}, \tag{24.2}$$

where $\overline{d_\nu}$ and $\overline{(\Delta d_\nu)^2}$ are sample mean and variance of the data points in section ν. We shall first assume that we have no prior knowledge about the values $\boldsymbol{\mu}$, the amounts of noise $\boldsymbol{\sigma}$ or the change point position m itself. This shall be expressed by the prior distributions

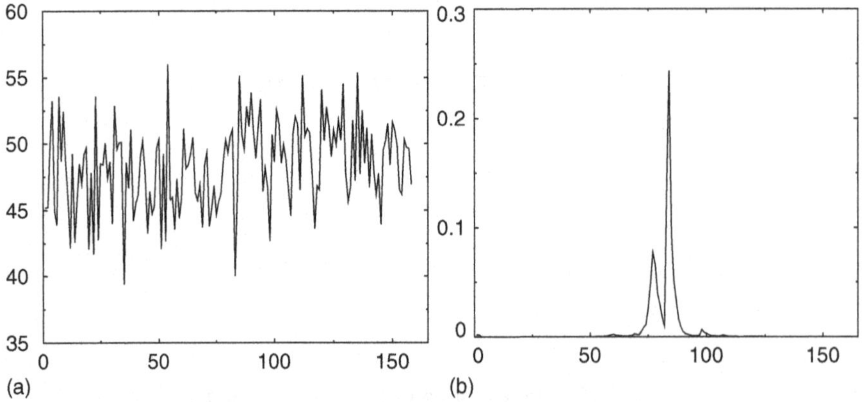

Figure 24.1 (a) Signal with $\mu_1 = 50$ for the first 80 points and $\mu_2 = 47$ for the remaining 78 points. The Gaussian noise is the same in the two regions and has $\sigma = 3$. (b) Unnormalized posterior distribution of the change point $p(m|\boldsymbol{d}, N, \mathcal{I})$.

$$p(\mu|\mathcal{I}) = \frac{1}{2W}, \qquad -W \leq \mu \leq W, \tag{24.3a}$$

$$p(\sigma|\mathcal{I}) = \frac{1}{2\log(B)}\,\frac{1}{\sigma}, \qquad 1/B \leq \sigma \leq B. \tag{24.3b}$$

These are regularized versions of the priors for a location parameter and a scale parameter. For $W \to \infty$ we obtain the improper location prior and for $B \to \infty$ Jeffreys' scale prior for a scale variable. The distribution that we want to find is $p(m|\boldsymbol{d}, N, \mathcal{I})$, which is obtained from Bayes' theorem:

$$p(m|\boldsymbol{d}, N, \mathcal{I}) = \frac{p(\boldsymbol{d}|m, N, \mathcal{I})\; p(m|N, \mathcal{I})}{p(\boldsymbol{d}|N, \mathcal{I})}. \tag{24.4}$$

Since we assume complete ignorance about the position of the single change point, the prior $p(m|N, \mathcal{I})$ is chosen to be uniform for $2 \leq m \leq N - 1$. We then need to calculate the marginal likelihood $p(\boldsymbol{d}|m, N, \mathcal{I})$. This is

$$p(\boldsymbol{d}|m, N, \mathcal{I}) = \int p(\boldsymbol{d}, \boldsymbol{\mu}, \boldsymbol{\sigma}|m, N, \mathcal{I})\, d\mu_1\, d\mu_2\, d\sigma_1\, d\sigma_2, \tag{24.5}$$

which, by introducing the above specified prior distributions, becomes

$$p(\boldsymbol{d}|m, N, \mathcal{I}) = \frac{1}{Z} \int\limits_{-W}^{W} d\mu_1 d\mu_2 \int\limits_{1/B}^{B} \frac{d\sigma_1}{\sigma_1}\frac{d\sigma_2}{\sigma_2}\; p(\boldsymbol{d}|\boldsymbol{\mu}, \boldsymbol{\sigma}, m, N). \tag{24.6}$$

We can now approximate the integrals over μ_1 and μ_2. Assume W to be finite but so large that the value of the μ-integrals is changed insignificantly if we replace the limits of integration $\pm W$ by $\pm\infty$. In other words, we assume that the standard errors for the sample estimates of μ_1 and μ_2, respectively, are much smaller than W, which is a sensible assumption if there are sufficient data on both sides of the change point. The integrals are then standard Gaussian and we obtain

$$p(\boldsymbol{d}|m, N, \mathcal{I}) = \frac{1}{Z} \frac{1}{\sqrt{N_1 N_2}} \prod_{\nu=1}^{2} \int_{1/B}^{B} \frac{d\sigma_\nu}{\sigma_\nu} \sigma_\nu^{-N_\nu+1} \exp\left\{ -\frac{\overline{(\Delta d_\nu)^2}}{2\sigma_\nu^2/N_\nu} \right\}. \tag{24.7}$$

The remaining integrals over σ are also standard if we choose B so large that we may replace the limits of integration by $0, \infty$ without altering the values of the integral. The integrals have the structure

$$\int_0^\infty \frac{d\sigma}{\sigma} \sigma^{-m+1} \exp\left(-\frac{\phi^2}{\sigma^2}\right) = \Gamma\left(\frac{m-1}{2}\right) (\phi^2)^{-\frac{m-1}{2}}. \tag{24.8}$$

The likelihood marginalized over $\mu_1, \mu_2, \sigma_1, \sigma_2$ then becomes

$$p(\boldsymbol{d}|m, N, \mathcal{I}) = \frac{1}{Z} \prod_{\nu=1}^{2} \frac{\Gamma\left(\frac{N_\nu-1}{2}\right)}{\sqrt{N_\nu}} \left(N_\nu \overline{(\Delta d_\nu)^2}\right)^{-\frac{N_\nu}{2}}. \tag{24.9}$$

Since the prior on the change point position was assumed flat:

$$p(m|\mathcal{I}) = \frac{1}{N-3} \boldsymbol{1}(1 < m < N), \tag{24.10}$$

the likelihood in equation (24.9) is also the unnormalized distribution $p(m|\boldsymbol{d}, N, \mathcal{I})$. This function is shown in the right-hand panel of Figure 24.1. The spiky structure of this function reflects the strong noise in the data. Nevertheless, the mean and standard deviation of this function yield $\overline{m} = 82.0 \pm 7.95$, which is in very good agreement with the 'true' change point position $m = 80$. We shall now briefly address the question of how these figures change as we supply more prior knowledge. Let us first assume that the noise level remains the same over the entire data range and therefore that only one common σ enters our likelihood function. In the calculation of the marginal likelihood the μ_1 and μ_2 integrations remain the same and the modification of equation (24.7) is

$$p(\boldsymbol{d}|m, N, \mathcal{I}) = \frac{1}{Z} \frac{1}{\sqrt{N_1 N_2}} \int_{1/B}^{B} \frac{d\sigma}{\sigma} \sigma_\nu^{-N+2} \exp\left\{ -\sum_i \frac{\overline{(\Delta d_\nu)^2}}{2\sigma^2/N_\nu} \right\} \tag{24.11}$$

which yields, according to equation (24.8),

$$p(\boldsymbol{d}|m, N, \mathcal{I}) \propto \frac{1}{\sqrt{m(N-m)}} \left(m\overline{(\Delta d_1)^2} + (N-m)\overline{(\Delta d_2)^2} \right)^{-\frac{N-2}{2}}. \tag{24.12}$$

Note the difference from the previous result in equation (24.9). The resulting probability density is almost identical to that in the right-hand panel of Figure 24.1. The reason is obvious, the sample sizes to estimate σ_1 and σ_2 are sufficiently large for reliable estimates and the 'true' variances of the two regions are indeed identical. In view of the large noise level, we likewise expect only a minor narrowing of the curves if the mean values μ_1 and μ_2 are known. The estimate of m based on equation (24.12) now becomes $m = 80.5 \pm 5.61$. This means that the function in equation (24.12) is slightly narrower than that in equation (24.9). We should recall, that the same data set has been used in both cases and the smaller variance originates from the fact that we have introduced the prior knowledge that $\sigma_1 = \sigma_2 = \sigma$. This suggests studying the effect of introducing still further prior knowledge. Suppose the noise levels σ_1 and σ_2 for the common noise level σ are known numerically. In this case we need only the marginal likelihood $p(\boldsymbol{d}|\sigma_1, \sigma_2, m, N)$, which we can pick from our previous calculations in equation (24.2). As expected, the resulting probability density is still narrower and yields $m = 80.6 \pm 4.91$. The still considerable uncertainty is due to the large ratio $\frac{\sigma}{\Delta\mu}$. It is worthwhile mentioning that the particular numbers for $\overline{m}$ and $\overline{(\Delta m)^2}$ depend of course on the particular realization of the signal in Figure 24.1. Other sequences with different seeds of the random generator used to generate the data will lead to different values. An average over a large sample of equivalent signals confirms, however, the trend exhibited by the present data.

24.1.2 Analysis of a tokamak magnetic diagnostic

A second, less artificial example for fixing a single change point arises in data from magnetic diagnostics of tokamaks. A tokamak is a toroidal magnetic confinement plasma device. A traditional method for studying plasma dynamics employs pick-up coils distributed poloidally around the torus, so-called '*Mirnov*' coils. Signals in these coils are induced by time variation of the magnetic flux created by particular structures in the toroidally rotating plasma. Such structures can be expanded in a Fourier series with terms of the form $\exp\{i(m\theta - n\phi)\}$, with m, n being the poloidal and toroidal mode numbers and θ, ϕ the poloidal and toroidal angular coordinates. Each term in the expansion is called an (m, n)-mode. The upper panel of Figure 24.2 shows a typical Mirnov signal which exhibits a transition from a mixed (4,3)/(3,2) mode to a pure (3,2) mode. This trace is the time derivative of the magnetic flux. The time variation of the flux itself is obtained by integration, which can conveniently be done by employing the Fourier transform technique; the function $f(t)$ is transformed from the time domain to Fourier space yielding $F(\omega)$. In Fourier space the indefinite integration reduces to a division by the frequency ω. Back-transformation to the time domain $F^{-1}(\omega)$ then yields the desired magnetic flux as a function of time. This is displayed in the lower panel of Figure 24.2. It is immediately obvious that the signal is composed of two regions with different amplitudes and different noise levels. To find the change point between the two regions is a genuine Bayesian problem. In the absence of a first principles theory for the time variation of the magnetic flux in this experiment we choose to represent the associated magnetic field $B_p(t)$ as a low-order polynomial:

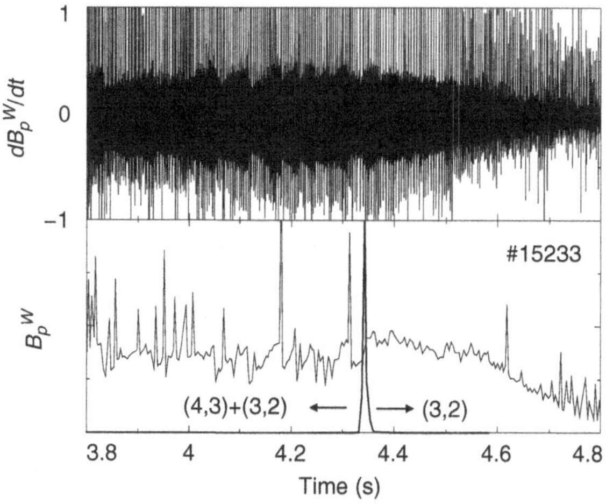

Figure 24.2 Mirnov coil signal from magnetic islands in a tokamak discharge (upper panel). The associated magnetic flux is shown in the lower panel. It results from a mixed mode (4,3) + (3,2). The (4,3) contribution dies away. The heavy trace spike shows the probability distribution of the change point between the two regions.

$$B_p(t) = \sum_{j=0}^{n_1} a_j^{(1)} t^j \ , \ t < T,$$

$$B_p(t) = \sum_{j=0}^{n_2} a_j^{(2)} t^j \ , \ t \geq T, \tag{24.13}$$

where T denotes the change point between the two domains. This model for the function $B_p(t)$ allows us then to formulate the likelihood of the problem. In the actual measurement the time variable is discrete and is denoted by t_i. The associated signal is d_i with noise variance σ_1^2 for $i \leq E$ and σ_2^2 for $i > E$, where E is the index of the change point channel and the total number of data is N. This yields the likelihood

$$p(\boldsymbol{d}|\boldsymbol{t}, \boldsymbol{\sigma}, \boldsymbol{n}, \boldsymbol{a}, E, \mathcal{I}) = \prod_{\nu=1}^{2} \left(\frac{1}{\sigma_\nu \sqrt{2\pi}} \right)^{N_\nu} \exp \left\{ - \sum_{i \in I_\nu} \frac{\left(d_i - \sum_{j=0}^{n_\nu} a_j^{(\nu)} t_i^j \right)^2}{2\sigma_\nu^2} \right\}.$$

From this likelihood we want to determine the change point channel number E. This requires the probability distribution $p(E|\boldsymbol{t}, \boldsymbol{d}, n_1, n_2, \mathcal{I})$. By Bayes' theorem

$$p(E|\boldsymbol{t}, \boldsymbol{d}, n_1, n_2, \mathcal{I}) = \frac{p(E|\mathcal{I}) \cdot p(\boldsymbol{d}|\boldsymbol{t}, E, n_1, n_2, \mathcal{I})}{p(\boldsymbol{d}|\boldsymbol{t}, n_1, n_2, \mathcal{I})}, \tag{24.14}$$

and this is related to the marginal likelihood $p(\boldsymbol{d}|\boldsymbol{t}, E, n_1, n_2, \mathcal{I})$ which we obtain from

$$p(\boldsymbol{d}|\boldsymbol{t}, E, n_1, n_2, \mathcal{I}) = \int d\sigma_1 d\sigma_2 d^N a \; p(\boldsymbol{d}, \sigma_1, \sigma_2, \boldsymbol{a}|\boldsymbol{t}, E, n_1, n_2, \mathcal{I}). \tag{24.15}$$

We assume flat priors on the expansion coefficients $\boldsymbol{a}$. This is debatable. Flat priors would not allow a model comparison. There could be cases where, in addition to the change point position E, the optimum expansion lengths n_1, n_2 were of interest. In this case one could not get away with flat improper priors since they could automatically select the simplest, e.g. a constant model. We are treating here a less demanding problem and shall be content with finding the change point E given the expansion orders n_1, n_2. Flat priors on $\boldsymbol{a}$ are then sufficient. We also want to get rid of the unknown variances. Here we need a truly uninformative prior for σ_1, σ_2. Even if a rough estimate of the variance of the signal dB_p/dt were initially available, this information is completely blurred in the integration step. With these assumptions the marginal likelihood reduces to

$$p(\boldsymbol{d}|\boldsymbol{t}, E, n_1, n_2, \mathcal{I}) = \int \frac{d\sigma_1}{\sigma_1} \frac{d\sigma_2}{\sigma_2} d^N a \; p(\boldsymbol{d}|\boldsymbol{t}, E, \sigma_1, \sigma_2, n_1, n_2, \boldsymbol{a}, \mathcal{I}). \tag{24.16}$$

Inspection of equation (24.16) reveals that the integral factors into an integral over $d\sigma_1 d\boldsymbol{a}^{(1)}$ and another over $d\sigma_2 d\boldsymbol{a}^{(2)}$. Both are of the same basic form. We outline the integration over $d\sigma_1 d\boldsymbol{a}^{(1)}$. Here it is convenient to go over to matrix notation. Let $\mathbf{M}$ be the matrix

$$\mathbf{M} = \left\{ \begin{array}{ccccc} 1 & t_1 & t_1^2 & \cdots & t_1^{n_1} \\ \vdots & & & & \\ 1 & t_E & t_E^2 & \cdots & t_E^{n_1} \end{array} \right\}. \tag{24.17}$$

The integral over $d\sigma_1 d\boldsymbol{a}^{(1)}$, which we denote by I_a, can then be written as

$$I_a = \left(\frac{1}{\sigma_1 \sqrt{2\pi}} \right)^E \cdot \int \frac{d\sigma_1}{\sigma_1} d^N a \; \exp\left\{ -\frac{1}{2\sigma_1^2} (\boldsymbol{d} - \mathbf{M}\boldsymbol{a})^T (\boldsymbol{d} - \mathbf{M}\boldsymbol{a}) \right\}. \tag{24.18}$$

For notational convenience, we have omitted the upper index of $\boldsymbol{a}^{(1)}$. The order of integration does not matter of course. It is convenient to perform the $\boldsymbol{a}$-integration first. To this end we need to express the argument of the exponential as a complete square plus a residue:

$$(\boldsymbol{d}_1 - \mathbf{M}\boldsymbol{a})^T (\boldsymbol{d}_1 - \mathbf{M}\boldsymbol{a}) = (\boldsymbol{a} - \boldsymbol{a}_0)\mathbf{Q}_1(\boldsymbol{a} - \boldsymbol{a}_0) + R_1. \tag{24.19}$$

Employing the standard result for integrals over multivariate Gaussians, given in Appendix A.4 [p. 602], we obtain

$$I_a = \left(\frac{1}{\sigma_1 \sqrt{2\pi}} \right)^{E-n_1} |\mathbf{Q}_1|^{-1/2} \cdot \exp\left\{ -\frac{1}{2\sigma_1^2} R_1 \right\}. \tag{24.20}$$

Comparison of coefficients that are quadratic, linear and independent of $\boldsymbol{a}$ in equation (24.19) yields

$$\mathbf{Q}_1 = \mathbf{M}_1^T \mathbf{M}_1, \quad R_1 = \boldsymbol{d}_1^T \left(\mathbb{1} - \mathbf{M}_1 (\mathbf{M}_1^T \mathbf{M}_1)^{-1} \mathbf{M}_1^T \right) \boldsymbol{d}_1. \tag{24.21}$$

The final integration over σ_1 is also of the previous standard form and yields finally

$$I_a = \left(\frac{1}{\pi}\right)^{\frac{E-n_1}{2}} |\mathbf{Q}_1|^{-1/2}\, \Gamma\left(\frac{E-n_1}{2}\right) R_1^{-\frac{E-n_1}{2}}. \tag{24.22}$$

Then the desired marginal likelihood becomes finally

$$p(\boldsymbol{d}|t, E, n_1, n_2, \mathcal{I}) = \pi^{-\frac{N-n_1-n_2}{2}} \prod_{\nu=1}^{2} |\mathbf{Q}_\nu|^{-1/2}\, \Gamma\left(\frac{N_\nu - n_\nu}{2}\right) R_\nu^{-\frac{N_\nu - n_\nu}{2}}. \tag{24.23}$$

The final step for our change point calculation is the assignment of a prior to the position E. We choose it flat. The range is from $n_1 + 1$ to $N - n_2 - 1$ since the minimum ranges which are necessary to define a polynomial of order n_i are n_{i+1} data points, therefore

$$p(E|\mathcal{I}) = \frac{1}{N - n_1 - n_2 - 2}. \tag{24.24}$$

This completes the analysis. For the treatment of the particular data in Figure 24.2 we have chosen $n_1 = n_2 = 4$, but $n_1 = n_2 = 5, 6$ gave an essentially identical result. The distribution $p(E|\boldsymbol{t}, \boldsymbol{d}, n_1, n_2, \mathcal{I})$ is shown in the lower panel of Figure 24.2 as the heavy trace. In view of the previous experience with the simpler change point problem in much noisier data, the very precise location of the change point here does not really come as a surprise.

24.2 Change points in a binary image

We turn now to a more complicated case. It requires a non-Gaussian likelihood and a number of change points greater than one. The data which we are going to consider consist of one line of the image shown in Figure 24.3. The figure shows a binary image obtained from phase-shifting speckle interferometry of an inclined microscopically rough surface. The determination of change points in this image is required as a first step for surface reconstruction from the interferogram. We shall limit our consideration to change point detection in a single line of the image. The change point problem does not lend itself to easy generalization in more than one dimension. The sequence of zeros and ones obtained in a single line is shown above the image. This one-dimensional function shows alternating regions dominated by 'zeros' or 'ones', respectively. The noise in this data set consists of 'ones' distorting the 'zero-dominated' regions and 'zeros' distorting the 'one-dominated' regions. Let $\boldsymbol{\xi}$ denote the positions of the change points and $[\xi_1, \xi_E]$ be the support interval of the function. Given an interval $\xi_r, \xi_r + 1$, the probability of having n_r zeros in a total of N_r data points is then binomial:

$$p(n_r|N_r, q, \boldsymbol{\xi}, \mathcal{I}) = \binom{N_r}{n_r} q^{n_r}(1-q)^{N_r - n_r}, \tag{24.25}$$

where q is the individual probability for a zero in this interval. The function in equation (24.25) has its maximum as a function of q at $q = n_r/N_r$, the fraction of zeros in

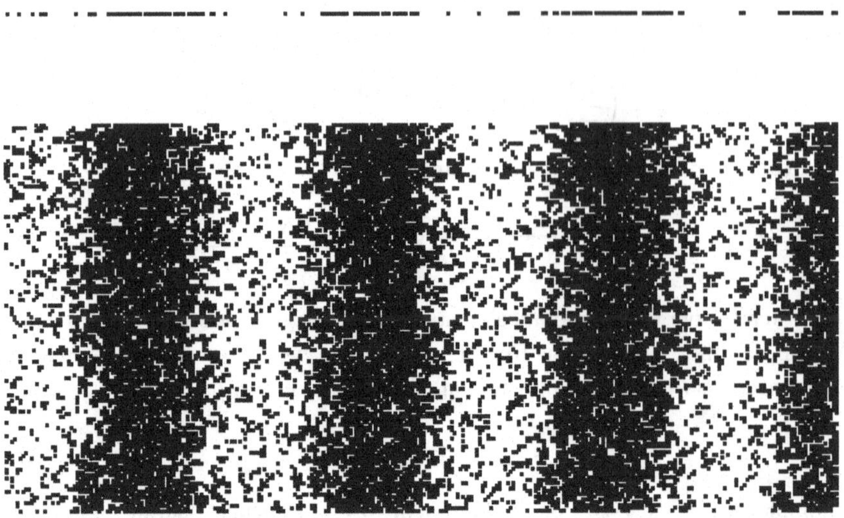

Figure 24.3 Binary image obtained from phase-shifting speckle interferometry of an inclined rough surface. The upper trace represents a single line of the image with functional values '1' assigned to the black dots and '0' assigned to the white dots.

N_r data points. This legitimates the interpretation of q as the signal-to-noise ratio in the present case. We assume that it is constant in the whole data range $[\xi_1, \xi_E]$, but unknown. Let us now consider the interval ξ_{r+1}, ξ_{r+2} which follows the above domain. The role of 'zeros' and 'ones' given q is reversed in this interval and the probability of finding n_{r+1} zeros becomes

$$p(n_{r+1}|N_{r+1}, q, \xi, \mathcal{I}) = \binom{N_{r+1}}{n_{r+1}} (1-q)^{n_{r+1}} q^{N_{r+1}-n_{r+1}}. \tag{24.26}$$

The two different cases of the likelihood given in equations (24.25) and (24.26) can be merged into a single expression if we introduce the auxiliary variables

$$\alpha_r = \begin{cases} n_r & \text{if } r \text{ is odd} \\ N_r - n_r & \text{if } r \text{ is even.} \end{cases} \tag{24.27}$$

Accordingly, $\beta_r = N_r - \alpha_r$. This yields the likelihood for the whole data set $(\boldsymbol{n}, \boldsymbol{N})$:

$$p(\boldsymbol{n}|\boldsymbol{N}, q, \boldsymbol{\xi}, E, \mathcal{I}) = \prod_{r=1}^{E-1} \binom{N_r}{\alpha_r} q^{\alpha_r} (1-q)^{\beta_r}. \tag{24.28}$$

The product can conveniently be split into

$$p(\boldsymbol{n}|\boldsymbol{N}, q, \boldsymbol{\xi}, E, \mathcal{I}) = q^{\sum_r \alpha_r} (1-q)^{\sum_r \beta_r} \cdot \prod_{r=1}^{E-1} \binom{N_r}{\alpha_r}. \tag{24.29}$$

We assume the number of change points E to be given in the sequel. The parameters which remain variable in equation (24.29) are then q and $\boldsymbol{\xi}$. Note that our notation is redundant in that a knowledge of $\boldsymbol{\xi}$ implies a knowledge of $\boldsymbol{N}$. We shall therefore simplify notation. Before we turn to the calculation of the change point positions we shall estimate the signal-to-noise ratio $\langle q \rangle$ by employing the likelihood equation (24.29):

$$p(q|\boldsymbol{n}, \boldsymbol{\xi}, E, \mathcal{I}) = \frac{p(q|\mathcal{I}) \cdot p(\boldsymbol{n}|q, \boldsymbol{\xi}, E, \mathcal{I})}{p(\boldsymbol{n}|\boldsymbol{\xi}, E, \mathcal{I})}. \tag{24.30}$$

Evaluation of $\langle q \rangle$ from equation (24.30) requires the choice of a prior $p(q|\mathcal{I})$ for q. We choose a two-parameter normalized beta distribution (see Section 7.5.2 [p. 100]):

$$p(q|a, b) = \frac{\Gamma(a+b)}{\Gamma(a) \cdot \Gamma(b)}\, q^{a-1}(1-q)^{b-1}. \tag{24.31}$$

The reason for this particular choice is that equation (24.31) is conjugate to the likelihood equation (24.29) and at the same time sufficiently flexible to accommodate two pieces of prior knowledge on q. The mean and variance of q from equation (24.31) alone are, according to Section 7.5.2 [p. 100], given by $a/(a+b)$ and $a \cdot b/(a+b+1)(a+b)^2$. Equation (24.31) does of course include the flat prior in $[0, 1]$ for $a = b = 1$. The flat prior is, however, not the one with the largest variance in the interval $[0, 1]$. If we rewrite the variance of the β distribution in terms of $x = b/a$ and $s = a + b$, we obtain

$$\text{var}(x, s|a, b) = \frac{x}{(1+x)^2} \cdot \frac{1}{s+1}. \tag{24.32}$$

This function peaks at $x = 1/2$ or $a = b$ or $\langle q \rangle = 1/2$ and the least informative prior, namely that with maximum variance, is obtained for $s \to 0$. For $s = 0$ the prior in equation (24.32) is not normalizable and it can be shown that it transforms into Jeffreys' scale prior in $[0, \infty]$ under the mapping $q = z/1+z$. We shall first calculate $\langle q \rangle$ and $\langle \Delta q^2 \rangle$ from equation (24.30) with a, b arbitrary and consider special cases afterwards:

$$p(\boldsymbol{n}|\boldsymbol{\xi}, E, \mathcal{I}) = \prod_{r=1}^{E-1} \frac{\Gamma(\alpha_r + \beta_r + 1)}{\Gamma(\alpha_{r+1})\Gamma(\beta_{r+1})} \cdot \frac{\Gamma(a+b)}{\Gamma(a)\Gamma(b)} \cdot \int_0^1 q^{\sum_r \alpha_r + a - 1} (1-q)^{\sum_r \beta_r + b - 1}\, dq, \tag{24.33}$$

$$p(\boldsymbol{n}|\boldsymbol{\xi}, E, \mathcal{I}) = \prod_{r=1}^{E} \frac{\Gamma(\alpha_r + \beta_r + 1)}{\Gamma(\alpha_r + 1)\Gamma(\beta_r + 1)} \frac{\Gamma(a+b)}{\Gamma(a)\Gamma(b)} \cdot \frac{\Gamma(a + \sum_r \alpha_r)\Gamma(b + \sum_r \beta_r)}{\Gamma(N + a + b)}. \tag{24.34}$$

Here N denotes the total number of data points, $N = \sum_r (\alpha_r + \beta_r)$. Moments $\langle q^n \rangle$ of the posterior equation (24.30) are then given by

$$\langle q^n \rangle = \int_0^1 dq q^n\, \frac{p(q|a, b)\, p(\boldsymbol{n}|q, \boldsymbol{\xi}, E, \mathcal{I})}{p(\boldsymbol{n}|\boldsymbol{\xi}, E, \mathcal{I})}. \tag{24.35}$$

The necessary integral can be derived from equation (24.33) by augmenting a to $a+n$. We find in particular

$$\langle q \rangle = \frac{a + \sum \alpha_r}{a+b+N}, \tag{24.36}$$

$$\langle \Delta q^2 \rangle = \frac{\left(a + \sum \alpha_r\right)}{(N+a+b)^2} \frac{\left(b + \sum \beta_r\right)}{(N+a+b+1)}. \tag{24.37}$$

If we now assume the least informative prior of the form given in equation (24.31), which means $a \to 0$ and $b \to 0$, we observe that the results of equations (24.36) and (24.37) remain well behaved. Moreover, the result for $\langle q \rangle$ is entirely in line with intuition; it is the fraction of N data points which is correctly described by equation (24.27). We now turn to the determination of the change point positions. From Bayes' theorem we have

$$p(\boldsymbol{\xi}|\boldsymbol{n}, E, \mathcal{I}) = \frac{p(\boldsymbol{\xi}|E, \mathcal{I}) \cdot p(\boldsymbol{n}|\boldsymbol{\xi}, E, \mathcal{I})}{p(\boldsymbol{n}|E)}. \tag{24.38}$$

Note that the denominator is the $\boldsymbol{\xi}$-integral of the numerator, and the PDF is therefore properly normalized. It remains to specify the prior distribution $p(\boldsymbol{\xi}|E, \mathcal{I})$. We assume here a flat prior on each element of $\boldsymbol{\xi}$ in $[\xi_1, \xi_E]$ and point out that the components of $\boldsymbol{\xi}$ take on only integer values. The number of ways to draw $E-2$ pivots from $N-2$ grid points (remember that the endpoints ξ_1, ξ_E are fixed) is

$$Z = \binom{N-2}{E-2} = \frac{(N-2)!}{(E-2)!(N-2-E+2)}, \tag{24.39}$$

hence

$$p(\boldsymbol{\xi}|E, \mathcal{I}) = \frac{(E-2)!(N-E)!}{(N-2)!}. \tag{24.40}$$

Though we are dealing with a fixed number of change points in the present example, it is instructive to abandon this assumption for a moment. We are then led to compare the priors for E and $E+1$ pivots

$$\frac{p(\boldsymbol{\xi}|E+1, \mathcal{I})}{p(\boldsymbol{\xi}|E, \mathcal{I})} = \frac{(E-1)!(N-E-1)!}{(E-2)!(N-E)!} = \frac{E-1}{N-E} \cong \frac{E}{N}, \tag{24.41}$$

which means that the likelihood in going from the proposition E to $E+1$ must at least increase by a factor of N/E in order to compensate for the decrease of the constant prior in order to obtain the same global probability or evidence of the data $p(\boldsymbol{n}|E)$. For numerical values of the change points and their variances we need to take moments $\langle \xi^n \rangle$ from equation (24.38). The necessary integration is a typical Monte Carlo problem which lends itself readily to treatment with the Metropolis–Hastings algorithm (see Chapter 30 [p. 537]). Note that for an application of the Metropolis–Hastings algorithm it is sufficient to know that the sampling density is normalizable. It is not required that the normalization is known. This means that we can leave the denominator $p(\boldsymbol{n}|E)$ undetermined in equation (24.38) for sampling the $\boldsymbol{\xi}$-space to evaluate $\langle \xi_k^n \rangle$. Results for the change point positions together

Table 24.1 Calculated change point positions for the data in the upper panel of Figure 24.3 for various choices of the prior on q

25.8	73.1	120.1	145.6	192.5	246.8	283.8
± 13.9	± 18.1	± 13.6	± 15.0	± 20.7	± 15.5	± 15.2
32.5	75.0	118.2	149.6	196.1	243.4	282.8
± 4.5	± 5.5	± 5.3	± 4.6	± 6.6	± 5.1	± 4.9
25.6	65.8	112.0	146.6	192.0	240.9	279.9
± 8.1	± 11.0	± 10.4	± 11.0	± 15.2	± 12.9	± 11.6
32.5	74.4	117.2	150.2	196.8	242.7	282.0
± 4.1	± 5.3	± 4.8	± 3.8	± 5.6	± 5.1	± 4.3

with confidence limits are given in the first line of Table 24.1 . The prior distributions which we have used so far were the least informative for the signal-to-noise ratio q as well as for the distribution of pivots. We shall now investigate the effect of more informative priors. Let us consider $p(q|\mathcal{I})$ first. If we make more data available, for example the whole binary image, then we can estimate from the preferentially black (white) regions along a vertical line the number of scattered in white (black) points. From equation (24.36) for $N \gg 1$ we know that the percentage of black points in a black segment is the signal-to-noise ratio $\langle q \rangle$. It fixes the ratio of a and b entering equation (24.31). From the variation of this fraction along different black or white sections we obtain an estimate of the variance in equation (24.37) (again for $N \gg 1$). This provides the second relation for a and b in equation (24.36). Choosing columns 21, 22, 23 of the image in Figure 24.3 for the determination of the signal-to-noise ratio yields $\langle q \rangle = 0.86 \pm 0.02$. From equation (24.32) we obtain the corresponding exponents $a = 41$ and $b = 246$. The choice of only three columns for the determination of the signal-to-noise ratio is arbitrary. Including more than three columns would make the prior information even more precise. However, even the prior information based on only three columns narrows down the posterior estimates of the change point positions considerably. Positions and confidence intervals are given in the second line of Table 24.1 . The second prior distribution, $p(\boldsymbol{\xi}|E)$, can also easily be made more informative. For the laser speckle data in Figure 24.3 it is obvious for it is extremely unlikely for two successive change points ξ_r, ξ_{r+1} to fall 'close' to each other. This assumption is of course general and still rather diffuse, so that it will apply to a much wider class of data than those in Figure 24.3. We model the idea that neighbouring pivots should 'repel' each other in the distribution

$$p(\xi_2, \ldots, \xi_{E-1}|\xi_1, \xi_E) = \frac{1}{Z} (\xi_2 - \xi_1)^n (\xi_3 - \xi_2)^n \ldots (\xi_E - \xi_{E-1})^n. \tag{24.42}$$

Though we do not need to know the normalization Z in the present context since we perform the ξ-integration by Monte Carlo techniques, it will be necessary in a later application. It is convenient to introduce the notation $z_i = \xi_i - \xi_1$. This yields

$$Z = \int_0^{z_3} dz_2 \int_0^{z_4} dz_3 \cdots \int_0^{z_E} dz_{E-1}\, z_2^n (z_3 - z_2)^n \cdots (z_E - z_{E-1})^n. \tag{24.43}$$

Now reverse the order of integrations:

$$Z = \int (z_E - z_{E-1})^n \cdots (z_4 - z_3)^n \cdot \int_0^{z_3} dz_2 (z_3 - z_2)^n z_2^n. \tag{24.44}$$

Substituting $\eta_i = z_i / z_{i+1}$ converts the inner integral to

$$\int_0^{z_3} dz_2 (z_3 - z_2)^n z_2^n = z_3^{2n+1} \int_0^1 d\eta (1 - \eta)^n \eta^n. \tag{24.45}$$

The integral on the right-hand side is the well-known beta function (see Section 7.5.2 [p. 100]) $B(n+1, n+1) = \Gamma(n+1)^2 / \Gamma(2n+2)$. The following integral over z_3 can be obtained by the same procedure; the arguments of the beta function however, become, different. Repeating these cycles until all integrals are done leads to

$$Z = \frac{(n!)^{E-1} (\xi_E - \xi_1)^{E(n+1)-(n+2)}}{\Gamma((E-1)(n+1))}. \tag{24.46}$$

It is of course instructive to calculate moments $\langle \xi^n \rangle$ from equation (24.38) with this more informative prior. The results, when only the more informative prior on the pivot points is included, are displayed in the third line of Table 24.1 and are seen to narrow down the variances in the first line appreciably. The last line in Table 24.1 presents finally the estimates of pivots and uncertainties using informative priors on both the pivot distribution given in equation (24.42) and the signal-to-noise ratio. As expected, this yields the most precise determination of the change point positions. The analysis of one line of the binary image in Figure 24.3 is thereby exhausted if we insist on the model of step-like changes.

24.3 Neural network modelling

But does a model of step-like changes really offer the best description of the data? It is by no means granted from an inspection of Figure 24.3 that homogeneous white stripes with occasional black points alternate with homogeneous black stripes with occasional white points. One might as well suspect some kind of smooth transition from black to white regions and vice versa. That this view is in fact supported by the data is revealed in Figure 24.4. The open circles in this figure are the column-wise sums of Figure 24.3 if we associate a '1' with a black point and a '0' with a white point in Figure 24.3. Note that the implicit assumption in taking column sums is that the stripes in Figure 24.3 are not only

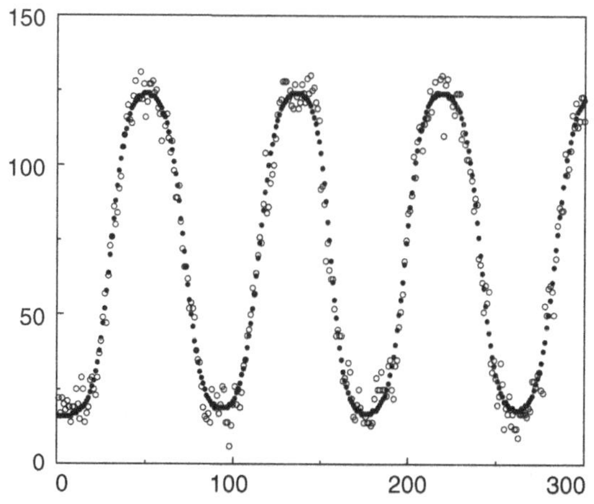

Figure 24.4 Open circles represent the column-wise sums of the binary image of Figure 24.3 if a '0' is associated with white spots and a '1' with black spots. The full dots are the fit to the data according to equations (24.47), (24.48) and (24.49).

parallel to each other but also strictly vertically oriented. At present these assumptions do not seem unreasonable. In contrast to our previous step function model of equation (24.25), we now need to formulate a model with locally varying signal-to-noise ratio $q(x)$ where x is the horizontal coordinate. Instead of equation (24.25) we then obtain the likelihood

$$L := p(n(x)|N, q(x), E, \mathcal{I}) = \binom{N}{n(x)} q(x)^{n(x)}(1 - q(x))^{N-n(x)}, \tag{24.47}$$

where N is the number of lines of the image in Figure 24.3, $n(x)$ the number of black dots in the column at position x, $q(x)$ the locally varying signal-to-noise ratio and E the number of change points that we consider to be given. By inspection of Figure 24.3 we find $E = 7$. Next we model the function $q(x)$. A convenient functional form for a single transition from '0' to '1' which contains the previous step function as a limiting case is given by a sort of Fermi function

$$\sigma(x) = \frac{1}{e^{v(1+ux)} + 1}. \tag{24.48}$$

This function takes on the value $\sigma = 1/2$ at $1 + ux = 0$. The width of the transition is given by $1/(uv)$. For $|v(1 + ux)| \gg 1$ and $ux < -1$, $\sigma(x)$ tends to one and describes a 'black' region, while for $ux > -1$ the function $\sigma(x)$ tends to zero and represents a 'white' region. The meaning of the two parameters u, v is thereby explained. The upper trace in Figure 24.5 displays the function σ for $v < 0$. It is obvious how to construct from such elements a function which exhibits multiple transitions between zero and one. Such a function is obtained by adding up a number of shifted σ functions with alternating sign, as shown in Figure 24.5. The explicit form is

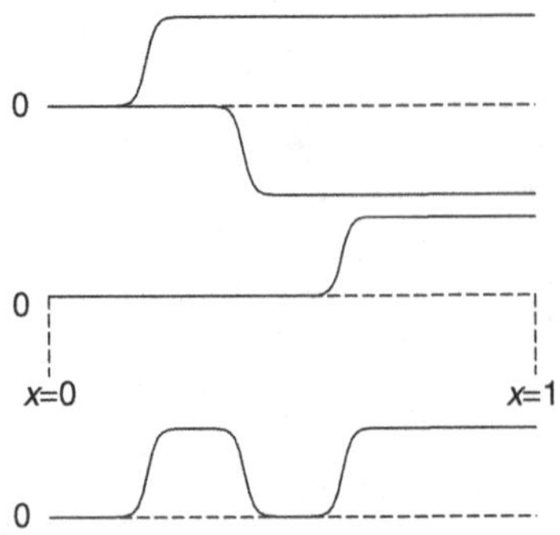

Figure 24.5 The upper trace shows a single sigma function as given in equation (24.48). Suitable combinations of shifted sigma functions of opposite sign are seen to combine to a function as in equation (24.49). See also the full dots in Figure 24.4.

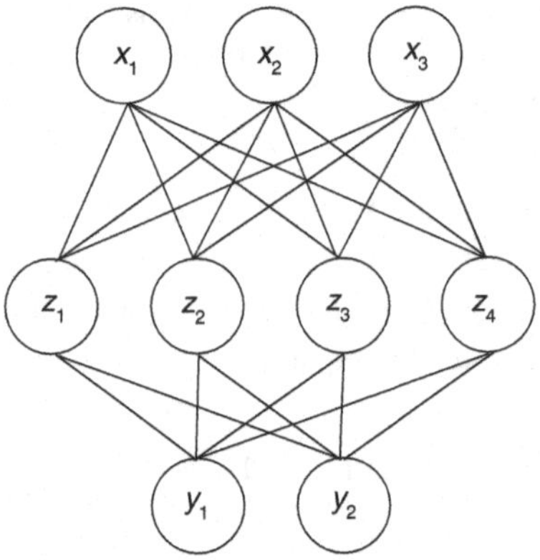

Figure 24.6 Schematic representation of a neural network with one hidden layer $\{z_1, \ldots, z_4\}$. $\{x_1, x_2, x_3\}$ are input and $\{y_1, y_2\}$ response variables. The intermediate signals z_i are sigma functions of the input variables of equation (24.48).

$$\tilde{q}(x) = \sum_i a_i \sigma(x|u_i, v_i), \tag{24.49}$$

with coefficients $a_i = (-1)^{i-1}$ for the special case at hand. $\tilde{q}(x)$ is a very simple special case of a neural network with one input x, one output q and one hidden layer. A graphical representation of a neural network with three input variables, two output variables and one hidden layer is shown in Figure 24.6. Each 'neuron' of the hidden layer is a σ function of the type in equation (24.48), however, with a more complicated argument. These hidden 'neurons' produce from the input variables x_1, x_2, x_3 the response variables $z_1, \ldots, z_4$. The outputs y_i are again formed from the new intermediate variables $z_1, \ldots, z_4$. In the simplest case, the operation $y_i(z)$ can be linear as assumed in equation (24.49) or, more complicated, a nonlinear mapping for which a convenient and common choice is again the 'neuron' in equation (24.48). The attractive feature of neural networks is that they can represent any multivalued function of several independent variables $\boldsymbol{x}$ to an accuracy which is only limited by the number of 'neurons' in the hidden layer. One last piece of information still needs to be incorporated in equation (24.49). $\tilde{q}(x)$ will tend to '0' and '1' in the respective white and black parts of the image. However, there are no entirely black and entirely white parts of the image, as can be seen from Figure 24.3 [p. 416]. In Figure 24.4 this is reflected in the fact that the column-wise sums vary between some finite level and an upper value somewhat less than N. This represents the noise in the image. We account for this noise by introducing a final parameter b and obtain

$$q(x) = (1-b) \cdot \sum_i a_i \sigma(x|u_i, v_i) + b \cdot \frac{1}{2}. \tag{24.50}$$

The likelihood for our analysis is thereby fully specified. The goal is of course the determination of the parameters $b, \boldsymbol{u}, \boldsymbol{v}$ which specify the model function $q(x)$. Application of Bayes' theorem then requires us to formulate appropriate prior probabilities. Formally,

$$p(\boldsymbol{u}, \boldsymbol{v}, b|\mathcal{I}) = p(\boldsymbol{u}|\mathcal{I}) \cdot p(\boldsymbol{v}|\boldsymbol{u}, \mathcal{I}) \cdot p(b|\not{\boldsymbol{v}}, \not{\boldsymbol{u}}, \mathcal{I}). \tag{24.51}$$

In the last probability the overall offset of the image b is logically independent of the parameters $\boldsymbol{u}, \boldsymbol{v}$, which specify the location and shape of the transition. Unless we have more precise independent knowledge about the numerical value of b, we assume a flat prior in the range $0 < b \leq b_{\max}$:

$$p(b|\mathcal{I}) = \frac{1}{b_{\max}} \boldsymbol{1}(0 < b \leq b_{\max}). \tag{24.52}$$

The prior distribution for $\boldsymbol{u}$ has been derived previously in Chapter 10 [p. 165]. The function $ux + 1 = 0$ is a special 'hyperplane' in two dimensions, for which we obtain

$$p(\boldsymbol{u}|\mathcal{I}) = r_0^N \prod_i \frac{1}{u_i^2}, \qquad u_i > r_0, \tag{24.53}$$

where N is the dimension of $\boldsymbol{u}$. Finally we are left with the determination of the conditional distribution $p(\boldsymbol{v}|\boldsymbol{u}, \mathcal{I})$. The key to this distribution is the minimum global slope of the function $q(x)$. The slope of a single $0 \to 1$ transition is proportional to $\sigma' = uv$. Consequently, the expression $\sum_i u_i^2 v_i^2$ is testable (measurable) information. The principle of maximum entropy assigns to this information the prior

$$p(\boldsymbol{v}|\boldsymbol{u}, \lambda, \mathcal{I}) = \frac{1}{Z} \exp\left\{-\lambda \sum_i u_i^2 v_i^2\right\}. \tag{24.54}$$

The normalization is

$$Z = \int_{-\infty}^{\infty} d^N v \exp\left\{-\lambda \sum_i u_i^2 v_i^2\right\} = \prod_i \left(\frac{2\pi}{\lambda u_i^2}\right)^{1/2}. \tag{24.55}$$

Merging the prior probabilities, we obtain

$$p(\boldsymbol{u}, \boldsymbol{v}, b|\lambda, \mathcal{I}) = \frac{1}{Z}\lambda^{N/2}\left|\prod_i \frac{1}{u_i}\right| \exp\left\{-\frac{\lambda}{2}\sum_i u_i^2 v_i^2\right\}, \tag{24.56}$$

with an unimportant normalization constant Z. The limits on u_i and b are omitted for notational simplicity.

The last point which needs consideration is the unknown hyperparameter λ. It is a nuisance parameter and a convenient way of dealing with it is marginalization. Since λ is a scale parameter, the appropriate prior distribution for marginalizing over λ is Jeffreys' scale prior. The integral is of the standard Gamma type, yielding

$$p(\boldsymbol{u}, \boldsymbol{v}|\mathcal{I}) = \frac{1}{Z'}\left|\prod_i \frac{1}{u_i}\right| \left(\sum u_i^2 v_i^2\right)^{-\frac{N}{2}}. \tag{24.57}$$

The problem is now fully specified and posterior estimates of the parameters can be derived from Bayes' theorem:

$$p(\boldsymbol{u}, \boldsymbol{v}, b|n(x), N, \boldsymbol{x}, E, \mathcal{I}) = \frac{p(\boldsymbol{u}, \boldsymbol{v}, b|\mathcal{I}) \cdot p(n(x)|\boldsymbol{u}, \boldsymbol{v}, b, N, \boldsymbol{x}, E, \mathcal{I})}{p(n(x)|N, \boldsymbol{x}, E, \mathcal{I})}. \tag{24.58}$$

Neither the likelihood of equation (24.47) nor the prior of equation (24.57) are of Gaussian type in this case. Posterior estimates of the parameters must therefore either rely on numerical integration or on the Gaussian approximation, which provides easy access to the evidence denominator in equation (24.58). Our calculations have employed the Gaussian approximation. The model function equation (24.47) along with the definition of $q(x)$ in equation (24.50), evaluated at the obtained posterior values of the parameters, is shown as full dots in Figure 24.4. They are seen to provide quite a convincing fit to the data. While marginalization of a nuisance parameter is certainly the orthodox Bayesian procedure, it may not always be possible analytically. It is therefore instructive to discuss an alternative approximate method of dealing with the hyperparameter λ. The basic idea is to use Bayesian probability theory to determine λ explicitly. This requires the PDF $p(\lambda|\boldsymbol{n}(x), N, E, \mathcal{I})$ for λ. This distribution results from Bayes' theorem:

$$p(\lambda|\boldsymbol{n}(x), N, E, \mathcal{I}) = \frac{p(\lambda|\mathcal{I})p(\boldsymbol{n}(\boldsymbol{x})|N, E, \boldsymbol{x}, \lambda, \mathcal{I})}{p(\boldsymbol{n}(\boldsymbol{x})|N, \boldsymbol{x}, E, \mathcal{I})}. \tag{24.59}$$

In the case of many and informative data the distribution of λ will be quite narrow and the mode can be regarded as a suitable posterior estimate for λ. A suitable prior $p(\lambda|\mathcal{I})$ is again a Jeffreys' scale prior. The PDF for λ is displayed in Figure 24.7 in a log–log and

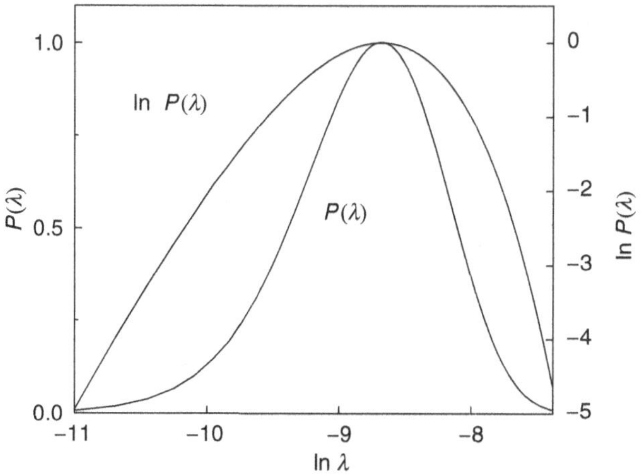

Figure 24.7 Probability distribution $p(\lambda)$ of the hyperparameter λ in equation (24.59) given the data in Figure 24.4.

a lin–log plot. We see that it does determine λ quite precisely. What about the likelihood? Assume that we had employed the prior in equation (24.56) instead of equation (24.57). The posterior corresponding to equation (24.58) would then be

$$p(\boldsymbol{u}, \boldsymbol{v}, b|n(\boldsymbol{x}), N, \boldsymbol{x}, E, \lambda^*, I) = \frac{1}{Z} p(\boldsymbol{u}, \boldsymbol{v}, b|\lambda^*, \mathcal{I})\; p(n(\boldsymbol{x})|\boldsymbol{u}, \boldsymbol{v}, b, N, \boldsymbol{x}, E, \mathcal{I}), \tag{24.60}$$

where we have replaced λ by the MAP solution, discussed before. The comparison of the posterior parameter estimates from equation (24.59) employing the prior in equation (24.58) and those obtained from equation (24.60) yields agreement to four decimal places for the image in Figure 24.4.

From the analysis discussed so far it is only a tiny step to a full two-dimensional change point analysis of the full binary image in Figure 24.3 [p. 416]. The generalization starts with equation (24.48), where the dependence from another independent variable y must be introduced to yield

$$\sigma(x, y) = \frac{1}{e^{v(1+ux+wy)} + 1}. \tag{24.61}$$

The location of the rim of the sigma function, defined by $\sigma(x, y) = 0$, is now given by $ux + wy + 1 = 0$, which is a straight line in two dimensions. The prior for the coefficients u, w has been derived in Chapter 10 [p. 165] and is

$$p(\boldsymbol{u}, \boldsymbol{w}|\mathcal{I}) = r_0^2 \prod_i \left(u_i^2 + w_i^2\right)^{-\frac{3}{2}}. \tag{24.62}$$

The prior probability $p(\boldsymbol{v}|\boldsymbol{u}, \boldsymbol{w}, \mathcal{I})$ should again characterize the smoothness of the image, which is characterized by the maximum gradient of $\sigma(x, y)$, which is achieved on the rim line. The gradient with respect to $\boldsymbol{z} := (x, y)^T$ is

$$\nabla_z \sigma(x, y) = -v \, \frac{e^{\Phi}}{(e^{\Phi} + 1)^2} \begin{pmatrix} u \\ w \end{pmatrix},$$

with $\Phi = v(1 + ux + wy)$, which is zero on the rim. So the global first derivative, controlling smoothness of the image reconstruction, and which we consider as testable information, is proportional to

$$\sum_i (\sigma_i')^2 = \sum_i v_i^2 (u_i^2 + w_i^2), \tag{24.63}$$

from which we obtain the Maxent prior distribution of $\boldsymbol{v}$ conditional on $\boldsymbol{u}, \boldsymbol{w}, \lambda$ as

$$p(\boldsymbol{v}|\boldsymbol{u}, \boldsymbol{w}, \lambda, \mathcal{I}) = \frac{\lambda^{N/2}}{Z} \exp\left\{ -\lambda \sum_i v_i^2 (u_i^2 + w_i^2) \right\} \prod_i (u_i^2 + w_i^2)^{1/2}. \tag{24.64}$$

This function can again be marginalized over λ by employing Jeffreys' prior. We introduce simultaneously the prior for $\boldsymbol{u}, \boldsymbol{w}$ to obtain finally

$$p(\boldsymbol{u}, \boldsymbol{w}, \boldsymbol{v}, b|\mathcal{I}) = \frac{1}{Z} \frac{\left(\prod_i (u_i^2 + w_i^2) \right)^{-1}}{\left(\sum_i v_i^2 (u_i^2 + w_i^2) \right)^{N/2}}. \tag{24.65}$$

Again, unimportant factors have been collected in the normalization constant Z. The likelihood of equation (24.47) [p. 421] remains meaningful even in the limit $N = 1$ and $n = 0, 1$. The log-likelihood can conveniently be written as

$$\ln(L) = \sum_{i,j} \Big\{ t_{ij} \ln\left[q(x_i, y_i, \boldsymbol{u}, \boldsymbol{v}, \boldsymbol{w}, b) \right] + (1 - t_{ij}) \ln\left[1 - q(x_i, y_i, \boldsymbol{u}, \boldsymbol{v}, \boldsymbol{w}, b) \right] \Big\}. \tag{24.66}$$

For black image points $t_{ij} = 1$ and white ones $t_{ij} = 0$ and the sum is over all image points. This completes the specification of the two-dimensional change point problem for the binary image. Posterior estimates have been obtained employing the Gaussian approximation again. As a result of our calculation, we show in Figure 24.8 the location of the level $\sigma = 1/2$ which is given by $ux + wy + 1 = 0$. We realize that none of these straight lines is strictly perpendicular and our previous treatment of the column-wise sums as a one-dimensional change point problem has therefore rested on an incorrect assumption which, after all, was not too bad. Analysis of an even more complicated image can be found in [215].

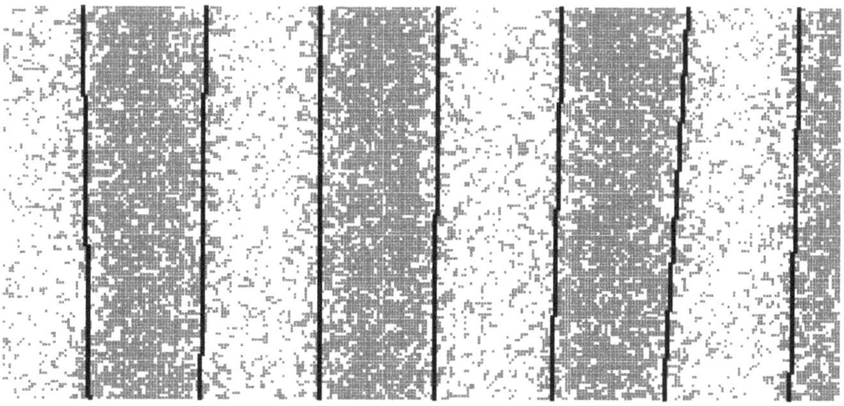

Figure 24.8 Results of a two-dimensional change point analysis of the binary image data. The heavy black traces are the locations where the sigma function in equation (24.61) takes on the value 1/2. The two-dimensional analysis reveals that the stripes in the image are not exactly parallel.

24.4 Thin film growth detected by Auger analysis

An interesting application of the change point analysis for a piecewise linear model arises in surface physics. The growth of thin metal films on metal substrates is usually monitored with Auger electron spectroscopy. The interest in thin films arises from the fact that their properties, e.g. magnetic, may differ radically from the bulk behaviour. Such a case is given by thin layers of iron on the (100) face of copper and the problem is to investigate the magnetic behaviour of the thin films as a function of thickness measured in atomic layers. The growth of iron thin films on Cu(100) proceeds layer by layer, as sketched schematically in Figure 24.9, which shows the $(n+1)$st layer with a fractional coverage θ on top of n complete layers. We shall consider the growth process as a function of evaporation time. The amount of iron on top of the copper substrate is derived from the size of the iron Auger electron signal. Let I_s be the Auger signal from an isolated single layer. The signal resulting from n layers is then given by

$$I_{(n)} = I_s + \alpha I_s + \cdots + \alpha^{n-1} I_s = I_s \frac{1-\alpha^n}{1-\alpha}, \tag{24.67}$$

where α is the attenuation coefficient for Auger electrons traversing one monolayer of iron. α is related to the inelastic mean free path λ via $\alpha = \exp\{-d/(\lambda \cos\vartheta)\}$, where ϑ is the angle of electron emission and d the depth below the surface where the electrons are created. By definition, $\alpha < 1$. The quantity

$$I_\infty := I_s/(1-\alpha)$$

is the intensity of an infinitely thick iron film. For the situation sketched in Figure 24.9 we expect a signal which is composed of a fraction θ originating from $(n+1)$ layers and a fraction $(1-\theta)$ originating from n layers:

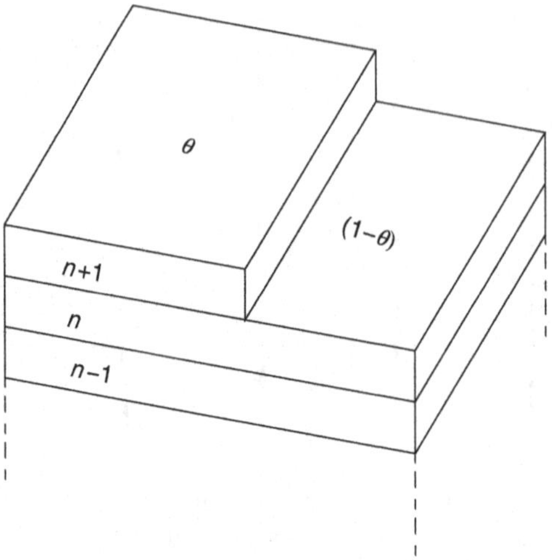

Figure 24.9 Scheme for metal-on-metal epitaxy. On top of n complete overlayers is the $(n+1)$st layer with fractional coverage θ.

$$I(t) = I_\infty \left(\theta(t) \left(1 - \alpha^{n+1}\right) + (1 - \theta(t))(1 - \alpha^n) \right). \tag{24.68}$$

Clearly, the derivative

$$\frac{dI(t)}{dt} = I_\infty \alpha^n (1 - \alpha) \frac{d\theta(t)}{dt} \tag{24.69}$$

is a decreasing function of n only since the evaporation rate $d\theta/dt$ is kept constant during the deposition process. For such conditions we find that

$$\theta(t_i) = \frac{t_i - \tau_n}{I_{n+1} - I_n}, \quad \tau_n \leq t_i \leq \tau_{n+1}, \tag{24.70}$$

with τ_n = time when n layers are completed and the nth change point occurs. Equation (24.68) defines our model for this film growth. The likelihood for the process is

$$p\left(\boldsymbol{d}|\boldsymbol{t}, \boldsymbol{\tau}, \alpha, \sigma, \mathcal{I}\right) = \left(\frac{1}{\sigma\sqrt{2\pi}}\right)^N \exp\left\{-\frac{1}{2\sigma^2} \sum_i \left(d_i - \frac{I\left(t_i, \boldsymbol{\tau}, \alpha\right)}{I_\infty}\right)^2\right\}. \tag{24.71}$$

For the kind of measurement we are dealing with, the assumption of a noise level σ independent of i is well justified. However, since σ is unknown we need to marginalize it with Jeffreys' scale prior as we did many times before. We end up with

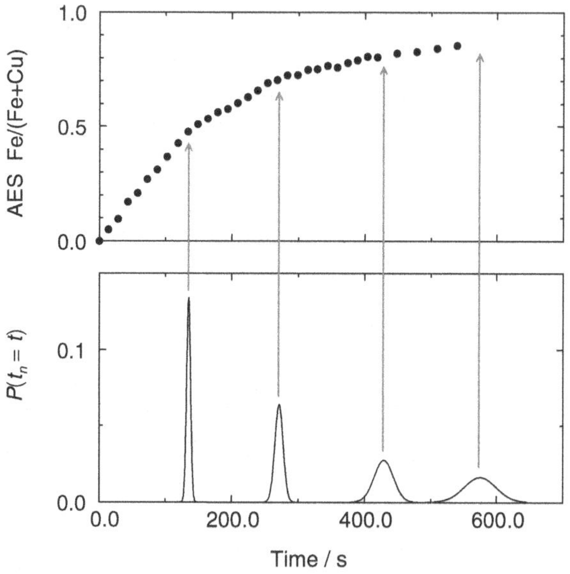

Figure 24.10 Auger signal from expitaxial iron films on a Cu(100) substrate. The piecewise linear sections follow the growth mechanism explained in Figure 24.9. The change point distributions in the lower panel indicate the completion of a monolayer.

$$p(\boldsymbol{d}|\boldsymbol{t}, \boldsymbol{\tau}, \alpha, \mathcal{I}) = \frac{1}{2}\left(\frac{1}{\pi}\right)^{N/2} \Gamma(N/2) \left[\sum_i \left(d_i - \frac{I\,(t_i, \boldsymbol{\tau}, \alpha)}{I_\infty}\right)^2\right]^{-\frac{N}{2}}. \tag{24.72}$$

The specification of the likelihood is thereby complete. Since we aim at the identification of change points in the measurements, we need to specify a prior on $\boldsymbol{\tau}$. While we have previously discussed uninformative equation (24.40) [p. 418] and weakly informative priors on change point positions, we can do better in the case of thin film growth. For an ideally stable apparatus the evaporation time for completion of n layers would be $\tau_n = n\overline{\tau}$. We relax the assumption of ideal stability by choosing

$$p\,(\boldsymbol{\tau}|\overline{\tau}, \delta, \mathcal{I}) = \prod_k \frac{1}{\delta_k\sqrt{2\pi}} \exp\left\{-\frac{1}{2\delta_k^2}\,(\tau_k - k\overline{\tau})^2\right\}. \tag{24.73}$$

An experimentally reasonable estimate for δ_k is $\delta_k = 0.1\tau_k$. This completes the specification of the posterior density of α and τ. Figure 24.10 shows in the upper panel experimental Auger intensities for thin films of iron on Cu(100) as a function of evaporation time. Numerical values for α, $\overline{\tau}$ turn out to be $\alpha = 0.462 \pm 0.015$ and $\overline{\tau} = 137 \pm 2$ s. The remarkable precision of these estimates seems surprising. Remember, however, that we needed only two parameters to model the Auger intensities. The lower panel of Figure 24.10 displays the change point distributions. The increasing width is due to the progressively less informative data. One may easily verify, however, that these posterior distributions are markedly narrower than the prior in equation (24.73) and therefore data dominated.

The layer-by-layer growth is not the only sensible model. Simultaneous multilayer growth, which has a different behaviour of Auger intensity versus evaporation time, is also frequently observed. A comparison of these two models, together with an in-depth discussion of various aspects of the experiment, can be found in [210].

25
Function estimation

The problem of estimating values of a function from a given set of data $\{y_i, x_i\}$ is a generalization of the well-known and much simpler problem of interpolation. In interpolation we infer the values of a function $f(x)$ which takes on the values $y_i = f(x_i)$ at the pivotal points $\{x_i\}$ at arguments x between the pivots, $x_k \leq x \leq x_{k+1}\ \forall k$. The interpolation problem becomes a function estimation problem if the data y_i which are regarded as samples from the function $f(x)$ at argument x_i are deteriorated by noise. In this case we must abandon the requirement $y_i = f(x_i)$ and require the function to pass through the given data $\{x_i, y_i\}$ in some sensible optimal way. We distinguish two categories of function estimation. In the first category the function $f(x)$ is a member of the class of functions $f(x|\boldsymbol{\theta})$ parametrized by a set of parameters $\boldsymbol{\theta}$. The simplest of these curves is a straight line passing through the origin of the coordinate system whose single parameter is the slope. Parametric function estimation amounts in this case to the determination of the single parameter 'slope'. This kind of function estimation is formally identical to the previously treated parameter estimation and regression. In this chapter we shall therefore only deal with the second category, nonparametric function estimation. In this case less specific assumptions about the function representing the data are made. We assume usually that the function shall be continuous or even continuously differentiable once or several times. But otherwise the data will determine the function $f(x)$ much more than would be the case if it were constrained to fall into a prescribed parametric family. A special case of function estimation is the problem of density estimation. In this problem, which has found considerable attention in the literature on statistics and data analysis, the function is further subject to the constraint that it be positive semidefinite.

In Section 25.1 we shall provide an example for function estimation in a problem of bioclimate research. A very general and flexible class of functions for nonparametric estimation is introduced in Section 13.1 [p. 216]. An application to the formulation of smoothness priors has been given in Section 13.2 [p. 219]. Section 25.2 [p. 439] is devoted to density estimation. Two data sets will be considered. In Section 25.2.1 [p. 440] we shall estimate the apparatus function of a Rutherford backscattering experiment from single isotope thin film scattering data. In Section 25.2.2 [p. 447] we give a density estimate of the Old Faithful geyser eruption data, a statistical benchmark data set [185].

25.1 Deriving trends from observations

The identification of changes in observational data relating to the climate change hypothesis remains a topic of paramount importance. It requires scientifically sound and rigorous methods. The kind of problem to be solved can be understood by reference to Figure 25.1. The open circles in this figure represent observations of the onset of cherry blossom since January 1st for the years 1896 until 2002 with missing data in the years 1946 and 1949. The scatter of the data is bewildering. It is caused by the natural climate variability as well as by uncertainties in the manmade and hence subjective observations of a lot of different people. It requires some time to acquaint oneself with the data, which at a very cursory glance do not seem to exhibit any trend at all. Certainly, over most of the data range the 'no trend' assumption is well justified. One might speculate, however, whether at the end of the 20th century a systematic variation of the cherry blossom data versus earlier blossom onset shows up. Considering the quality (scatter) of the data we shall try nonparametric function estimation with a very simple function [49]. This will be composed of two continuously matched linear segments, where the first segment models the constant over most of the data range and the second segment describes a possible change of $f(x)$ from late in the 20th century to present. The immediately arising question is then, at which year of observation should these two segments match? Clearly this is a special case of the previously, more generally treated change point problem (see Chapter 24 [p. 409]). Let E be the variable denoting the matching year. Our first problem is then to find the probability distribution of E given the observational data d, day of blossom onset, as a function of x, year of observation. This distribution is obtained from Bayes' theorem:

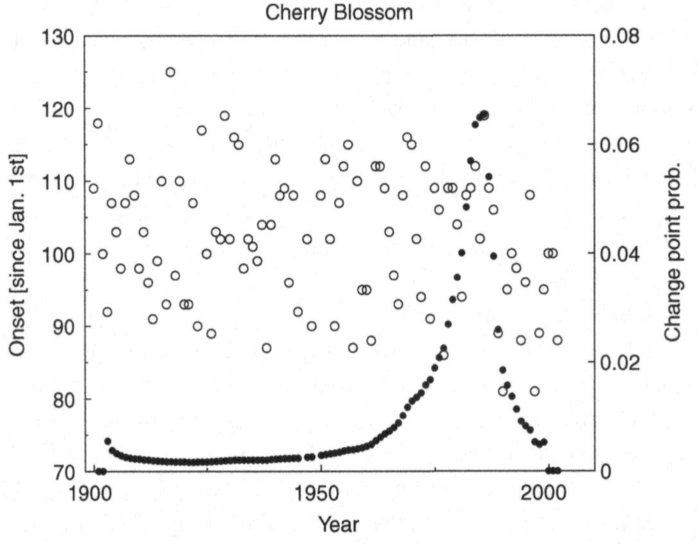

Figure 25.1 Open circles show the onset of cherry blossom in the days after January 1st at Geisenheim (Germany) as a function of the year of observation. Full dots denote the probability for the position of a matching point of two straight-line segments employed to model the data.

$$p(E|\boldsymbol{d}, \boldsymbol{x}, \mathcal{I}) = \frac{p(E|\boldsymbol{x}, \mathcal{I}) \cdot p(\boldsymbol{d}|\boldsymbol{x}, E, \mathcal{I})}{p(\boldsymbol{d}|\boldsymbol{x}, \mathcal{I})} \tag{25.1}$$

in terms of the prior distribution $p(E|\boldsymbol{x}, \mathcal{I})$ and the marginal likelihood $p(\boldsymbol{d}|\boldsymbol{x}, E, \mathcal{I})$. The prior distribution $p\,(E|x, \mathcal{I})$ will be chosen uninformative in the sense that we attach equal weight to all possible choices of E. If the time series has N elements then the number of possible change point positions is $N - 2$ and accordingly $p\,(E|\mathcal{I}) = 1/(N - 2)$. The marginal likelihood is derived from the full likelihood $p(\boldsymbol{d}|\boldsymbol{x}, E, \sigma, \boldsymbol{f}, \mathcal{I})$, where σ is the (unknown) true variance of the data and $\boldsymbol{f}$ the three-component vector of design ordinates at the beginning (x_1) and end (x_N) of the data range and at the change point. This two-segment polygon, which we choose to represent the data, is then for a given change point position x_E

$$\varphi(x|x_E, \boldsymbol{f}) = \begin{cases} f_1 \dfrac{x_E - x}{x_E - x_1} + f_2 \dfrac{x - x_1}{x_E - x_1}, & \text{for } x \le x_E \\[2ex] f_2 \dfrac{x_N - x}{x_N - x_E} + f_3 \dfrac{x - x_E}{x_N - x_E}, & \text{for } x \le x_E \end{cases} \tag{25.2}$$

and the full likelihood becomes

$$p(\boldsymbol{d}|\boldsymbol{x}, E, \sigma, f, \mathcal{I}) = \left(\frac{1}{\sigma\sqrt{2\pi}}\right)^N \exp\left\{-\frac{\|\boldsymbol{d} - \mathbf{A}\boldsymbol{f}\|^2}{2\sigma^2}\right\}. \tag{25.3}$$

We have introduced the matrix $\mathbf{A}$ whose ith row contains the coefficients of $\boldsymbol{f}$ (equation (25.2)) for the year $x = x_i$. The required marginal likelihood is obtained by marginalizing over σ and $\boldsymbol{f}$:

$$p(\boldsymbol{d}|\boldsymbol{x}, E, \mathcal{I}) = \int p(\sigma|\mathcal{I}) \cdot p(\boldsymbol{f}|\boldsymbol{x}, E, \mathcal{I}) \cdot p(\boldsymbol{d}|\boldsymbol{x}, E, \sigma, \boldsymbol{f}, I)\, d\sigma\, d^3 f. \tag{25.4}$$

Uninteresting conditions, which are logically independent from the variables σ and $\boldsymbol{f}$, have been suppressed in the respective prior distributions. Only faint prior knowledge is available about σ. We exaggerate the lack of knowledge slightly by choosing for $p(\sigma|\mathcal{I})$ a normalized form of Jeffreys' scale prior

$$p(\sigma|\beta, \mathcal{I}) = \frac{1}{2\ln\beta} \cdot \frac{1}{\sigma}, \quad \frac{1}{\beta} \le \sigma \le \beta. \tag{25.5}$$

From similar considerations that cherry blossom onset has never been observed prior to d_{min} or later than d_{max}, we may formulate a weakly informative prior

$$p(\boldsymbol{f}|\mathcal{I}) = \left(\frac{1}{2\gamma}\right)^3 \prod_{\nu=1}^{3} \boldsymbol{1}(d_{\text{min}} \le f_\nu \le d_{\text{max}}), \quad 2\gamma = d_{\text{max}} - d_{\text{min}}. \tag{25.6}$$

Since the limits $d_{\text{min/max}}$ can easily be included at the end, we will omit the Heaviside functions henceforward. With these prior distributions the marginal likelihood becomes

$$p(\boldsymbol{d}|\boldsymbol{x}, E, \mathcal{I}) = \frac{1}{Z} \int_{1/\beta}^{\beta} \frac{d\sigma}{\sigma} \int_{d_{\text{min}}}^{d_{\text{max}}} d^3 f\; p(\boldsymbol{d}|\boldsymbol{x}, E, \sigma, \boldsymbol{f}, \mathcal{I}). \tag{25.7}$$

The normalization Z will be restored at the end of the calculation. Exact evaluation of this integral is complicated due to the finite limits on σ and $\boldsymbol{f}$. Suppose, however, that the likelihood is sufficiently localized as a function of $\boldsymbol{f}$ within the limits $d_{\max}$ and $d_{\min}$, then we may replace the finite limits by infinite limits without affecting the value of the definite integral. An exactly analogous consideration applies to the σ-integral. For a suitably chosen, sufficiently large β the value of the double integral becomes independent of β such that we can replace the limits of integration by $0 \leq \sigma \leq \infty$. We then have

$$p(\boldsymbol{d}|\boldsymbol{x}, E, \mathcal{I}) = \frac{1}{Z'} \int_0^\infty \frac{d\sigma}{\sigma} \sigma^{-N} \int_{-\infty}^{\infty} d^3 f \exp\left\{-\frac{\|\boldsymbol{d} - \mathbf{A}\boldsymbol{f}\|^2}{2\sigma^2}\right\}. \tag{25.8}$$

The argument of the inner integration can be rewritten as

$$\|\boldsymbol{d} - \mathbf{A}\boldsymbol{f}\|^2 = (\boldsymbol{f} - \boldsymbol{f}_0)^T \mathbf{Q} (\boldsymbol{f} - \boldsymbol{f}_0) + R(E), \tag{25.9a}$$

$$\mathbf{Q} := \mathbf{A}^T \mathbf{A}, \tag{25.9b}$$

$$\boldsymbol{f}_0 := \mathbf{Q}^{-1} \mathbf{A} \boldsymbol{d}, \tag{25.9c}$$

$$R(E) := \boldsymbol{d}^T \big(\mathbf{1} - \mathbf{A}^T \mathbf{Q}^{-1} \mathbf{A}\big) \boldsymbol{d}. \tag{25.9d}$$

The remaining integral is of standard Gaussian type and yields

$$\int_{-\infty}^{\infty} d^3 f \exp\left\{-\frac{1}{2\sigma^2} \|\boldsymbol{d} - \mathbf{A}\boldsymbol{f}\|^2\right\} = \frac{(2\pi)^{3/2}\,\sigma^3}{\sqrt{|\mathbf{Q}|}} \exp\left\{-\frac{R(E)}{2\sigma^2}\right\}. \tag{25.10}$$

The outer σ-integral is then also elementary and we obtain

$$p(\boldsymbol{d}|\boldsymbol{x}, E, \mathcal{I}) = \frac{1}{Z''} \big(R(E)\big)^{-\frac{N-3}{2}}. \tag{25.11}$$

At this point we can also determine the posterior PDF for the noise level, which will be relevant later on:

$$\begin{aligned} p(\sigma|\boldsymbol{d}, \boldsymbol{x}, E, \mathcal{I}) &\propto p(\boldsymbol{d}|\sigma, \boldsymbol{x}, E, \mathcal{I})\; p(\sigma|\cancel{\boldsymbol{x}}, \cancel{E}, \mathcal{I}) \\ &\propto \int d^3 f p(\boldsymbol{d}|\boldsymbol{f}, \sigma, \boldsymbol{x}, E, \mathcal{I})\; p(\boldsymbol{f}|\sigma, \boldsymbol{x}, E, \mathcal{I}). \end{aligned}$$

With the previous results we find

$$p(\sigma|\boldsymbol{d}, \boldsymbol{x}, E, \mathcal{I}) = \frac{1}{Z} \sigma^{-N+2} e^{-\frac{R(E)}{2\sigma^2}}.$$

We will need the second moment $\langle \sigma^2 \rangle_E$, which still depends on E. The moments are obtained via the substitution $t = R(E)/2\sigma^2$, and we find

$$\langle \sigma^2 \rangle_E = \frac{R(E)}{N-5}. \tag{25.12}$$

If we draw further on the previously derived results (Section 21.1.5 [p. 344]) from singular-value decomposition (here of matrix $\mathbf{A}$), then we obtain expressions for $\boldsymbol{f}_0$, $|\mathbf{Q}|$ and $R(E)$. Let $\mathbf{A} = \sum_i \lambda_i \boldsymbol{u}_i \boldsymbol{v}_i^T$ be the singular-value decomposition of matrix $\mathbf{A}$, then

$$|\mathbf{Q}| = \prod_i \lambda_i^2, \quad \boldsymbol{f}_0 = \sum_i \frac{1}{\lambda_i} \boldsymbol{v}_i (\boldsymbol{u}_i^T \boldsymbol{d}), \quad R = \boldsymbol{d}^T \boldsymbol{d} - \sum_k (\boldsymbol{u}_k^T \boldsymbol{d})^2. \tag{25.13}$$

This completes the calculation of the marginal likelihood. The probability distribution for the change point position $p(E|\boldsymbol{d}, \boldsymbol{x}, \mathcal{I})$ then follows from equation (25.1). The result is shown in Figure 25.1 as full dots. Unlike our previous experience with change point problems, this distribution is rather diffuse and even the optimum change point choice has only a probability of about 6%. Function estimation from the data in Figure 25.1 in terms of a two-segment polygon or triangular function, which amounts to a maximum likelihood solution of the problem, is therefore entirely unsatisfactory. Instead, every possible change point has a finite probability and the final function estimate must incorporate all these choices according to their respective probabilities.

Our aim is now to make predictions for the beginning date D of the cherry blossom in year z. z may, but need not, coincide with years of previous observations. It must not even lie in the range which is covered by previous observations. The following analysis is similar to that already encountered in Section 21.1.4 [p. 342]. We want to find the PDF:

$$p(D|z, \boldsymbol{d}, \boldsymbol{x}, \mathcal{I}) = \sum_E \int d^3 f \, d\sigma \; p(\boldsymbol{f}, \sigma, E|\boldsymbol{d}, \boldsymbol{x}, \mathcal{I}) \cdot p(D|\boldsymbol{f}, E, \sigma, z, \boldsymbol{d}, \boldsymbol{x}, \mathcal{I}). \tag{25.14}$$

The second term is the probability density for the target datum D, given $\boldsymbol{f}$. Looking back at the model function in equation (25.2) we see that this term is logically independent of σ and $\boldsymbol{d}$ and becomes the delta function $\delta(D - \varphi(z|E, \boldsymbol{f}))$. Expectation values of D^k are therefore given by

$$\langle D^k \rangle = \sum_E \int d\sigma \int d^3 f \; p(\boldsymbol{f}, \sigma, E|\boldsymbol{d}, \boldsymbol{x}, \mathcal{I}) \, \varphi^k(z|E, \boldsymbol{f}). \tag{25.15}$$

The first term under the integral is the posterior distribution of the parameters. The product rule yields

$$p(\boldsymbol{f}, \sigma, E|\boldsymbol{d}, \boldsymbol{x}, \mathcal{I}) = p(\boldsymbol{f}|\sigma, E, \boldsymbol{d}, \boldsymbol{x}, \mathcal{I}) \, p(\sigma|E, \boldsymbol{d}, \boldsymbol{x}, \mathcal{I}) \, P(E|\boldsymbol{d}, \boldsymbol{x}, \mathcal{I}). \tag{25.16}$$

The posterior PDF for $\boldsymbol{f}$ follows Bayes' theorem:

$$p(\boldsymbol{f}|\sigma, E, \boldsymbol{d}, \boldsymbol{x}, \mathcal{I}) \propto p(\boldsymbol{d}|\boldsymbol{f}, \sigma, E, \boldsymbol{x}, \mathcal{I}) \, p(\boldsymbol{f}|\sigma, E, \boldsymbol{x}, \mathcal{I}).$$

Within the limits $d_{\text{min/max}}$ the prior for f_i is uniform and since the likelihood constraint appeared to be much more restrictive than the prior, we ignore the limits and obtain with equations (25.3) and (25.9)

$$p(\boldsymbol{f}|\sigma, E, \boldsymbol{d}, \boldsymbol{x}, \mathcal{I}) = \mathcal{N}(\boldsymbol{f}|\boldsymbol{f}_0, \mathbf{C}), \tag{25.17}$$

$$\mathbf{C} := \sigma^2\, \mathbf{Q}^{-1}. \tag{25.18}$$

In summary, the posterior PDF for $\boldsymbol{f}$ is Gaussian with mean $\boldsymbol{f}_0$ and covariance matrix $\mathbf{C}$. Remember that $\mathbf{Q} = \mathbf{A}^T\mathbf{A}$ depends on E, because $\mathbf{A}$ does. Now we are prepared to compute the moments $\langle D^k \rangle$ by rewriting equation (25.15) based on equations (25.16) and (25.17) as

$$\begin{aligned}\langle D^k \rangle = \sum_E P(E|\boldsymbol{d}, \boldsymbol{x}, \mathcal{I}) \int d\sigma p(\sigma|E, \boldsymbol{d}, \boldsymbol{x}, \mathcal{I}) \\ \ldots \int d^3 f\, \varphi^k(z|E, \boldsymbol{f})\, \mathcal{N}(\boldsymbol{f}|\boldsymbol{f}_0, \mathbf{C}).\end{aligned} \tag{25.19}$$

A convenient representation of our model function $\varphi(z|E, \boldsymbol{f})$ is obtained if we introduce the coefficient vector $\boldsymbol{b}(E, z)$. Then

$$\varphi(z|\boldsymbol{f}, E) = \boldsymbol{b}^T(E, z)\, \boldsymbol{f}. \tag{25.20}$$

Recalling $\boldsymbol{f}_0 = \mathbf{Q}^{-1}\mathbf{A}\boldsymbol{d}$, then the first moment is readily obtained as

$$\begin{aligned}\langle D \rangle &= \sum_E P(E|\boldsymbol{d}, \boldsymbol{x}, \mathcal{I}) \underbrace{\int d\sigma p(\sigma|E, \boldsymbol{d}, \boldsymbol{x}, \mathcal{I})}_{=1}\, \boldsymbol{b}^T(E, z)\, \mathbf{Q}^{-1}\mathbf{A}\boldsymbol{d} \\ &= \sum_E P(E|\boldsymbol{d}, \boldsymbol{x}, \mathcal{I})\, \boldsymbol{b}^T(E, z)\, \mathbf{Q}^{-1}\mathbf{A}\, \boldsymbol{d}.\end{aligned}$$

This is a rather transparent and intuitively pleasing result. Predictions are obtained as the sum over all possible model functions weighted with their respective posterior probability. The result of this calculation for the cherry blossom data is shown in Figure 25.2. The estimated function appears to be rather smooth and is quite different from a two-segment polygon which would be the maximum likelihood solution of the problem. In fact, the estimated function is still linear between two support data points with slope varying from interval to interval. Therefore it does not come as a surprise that the extrapolation is also linear. Figure 25.2 also shows the uncertainty range of the prediction, which we shall calculate now. What is needed for this purpose is the second moment of D. From equations (25.19) and (25.20) we get

$$\begin{aligned}\langle(\Delta D)^2\rangle &= \sum_E P(E|\boldsymbol{d}, \boldsymbol{x}, \mathcal{I}) \int d\sigma p(\sigma|E, \boldsymbol{d}, \boldsymbol{x}, \mathcal{I})\, \boldsymbol{b}^T(E, z)\, \underbrace{\left\langle \Delta \boldsymbol{f} \Delta \boldsymbol{f}^T \right\rangle}_{\mathbf{C}}\, \boldsymbol{b}(E, z) \\ &\overset{(25.17)}{=} \sum_E P(E|\boldsymbol{d}, \boldsymbol{x}, \mathcal{I}) \underbrace{\left(\int d\sigma p(\sigma|E, \boldsymbol{d}, \boldsymbol{x}, \mathcal{I})\, \sigma^2 \right)}_{\langle \sigma^2 \rangle_E} \boldsymbol{b}^T(E, z)\, \mathbf{Q}^{-1}\, \boldsymbol{b}(E, z).\end{aligned}$$

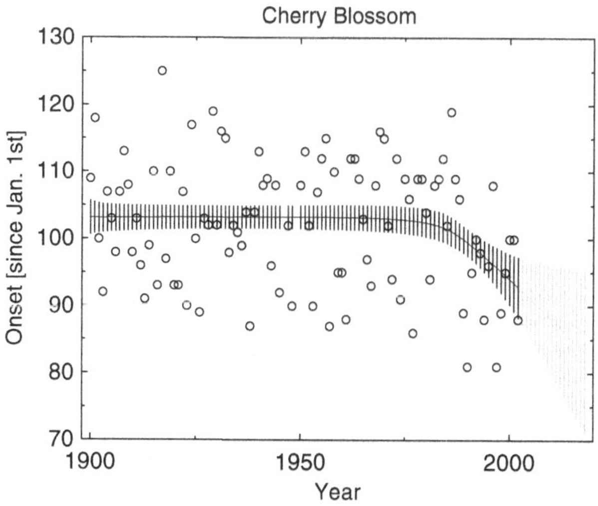

Figure 25.2 The trend of cherry blossom onset as a function of the year of observation as estimated from the observed data. Note the widening of the confidence interval for prediction at the ends of the data range and in particular in the extrapolation region.

So eventually we find

$$\langle (\Delta D)^2 \rangle \overset{(25.12)}{=} \sum_E P(E|\boldsymbol{d}, \boldsymbol{x}, \mathcal{I}) \, \frac{R(E)}{N-5} \, \boldsymbol{b}^T(E, z) \, \mathbf{Q}^{-1} \, \boldsymbol{b}(E, z). \tag{25.21}$$

So the second moment is also obtained from a sum of the moments of a particular change point choice E weighted with the posterior probability of that particular change point choice. This completes the analysis of the problem. Figure 25.2 shows a widening of the uncertainty band towards the ends of the data range and even more so in the extrapolated region. This is not at all surprising. The predictions must become progressively more uncertain as the year of observation, for which the prediction is made, moves away from the range where the already observed data lie. The behaviour is comparable to the previously investigated case (see Figure 21.2 [p. 344]).

Another interesting quantity in this problem is the rate of change in blossom onset as a function of the year of observation. The function to be estimated is then the derivative $\varphi'(z|\boldsymbol{f}, E)$, defined in equation (25.2). This can be expressed as

$$\varphi'(z|\boldsymbol{f}, E) = \boldsymbol{a}^T(E) \, \boldsymbol{f}. \tag{25.22}$$

The necessary integrations for $\langle \varphi' \rangle$ and $\langle \varphi'^2 \rangle$ are even simpler in this case, particularly for the second moment. The result is, however, interesting. Looking at the data in Figure 25.1, the attempt to estimate a trend looks rather daring, let alone its derivative. The rate of change shown in Figure 25.3 is, however, quite a smooth function with no change over most of the 20th century and a rapid change in the period 1980–1990, with a maximum rate of −0.6 days per year in 2002. A variant of the foregoing treatment has also been

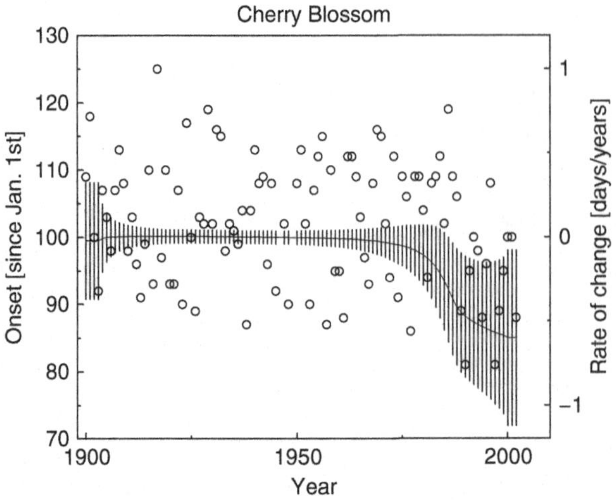

Figure 25.3 The derivative of the trend in days/years as estimated from the observed data.

applied in an analysis of Caribbean hurricane data [48]. The generalization of the previous model to n change points with $n > 1$ is straightforward and most of the previous analysis remains valid. The model function in equation (25.2) is now replaced by

$$\varphi_K(x|\boldsymbol{E}, \boldsymbol{f}, \mathcal{I}) = f_K \frac{x_{K+1} - x}{x_{K+1} - x_K} + f_{K+1} \frac{x - x_K}{x_{K+1} - x_K} \tag{25.23}$$

for $x_K \leq x \leq x_{K+1}$ and the vector $\boldsymbol{E}$ with components $E_K = x_{K+1}$ for $K = 1, \ldots, N-2$. The analysis following equation (25.2) remains unchanged except for the number of columns in matrix $\mathbf{A}$, which is now $(n+2)$, and E becomes an n-component vector. The number of ways to choose the n change points from $x_2, \ldots, x_{N-1}$ is

$$Z_n = \binom{N-2}{n}. \tag{25.24}$$

For sufficiently large N and moderate n the leading term in equation (25.24) is $N^n/n!$. For the example presented in Figure 25.4 we have $N = 149$ and $n = 4$, leading to $Z_n \approx 2\times 10^7$. A deterministic evaluation of the sum over change point positions as in equation (25.15) becomes therefore prohibitive even for relatively small n. The way out of this problem is to replace the deterministic exact sum in equation (25.15) by a Monte Carlo approximation. This consists of averaging over Z_{MC} random change point configurations instead of the $(N-2)$ single change point positions. An appropriate generator for the change point configurations uses the following three steps:

$$(1)\ x_1 = 0, \quad x_i = -\ln(1 - U(0,1)), \quad i = 2, \ldots, (n+2); \tag{25.25}$$

$$(2)\ y_i = \frac{x_i}{\sum_{i=1}^{n+2} x_i}; \tag{25.26}$$

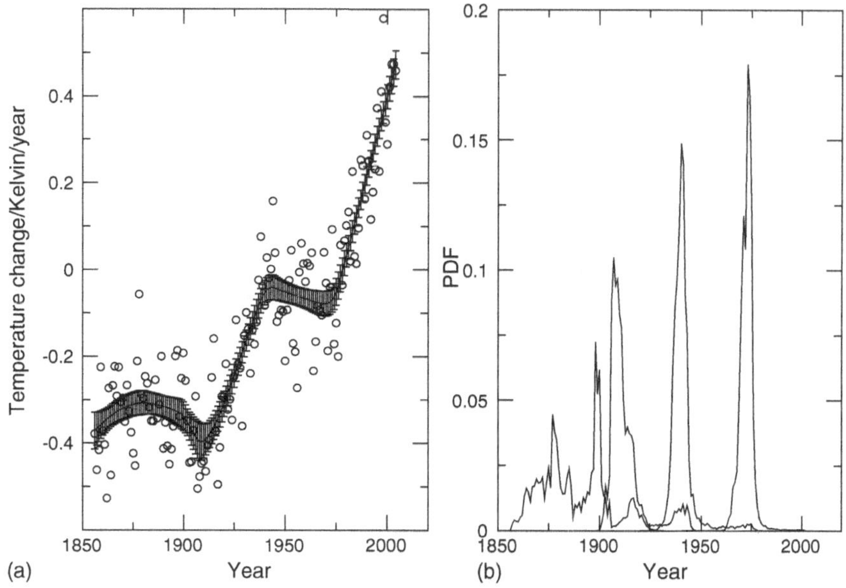

Figure 25.4 (a) Data on the earth's surface temperature variations in the last 150 years as open circles. The continuous trace is the nonparametric function estimation of these data with error bars. (b) Distributions of the four change points used to set up the model.

$$(3)\quad z_k = \sum_{i=1}^{k} y_i, \quad k = 1, \dots, (n+2). \tag{25.27}$$

$U(0, 1)$ is a uniform random number > 0 and < 1 and z_k an ordered set of change points $k = 1, \dots, n+1$ with fixed endpoints $z_1 = 0$ and $z_{n+2} = 1$. The z_k are finally matched to the actual data range $(x_1, \dots, x_N)$ by appropriate scaling and translation. Figure 25.4 shows an example of the multiple change point approach to observations of the global surface air temperature anomaly since 1850 (http://www.cru.uea.ac.uk/, see also [108]). The left panel shows the data as open circles and the fit function with error bars. Please note that the fit function shows also a curved section apart from the linear ones. It is instructive to calculate the probability distributions of the four change points chosen in this calculation. These distributions are obtained by averaging the configurations weighted by their respective probabilities. The result is shown in the right panel of Figure 25.4. We realize that the first change point has a rather smeared distribution and this accounts for the curved section of the function in the left panel of Figure 25.4, while the other three distributions are fairly localized in line with the corresponding linear sections of the fit function.

25.2 Density estimation

Density reconstruction is a special case of function estimation where the target function is positive semidefinite and normalizable. We shall illustrate the process with the

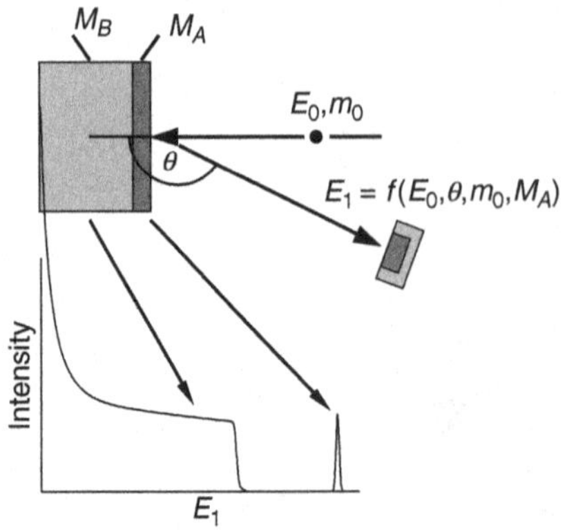

Figure 25.5 Schematic of an RBS experiment. Source: Courtesy of M. Mayer.

apparatus response function of an accelerator experiment called Rutherford backscattering. As a second example, particularly interesting because of the likelihood involved, we apply an estimation procedure to 107 data for the eruption time of the Old Faithful geyser in Yellowstone National Park [183].

25.2.1 RBS apparatus function

Rutherford backscattering (RBS) is a surface analytical technique whose importance derives from its quantitative nature. We discuss the principle by reference to Figure 25.5. Consider a beam of ions of mass m_0, usually either protons or helium nuclei, with well-defined energy E_0 in the lower MeV range impinging on a target with atomic mass number M_B covered by a thin film overlayer of atoms of mass number M_A. We shall assume – in line with the subsequent data – that M_A is very much larger than M_B. Ions undergoing an elastic Coulomb collision with deflection by an angle θ are recorded in an energy dispersive solid-state detector. Since the projectile–target interaction is Coulombic, the scattering cross-section is the quantitatively known Rutherford scattering cross-section. All backscattered ions have $E < E_0$. The final state energy E depends on E_0, m_0, $M_{A,B}$, θ and is larger for scattering from a target with mass M_A than for mass M_B since $M_A > M_B$. If the overlayer M_A in Figure 25.5 is monatomic then the incident beam energy is not attenuated by energy loss prior to the collision and the energy distribution for scattering from A atoms is a narrow peak. Scattering from the substrate material, however, produces a step function. The rising edge is due to scattering from the first substrate layer, while particles which have penetrated to some depth into material B before undergoing large-angle scattering lose energy on their way to the scattering centre and after scattering on their way out to the detector. These scattered particles contribute therefore to the spectrum at progressively lower energies. The overall rising signal towards lower final state energy reflects the well-known $1/E^2$ dependence of the Rutherford cross-section.

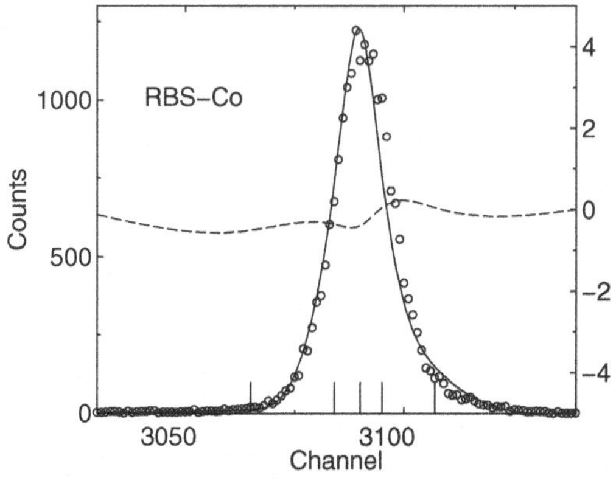

Figure 25.6 RBS spectrum of a thin cobalt overlayer on a silicon substrate (open circles). The maximum likelihood reconstruction of the underlying density is shown as a solid line. The dashed curve is the difference of the maximum likelihood result and an estimate from a posterior incorporating the smoothness prior given in equation (13.23) [p. 221] (right-hand scale). Tick marks on the abscissa show the pivot positions.

Figure 25.6 shows the surface part of an RBS spectrum obtained by scattering 2.6-MeV He^+ ions from a silicon target with a monatomic overlayer of cobalt. Cobalt is isotopically pure and has mass number $M_A = 59$. Silicon is lighter and has mean mass number $M_B = 28$ with small admixtures of masses 29 and 30. Figure 25.6 shows only the spectrum due to scattering off the cobalt overlayer. For an ideal experiment it should be an infinitely sharp spike. The finite width derives from various instrumental effects. The most important contributions are those from the detector angular and energy resolution and the energy width of the incident beam. The data in Figure 25.6 constitute therefore the apparatus transfer function which characterizes the difference between an ideal and a real-world experiment. We shall now derive a density function based on these experimental data. Let d_i^* be the number of counts measured in channel i. Assume that the target function can be represented as a cubic spline with design support ordinates $\boldsymbol{f}(\boldsymbol{\xi})$ at pivots $\boldsymbol{\xi}$. Our model equation is then

$$d_i^* = (\mathbf{A}^* \boldsymbol{f})_i + \delta_i, \tag{25.28}$$

where $\mathbf{A}^*$ is the spline coefficient matrix and δ_i represents noise on channel i. We assume $\langle \delta_i \rangle = 0$ and $\langle \delta_i^2 \rangle = \sigma_i^2$. The likelihood function for this model reads

$$p(\boldsymbol{d}^* | \mathbf{A}^*, \boldsymbol{\sigma}, \boldsymbol{f}, \mathcal{I}) = \frac{1}{\prod_i \sigma_i \sqrt{2\pi}} \exp\left\{ -\frac{1}{2} \left(\boldsymbol{d}^* - \mathbf{A}^* \boldsymbol{f}\right)^T \mathbf{C}^{-1} \left(\boldsymbol{d}^* - \mathbf{A}^* \boldsymbol{f}\right) \right\}.$$

The pivot points $\boldsymbol{\xi}$ have been included in the background information $\mathcal{I}$. In this experiment the noise is uncorrelated and the covariance matrix is diagonal, $\mathbf{C} = \text{diag}\{\sigma_i^2\}$, containing the variances. For convenience of notation we introduce $\boldsymbol{d} = \mathbf{C}^{-1/2} \boldsymbol{d}^*$ and $\mathbf{A} = \mathbf{C}^{-1/2} \mathbf{A}^*$, which simplifies the likelihood to

$$p(\boldsymbol{d}|\boldsymbol{f},\mathcal{I}) = \left(\frac{1}{2\pi}\right)^{N/2} \exp\left\{-\frac{1}{2}\|\boldsymbol{d} - \mathbf{A}\boldsymbol{f}\|^2\right\}. \tag{25.29}$$

In order to keep the notation readable, we consider $\mathbf{A}$ as part of $\mathcal{I}$. This likelihood is only approximate since we are dealing with a counting experiment which obeys the Poisson distribution. As has been shown earlier, the Poisson distribution can be better approximated by a Gaussian the larger the number of counts is. Hence, our approximation becomes poor in the wings of the data in Figure 25.6. On the contrary, even if the relative approximation error may be large for these data, it is acceptably small on an absolute scale since the maximum of the data lies around 1200 counts.

So far we have only formulated the likelihood. In order to derive the posterior density $p(\boldsymbol{f}|\mathbf{A},\boldsymbol{d},\mathcal{I})$ from Bayes' theorem we must specify a prior distribution on $\boldsymbol{f}$. If we take the same simple choice of a flat prior as in Section 25.1 [p. 432], then the posterior solution for $\boldsymbol{f}$ is identical to the maximum likelihood solution. It is displayed in Figure 25.6 as a continuous curve. The number of pivots chosen for this reconstruction was seven. Their distribution is shown by the tick marks on the abscissa of Figure 25.6. They were chosen by common sense such that their spacing is narrower where the data peak and wider where there is only small variation. We shall return to the problem of pivot choice later.

A flat prior on $\boldsymbol{f}$ is the simplest possible choice but certainly not the most sensible one. We might be interested in a smooth reconstruction, and suitable priors for $\boldsymbol{f}$ which encode such a behaviour are given for example in equations (13.23) and (13.27)[p. 221]. For definiteness we choose the first one. This is

$$p(\boldsymbol{f}|\mu,E,\mathcal{I}) = \left(\frac{\mu}{2\pi}\right)^{E/2} \varepsilon|\det{}'\mathbf{P}_{f''}|^{1/2} \exp\left\{-\frac{\mu}{2}\boldsymbol{f}^T\hat{\mathbf{P}}_{f''}\boldsymbol{f}\right\}, \tag{25.30a}$$

$$\hat{\mathbf{P}}_{f''} = \mathbf{P}_{f''} + \varepsilon\mathbf{1}, \tag{25.30b}$$

where $\mathbf{P}_{f''}$ is the matrix representing $\int f''^2 dx$ and μ an unknown scaling parameter for the overall smoothness that we want to impose. E is the number of pivots and hence the dimension of the vector $\boldsymbol{f}$. $\det{}'\mathbf{P}_{f''}$ is the product of the nonvanishing eigenvalues of matrix $\mathbf{P}_{f''}$. ε is the regularization variable introduced in Section 13.2 [p. 219]. An order of magnitude estimate of E can be obtained as follows. We need two parameters to specify the endpoints of the reconstruction interval, one parameter each for position, width and height of the signal and at least one parameter to describe the asymmetry of the data. So we expect E to be greater than six. In fact, explicit calculations will be limited here to $E = 7$. The question of how to treat this variable in more detail will be addressed later. Analogous to the previous procedure in equation (25.9) [p. 434], we rewrite

$$\|\boldsymbol{d} - \mathbf{A}\boldsymbol{f}\|^2 = (\boldsymbol{f} - \boldsymbol{f}_0)^T\mathbf{Q}(\boldsymbol{f} - \boldsymbol{f}_0) + R_L, \tag{25.31}$$

where f_0 $\mathbf{Q}$ and R_L follow from a singular-value decomposition of matrix $\mathbf{A}$. The posterior is obtained upon combining equations (25.29) and (25.30):

$$p(\boldsymbol{f}|\boldsymbol{d}, \mu, E, \mathcal{I}) \propto p(\boldsymbol{d}|\boldsymbol{f}, \mu, E, \mathcal{I}) p(\boldsymbol{f}|\boldsymbol{d}, \mu, E, \mathcal{I}) = \frac{1}{Z} \exp\left\{-\frac{\Phi}{2}\right\}, \tag{25.32a}$$

$$\begin{aligned} \Phi &:= (\boldsymbol{f} - \boldsymbol{f}_0)^T \mathbf{Q}(\boldsymbol{f} - \boldsymbol{f}_0) + R_L + \mu\, \boldsymbol{f}^T \hat{\mathbf{P}}_{f''} \boldsymbol{f} \\ &= (\boldsymbol{f} - \boldsymbol{f}_\mu)^T \mathbf{H}(\boldsymbol{f} - \boldsymbol{f}_\mu) + R_H, \end{aligned} \tag{25.32b}$$

and the newly introduced variables $\mathbf{H}$, R_H and $\boldsymbol{f}_\mu$ follow from comparison of coefficients in powers of $\boldsymbol{f}$:

$$\mathbf{H} = \mathbf{Q} + \mu \hat{\mathbf{P}}_{f''}, \tag{25.33a}$$

$$\boldsymbol{f}_\mu = \mathbf{H}^{-1} \mathbf{Q} \boldsymbol{f}_0, \tag{25.33b}$$

$$R_H = R_L + \boldsymbol{f}_0^T \mathbf{Q} \boldsymbol{f}_0 - \boldsymbol{f}_0^T \mathbf{Q}^T \mathbf{H}^{-1} \mathbf{Q} \boldsymbol{f}_0. \tag{25.33c}$$

The mode of the posterior distribution $\boldsymbol{f}_\mu$ has been shifted from the maximum of the likelihood $\boldsymbol{f}_0$ by an amount depending on the as yet unknown hyperparameter μ.

Remarks on hyperparameters

The present example is well suited as a case study for a common treatment of hyperparameters.

Evidence approximation

We start out with the so-called 'evidence approximation'. The approach we have followed so far was to introduce a prior $p(\boldsymbol{f}|\mu, E, \mathcal{I})$, yet depending on a 'hyperparameter'. The obvious way to handle hyperparameters is marginalization, resulting in the prior

$$p(\boldsymbol{f}|E, \mathcal{I}) = \int d\mu\; p(\boldsymbol{f}|\mu, E, \mathcal{I})\; p(\mu). \tag{25.34}$$

In many cases, however, $p(\boldsymbol{f}|\mu, E, \mathcal{I})$ has a simple structure that allows an analytic treatment of posterior probabilities, which is not possible with $p(\boldsymbol{f}|E, \mathcal{I})$. As in the present case, the posterior in equation (25.32) is Gaussian. In such cases it might be advantageous to postpone the marginalization of the hyperparameter to the very end, which then reads

$$p(\boldsymbol{f}|\boldsymbol{d}, E, \mathcal{I}) = \int d\mu\; p(\boldsymbol{f}|\mu, \boldsymbol{d}, E, \mathcal{I})\; p(\mu|\boldsymbol{d}, E, \mathcal{I}). \tag{25.35}$$

In many cases this is the preferable approach, in combination with a numerical evaluation of the marginalization integral in equation (25.35).

However, in many cases the posterior PDF $p(\mu|\boldsymbol{d}, E, \mathcal{I})$ is sharply peaked at $\hat{\mu}$, while $p(\boldsymbol{f}|\mu, \boldsymbol{d}, E, \mathcal{I})$ has a fairly weak μ-dependence. Then we can replace $p(\boldsymbol{f}|\mu, \boldsymbol{d}, E, \mathcal{I})$ by $p(\boldsymbol{f}|\hat{\mu}, \boldsymbol{d}, E, \mathcal{I})$, take it out of the integral and the remaining integral is unity. This defines the so-called

EVIDENCE APPROXIMATION

$$p(\boldsymbol{f}|\boldsymbol{d}, E, \mathcal{I}) = p(\boldsymbol{f}|\hat{\mu}, \boldsymbol{d}, E, \mathcal{I}), \tag{25.36a}$$

$$\hat{\mu}: \quad p(\hat{\mu}|\boldsymbol{d}, E, \mathcal{I}) \geq p(\mu|\boldsymbol{d}, E, \mathcal{I}), \qquad \forall \mu. \tag{25.36b}$$

We will apply the evidence approximation to the present problem. To this end we need the PDF for μ, which we obtain through Bayes' theorem:

$$\begin{aligned} p(\mu|\boldsymbol{d}, E, \mathcal{I}) &\propto p(\boldsymbol{d}|\mu, E, \mathcal{I})\; p(\mu|\not{E}, \mathcal{I}) \\ &\propto \int d^N f\; p(\boldsymbol{d}|\boldsymbol{f}, \mu, E, \mathcal{I})\; p(\boldsymbol{f}|\mu, E, \mathcal{I})\; p(\mu). \end{aligned}$$

Naturally, the smoothness parameter is logically independent of the expansion order E. The first two factors are the same as those in equation (25.32a). In addition to $\exp(-\Phi)$ we have to include here the μ-dependent prefactor of equation (25.30) as well. Then we find

$$p(\mu|\boldsymbol{d}, E, \mathcal{I}) \propto \mu^{\frac{E}{2}}\; p(\mu) \int d^N f\; \exp\left\{-\frac{\Phi}{2}\right\}.$$

The proportionality constant follows from normalization. Next we determine the remaining integral. To this end we invoke equation (25.32b):

$$\begin{aligned} p(\mu|\boldsymbol{d}, E, \mathcal{I}) &\propto \mu^{\frac{E}{2}}\; e^{-R_H} \int d^N f\; \exp\left\{-\frac{1}{2}\left(f - f_{\mu}\right)^T \mathbf{H} \left(f - f_{\mu}\right)\right\} \\ &\propto p(\mu)\; \mu^{\frac{E}{2}}\; e^{-\frac{R_H}{2}}\; |\mathbf{H}|^{-\frac{1}{2}}. \end{aligned}$$

Then we find, apart from uninteresting constants (C, C'),

$$\begin{aligned} 2\,\ln\left[p(\mu|\boldsymbol{d}, E, \mathcal{I})\right] &= \ldots \\ &= C + 2\ln(p(\mu)) + E\ln(\mu) + \boldsymbol{f}_0^T \mathbf{Q}^T \mathbf{H}^{-1}(\mu) \mathbf{Q} \boldsymbol{f}_0 - \ln\left(\left|\mathbf{Q} + \mu \hat{\mathbf{P}}_{f''}\right|\right) \\ &= C' + 2\ln(p(\mu)) + E\ln(\mu) + \boldsymbol{f}_0^T \mathbf{Q}^T \mathbf{H}^{-1}(\mu) \mathbf{Q} \boldsymbol{f}_0 - \ln\left(\left|\tilde{\mathbf{Q}} + \mu \mathbf{1}\right|\right), \end{aligned} \tag{25.37}$$

$$\tilde{\mathbf{Q}} := \hat{\mathbf{P}}_{f''}^{-\frac{1}{2}} \mathbf{Q} \hat{\mathbf{P}}_{f''}^{-\frac{1}{2}}.$$

The μ-dependence has been made explicit. Next we seek the MAP solution for μ, that is we determine the root of the derivative of $\ln\left[p(\mu|\boldsymbol{d}, E, \mathcal{I})\right]$ w.r.t. μ. Along with

$$\frac{d}{d\mu}\mathbf{H}^{-1}(\mu) = \frac{d}{d\mu}\left(\mathbf{Q} + \mu \hat{\mathbf{P}}_{f''}\right)^{-1} = -\mathbf{H}^{-1} \hat{\mathbf{P}}_{f''} \mathbf{H}^{-1}$$

and equation (A.15) [p. 602], we obtain the condition

$$
\begin{aligned}
2\frac{d\ln[p(\mu|\boldsymbol{d},E,\mathcal{I})]}{d\mu} &= \dots \\
&= 2\frac{d\ln(p(\mu))}{d\mu} + \frac{E}{\mu} - \underbrace{\boldsymbol{f}_0^T\mathbf{Q}^T\mathbf{H}^{-1}}_{\boldsymbol{f}_\mu^T}\hat{\mathbf{P}}_{f''}\underbrace{\mathbf{H}^{-1}\mathbf{Q}\boldsymbol{f}_0}_{\boldsymbol{f}_\mu} - \sum_{i=1}^{E}\frac{d\ln(\mu+\lambda_i)}{d\mu} \\
&= 2\frac{d\ln(p(\mu))}{d\mu} + \frac{E}{\mu} - \boldsymbol{f}_\mu^T\hat{\mathbf{P}}_{f''}\boldsymbol{f}_\mu - \sum_{i=1}^{E}\frac{1}{\mu+\lambda_i} \stackrel{!}{=} 0,
\end{aligned}
\tag{25.38}
$$

where λ_i are the eigenvalues of $\tilde{\mathbf{Q}}$, which is independent of μ. The condition for the MAP solution therefore reads

$$
\mu\, \boldsymbol{f}_\mu^T\hat{\mathbf{P}}_{f''}\boldsymbol{f}_\mu = -\sum_{i=1}^{E}\frac{\mu}{\mu+\lambda_i} + E + 2\mu\frac{d\ln(p(\mu))}{d\mu}.
$$

Then

$$
\mu\, \boldsymbol{f}_{\hat{\mu}}^T\hat{\mathbf{P}}_{f''}\boldsymbol{f}_{\hat{\mu}} = E\left[\overline{\lambda} + \frac{2\mu}{E}\frac{d\ln(p(\mu))}{d\mu}\right], \tag{25.39a}
$$

$$
\overline{\lambda} := \frac{1}{E}\sum_{i}^{E}\frac{\lambda_i}{\hat{\mu}+\lambda_i}. \tag{25.39b}
$$

This result is very akin to that obtained in the frame of 'quantified Maxent' in Chapter 12 [p. 201]. For a uniform, or almost uniform, prior the second term in the square brackets is negligible and the condition for the MAP solution simply reads

$$
\hat{\mu}\, \boldsymbol{f}_{\hat{\mu}}^T\hat{\mathbf{P}}_{f''}\boldsymbol{f}_{\hat{\mu}} = E\ \overline{\lambda}. \tag{25.40}
$$

As discussed in Chapter 12 [p. 201], $\overline{\lambda}$ corresponds to the percentage of 'good data'. Figure 25.7 shows the log-evidence of equation (25.37) (for a uniform prior) together with the left and right-hand sides of equation (25.40) as a function of $\ln(\mu)$. We realize that the evidence procedure fixes $\hat{\mu}$ quite well. Moreover, we can tell from the dashed line, which represents the right-hand side of equation (25.40), that $\overline{\lambda} = 1$. If we use $\hat{\mu}$ in equation (25.36), or rather in equation (25.33), in order to find posterior values $\boldsymbol{f}_{\hat{\mu}}$, we obtain a new estimate of the reconstruction of the apparatus function. Since it falls very close to the maximum likelihood solution, we have chosen to display the difference of the two solutions in Figure 25.6 as a dashed curve. The influence of the smoothness prior is quite small in this case, the reason being that we have many and quite accurate data to determine $\boldsymbol{f}$.

Marginal prior

In the previous section we have first determined the posterior PDF for $\boldsymbol{f}$ as a function of the hyperparameter μ and then we have approximated the μ-marginalization by the evidence approximation. The alternative approach starts out directly from the marginal prior

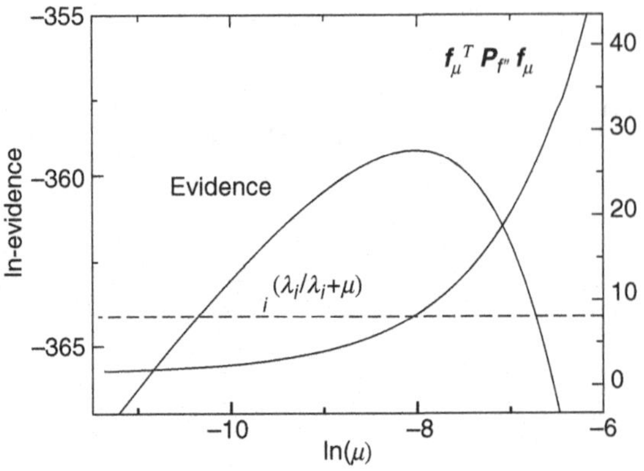

Figure 25.7 The log-evidence of equation (25.37) as a function of $\ln(\mu)$ (left-hand scale). Also shown are both sides of equation (25.40) (right-hand scale).

in equation (25.34). In the present case the latter can easily be determined. The calculation of this integral requires specifying a prior distribution for μ. Complete ignorance about the scale parameter μ leads to the choice of Jeffreys' prior $1/\mu$. The marginal prior is then

$$p(\boldsymbol{f}|E,\boldsymbol{\xi},\mathcal{I}) \propto \int_0^\infty \frac{d\mu}{\mu} \mu^{E/2} \exp\left\{-\frac{\mu}{2}\boldsymbol{f}^T \mathbf{P}_{f''}\boldsymbol{f}\right\} \propto \{\boldsymbol{f}^T \mathbf{P}_{f''}\boldsymbol{f}\}^{-\frac{E}{2}}.$$

This prior has a singularity at $\boldsymbol{f} = 0$. This is the price to be paid for pretending complete ignorance. For sufficiently informative likelihoods we shall ignore this posterior solution, which corresponds to infinite smoothness, and consider the posterior only in the vicinity of the maximum likelihood solution. The log-posterior is then

$$\ln p(\boldsymbol{f}|\mathbf{A}, E, \boldsymbol{\xi}, \mathcal{I}) = C - \frac{1}{2}(\boldsymbol{f} - \boldsymbol{f}_0)^T \mathbf{Q}(\boldsymbol{f} - \boldsymbol{f}_0) + R_L - \frac{E}{2}\ln \boldsymbol{f}^T \mathbf{P}_{f''}\boldsymbol{f}. \quad (25.41)$$

Let us have a look at the MAP solution, which we obtain from the root of the gradient w.r.t. $\boldsymbol{f}$, resulting in

$$0 = \mathbf{Q}\boldsymbol{f} - \mathbf{Q}\boldsymbol{f}_0 - \frac{E}{\boldsymbol{f}^T \mathbf{P}_{f''}\boldsymbol{f}} \mathbf{P}_{f''}\boldsymbol{f},$$

$$\hat{\boldsymbol{f}} = \mathbf{H}^{-1}(\tilde{\mu})\mathbf{Q}\boldsymbol{f}_0,$$

$$\tilde{\mu} := \frac{E}{\boldsymbol{f}^T \mathbf{P}_{f''}\boldsymbol{f}}.$$

The self-consistent solution defines an effective μ, which, interestingly, is the same result if we replace $\overline{\lambda}$ by one. We can even go one step further and replace $\boldsymbol{f}_\mu$ by $\boldsymbol{f}_0$ to obtain

$$\mu \approx \frac{E}{\boldsymbol{f}_0^T \mathbf{P}_{f''}\boldsymbol{f}_0}. \quad (25.42)$$

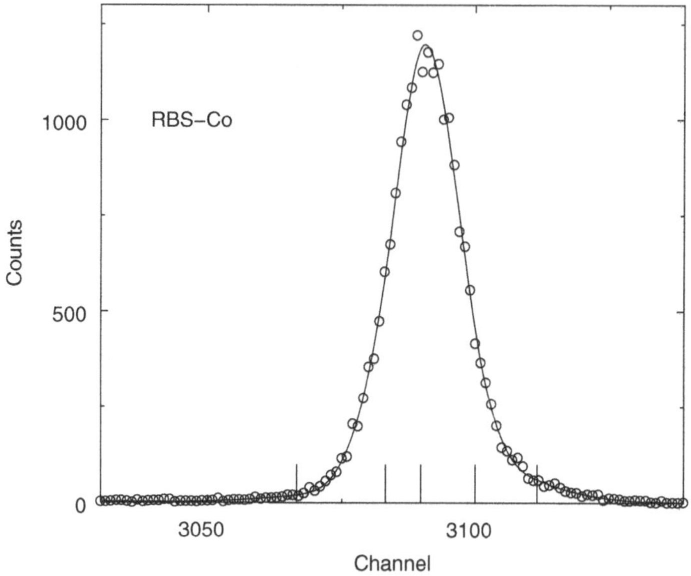

Figure 25.8 Reconstruction with optimum choice of pivot positions. Compare the clearly improved fit (especially at the high-energy side) to that of Figure 25.6 [p. 441].

The numerical value of μ from equation (25.42) is $\ln(\mu) = -8.039$, which compares well with $\ln(\mu) = -8.049$ from the evidence approximation.

Further details on hyperparameters can be found in [69, 199]. Having concentrated on the determination of the hyperparameter, we have slighted the problem that the density reconstruction in Figure 25.6 [p. 441] is far from being satisfactory. A better fit can of course be achieved by an increase of the number of pivots. If we insist, however, on a parsimonious reconstruction then an alternative way is to optimize the position of the pivots which was up to now chosen more or less arbitrarily. In fact, this optimization meets with considerable success, as can be seen in Figure 25.8. We want to point out, however, that the solution in Figure 25.8 is an approximation to the genuine Bayesian solution which would require us to marginalize over the position of the pivots (as was done in Section 25.1 [p. 432]) and over their number. This requires considerable effort, which should be invested if the reconstruction is the problem solution one is aiming at. If, in contrast, the reconstruction will be used frequently in further calculations, then a representation in terms of a few parameters may be the most economical approach.

25.2.2 Modelling Old Faithful data

Old Faithful is the name of a famous geyser in Yellowstone National Park. A table of observed eruption time durations has become a benchmark data set for density estimation [183, 185]. The goal is to estimate the PDF $\rho(t)$ for eruption lasting for t minutes. The data set which we use here is reproduced in Table 25.1 and comprises 109 observations.

Table 25.1 Eruption durations in minutes of 109 eruptions of Old Faithful geyser

1.67	1.67	1.68	1.73	1.73	1.75	1.77	1.77	1.80	1.80
1.82	1.83	1.83	1.83	1.85	1.85	1.88	1.90	1.90	1.90
1.95	1.97	1.97	2.00	2.03	2.25	2.27	2.33	2.50	2.72
2.93	2.93	3.10	3.20	3.33	3.43	3.43	3.50	3.50	3.50
3.52	3.58	3.58	3.67	3.68	3.70	3.70	3.72	3.73	3.73
3.75	3.77	3.80	3.80	3.83	3.87	3.92	3.92	3.93	3.95
4.00	4.00	4.00	4.00	4.00	4.03	4.03	4.05	4.07	4.08
4.08	4.10	4.10	4.12	4.13	4.13	4.18	4.20	4.25	4.25
4.25	4.28	4.33	4.33	4.35	4.37	4.40	4.42	4.43	4.50
4.50	4.50	4.53	4.57	4.58	4.58	4.60	4.60	4.62	4.62
4.63	4.63	4.65	4.70	4.73	4.83	4.93	6.10	6.20	

Source: Adapted from [185].

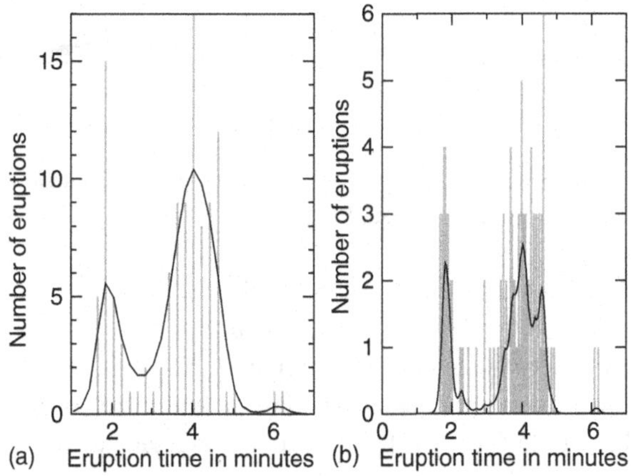

Figure 25.9 The data range was subdivided into (a) 30 bins and (b) 150 bins. The solid line is an interpolation of pointwise reconstructions according to equation (25.46).

There are no observations for durations less than a minute or longer than seven minutes. We shall limit our considerations to this range. First we bin the data in intervals of width δ centred about equidistant points t_i. This yields a histogram displaying the number of eruptions n_i having a duration $t \in \Delta T_i := (t_i - \delta/2, t_i + \delta/2)$, as in Figure 25.9. The probability for an event in T_i is

$$P_i = \int_{\Delta T_i} \rho(t)\, dt.$$

If the total count of events is N, we expect on average $\rho_i = N P_i$ events in ΔT_i. For sufficiently small intervals ΔT_i the distribution becomes Poissonian as outlined in Chapter 9 [p. 147] and the probability for n_i events in ΔT_i, corresponding to a mean count ρ_i, is

$$P(n_i|\rho_i, \mathcal{I}) = \frac{\rho_i^{n_i}}{n_i\,!}\, e^{\rho_i} = \frac{1}{n_i\,!} \exp\left\{-\rho_i + n_i \ln \rho_i\right\}. \tag{25.43}$$

Then the joint likelihood for all data in the independent bins is

$$P(\boldsymbol{n}|\boldsymbol{\rho}, \mathcal{I}) = \prod_i P(n_i|\rho_i) = \frac{1}{\prod_i n_i\,!} \exp\left\{-\sum_i \rho_i + \sum_i n_i \ln \rho_i\right\}. \tag{25.44}$$

This is yet another of the rare cases where the likelihood is not of Gaussian type. From equation (25.44) we derive immediately the maximum likelihood estimate for $\boldsymbol{\rho}$:

$$\frac{d}{d\rho_k}\left(-\sum_i \rho_i + \sum_i n_i \ln \rho_i\right) = -1 + \frac{n_k}{\rho_k}, \tag{25.45}$$

to be $\boldsymbol{\rho} = \boldsymbol{n}$. So the histogram itself turns out to be the maximum likelihood estimator for the density. A glance at Figure 25.9 shows that this result is not very attractive. It is spiky and depends strongly on the chosen binning, as one can see from the pronounced difference of the results in the two panels. This dependence on binning is not unexpected, since binning inevitably goes along with smoothing. So, while we shall not be able to obtain results entirely independent of the chosen binning, a less pronounced sensitivity would be very desirable. A first attempt to achieve this will be to impose the requirement of smoothness on the posterior density for ρ. Remember that the maximum likelihood solution is the Bayesian posterior for a flat, $\boldsymbol{\rho}$-independent prior. The appropriate prior for the present problem is the Fisher information, since it combines the smoothness with the positivity requirement. Details have been discussed in Section 13.4 [p. 221]. The appropriate procedure is to replace ρ_i in the likelihood equation (25.44) by Ψ_i^2 and use the global first derivative prior in equation (13.27) [p. 221] to set up the posterior distribution of $\boldsymbol{\Psi}$:

$$\ln P\left(\boldsymbol{\Psi}|\boldsymbol{n}, E, \mu, \mathcal{I}\right) = C + \frac{E}{2}\ln\mu - \frac{\mu}{2}\boldsymbol{\Psi}^T \mathbf{P}_{f'}\boldsymbol{\Psi} - \sum_{i=1}^{N} \Psi_i^2 + 2\sum_{i=1}^{N} n_i \ln \Psi_i. \tag{25.46}$$

We have suppressed in equation (25.46) all terms not involving μ and $\boldsymbol{\Psi}$. Furthermore, for pointwise reconstruction of the density we have $E = N$, the dimension of $\boldsymbol{\Psi}$ equal to that of $\boldsymbol{n}$. Figure 25.9(a) shows the MAP solution of equation (25.46) for 30 data bins together with the data which constitute also the least-squares solution of the problem. The density reconstruction looks quite acceptable. But what about a dependence on the binning which of course goes along with data smoothing? Figure 25.9(b) shows a similar solution of equation (25.46) with 150 data bins. It looks much less satisfactory. The essential change from (a) to (b) is not only the change in the number of data from 30 to 150. At the same time we have changed the number of variables from 30 to 150 in the density reconstruction. If we are prepared to accept a solution similar to that in the left-hand panel then the number of parameters necessary to characterize such a function can be estimated similarly to the RBS apparatus function in Section 25.2.1 [p. 440]. We would probably need to characterize each peak by position, width and height. Together with the two endpoints of the data range this amounts to eight parameters, which we choose to represent a cubic spline function

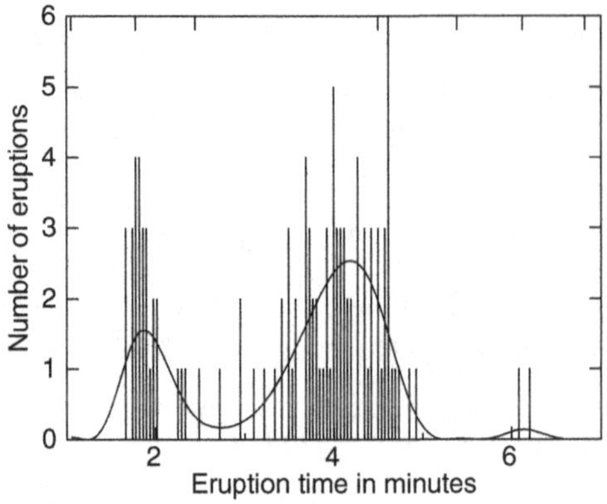

Figure 25.10 Reconstruction of the density according to equation (25.47). The optimum number of spline knots turned out to be seven and their position is indicated as tick marks on the upper frame.

pivoted at the knots $\boldsymbol{\xi}$. The log-posterior from which we determine the support functional values $\boldsymbol{\Psi}$ is

$$\ln P(\boldsymbol{\Psi}|\boldsymbol{n}, \boldsymbol{t}, \boldsymbol{\xi}, E, \mu, \mathcal{I}) = C + \frac{E}{2} \ln \mu - \frac{\mu}{2} \boldsymbol{\Psi}^T \mathbf{P}_{f'} \boldsymbol{\Psi} - \sum_{i=1}^{N} \Psi_i^2 + 2 \sum_{i=1}^{N} n_i \ln \Psi_i, \tag{25.47}$$

where N is again the number of data bins and E the dimension of $\boldsymbol{\Psi}$ and $\boldsymbol{\xi}$. Figure 25.10 shows the density reconstructions for 150 data bins and seven pivots. The number of knots was determined by evidence analysis. The optimum number of knots turned out to be seven instead of the initially estimated eight. From the previous section we know already that the positioning of the pivots can have a strong influence on the quality of the reconstructions. The pivot positions were therefore varied in a random fashion to maximize the evidence. The resulting pivot positions are indicated as tick marks at the upper frame of Figure 25.10. The reconstruction looks quite reasonable and compares favourably with that in Figure 25.9(a). It is, however, more acceptable because the finer data binning has reduced considerably the effect of data smoothing which is inevitably connected with binning.

26
Integral equations

If we face the problem to determine a function $f(x)$ from an equation where it also appears under an integral sign, then we are dealing with the problem of solving an integral equation. A particularly important kind of integral equation arises if the unknown function appears linearly and only under the integral sign. This class of integral equations is called linear of the first kind:

$$h(y) = \int_{-\infty}^{\infty} g(x, y) f(x) dx. \tag{26.1}$$

The so-called 'integral kernel' $g(x, y)$ is assumed to be known, as well as the left-hand side $h(y)$. The function $f(x)$ is to be reconstructed. Expressions of the form given in equation (26.1) arise in many situations in the natural sciences. Probably the most prominent example is the 'Fourier transform'. Here the time domain function $f(t)$ is related to the frequency domain function $A(\omega)$ through

$$A(\omega) = \int_{-\infty}^{\infty} f(t) \cdot e^{i\omega t} dt, \tag{26.2}$$

and $f(t)$ is given by the inverse transform

$$f(t) = \frac{1}{2\pi} \int_{-\infty}^{\infty} A(\omega) e^{-i\omega t} dw. \tag{26.3}$$

An important aspect of Fourier inversion is that this solution is also readily obtainable if the function under the integral is only given numerically. For the special case of an equidistantly spaced grid and a number of data points equal to a power of two, the famous efficient and stable fast finite Fourier transform algorithm solves the problem. The case of the Fourier transform is exceptional. For many integral equations closed-form solutions are not known or, if so, their numerical evaluation poses serious problems. An example for this case is the Laplace transform. Other limitations arise if $h(y)$ in equation (26.1) is only known numerically with finite resolution on a compact support. If it stems in addition from an experimental observation, it will always be corrupted by noise. This means that we should not insist on exact equality in equation (26.1) but allow for some slip which accounts for the noise. The solution of linear integral equations of the first kind is therefore addressed in this chapter by optimization employing the Bayesian framework. Section 26.1 discusses Abel's integral equation which is of importance in several fields of science for problems with cylindrical symmetry. In Section 26.2 we deal with the inversion of the Laplace transform, a particularly nasty problem. Kramers–Kronig inversion is discussed in Section 26.3 [p. 459]. The chapter concludes with the inversion of equation (26.1) for the special case that $g(x, y) = g(y - x)$. In this case the integral is called a convolution of the spectral function $f(x)$ by an apparatus response function $g(y - x)$ yielding the observable signal $h(y)$. Recovering the spectral function $f(x)$ from the measured $h(y)$ is a well-known evergreen of data analysis.

While all the above problems require the solution of linear integral equations, the procedures which we shall adopt in the following can also be (and have been) applied to the solution of nonlinear equations which arise for example in two-electron spectroscopy, like Auger analysis or appearance potential spectroscopy [79, 129].

26.1 Abel's integral equation

'Abel's integral equation' results if an experiment is considered which measures 'line integrals' of a density with cylindrical symmetry. The general case is depicted in Figure 26.1 and examples of the measured quantity are the line integrated absorption or index of refraction. Refraction index measurements have been used for many years in toroidal

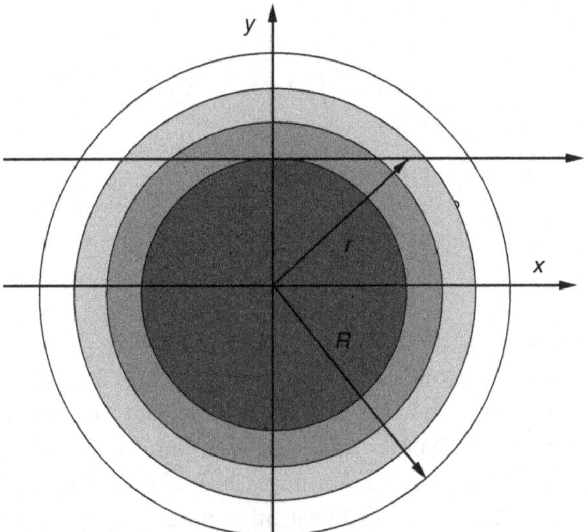

Figure 26.1 Geometry of a problem leading to Abel's integral equation. The measurement yields the line integral of a quantity at distance y from the x-axis through a medium of cylindrical symmetry.

magnetic confinement fusion devices for the determination of the electron density distribution in the hot fusion plasma. Let $\rho(r = \sqrt{x^2 + y^2})$ be the quantity of interest. The measured signal $d(y)$ is then the line integral of $\rho(r(x))$ along the viewing chord, which has been chosen parallel to the x-axis in Figure 26.1:

$$d(y) = \int_{-\sqrt{R^2-y^2}}^{\sqrt{R^2-y^2}} \rho\left(\sqrt{x^2 + y^2}\right) dx. \tag{26.4}$$

The finite limits of integration assume that ρ has decreased to negligibly low values at $r = R$. Substitute $r^2 = x^2 + y^2$ to obtain

$$d(y) = 2 \int_{y}^{R} \frac{r}{\sqrt{r^2 - y^2}} \rho(r) dr. \tag{26.5}$$

We notice that the integrand becomes singular at $r = y$. The singularity is, however, harmless since all integrals for arbitrary powers of r exist. In view of the success of the spline approach in function estimation we model the unknown density function $\rho(r)$ as a cubic spline. In other words, the radial symmetric density $\rho(r)$ is presented by its values at some suitable pivot points r_j, for $j = 1, \ldots, N$. For arbitrary but fixed y, the integral is given in equation (26.5). This implies that $d(y)$ is a linear combination of integrals of the form

$$I_n = 2 \int_{r_j}^{r_{j+1}} \frac{r^n}{\sqrt{r^2 - y^2}} dr, \quad n = 1, \ldots, 4, \tag{26.6}$$

where only those pivot points with $r_j > y$ contribute to $d(y)$. For a cubic spline basis in equation (26.5), I_n is required for $n = 1, \ldots, 4$.

For $y > 0$ the integrals are

$$\begin{aligned}
I_1 &= 2 \cdot \sqrt{r^2 - y^2}, \\
I_2 &= r \cdot \sqrt{r^2 - y^2} + y^2 \ln\left(r + \sqrt{r^2 - y^2}\right), \\
I_3 &= 2y^2 \cdot \sqrt{r^2 - y^2} + \frac{2}{3}\left(r^2 - y^2\right)^{3/2}, \\
I_4 &= \frac{r}{2} \cdot \sqrt{r^2 - y^2}\left(r^2 + \frac{3}{2} y^2\right) + \frac{3}{4} y^4 \ln\left(r + \sqrt{r^2 - y^2}\right).
\end{aligned} \tag{26.7}$$

The central chord $y = 0$ needs special attention:

$$I_n = \int_{0}^{r_1} \frac{r^n}{r} dr = \frac{1}{n} \cdot r_1^n, \quad \forall n. \tag{26.8}$$

If we now refer to the explicit form of the spline function $S_j(r)$ on the interval r_j, r_{j+1} of equation (13.7) [p. 217] and define the functions

$$p_0(r_j) := \frac{r_{j+1} I_1 - I_2}{h_{j+1}},$$
$$p_1(r_j) := \frac{I_2 - r_j I_1}{h_{j+1}},$$
$$u_0(r_j) := \frac{-I_4 + 3r_{j+1} I_3 + I_2\left(h_{j+1}^2 - 3r_{j+1}^2\right) + I_1\left(r_{j+1}^3 - h_{j+1}^2 r_j - h_{j+1}^3\right)}{6h_{j+1}},$$
$$u_1(r_j) := \frac{I_4 - 3r_j I_3 + I_2\left(2r_j^2 - h_{j+1}^2\right) + I_1\left(h_{j+1}^2 - r_j^2\right) r_j}{6h_{j+1}},$$

we obtain

$$\int_{r_j}^{r_{j+1}} \frac{r S_j(r) dr}{\sqrt{r^2 - y^2}} = \sum_{\alpha=0}^{1} \left(u_\alpha(r_j) \sum_k L_{j+\alpha,k} f_k + p_\alpha(r_j) \cdot f_{j+\alpha} \right). \tag{26.9}$$

$L_{j,k}$ are the elements of the matrix $\mathbf{L}$ defined in equation (13.11) [p. 218], which allows us to express the second derivatives of the spline by the support functional values. The right-hand side of equation (26.9) can obviously be written as the scalar product of a vector $\boldsymbol{a}_j^T$ and $\boldsymbol{f}$. Let $\mathbf{A}$ be the matrix with row vectors $\boldsymbol{a}_j^T$. Our model is then completely specified and the task is to recover $\rho(r) = \{S_j(r)\}$ from the equation

$$d(y_i) = (\mathbf{A}\boldsymbol{f})_i + \epsilon_i. \tag{26.10}$$

Assuming as usual that $\langle \epsilon_i \rangle = 0$ and $\langle \epsilon_i^2 \rangle = \sigma_i^2$, we obtain the Gaussian likelihood

$$p(\boldsymbol{d}|\mathbf{A}, \boldsymbol{f}, \boldsymbol{y}, \boldsymbol{\sigma}, \mathcal{I}) = \frac{1}{\prod_i \sigma_i \sqrt{2\pi}} \exp\left\{ -\frac{1}{2} \sum_i \frac{(d_i - (\mathbf{A}\boldsymbol{f})_i)^2}{\sigma_i^2} \right\}. \tag{26.11}$$

This is exactly the same form that we met in equation (25.3) [p. 433], where we considered function estimation. So what is the problem here? Our aim is to infer the vector $\boldsymbol{f}$ from the vector of measured data $\boldsymbol{d}$, given errors and mapping matrix $\mathbf{A}$. Assuming for a moment a uniform prior on $\boldsymbol{f}$, this problem is identical to the earlier similar regression problem in equation (21.41) [p. 345]. In Section 12.7.1 [p. 211] we have already discussed 'ill-posed inversion' problems and the reason why some inversions are ill behaved. The reason was the fact that the eigenvalues of the kernel matrix $\mathbf{A}$ span several orders of magnitude, which results in severe noise amplification. This holds if only the likelihood is considered, which corresponds to a uniform prior on $\boldsymbol{f}$. However, while the prior did not show much influence on the result in function estimation, it makes all the difference in the present solution of ill-posed integral equations. In purely technical terms an appropriate prior distribution will enlarge the small singular values of the matrix $\mathbf{A}$ such that the solution becomes considerably less sensitive to data corruption by noise. In order to illustrate the various aspects of the problem, data were generated from the density

$$\rho(r) = \left\{ \left(\frac{r}{r_0}\right)^2 + 0.2 \right\} \cdot \exp\left\{ -\left(\frac{r}{r_0}\right)^2 \right\} \tag{26.12}$$

in the range $0 \leq y/r_0 \leq 3$. The line integrals were then superimposed by white noise of variance $(0.03)^2$, which corresponds to a 'data' precision of 3% for the largest line integrals and much less for the outer chords.

We first address the 'unique' problem where the number of reconstruction points equals that of the data points, 30 each. The singular values of $\mathbf{A}$ spanned approximately four orders of magnitude and the straightforward least-squares solution (uniform prior on $\boldsymbol{f}$) covered the range from −71 to +127 while the known true solution is positive definite with a maximum around 0.45, in other words, it results in complete nonsense. We shall now introduce a prior on $\boldsymbol{f}$ which incorporates our prior knowledge that particle densities are smooth and positive. We choose equation (13.27) [p. 221], which enforces a smooth solution and compare the performance to the Fisher information prior in Section 13.4 [p. 221] which enforces both positivity and smoothness. The results are depicted in Figure 26.2.

Panel (a) shows the result for the global first derivative prior, while panel (b) was obtained by employing the Fisher information prior. We notice that there is very little difference between the two solutions. The only point worth mentioning is that the confidence intervals in case of the Fisher information prior, compared with the global first derivative prior, are smaller where the function is small and larger where the function is large. The effect is, though small, not unexpected since the Fisher information results from weighting the first derivative squared with the inverse of the function itself. The use of equal numbers of data and reconstruction points seems rather special. Figure 26.3 shows what happens when the number of reconstruction points is larger or smaller than the number of data. The Fisher

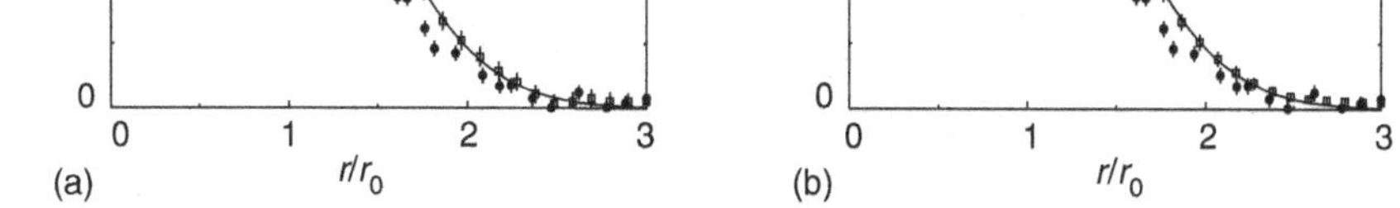

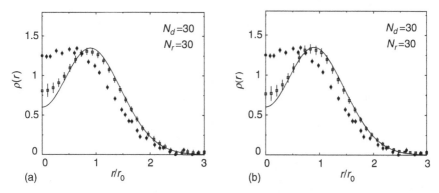

Figure 26.2 The solution of Abel's integral equation is given as full squares. The continuous line depicts the 'true solution' from which the data (full dots) were generated. The reconstructed density is represented by full squares. (a) The reconstruction employing the global first derivative prior, (b) obtained for the Fisher information prior.

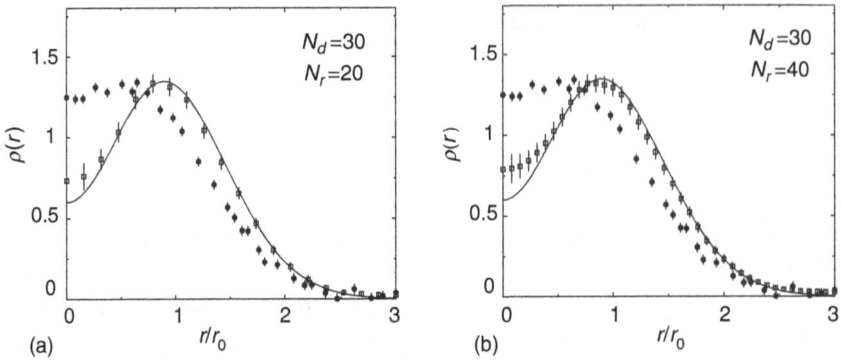

Figure 26.3 Density reconstructions for (a) $N_d > N_r$ and (b) $N_d < N_r$, based on the Fisher information prior. Full dots with error bars display input data. The continuous curve is the true density of equation (26.12) and full squares with error bars show the reconstructed density along with confidence intervals.

information prior was used for these calculations. Again we do not find much difference between the two cases. Remember also that the reconstruction points are support values of a spline function which permits interpolation on even finer scales. In fact, we expect that no more than about five or six reconstruction points are necessary to completely define the density function. This remark is important since the computation time increases roughly as the third power of the number of reconstruction points. The spline representation of the unknown function therefore also has important economic aspects.

26.2 The Laplace transform

We have already encountered the Laplace inversion problem for quantum Monte Carlo data in Section 12.7 [p. 210]. In general, the Laplace integral transform of a function $f(x)$ is defined as

$$F(s) = \int_0^\infty f(x)e^{-sx}dx. \tag{26.13}$$

There exists an analytic formula for inversion:

$$f(x) = \frac{1}{2\pi i}\int_{c-i\infty}^{c+i\infty} F(s)e^{sx}ds. \tag{26.14}$$

The constant c in the integration limits must be chosen so large that all singularities of $F(s)$ lie to the left of the line $c - i\infty$, $c + i\infty$. While the calculation of the 'direct' transform in equation (26.13) is easily possible even if $f(x)$ is given only numerically, it seems clear that the evaluation of the inverse formula can become very complicated, if not impossible, even when the functional form of $F(s)$ is known. In that case one can proceed as follows: expand

$F(s)$ in terms of basis functions $\phi_i(s)$ for which the inverse transformation is known. Then $f(x)$ is given by the linear combination of the inverse transforms $\phi_i(x)$. Extensive tables of Laplace transforms are available for this route [46]. This approach, however, is not applicable if $F(s)$ is only given for a discrete set of points s_i and if $F(s_i)$ is even corrupted by noise. Then we are facing an ill-posed inversion problem, which can only be solved consistently in the Bayesian framework.

As outlined in Chapter 12 [p. 201], a major problem in theoretical many-body physics is to infer a spectral function $f(x)$ from its computed Green's function $d(s)$. The latter is related to the spectral function via a Laplace transform

$$d(s) = \int_0^\infty f(x) \cdot e^{-sx} dx, \tag{26.15}$$

and since $d(s)$ is computed (mostly via quantum Monte Carlo methods) it is deteriorated by numerical noise. We shall proceed with the solution as we did in the previous case. In Chapter 12 [p. 201] we have used the 'entropic prior' for a form-free reconstruction. In many cases, however, the result exhibits spurious features, which result from the fact that the form-free reconstruction is performed with higher resolution than supported by the data. Along with the missing correlation in the entropic prior, noise may still be enhanced. In order to incorporate local smoothness, we will therefore represent $f(x)$ by a spline function. Sharp structures can still be resolved, if they are supported by the data, by the adequate choice of pivot points. For the spline representation we need the integrals

$$I_n(x_j) = \int_{x_j}^{x_{j+1}} x^n e^{-sx} dx, \qquad n = 0, \ldots, 3. \tag{26.16}$$

They can best be determined in this context via a recursion relation which is obtained through integration by parts:

$$I_n(x_j) = \frac{1}{s}\left\{x_j^n e^{-sx_j} - x_{j+1}^n e^{sx_{j+1}}\right\} + \frac{n}{s} I_{n-1}(x_j), \tag{26.17a}$$

$$I_0(x_j) = \frac{1}{s}\left\{e^{-sr_j} - e^{sr_{j+1}}\right\}. \tag{26.17b}$$

The rest of the analysis develops in close analogy to the solution of Abel's integral equation; after all, the two differ only in the shape of their positive definite kernels. Illustrative calculations were performed with the spectral function

$$f(x|\gamma, x_0) = \frac{\gamma^3}{4\pi} \frac{1}{\{(x - x_0)^2 + \gamma^2/4\}^2}, \tag{26.18}$$

with $\gamma = 0.8$, $x_0 = 1.2$ and data restricted to the range $s \leq 3$. The number of data was chosen to be $N_d = 75$ and the number of reconstruction pivots was fixed at $N_r = 24$. The posterior PDF for f follows from Bayes' theorem:

$$p(\boldsymbol{f}|\boldsymbol{d},\mathcal{I}) = \frac{p(\boldsymbol{f}|\mu,\mathcal{I})p(\boldsymbol{d}|\boldsymbol{f},\mathcal{I})}{p(\boldsymbol{d}|\mathcal{I})}. \tag{26.19}$$

As shown in Chapter 12 [p. 201], the Laplace kernel is particularly ill-posed and a uniform prior results in a completely useless reconstruction. Even if local smoothness is enforced by the spline basis, the kernel still turns out to span more than six orders of magnitude, resulting in an entirely useless reconstruction $\boldsymbol{f}$. This is not unexpected but the effect is much more pronounced than in the Abel inversion. An appropriate prior for the problem is of course again the Fisher information and it does in fact provide a satisfactory solution of the problem, as can be seen in Figure 26.4(a). The small vertical tick marks show the data in a semi-logarithmic representation. Note that the influence of the spectral function $f(x)$ is buried in the deviation of the data from a straight line! The open circles with error bars are the reconstruction of $f(x)$ obtained by treating the hyperparameter μ in the prior in evidence approximation. The thin continuous curve is the true spectral function of equation (26.18). The reconstruction is found to be satisfactory but not perfect. The question arises whether another prior performs better. The right-hand panel in Figure 26.4 shows the reconstruction using the global first derivative squared as a prior. The reconstruction is considerably worse than for the Fisher information, in particular it becomes negative at the beginning and end of the support.

Let us now test the performance of the form-free QME, outlined in Chapter 12 [p. 201], for the inversion of the Laplace transform. The results are depicted in Figure 26.5, for which the hyperparameter α has been treated in evidence approximation. The reconstruction is of course restricted to the domain of positive values for the components of $\boldsymbol{f}$. This brings the solution much closer to Figure 26.4(a). However, we realize that the uncertainties in the reconstruction are considerably larger than for the solution employing the Fisher information prior. The effect is due to the missing smoothness, or rather correlation,

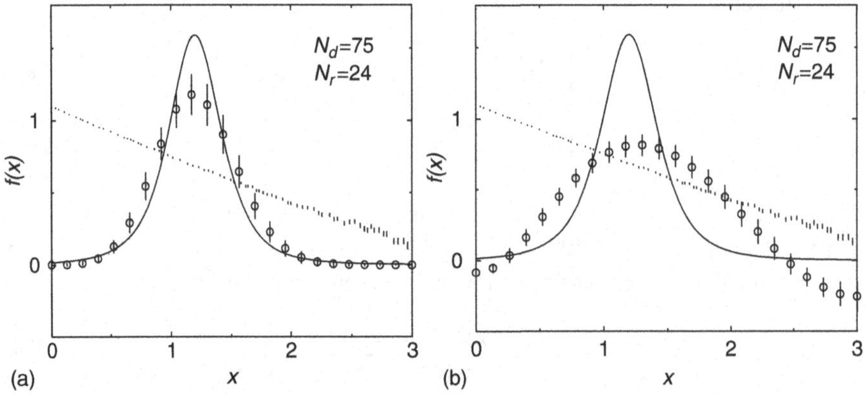

Figure 26.4 The Laplace transform of the continuous curve is shown in a semi-log plot, marked by small vertical bars, which represent the errors of the data. Reconstruction of the source function is shown by open circles with error bars. (a) Fisher information prior, (b) first derivative prior.

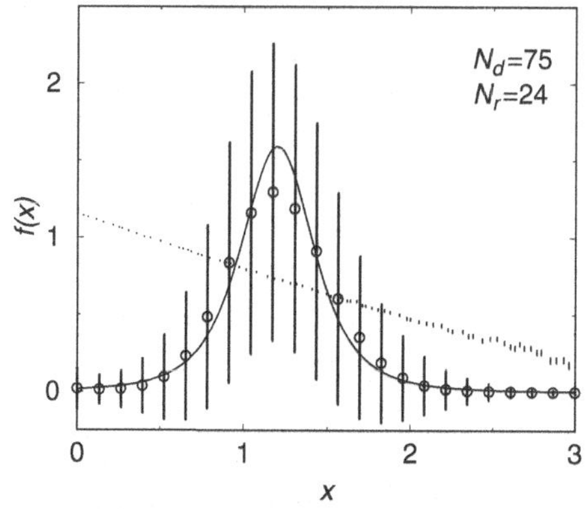

Figure 26.5 Form-free inversion of the Laplace transform employing the traditional QME prior with a uniform default model in Bayes' theorem.

constraint in the entropic prior. This may be an advantage in other problems. For the present task it is definitely a disadvantage since it is part of our prior knowledge that the reconstruction is a smooth function. This property is encoded in the priors of Chapter 13 [p. 215] and the present example shows their superior performance in appropriate problems.

Nevertheless, the entropic prior has contributed much to the dissemination of Bayesian procedures in the natural sciences. The authors confess that they were also enthusiasts of the entropic prior for several years, until some of its limitations gradually became known.

26.3 The Kramers–Kronig relations

Consider a complex function $h(\omega) = f(\omega) + ig(\omega)$ of the complex variable ω. The functions $f(\omega)$ and $g(\omega)$ are real and $h(\omega)$ shall be analytic in the upper half of the complex plane. Moreover, it shall decay faster than $|\omega|^{-1}$ for $|\omega| \to \infty$, to ensure that the integral along an infinite semicircle in the upper half-plane is zero. In this case the following relation follows from Cauchy's residue theorem:

KRAMERS–KRONIG RELATION

$$h(\omega) = \frac{1}{i\pi}\mathcal{P}\int_{-\infty}^{\infty} \frac{h(x)dx}{x-\omega}, \tag{26.20}$$

where $\mathcal{P}$ indicates the principal value. If, in addition, $h(-x) = h^*(x)$, that is $f(x)$ is an even and $g(x)$ an odd function, then the relations are

Kramers–Kronig relations for odd $g(x)$

$$f(\omega) = \frac{2}{\pi}\mathcal{P}\int_0^\infty \frac{xg(x)dx}{x^2-\omega^2}, \tag{26.21a}$$

$$g(\omega) = -\frac{2\omega}{\pi}\mathcal{P}\int_0^\infty \frac{f(x)dx}{x^2-\omega^2}. \tag{26.21b}$$

The Kramers–Kronig relations make it possible to find the real part of the response function of a linear passive system if the imaginary part is known and vice versa. They constitute an important tool in the analysis of optical measurements on solids such as absorption, refraction and reflection. We shall address the problem of calculating the absorption $g(\omega)$ given a measured, noisy, truncated data set on $f(\omega)$. The procedure follows closely the lines that we pursued in the analysis of Abel's and Laplace's integral equation. We shall model again the unknown function $g(x)$ in a spline basis and need therefore to attend to the principal value of the integral in equation (26.21a). This principal value can be tackled as follows:

$$\begin{aligned} f(\omega) &= \frac{2}{\pi}\mathcal{P}\int_0^\infty \frac{xg(x)-\omega g(\omega)}{x^2-\omega^2}dx + \frac{2\omega g(\omega)}{\pi}\mathcal{P}\int_0^\infty \frac{dx}{x^2-\omega^2} \\ &= \frac{2}{\pi}\int_0^\infty \frac{xg(x)-\omega g(\omega)}{x^2-\omega^2}dx + \frac{2\omega g(\omega)}{\pi}\mathcal{P}\int_0^\infty \frac{dx}{x^2-\omega^2}, \end{aligned} \tag{26.22}$$

where the first integral is now regular and the principal value has been shifted to an elementary integral

$$\mathcal{P}\int_0^\infty \frac{dx}{x^2-\omega^2} = 0, \tag{26.23}$$

which is shown to vanish by elementary integration. We will be interested in a discrete mesh of ω_i values. The integrals to be considered depend on whether the zero in the denominator of equation (26.22) is contained in the pivotal interval $[x_l, x_{l+1}]$ or not. For $x_l < \omega_i < x_{l+1}$, equation (26.22) becomes

$$\begin{aligned} \frac{\pi}{2}f(\omega_i) &= \int_0^{x_l} \frac{xg(x)}{x^2-\omega_i^2}dx + \int_{x_l}^{x_{l+1}} \frac{xg(x)-\omega_i g(\omega_i)}{x^2-\omega_i^2}dx + \int_{x_{l+1}}^{x_N} \frac{xg(x)}{x^2-\omega_i^2}dx \\ &\quad - \omega_i g(x_i)\left\{\int_0^{x_l} \frac{dx}{x^2-\omega_i^2} + \int_{x_{l+1}}^{x_N} \frac{dx}{x^2-\omega_i^2}\right\}. \end{aligned} \tag{26.24}$$

Since we have assumed that $g(x)$ is a spline function, the required integrals not containing the singularity are of the form

$$I_n(\omega_i) = \int_a^b \frac{x^{n+1}}{x^2 - \omega_i^2} dx := F_n(b|\omega_i) - F_n(a|\omega_i), \tag{26.25}$$

which can be determined by elementary integration, resulting in

$$F_1(a|\omega_i) = \frac{1}{2} \log\left(a^2 - \omega^2\right), \tag{26.26a}$$

$$F_2(a|\omega_i) = a - \omega \tanh^{-1}\left(\frac{a}{\omega}\right), \tag{26.26b}$$

$$F_3(a|\omega_i) = \frac{\omega^2}{2} \log\left(a^2 - \omega^2\right) + \frac{a^2}{2}, \tag{26.26c}$$

$$F_4(a|\omega_i) = \frac{a^3}{3} - \omega^3 \tanh^{-1}\left(\frac{a}{\omega}\right) + a\omega^2. \tag{26.26d}$$

For the singularity-containing interval a second type of integral is needed:

$$I_n = \int_{x_l}^{x_{l+1}} \frac{x^{n+1} - \omega_i^{n+1}}{(x + \omega_i)(x - \omega_i)} dx. \tag{26.27}$$

The numerator of the integrand can be written as $(x - \omega_i)P_n(x, \omega_i)$, where P_n is a polynomial of order n. The resulting expression for the integral

$$I_n = \int_{x_l}^{x_{l+l}} \frac{P_n(x, \omega_i)}{(x + \omega_i)} dx \tag{26.28}$$

has no singularity any more. This concludes the formulation of the model equations, since the last term in equation (26.24) is trivial.

An important example of a function $h(w)$ whose real and imaginary parts obey the Kramers–Kronig relations is provided by the response of a linear damped oscillator of resonance frequency ω_0 and damping γ subject to a harmonic excitation at frequency ω:

$$h(\omega) = \frac{\omega_0^2 - \omega^2 + i\omega\gamma}{(\omega_0^2 - \omega^2)^2 + \omega^2\gamma^2}. \tag{26.29}$$

Data were generated from the real part of equation (26.29) in the range $0 \le \omega \le 1$ with $\gamma = 0.1$ and $\omega_0 = 0.4$. Noise of variance 0.75 was then added to simulate a measurement. The data we use are displayed in Figure 26.6. Since the absorption to be reconstructed is strictly positive definite, the appropriate prior for reconstruction is the Fisher information. The hyperparameter μ was determined by employing the evidence approximation. Results are presented for $N_d = 18$ data points and $N_r = 38$ reconstruction points in

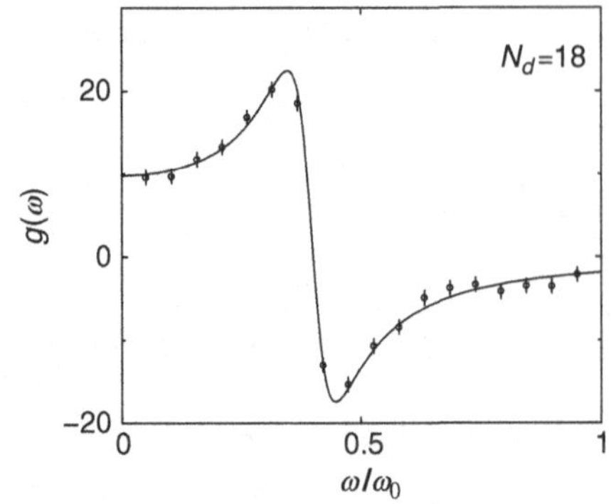

Figure 26.6 Data (full circles) for the Kramers–Kronig problem were generated from the dispersion part of equation (26.29).

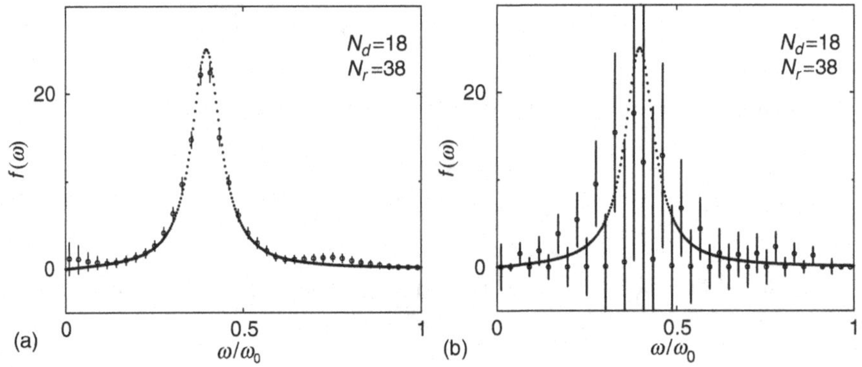

Figure 26.7 (a) The reconstruction (circles with error bars) of the absorption from the dispersion data in Figure 26.6 using the Fisher prior. (b) The performance of QME for the same inversion problem. In both panels exact results are marked by continuous dotted lines.

Figure 26.7(a). The circles with error bars denote the reconstructed values of the absorption with associated confidence limits. As in previous cases, the true solution of the inversion problem is known exactly as the imaginary part of equation (26.29) and is depicted as well. The overall agreement between reconstruction and exact solution is quite satisfactory and the reconstruction is also smooth. The rise of the reconstruction as ω tends to zero is unexpected. In this region the data information decreases progressively, which is also reflected in the enlarging uncertainty of the reconstruction. It was tempting to try QME on this problem since by suitable choice of the invariant measure, constant and sufficiently small, such a behaviour should be suppressible. We gave this idea a try with the interesting experience displayed in Figure 26.7(b). The point-to-point oscillations of the reconstruction render it entirely useless. Yet an explicit calculation of the entropies associated with the reconstruction, and that of the true solution, proved that the QME solution indeed has a greater entropy than the solution obtained by the Fisher prior. The peculiar unacceptable behaviour is due to the permutation invariance of the entropic prior, which ignores any possible correlations. As a matter of fact, this is the only case known to the authors where QME fails in such a dramatic way, and it does so in the present case only if we formulate an underdetermined problem where the number of reconstruction points is much larger than the number of data points. What remains, however, is that even on a diluted reconstruction grid the performance of the Fisher information prior is superior: it provides the smoother reconstruction and the smaller uncertainties.

26.4 Noisy kernels

The common property of the integral equations considered in the preceding three sections is that the kernel of the integral was known exactly. In the important case of convolutions, which we shall address in the next section, this is usually not the case. What we shall assume, however, is that the kernel has been measured. It is clear that in such a situation the associated measurement errors of the kernel will contribute to the uncertainty of the reconstructed function in addition to the errors in the left-hand side of equation (26.1) [p. 451].

Consider again the multivariate linear problem

$$p(\boldsymbol{d}|\mathbf{A}, \boldsymbol{f}, \boldsymbol{\sigma}, \mathcal{I}) = \prod_i \frac{1}{\sigma_i\sqrt{2\pi}} \exp\left\{-\frac{1}{2}\sum_i \frac{\left(d_i - \sum_k a_{ik} f_k\right)^2}{\sigma_i^2}\right\}, \tag{26.30}$$

where a_{ik} are the elements of the exact transfer matrix $\mathbf{A}$. We assume now that this is not known. Instead, the elements of a matrix $\tilde{\mathbf{A}}$ are assumed to be available as a result of a measurement. Hence

$$p(\mathbf{A}|\tilde{\mathbf{A}}, \boldsymbol{\delta}) = \prod_{i,j} \frac{1}{\delta_{ij}\sqrt{2\pi}} \exp\left\{-\frac{1}{2}\sum_{i,j} \frac{\left(a_{ij} - \tilde{a}_{ij}\right)^2}{\delta_{ij}^2}\right\}. \tag{26.31}$$

Before we can use Bayes' theorem to infer the unknown functional values $\boldsymbol{f}$ in equation (26.30), we must remove the M unknown elements a_{ik} of the exact transfer matrix $\mathbf{A}$ from equation (26.30) by marginalization, employing the probability distribution of equation (26.31) to obtain a marginal likelihood in terms of the M known approximate matrix elements $\tilde{a}_{ij}$:

$$\begin{aligned} p(\boldsymbol{d}|\tilde{\mathbf{A}}, \boldsymbol{f}, \boldsymbol{\sigma}, \boldsymbol{\delta}) &= \int d^M A \; p(\mathbf{A}, \boldsymbol{d}|\tilde{\mathbf{A}}, \boldsymbol{f}, \boldsymbol{\sigma}, \boldsymbol{\delta}) \\ &= \int d^M A \; p(\mathbf{A}|\tilde{\mathbf{A}}, \boldsymbol{\delta}) \; p(\boldsymbol{d}|\mathbf{A}, \boldsymbol{f}, \boldsymbol{\sigma}). \end{aligned} \tag{26.32}$$

Since $p(\mathbf{A}|\tilde{\mathbf{A}}, \boldsymbol{\delta})$ is also quadratic in the $\{a_{ij}\}$ we can express the argument of the exponential under the integral in equation (26.32),

$$\sum_i \frac{\left(d_i - \sum_k a_{ik} f_k\right)^2}{\sigma_i^2} + \sum_{i,j} \frac{\left(a_{ij} - \tilde{a}_{ij}\right)^2}{\delta_{ij}^2} = (\boldsymbol{a} - \boldsymbol{a}_0)^T \mathbf{Q} (\boldsymbol{a} - \boldsymbol{a}_0) + R, \tag{26.33}$$

as a complete square with residue R. The vector $\boldsymbol{a}$ in equation (26.33) contains all the elements of matrix $\mathbf{A}$ and thus has $M = I \cdot J$ components if the summation over i is understood to run from $i = 1, \dots, I$ and that over j from $j = 1, \dots, J$. $\boldsymbol{a}_0$ is the minimum of equation (26.33) and therefore also the mode of the integrand in equation (26.32). The distribution in equation (26.32) is now readily obtained in terms of $\boldsymbol{a}_0$, $\mathbf{Q}$ and R:

$$p(\boldsymbol{d}|\tilde{\mathbf{A}}, \boldsymbol{f}, \boldsymbol{\sigma}, \boldsymbol{\delta}) = \frac{1}{Z}|\mathbf{Q}|^{1/2} \exp\left\{ -\frac{R}{2}\right\}, \tag{26.34}$$

where all the dependencies on $\boldsymbol{f}$ and $\tilde{\mathbf{A}}$ are now hidden in $\mathbf{Q}$ and R. We shall postpone the problem of calculating $\mathbf{Q}$ and focus first on the evaluation of R. Since $\boldsymbol{a}_0$ is the minimum of equation (26.33), its components satisfy

$$\frac{\partial}{\partial a_{mn}} \left\{ \frac{1}{2} \sum_i \frac{(d_i - \sum_k a_{ik} f_k)^2}{\sigma_i^2} + \frac{1}{2} \sum_{i,j} \frac{(a_{ij} - \tilde{a}_{ij})^2}{\sigma_{ij}^2} \right\} = 0, \tag{26.35}$$

which transforms into

$$\frac{a_{mn}^0}{\delta_{mn}^2} = \frac{\tilde{a}_{mn}}{\delta_{mn}^2} + \frac{d_m - \sum_k a_{mk}^0 f_k}{\sigma_m^2} \cdot f_n. \tag{26.36}$$

We multiply this equation by $f_n \delta_{mn}^2$, sum over n and subtract d_m on both sides. Reordering the terms finally yields

$$d_m - \sum_n a_{mn}^0 f_n = \frac{d_m - \sum_k \tilde{a}_{mn} f_k}{1 + \sum_n \delta_{mn}^2 f_n^2 / \sigma_m^2}. \tag{26.37}$$

We abbreviate $\sum_n \delta_{mn}^2 f_n^2 = S_m^2$ and obtain for the first term on the left-hand side of equation (26.33):

$$\frac{1}{2} \sum_i \Big(d_i - \sum_k a_{ik} f_k \Big)^2 = \frac{1}{2} \sum_m \left\{ \frac{d_m - \sum_n \tilde{a}_{mn} f_n}{1 + S_m^2/\sigma_m^2} \right\}^2 \cdot \frac{1}{\sigma_m^2}. \tag{26.38}$$

Next we calculate the second term on the left-hand side of equation (26.33) from equation (26.36):

$$a_{mn}^0 - \tilde{a}_{mn} = \frac{d_m - \sum_k \tilde{a}_{mk} f_k}{\sigma_m^2 + S_m^2} \delta_{mn}^2 f_n. \tag{26.39}$$

This must now be squared, divided by δ_{mn}^2 and summed over m and n to yield

$$\frac{1}{2} \sum_{m,n} \frac{(a_{mn}^0 - \tilde{a}_{mn})^2}{\delta_{mn}^2} = \frac{1}{2} \sum_m \left\{ \frac{d_m - \sum_k \tilde{a}_{mk} f_k}{\sigma_m^2 + S_m^2} \right\}^2 \sum_n \delta_{mn}^2 f_n^2. \tag{26.40}$$

This is very similar to the right-hand side of equation (26.38) and can easily be combined with equation (26.40) to obtain the marginal prior distribution

$$p(\boldsymbol{d}|\tilde{\mathbf{A}}, \boldsymbol{f}, \boldsymbol{\sigma}, \boldsymbol{\delta}) = \frac{1}{Z} \exp\left\{ -\frac{1}{2} \sum_m \frac{(d_m - (\tilde{\mathbf{A}}\boldsymbol{f})_m)^2}{\sigma_m^2 + \sum_n \delta_{m,n}^2 f_n^2} \right\}. \tag{26.41}$$

We find that its form becomes identical to equation (26.30) if we replace σ_m^2 by

$$\sigma_{m,\text{eff}}^2 = \sigma_m^2 + \sum_n \delta_{mn}^2 f_n^2. \tag{26.42}$$

Note that this result is very reasonable and identical to what one would have obtained by employing the elementary law of error propagation.

We can now bypass the still pending calculation of matrix $\mathbf{Q}$ in equation (26.33) by normalizing equation (26.41). Compare equations (26.41) and (26.42) to equation (26.30) to find the normalized marginal likelihood

$$p(\boldsymbol{d}|\tilde{\mathbf{A}}, \boldsymbol{f}, \boldsymbol{\sigma}, \boldsymbol{\delta}) = \prod_m \left(2\pi\sigma_{m,\text{eff}}^2\right)^{-\frac{1}{2}} \exp\left\{-\frac{1}{2}\sum_m \frac{(d_m - (\tilde{\mathbf{A}}\boldsymbol{f})_m)^2}{\sigma_{m,\text{eff}}^2}\right\}. \tag{26.43}$$

26.5 Deconvolution

In Section 25.2.1 [p. 440] we discussed a Rutherford backscattering experiment on an ideal target. The target was ideal insofar as it was monatomic and hence did not suffer from energy loss resulting in additional energy broadening, and it was assumed to consist of cobalt atoms which are known to occur only with a single nuclear mass, that is isotopically pure. Figure 25.6 [p. 441] shows the result of an experiment on such an ideal target. The final state energy is distributed over several detection channels instead of a delta function, as expected from the ideal version of the experimental apparatus sketched in Figure 25.5 [p. 440]. These data comprise all deviations of the experiment from the unavailable ideal, and are called the apparatus response function, or apparatus transfer function of the experiment. Let $g(y)$ denote the apparatus response as a function of energy y. Consider now the presence of other elements in the target with masses adjacent to cobalt. The cobalt signal is now $f(x = 0) \cdot g(y)$, where $f(0)$ is the cobalt fraction in the mixture target. Target atoms with higher mass $M = M_{\text{Co}} + \Delta M$ will produce an energy distribution of the same shape as cobalt, however, shifted to higher energy $g(y - x^+)$ with amplitude $f(x^+)$. Target atoms with smaller mass will be reflected in $g(y - x^-)$ and amplitude $f(x^-)$. The total scattering signal is obtained from adding up all these contributions to $h(y)$, with $h(y)$ given by

$$h(y) = \int_{-\infty}^{\infty} f(x) g(y - x) dx. \tag{26.44}$$

The interesting problem is now to infer the mass distribution $f(x)$ from a measurement on the mixture target $h(y)$ and a measurement of the apparatus transfer function $g(y)$. This process is called deconvolution and fits into the general form of integral equations treated so far in this chapter.

A convolution shows up in most experiments due to the finite resolution of the apparatus. A particularly challenging problem occurs in inverse photoemission, where the resolution is much less than in photoemission and where in addition the presence of the Fermi–Dirac distribution suppresses information from the occupied part of the energy bands. According to conventional wisdom, inverse photoemission only provides information about the unoccupied bands. It has been demonstrated in [208] that the application of Bayesian probability theory in the frame of QME is capable of also recovering occupied parts of the bands and increasing the resolution by a factor of 5. In this application it was possible to recover the temperature dependence of the exchange splitting of the nickel bands below T_C, resulting in a band that crosses the Fermi level. Without the Bayesian reconstruction, the experimental raw data are localized way above the Fermi level and give no clue to the energy splitting.

Let us now turn back to some general remarkable properties of equation (26.44). By a transform of variables $y - x = z$ we find that

$$h(y) = \int_{-\infty}^{\infty} f(x)g(y-x)dx = \int_{-\infty}^{\infty} f(y-z)g(z)dz. \tag{26.45}$$

Equation (26.44) also obeys an interesting moment theorem, discussed already in equation (7.69) [p. 131]:

$$M_n(h) = \sum_{k=0}^{n} \binom{n}{k} M_k(f) M_{n-k}(g), \tag{26.46}$$

where $M_n(F)$ stands for the nth moment of the density $F \in \{h, f, g\}$. Without loss of generality we assume that the apparatus function is normalized ($M_0(g) = 1$) and centred ($M_1(g) = 0$). Moreover, we normalize the measured distribution h to one, i.e. $M_0(h) = 1$. Then the first three moments of h are

$$1 = M_0(f), \tag{26.47a}$$

$$M_1(h) = M_1(f), \tag{26.47b}$$

$$M_2(h) = M_2(f) + M_2(g). \tag{26.47c}$$

Apparently, the reconstructed density is also normalized ($M_0(f) = 1$), in other words, the amplitude of the measured signal is uniquely given by the abundance of target atoms with mass M. This ensures 'mass' conservation. From equation (26.47b) we then find that the centre of mass of h is equal to the centre of mass of function f. Finally, equation (26.47c) reveals that the second moment of the convolution is the sum of the second moments of the individual contributions, and due to equation (26.47b) this implies

$$\sigma_h^2 = \sigma_f^2 + \sigma_g^2, \tag{26.48}$$

that is, the variances of function f and function g add up under convolution.

In Section 7.8.5 [p. 129] we have shown that the characteristic function (Fourier transform) of a convolution $\Phi_h(\omega)$ is the product of the characteristic function of the constituents $\Phi_f(\omega)$ and $\Phi_g(\omega)$, respectively. This property points to a possible solution for the deconvolution problem:

$$\Phi_f(\omega) = \frac{\Phi_h(\omega)}{\Phi_g(\omega)}\,, \tag{26.49a}$$

$$f(x) = \frac{1}{2\pi}\int_{-\infty}^{\infty} \frac{\Phi_h(\omega)}{\Phi_g(\omega)} \cdot e^{-i\omega x} d\omega. \tag{26.49b}$$

This formal solution is, however, of minor practical importance since $h(\omega)$ and $g(\omega)$ will never be available on the whole interval $(-\infty, \infty)$ and moreover they will never be available exactly since they result from finite effort experiments and are hence affected by noise. The only sensible approach to deal with truncated noisy data sets is the probabilistic one at the expense that we no longer require an exact solution of equation (26.44) but require instead an optimal one on the basis of the available information. The available information will in most cases be discrete. Let $h(x)$ be measured on an equidistant grid $\{x_i\}$, with $h_i := h(x_i)$. We assume that $g(x_i)$ is also available on the same grid. In view of equation (26.47b) we know that both f and g are restricted to the interval

$$I := [M_1(f) - \kappa\ \sigma_h, M_1(f) + \kappa\ \sigma_h], \tag{26.50}$$

where κ has to be adjusted appropriately, e.g. $\kappa = 2$ will do in most cases. Clearly, the grid points are in I, otherwise we have to increase κ. Then we extend the equidistant grid such that it covers the entire interval I and renumber the grid points in increasing order. Then we can utilize equation (26.45) and our model equation becomes

$$h_k = \sum_{i=1}^{k-1} f_{k-i} g_i + \epsilon_k. \tag{26.51}$$

By construction (equation (26.50)) it is guaranteed that boundary effects are negligible. As before we will invoke cubic spline functions supported at some selected pivots $\boldsymbol{\xi}$ and represent $\boldsymbol{f}$ as the square of a spline function

$$f_k = \Psi^2(x_k|\boldsymbol{\xi}, \boldsymbol{\varphi}) := \Psi_k^2(\boldsymbol{\xi}, \boldsymbol{\varphi}), \tag{26.52}$$

which enforces positivity and is a convenient representation for the Fisher information prior. Here $\{\varphi_j\}$ are the functional values at the pivot points $\{\xi_j\}$. This completes the formulation of our model equation:

$$h_k = \sum_{i=1}^{k-1} \Psi_{k-1}^2(\boldsymbol{\xi}, \boldsymbol{\varphi})\ g_i + \epsilon_k. \tag{26.53}$$

Assuming that $\langle \epsilon_k \rangle = 0$ and $\langle \epsilon_k^2 \rangle = \sigma_k^2$ finally yields the likelihood

$$p(\boldsymbol{h}|\boldsymbol{g}, \boldsymbol{\xi}, \boldsymbol{\varphi}, \boldsymbol{\sigma}) = \frac{1}{Z} \exp\left\{ -\frac{1}{2} \sum_k \frac{\left(h_k - \sum\limits_{i=1}^{k-1} \Psi_{k-i}^2(\boldsymbol{\xi}, \boldsymbol{\varphi})\ g_i \right)^2}{\sigma_k^2} \right\}. \tag{26.54}$$

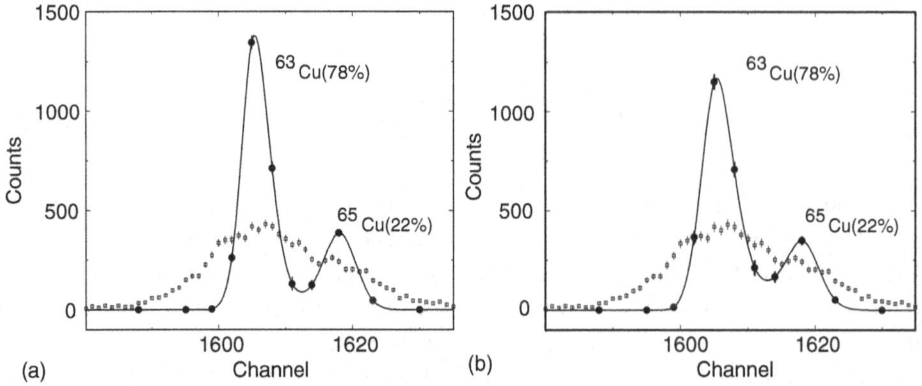

Figure 26.8 Full squares with error bars represent RBS measurements on a thin copper layer on a silicon substrate. The deconvolution is depicted as a continuous curve. The heavy dots show the reconstructed support functional values, defining the spline anchor points. Decreasing the precision of the apparatus function measurement by three leads to the result in (b).

The exponent of the likelihood is of fourth order in $\{\varphi_i\}$. We now form the posterior PDF by combining equation (26.54) with the global first derivative prior of equation (13.27) [p. 221] on $\boldsymbol{\varphi}$. Remember that this is equivalent to employing the Fisher information prior on $\boldsymbol{f}$ with the important advantage that $\boldsymbol{f}$ is positive semidefinite everywhere and not only at the pivotal points. As an example for the application of the developed formalism we consider deconvolution of a Rutherford backscattering spectrum obtained from a thin layer of copper on a silicon substrate. Copper exists with isotopic mass numbers 63 and 65. The relative abundances are 69% and 31%, respectively. The data are shown as full squares with error bars in Figure 26.8. The apparatus function is derived from the cobalt data and their spline representation in Figure 25.8 [p. 447]. Cobalt has mass number 59 and is isotopically pure. However, if we assume a constant resolution of the RBS experiment, then the energy width of the cobalt spectrum would be smaller than the width of a spectrum from an atom with fictitious mass $(0.69 \times 63 + 0.31 \times 65) = 63.62$ by a factor of 59/63.62. After stretching the cobalt spectrum relative to its centre of mass by the inverse of this factor, it can be used as the apparatus response function in the deconvolution problem. The left-hand panel of Figure 26.8 shows the deconvolution result using the smooth approximation to the apparatus function. The heavy dots with error bars are the squares of the reconstruction variables $\{\varphi_k\}$ taken at the mode of the posterior distribution and the continuous curve is the spline interpolation in between. We notice an important resolution enhancement. If we assume that the two peaks associated with copper masses 63 and 65, respectively, have approximately the same shape, then the ratio of their amplitudes is representative for the ratio of their areas and yields isotopic abundance of 78% for ^{63}Cu and 22% for ^{65}Cu. The corresponding literature values are 69% and 31%, respectively, which are close to but not identical with the result of the deconvolution.

The use of the spline smoothed cobalt data as the apparatus function is of course an unnecessary approximation. We have shown in Section 26.4 how to employ the apparatus response function 'as measured', including the impact of measurement errors. The respective calculation leads to results which are hardly distinguishable from Figure 26.8(a), the reason being that the cobalt data have about three times as many counts in the maximum as the copper data. They are therefore much more precise and the influence of their measurement errors on the deconvolution results remains negligible. Figure 26.8(b) shows what would happen if the cobalt data had at maximum only one-third as many counts as the copper data. In this fictitious case we see an appreciable loss of resolution compared with Figure 26.8(a), which is reflected in attenuation of the peaks and filling up of the valley between them. As is to be expected, the error bars on the reconstruction have also increased somewhat.

This concludes Section 26.5 on deconvolution. We believe that it was appropriate to present the topic in somewhat more detail due to its ubiquitous importance for every measurement in the experimental sciences.

27

Model selection

Interpretation of a set of experimental measurements requires us to choose a model which will fit the data to some extent. The model choice may not be unique, but model M_k may be just one possible choice out of a family of models $\{M_i\}$. In this case we are finally faced with the problem of deciding which member M_k of the family represents the most probable physics hidden in the measured data. It is immediately clear that the choice of the 'best' model is a topic of paramount importance since it will most probably influence the direction of further research. The traditional tendency is to equate best fit to best model. This is certainly incorrect, as will be explained by reference to the fit of an orthogonal expansion to the experimental and hence noisy representation of a function. Consider N data points d_i corresponding to N different values x_i of an input variable. The theoretical model function $f(x)$ shall be a linear combination of E orthonormal basis functions $\phi_k(x)$, then the theoretical prediction based on this model reads

$$f(x_i) = \sum_{k=1}^{E} a_k \, \phi_k(x_i) := \sum_{k=1}^{E} \Phi_{i,k} \, a_k .$$

We make the fairly weak assumption that the column vectors in the matrix Φ are still linearly independent. Otherwise we eliminate linearly dependent vectors from the set. The unknowns to be determined are the expansion coefficients $a_1, \ldots, a_E$. Apparently the expansion becomes complete when E approaches N. In the limit $E = N$ the expansion will reproduce exactly every measured data point. What we have found is a model that describes both signal plus noise. This is not of interest at all, what we really want is a representation which is optimal in the sense that it solely describes the physical model, avoiding the fitting of uninteresting noise. An expansion affording this requirement will be a truncated and hence more parsimonious one.

The philosophy just described is very old and is known as Ockham's principle: Employ a model as complicated as necessary to describe the essentials of the data and as simple as possible to avoid the fitting of noise [22, 80]. General considerations concerning Ockham's razor have been given already in Section 3.3 [p. 43]. Ockham's principle can be accessed in quantitative terms by use of Bayesian probability theory. We have encountered this topic in various contexts and differing guises already, in Chapter 17 [p. 255] and

Section 3.2.4 [p. 38]. In this section we want to apply these ideas to real-world problems. The quantity we want to know is $p(E|\boldsymbol{d}, N, \mathcal{I})$, the probability of expansion order E of a system of functions specified by $\mathcal{I}$ when applied to characterize a given set of data $\boldsymbol{d}$. From Bayes' theorem

$$P(E|\boldsymbol{d}, N, \mathcal{I}) = \frac{P(E|N, \mathcal{I})\ p(\boldsymbol{d}|E, N, \mathcal{I})}{p(\boldsymbol{d}|N, \mathcal{I})}. \tag{27.1}$$

This probability is given by a prior probability $p(E|N, \mathcal{I})$ and the marginal likelihood

$$p(\boldsymbol{d}|E, N, \mathcal{I}) = \int d^E a\ p(\boldsymbol{a}|E, N, \mathcal{I})\ p(\boldsymbol{d}|\boldsymbol{a}, E, N, \mathcal{I}), \tag{27.2}$$

which we obtain from the full likelihood by marginalizing over the possible values of the expansion coefficients $\boldsymbol{a}$. Noting further that

$$p(\boldsymbol{d}|N, \mathcal{I}) = \sum_E P(E|N, \mathcal{I})\ p(\boldsymbol{d}|E, N, \mathcal{I}), \tag{27.3}$$

we see when returning to equation (27.1) that $p(E|\boldsymbol{d}, N, I)$ equals one regardless of what the prior and the likelihood are if we consider a single, isolated model, which means $P(E|N, \mathcal{I}) = \delta_{E,E_0}$. The question of what the probability of a given model is does not make sense at all. What makes sense, however, is the question of whether $P(E_1|\boldsymbol{d}, N, \mathcal{I})$, is greater or smaller than $P(E_2|\boldsymbol{d}, N, \mathcal{I})$, or more generally which expansion order $E_k \in \{1, \ldots, N\}$ is the most probable one. This meaningful question may be answered by considering the odds ratio

$$\frac{P(E_j|\boldsymbol{d}, N, \mathcal{I})}{P(E_k|\boldsymbol{d}, N, \mathcal{I})} = \frac{P(E_j|N, \mathcal{I})}{P(E_k|N, \mathcal{I})} \frac{p(\boldsymbol{d}|E_j, N, \mathcal{I})}{p(\boldsymbol{d}|E_k, N, \mathcal{I})}, \tag{27.4}$$

of two alternatives of interest. The odds ratio factors into two terms, the prior odds $P(E_j|N, \mathcal{I})/P(E_k|N, \mathcal{I})$ and the Bayes factor $p(\boldsymbol{d}|E_j, N, \mathcal{I})/p(\boldsymbol{d}|E_k, N, \mathcal{I})$. The prior odds specifies our preference between E_j and E_k before we have made use of the new data $\boldsymbol{d}$. The prior odds represents expert knowledge which may stem from earlier experiments, related facts from other sources, etc. The prior odds may favour model E_j over E_k even if the marginal likelihood $p(\boldsymbol{d}|E_j, N, \mathcal{I})$ is smaller than $p(\boldsymbol{d}|E_k, N, \mathcal{I})$. Most often, however, the prior odds is chosen to be one, which expresses complete indifference instead of preference. In fact, this status of indifference is the main driving force for the design and performance of new experiments in the natural sciences.

We shall now investigate the Bayes factor a little further. In the case of well-conditioned (informative) data, the likelihood $p(\boldsymbol{d}|\boldsymbol{a}, E, N, \mathcal{I})$ will exhibit a well-localized maximum $\hat{\boldsymbol{a}}$ as a function of $\boldsymbol{a}$. Let us assume in particular that the prior in equation (27.2) is comparatively diffuse. A sensible approximation to the marginal likelihood is then obtained by pulling the prior in equation (27.2) with $\boldsymbol{a} = \hat{\boldsymbol{a}}$ out of the integral:

$$p(\boldsymbol{d}|E, N, \mathcal{I}) \approx p(\hat{\boldsymbol{a}}|E, N, \mathcal{I}) \int d\boldsymbol{a}\ p(\boldsymbol{d}|\boldsymbol{a}, E, N, \mathcal{I}). \tag{27.5}$$

The integral may further be characterized by

$$p(\boldsymbol{d}|E, N, \mathcal{I}) \approx p(\hat{\boldsymbol{a}}|E, N, \mathcal{I})\ p(\boldsymbol{d}|\hat{\boldsymbol{a}}, E, N, \mathcal{I})\ (\delta a)^E, \tag{27.6}$$

where $(\delta a)^E$ is an effective volume occupied by the likelihood in parameter space. In order to highlight the key idea, we assume a uniform and normalized prior. Then

$$p(\hat{\boldsymbol{a}}|E, N, \mathcal{I}) = 1/(\Delta a)^E, \tag{27.7}$$

where $(\Delta a)^E$ is the prior volume in parameter space. Merging equations (27.6) and (27.7) we obtain, for the logarithm of the marginal likelihood,

$$\ln p(\boldsymbol{d}|E, N, \mathcal{I}) \approx \ln p(\boldsymbol{d}|\hat{\boldsymbol{a}}, E, N, \mathcal{I}) + E \ln \frac{\delta a}{\Delta a}. \tag{27.8}$$

Since, according to our assumption, the prior is more diffuse than the likelihood, $\delta a < \Delta a$, we arrive at the important result that the marginal likelihood is approximately equal to the maximum likelihood penalized by $-E \ln(\Delta a/\delta a)$. This penalty may be detailed further since δa is certainly a function of the number of data. In fact, in the limiting case of the number of data approaching infinity, δa will vanish and the likelihood collapses to a multidimensional unnormalized delta function in $\boldsymbol{a}$-space. The approximate dependence of the width of the likelihood as a function of the number of data can be guessed from previously discussed examples (see, for example, equation (21.7) [p. 335]) to be inversely proportional to the square root of the number of data, hence if σ is the standard deviation of a single datum then $\sigma/\sqrt{N}$ is approximately the width of the likelihood along one direction in parameter space. Then our final estimate for the marginal likelihood becomes

$$\ln p(\boldsymbol{d}|E, N, \mathcal{I}) \approx \ln p(\boldsymbol{d}|\hat{\boldsymbol{a}}, E, N, \mathcal{I}) - \frac{E}{2} \ln N\left(\frac{\Delta a}{\sigma}\right)^2. \tag{27.9}$$

Equation (27.9) is a form of the Bayesian information criterion [125], which measures the information content of the data with respect to a model specified by the set of parameters $\boldsymbol{a}$ regardless of the numerical values of the parameters. This information is obtained from the maximum likelihood of the problem penalized by a term which accounts for the complexity of the model represented by E and the precision of the likelihood given by the number of data N. We see that Ockham's razor is an integral part of Bayesian probability theory. Note that our hand-waving derivation is far from being rigorous. In the ensuing examples, the Bayesian expressions will be evaluated rigorously and we will see that the key features are still the same.

For the initially considered example of an orthogonal expansion we can even push the formal evaluation of the Bayes factor using equation (27.9) one step further. An orthogonal expansion constitutes an example for the important type of nested models. A family of models is called 'nested' if the model of (expansion) order E_j is completely contained in the model of (expansion) order E_k for $E_k > E_j$. The prominent property of such a model family is that the likelihood is a monotonic function of the expansion order. A more complicated model provides a fit at least as good as a simpler one but most often even better. For the log of the Bayes factor $B_E := p(\boldsymbol{d}|E+1, N, \mathcal{I})/p(\boldsymbol{d}|E, N, \mathcal{I})$ for two consecutive models of a family of nested models we find

$$\ln B_E \approx \ln\left[\frac{p(\boldsymbol{d}|\hat{\boldsymbol{a}}_{E+1}, E+1, N, \mathcal{I})}{p(\boldsymbol{d}|\hat{\boldsymbol{a}}_E, E, N, \mathcal{I})}\right] - \frac{1}{2}\ln N\left(\frac{\Delta a}{\sigma}\right)^2. \tag{27.10}$$

The conclusion is that expansion order $E + 1$ is accepted only if the corresponding log-likelihood is sufficiently larger than that of model E, such that the penalty term $-1/2 \ln N$ $(\Delta a/\sigma)^2$ is overcompensated. Equation (27.10) allows us to draw a final conclusion with validity beyond the special case used to set up equation (27.10). The penalty term in equation (27.10) depends also on the prior width and the penalty is a monotonic function of Δa. Hence, if we choose the prior to be sufficiently uninformative, $\Delta a \to \infty$, then the simpler model (and by iteration of the argument, the simplest member of the model family) will win the competition. This means that improper priors, though in many cases acceptable in parameter estimation, are entirely useless in model selection problems. Model selection depends critically on the prior employed in the analysis and due care should be exercised in order to choose it to be as informative as possible. The deeper meaning of Ockham's razor has been discussed in Section 3.3 [p. 45].

27.1 Inelastic electron scattering

Cross-sections and rate coefficients for electron impact ionization of process gases such as methane and hydrogen constitute the necessary database for modelling low-temperature glow discharges. The interest in the latter stems from their application in the formation of diamond and diamond-like carbon films [102]. Experimental data on electron impact ionization are available for many gaseous species, however, they always span only a limited interval on the energy scale. The calculation of rate coefficients requires, in contrast, the availability of cross-section data for all energies. While rate coefficients for low temperatures are mainly determined by the threshold behaviour of the cross-section, for high temperatures such as in the boundary of magnetic confinement fusion plasmas the high-energy cross-section functional behaviour becomes important. The asymptotic form of cross-sections for inelastic electron scattering is fortunately known from theory and this knowledge can be used to set up a parametric family of functions for which the optimum parameters are then estimated on the basis of the available experimental data. Atomic collision theory distinguishes between two cases for the asymptotic behaviour of inelastic scattering cross-sections. In the case that the final state considered in the collision can also be reached by photon absorption from the initial (ground) state ('optically allowed'), the cross-section $\sigma(E)$ as a function of energy E behaves as

$$\sigma(E) \underset{E\to\infty}{\to} \sim \frac{1}{E}\ln\frac{E}{c}, \tag{27.11}$$

with an adjustable parameter c. If optical transitions between the initial and final state are forbidden, then

$$\sigma(E) \underset{E\to\infty}{\to} \sim \frac{1}{E}. \qquad (27.12)$$

Equation (27.11) must of course still be modified to provide a sensible representation of the cross-section for all energies. One obvious modification is the choice of c. If we choose c to be equal to the threshold energy for ionization, then $\sigma(E)$ tends to zero as the threshold energy is approached from $E > E_0$. We introduce as a new variable the reduced energy $\mathcal{E} = E/E_0$, describe the threshold behaviour by an additional factor $(\mathcal{E}-1)^\alpha$ and introduce a constant β which tunes the transition to the asymptotic region:

$$\sigma(\mathcal{E}) = \sigma_0 \frac{(\mathcal{E}-1)^\alpha}{\beta + \mathcal{E}(\mathcal{E}-1)^\alpha} \ln \mathcal{E}. \qquad (27.13)$$

A similar modification is required for equation (27.12). For the region close to the threshold $\mathcal{E} = 1 + \Delta\mathcal{E}$ we notice that equation (27.13) becomes proportional to $(\Delta\mathcal{E})^{\alpha+1}$. The appropriate modification of equation (27.13) is then

$$\sigma(\mathcal{E}) = \sigma_0 \frac{(\mathcal{E}-1)^{\alpha+1}}{\beta + \mathcal{E}(\mathcal{E}-1)^{\alpha+1}}. \qquad (27.14)$$

σ_0, α and β now have exactly the same meaning in both cases. The prior $p(\sigma_0, \alpha, \beta|\mathcal{I})$ for a Bayesian estimate of the parameters is therefore also the same in both cases. Assuming then that we adopt a prior which is uniform over some sensible, positive, sufficiently large range implies that the Bayes factor for comparison of the two models in equations (27.13) and (27.14) is reduced to the ratio of maximum likelihoods. This in turn means that in the present case a better fit means a more probable model, since the expense of volume in parameter space is model independent.

Figure 27.1 shows data and fits for electron impact ionization of H_2. The ionization channel leading to H_2^+ is classified as an optically allowed transition with log(odds) = +7.6, while the dissociative ionization channel leading to an atomic ion H^+ turns out to be an optically forbidden transition with log (odds) = +64.5. The analysis has therefore made a model choice with very high selectivity and provides good reason to employ equations (27.13) and (27.14) for the calculation of rate coefficients for arbitrary temperatures. This example constitutes the simplest possible case of Bayesian model selection, since we have also taken the prior odds equal to one.

27.2 Signal–background separation

An evergreen in physical data analysis is the separation of a signal from the background on which it resides. A probabilistic answer to this problem has been given in Chapter 23 [p. 396]. The present alternative approach will be applied to the data shown in Figure 23.3 [p. 404]. The traditional procedure chooses a background model and defines the signal as the difference between measured data and simulated background. The result of such an operation

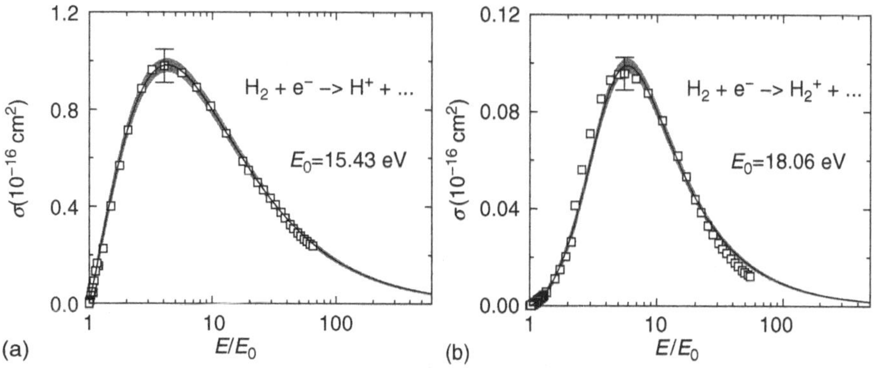

Figure 27.1 Cross-sections for direct and dissociative electron impact ionization of molecular hydrogen. Open squares are experimental data, with common uncertainty indicated at the cross-section maximum. The continuous curves are the fits employing equations (27.13) and (27.14), respectively. The shaded band represents the confidence interval of the fit.

can turn out to be disturbing. In case of counting experiments the signal may contain negative counts, an unacceptable result which has been tolerated by generations of physicists. A Bayesian analysis of the problem resolves this unsatisfactory situation. Let

$$d_i^* = f_i^* + b_i^* + \epsilon_i \tag{27.15}$$

be our data description (d_i^*) in terms of a signal (f_i^*) and background (b_i^*) measured at x_i. Assuming that the noise is characterized by $\langle \epsilon_i \rangle = 0$ and $\langle \epsilon_i^2 \rangle = \sigma_i^2$, we obtain the likelihood

$$p(\boldsymbol{d}|\boldsymbol{f}, \boldsymbol{b}, \boldsymbol{\sigma}, \mathcal{I}) = \prod_i \frac{1}{\sigma_i \sqrt{2\pi}} \exp\left\{ -\frac{1}{2} \sum_i^N \frac{(d_i - f_i - b_i)^2}{\sigma_i^2} \right\}. \tag{27.16}$$

We simplify notation by the definitions

$$d_i := d_i^*/\sigma_i, \quad f_i := f_i^*/\sigma_i, \quad b_i := b_i^*/\sigma_i \tag{27.17}$$

to obtain

$$p(\boldsymbol{d}|\boldsymbol{f}, \boldsymbol{b}, \mathcal{I}) = \left(\frac{1}{\sqrt{2\pi}} \right)^N \exp\left\{ -\frac{1}{2} \sum_i^N (d_i - f_i - b_i)^2 \right\}. \tag{27.18}$$

Next we define our background model as

$$b_i = \sum_k^E A_{ik} c_k^* \tag{27.19}$$

and choose $A_{ik} = \phi_k(x_i)$. We shall use Legendre polynomials for ϕ_k on the support of $\boldsymbol{x}$ in the subsequent analysis. Equation (27.18) may conveniently be rewritten in vector/matrix notation as

$$p(\boldsymbol{d}|\boldsymbol{f}, \mathbf{A}, \boldsymbol{c}^*, \mathcal{I}) = (2\pi)^{-N/2} \exp\left\{-\frac{1}{2}\|\boldsymbol{d} - \boldsymbol{f} - \mathbf{A}\boldsymbol{c}^*\|^2\right\}. \tag{27.20}$$

The kth column of $\mathbf{A}$ is then the $(k-1)$th-order polynomial evaluated at $\boldsymbol{x}$. For matrix $\mathbf{A}$ we introduce its singular-value decomposition defined in Section 21.1.5 [p. 344] and find

$$\mathbf{A}\boldsymbol{c}^* = \mathbf{U}\mathbf{W}\mathbf{V}^T\boldsymbol{c}^*. \tag{27.21}$$

Along with the definition $\boldsymbol{c} := \mathbf{W}\mathbf{V}^T\boldsymbol{c}^*$, we finally obtain

$$p(\boldsymbol{d}|\boldsymbol{f}, \mathbf{U}, \boldsymbol{c}, \mathcal{I}) = (2\pi)^{-N/2} \exp\left\{-\frac{1}{2}\|\boldsymbol{d} - \boldsymbol{f} - \mathbf{U}\boldsymbol{c}\|^2\right\}. \tag{27.22}$$

Matrix $\mathbf{U}$ has E columns and N rows and the column vectors are mutually orthogonal and normalized to one. Removal of the background from the data is now achieved by marginalizing $\boldsymbol{c}$:

$$p(\boldsymbol{d}|\boldsymbol{f}, \mathbf{U}, \mathcal{I}) = (2\pi)^{-N/2} \int d^E c \; p(\boldsymbol{d}|\boldsymbol{f}, \mathbf{U}, \boldsymbol{c}, \mathcal{I}) \; p(\boldsymbol{c}|\mathbf{U}, \mathcal{I}). \tag{27.23}$$

A prior for the expansion coefficients of a multivariate linear problem has already been derived in Section 21.1.5 [p. 344] and equation (21.68) [p. 349]. In terms of the present notation it reads

$$p(\boldsymbol{c}|\mathbf{U}, \mathcal{I}) = \frac{1}{2} \frac{\Gamma(\frac{E}{2})}{\left|\mathbf{U}^T\mathbf{U}\right|^{\frac{1}{2}}} \left(\pi \boldsymbol{c}^T\mathbf{U}^T\mathbf{U}\boldsymbol{c}\right)^{-\frac{E}{2}}, \tag{27.24}$$

which simplifies due to the orthogonality of the column vectors of $\mathbf{U}$, to

$$p(\boldsymbol{c}|\mathbf{U}, \mathcal{I}) = \frac{\Gamma\left(\frac{E}{2}\right)}{2(\pi\|\boldsymbol{c}\|)^{\frac{E}{2}}}. \tag{27.25}$$

Exact evaluation of equation (27.23) is impossible. A useful approximate evaluation of equation (27.23) relies on the fact that $p(\boldsymbol{c}|\mathbf{U}, \mathcal{I})$ is a rather smoothly varying function while the likelihood $p(\boldsymbol{d}|\boldsymbol{f}, \mathbf{U}, \boldsymbol{c})$ is usually strongly peaked at the maximum likelihood values of $\boldsymbol{c}$. Since equation (27.25) is in this approximation replaced by a constant, it can be taken out of the integrand resulting in

$$p(\boldsymbol{d}|\boldsymbol{f}, \mathbf{U}, \mathcal{I}) = (2\pi)^{-\frac{N}{2}} \frac{\Gamma(\frac{E}{2})}{(2\pi\|\boldsymbol{c}^{\mathrm{ML}}\|)^{\frac{E}{2}}} \int d^E c \exp\left\{-\frac{1}{2}\|\boldsymbol{d} - \boldsymbol{f} - \mathbf{U}\boldsymbol{c}\|^2\right\}. \tag{27.26}$$

The exponent of the integrand is a quadratic functional in $\boldsymbol{c}$ which we rewrite as

$$\|\boldsymbol{d} - \boldsymbol{f} - \mathbf{U}\boldsymbol{c}\|^2 = \left(\boldsymbol{c} - \boldsymbol{c}^{\mathrm{ML}}\right)^T \left(\boldsymbol{c} - \boldsymbol{c}^{\mathrm{ML}}\right) + R. \tag{27.27}$$

Comparison of the coefficients of $\boldsymbol{c}$ and $\boldsymbol{c}^T\boldsymbol{c}$ yields

$$\boldsymbol{c}^{\mathrm{ML}} = \mathbf{U}^T(\boldsymbol{d} - \boldsymbol{f}), \quad R = (\boldsymbol{d} - \boldsymbol{f})^T(\mathbb{1} - \mathbf{U}\mathbf{U}^T)(\boldsymbol{d} - \boldsymbol{f}), \tag{27.28}$$

where we have used the fact that $\mathbf{U}^T\mathbf{U} = \mathbb{1}$. Now $\mathbf{P} := \mathbf{U}^T\mathbf{U}$ is an operator projecting into the subspace spanned by the column vectors in $\mathbf{U}$, that describe the background. The integral in equation (27.26) becomes $(2\pi)^{E/2}\exp\{-R/2\}$, yielding the marginal likelihood

MARGINAL LIKELIHOOD FOR THE SIGNAL AFTER BACKGROUND MARGINALIZATION

$$p(\boldsymbol{d}|\boldsymbol{f}, \mathbf{U}, \mathcal{I}) = \frac{1}{Z}\exp\left\{-\frac{1}{2}(\boldsymbol{d} - \boldsymbol{f})^T(\mathbb{1} - \mathbf{P})(\boldsymbol{d} - \boldsymbol{f})\right\}. \tag{27.29}$$

The maximum likelihood solution of equation (27.29) is

$$(\mathbb{1} - \mathbf{P})\boldsymbol{d} = (\mathbb{1} - \mathbf{P})\boldsymbol{f}. \tag{27.30}$$

We decompose the signal $\boldsymbol{f}$ into

$$\boldsymbol{f} = \mathbf{P}\boldsymbol{f} + (\mathbb{1} - \mathbf{P}) := \boldsymbol{f}_{\parallel} + \boldsymbol{f}_{\perp},$$

that is, into the part $\boldsymbol{f}_{\parallel}$ that is describable by the background basis and the orthogonal rest $\boldsymbol{f}_{\perp}$. Insertion into equation (27.30) yields

$$(\mathbb{1} - \mathbf{P})\boldsymbol{d} = \underbrace{(\mathbb{1} - \mathbf{P})\boldsymbol{f}_{\parallel}}_{=\mathbf{0}} + (\mathbb{1} - \mathbf{P})\boldsymbol{f}_{\perp} = \boldsymbol{f}_{\perp}.$$

The signal part $\boldsymbol{f}_{\perp}$ is therefore fixed by the data

$$\boldsymbol{f}_{\perp} = (\mathbb{1} - \mathbf{P})\boldsymbol{d}, \tag{27.31}$$

while the part $\boldsymbol{f}_{\parallel}$ is not determined by the marginal likelihood but follows from the prior. As a first step, we could set $\boldsymbol{f}_{\parallel} = \mathbf{0}$ because that part of the data is part of the background and we ignore the prior for $\boldsymbol{f}$, in other words we use a uniform prior. Reconstruction then reads

$$\boldsymbol{f} = (\mathbb{1} - \mathbf{P})\boldsymbol{d}.$$

The corresponding result is depicted in Figure 27.2(a). It has slightly negative parts, which is of course unacceptable since the signal from a counting experiment cannot be negative. The origin of this deficiency is a missing prior. In this application we use the entropic prior (see Chapter 12 [p. 201]) as the Fisher prior is inadequate because of its inherent smoothing property, which is detrimental in the present case of very sharp structures. Figure 27.2(b) shows the posterior solution of equation (27.29) using the Shannon entropy as a prior for $\boldsymbol{f}$. As required, the signal obtained is now strictly positive. Along the lines outlined in the previous section, it is also possible to determine the probability $P(E|\boldsymbol{d}, \mathbf{U}, \mathcal{I})$ for the

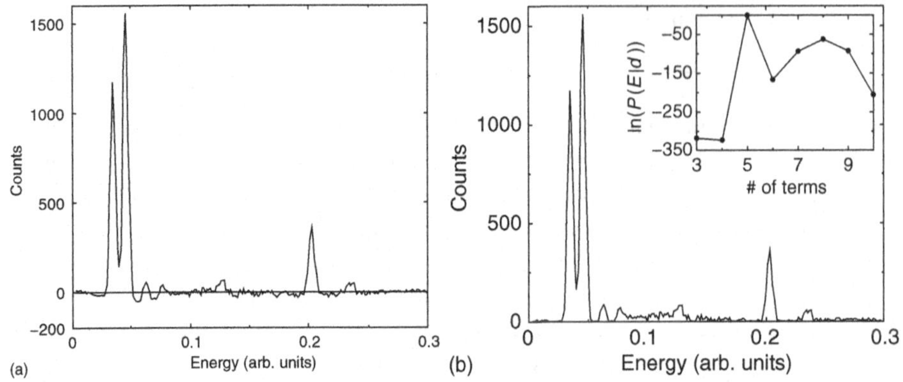

Figure 27.2 ML recovery of the signal in (a) shows unacceptable negative events. MAP solution based on the entropic prior leads to a strictly positive reconstruction (b). The inset shows $\ln(P(E|\boldsymbol{d}, \mathcal{I}))$ for the expansion order of the Legendre polynomials describing the background.

expansion order E. Here we have used Legendre polynomials, which are particularly suitable to describe the smooth background. The inset of Figure 27.2(b) shows the logarithm of $P(E|\boldsymbol{d}, \mathbf{U}, \mathcal{I})$. We realize that stepping from order four to order five leads to an appreciable decrease of the evidence. Please note the logarithmic scale. Order six is obviously unnecessary to improve the background description and the Ockham penalty increases strongly in this step. There is much more to be said about the signal–background separation. First of all the data are really Poissonian and not Gaussian, which becomes important in those parts where the signal and/or background counts are small. It is a straightforward application of the rules of Bayesian probability theory to use the Poisson distribution. As a matter of fact, the numerical effort is the same. In many applications it is also of great interest to analyse the various peaks in the experimental spectrum and to assign probabilities to what extent they are real or part of the background or even spurious noise-induced structures. These aspects have been analysed in [70, 158, 171, 211].

27.3 Spectral line broadening

The light emitted by a radiating atom at rest is not strictly monochromatic. The range of energies spanned by a spectral line can be estimated from the lifetimes of the levels involved in the transition using the uncertainty relation $\Delta t \Delta E \geq h/2\pi$. Moreover, the distribution of energies is not uniform but follows a Lorentzian if external perturbations can be neglected. The external perturbations which we consider here are the thermal motion of the atoms embedded in a gaseous background and the collisional interaction of the radiative atoms with the surrounding gas. The thermal motion of the atoms results in a Doppler shift of the observed radiation depending on the velocity component in the direction of observation and leads to a Gaussian line profile assuming that the motional broadening is very large compared with the natural line width. The collisional perturbation of the radiating atoms leads to a more or less drastic phase change of the emitted light wave, which transforms to a Lorentzian energy distribution of the emitted light. Usually both effects are present at the same time, leading to the so-called Voigt profile which consists of a short-range Gaussian kernel with long-range Lorentzian wings. In special situations like a hot dilute gas or a cold dense gas, Gaussian and Lorentzian line shapes occur and an analysis of the line shape provides access to the physical conditions in the environment of the radiating atom. The question of whether spectral observations of line emissions are fitted better by a Gaussian than by a Lorentzian is therefore a model selection problem of considerable relevance, in particular if the precision of the data is limited. The data which we are going to use in this model comparison exercise were generated from a Gaussian distorted by an energy-independent background:

$$d_G(E) = A \exp\left\{-\frac{1}{2}\left(\frac{E-E_0}{\sigma}\right)^2\right\} + B. \tag{27.32}$$

For fixed values of the four parameters A, σ, B, E_0 and a set of energies $\boldsymbol{E} = \{E_1, \ldots, E_N\}$ we have computed Poisson deviates according to

$$p(n_i|d_G, E_i, \mathcal{I}) = e^{-d_G(E_i)} \frac{\left[d_G(E_i)\right]^{n_i}}{n_i!}. \tag{27.33}$$

Such artificial data for two different runs of the random number generator are shown in Figure 27.3, denoted as data sets I and II.

For the analysis of the data sets we choose two alternative hypotheses: (a) the line broadening is Gaussian with background, as described by equation (27.32) and according to which the data sets were generated, or (b) the line broadening is Lorentzian without background. The model function for the latter case is

$$d_L(E) = \frac{A\sigma^2}{\alpha(E-E_0)^2 + \sigma^2}. \tag{27.34}$$

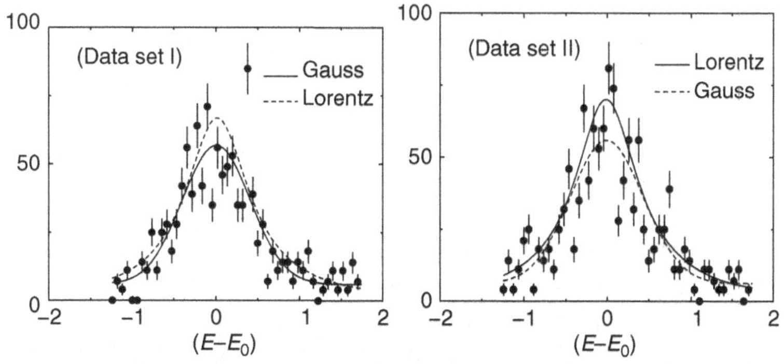

Figure 27.3 Artificial Poissonian counts (dots plus error bars) for a Gaussian spectral line (equation (27.33)). Model functions are Gaussian for Doppler broadening and Lorentzian for collisional broadening. The superior (subordinate) model function is displayed as a solid (dashed) line.

This particular parametrization of the Lorentzian has been chosen by the requirement that A and σ should have the same meaning as in equation (27.32). Firstly, in both cases the maximum is given by A. Since the Lorentzian has infinite variance, it is reasonable to compare the half width at half maximum (HWHM). For a Gaussian it is HWHM $= \sigma\sqrt{\ln(4)}$, while in the Lorentzian case we have HWHM $= \sigma/\sqrt{\alpha}$. As the two shall be the same, we find $\alpha = 1/\ln(4)$. The advantage of this particular choice of parametrization is that the priors on A, E_0 and σ cancel upon calculating the Bayes factor as long as they are uniform. Therefore only the prior for the background B is relevant for model comparison. If B_0 is a point estimate of the background then the distribution of B is, by virtue of the principle of maximum entropy,

$$p(B|B_0) = \frac{1}{B_0}\, e^{-\frac{B}{B_0}}. \tag{27.35}$$

The posterior distribution of the parameters for the Lorentzian model function is then

$$p(A, E_0, \sigma|\boldsymbol{n}, \mathcal{I}) = \frac{p(A, E_0, \sigma|\mathcal{I})}{Z_L} \prod_k \frac{\left[d_L(E_k)\right]^{n_k}}{n_k\,!} e^{-d_L(E_k)}, \tag{27.36}$$

where Z_L is the evidence for the model. The set of energies $\boldsymbol{E}$ is hidden in the background information $\mathcal{I}$. In the Gaussian case we have

$$p(A, E_0, \sigma, B|\boldsymbol{n}, \mathcal{I}) = \frac{p(A, E_0, \sigma|\mathcal{I})}{Z_G} \frac{1}{B_0} e^{-\frac{B}{B_0}} \prod_k \frac{\left[d_G(x_k)\right]^{n_k}}{n_k\,!} e^{-d_G(E_k)}. \tag{27.37}$$

Both posterior distributions are manifestly non-Gaussian and exact calculation of the model evidences is therefore not possible. We employ the Gaussian approximation separately for the two models. Let $\boldsymbol{\lambda}$ be the vector of parameters, then

$$\ln p(\boldsymbol{\lambda}|\boldsymbol{n}, \mathcal{I}) \approx \ln p(\boldsymbol{\lambda}_0|\boldsymbol{n}, \mathcal{I}) + \frac{1}{2}(\boldsymbol{\lambda} - \boldsymbol{\lambda}_0)^T \;\nabla\nabla^T \ln p\big|_{\boldsymbol{\lambda}=\boldsymbol{\lambda}_0} (\boldsymbol{\lambda} - \boldsymbol{\lambda}_0), \tag{27.38}$$

where the gradients are understood with respect to $\boldsymbol{\lambda}$. The posterior mode $\boldsymbol{\lambda}_0$ is given by $\nabla_{\boldsymbol{\lambda}} \ln p(\boldsymbol{\lambda}|\boldsymbol{n}, \boldsymbol{E}, \mathcal{I}) = 0$ and can be determined iteratively. The evidence follows from a Gaussian integral and results in

$$\ln p(\boldsymbol{n}|\mathcal{I}) = \ln p(\boldsymbol{\lambda}_0|\boldsymbol{n}, \mathcal{I}) + \frac{N_p}{2} \ln 2\pi - \frac{1}{2} \ln\left(\left|\nabla\nabla^T \ln p\big|_{\boldsymbol{\lambda}=\boldsymbol{\lambda}_0}\right|\right). \tag{27.39}$$

Here N_p stands for the number of parameters in the two models, i.e. $N_p = 4$ (3) in the Gaussian (Lorentzian) case. The Gaussian approximation for calculating the evidence is problematic in two respects in this case: First of all, it is by no means clear that the shape of the posterior resembles a Gaussian sufficiently closely and second, the range of the parameters A, σ and B is restricted to the positive domain, while we have integrated over the entire real axis. We have therefore also computed the evidence numerically without approximations. The results are given in Table 27.1. It contains a comparison of the

Table 27.1 Log evidences for the Gaussian and Lorentzian model for two different data sets (see Figure 27.3)

	Data set I		Data set II	
Model	GA	NI	GA	NI
Gauss	−115.89	−115.94	−120.00	−119.97
Lorentz	−117.69	−117.67	−116.73	−116.72

GA: Gaussian approximation
NI: Numerical integration

Table 27.2 Meaning of log-odds numbers

Range of $\ln(o)$	Meaning
< 1	hardly significant
$1.0, \ldots, 2.5$	positive evidence
$2.5, \ldots, 5.0$	strong evidence
> 5	overwhelming evidence

evidence values for the two models for two different data sets, generated from the same model function (equation (27.32)) by two different seeds of the random generator for the Poisson deviates. By inspection of Table 27.1 we find that for data set I the Gauss model wins the competition in the analysis, while for data set II the Lorentzian model is more probable. The more probable model function is shown as a solid line in Figure 27.3. The exact values of the evidence are also given, obtained by numerical integration of the posterior (see Chapter 29 [p. 509]), and we find that the Gaussian approximation is quite accurate in this case. It has been suggested in the literature [113] that the scheme displayed in Table 27.2 provides a suitable guideline for interpreting log-odds values. It might be disturbing that we find contradictory results although the true model is the same in both cases. The lesson to be learned is that the answer to a probabilistic analysis always refers to the data. The romantic desire to infer the 'truth', in this case the data-generating function from a given data set, is frustrated as a matter of principle in any real-world problem.

27.4 Adaptive choice of pivots

In Chapters 25 and 26 we have addressed the problem of reconstructing a function from a given set of data. While in Chapter 25 [p. 431] the data themselves were noisy samples of the function to be estimated, in Chapter 26 [p. 451] the data were definite integrals over a kernel times the unknown function. The basic problem is the same in both cases; the complications, however, are widely different. If we employ pointwise reconstructions of the unknown function then the dimension of the data space is equal to the dimension of the function space. Such a high-resolution reconstruction, which allows for considerable point-to-point changes of the function, is in many cases not necessary. In fact, in most cases the function will be slowly varying in some ranges of the independent variable and be highly structured in others. We have used spline functions in such cases with pivots of varying density. Our previous treatment has assumed that the number of pivots for the spline and their position are given. We shall now abandon both of these assumptions and determine the number of pivots by model selection and marginalize over their position. It is clear that there exists an optimum model. Increasing the number of spline pivots will of course decrease the misfit but increase the model complexity. The maximum evidence and hence the optimum model will emerge from a trade-off between best fit and model complexity. We shall discuss the process with a renewed density estimation of the Old Faithful data. From equation (25.44) [p. 449] the likelihood for having n_i eruptions in the observation interval Δt_i is

$$p(\boldsymbol{n}|\boldsymbol{\rho}, \mathcal{I}) = \frac{1}{\prod_i n_i\,!} \exp\left\{ -\sum_i \rho_i + \sum_i n_i \, \ln \rho_i \right\}. \tag{27.40}$$

For the density function ρ we choose

$$\rho_i = \Psi^2(t_i, E, \boldsymbol{\xi}, \boldsymbol{f}), \tag{27.41}$$

where Ψ is a spline function with E pivots at positions $\boldsymbol{\xi}$ and associated support functional values $\boldsymbol{f}$. Our aim is then to calculate the marginal likelihood $p(\boldsymbol{n}|E, \boldsymbol{t}, \mathcal{I})$ which enters the Bayes factor for model comparison. The marginal likelihood is obtained from the full likelihood by marginalization:

$$p(\boldsymbol{n}|E, \mathcal{I}) = \sum_{\boldsymbol{\xi}} \int d^E f \; p(\boldsymbol{n}, \boldsymbol{\xi}, \boldsymbol{f}|E, \mathcal{I}). \tag{27.42}$$

The sum over $\boldsymbol{\xi}$ indicates that we are choosing to vary $\boldsymbol{\xi}$ only in discrete steps and choose as the possible components of $\boldsymbol{\xi}$ the grid $\{t_i\}$ on which the data are available. The information about these grid points is part of the background information $\mathcal{I}$. Continuous variation of the components of $\boldsymbol{\xi}$ is of course also possible; the finite sum then has to be replaced by an integral. We expand the integrand of equation (27.42) using the product rule

$$p(\boldsymbol{n}|E, \mathcal{I}) = \sum_{\boldsymbol{\xi}} p(\boldsymbol{\xi}|E, \mathcal{I}) \int d^E f \; p(\boldsymbol{f}|\boldsymbol{\xi}, E, \mathcal{I}) \; p(\boldsymbol{n}|\boldsymbol{f}, \boldsymbol{\xi}, E, \mathcal{I}). \tag{27.43}$$

For the execution of the $\boldsymbol{f}$-integral we choose the minimum global slope prior in equation (13.1a) [p. 215], which amounts to using the Fisher information prior (see equation (13.1c) [p. 215]) on $\rho = \Psi^2$ and evaluating the integral in a Gaussian approximation. The final step is then the summation over pivot positions $\boldsymbol{\xi}$. We assume that ξ_1, ξ_E remain fixed

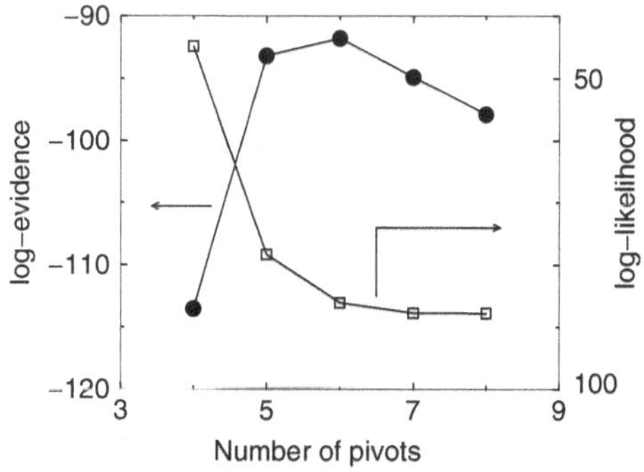

Figure 27.4 Evidence for the number of pivot points and likelihood for estimation of Old Faithful data as a function of the number of pivots, demonstrating Ockham's razor in action.

at the boundaries of the support. An appropriate prior for $\boldsymbol{\xi}$ was already met in equation (24.42) [p. 419]:

$$p(\xi_2, \ldots, \xi_{E-1}|\xi_1, \xi_E, \mathcal{I}) = \frac{(\xi_2 - \xi_1)^n \ldots (\xi_E - \xi_{E-1})^n}{(n\,!)^{E-1}(\xi_E - \xi_1)^{E(n+1)-(n+2)}}\,\Gamma((E-1)(n+1)).$$

The maximum value of this prior is obtained for equidistant spacing of the pivots. This is indeed uninformative. As long as we do not know anything about the function we cannot assign pivot spacing according to the rate of variation of the function. The parameter n adjusts for how strongly we insist on the uninformative situation. The summation can now be performed. In principle, however, it meets with a technical difficulty. If we have only one pivot in between the endpoints ξ_1, ξ_E of the support, the summation will involve exactly $n_g - 2$ terms where n_g is the number of points on the data grid (e.g. the dimension of $\boldsymbol{t}$). For two pivots the summation will already involve $O(n_g^2)$ terms and we realize that deterministic evaluation of the sum amounts to $O(n_g^{(E-2)})$ operations, which is prohibitive already for moderate E. Monte Carlo methods, as will be discussed in Chapter 30 [p. 537], come to the rescue. We skip the details of $\boldsymbol{\xi}$-marginalization here and return to the problem in Chapter 29 [p. 509]. The results for the Old Faithful density estimation problem are shown in Figure 27.4. The full dots display the evidence as a function of the number of pivots N_p. It is seen to pass through a pronounced maximum at $N_p = 6$. The open squares in Figure 27.4 show the expectation value of the likelihood, which is a monotonously increasing function of the number of pivots equivalent to a monotonously decreasing misfit. The expectation value of the likelihood has been calculated from the predictive distribution for the present problem, which is similar to the one in Section 25.1 [p. 432]. Finally we want to use the predictive distribution to estimate the density. The result is shown in Figure 27.5 as full dots. The thin continuous curve is the prediction based on the most probable pivot position, shown as tick marks at the upper frame boundary. The difference between the two solutions is shown in the lower panel. It is small but non-negligible. The present result should be compared with Figure 25.10 [p. 450] in order to realize that the earlier result clearly exhibits overfitting.

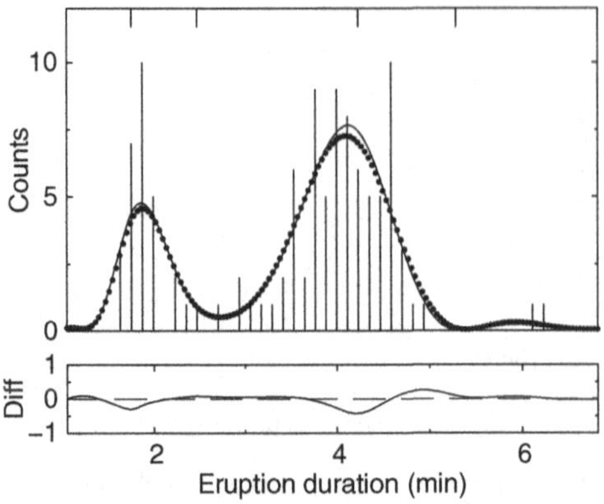

Figure 27.5 Density estimation of Old Faithful data for the optimum number of six pivots marginalized over their positions (full dots). The most likely pivot position is marked by ticks at the upper frame and the corresponding density function is shown as the thin continuous curve.

27.5 Mass spectrometry

Mass spectrometry is a common technique in plasma, surface and vacuum physics for the analysis of the composition of gas mixtures. Typical low-resolution quadrupole instruments offer in principle the possibility of identifying the constituents of a gas mixture with very high sensitivity. Neutral molecules are ionized by electron impact in an ion source at electron energies of 50–100 eV. This choice of impact energy provides an optimum ionization efficiency, but it also introduces a serious complication. Consider a simple diatomic molecule AB. Electron impact ionization will not only produce the parent molecular ion AB^+ but also A^+ and B^+ fragment ions from dissociative ionization. Also, doubly charged ions A^{++} and B^{++} may occur. The quadrupole filter will transmit these doubly charged ions at mass tuning $A/2$ and $B/2$, respectively, since the filter is sensitive to the mass-to-charge ratio rather than to mass directly. Even a molecule as simple as AB can therefore produce, under quite common conditions, signals in as many as five different mass channels. This mass distribution is called the 'cracking pattern' of the molecule. Cracking patterns become progressively more complicated as the number of atomic constituents of a polyatomic molecule increases. Quantitative detection of the constituents of a gas mixture is therefore only simple and straightforward if the cracking patterns of the species do not overlap. This is a rare and unimportant case in practical situations. Cracking patterns overlap nearly always and mixture signals must be disentangled. A simple example is shown in Figure 27.6. Here the mixture results from thermal decomposition of the molecule $CH_3 - N = N - CH_3$. The expected pyrolysis products are the molecule N_2 and the radical CH_3. The former would give rise to mass signals in channels 14 for N^+ and 28 for N_2^+ while CH_3 shows up in channels 12–15. We realize immediately that this cannot be the whole story. Appreciable signals are also detected in channels 24–27, 29, 30. These must stem from molecules which are reaction products of the main species CH_3 and N_2 in

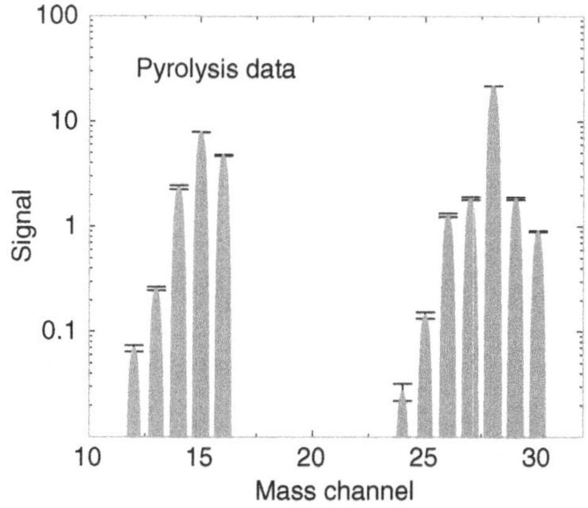

Figure 27.6 Product mass spectrum from the thermal decomposition of $CH_3-N=N-CH_3$.

Table 27.3 Mass signals from five calibration gases and from the mixture arising from $(CH_3)_2N_2$ pyrolysis. The values in brackets indicate the uncertainty of the corresponding last digits

Mass	N_2	CH_4	C_2H_2	C_2H_4	C_2H_6	Mixture
12	—	0.142(12)	0.193(15)	0.072(5)	—	0.069(5)
13	—	0.435(2)	1.27(5)	0.135(5)	0.042(5)	0.257(1)
14	1.17(10)	1.64(5)	—	0.342(5)	0.28(1)	2.33(10)
15	—	19.96(20)	—	—	0.57(1)	7.86(5)
16	—	24.87(20)	—	—	—	4.74(5)
24	—	—	1.26(5)	0.239(1)	—	0.027(5)
25	—	—	5.95(10)	0.829(20)	0.26(1)	0.145(10)
26	—	—	32.86(20)	8.23(10)	2.12(5)	1.29(5)
27	—	—	—	7.98(10)	3.62(7)	1.86(5)
28	38.46(20)	—	—	14.65(10)	12.1(1)	21.73(10)
29	0.4(1)	—	—	—	3.87(6)	1.85(4)
30	—	—	—	—	2.50(5)	0.907(10)

collision with other molecules on the surface of the reactor or in the gas phase. Candidate species are $CH_4, C_2H_6, C_2H_4, C_2H_2$ and our primary aim is to determine the subset of this group which explains the data best. The data are also reproduced in numerical form in Table 27.3 [179]. Disentanglement of the spectrum in Figure 27.6 requires us to determine the constituent concentrations $\boldsymbol{x}^*$ from the model relation

$$\boldsymbol{d}^* \to \mathbf{C}^*\boldsymbol{x}^* + \boldsymbol{\epsilon}^*, \qquad (27.44)$$

where $\boldsymbol{d}^*$ is the vector of detected mass signals, $\boldsymbol{\epsilon}^*$ the associated measurement uncertainty and $\mathbf{C}^*$ the cracking matrix. The kth column vector of $\mathbf{C}^*$ is equal to the cracking pattern associated with species k of concentration x_k^*. The above relation is not an equation since the dimension of the right-hand side is not the same as that of $\boldsymbol{d}^*$, even if $\mathbf{C}^*$ contains only dimensionless numbers. We eliminate this problem by introducing normalized quantities. Thus $\boldsymbol{d}$ is normalized to unit length and the columns of $\mathbf{C}$ are also normalized to unit length. This implies normalization of $\boldsymbol{x}$ to unit length; $\boldsymbol{\epsilon}$ is the rescaled vector $\boldsymbol{\epsilon}^*$. We arrive at the model equation

$$\boldsymbol{d} = \mathbf{C}\,\boldsymbol{x} + \boldsymbol{\epsilon}. \tag{27.45}$$

We assume that $\langle \epsilon_i \rangle = 0$ and $\langle \epsilon_i^2 \rangle = \sigma_i^2$, and obtain the likelihood

$$p(\boldsymbol{d}|\mathbf{C}, \boldsymbol{x}, \boldsymbol{\sigma}, \mathcal{I}) = \prod_i \left(\frac{1}{\sigma_i \sqrt{2\pi}} \right) \exp\left\{ -\frac{1}{2\sigma_i^2} [d_i - \sum_k c_{ik} x_k]^2 \right\}. \tag{27.46}$$

This is a good opportunity to comment on the importance of measurement errors in Bayesian analysis. The measurement uncertainties in the mixture measurement are indicated in Figure 27.6, and are seen to be widely different. Great care has been taken to obtain as accurate estimates as possible in this experiment. In order to see what information is contained in the uncertainties, imagine for a moment a constant relative error. In this hypothetical case all the terms in the exponential in equation (27.46) would carry equal weight and the information contained in the spectrum equation (27.6) would be reduced to the channel numbers where a signal is detected, its size would be irrelevant and indeed a dramatic loss of information. Up to now we have assumed that we know the cracking patterns of the species contributing to the mixture signal. This is nearly always an unjustified assumption, since the cracking pattern of a molecule is not an intrinsic molecular property but depends on the particular mass spectrometer employed and its tuning. What is known in the case of stable neutrals are cracking patterns from previous measurements published in tables on mass spectrometry. This valuable prior knowledge is available as point estimates and can be obtained from published tables [34]. Data relevant for the present analysis are reproduced in the first five columns of Table 27.4. No prior information of comparable quality is available for the radical CH_3.

We would certainly reject the proposal to assign equal cracking coefficients to CH_3, CH_2, CH and C. General experience in mass spectrometry says that at least for small molecules, the cracking coefficient should decrease strongly from CH_3 to C. The rate of this decrease can be approximated by reference to CH_4, for which data are available. Two choices suggest themselves: select either the first four or the last four cracking coefficients of CH_4 as an estimate for CH_3. For the present analysis we took the first four, which are shown after renormalization in the last column of Table 27.4. The point estimates c_{jk}^0 constitute rather weak prior knowledge and the associated prior density is, according to the maximum entropy principle,

$$p(c_{jk}|c_{jk}^0, \mathcal{I}) = \frac{1}{Z(\lambda_{jk})} \exp\{-\lambda_{jk}\, c_{jk}\}. \tag{27.47}$$

If the support for the variable c_{jk} were $[0, \infty)$, then λ_{jk} would be its inverse expectation

Table 27.4 Prior point estimates of cracking coefficients

Mass	N_2	CH_4	C_2H_2	C_2H_4	C_2H_6	CH_3
12	—	0.010	0.011	0.006	—	0.032
13	—	0.032	0.039	0.011	0.005	0.067
14	0.049	0.067	—	0.021	0.012	0.407
15	—	0.402	—	—	0.017	0.494
16	—	0.489	—	—	—	—
24	—	—	0.042	0.012	—	—
25	—	—	0.153	0.043	0.016	—
26	—	—	0.755	0.241	0.101	—
27	—	—	—	0.252	0.150	—
28	0.951	—	—	0.414	0.469	—
29	—	—	—	—	0.104	—
30	—	—	—	—	0.126	—

value $1/c^0_{jk}$. But the support of the variable is [0, 1] in the present case. Therefore, λ_{jk} has to be determined as in the die problem in Section 11.2.2 [p. 186], where we had to fix the normalization and determine the Lagrange parameter matching the constraint for the mean. Here λ_{jk} plays the role of the Lagrange parameter, a scale parameter which must be determined from the equations

$$Z(\lambda_{jk}) = \frac{1 - \exp(-\lambda_{jk})}{\lambda_{jk}}, \tag{27.48a}$$

$$\langle c_{jk} \rangle = \frac{1}{\lambda_{jk}} + \frac{1}{1 - \exp(\lambda_{jk})} \stackrel{!}{=} c^0_{jk}, \tag{27.48b}$$

where c^0_{jk} is the corresponding entry in Table 27.4. Examples for the prior distribution in equation (27.47) for three different values of c^0_{jk} are displayed in Figure 27.7. The result is analogous to that of the Maxent solution for the PMF of the die problem in Figure 11.2 [p. 187].

For quantitative analysis, the cracking patterns of as many stable components as possible, supposed to be included in the mixture, should be determined from a calibration experiment combined with the independent measurement on the mixture. Such calibration measurements do not provide exact cracking patterns. Rather, they give an approximation to the exact pattern corrupted by noise. If we assume that this noise is comparable to the noise on the mixture measurement, it cannot be neglected. On the contrary, both sources of uncertainty will contribute about equal amounts to the uncertainty of constituent concentrations. A calibration measurement is of course just a special case of a mixture measurement. The 'mixture' has just one component, which is known. The model equation for calibrating species j is then

$$\boldsymbol{d}_j = \boldsymbol{c}_j + \boldsymbol{s}_j, \tag{27.49}$$

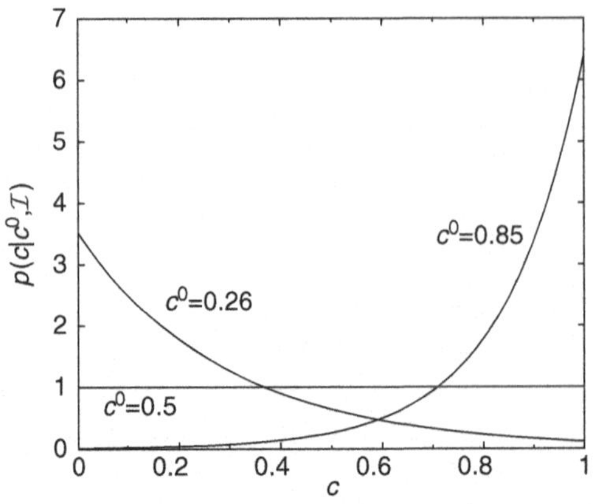

Figure 27.7 The exponential prior of equation (27.47) for three different expectation values c_{jk}^0 (see equation (27.48)).

where $\boldsymbol{d}_j$ is the normalized measured data vector, $\boldsymbol{c}_j$ the jth column vector of cracking matrix $\mathbf{C}$ and $\boldsymbol{s}_j$ the rescaled measurement uncertainty on $\boldsymbol{d}_j$. The corresponding likelihood is

$$p(\boldsymbol{d}_j|\boldsymbol{c}_j, \boldsymbol{s}_j, \mathcal{I}) = \prod_k \frac{1}{s_{jk}\sqrt{2\pi}} \exp\left\{-\frac{1}{2}\frac{(D_{jk} - c_{jk})^2}{s_{jk}^2}\right\}. \tag{27.50}$$

Of course, the data $\boldsymbol{d}_j$ are treated on the same basis as the data $\boldsymbol{d}$. It may be advantageous to make a formal difference. We combine equations (27.47) and (27.50) in cases where calibration measurements are available to form a new highly informative prior density. The product of equations (27.47) and (27.50) is again a Gaussian for $c_{jk} > 0$ and zero otherwise. If we make the realistic assumption that the calibration measurement $\boldsymbol{d}_j$ is vastly more precise and hence informative than equation (27.47) but has more or less the same point estimate, then the truncation can be ignored and the range of the new prior be extended to $(-\infty, \infty)$ for computational convenience:

$$\begin{aligned}\tilde{p}(c_{jk}|c_{jk}^0, D_{jk}, s_{jk}, \mathcal{I}) = \frac{1}{Z(\mu_{jk})} \exp\left\{-\frac{\lambda_{jk}}{2}(2D_{jk} - \lambda_{jk}\, s_{jk}^2)\right\}\\ \times\, p(c_{jk}|c_{jk}^0, D_{jk}, s_{jk}, \mathcal{I}),\end{aligned} \tag{27.51}$$

where we separate as a prior normalized in c_{jk} the PDF

$$p(c_{jk}|c_{jk}^0, D_{jk}, s_{jk}, \mathcal{I}) = \frac{1}{s_{jk}\sqrt{2\pi}} \exp\left\{-\frac{1}{2}\frac{\left(c_{jk} - D_{jk} + \lambda_{jk}\, s_{jk}^2\right)^2}{s_{jk}^2}\right\} \tag{27.52}$$

and treat the remaining factor as a scale factor of the posterior. The effect of combining equations (27.47) and (27.50) is a shift of the Gaussian by a fraction $s_{jk}\lambda_{jk}$ of its standard deviation s_{jk}. Since $s_{jk}\mu_{jk}$ is approximately the relative precision of the calibration measurement, which is of the order of 1%, the shift is indeed very small and hence the calibration measurement is only modified to a negligible extent by the prior drawn from the tables.

The final ingredient to fully specify the inference problem is the prior distribution on concentrations $\boldsymbol{x}$. Our prior knowledge about the components of $\boldsymbol{x}$ will usually be very foggy, but in many cases also more informative than a flat prior. If our prior estimate on x_k is x_{k0}, then the appropriate prior is again an exponential:

$$p(x_k|x_{k0}, \mathcal{I}) = \frac{1}{Z(x_{k0})} \exp\{-\nu_k \, x_k\}, \tag{27.53}$$

on the support [0, 1]. The Lagrange parameter ν_k must be determined such that $\langle x_k \rangle = x_{k0}$ (see equation (27.48)). In the present analysis we use a flat prior. All the probability distributions needed to specify the posterior are now available and we are ready to calculate expectation values of any function $f(\mathbf{C}, \boldsymbol{x})$ depending on the cracking coefficients $\mathbf{C}$ and/or on the concentrations $\boldsymbol{x}$:

$$\begin{aligned}\langle f(\mathbf{C}, \boldsymbol{x}) \rangle = \frac{1}{Z} \int_0^1 d^{N_x}x \int_0^1 d^{N_C}C \; f(\mathbf{C}, \boldsymbol{x}) \\ \times \; p(\boldsymbol{x}|\boldsymbol{x}_0, \mathcal{I}) \; p(\mathbf{C}|\mathbf{C}^0, \mathbf{D}, \mathbf{s}, M, \mathcal{I}) \; p(\boldsymbol{d}|\mathbf{C}, \boldsymbol{x}, \boldsymbol{\sigma}, M, \mathcal{I}).\end{aligned} \tag{27.54}$$

Here N_x and N_C are the number of elements in $\boldsymbol{x}$ and $\mathbf{C}$, respectively. A variable M has been split off from the general background information $\mathcal{I}$ to express explicitly that we shall consider different models in the analysis. The normalization constant Z of the posterior in equation (27.54) is the marginal likelihood which enters the Bayes factor in model comparison:

$$Z = p(\boldsymbol{d}|\boldsymbol{\sigma}, \mathbf{D}, \mathbf{s}, \mathbf{C}_0, M, \mathcal{I}) \tag{27.55}$$

$$= \int_0^1 d^{N_x}x \int_0^1 d^{N_C}C \; p(\boldsymbol{x}|\boldsymbol{x}_0, \mathcal{I}) \; p(\mathbf{C}|\mathbf{C}^0, \mathbf{D}, \mathbf{s}, M, \mathcal{I}) \; p(\boldsymbol{d}|\mathbf{C}, \boldsymbol{x}, \boldsymbol{\sigma}, M, \mathcal{I}). \tag{27.56}$$

The integrals in equations (27.54) and (27.56) can only be carried out numerically, since, even if a Gaussian approximation to the integrand exists (which is by no means guaranteed), the limits of integration are finite and the transition to infinite limits will no longer be an acceptable approximation in general. Chapter 29 [p. 509] is devoted to the topic of numerical integration and we shall quote only the results here. The results for parameter estimation are shown in Table 27.5. The most interesting entry for model M_1 is the number for C_2H_2. It has a much larger relative uncertainty than the other entries in this row. The question arises immediately of whether this molecule must really be included in the candidate list. Deleting C_2H_2 constitutes our model M_2. The only change which occurs is in the concentration of C_2H_4, while all others remain unaltered. Even the change in the concentration of C_2H_4 stays in the limits of its uncertainty under model M_1. Reduced complexity seems to explain the data equally well. M_3 even goes one step further and deletes both

Table 27.5 Species concentrations x for three different models

Model	CH_3	CH_4	C_2H_6	N_2	C_2H_4	C_2H_2
M_1	0.126(5)	0.208(2)	0.220(2)	0.395(3)	0.045(5)	0.001(1)
M_2	0.126(5)	0.208(2)	0.220(2)	0.395(3)	0.048(3)	—
M_3	0.126(5)	0.209(2)	0.234(2)	0.409(3)	—	—

C_2H_2 and C_2H_4. This time the changes in the other estimates are no longer negligible and M_3 is probably no longer appropriate to explain the data satisfactorily. The quantitative measure for ranking the different models is the posterior odds, which we take with reference to model M_2. For equal prior model probabilities the odds ratio reduces to the Bayes factor, which yields on a log scale $-4.1, 0, -67.9$ for models M_1, M_2, M_3, respectively. Such a pattern was already met in Figure 27.4 [p. 483]. Simplifying the optimum model results in a drastic loss in evidence due to a strong increase of the misfit to the data, while added complexity improves the misfit marginally and is overridden by the Ockham factor.

28

Bayesian experimental design

The previous examples on parameter estimation and model comparison have demonstrated the benefits of Bayesian probability theory for quantitative inference based on prior knowledge and measured data. However, Bayesian probability theory is not a magic black box capable of compensating for badly designed experiments. Information absent in the data cannot be revealed by any kind of data analysis. This immediately raises the question of how the information provided by a measurement can be quantified and, in a next step, how to optimize experiments to maximize the information gain. Here one of the very recent areas of applied Bayesian data analysis is entered: *Bayesian experimental design* is an increasingly important topic driven by progress in computer power and algorithmic improvements [132, 214]. So far it has been implicitly assumed that there is little choice in the actual execution of the experiment, in other words, the data to be analysed were assumed to be given. While this is the most widespread use of data analysis, the *active* selection of data holds great promise to improve the measurement process. There are several scenarios in which an active selection of the data to be collected or evaluated is obviously very advantageous, for example:

- Expensive and/or time-consuming measurements, thus one wants to know where to look next to learn as much as possible – or when to stop performing further experiments.
- Design of a future experiment to obtain the best performance (information gain) within the scheduled experimental scenarios.
- Selection of the most useful data points from a huge amount of data.
- Optimal (most informative) combination of different experiments for the quantity of interest.

These ideas emphasize the importance that inference has to be independent of the experimental design, as discussed for the stopping criterion in Section 20.2 [p. 325].

28.1 Overview of the Bayesian approach

The theory of frequentist experimental design dates back to the 1970s [29, 66]. At about the same time, the Bayesian approach was put forward by the influential review of Lindley [128]. His decision-theoretic approach involves the specification of a suitable utility function $U(\boldsymbol{y}, \eta)$ which depends on the result, that is measured data $\boldsymbol{y}$ of an experiment and the design parameters η. Design parameters are understood as the parameters of an experiment

which
are accessible and adjustable. Examples are the point in time for the next measurement or the analysis beam energy. The utility function has to be defined with respect to the goals of the experiment and cannot be derived from first principles. It may contain considerations about the cost of an experiment or the value of a reduced uncertainty of a parameter estimation. For a discussion about the formal requirements for utility functions, see e.g. [15]. The experimental design decision η^* (e.g. where to measure next) which maximizes the chosen utility function $U(\boldsymbol{y}, \eta)$ is the optimal design. However, the data L data points $\boldsymbol{y}$ are uncertain before the actual measurement due to statistical or systematic uncertainties and incomplete knowledge about the parameters $\boldsymbol{\theta}$ of the physical system. Therefore, the Bayesian experimental design has to take into account all possible data sets and the utility function has to be marginalized over the data space, which results in the expected utility

$$\mathrm{EU}(\eta) = \int d^L y \, p(\boldsymbol{y}|\eta, \mathcal{I}) \, U(\boldsymbol{y}, \eta) \,. \tag{28.1}$$

Apparently, the expected utility is the integral over all possible data weighted by the probability of the data under the design decision η and the utility of the corresponding data. In general, the required predictive distribution $p(\boldsymbol{y}|\eta, \mathcal{I})$ is not immediately accessible if there are unknown parameters $\boldsymbol{\theta}$. In that case we have to invoke the marginalization rule

$$p(\boldsymbol{y}|\eta, \mathcal{I}) = \int d^n\theta \, p(\boldsymbol{y}|\boldsymbol{\theta}, \eta, \mathcal{I}) \, p(\boldsymbol{\theta}|\mathcal{I}) \,. \tag{28.2}$$

The number of parameters is denoted by n. Substituting $p(\boldsymbol{y}|\eta, \mathcal{I})$ from equation (28.1) in equation (28.2) yields

$$\mathrm{EU}\left(\eta^*\right) = \max_{\eta} \int d^L y \int d^n\theta \, p(\boldsymbol{y}|\theta, \eta, \mathcal{I}) \, p(\boldsymbol{\theta}|\mathcal{I}) \, U(\boldsymbol{y}, \eta) \tag{28.3}$$

for the best experimental design decision. Therefore, the expected utility can be expressed in terms of likelihood and prior distributions combined with a suitable utility function $U(\boldsymbol{y}, \eta)$. The evaluation, however, requires nested integrations. Only for very few cases (almost always involving Gaussian likelihoods and linear models) can the integration over parameter space and data space be performed analytically.

28.2 Optimality criteria and utility functions

The most widely used optimality criteria for experimental design are derived from various desirable properties of parameter estimates of linear models. Minimizing the average variance of the best estimates of the regression coefficients by minimizing the trace of the variance–covariance matrix is called A-optimality [66, 197]. The sometimes harmful neglect of the parameter covariances in A-optimality motivates D-optimality, where the determinant of the variance–covariance matrix is minimized [197]. Other optimality criteria focus on the variance of predictions instead of the variance of the parameter estimates. For an overview of the various optimality criteria, see e.g. [7, 172] and the relationship between frequentist and Bayesian optimality criteria as discussed in [28, 39]. However, the focus on the best estimate $\boldsymbol{\theta}^*$ only as a basis for experimental design does not take into account the full information content of the probability distribution of the parameters in nonlinear settings. The suggestion of Lindley [127] to use the information gain of an experiment as utility function has therefore been followed by several authors [14, 40, 41, 132, 134, 198]. The information gain is given by the expected Kullback–Leibler divergence between the posterior distribution $p(\boldsymbol{\theta}|\boldsymbol{y}, \eta, \mathcal{I})$ and the prior distribution $p(\boldsymbol{\theta}|\mathcal{I})$:

$$U_{\mathrm{KL}}(\boldsymbol{y}, \eta) = \int d^n\theta \, p(\boldsymbol{\theta}|\boldsymbol{y}, \eta, \mathcal{I}) \log \frac{p(\boldsymbol{\theta}|\boldsymbol{y}, \eta, \mathcal{I})}{p(\boldsymbol{\theta}|\mathcal{I})}. \tag{28.4}$$

For the standard Gaussian linear regression model this utility function yields the same results as using a Bayes D-optimality criterion for design [28]. A discussion of some of the properties of the Kullback–Leibler divergence as utility function in experimental design can be found in [135]. From a theoretical point of view the decision-theoretic formulation of experimental design is well understood. Nevertheless, nonlinear experimental design methods received only little attention within and outside the physics community (see, for example, statements about lack of real applications in [28, 86, 205]) until Loredo [133] published an illustrative example about optimization of observation times in astronomy, highlighting the potential and feasibility of nonlinear Bayesian experimental design.

28.3 Examples

28.3.1 Adaptive exploration for extrasolar planets

The search for extrasolar planets is one of the foci of astrophysical research programmes – supported by space-based missions like Kepler [119] or by ground instruments like HARPS (High Accuracy Radial velocity Planet Searcher). But the necessary high-accuracy measurements are time-consuming, seriously restricting the number of stars that can be examined in search of extrasolar planets. Observation time is thus a precious resource that must be carefully allocated. Therefore, observations of stars with companions should be scheduled optimally to determine the orbital parameters with the fewest number of observations. In [132] the problem of determining the best time for the next measurement of the radial velocity (RV) of a star known to have a single planetary companion is addressed. The time-dependent radial velocity is a nonlinear function given by

$$\mathrm{v}(t; \tau, e, K) := \mathrm{v}_0 + K\left(e \cos\omega + \cos\left[\omega + \upsilon(t)\right]\right), \tag{28.5}$$

where the true anomaly $\upsilon(t)$ can be computed by the solution of a system of two nonlinear equations for the eccentric anomaly (an angular parameter that defines the position of a body that is moving along an elliptic Kepler orbit [150, 219]):

$$E(t) - e \sin(E(t)) = \frac{2\pi t}{\tau} - M_0 \tag{28.6}$$

and

$$\tan\frac{\upsilon(t)}{2} = \sqrt{\frac{1+e}{1-e}}\tan\frac{E(t)}{2}. \tag{28.7}$$

The six parameters of the model are the orbital period τ, the orbital eccentricity e, the velocity amplitude K, the centre-of-mass velocity of the system v_0, the mean anomaly at $t = 0$, M_0 and the argument of the pericentre ω. The orbital eccentricity e describes the amount by which an orbit deviates from a perfect circle: $e = 0$ is a perfectly circular orbit and $e = 1$ corresponds to an open parabolic orbit. In [132] the treatment is simplified, taking into account only three of the six parameters and assuming Gaussian additive noise ϵ with a standard deviation σ, so that the measured datum d_i at time t_i is given by

$$d_i = v(t_i; \tau, e, K) + \epsilon_i. \tag{28.8}$$

For parameter values $\tau = 800$ d, $e = 0.5$, $K = 50$ m s^{-1} and $\sigma = 8$ m s^{-1} a vector $\boldsymbol{y}$ of 10 simulated observations was computed.

In Figure 28.1 the 10 data points with error bars are displayed together with the true velocity curve. The posterior distribution of the parameters is given by Bayes' theorem as

$$p(\tau, e, K|\boldsymbol{y}, \mathcal{I}) \propto p(\boldsymbol{y}|\tau, e, K, \mathcal{I})\, p(\tau, e, K|\mathcal{I}), \tag{28.9}$$

where $p(\tau, e, K|\mathcal{I})$ is the prior probability density for the orbital parameters and $(\mathbf{d}|\tau, e, K, \mathcal{I})$ the Gaussian likelihood function. For the data set displayed in Figure 28.1, rejection sampling was used to draw independent samples from the posterior distribution. From a sample of size $N = 100$ only the τ and e values are plotted in Figure 28.2, corresponding to the marginal PDF $(\tau, e|\boldsymbol{y}, \mathcal{I})$. The distribution is roughly located at the true values $(\tau, e) = (800, 0.5)$ but with an asymmetric shape and still significant uncertainty. Based on the posterior distribution in equation (28.9) the predictive distribution $p(\eta|t, \boldsymbol{y}, \mathcal{I})$ for

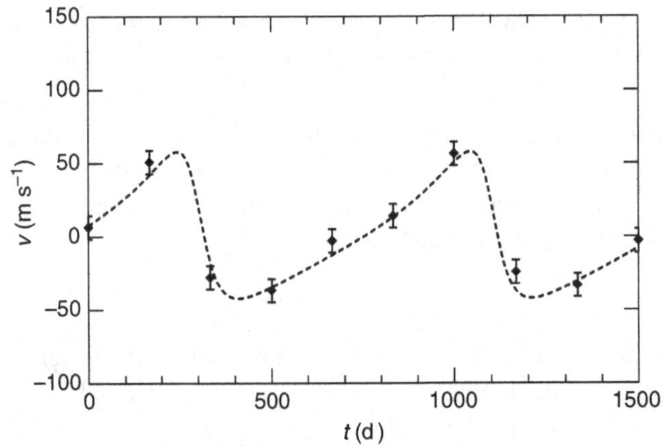

Figure 28.1 From this true velocity curve (dashed line) 10 simulated observations distorted with Gaussian noise have been generated (dots plus error bars). Data are taken from [131].

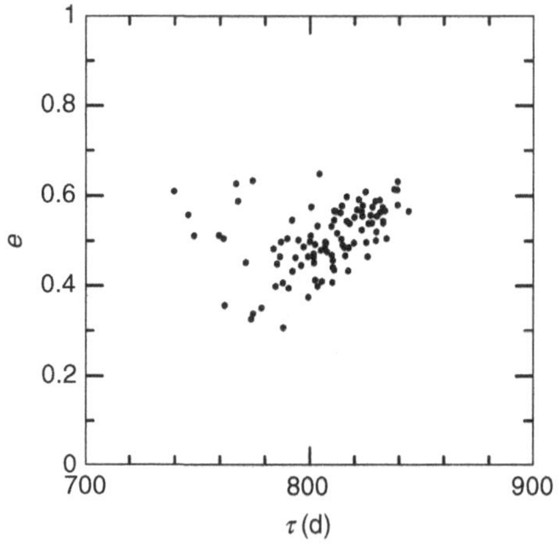

Figure 28.2 Posterior samples drawn from the marginal posterior PDF $p(\tau, e|\boldsymbol{d}, \mathcal{I})$ (published in [131]).

a datum η at a future time t can be computed. For given values of (τ, e, K) the predictive probability density for η is a Gaussian centred at $v\,(t;\,\tau, e, K)$. The predictive distribution $p\,(\eta|t, \tau, e, K, \boldsymbol{y}, \mathcal{I})$ is thus given by the product of the Gaussian likelihood for η and the posterior distribution $p\,(\tau, e, K\,|\,\boldsymbol{y}, \mathcal{I})$. To account for the parameter uncertainty, the model parameters have to be marginalized. Use of the posterior samples circumvents the time-consuming integration over the parameter space, since $p\,(\eta|t, \boldsymbol{y}, \mathcal{I})$ can be expressed as

$$
\begin{aligned}
p\,(\eta|t, \boldsymbol{y}, \mathcal{I}) &= \int d\tau \int de \int dK\, p\,(\tau, e, K\,|\,\boldsymbol{y}, \mathcal{I}) \\
&\quad \times \frac{1}{\sqrt{2\pi\sigma^2}} \exp\left(-\frac{1}{2}\frac{[\eta - v\,(t;\,\tau, e, K)]^2}{\sigma^2}\right) \\
&\approx \frac{1}{N} \sum_{\tau_i, e_i, K_i} \frac{1}{\sqrt{2\pi\sigma^2}} \exp\left(-\frac{1}{2}\frac{[\eta - v\,(t;\,\tau_i, e_i, K_i)]^2}{\sigma^2}\right).
\end{aligned}
$$

The last line gives a Monte Carlo estimate of the predictive distribution using the independent samples from the posterior distribution. In Figure 28.3 the predicted velocity $v\,(t;\,\tau, e, K)$ is displayed for the first 15 sampled parameter values $(t;\,\tau_i, e_i, K_i)$. The spread of the velocity functions represents the uncertainty in the predictive distribution. The uncertainty is largest where the velocity changes most quickly, and is slowly increasing with time since predictions with different periods fall increasingly out of synchronization. Once the predictive distribution is available, the expected utility can be computed as a function of time using equation (28.1):

$$
\mathrm{EU}\,(t) = \int d\eta\, p\,(\eta|t, \boldsymbol{y}, \mathcal{I})\, U\,(\eta, t)\,. \tag{28.10}
$$

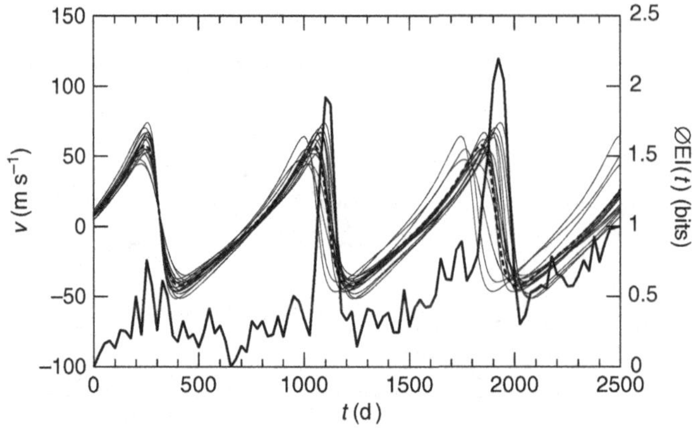

Figure 28.3 Fifteen predicted velocity curves (thin lines) compared with the true velocity curve (dashed line) used for the simulation of the data. Their spread is a measure of the uncertainty of the predicted velocity. The expected information gain (right axis) for a further measurement as a function of time is indicated by the thick solid line (data taken from [131]). Note that the positions of largest uncertainty and largest information gain coincide.

For the present problem where the width of the noise distribution does not depend on the underlying signal, it can be shown that the expected information gain is equal to the entropy of the predictive distribution [180]:

$$\mathrm{EU}\,(t) = \int d\eta\; p\,(\eta|t, \boldsymbol{y}, \mathcal{I}) \log\left(p\,(\eta|t, \boldsymbol{y}, \mathcal{I})\right). \tag{28.11}$$

This equality is saving one (possibly high-dimensional) integration over the parameter space otherwise needed for the computation of the information-based utility function. Thus the best sampling time is the time at which the entropy (uncertainty) of the predictive distribution is largest. The thick line in Figure 28.3 shows the estimate of EU$\,(t)$ using base-2 logarithms so that the relative information gain is measured in bits (ordinate on the right-hand side). It is largest near the periastron crossings, thus recommending an additional observation at the maximum, i.e. at $t = 1925$ d. Incorporating the new noisy datum measured at $t = 1925$ d into the posterior yields a significantly reduced uncertainty in the period estimate, as a comparison of the posterior distributions without (see Figure 28.2) and with the new data point (Figure 28.4) reveals. The optimization procedure can be repeated and the well-chosen data points yield an increase in precision exceeding the rule-of-thumb $\sqrt{n}$ dependence often seen for random sampling [132].

28.3.2 *Optimizing NRA measurement protocols*

Nuclear reaction analysis (NRA) is a well-known technique for depth profiling of light elements in the near-surface region of solids (up to depths of several micrometres) using ion beams with energies in the MeV range. NRA measurements yield quantitative information

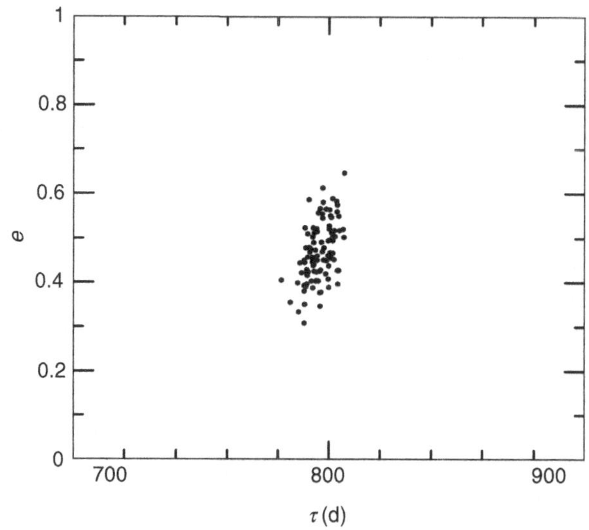

Figure 28.4 Same as Figure 28.2 but with one additional measurement at the optimal time. Comparison with Figure 28.2 reveals that the parameters have significantly higher precision (from [131]).

on the isotopic depth distribution within the target and are highly sensitive. For an introduction to NRA for material analysis, see e.g. [5, 201]. The basic principle of NRA is straightforward: the sample is subjected to an energetic ion beam with an initial energy E_i^0 at an angle of incidence ϕ which reacts with the species of interest. The products of the reaction are measured under a specified angle θ. The measured total signal count, d_i, depends (in the limit of small concentrations) linearly on the concentration profile $c(x)$ of the species in depth x:

$$d_i := d\left(E_i^0\right) = \mu N_i \int_0^{x(E_i^0)} dx\, \sigma\left(E\left(x, E_i^0\right)\right)\, c\,(x) + \epsilon_i. \tag{28.12}$$

The additional parameters are the total number of impinging ions N_i, the energy-dependent cross-section of the reaction $\sigma\,(E)$, the efficiency of the detection and the geometry of the setup, μ. The measurement uncertainty ϵ_i is approximated by a Gaussian distribution $\mathcal{N}\,(0, \sigma_i)$. Repeated measurements with different initial energies E_i^0 provide increasing information about the depth profile of the species of interest. The optimization of NRA measurements of deuterium profiles for the weakly resonant nuclear reaction $D + {}^3He \rightarrow p + {}^4He + 18.352$ MeV [5] has been studied in [216]. The high Q-value of 18.352 MeV provides an analysis depth for deuterium of several micrometres even in high-Z materials such as tungsten. Therefore, the reaction is commonly used to study the hydrogen isotope retention in plasma-facing components of fusion experiments [195]. However, measurements are time-consuming and the extraction of the concentration depth profile from the measured data is an ill-conditioned inversion problem due to the very broad cross-section of the D $(^3He, p)\,^4He$ reaction [144]. Therefore the experimental setup, that is the choice of analysis energies, should be optimized to provide a maximum of information about the depth profile. To evaluate equation (28.12), the energy $E\left(x;\,E_i^0\right)$of the incident particle on its path through the sample for a given initial energy E_i^0 is required. The energy loss of the impinging ^3He ion in the sample is determined by the *stopping power* $S(E)$ of the sample:

$$\frac{dE}{dx} = -S(E), \tag{28.13}$$

which can be solved to get the depth-dependent energy $E(x;\,E_i^0)$ for different initial energies E_i^0. Parametrizations and tables of $S(E)$ for different elements are given in [201]. Since the amount of hydrogen in the sample is usually well below 1% (with the exception of a very thin surface layer), the influence of the hydrogen concentration on the stopping power can be neglected in most cases. A parametrization for the cross-section $\sigma\,(E)$ [3] is provided in [216]. A tungsten sample $\left(\rho = 19.3\,\mathrm{g\,cm^{-3}}\right)$ with a (high) surface concentration of 12% deuterium, followed by an exponentially decaying deuterium concentration profile down to a constant background level described by

$$c(x) = a_0 \cdot \exp\left(-\frac{x}{a_1}\right) + a_2 \tag{28.14}$$

has been used. The parameter values are $a_0 = 0.1, a_1 = 395$ nm and $a_2 = 0.02$. The corresponding mock data for a set of initial energies E^0={500, 700, 1000, 1300, 1600, 2000, 2500, 3000} keV is shown in Figure 28.5. The variations in the detected yields reflect the interplay of the increasing range of the ions with increasing energy and the reduced cross-section at higher energies modulated with the decreasing deuterium concentration

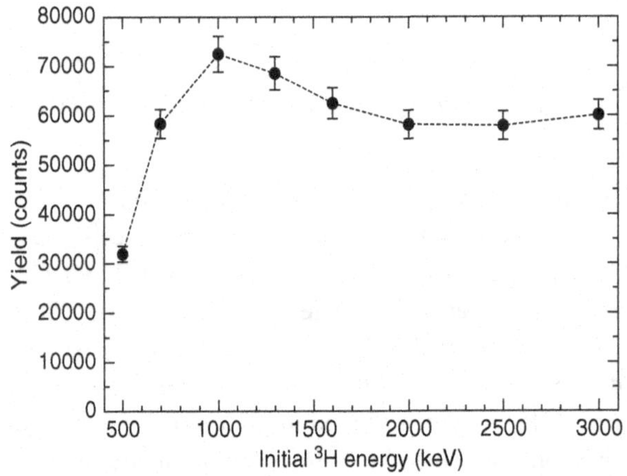

Figure 28.5 Simulated yield data of a D(^{3}He, p)^{4}He nuclear reaction analysis of a tungsten sample with an exponentially decaying deuterium concentration profile. The varying intensity reflects the interplay of energy-dependent cross-section, ion beam range and concentration profile.
Source: Adapted from [216].

at larger depths. The increase of the signal by raising the initial energy from 2500 keV to 3000 keV is caused by the constant deuterium background of 2%. The time which would be needed to obtain the eight data points is around one working day, taking into account necessary calibration measurements. The uncertainty of the measurement is given by a Poisson statistic. However, fluctuations in the beam current measurements are very often the dominating factor, affecting the prefactor N_i in equation (28.12). An accuracy of up to 3% can be achieved in favourable circumstances (e.g. by using the number of Rutherford-scattered ^{3}He ions on a thin gold coating on top of the sample as reference). Therefore, a realistic estimate of the measurement uncertainty was assumed to be $\sigma_i = \max\left(5\%d_i, \sqrt{d_i}\right)$. In [216] the best experimental design for a linear setting (assuming a piecewise linear concentration profile) compared very favourably with the established experimental technique of an equidistant choice of the beam energies. For the nonlinear design the Kullback–Leibler divergence was optimized. In the experimental design approach for the Kepler orbit measurements, the computational effort could be reduced to exploit the maximum entropy sampling. Instead, in the present case the data-dependent uncertainty $\sigma_i = \max\left(5\%d_i, \sqrt{d_i}\right)$ requires an additional parameter space integration to compute the information gain of a measurement using the Kullback–Leibler divergence (equation (28.4) [p. 493]), increasing computation time. However, the optimal next accelerator energy can only be computed after the result of the previous measurement is available. Therefore, long computing times are not compatible with an efficient operation. To circumvent this problem the posterior sampling method [130] was used, reducing the computation of the next measurement energy to less than 5 minutes using a standard PC with 2 GHz CPU [216]. In Figure 28.6 three cycles of nonlinear Bayesian experimental design are shown. After a first measurement at 500 keV, a sample of the posterior distribution of $\{a_1, a_2\}$ is depicted in (a). The single measurement does not allow us to distinguish between a large decay constant a_1 and a low constant offset a_2 or vice versa. The EU, plotted in (A), now favours a measurement at the other end of the energy range (the maximum of the utility function is encircled). After a measurement with 3 MeV ^{3}He the 'area' of the posterior distribution is significantly reduced. In (b), the background concentration is below 3% but the decay length is still quite undetermined. The new EU in subplot (B) now has a maximum at 1500 keV, still with a pretty high EU. Performing a measurement with 1500 keV localizes the posterior distribution around the true, in general unknown, value of $a_1 = 395$ nm and $a_2 = 0.02$, as depicted in (c). The next measurement, displayed in (C), should be performed at 1200 keV but the EU is significantly lower than before. Subsequent measurements are predominantly improving the statistics; a second measurement at 3 MeV provides nearly the same information.

28.3.3 Autonomous exploration

Most present-day advanced remote science operations use semi-automated systems that can carry out basic tasks such as locomotion and directed data collection, but require human intervention when it comes to deciding where to go or which experiment to perform.

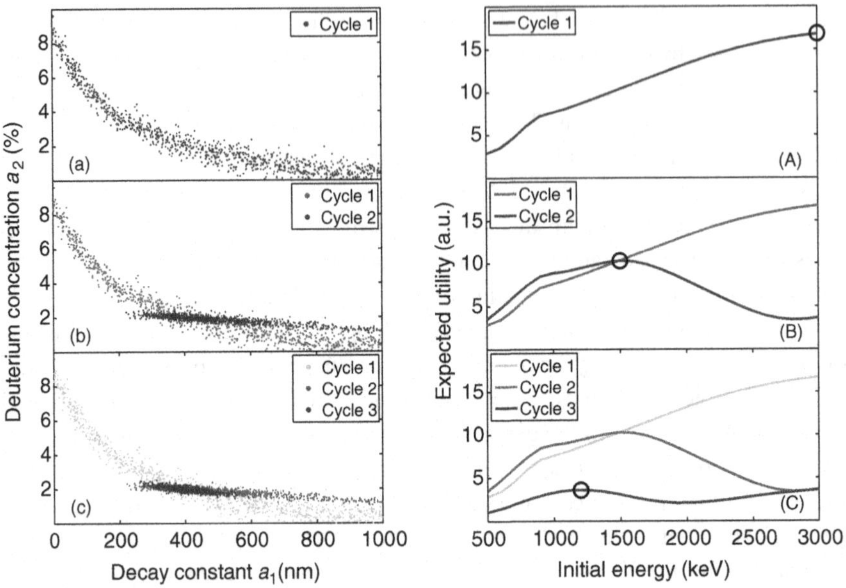

Figure 28.6 Three cycles of the experimental design process. In the first cycle a sample of size 1000 is drawn from the posterior distribution $p\,(a_1, a_2|\mathbf{d}, \eta)$. The result is depicted in (a). The corresponding EU is plotted as function of energy in (A) and the maximum is indicated by a circle. Its position defines the suggested next measurement energy. Performing that measurement yields the posterior samples (black dots in (b)). The previous posterior samples are retained in grey. The new EU as a function of beam energy is given in (B). Its maximum position defines the next measurement energy. Adding a new measurement with that energy results in the posterior sample displayed in (c). The previous posterior samples are also given in different shades. The posterior volume has decreased significantly. Therefore the best EU for the next measurement (see (C)) is lower than before.
Source: Adapted from [216].

However, for many applications, instruments that can both act and react with minimal human intervention would be advantageous. In [117] a simple robot is described which collects data in an automated fashion, and based on the results, decides which new measurement to take, thus pursuing the learning cycle of observation, inference and hypothesis refinement.

The experimental problem addressed is the localization (x, y-position) and characterization (radius) of a white disc on a black plane using a LEGOR® robot arm equipped with a light sensor capable of **noisy point** measurements only. This toy problem can be considered as a crude representation of a landmine search problem [85]. The parameter vector of the disc consists of the disc centre coordinates $\boldsymbol{x}_0$ and the disc radius r_0:

$$\boldsymbol{c} = \{\boldsymbol{x}_0, r_0\}\,. \tag{28.15}$$

The data vector is given by a set of N light measurements

$$\mathbf{d} = \{d_1, d_2, \ldots, d_\mathrm{N}\} \tag{28.16}$$

recorded at positions

$$\mathbf{X} = \{\boldsymbol{x}_1, \ldots, \boldsymbol{x}_\mathrm{N}\}\,. \tag{28.17}$$

When the positions are assumed to be known with certainty, the posterior probability for the disc parameters is given by

$$p(\boldsymbol{c}|\mathbf{d}, \mathbf{X}, \mathcal{I}) = \frac{p(\mathbf{d}|\boldsymbol{c}, \mathbf{X}, \mathcal{I})\, p(\boldsymbol{c}|\mathcal{I})}{p(\mathbf{d}|\mathcal{I})}. \tag{28.18}$$

A uniform prior probability is assigned for the disc parameters:

$$p(\boldsymbol{c}|\mathcal{I}) = \left(\frac{1}{x_{\max} - x_{\min}}\right)\left(\frac{1}{y_{\max} - y_{\min}}\right)\left(\frac{1}{r_{\max} - r_{\min}}\right) \tag{28.19}$$

with $r_{\min} = 1$ cm and $r_{\max} = 15$ cm. The associated Heaviside functions are omitted as they have no impact since the likelihood is sharply peaked. The likelihood function for one measurement d_i taken at $\boldsymbol{x}_i$ can be written as

$$\begin{aligned} p(d_i|\boldsymbol{c}, \boldsymbol{x}_i, \mathcal{I}) &= p(d_i|\{\boldsymbol{x}_0, r_0\}, \boldsymbol{x}_i, \mathcal{I}) \\ &= \begin{cases} \mathcal{N}(d_W, \sigma), & \text{if } \|\boldsymbol{x}_i - \boldsymbol{x}_0\|^2 \le r_0^2 \\ \mathcal{N}(d_B, \sigma), & \text{if } \|\boldsymbol{x}_i - \boldsymbol{x}_0\|^2 > r_0^2 \end{cases}. \end{aligned} \tag{28.20}$$

The mean value μ of a light measurement on the white (black) disc is d_W (d_B). The uncertainty of the intensity measurement is given by a Gaussian distribution $\mathcal{N}(\mu, \sigma)$ with uncertainty σ, centred around the mean value μ. The information gain (Shannon entropy) of a measurement has been taken as utility function. As the noise level is independent of the sampling location, the maximum entropy sampling [180] can be used for an efficient computation of the expected utility based on posterior samples:

$$(\hat{x}_e, \hat{y}_e) = \arg\min_{(x_e, y_e)} \int dd_e\; p(d_e|\mathbf{d}, \boldsymbol{x}_e, \mathcal{I})\, \log p(d_e|\mathbf{d}, \boldsymbol{x}_e, \mathcal{I})\,. \tag{28.21}$$

For an efficient computation of posterior samples, the nested sampling algorithm has been used. Nested sampling is described in Chapter 31 [p. 572]. In order to find the next measurement position, a grid on the space of possible measurement locations is considered and equation (28.21) is only computed at the grid points. The alignment of this grid is randomly jittered so that a greater variety of points can be considered during the measurement process. In Figure 28.7(A) the initial stage of the inference process is displayed; a first measurement has been taken (indicated by the black mark in the upper right part of panel (A)). The white disc has not yet been located. For this reason there are large regions of the measurement space that are potentially equally informative, indicated by the homogeneous areas in Figure 28.7(B), where the entropy gain of a further measurement at that location is displayed. Locations with a darker colour provide less information (e.g. already measured locations). After several iterations, the robot will eventually find a white area belonging to the disc, thus immediately constraining the possible parameter space considerably. In Figure 28.7(C) the set of circles in agreement with all measurements is now already

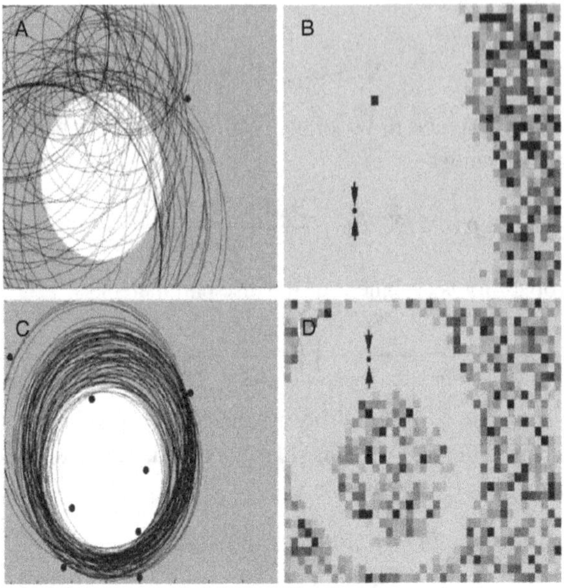

Figure 28.7 Panels (A/C) show the area to be searched together with the position of the white disc to be detected. In panel (A) a first measurement has been marked by a black dot. Since the background was detected, a number of possible positions of the white disc can immediately be excluded. This is visualized by 150 circles sampled from the updated posterior. These represent possible disc positions and sizes consistent with the measurement(s). Panel (C) displays the situation after several measurements. The scatter of possible disc positions is already fairly small. Panels (B/D) visualize the selection algorithm for the location of the next measurement. They display the information gain for each measurement location (dark colours indicate uninformative locations) and the location with the highest information gain is selected for the next measurement (indicated by arrows). Further details are given in the text.
Source: Graph adapted from [117].

constrained to the vicinity of the true disc location. The measurement with the highest expected utility (indicated by two arrows in Figure 28.7(D)) is in the region with the highest scattering of the posterior samples. It is essentially asking a binary question that rules out half of the models. This results in a rapid convergence, significantly reducing the number of necessary measurements. These binary questions are not hard-wired into the system but are a natural consequence of the selection of the most informative measurement [117].

28.3.4 Optimizing interferometric diagnostics

In a series of papers, Dreier *et al.* [54–57, 71] studied the design of a multichannel interferometer at the Wendelstein 7-X stellarator with respect to beam-line configuration, number of beam-lines and joint evaluation with other diagnostics. In [57] the impact of technical boundary conditions on the measurement of plasma electron density distributions using a four-channel, two-colour interferometer is investigated. For the interferometry system at

W7-X three entrance ports into the vacuum vessel are reserved, allowing different beam-line configurations from vertical to horizontal optical paths [121]. Because no opposite ports are available, the probing beams have to be reflected by corner cube retro-reflectors mounted at the opposite wall. These reflectors have to fit to the structure of the in-vessel components. In combination with other constraints (e.g. limited port size), the number of realizable beam-lines is 101. One of the physical questions to be addressed in the W7-X stellarator is the variation of the plasma density profiles at various confinement regimes (H-mode and high-density H-mode (HDH)). Maximizing the expected utility using the information gain of measurements (Kullback–Leibler divergence) as utility function yielded an optimal design (Figure 28.8, right panel) with an expected utility of $\text{EU} = 28.3 \pm 0.2$ bit. The best design taking technical boundary conditions into account (Figure 28.8, left panel) leads to an expected utility of only $\text{EU} = 8.53 \pm 0.01$ bit. A comparison of the two different designs reveals the reason for the large difference of the expected information gains. In the unconstrained design (Figure 28.8, right panel) two lines of sight are localized at the very edge of the plasma, additionally passing the plasma on a very long path. This provides a good signal-to-noise ratio and at the same time a high sensitivity to small shifts in the position of the plasma edge. In both aspects the design where the port system had to be taken into account is inferior. Therefore, the expected utility of modifications of the experimental setup has been investigated, yielding a different design suggestion based on an out-of-plane setup [57].

Other applications of (nonlinear) Bayesian experimental design are optimized material testing schemes [175], filter design for Thomson scattering diagnostics [68], optimized experiment proposals based on scaling laws [168] or the optimal design of heart defibrillators [32].

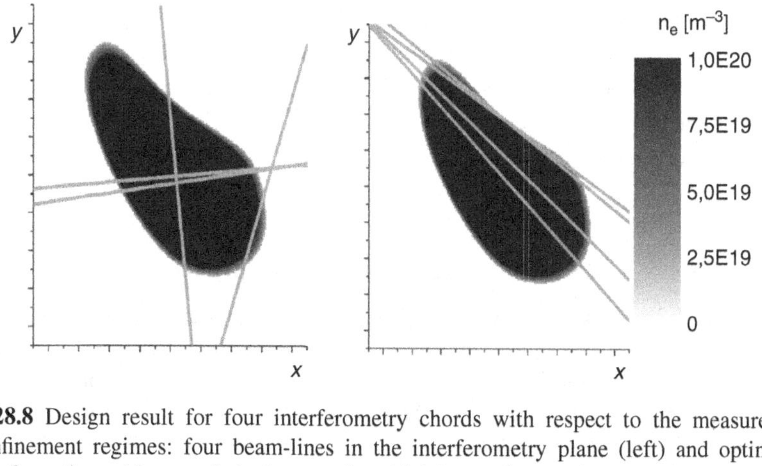

Figure 28.8 Design result for four interferometry chords with respect to the measurement of high confinement regimes: four beam-lines in the interferometry plane (left) and optimal four-beam configuration without technical constraints (right) superimposed onto a cross-section of the W7-X plasma.
Source: Adapted from [56].

28.4 N-step-ahead designs

All the design policies that have been considered so far are greedy, selecting the best action using a one-step-ahead approach (i.e. the best action is chosen as if the next measurement would be the last one). In practice, however, most experimental design optimizations are used in a repetitive manner. This may lead to less than optimal designs, as can be demonstrated with a simple example. The interval $[1, 4]$ has to be segmented with two support points such that the largest segment is minimized (e.g. for efficient regression). A one-step-ahead algorithm selects $x_1 = 2.5$ as best segmentation value for the first support point but in the subsequent optimization no placement of the second support point can reduce the size of the largest segment below 2.5. A two-step-ahead algorithm would position the support points instead at $x_1 = 2$ and $x_2 = 3$, achieving an upper limit on the segment length of $s = 1$.

N-step-ahead designs, also known as *full-sequential designs*, correspond to stochastic dynamic programming problems (SDP) [11]. The dependence of later experiments on previous actions and observations (here for a two-step-ahead design)

$$\begin{aligned}\mathrm{EU}(\eta_1) = \max_{\eta_1} \Bigg\{ \int d^L y_1\, p\left(\boldsymbol{y}_1 | \eta_1, \mathcal{I}\right) \\ \times \max_{\eta_2} \left[\int d^L y_2\, p\left(\boldsymbol{y}_2 | \boldsymbol{y}_1, \eta_1, \eta_2, \mathcal{I}\right) U\left(\boldsymbol{y}_1, \boldsymbol{y}_2, \eta_1, \eta_2\right) \right] \Bigg\}\end{aligned} \tag{28.22}$$

introduces a feedback of information. This feedback, leading to the repeated embedded maximizations and integrations in equation (28.22), is the reason for the extreme difficulty of full-sequential designs with $N > 1$. Approximate solution methods of equations of similar structure are discussed in the areas of feedback control [82, 169] and partially observable Markov decision processes (POMDP) [109, 110, 154]. The computational complexity of SDP has so far mostly precluded the use of full-sequential designs in experimental design. In contrast, there is some evidence that in many cases the largest benefit is already provided by the step from $N = 1$ to $N = 2$ (see [169] and references cited therein). The emerging computer power widens the range of models for which this limited increase in the prediction horizon is feasible.

28.5 Experimental design: Perspective

In the preceding sections the Bayesian approach to experimental design was illustrated with several examples, all of them focusing on the best strategy for parameter estimation $p(\boldsymbol{\theta} | \boldsymbol{y}, M_1)$ for a given model M_1. In contrast, the closely related approach of experimental design for model identification [204], i.e. the selection of measurements which best discriminate between a set of models M_k, $k = 1, \dots, K$ has so far only rarely been applied for nonlinear models, most likely due to the increased numerical complexity of computing $p(M_k | \boldsymbol{y})$. Some further aspects of experimental design which are in the focus of current research are listed below.

- The optimization procedure assumes that the model is correct. This may (especially for linear models) lead to design suggestions which appear strange and are not robust with respect to minor deviations from the model. This is reflected in optimized designs which suggest measurements only at the endpoints of the design interval [39] or repeated measurements with the same settings (sometimes referred to as *thinly supported designs*). Averaging the expected utility over a set of plausible models (the mixture approach) may provide more robust designs [28].
- The majority of experimental design techniques focus either on the estimation of parameters of a given model or on model identification. For both cases, appropriate utility functions are known [172]. Relatively little work has been done to develop experimental design criteria to jointly improve parameter estimates and model identification. Some ideas are given in [17, 30], but these ideas still wait to be tested in physics applications.
- The applicability of Bayesian experimental design depends on the feasibility of the necessary integrations. Several efficient algorithms have already been proposed [146–148, 180], but the special structure of (sequential) expected utility computation still provides possibilities for further optimization. This is an active area of research [25, 149].
- Once a joint environment–sensor model is created, the act of calibration becomes another potential experiment. In such a system, the instrument can decide to interact with either the environment via measurements or itself via calibration, giving rise to an instrument that actively self-calibrates during an experiment [117, 202]. The potential of these and similar ideas still waits to be explored.

Another noteworthy application of experimental design is the optimization of computer experiments. Elaborate computer simulations are progressively used in scientific research. As surrogates for physical systems, computer simulations can be subjected to experimentation, the goal being to predict how the real system would behave or to validate the computer model. Complex simulation models often require long running times, thus severely limiting the size and scope of computer simulations. A frequently used approach to circumvent these restrictions is based on fitting a cheaper predictor or surrogate model (e.g. a response surface model [151]) of the simulation code output $y(t)$ to the input data t. The predictor is then used for parameter studies instead of the original computer code. The experimental design is concerned with the best prediction of the simulation code output $y(t)$ using an optimized selection of sites $\{t_1, t_2, \ldots, t_n\}$ [37, 116, 176] and the efficient identification of the most relevant input parameters [178]. Bayesian approaches based on Gaussian processes [152] may require a far smaller number of model runs than standard Monte Carlo approaches [156]. A transfer of ideas from control theory for dynamic model systems [98] and correlated, multidimensional response variables may provide further progress.

PART VI

PROBABILISTIC NUMERICAL TECHNIQUES

29
Numerical integration

In the preceding sections we have frequently employed the saddle-point approximation (Gaussian approximation) to problems which are nonlinear in the parameters. Two requirements have been more or less tacitly assumed in doing so: the first assumption is that the true posterior resembles relatively closely a multivariate Gaussian. This is frequently the case, but of course not necessarily generally so. The second assumption is that the limits of integration and hence the support of the parameters is $(-\infty, \infty)$. This latter assumption can be relaxed to the requirement that the parameter support and position and width of the posterior are such that extending the limits of integration causes negligible approximation error to the integral. This assumption is, however, frequently not justified. Since analytic integration of a multivariate Gaussian within finite limits is not possible, the only solution to the problem is numerical integration. This does not sound like a complicated task, but we shall see in the sequel that things are different in many dimensions and frequently people discussing the topic are misled by a dimension fallacy.

The evaluation of multidimensional integrals of arbitrary posterior distributions, which would of course also include multimodal posteriors, can become rather complicated. For pedagogical reasons, we shall pursue a much simpler problem and limit the discussion to problems where a Gaussian approximation exists and discuss how to account by numerical integration for approximation errors introduced by differences between the true shape of the posterior and its Gaussian approximation and by finite limits of integration. We shall first investigate deterministic numerical integration, and then proceed in a heuristic manner to Monte Carlo methods which include simple sampling, importance sampling and MCMC in its simplest form. We will return to Monte Carlo methods in Chapter 30 [p. 537] in a mathematically rigorous way.

29.1 The deterministic approach

The general Bayesian integration problem is the evaluation of expectation values

$$\langle g \rangle = \int g(\boldsymbol{\mu})\ p(\boldsymbol{\mu}|\boldsymbol{d}, \mathcal{I})\ d^E\mu, \tag{29.1}$$

where $p(\boldsymbol{\mu}|\boldsymbol{d}, \mathcal{I})$ is the posterior distribution of the parameters and $g(\boldsymbol{\mu})$ is some arbitrary function of $\boldsymbol{\mu}$. Frequently $g(\boldsymbol{\mu}) = \boldsymbol{\mu}$ or $g(\boldsymbol{\mu}) = (\boldsymbol{\mu} - \boldsymbol{\mu}_0)^2$, where $\boldsymbol{\mu}_0$ is the posterior mean or the posterior mode. The case causing real trouble occurs when the normalization of the PDF is not known and when we are interested in it for reasons which have been discussed before. In this case the normalization of the posterior PDF is called 'evidence' and is a measure for the plausibility of the model assumptions in light of the data

$$\mathcal{E} := p(\boldsymbol{d}|\mathcal{I}) = \int d^E\mu \; p(\boldsymbol{d}|\boldsymbol{\mu}, \mathcal{I}) \; p(\boldsymbol{\mu}|\mathcal{I}).$$

The evidence basically indicates how likely the measured data are for a given problem, specified by the background information $\mathcal{I}$, regardless of the model parameters. For example, if $\mathcal{I}$ tells us that the data stem from optical experiments in the visible regime, then the evidence will be restricted to wavelengths in the range of 390–680 nm. The evidence drops rapidly to zero for data values outside this interval. In statistical physics, the analogous integral is the partition function. As a matter of fact, the evidence integral can be considered as the prior mean of the likelihood. For notational simplicity we abbreviate the integral by

$$\mathcal{E} = \int d^E\mu \; g(\boldsymbol{\mu}). \tag{29.2}$$

Moreover, we start out with the assumption that $g(\boldsymbol{\mu})$ is unimodal in $\boldsymbol{\mu}$, with $\boldsymbol{\mu}_0$ being the mode. A widespread approximation in the field of statistical inference is the 'saddle-point approximation' which is also referred to as 'Gaussian approximation' and is introduced in Appendix A.2 [p. 598]. Taylor expansion of the logarithm of $g(\boldsymbol{\mu})$ around $\boldsymbol{\mu}_0$ up to second order yields

$$g(\boldsymbol{\mu}) \approx g(\boldsymbol{\mu}_0) \; \exp\left\{-\frac{1}{2}(\boldsymbol{\mu} - \boldsymbol{\mu}_0)^T \mathbf{H}(\boldsymbol{\mu} - \boldsymbol{\mu}_0)\right\}, \tag{29.3}$$

where $\mathbf{H}$ is the Hessian evaluated at $\boldsymbol{\mu}_0$. In this approximation, assuming infinite integration limits to be justified, the evidence is given by

$$\mathcal{E} = \int d^E\mu \; g(\boldsymbol{\mu}) = g(\boldsymbol{\mu}_0)(2\pi)^{E/2} \; |\mathbf{H}|^{-1/2}. \tag{29.4}$$

For later discussion it is worthwhile recalling that the covariance matrix of the multivariate Gaussian PDF is $\mathbf{C} = \mathbf{H}^{-1}$, i.e. the evidence integral is proportional to $\sqrt{|\mathbf{C}|} = \prod \tilde{\sigma}_i$, where the variances $\tilde{\sigma}_i^2$ are the eigenvalues of the covariance matrix. As mentioned before, the Gaussian approximation, as widespread as it may be, is based on two crucial assumptions: (a) the integration limits can be extended to infinity and (b) the shape of the integrand can be approximated with sufficient accuracy by a multivariate normal PDF. If one of these assumptions does not hold, the results cannot be trusted and one should better

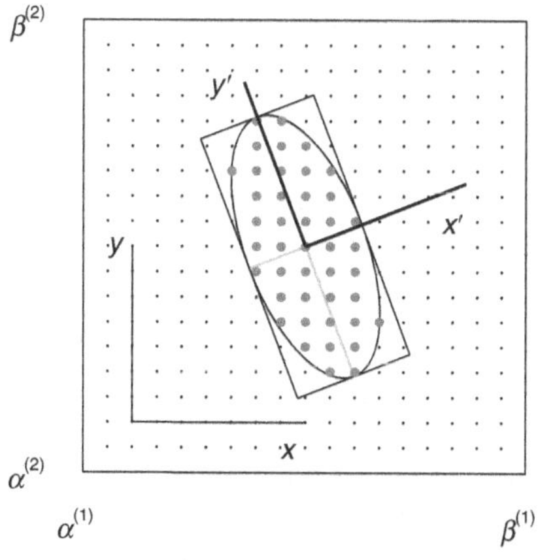

Figure 29.1 Contour plot of the significant part of a two-dimensional Gaussian function.

invoke numerical integration techniques. For these, we have to introduce finite integration intervals $I_i := (\alpha_i, \beta_i)$ in parameter space:

$$\mathcal{E} = \int_{\alpha_1}^{\beta_1} d\mu_1 \ldots \int_{\alpha_E}^{\beta_E} d\mu_E \; g(\boldsymbol{\mu}). \tag{29.5}$$

The intervals follow either from prior restrictions (for example, positivity) or are ad-hoc limitations, chosen such that accurate numerical results are obtained. To illustrate the situation, we start out with a representative two-dimensional example as depicted in Figure 29.1.

The upright rectangle defines the support area defined by the prior or ad-hoc restrictions. The common-sense approach would fill in a grid of suitable pivot points. For simplicity and given the lack of better knowledge, we use the same number of pivot points in both directions (17 in this example). In general, a numerical approximation to the evidence is then

$$g(\boldsymbol{\mu}) \approx h^2 \sum_{i,k=1}^{p} g(\boldsymbol{\mu} = \mu_{ik}), \tag{29.6}$$

where h is the distance between neighbouring pivot points. The computational effort involved in equation (29.6) consists of $N = p^2$ evaluations of f. In the example of Figure 29.1 the number of pivot points is $N = 17^2$. Now assume that the ellipse in Figure 29.1 was chosen such that f can safely be neglected outside the ellipse. In the Gaussian approximation this region can be defined by $\Delta\boldsymbol{\mu}^T \mathbf{H} \Delta\boldsymbol{\mu} = \text{const.}$, with the constant being chosen such that the integrand is negligibly small. We see immediately that most of the evaluations of f take place at coordinates where the function contributes negligibly to the sum in equation (29.6), which is given to sufficient precision by extending it only over the functional values of f at the solid dots inside or on the boundary of the ellipse. By simple counting we find that only 15% 'intelligent' function evaluations of the brute-force lot are necessary. The point here is that the ellipses are in general arbitrarily oriented and stretched with respect to the problem, dictated by the $\boldsymbol{\mu}$-coordinate system. In order to obtain the evidence more efficiently, by numerical integration we would be inclined to choose an adjusted support area. Introduce a reference frame centred at the mode μ_0 with axes given by the eigenvectors of the covariance matrix or the Hessian, respectively. The edge size of the support area along the new axes would be some constant times the standard deviation $\tilde{\sigma}_i$. The resulting support area is depicted as a tilted rectangle in Figure 29.1. The number of brute-force integrand evaluations is then proportional to the area of the rotated rectangle, which is (2 const. $\tilde{\sigma}_1$) (2 const. $\tilde{\sigma}_2$). This has to be compared with the area of the ellipse, which is (const. $\tilde{\sigma}_1$) (const. $\tilde{\sigma}_2$) π, and we see that now the fraction of 'intelligent' integrand evaluations is $\pi/4 = 78.5\%$. So far so good, but Figure 29.1 makes another problem obvious. With the initially assumed grid, the number of pivot points in the x'-direction differs significantly from that in the y'-direction. The two should, however, be the same, since the variation of the function is by construction of the ellipse the same in both directions. The consequence is an unnecessarily large number of pivot points for a given accuracy. Say we need L pivot points along the principal axis of $\mathbf{C}$. The spacing of the pivot points along the axis corresponding to $\tilde{\sigma}_{\text{min}}$ is $\tilde{\sigma}_{\text{min}}/L$. So, in order to achieve the desired accuracy, the number of pivot points within the ellipse is

$$N = \frac{\tilde{\sigma}_1 \tilde{\sigma}_2 \pi}{(\tilde{\sigma}_{\text{min}}/L)^2} = L^2 \, \pi \, \frac{\tilde{\sigma}_{\text{max}}}{\tilde{\sigma}_{\text{min}}}.$$

The problem can be solved by an additional stretching transformation. The result would be a circle, which has of course no longer a preferred direction. The two transformations which we have just discussed by reference to Figure 29.1 can be obtained in a single step by choosing in the Gaussian PDF

$$\Delta\boldsymbol{\mu}^T \mathbf{H} \Delta\boldsymbol{\mu} = \boldsymbol{z}^T \boldsymbol{z}. \tag{29.7}$$

Since $\mathbf{H}$ is a real symmetric positive definite matrix, the square root exists and also $\mathbf{H}^{1/2} = (\mathbf{H}^{1/2})^T$. Hence

$$\mathbf{H}^{1/2} \, \Delta\boldsymbol{\mu} = \boldsymbol{z}, \qquad \boldsymbol{\mu} = \boldsymbol{\mu}_0 + \mathbf{H}^{-1/2} \boldsymbol{z}, \tag{29.8}$$

and a uniform grid for the components of $\boldsymbol{z}$ is transformed to the appropriately oriented and compressed grid for $\boldsymbol{\mu} = (x, y)$. In a more general unimodal PDF we would replace the Hessian by the inverse of the covariance matrix $\mathbf{H} \to \mathbf{C}^{-1}$, i.e. $\boldsymbol{\mu} = \boldsymbol{\mu}_0 + \mathbf{C}^{1/2} \boldsymbol{z}$.

The square root of $\mathbf{H}$ (and similarly for $\mathbf{C}$) can be obtained by the previously discussed singular-value decomposition in equation (21.47) [p. 346]. For a real symmetric matrix this reads

$$\mathbf{H}^{-1/2} = \mathbf{V} \mathbf{D}^{-1/2} \mathbf{V}^T. \tag{29.9}$$

Note that now in the transformed system $\boldsymbol{z}$ the fraction of intelligent function evaluations is equal to the volume of the 2-D sphere of radius R divided by the volume of the 2-D cube with side length $2R$. This is still equal to $\pi/4$, as it was prior to the stretching transformation, but for a given accuracy the total number of pivot points has been reduced. For the desired accuracy the spacing between the pivot points in the z representation is now given by $1/L$, hence the number of points within the relevant unit circle is

$$N = \frac{\pi}{(1/L)^2} = \pi L^2.$$

Compared with the result prior to the stretching transformation, this amounts to a reduction by $\tilde{\sigma}_{\min}/\tilde{\sigma}_{\max}$. This is in most cases a significant improvement. Along with the yield fraction of $\pi/4 = 78.5\%$, we could be satisfied and call it a day.

But, here comes part I of the dimension fallacy. Consider dimension three. The volume of the sphere V_s is now $4\pi R^3/3$ and that of a cube $V_c = (2R)^3$. The ratio $V_s/V_c = \pi/6$ has decreased to 0.52. This raises the question of whether even higher dimensions have progressively less favourable ratios. This is indeed the case. Consider an E-dimensional sphere. Its volume is given by

$$V_E^s(R) = \pi^{E/2} R^E / \Gamma\left(\frac{E}{2} + 1\right), \tag{29.10}$$

while the volume of the corresponding cube is

$$V_E^c(R) = 2^E R^E. \tag{29.11}$$

We want to restrict the following discussion to the case where the cutoff length R is chosen such that the integrand outside the hypersphere can safely be ignored. The yield ratio q_E of intelligent pivot points is then

$$q_E = \frac{\pi^{E/2}}{2^E \Gamma\left(\frac{E+2}{2}\right)}. \tag{29.12}$$

For $E = 8$, for example, we find $q_8 = 0.016$ and see that equal subdivision in all components of z leads to a drastic loss of efficiency, since most evaluations of the integrand take place for arguments where it has decreased to negligible values.

For the sake of clarity, we will invoke Stirling's formula as discussed in Section 6.1 [p. 83]. Note also the discussion on the accuracy. Here we are interested merely in relative accuracy, therefore

$$\Gamma\left(\frac{E+2}{2}\right) \approx \sqrt{2\pi}\, e^{-\frac{E}{2}} \left(\frac{E}{2}\right)^{\frac{E+1}{2}} = \sqrt{\pi\, E} \left(\frac{E}{2e}\right)^{\frac{E}{2}}. \tag{29.13}$$

Collecting terms we obtain, for the number of integrand evaluations $Z_{p,E}$ for an E-dimensional problem with p pivot points in one direction,

$$Z_{p,E} = p^E\, q_E \approx \frac{1}{\sqrt{\pi E}} \left(\frac{\pi\, e\, p^2}{2\, E}\right)^{\frac{E}{2}}. \tag{29.14}$$

The cure to get rid of the unintelligent pivot points is very simple. Choose a vector z on the E-dimensional equidistantly chosen grid in z-space and accept it only if $z^T z$ is

Table 29.1 Logarithm of the integrand evaluations $\log_{10}(Z_{p,E})$ as function of pivot number p and dimension E for the optimized deterministic numerical integration (OI), or by the brute-force approach (BF)

	7		9		11		13		15		17	
$E\backslash p$	OI	BF	OI	BF	OI	BF	OI	BF	OI	BF	OI	BF
3	2.3	2.5	2.6	2.9	2.9	3.1	3.1	3.3	3.3	3.5	3.4	3.7
6	4.0	5.1	4.6	5.7	5.2	6.2	5.6	6.7	6.0	7.1	6.3	7.4
9	5.4	7.6	6.4	8.6	7.2	9.4	7.8	10.0	8.4	10.6	8.9	11.1
12	6.7	10.1	8.0	11.5	9.0	12.5	9.9	13.4	10.6	14.1	11.3	14.8
15	7.7	12.7	9.4	14.3	10.7	15.6	11.8	16.7	12.7	17.6	13.5	18.5

smaller than some predefined constant R^2_{max}. This limits the evaluation of the posterior to the 'intelligent' volume. Given a number p of pivot points along one component of z, we then obtain a computational effort of $p^E q_E$ which is much less than the brute-force number of p^E which rises exponentially. Table 29.1 shows the **logarithm** of the number of integrand evaluations as a function of dimension E and pivot variable p. Note that the maximum number of pivots along each axis is p. The table shows clearly that the computational effort eventually becomes prohibitive for $pE \gtrsim 20$. In other words, since p is dictated by the desired accuracy, the admissible dimensions are limited by $a \approx 20/p$. Though the range of application for deterministic numerical integration can be extended to much larger sizes than commonly implied for certain integrands by the actions described above, other methods must be considered for problems with large dimensions.

There is, however, an interesting feature in equation (29.14). For large values of E, specifically for $E > 2\pi e N^2$, the increasing behaviour turns into a strongly decreasing

$$Z_{p,E} = E^{-E/2} \tag{29.15}$$

dependence on the dimension. But that does not necessarily mean that eventually one function call suffices to evaluate an infinite-dimensional integral. There are two additional effects coming into play. Firstly, the decline in accuracy with increasing dimension and secondly, the increasing cutoff R necessary to cover the important volume for the integral.

The latter point can best be illustrated if we consider a spherical symmetric Gaussian PDF. In E dimensions, the evidence integral reads

$$\mathcal{E} \propto \int d^E\mu \, e^{-\frac{\mu^2}{2\sigma^2}} \propto \int dr \, r^{E-1} e^{-\frac{r^2}{2\sigma^2}}.$$

The radial part of the integrand has a maximum at $r^* = \sigma\sqrt{E-1}$. The first point is a bit more subtle and shall be discussed next. To this end, we will analyse the question of whether the spacing h of the grid depends on the dimension E of the problem. The answer certainly depends on the detailed structure of the integrand. The most reasonable assumption for this question is that the integrand has the same structure in all spatial directions, or rather if we assume a tensor product form:

$$g(\boldsymbol{\mu}) = \prod_{i=1}^{E} g_1(\mu_i),$$

which means increasing the dimension adds additional identical factors. The multivariate normal PDF is an example, with

$$g_1(\mu_i) = e^{-\mu_i^2/2}.$$

We integrate numerically over the hypercube with edges ranging from $-r_c$ to r_c with tensor product pivot points resulting in the approximate integral

$$\tilde{\mathcal{E}}_E = h^E \sum_{i_1 \ldots i_E} g(\mu_{i_1}, \ldots, \mu_{i_E}) = \left(h \sum_{i=1} g(\mu_i) \right)^E := \tilde{\mathcal{E}}_1^E.$$

Let the exact answer be $\mathcal{E}_E = \mathcal{E}_1^E$ (in the Gaussian example, $\mathcal{E}_1 = \sqrt{2\pi}$). The relative error is given by

$$\varepsilon_E := \left| \frac{\tilde{\mathcal{E}}_1^E - \mathcal{E}_1^E}{\mathcal{E}_1^E} \right| = \left| \left(\frac{\tilde{\mathcal{E}}_1}{\mathcal{E}_1} \right)^E - 1 \right| = \left(1 + \varepsilon_1 \right)^E - 1 = e^{E \ln(1+\varepsilon_1)} - 1. \tag{29.16}$$

For very small relative error $\varepsilon_1 \ll 1$ and moderate E such that $E\varepsilon_1 \ll 1$, we have $\varepsilon_E \approx e^{E\varepsilon_1} - 1 \approx E\varepsilon_1$, i.e. the relative error increases linearly with the dimension E. For sufficiently large E or for $\varepsilon_1 \not\ll 1$, respectively, we obtain, however, an exponential increase according to (29.16). In order to ensure a fixed relative error $\varepsilon_E = \varepsilon \ll 1$, (29.16) yields the condition

$$\varepsilon_1 \approx \frac{\varepsilon}{E}.$$

Hence, for a numerical integration scheme of order $\varepsilon_1 = O(h^2)$, the spacing scales like $h \propto \frac{1}{\sqrt{E}}$ and the number of pivot points per spatial dimension has to increase like $\sqrt{E}$. Consequently, the counterintuitive behaviour discussed in (29.15) vanishes if we invoke the property $p \propto \sqrt{E}$ in (29.14). What is left is an exponential increase with dimension E.

29.2 Monte Carlo integration

The bottom line of the previous discussion is that in most realistic cases deterministic numerical approaches are highly inefficient and we have to resort to stochastic approaches. Here we will give a first heuristic introduction to the stochastic evaluation of high-dimensional integrals and postpone the more thorough mathematical treatment to Chapter 30 [p. 537].

29.2.1 Simple sampling

The integral in equation (29.2) can also be written in the form

$$\langle f \rangle = \frac{1}{V_I} \int_{V_I} d^E\mu \; f(\boldsymbol{\mu}), \tag{29.17}$$

with $f(\boldsymbol{\mu}) = V_I \; g(\boldsymbol{\mu})$. So far, we have approximated the mean value $\langle f \rangle$ by summation over a regular grid:

$$\langle f \rangle \approx \frac{1}{N} \sum_{i,k=1}^{p} f(\boldsymbol{\mu}_{ik}), \tag{29.18}$$

where $N = p^2$ is the total number of integration points. In one dimension we know that – if the endpoints are assumed to be zero – this scheme amounts to the well-known trapezoidal rule, which over several equidistant intervals with step width h has an integration error proportional to h^2. It is not at all simple to estimate the precise value of the approximation error in realistic many-dimensional problems. The regular sampling of the mean value in equation (29.18) is by no means the only possible way. In view of equation (29.17), an equally acceptable procedure would approximate $\langle f \rangle$ by a sample mean

$$\langle f \rangle \approx \overline{f} := \frac{1}{N} \sum_{i,k=1}^{p} f(x_i, y_k), \tag{29.19}$$

where the two components of the sampling points $\boldsymbol{\mu}_{ik} = (x_i, y_k)^T$ are chosen from a uniform random number generator $U(0, 1)$ via

$$\begin{aligned} x_i &= \alpha_1 + (\beta_1 - \alpha_1)U(0, 1), \\ y_k &= \alpha_2 + (\beta_2 - \alpha_2)U(0, 1), \\ &\text{accept if} \quad x_i^2 + y_k^2 \leq R_{\max}^2. \end{aligned} \tag{29.20}$$

This is a simple version of the *rejection method*, due to John von Neumann [213]. In the last step we have accounted for the fact that the integrand is negligible outside a circle of radius R in parameter space. The advantage of the stochastic approach is that we can immediately give a numerical error estimate for the integral based on the central limit theorem equation (8.12) [p. 145]:

$$\langle (\Delta \overline{f})^2 \rangle = \frac{\sigma_f^2}{N}. \tag{29.21}$$

The variance σ_f^2 of $f(\boldsymbol{\mu})$ can be estimated by the sample variance v_f^2:

$$v_f^2 = \frac{1}{N-1} \sum_{i,k} \left(f(\boldsymbol{\mu}_{ik}) - \overline{f} \right)^2,$$

and we see that the precision (standard error) of the estimated integral is proportional to $1/\sqrt{N}$. Precision is doubled with a fourfold increase of samples. This is a poor performance in one dimension. However, equation (29.21) holds **independent** of the dimension E of the integral. If applied carefully, Monte Carlo integration can be constructed such that the variance σ_f^2 depends only weakly on the dimension.

So far we have assumed that the time-consuming part in simple sampling is the evaluation of the posterior density. We have not yet paid attention to the random number generation, which becomes ever more important as the dimension E increases. Consider an intermediate $E = 15$. The fraction q_E of samples generated in a hypercube of dimension 15 falling into the largest hypersphere contained entirely in the hypercube is, following equation (29.12), $q_{15} \approx 10^{-5}$, meaning that creating random numbers on the hypercube and ignoring those outside the hypersphere becomes terribly inefficient. It is therefore of interest to pursue the direct approach to generating such random vectors. Needless to say, this algorithm to generate random numbers uniformly distributed over the volume of a hypersphere of radius R outperforms the rejection method by far and is essential for employing simple sampling already at moderate dimensions.

29.2.2 Random number generators

The integral in equation (29.17) is a special case of the general type of integrals we are interested in, namely

$$\langle f \rangle := \int d^E\mu \; f(\boldsymbol{\mu}) \; p(\boldsymbol{\mu}).$$

For Monte Carlo simulations in general, we have to be able to generate random numbers according to a given PDF $p(\boldsymbol{\mu})$. We will start out with some simple yet relevant situations.

Inverse transform sampling

A fairly general procedure to generate scalar random numbers x according to a given PDF $p_x(x)$ is by generating random numbers y for another simpler PDF $p_y(y)$, for which a random number generator is known, and to do some more or less complex manipulation with these numbers. In other words, we have the functional relation

$$x = G(y),$$

and we know how to draw samples from $p_y(y)$. We assume that $G(y)$ is a monotonically increasing function. Then, according to equation (7.51) [p. 122], the transformation reads

$$p_y(y) = p_x(x) \left| \frac{dx}{dy} \right| = p_x(G(y)) \frac{dG(y)}{dy}.$$

The sought-for PDF $p_x(x)$ and the input PDF $p_y(y)$ can be expressed in terms of the corresponding CDFs F_x and F_y, respectively, resulting in

$$\frac{dF_y}{dy} \stackrel{!}{=} \left.\frac{dF_x}{dx}\right|_{x=G(y)} \frac{dG}{dy} = \frac{dF_x(G(y))}{dy}.$$

The solution is $F_y(y) = F_x(G(y))$. There is no additional constant, since both CDFs have to vary between 0 and 1. Clearly, the sought-for function $G(y)$ is

$$G(y) = F_x^{-1}(F_y(y)),$$

provided the CDF F_x has an inverse. Cumulative distribution functions are integrals over non-negative PDFs. Consequently, they are in general monotonically increasing functions. They are not necessarily strictly increasing. Here we restrict the discussion to strictly increasing functions. In this case, F^{-1} exists and is itself strictly increasing. The most elementary input PDF is the standard uniform distribution $p(y) = \text{const.}$ on the interval $(0, 1]$, for which $F_y(y) = y$, and we obtain the familiar results:

UNIVARIATE RANDOM NUMBER GENERATOR FOR $p_x(x)$

- let $F_x(x)$ be the CDF of $p_x(x)$,
- draw a random number y from the uniform PDF $U(0, 1)$,
- calculate $x = F_x^{-1}(y)$.

This procedure shall be illustrated in terms of a few examples.

Power law PDF

As a first example we consider $p_x(x) = \frac{E\,x^{E-1}}{R^E}$. The corresponding CDF reads

$$F_x(x) = \left(\frac{x}{R}\right)^E,$$

the inverse of which yields

$$F_x^{-1}(y) = R\,y^{1/E}.$$

Exponential distribution

Another important example are exponential random numbers

$$p_x(x) = \lambda\,e^{-\lambda x}, \tag{29.22}$$

with CDF

$$F_x(x) = 1 - e^{-\lambda x}.$$

Apparently, the inverse function is

$$F_x^{-1}(u) = -\frac{1}{\lambda} \ln(1-u). \tag{29.23}$$

Hyperbolic Cauchy distribution

Next we compute the transformation needed to draw random numbers from the hyperbolic Cauchy distribution

$$p_x(x) = \frac{2}{a} \frac{1}{\left(e^{x/a} + e^{-x/a}\right)^2}. \tag{29.24}$$

The corresponding CDF reads

$$F_x(x) = \frac{1}{2}\left(1 + \tanh\left(\frac{x}{a}\right)\right),$$

and the inverse function is readily obtained as

$$F_x^{-1}(u) = \frac{a}{2} \ln\left(\frac{u}{1-u}\right). \tag{29.25}$$

Cauchy distribution

Finally, we compute the transformation needed to draw random numbers from the Cauchy distribution:

$$p_x(x) = \frac{\gamma}{\pi} \frac{1}{z^2 + \gamma^2}. \tag{29.26}$$

The corresponding CDF is

$$F_x(x) = \frac{1}{2} + \frac{\arctan\left(\frac{x}{\gamma}\right)}{\pi}.$$

The inverse transform reads

$$F_x^{-1}(u) = \frac{\gamma}{\tan(u\pi)}.$$

Note that random number generators for the standard uniform distribution $U(0,1)$ usually include either the value 0 or 1. These values must be skipped in all aforementioned cases in order to avoid numerical singularities.

Uniform random numbers on a hypersphere

In some applications random vectors $\boldsymbol{x} = (x_1, \ldots, x_E)$ are required which are uniformly distributed on the surface of the E-dimensional hypersphere ($\boldsymbol{x}^2 = R^2$). The first intuitive guess would be to generate random numbers uniformly distributed in a hypercube, i.e $y_i = 2U(0,1) - 1$, and to enforce the desired length by $x_i = y_i \; R/\|\boldsymbol{y}\|$. But due to the spatial structure of the hypercube, the so-generated PDF is not rotational invariant, as can be seen in what follows. The resulting PDF $p(\boldsymbol{x}|\mathcal{I})$ is obtained upon introducing the sample of uniform random numbers y_i by means of the marginalization rule

$$\begin{aligned} p(\boldsymbol{x}|\mathcal{I}) &= \int d^E y\; p(\boldsymbol{x}|\boldsymbol{y},\mathcal{I}) p_y(\boldsymbol{y}) \\ &= \int d^E y \prod_i \delta\Big(x_i - y_i \frac{R}{\|\boldsymbol{y}\|}\Big) p_y(\boldsymbol{y}). \end{aligned}$$

The evaluation is simplified if we introduce an auxiliary variable s by the dummy integral $\int_0^\infty ds\; \delta(s - \|\boldsymbol{y}\|/R)$, resulting in

$$\begin{aligned} p(\boldsymbol{x}|\mathcal{I}) &= \int_0^\infty ds \int d^E y \prod_i \delta\Big(x_i - y_i \frac{1}{s}\Big) p_y(\boldsymbol{y})\; \delta(s - \|\boldsymbol{y}\|/R) \\ &= \int_0^\infty ds\; s^E\; p_y(s\boldsymbol{x})\; \delta(\frac{s}{R}(R - \|\boldsymbol{x}\|)) \\ &= R \int_0^\infty dt\; t^E p_y(t\boldsymbol{x})\; \delta(t(R - \|\boldsymbol{x}\|)). \end{aligned}$$

The delta function carries over to a delta function $\delta(\|\boldsymbol{x}\| - R)$. For the contribution of $\|\boldsymbol{x}\| = R$ we introduce $\boldsymbol{x} = r\hat{\boldsymbol{x}}$, with $\|\hat{\boldsymbol{x}}\| = 1$, and we integrate over r in the infinitesimal vicinity of $r = R$:

$$\begin{aligned} \int_{R-\varepsilon}^{R+\varepsilon} p_y(r\hat{\boldsymbol{x}})\; dr &= R \int_0^\infty dt\; t^E p_y(tR\hat{\boldsymbol{x}}) \int_{R-\varepsilon}^{R+\varepsilon} \delta(t(R - r))\; dr \\ &= R \int_0^\infty dt\; t^{E-1} p_y(tR\hat{\boldsymbol{x}}). \end{aligned}$$

Hence, the functional dependence of $p_y(r\hat{\boldsymbol{x}})$ on $\|\boldsymbol{x}\|$ is a delta function and the final result reads

$$\begin{aligned} p(\boldsymbol{x}|\mathcal{I}) &= R\; \delta(R - \|\boldsymbol{x}\|) \int_0^\infty dt\; t^{E-1} p(t\boldsymbol{x}|\mathcal{I}) \qquad (29.27) \\ &= R\; \delta(R - \|\boldsymbol{x}\|) \int_0^\infty dt\; t^{E-1} \prod_i \frac{1}{2}\, \mathbf{1}(-1 \le t x_i \le 1) \\ &= R\; \delta(R - \|\boldsymbol{x}\|)\; 2^{-E} \int_0^\infty dt\; t^{E-1} \prod_i \mathbf{1}\Big(-\frac{1}{x_i} \le t \le \frac{1}{x_i}\Big) \end{aligned}$$

$$= R\,\delta(R - \|\boldsymbol{x}\|)\,2^{-E} \int\limits_0^{\frac{1}{\max_i(x_i)}} dt\; t^{E-1}$$

$$= R\,\delta(R - \|\boldsymbol{x}\|)\,2^{-E}\,\frac{\left(\max_i(x_i)\right)^{-E}}{E}.$$

In order to illustrate the behaviour, we compare the outcome for a vector $\boldsymbol{x}$ of unit length in one case (a) pointing in one of the coordinate axes, i.e. $x_i = \delta_{i,i_0}$ and in the other case (b) with equal contributions in all coordinates, i.e. $x_i = 1/\sqrt{E}$. In these two cases we have:

$$\text{(a)}\quad \left(\max_i x_i\right)^{-E} = 1,$$

$$\text{(b)}\quad \left(\max_i x_i\right)^{-E} = E^{E/2},$$

which reflects the significantly higher probability mass contributing if $\boldsymbol{x}$ points into the corners of the hypercube compared with the case if the vector is perpendicular to one of the faces of the hypercube.

The rotational symmetry can be restored by using a Gaussian PDF for the individual random variables y_i, since the tensor product only depends on $\|\boldsymbol{y}\|$. Inserting the Gaussian PDF $p_y(y_i|\mathcal{I}) = \frac{1}{\sqrt{\pi}}e^{-\frac{1}{2}y_i^2}$ in (29.27), we obtain

$$p(\boldsymbol{x}|\mathcal{I}) = R\,\delta(R - \|\boldsymbol{x}\|) \int\limits_0^\infty dt\; t^{E-1} p_y(t\boldsymbol{x}|\mathcal{I})$$

$$= R\,\delta(R - \|\boldsymbol{x}\|) \int\limits_0^\infty dt\; t^{E-1} (2\pi)^{-E/2} e^{-\frac{t^2\|\boldsymbol{x}\|^2}{2}}$$

$$= R\,\delta(R - \|\boldsymbol{x}\|)\,(2\pi)^{-E/2} \int\limits_0^\infty dt\; t^{E-1} e^{-\frac{t^2R^2}{2}}$$

$$= R^{-(E-1)}\,\delta(R - \|\boldsymbol{x}\|)\,(2\pi)^{-E/2} \int\limits_0^\infty dt\; t^{E-1} e^{-\frac{t^2}{2}}$$

$$= \frac{1}{S(R)}\,\delta(R - \|\boldsymbol{x}\|),$$

$$S(R) := R^{E-1}\frac{2\pi^{E/2}}{\Gamma\left(\frac{E}{2}\right)}.$$

The normalization is correctly given by the surface area of the hypersphere $S(R)$ of radius R. Summarizing, the algorithm to produce E-dimensional random numbers, uniformly distributed over a hypersphere of radius R, is

Uniform random numbers on the surface of a hypersphere of radius R

- Generate a vector $\boldsymbol{y}$, whose components y_i $(i = 1, \dots, E)$ are drawn from a normal PDF with zero mean and unit variance.
- Normalize appropriately by $\boldsymbol{x} = \boldsymbol{y}\,\frac{R}{\|\boldsymbol{y}\|}$.

Uniform random numbers inside a hypersphere

Along the lines discussed before, the obvious algorithm to generate random numbers uniformly spread within a hypersphere of radius R is

Uniform random numbers inside a hypersphere of radius R

- Generate a vector $\boldsymbol{y}$, whose components y_i $(i = 1, \dots, E)$ are drawn from a normal PDF with zero mean and unit variance.
- Draw a random number r from a suitable PDF $p_r(r|R)$.
- Normalize accordingly $\boldsymbol{x} = \boldsymbol{y}\,\frac{r}{\|\boldsymbol{y}\|}$.

The spherical symmetry is guaranteed, as discussed before, and the radial part has to be accounted for by the proper choice of $p_r(r)$. The resulting PDF is obtained by introducing the random numbers $\boldsymbol{y}$ and r of the algorithm through the marginalization rule

$$
\begin{aligned}
p(\boldsymbol{x}|\mathcal{I}) &= \int d^E y \int_0^\infty dr\; p(\boldsymbol{x}|\boldsymbol{y}, r, \mathcal{I})\; p(\boldsymbol{y}|\mathcal{I})\; p_r(r) \\
&= (2\pi)^{-E/2} \int d^E y\; e^{-\frac{\|\boldsymbol{y}\|^2}{2}} \int_0^\infty dr\; p_r(r) \prod_i \delta\left(x_i - y_i \frac{r}{\|\boldsymbol{y}\|}\right).
\end{aligned}
$$

Next we substitute $r = s\|\boldsymbol{y}\|$ and obtain

$$
\begin{aligned}
p(\boldsymbol{x}|\mathcal{I}) &= (2\pi)^{-E/2} \int d^E y\; e^{-\frac{\|\boldsymbol{y}\|^2}{2}} \int_0^\infty ds\; \|\boldsymbol{y}\|\; p_r(s\|\boldsymbol{y}\|) \prod_i \delta(x_i - y_i s) \\
&= (2\pi)^{-E/2} \int_0^\infty ds \int d^E y\; e^{-\frac{\|\boldsymbol{y}\|^2}{2}}\; \|\boldsymbol{y}\|\; p_r(s\|\boldsymbol{y}\|) \prod_i \frac{\delta(y_i - \frac{x_i}{s})}{s} \\
&= (2\pi)^{-E/2}\; \|\boldsymbol{x}\|\; p_r(\|\boldsymbol{x}\|) \int_0^\infty ds\; e^{-\frac{\|\boldsymbol{x}\|^2}{2s^2}}\; s^{-(E+1)}.
\end{aligned}
$$

This integral has been computed in equation (7.27) [p. 108] and yields

$$p(\boldsymbol{x}|\mathcal{I}) = \pi^{-E/2} \frac{\|\boldsymbol{x}\|^{-(E-1)}}{2} \Gamma\left(\frac{E}{2}\right) p_r\left(\|\boldsymbol{x}\|\right).$$

In order to obtain a uniform $p(\boldsymbol{x}|\mathcal{I})$ within a hypersphere of radius R, we need the PDF $p_r(r) \propto r^{E-1}\, \boldsymbol{1}(r \leq R)$. Along with the normalization, the correct radial PDF is

RADIAL PDF FOR UNIFORM RANDOM NUMBERS IN A HYPERSPHERE

$$p_r(r) = \frac{E}{R^E}\ \boldsymbol{1}(r \leq R)\ \ r^{E-1}.$$

The radial random number r can be generated via inverse transform sampling, as outlined before. The resulting PDF $p(\boldsymbol{x}|\mathcal{I})$ has the correct normalization

$$p(\boldsymbol{x}|\mathcal{I}) = \boldsymbol{1}(r \leq R)\ \pi^{-E/2}\, \Gamma\left(\frac{E}{2}+1\right) \frac{1}{R^E} \overset{(29.10)}{=} \frac{1}{V_E^s(R)}\ \boldsymbol{1}(\boldsymbol{x} \leq R).$$

29.2.3 Importance sampling

We shall now investigate the efficiency of simple sampling as considered in Section 29.2.1 [p. 516]. Remember that we limit our considerations in this chapter to posteriors, which allow for an approximate representation by the multivariate Gaussian in equation (29.3) [p. 510]. The evidence is then approximated by

$$\mathcal{E} = \langle f \rangle = \int f(\boldsymbol{\mu})\, p(\boldsymbol{\mu}) \approx f(\boldsymbol{\mu}_0) \int d^E\mu\ \exp\left\{-\frac{1}{2}\Delta\boldsymbol{\mu}^T \mathbf{H} \Delta\boldsymbol{\mu}\right\}\ p(\boldsymbol{\mu}),$$

where $p(\boldsymbol{\mu})$ is a uniform PDF in the hypersphere of volume V_I. The uncertainty of simple sampling is, according to equation (29.21) [p. 516], given by $\sigma_f/\sqrt{N}$. In the case of a Gaussian integrand, the variance can be determined easily:

$$\begin{aligned}
\sigma_f^2 &= \langle f^2 \rangle - \langle f \rangle^2, \\
\langle f^n \rangle &= \int d^E\mu\, f^n(\boldsymbol{\mu}) \\
&\approx \Big(f(\boldsymbol{\mu}_0)\Big)^n \int d^E\mu\ \exp\left\{-\frac{n}{2}\Delta\boldsymbol{\mu}^T \mathbf{H} \Delta\boldsymbol{\mu}\right\}\ p(\boldsymbol{\mu}) \\
&:= \Big(f(\boldsymbol{\mu}_0)\Big)^n \frac{1}{V_I} \int_{V_I} d^E\mu\ \exp\left\{-\frac{n}{2}\Delta\boldsymbol{\mu}^T \mathbf{H} \Delta\boldsymbol{\mu}\right\}.
\end{aligned}$$

As before, we assume that the volume V_I covers the essential region of the integrand and we can safely extend the integration limit to infinity. Then

$$\begin{aligned}\langle f^n\rangle &\approx \Big(f(\boldsymbol{\mu}_0)\Big)^n \frac{1}{V_I}\int d^E\mu\ \exp\Big\{-\frac{n}{2}\Delta\boldsymbol{\mu}^T\mathbf{H}\Delta\boldsymbol{\mu}\Big\}\\ &= \Big(f(\boldsymbol{\mu}_0)\Big)^n \frac{1}{V_I}\left|\det(\mathbf{H}^{-1/2})\right|\Big(2\pi\Big)^{E/2} n^{-E/2}\\ &= \Big(f(\boldsymbol{\mu}_0)\Big)^n \frac{\langle f\rangle}{f(\boldsymbol{\mu}_0)} n^{-E/2}.\end{aligned}$$

The variance then reads

$$\sigma_f^2 = \langle f\rangle^2\left[\frac{f(\boldsymbol{\mu}_0)}{2^{E/2}\langle f\rangle} - 1\right].$$

This expression has a simple geometrical meaning. We consider

$$\frac{\langle f\rangle}{f(\boldsymbol{\mu}_0)} = \frac{1}{V_I}\int_{V_I} d^E\mu\ \exp\Big\{-\frac{n}{2}\Delta\boldsymbol{\mu}^T\mathbf{H}\Delta\boldsymbol{\mu}\Big\} := \frac{1}{V_I}\int_{V^*} d^E\mu\ 1 = \frac{V^*}{V_I}.$$

Here V^* is a volume over which the integral has to be extended if we replace the integrand by a constant. Owing to the tensorial structure of the problem we have $V_*/V_I = \Big(L_*/L_I\Big)^E$, where L_I is the linear size covered by V_I and L_* is related to the width of the Gaussian. So we find for the variance

$$\sigma_f^2 \approx \langle f\rangle^2\left[\left(\frac{L_I}{\sqrt{2}L_*}\right)^E - 1\right].$$

The ratio $\frac{L_I}{\sqrt{2}L_*}$ will clearly be greater than one and we observe that the relative uncertainty $\sigma_f/\langle f\rangle$ increases exponentially with increasing dimension E, which is the **second dimension fallacy**. In connection with equation (29.21) [p. 516], we have claimed that MC integration can be constructed such that the variance depends only weakly on the dimension. This obviously does not apply to simple sampling.

Next we will study the origin of the dimensional deterioration in more detail and present 'importance sampling' as a way to reduce the variance of the sample mean. To this end we transform the evidence integral based on equation (29.8) [p. 512]:

$$\mathcal{E} \approx f(\boldsymbol{\mu}_0)\ |H|^{-1/2}\int dz\ \exp\Big\{-\frac{1}{2}z^Tz\Big\} \qquad (29.28)$$

and introduce spherical coordinates:

$$\int dz\ \exp\Big\{-\frac{1}{2}z^Tz\Big\} = \Omega_E\int dr\ r^{E-1}\exp\Big\{-\frac{1}{2}r^2\Big\}. \qquad (29.29)$$

The prefactor Ω_E stands for the solid angle in E dimensions. The integrand of the r-integral is shown in Figure 29.2 for $E = 4, 8, 12$. The shapes are very similar and the influence of

Table 29.2 Characteristics of the function $f(r) = r^{E-1}\exp\{-r^2/2\}$

E	$\langle r\rangle$	$\langle(\Delta r)^2\rangle$	V_-/V_+
4	1.88	0.466	0.048
8	2.74	0.484	0.016
12	3.39	0.489	0.007

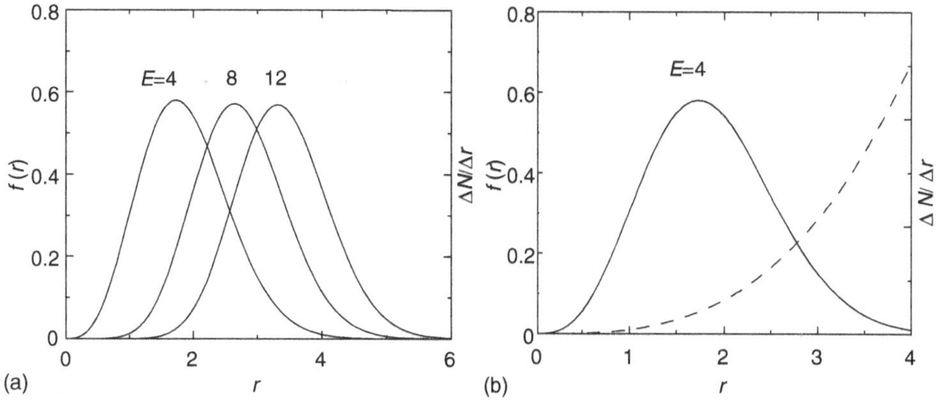

Figure 29.2 (a) The integrand in equation (29.29) for dimensions $E = 4, 8, 12$. (b) The integrand for $E = 4$ and the number of sampling points ΔN in a spherical shell of thickness Δr in the case of simple sampling (see equation (29.32)).

dimension is a displacement, the mode being $\hat{r} = \sqrt{E-1}$. It is equally simple to calculate the first two moments of the distribution:

$$\langle r\rangle = \sqrt{2}\,\frac{\Gamma(\frac{E+1}{2})}{\Gamma(\frac{E}{2})}, \quad \langle r^2\rangle = E. \tag{29.30}$$

An approximate evaluation of $\langle r\rangle$ for large E, employing Stirling's formula for the Gamma function, yields after an elementary but tedious calculation

$$\langle(\Delta r)^2\rangle = \frac{1}{2} - \frac{1}{8E}, \tag{29.31}$$

which falls close to the exact result, given in Table 29.2, even for $E = 4$.

Table 29.2 lists for a few values of E the values of $\langle r\rangle$ and $\langle(\Delta r)^2\rangle$. For large E, we observe $\langle(\Delta r)^2\rangle \to 0.5$, that is the width of the distribution $r^{E-1}\exp(-r^2/2)$ is nearly independent of dimension E. Also given in the table is the ratio V_-/V_+, where $V_\pm$ is the volume of a sphere with radius $\langle r\rangle \pm \langle(\Delta r)^2\rangle^{1/2}$. We see that the sampling effort for small radii becomes progressively negligible in comparison with that at larger radii. In fact, the number of sampling points in a spherical shell of thickness Δr is

$$\Delta N = N \frac{E\, \pi^{E/2} r^{E-1}}{\Gamma\left(\frac{E+2}{2}\right)} \Delta r. \tag{29.32}$$

This function is shown in arbitrary units for $E = 4$ in Figure 29.2. This figure disguises the second dimension fallacy. The power-law increase r^{E-1} is present in both the integrand as well as in the density of pivot points, but the uniform sampling has the highest density at radii which no longer contribute to the value of the integral due to the Gaussian part. Considerable savings could therefore be achieved if we were able to sample with a density proportional to $r^{E-1}\exp(-r^2/2)$ instead of only r^{E-1}. Remember that

$$\Omega_E\, r^{E-1} \exp(-r^2/2) = \exp\left\{-\frac{1}{2}\sum_i z_i^2\right\} = \prod_i \exp\left\{-\frac{1}{2} z_i^2\right\}. \tag{29.33}$$

We can therefore calculate the required distribution if we can produce random numbers from a zero mean, unit variance, one-dimensional Gaussian distribution ($\phi(z)$). Fortunately, fast and simple algorithms are available which afford this. In order to compensate for the non-uniform sampling, we rewrite the evidence as

$$\mathcal{E} = \langle f \rangle = |\mathbf{H}|^{-1/2} \frac{1}{V_I} \int d^E z \underbrace{\frac{f(\boldsymbol{\mu}(z))}{\phi(z)}}_{:=\tilde{f}(z)} \phi(z). \tag{29.34}$$

The Monte Carlo approximation to this integral then becomes

$$\mathcal{E} \approx \frac{|\mathbf{H}|^{-1/2}}{V_I} \frac{1}{N} \sum_{j=1}^{N} \tilde{f}(\boldsymbol{\mu}(z_j)). \tag{29.35}$$

The key idea of importance sampling is to replace the observable f by a different $\tilde{f}$, which has a considerably smaller variance, and to transfer as much variability as possible into a probability density $\phi(z)$. In this section we restrict the discussion to densities $\phi(z)$ for which random numbers can be drawn by standard methods, e.g. by inverse transform sampling. Importance sampling can also be combined with Markov chain methods, which allow us to utilize densities that are intractable by standard random number generators. By contrast to the approaches in equations (29.1) [p. 509] and (29.3) [p. 510], we no longer need to define a finite support for the integration unless the prior requires us to do so. This is a very welcome simplification. Note that if $f(\boldsymbol{\mu})$ happens to be a Gaussian, all the terms under the summation are equal to a constant and a single term will apparently give the exact value of the integral. This is no longer true if one or more variables z_i are confined to a compact support due to the prior.

However, consideration of the artificial limiting case, in which the whole job can be done analytically, gives us a clue to the efficiency of importance sampling. The error estimate remains of course as given in equation (29.21) [p. 516]. It is clear that the Monte Carlo estimate of the evidence by importance sampling will be more efficient the more the posterior distribution resembles a Gaussian function. If a Gaussian approximation to the posterior exists at all, this will fit best in the vicinity of the mode. Larger discrepancies will evolve as we move further out into the wings of the posterior. This will cause severe problems if the sampling PDF ϕ decays faster than the original integrand f. In this case the variance of the new observable $\tilde{f} = f/\phi$ can even become infinite. An obvious example is given if ϕ is also a Gaussian. We assume for the sake of simplicity that $f(\boldsymbol{\mu})$ is a zero mean, unit variance Gaussian and for ϕ we use a zero mean Gaussian with variance σ_ϕ^2 in each direction. According to equation (29.34), the expectation value is independent of ϕ but the second moment changes. Without loss of generality we suppress the prefactors in equation (29.34) and set $f(\boldsymbol{\mu}_0) = 1$ as these factors do not enter the relative accuracy. In other words, we use the following posterior function $f(\boldsymbol{z})$:

$$f(\boldsymbol{z}) = \prod_{i=1}^{E} \left(\frac{1}{\sqrt{2\pi}} e^{-\frac{1}{2} z_i^2} \right).$$

The expectation values are then

$$\begin{aligned}
\langle \tilde{f} \rangle &= \langle f \rangle = 1, \\
\langle \tilde{f}^2 \rangle &= \int d^E z \left(\frac{f(\boldsymbol{\mu}(\boldsymbol{z}))}{\phi(\boldsymbol{z})} \right)^2 \phi(\boldsymbol{z}) \\
&\propto \int d^E z \, \exp\left\{ -(1 - \frac{1}{2\sigma_\phi^2}) \, z^2 \right\} \\
&\propto \begin{cases} \left(2\pi \frac{\sigma_\phi^2}{2\sigma_\phi^2 - 1} \right)^{E/2} & \text{for } \sigma_\phi^2 > 1/2 \\ \infty & \text{otherwise.} \end{cases}
\end{aligned}$$

We see that we have severe problems if ϕ declines faster than the Gaussian posterior. In contrast, if the decay is much slower than that of the Gaussian, the power law increase due to the volume factor of the hypersphere shells cannot be overcome.

Next we will discuss two possible choices for the importance sampling density ϕ that we will use in the next section for numerical examples. Remember that we have used the Gaussian approximation to the posterior to find a transformation $\boldsymbol{\mu} \leftrightarrow \boldsymbol{z}$. If the posterior is really Gaussian, then the new variables $\boldsymbol{z}$ are uncorrelated and the importance sampling density has a product form

$$\phi(\boldsymbol{z}) = \prod_i \phi(z_i). \tag{29.36}$$

We can use any one-dimensional positive normalizable distribution from which we can draw random numbers. Assuming that the posterior is indeed Gaussian, we can easily compute the moments of $\tilde{f}$:

$$\begin{aligned}\langle \tilde{f} \rangle &= \langle f \rangle = 1,\\ \langle \tilde{f}^2 \rangle &= \prod_{i=1}^{E} dz_i \left(\int \frac{\tilde{f}(z_i)}{\phi(z_i)} \right)^2 \phi(z_i)\\ &= \left(\int dz \, \frac{(\tilde{f}(z))^2}{\phi(z)} \right)^E .\end{aligned}$$

Eventually, the variance that determines the Monte Carlo error reads

$$\sigma_{\tilde{f}}^2 = \left(\frac{1}{2\pi} \int dz \, \frac{e^{-z^2}}{\phi(z)} \right)^E - 1. \tag{29.37}$$

In view of the above discussion, functions ϕ with heavier tails are of interest and two alternatives are of considerable practical importance. The first is the hyperbolic Cauchy distribution, which is given by

$$\phi(z|a) = \frac{2}{a} \, \frac{1}{(e^{z/a} + e^{-z/a})^2}. \tag{29.38}$$

The wings of this function decay exponentially as $\exp\{-2|z/a|\}$ and they are able to suppress the power-law increase of the volume factor (r^{E-1}). At the same time the decay is still slower than Gaussian to avoid an increase of the variance over the simple sampling case. In this case the variance of the sample mean, given in equation (29.37), can be computed analytically. A straightforward calculation yields

$$\sigma_{\tilde{f}}^2 = \left(\frac{a}{2\sqrt{\pi}} \left(1 + e^{\frac{1}{a^2}} \right) \right)^E - 1.$$

The variance is minimized with respect to a for $a = 1.1634$ and the minimal variance yield

$$\sigma_{\tilde{f}}^2 = 1.0152^E - 1 \approx e^{0.0152\,E} - 1. \tag{29.39}$$

We also obtain a very weak (initially linear) dependence on the dimension E.

Alternatively, the shape parameter a can be chosen such that the area between $\phi(z|a)$ and a zero mean, unit variance Gaussian becomes minimal. By numerical integration of

$$\left| \phi(z|a) - \frac{1}{\sqrt{2\pi}} \exp\{-z^2/2\} \right| \tag{29.40}$$

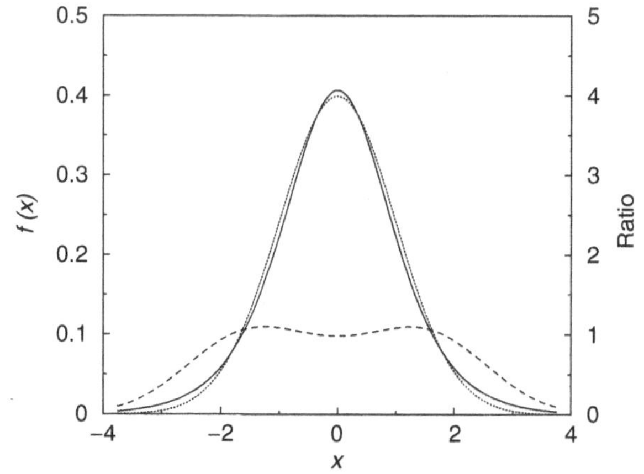

Figure 29.3 Comparison of a Gaussian (dotted) to a hyperbolic Cauchy distribution (continuous). Their ratio (dashed) tends to zero for large values of the argument.

as a function of a we find that the optimum value is $a = 1.23$, which is close to the parameter that minimizes the variance. Random numbers x_i from the distribution in equation (29.38) are then obtained via inverse transform sampling, as described in Section 29.2.2 [p. 519]. In Figure 29.3 we compare a Gaussian as a posterior candidate to a hyperbolic Cauchy distribution and display also their ratio. The latter shows beautifully that the wings are also well described. A typical example that does not improve so well over simple sampling is Student's t-distributions, which we have met already in equation (7.31a) [p. 109]. They exhibit only a power-law decay and are therefore not able to suppress the power-law increase due to the volume of the hyperspheres. A suitable distribution to meet this case is the Cauchy distribution

$$p(z|\gamma) = \frac{\gamma}{\pi} \frac{1}{z^2 + \gamma^2}. \tag{29.41}$$

In this case the variance can also be computed analytically, and we obtain

$$\sigma_{\bar{f}}^2 = \left(\frac{\sqrt{\pi}}{4\gamma} (1 + 2\gamma^2) \right)^E - 1.$$

The minimum is obtained for $\gamma = 1/\sqrt{2} = 0.707$, resulting in

$$\sigma_{\bar{f}}^2 = 1.253^E - 1, \tag{29.42}$$

which exhibits a much more pronounced E dependence than the hyperbolic Cauchy function. The numerical example in the next section will corroborate this observation. Alternatively, we minimize again the modulus of the area between $p(z|\gamma)$ and a zero mean, unit variance Gaussian and obtain $\gamma = 0.73$, which is very close to the optimal value as far as the variance is concerned. Random numbers x_i from the distribution in equation (29.41) are again obtained via inverse transform sampling, as described in Section 29.2.2 [p. 519].

29.2.4 An illustrative example

We shall now demonstrate the performance of the various methods described so far. As a test integrand we choose a multivariate Gaussian because in this case the numerical approximations to the integral can be compared to the exact answer:

$$I = \int d^E x \exp\left\{-\frac{1}{2}(\boldsymbol{x}-\boldsymbol{x}_0)^T \mathbf{Q}(\boldsymbol{x}-\boldsymbol{x}_0)\right\} = \frac{(2\pi)^{E/2}}{\sqrt{|\mathbf{Q}|}}. \tag{29.43}$$

The matrix $\mathbf{Q}$ must be real symmetric positive definite and can be constructed with known properties from any system of orthonormal vectors $\{\boldsymbol{b}_i\}$ via

$$\mathbf{Q} = \sum_i \lambda_i \boldsymbol{b}_i \boldsymbol{b}_i^T. \tag{29.44}$$

The determinant of $\mathbf{Q}$ is then $\prod_i \lambda_i$ and by choice of the ratio between largest and smallest eigenvalues $\lambda_{\max}/\lambda_{\min}$, the multidimensional ellipsoid defined by $\mathbf{Q}$ can be tuned to any desired elongation. The numerical data cited below were obtained with an eigenvalue spectrum set to $\ln \lambda_i = (i-1)\ln(10^{-4})/(E-1)$, a very much elongated ellipsoid indeed since the largest and smallest half-axes differ by a factor of 100.

For application of the deterministic approach and Monte Carlo simple sampling it is further necessary to define a compact support such that the finite range has only negligible effect on the accuracy of the integral. We derive an estimate for the cutoff radius r_c from the properties of the function $r^{E-1}\exp(-r^2/2)$. The maximum of this function lies at $r_{\max} = \sqrt{E-1}$. If we choose the boundary of the support such that the function has decayed to less than 2% of its maximum value, we find that the cutoff radius obeys approximately the relation

$$r_c = \sqrt{E-1} + 2 + \frac{2}{E+3}. \tag{29.45}$$

The most important problem of deterministic numerical integration is the dimensional curse. Although we have been able to reduce the effort according to equation (29.14) [p. 513] significantly compared with the general exponential dependence on dimension E for the type of integrands considered here, the choice of $p = 8$ pivots requires for dimension $E = 10$ already $Z_{p,E} > 5 \times 10^5$ integrand evaluations. What happens for a larger dimension and/or number of pivots was shown in Table 29.1 [p. 514].

So far, we have avoided raising the question of how Monte Carlo methods perform in this respect. We anticipate the answer to this question and then test its validity. Experience shows that Monte Carlo methods of integration exhibit a very weak dependence on the dimension of the integral, given a certain precision ΔI. In fact, this dependence is frequently found to be close to linear. If this was true, then a given precision ΔI of the integrand would imply that the number of necessary integrand samples was approximately $N \approx \text{const.}(E/\Delta I)^2$. The data displayed in Figure 29.4 have been obtained with $N = 200E^2$. The full dots in Figure 29.4(a) show the calculated relative Monte Carlo error in simple sampling for dimensions ranging from 4 to 26. The open circles show the

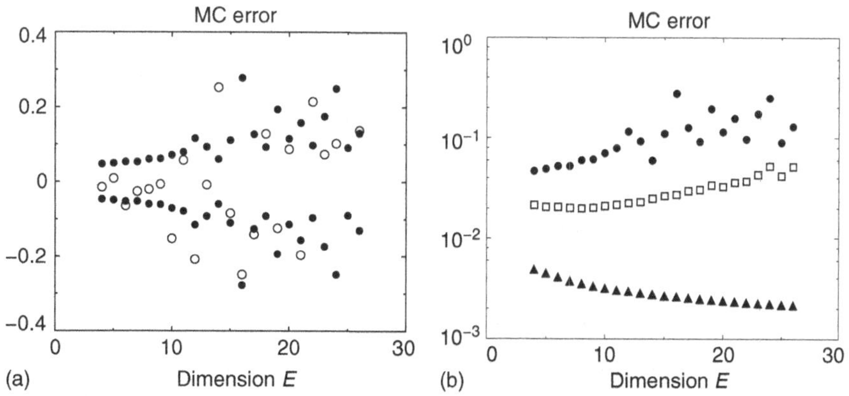

Figure 29.4 (a) Full dots display the calculated Monte Carlo error for simple sampling and open circles show the difference between the true value of the integral and the Monte Carlo error. (b) We compare the relative Monte Carlo error of simple sampling (full dots) with results obtained by importance sampling employing Cauchy random numbers (open squares) and hyperbolic Cauchy random numbers (triangles).

difference between the numerical value of the integral, which includes also the finite support approximation, and the exact number of equation (29.43). The right panel compares, on a logarithmic scale, the relative Monte Carlo errors in simple sampling (full dots) with those obtained using importance sampling either employing Cauchy random numbers (open squares) or hyperbolic Cauchy random numbers (triangles). Note the impressive gain in precision when moving from simple sampling to importance sampling. The different behaviour of the Cauchy and the hyperbolic Cauchy random numbers corroborates the different values for the variances given in equations (29.39) and (29.42). Moreover, we see that in the case of Cauchy random numbers, the variance increases faster than E^2. In conclusion, we find that due to the dimensional curse in deterministic numerical integration its application is limited to dimensions smaller than 10. Monte Carlo methods are much more efficient already at $E = 10$ and hold the promise to perform satisfactorily even at much higher dimensions, with the additional advantage of a rigorous assessment of the integration error.

29.3 Beyond the Gaussian approximation

Our assumptions that the posterior density has approximately a Gaussian shape cover a large number of practical data analysis problems, however, not all of them. The main application for Monte Carlo integration turns out to be the generation of random numbers from general multivariate distributions. This problem can in principle be reduced to the generation of random numbers from one-dimensional distributions, which we assume to be always possible whatever the density looks like. By repeated use of the product rule we obtain

$$\phi(\boldsymbol{x}) := p(\boldsymbol{x}|\mathcal{I}) = p(x_1|\mathcal{I})\; p(x_2|x_1,\mathcal{I}) \ldots p(x_n|x_1,\ldots,x_{n-1},\mathcal{I}). \tag{29.46}$$

This looks simple enough. However, it implies that all the one-dimensional conditional distributions can be calculated exactly. This is only possible in very rare, exceptional cases such as for example a multivariate Gaussian on infinite support. The discovery of Metropolis *et al.* [142] that random vectors from a given multivariate density can be obtained by a very simple algorithm is therefore of paramount importance. Originally it was formulated to compute integrals that correspond to posterior expectation values. Evidence integrals require additional considerations. The method consists of exploring the space of the posterior density $\phi(\boldsymbol{x})$ by a random walk. Let $\boldsymbol{x}^{(i)}$ be the coordinate of the ith step in this exploration. A new candidate $\boldsymbol{y}$ for step $(i+1)$ is then generated from a 'proposal' density $\rho(\boldsymbol{y}|\boldsymbol{x}^{(i)},\mathcal{I})$. A simple form of a proposal distribution would consist of local uniform changes

$$\rho(\boldsymbol{y}|\boldsymbol{x},\mathcal{I}) = \prod_i \frac{1}{2\delta}\, \boldsymbol{1}(x_i-\delta \le y_i \le x_i+\delta)\,. \tag{29.47}$$

If $\phi(\boldsymbol{y})$ is greater than $\phi(\boldsymbol{x}^{(i)})$ then $\boldsymbol{y}$ is accepted as the $(i+1)$th step $\boldsymbol{x}^{(i+1)}$ of the random walk. If $\phi(\boldsymbol{y}) < \phi(\boldsymbol{x}^{(i)})$ then a decision must be made whether $\boldsymbol{x}^{(i+1)} = \boldsymbol{x}^{(i)}$ or $\boldsymbol{x}^{(i+1)} = \boldsymbol{y}$. The rule for this decision is that $\boldsymbol{x}^{(i+1)} = \boldsymbol{y}$ if $\varepsilon < \phi(\boldsymbol{y})/\phi(\boldsymbol{x}^{(i)})$, where ε is a random number from $U(0,1)$. Interestingly, the norm of the density $\phi(\boldsymbol{x})$ does not enter into the above decision criterion and is therefore immaterial for the generation of the random walk. An implicitly assumed side condition for this simple algorithm is that the proposal distribution ρ is symmetric, $\rho(\boldsymbol{y}|\boldsymbol{x},\mathcal{I}) = \rho(\boldsymbol{x}|\boldsymbol{y},\mathcal{I})$, in order to ensure detailed balance. Details will be discussed in Section 30.3.1 [p. 544]. This criterion is of course fulfilled by equation (29.47) and holds in general for distributions symmetric in $\boldsymbol{x}$ and $\boldsymbol{y}$. An extension of the algorithm to asymmetric distributions has been proposed by Hastings. For proposal densities with $\rho(\boldsymbol{x}|\boldsymbol{y},\mathcal{I}) \ne \rho(\boldsymbol{y}|\boldsymbol{x},\mathcal{I})$ the above decisions must be made for the product $\phi(\boldsymbol{y})\; \rho(\boldsymbol{x}|\boldsymbol{y},\mathcal{I})$ instead of $\phi(\boldsymbol{y})$ in order to ensure detailed balance.

The sequence of random vectors $\boldsymbol{x}^{(i)}, \boldsymbol{x}^{(i+1)}, \ldots$ is called a Markov chain. That is the reason why the Metropolis–Hastings approach is also called Markov chain Monte Carlo. MCMC is however not restricted to the Metropolis–Hastings recipe. The general approach will be discussed in Section 30.3.1 [p. 544]. The characteristic of a Markov chain is that each configuration $\boldsymbol{x}^{(i)}$ depends explicitly only on its predecessor $\boldsymbol{x}^{(i-1)}$. This implies that the chain has no memory but it is still correlated. In fact, if δ in equation (29.47) is chosen to be sufficiently small then $\boldsymbol{x}^{(i+1)}$ will be very near to $\boldsymbol{x}^{(i)}$. Imagine that $\boldsymbol{x}^{(i)}$ is close to the maximum of the distribution $\phi(\boldsymbol{x})$. Consider then the opposite choice of very large δ. The density $\phi(\boldsymbol{y})$ will very likely be so small that it has a very low chance of being accepted and state $\boldsymbol{x}^{(i)}$ may be perpetuated for several steps. Exploration of the whole density will then clearly be very slow. The choice of the step width δ in equation (29.47) is therefore obviously of some importance. As a very coarse rule of thumb, an acceptance rate of about 50% has been established as a criterion for a suitable choice of δ.

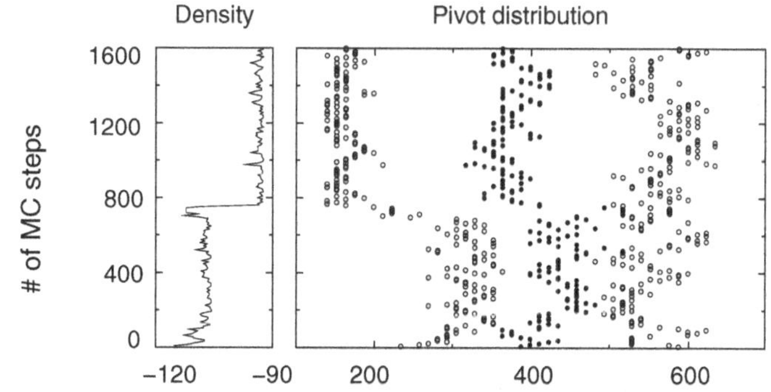

Figure 29.5 Evolution of density and pivot distribution in a Metropolis random walk.

In order to demonstrate the Metropolis algorithm in action, we return to the evaluation of the $\boldsymbol{\xi}$-summation in equation (27.43) [p. 482]. Assume a certain pivot distribution $\boldsymbol{\xi}_k$ in the kth step of the Metropolis walk with associated $p_k = p(\boldsymbol{\xi}_k, \boldsymbol{n}|E, \boldsymbol{t}, \mathcal{I})$. A new configuration $\boldsymbol{\xi}_p^*$ is then proposed by adding uniform deviates ranging from $-\delta$ to δ to $\boldsymbol{\xi}_k$, resulting in $\boldsymbol{\xi}_p^*$. The update is done componentwise. The proposal values are then rounded to integers and ordered to form an ascending series. If all intervals $\xi^*_{p,k+1} - \xi^*_{p,k}$ are greater than zero, then $\boldsymbol{\xi}_p = \boldsymbol{\xi}_p^*$ and the density $p_p = p(\boldsymbol{\xi}_p, \boldsymbol{n}_p|E, \boldsymbol{t}, \mathcal{I})$ is calculated. Comparison of p_k and p_p subject to the Metropolis prescription then decides whether the proposed configuration should become $\boldsymbol{\xi}_{k+1}$ and $p_{k+1} = p_p$ or whether $\boldsymbol{\xi}_{k+1} = \boldsymbol{\xi}_k$ and $p_{k+1} = p_k$ persist for the next step in the random walk. Figure 29.5 displays results for the log density and the pivot position from a sequence of 1600 Metropolis steps. Only every eighth sample is shown. The walk starts at the bottom with an equidistant pivot distribution and takes some time to equilibrate to new 'average' positions of ξ_2, ξ_3, ξ_4. The equilibration is also seen in the density, which is shown in the left panel. For every block of 200 steps the fraction of accepted moves is calculated. If the ratio is too large then δ is increased, while δ is decreased if the ratio is too small. After about 700 steps we notice an important jump in the density to considerably higher values, which equilibrates quickly at the new higher level. This jump tells us that the problem has at least two different maxima of the density corresponding to markedly differing pivot distributions. Note that these maxima are separated by a density minimum which can only be traversed because the Metropolis algorithm also accepts moves of the variables to regions of lower density, albeit with probability smaller than one. The first 1200 steps of the random walk in Figure 29.5 are discarded and considered as a 'warm up' of the algorithm. Samples of the sum in equation (27.43) [p. 482] are only derived from the last 400 steps. The importance of equilibration will be analysed in Section 30.5 [p. 560].

The jump in the density near step 700 can be understood by reference to Figure 27.5 [p. 484]. For equidistantly spaced pivots, and this was the initial configuration chosen above, the second pivot falls in the region of the minimum between the two peaks. Such a configuration cannot even approximately fit the left peak. Similarly, a relative best fit occurs when the pivots are more or less symmetrically distributed around the larger peak. The jump in density, equivalent to a dramatic increase in the goodness of fit, occurs only after the leftmost pivot has accidentally been placed to the left of the smaller peak.

The proposal distribution of equation (29.47) is by no means the only possible choice. Attractive properties are offered by

$$\rho(\boldsymbol{y}|\boldsymbol{x},\mathcal{I}) = \prod_i \frac{\gamma_i}{\pi} \frac{1}{(x_i - y_i)^2 + \gamma_i^2}. \tag{29.48}$$

Every term in the product in equation (29.48) is a normalized Cauchy distribution with half of its area in the range $|x| < \gamma$ and the other half spread out from $|x| = \gamma$ to infinity. This property is of course very welcome to combine the detailed exploration of highly structured regions of $\phi(\boldsymbol{x})$ with large moves to remote places. The scale of the moves set by γ_i will differ considerably in real-world problems and therefore needs individual adjustment. This may be understood by reference to Figure 29.1 [p. 511]. In order to cover the ellipse 'uniformly' we should use step widths for the x' and y' coordinates proportional to the half-axis of the ellipse in the corresponding directions.

A valuable advantage of importance sampling is the rigorous error estimate that we obtain. In the present example we have not assumed knowledge about the posterior, neither shape nor mode nor covariance. This could be included to some extent to initiate the random walk and to tailor the proposal distribution. If such information is available, it is highly recommended to use it.

A disadvantageous part of MCMC is the correlation between the samples. This correlation implies that we can no longer use the simple error estimate of equation (29.21) [p. 516] since this is based on the assumption that the samples in equation (29.35) [p. 526] are independent. This topic will be discussed in detail in Section 30.3.1 [p. 544]. It may be clear by now that MCMC is not a panacea and should be used with great care, and only if deterministic numerical integration or exact sampling based on importance sampling cannot be applied. MCMC requires careful attention to the equilibration of the random walk, acceptance adjustment and correlation in order to perform successfully, and all three items require a lot of experience.

Dependence of the statistical error on the dimension E

To conclude this section on Monte Carlo integration, we would like to study the dependence of the MCMC uncertainty on the dimension of the underlying problem. In order to get a qualitative idea and to be able to proceed analytically, we assume that the posterior PDF as well as the observable have a product form, i.e.

$$p(\boldsymbol{x}|\mathcal{I}) = \prod_{i=1}^{E} p_1(x_i|\mathcal{I}) \; ; \quad f(\boldsymbol{x}) = \prod_{i=1}^{E} f_1(x_i),$$

where the functions p_1 and f_1 are independent of i. The expectation value of the observable is then given by

$$\langle f \rangle = \int f(\boldsymbol{x}) p(\boldsymbol{x}|\mathcal{I}) d^E x = \prod_{i=1}^{E} \int p_1(x_i|\mathcal{I}) f_1(x_i) dx_i := \langle f \rangle_1^E. \tag{29.49}$$

The Monte Carlo estimate for $\langle f \rangle$ is the sample mean

$$S := \frac{1}{N} \sum_{i=1}^{N} f(\boldsymbol{\xi}_i).$$

We assume that the random numbers are uncorrelated, either because we can draw exact samples or skip enough Markov steps in MCMC. According to equation (29.21) [p. 516], the variance of the sample mean reads

$$\langle (\Delta S)^2 \rangle = \frac{\sigma_f^2}{N},$$

and σ_f^2 can be computed easily:

$$\begin{aligned}
\sigma_f^2 &= \int f(\boldsymbol{x})^2 p(\boldsymbol{x}|\mathcal{I}) d^E x - \langle f \rangle^2 \\
&= \left(\int f_1(x)^2 p(x|\mathcal{I}) dx \right)^E - \langle f \rangle^2 \\
&= \Bigg(\langle f \rangle_1^2 + \underbrace{\int (\Delta f_1(x))^2 p(x|\mathcal{I}) dx}_{\sigma_{f,1}^2} \Bigg)^E - \langle f \rangle^2 \\
&\overset{(29.49)}{=} \langle f \rangle^2 \left[\left(1 + \frac{\sigma_{f,1}^2}{\langle f \rangle_1^2} \right)^E - 1 \right].
\end{aligned}$$

The relative standard deviation $\varepsilon_f := \sigma_f / \langle f \rangle$ is therefore related to that for one dimension $\varepsilon_{f,1} := \sigma_{f,1} / \langle f \rangle_1$:

$$\varepsilon_f^2 = \left(1 + \varepsilon_{f,1}^2 \right)^E - 1.$$

It is apparently reasonable to assume that $\varepsilon_{f,1}^2 \ll 1$, which yields

$$\varepsilon_f^2 = e^{E \ln(1+\varepsilon_{f,1}^2)} - 1 \approx e^{E \varepsilon_{f,1}^2} - 1.$$

If in addition $E \varepsilon_{f,1}^2 \ll 1$, then the relative standard deviation of the sample mean S, that is the relative statistical error, is given by

$$\varepsilon_S := \frac{\sqrt{\langle (\Delta S)^2 \rangle}}{\langle f \rangle} = \frac{\sqrt{E}}{\sqrt{N}} \varepsilon_{f,1}.$$

In this case, the sample size has to increase linearly with the dimension of the problem in order to retain the relative accuracy. Otherwise we find an exponential dependence on E.

Another common situation is given by the additive observable

$$f(\boldsymbol{x}) = \sum_{i=1}^{E} f_1(x_i).$$

In this case, if we still use the same posterior PDF as before, we readily see along the lines outlined above that

$$\langle f \rangle = E \langle f \rangle_1,$$

and the variance is

$$\begin{aligned}\sigma_f^2 &= \sum_{i=1}^{E} \int f_1(x_i)^2 p_1(x_i) dx_i + \sum_{\substack{i,j=1 \\ i \neq j}}^{E} \int f_1(x_i) f_1(x_j) p_1(x_i) p_1(x_j) dx_i dx_j - \langle f \rangle^2 \\ &= E \langle f^2 \rangle_1 + E(E-1) \langle f \rangle_1^2 - E^2 \langle f \rangle_1 = E \sigma_{f,1}^2.\end{aligned}$$

Then the relative error reads

$$\varepsilon_f = \frac{\sigma_f}{\langle f \rangle} = \frac{\sigma_{f,1}}{\sqrt{E} \langle f \rangle_1} = \varepsilon_{f,1} \frac{1}{\sqrt{E}}.$$

Hence, in this case the relative statistical error of the sample mean is given by

$$\varepsilon_S = \frac{1}{\sqrt{EN}} \varepsilon_{f,1}.$$

This is a very reasonable result, as each dimension has the same mean $\langle f \rangle_1$ and the dimension acts like repeated measurements.

30

Monte Carlo methods

In the previous chapter we have given a heuristic introduction to numerical integration. The goal of the present chapter is twofold. On the one hand we will present Markov chain Monte Carlo as a tool to compute multidimensional integrals occurring in data analysis problems or in statistical physics. On the other hand we will use the rules of probability theory to study some of the important properties of MCMC. For proofs and additional material not covered here, we refer the reader to [16, 74, 84].

30.1 Simple sampling

The most basic MC idea is the following. Let's assume we are interested in the evaluation of the integral

$$q := \int_0^1 g(t)dt, \qquad \text{with } 0 \leq g(t) \leq 1 \; \forall t \in [0, 1]. \tag{30.1}$$

We generate points (t, y), where both coordinates are uniform random numbers in the unit interval $(0, 1]$ and we count the events E where points lie below the curve defined by $g(t)$. The probability for these events $P(E|\mathcal{I})$ is equal to the ratio of the corresponding areas. Since the unit square has size 1, the ratio is the sought-for integral $P(E|\mathcal{I}) = q$. So far, we have just defined a Bernoulli experiment and the probability that n points of N lie below the curve is given by the ubiquitous binomial distribution, introduced in Section 4.2 [p. 54]:

$$P(n|N, q, \mathcal{I}) = \binom{N}{n} q^n (1 - q)^{N-n}.$$

30.1.1 The Bayesian approach

In order to infer q, given the experimental value n, we invoke Bayes' theorem:

$$p(q|n, N, \mathcal{I}) = \frac{P(n|q, N, \mathcal{I})p(q|N, \mathcal{I})}{Z}.$$

In order to keep the discussion transparent, we assume a beta prior

$$p(q|N,\mathcal{I}) = \frac{1}{B(a,b)} q^{a-1}(1-q)^{b-1},$$

with appropriate values for a and b representing mean and variance of q according to our prior knowledge. An uninformative, uniform prior corresponds to $a = b = 1$. We end up with the beta PDF for the posterior

$$p(q|n,N,\mathcal{I}) = p_\beta(q|\alpha,\rho); \qquad \text{with } \alpha = a+n, \quad \rho = b+N-n$$

and we characterize the result by

MEAN VARIANCE AND RELATIVE ACCURACY OF BETA PDF

$$\langle q \rangle = \frac{n+a}{N+a+b}, \tag{30.2a}$$

$$\text{var}(q) = \frac{\alpha\rho}{(\alpha+\rho)^2(\alpha+\rho+1)}, \tag{30.2b}$$

$$\varepsilon_{rel} := \frac{\text{std}(q)}{\langle q \rangle} = \sqrt{\frac{\rho}{\alpha(\alpha+\rho+1)}}. \tag{30.2c}$$

The last equation defines the relative uncertainty. In order to separate the wood from the trees we use the special case $q = \frac{1}{2}$, implying $a = b$. The results for the bare prior, corresponding to $N = n = 0$, read

$$\langle q \rangle^{\text{prior}} = \frac{1}{2}; \qquad \varepsilon_{rel}^{\text{prior}} = \frac{1}{\sqrt{2a+1}}, \tag{30.3}$$

while for the posterior we have

$$\langle q \rangle = \frac{n+a}{N+2a}; \qquad \varepsilon_{rel} = \sqrt{\frac{a+N-n}{(a+n)(2a+N+1)}}. \tag{30.4}$$

For the symmetric case under consideration the experimental value n will be close to $N/2$. So, we can estimate the relative uncertainty by

$$\varepsilon_{rel} \approx \sqrt{\frac{a+N/2}{(a+N/2)(2a+N+1)}} = \frac{1}{\sqrt{2a+N+1}}.$$

There are two interesting limiting cases. On the one hand, there is the 'scientific novice', who has no clue whatsoever about the possible result. His or her prior knowledge is best encoded by a uniform PDF. On the other hand, there is the expert, who is an old hand at science. His or her prior knowledge is adequately described by an *expert* prior with $a \gg 1$. The *novice* prior yields

$$\langle q \rangle^{\text{prior}} = \frac{n+1}{N+2}; \qquad \varepsilon_{rel}^{\text{prior}} = \frac{1}{\sqrt{3}}, \tag{30.5}$$

$$\varepsilon_{rel} = \sqrt{\frac{N/2+1}{(N/2+1)(N+3)}} = \frac{1}{\sqrt{N+3}}. \tag{30.6}$$

The *expert* prior, in contrast, results in

$$\langle q \rangle = \frac{n+a}{N+2a}; \qquad \varepsilon_{rel}^{\text{prior}} \approx \frac{1}{\sqrt{2a}},$$

$$\varepsilon_{rel} = \sqrt{\frac{a+N/2}{(a+N/2)(2a+N)}} \approx \frac{1}{\sqrt{2a+N}} = \frac{1}{\sqrt{\frac{1}{\left(\varepsilon_{rel}^{\text{prior}}\right)^2}+N}}.$$

The information gain by the experiment is reflected in the reduced posterior uncertainty. For a sensible reduction in the latter case, we need

$$N > \frac{1}{\left(\varepsilon_{rel}^{\text{prior}}\right)^2} = 2a.$$

Hence, the experienced scientist needs significantly 'better' experiments to improve his/her state of knowledge. Moreover, we see that it takes much more data to convince the expert that q is different from his/her prior knowledge. As long as $a \gg N$ the mean is still $\langle q \rangle \approx 1/2$, which was the expert's conjecture.

30.1.2 The frequentist approach

An alternative approach, commonly used in frequentist statistics, to evaluate the random experiment is to invent an estimator for the sought-for quantity. In the present case one is prompted to use the relative frequency

$$q^* = \frac{n}{N}.$$

Since n is obtained by a Bernoulli experiment, it is binomial and we get

FREQUENTIST ESTIMATE

$$\langle q^* \rangle = \frac{\langle n \rangle}{N} = q, \tag{30.7a}$$

$$\text{std}(q^*) = \frac{\text{std}(n)}{N} = \sqrt{\frac{q(1-q)}{N}},$$

$$\varepsilon_{rel} = \sqrt{\frac{(1-q)}{qN}}. \tag{30.7b}$$

For the symmetric case $q = 1/2$ the latter simplifies to

$$\varepsilon_{rel} = \frac{1}{\sqrt{N}}.$$

Since the frequentist approach disdains prior knowledge, the result has to be compared with that for the *novice* prior in equation (30.6). In the Bayesian reasoning, based on quadratic costs, one would use the posterior mean $\langle q \rangle = \frac{n+1}{N+2}$ as estimator. The results are roughly the same. The relative uncertainty is slightly better in the Bayesian case. But the frequentist statistician would discredit the Bayesian estimator for being biased.

30.1.3 Bayes, or not biased: That is the question

At first glance, one might agree with the frequentist point of view, since an estimator that yields on average a wrong result seems useless. However, the Bayesian approach is consistent and free from ambiguities. The seeming contradiction has a subtle origin.

In the frequentist approach one assumes a given data analysis problem, in our case the determination of q where the parameters are not random in any way but fixed and as yet unknown. In order to decide whether an estimator is biased or not, the random experiment is repeated ad infinitum for the same value of q. To verify experimentally that the estimator is unbiased, one has to repeat the experiment with the same sample size and for one and the same value for the parameter q over and over again. In doing so, we produce a set of samples $\{n_1, n_2, \ldots, n_L\}$ with $L \to \infty$. The sample mean of the frequentist estimator is

$$\langle q^* \rangle = \lim_{L\to\infty} \frac{1}{L} \sum_{\nu=1}^{L} \frac{n_\nu}{N} = q,$$

and therefore unbiased. The same situation would be dealt with in Bayesian logic by computing the posterior PDF $p(q|\{n_1, \ldots, n_L\}, N, \mathcal{I})$. Again using Bayes' theorem and the fact that the random variables n_ν are uncorrelated, we end up with a beta PDF

$$\begin{aligned} p(q|\{n_1, \ldots, n_L\}, N, \mathcal{I}) &= p_\beta(q|\alpha', \rho') \\ \text{with} \qquad \alpha' = a + \sum_\nu n_\nu, & \qquad \rho' = b + N\,L - \sum_\nu n_\nu. \end{aligned}$$

In the limit $L \to \infty$ the data overrule the prior completely and also in the Bayesian approach we end up with the unbiased result $\langle q \rangle = q$.

But the frequentist way of reasoning is hypothetical and unrealistic. Not all experiments are dedicated to determining one and the same parameter value over and over again. The realistic situation is that we use the same experimental apparatus and the same data analysis tool to analyse different probes with different parameter values. The Bayesian goal, in the case of a quadratic cost function, is to minimize the mean square deviation between estimator and true value, averaged over all parameter values as they occur in nature. Since the Bayesian approach is consistent and the quadratic costs are minimized by the posterior mean, there is no better estimator in this respect. We will illustrate this point guided by the present simple example.

Quadratic costs

The procedure is as follows. We generate a value q at random according to a certain prior PDF that encodes the distribution of q values occurring in nature. For a given value q and sample size N, we draw a random integer n_q from the binomial probability $P(n|q, N, \mathcal{I})$. Next we determine the estimator $q^* = \frac{n_q+c}{N+d}$, with $c = d = 0$ in the frequentist case and $c = 1, d = 2$ in the Bayesian case. Finally, the discrepancy with the true value is measured by the quadratic cost function

$$\Delta^2 := \left(q - \frac{n_q + c}{N + d}\right)^2 .$$

We are interested in the mean costs, i.e.

$$\begin{aligned}\langle\Delta^2\rangle &= \int_0^1 dq p(q) \sum_{n_q=0}^{N} P(n_q|q, N, \mathcal{I}) \left(q - \frac{n_q + c}{N + d}\right)^2 \\ &= \frac{1}{(N + d)^2}\left(\langle(\Delta q)^2\rangle(d^2 - N) + N\langle q\rangle(1 - \langle q\rangle) + \left(c - d\langle q\rangle\right)^2\right).\end{aligned}$$

Particularly for the *novice* prior ($a = b = 1$) we have

$$\langle q\rangle = \frac{1}{2}, \qquad \langle(\Delta q)^2\rangle = \frac{1}{12}$$

and hence

$$\langle\Delta^2\rangle = \frac{1}{(N + d)^2}\left(\frac{d^2}{12} + \frac{N}{6} + \left(c - \frac{d}{2}\right)^2\right).$$

For the frequentist estimator ($c = d = 0$) we get $\langle\Delta^2\rangle_0 = \frac{1}{6N}$, while the mean costs for the Bayesian estimator $\langle\Delta^2\rangle_B = \frac{1}{6(N+2)}$ are slightly smaller, corroborating that the Bayesian estimator is superior over the unbiased one.

30.1.4 A hypothetical application

Here we will assess the power of simple sampling by a hypothetical application. A scientist, cast away on a lonesome island, for some weird reason desperately needs the number π with relative uncertainty 10^{-5}. Since he is familiar with probability theory he decides to determine it through a random counting experiment. He draws a large circle in the sand entirely enclosed by a square. He starts throwing little stones at random in the direction of the square. The size of its edges is defined as of unit length. Ignoring those stones that land outside the unit square, the (conditional) probability that a stone lands inside the circle is given by the area of the circle. So, in principle, the goal is the evaluation of an integral given in equation (30.1) and the manual stochastic evaluation corresponds to simple sampling. The probability for a stone landing inside the circle is

$$q = \frac{\pi}{4}.$$

Since the prior knowledge is weak compared with the desired accuracy, the likelihood will take the leading part and the relative uncertainty is given by equation (30.7), which for N random throws yields

$$\varepsilon_{rel} = \frac{1}{\sqrt{N}} \underbrace{\sqrt{\frac{4-\pi}{\pi}}}_{\approx 1/2}.$$

We infer the sample size N necessary for the desired accuracy:

$$N \geq \frac{1}{(2\varepsilon_{rel})^2} \approx 3 \times 10^9.$$

How long would it take to reach the accuracy? Let's assume that throwing a stone along with the necessary bookkeeping takes on average 5 seconds. In total it would take the lonesome scientist roughly 95 years to determine π with the desired accuracy. This utterly simple example illustrates that brute-force Monte Carlo is not very efficient. In this case it might help to use a computer to throw virtual stones but as soon as the problem gets mega-dimensional, high-performance computers are also rapidly in the same situation as the poor scientist. The reason is simple: the relative uncertainty is given by

$$\varepsilon_{rel} = \sqrt{\frac{(1-q)}{qN}},$$

irrespective of the spatial dimension of the problem. The probability q, that is the ratio of volumes, however, depends exponentially ($ce^{-\beta d}$) on the dimension d, with some positive parameter β. Hence q is exponentially small, resulting in an exponentially increasing

$$\varepsilon_{rel} = \frac{e^{\beta d/2}}{\sqrt{cN}},$$

resulting in an exponentially increasing sample size for fixed accuracy.

30.2 Variance reduction

One of the central applications of Monte Carlo techniques is the evaluation of integrals of the form

$$\langle f \rangle = \int \underbrace{f(\boldsymbol{x})\, \rho(\boldsymbol{x})}_{:=g(\boldsymbol{x})} \, d^L x \; := I,$$

where $\boldsymbol{x}$ is an L-dimensional vector and $d^L x$ the corresponding integration measure. We have seen that simple sampling is in general not adequate for such high-dimensional problems, due to the increasing variance. We therefore need some means to reduce the variance. In data analysis or classical statistical physics the integrand $g(\boldsymbol{x})$ is by definition split into two factors $f(\boldsymbol{x})$ and $\rho(\boldsymbol{x})$, where the latter is a probability density function. But even if this is not a priori the case, each function $g(\boldsymbol{x})$ can always be split in the aforementioned way. In the context of reweighting, we will look at this aspect more generally.

The key idea of Monte Carlo methods is to generate samples of vectors $\{\boldsymbol{x}^{(1)}, \ldots, \boldsymbol{x}^{(N)}\}$ according to the target PDF $\rho(\boldsymbol{x})$ and to estimate the integral I by the sample mean

$$I^* := \frac{1}{N} \sum_{i=1}^{N} f(\boldsymbol{x}^{(i)}).$$

The standard deviation of the sample mean in the case of uncorrelated events is given by the standard error, outlined above, namely

$$\text{std}(I^*) := \frac{\sigma_f}{\sqrt{N}}; \qquad \sigma_f^2 := \int \left(f(\boldsymbol{x}) - \langle f \rangle\right)^2 \rho(\boldsymbol{x})\, d^L x.$$

Although the standard error can be viewed as a special application of the central limit theorem, it is also valid for small sample sizes N, as we have seen in Section 19.1.1 [p. 284]. Obviously, it is extremely advantageous if the variation of $g(\boldsymbol{x})$ is mainly due to $\rho(\boldsymbol{x})$ and $f(\boldsymbol{x})$ varies very little in the range of the peaks of $\rho(\boldsymbol{x})$, as the latter determines σ_f. If that is not the case from the outset one can apply the remedy of 'reweighting', which is closely related to *importance sampling* as discussed in Section 29.2.3 [p. 523].

Reweighting: If a positive function $\tilde{g}(\boldsymbol{x})$ is available, for which the integral $\int \tilde{g}(\boldsymbol{x}) d^L x = Z$ is known analytically, and that 'closely resembles' the integrand $g(\boldsymbol{x})$ in a sense that will soon become clear, then we can rewrite the integral as

$$I = \int \underbrace{\frac{Z g(\boldsymbol{x})}{\tilde{g}(\boldsymbol{x})}}_{\tilde{f}(\boldsymbol{x})} \underbrace{\frac{\tilde{g}(\boldsymbol{x})}{Z}}_{:=\tilde{\rho}(\boldsymbol{x})} d^L x.$$

The central goal of reweighting is achieved if the new PDF, $\tilde{\rho}(\boldsymbol{x})$, implies a significantly smaller $\sigma_{\tilde{f}}$. This is not the case if $\|\tilde{g}(\boldsymbol{x})\|$ falls off more rapidly than $\|g(\boldsymbol{x})\|$ for large $\|\boldsymbol{x}\|$. Reweighting is also most useful in the case of parameter studies. Let's assume we are interested in the parameter dependence of an observable

$$\langle O \rangle_a := \int O(\boldsymbol{x})\, \rho(\boldsymbol{x}|a)\, d^L x,$$

where a is the parameter under consideration. The expectation value shall be determined by MC means. That is, we generate a sample $\{\boldsymbol{x}^{(1)}, \ldots, \boldsymbol{x}^{(N)}\}$ drawn from the PDF $\rho(\boldsymbol{x}|a)$ and compute the sample mean

$$\overline{O}_a := \frac{1}{N} \sum_{i=1}^{N} O(\boldsymbol{x}^{(i)}).$$

In order to study the parameter dependence, we would have to repeat the procedure for all parameter values we are interested in. This can be quite CPU time-consuming and on top of that the parameter dependence is masked by the statistical noise. We can apply the reweighting idea to overcome both problems. To this end we write the expectation value for a parameter value a' as

$$\langle O \rangle_{a'} := \int \underbrace{O(\boldsymbol{x})\,\frac{\rho(\boldsymbol{x}|a')}{\rho(\boldsymbol{x}|a)}}_{:=\tilde{O}(\boldsymbol{x}|a,a')}\,\rho(\boldsymbol{x}|a)\,d^L x.$$

This expression implies that we can use the same sample, drawn from the PDF for the parameter value a, to estimate the value for a' by the sample mean

$$\overline{O}_{a'} := \frac{1}{N}\sum_{i=1}^{N} \tilde{O}(\boldsymbol{x}^{(i)}|a,a').$$

Reweighting in this context is only applicable if the two PDFs for the parameter values a' and a, respectively, do not differ too much. Otherwise the standard error of $\overline{O}_{a'}$ will be large.

30.3 Markov chain Monte Carlo

Once the variance has been reduced by a proper splitting of the integrand into probability density $\tilde{p}(\boldsymbol{x})$ and observable $\tilde{f}(\boldsymbol{x})$, we still need a way to sample from $\tilde{p}(\boldsymbol{x})$. This combination of variance reduction and sampling from the PDF that contains the dominant part of the variation of the integrand is often referred to as *importance sampling*, as already introduced in the previous chapter. In general it will not be possible to draw an exact and uncorrelated sample from $\tilde{p}(\boldsymbol{x})$, as it will neither have a tensor structure nor be akin to a Gaussian. The most widely used approach for this endeavour is Markov chain Monte Carlo. MCMC has been discussed in many review articles and books. At first glance, the MCMC algorithm seems embarrassingly trivial and for simple problems it works strikingly well. However, when it comes to real-world problems, it is essential to know the possible pitfalls. In order to reveal them we will analyse MCMC following the rules of probability theory.

30.3.1 Metropolis–Hastings algorithm

For concreteness we describe the MCMC method in terms of the Metropolis–Hastings algorithm. Other algorithms can be found in [16, 74, 84]. We restrict the discussion to problems with discrete states $\mathcal{I} = \{\boldsymbol{\xi}_1, \boldsymbol{\xi}_2, \ldots, \boldsymbol{\xi}_{\mathcal{N}}\}$. Continuous problems can always be cast into a discrete version by suitable discretization. A sample shall be drawn from the probability mass function (PMF) $\rho_j := \rho(\boldsymbol{\xi}_j)$. Initially for discrete time $\tau = 0$ a state is generated according to some initial PMF p_j^0, which could even be a Kronecker delta δ_{j,j_0}, if the Markov processes always start out from one and the same state $\boldsymbol{\xi}_{j_0}$. At some later time $\tau \geq 0$ the Markov chain contains already the elements $\{\boldsymbol{x}_0, \dots, \boldsymbol{x}_\tau\}$. The next element of the Markov chain, i.e. $\boldsymbol{x}_{\tau+1}$, is generated according to the following recipe.

- Create a trial state $\boldsymbol{x}_p$ according to a proposal distribution $q(\boldsymbol{x}_p|\boldsymbol{x}_\tau)$, which is typically restricted to the close vicinity of $\boldsymbol{x}_\tau$.
- The trial state $\boldsymbol{x}_p$ is accepted with acceptance probability

$$\alpha(\boldsymbol{x}_p, \boldsymbol{x}_\tau) = \min\left(1, \frac{P(\boldsymbol{x}_p)\, q(\boldsymbol{x}_\tau|\boldsymbol{x}_p)}{P(\boldsymbol{x}_\tau)\, q(\boldsymbol{x}_p|\boldsymbol{x}_\tau)}\right). \tag{30.8}$$

  The q-dependent terms are due to Hastings and allow for asymmetric proposal distributions, i.e. in case $q(\boldsymbol{\xi}_i|\boldsymbol{\xi}_j) \neq q(\boldsymbol{\xi}_j|\boldsymbol{\xi}_i)$.
- The next element in the Markov chain is chosen by

$$\boldsymbol{x}_{\tau+1} = \begin{cases} \boldsymbol{x}_p & \text{if the proposal is accepted} \\ \boldsymbol{x}_\tau & \text{if the proposal is rejected.} \end{cases} \tag{30.9}$$

30.3.2 The Markov matrix

The recipe presented in the previous section defines a special Markov process and serves as a concrete example. Most of the following discussion, however, will be valid generally. Restrictions on the Metropolis–Hastings algorithm will be pointed out explicitly. The Markov chain is governed by a Markov matrix $\mathbf{M}$, with components

$$M_{ji} := P(\boldsymbol{x}_{\tau+1}{=}\boldsymbol{\xi}_j|\boldsymbol{x}_\tau{=}\boldsymbol{\xi}_i, \mathcal{I}), \tag{30.10}$$

which is homogeneous (i.e. independent of 'time' τ). The background information $\mathcal{I}$ contains the details of the Markov process, for example whether Metropolis–Hastings, Glauber [84] or any other dynamics is used and which proposal distribution is employed. Since the Markov matrix corresponds to a conditional probability, it fulfils the normalization rule

$$\sum_{j=1}^{\mathcal{N}} M_{ji} = 1. \tag{30.11}$$

We will show in the sequel that the Markov process converges under some general conditions towards a unique invariant distribution, which satisfies the equation

$$\sum_{i=1}^{\mathcal{N}} M_{ji}\, \rho_i = \rho_j\,. \tag{30.12}$$

The crucial question, however, remains of how to construct a Markov matrix $\mathbf{M}$ that has a predefined invariant distribution $\boldsymbol{\rho}^0$. A sufficient but not necessary condition to this end is the detailed balance relation

$$M_{ij}\rho_j^0 = M_{ji}\rho_i^0. \tag{30.13}$$

Here $\boldsymbol{\rho}^0$ is the desired PMF for which we want to generate a sample by MCMC means. Summing both sides of the detailed balance condition over index j, along with the normalization equation (30.11) yields

$$\sum_{j=1}^{\mathcal{N}} M_{ij}\rho_j^0 = \rho_i^0,$$

showing that $\boldsymbol{\rho}^0$ is indeed an invariant distribution of the Markov matrix. We still have to prove that there is only one invariant distribution and that the Markov process converges towards it. In general, it is straightforward to decide whether the detailed balance condition is fulfilled. In particular, the Metropolis–Hastings algorithm, outlined in the previous section, satisfies the detailed balance. According to the recipe, we decompose the Markov matrix element M_{ji}, which is the probability of moving in one time step from state $\boldsymbol{\xi}_i$ to $\boldsymbol{\xi}_j$, into the proposal probability $q(\boldsymbol{\xi}_k|\boldsymbol{\xi}_i)$ to offer a state $\boldsymbol{\xi}_k$ starting from state $\boldsymbol{\xi}_i$ and the acceptance probability $\alpha(\boldsymbol{\xi}_k|\boldsymbol{\xi}_i)$ to accept state $\boldsymbol{\xi}_k$ coming from state $\boldsymbol{\xi}_i$. These elements can be introduced by the marginalization rule

$$\begin{aligned} M_{ji} &= P(\boldsymbol{x}_{\tau+1}{=}\boldsymbol{\xi}_j|\boldsymbol{x}_\tau{=}\boldsymbol{\xi}_i,\mathcal{I}) \\ &= \sum_{k=1}^{\mathcal{N}} P(\boldsymbol{x}_{\tau+1}{=}\boldsymbol{\xi}_j|\boldsymbol{x}_p{=}\boldsymbol{\xi}_k,\boldsymbol{x}_\tau{=}\boldsymbol{\xi}_i,\mathcal{I})\; P(\boldsymbol{x}_p{=}\boldsymbol{\xi}_k|\hat{x}_\tau{=}\boldsymbol{\xi}_i,\mathcal{I}) \\ &= \sum_{k=1}^{\mathcal{N}} P(\hat{x}_{\tau+1}{=}\boldsymbol{\xi}_j|\boldsymbol{x}_p{=}\boldsymbol{\xi}_k,\hat{x}_\tau{=}\boldsymbol{\xi}_i,\mathcal{I})\; q(\boldsymbol{\xi}_k|\boldsymbol{\xi}_i)\,. \end{aligned} \tag{30.14}$$

According to the rules underlying the Markov process, the first factor represents the acceptance probability. To begin with, we consider the case $\boldsymbol{\xi}_j \neq \boldsymbol{\xi}_i$. In this situation, the proposed state $\boldsymbol{\xi}_k$ has to be equal to $\boldsymbol{\xi}_j$ and it has to be accepted, i.e. $P(\hat{x}_{\tau+1}=\boldsymbol{\xi}_j|\boldsymbol{x}_p=\boldsymbol{\xi}_k,\hat{x}_\tau{=}\boldsymbol{\xi}_i,\mathcal{I}) = \delta_{kj}\alpha(\boldsymbol{\xi}_j|\boldsymbol{\xi}_i)$. This results in

$$M_{ji} = q(\boldsymbol{\xi}_j|\boldsymbol{\xi}_i)\,\alpha(\boldsymbol{\xi}_j|\boldsymbol{\xi}_i) \qquad \text{for } j \neq i. \tag{30.15}$$

The diagonal element follows readily from the normalization equation (30.11):

$$M_{ii} = 1 - \sum_{k\neq i} q(\boldsymbol{\xi}_k|\boldsymbol{\xi}_i)\,\alpha(\boldsymbol{\xi}_k|\boldsymbol{\xi}_i). \tag{30.16}$$

So, the final result is

Markov matrix for Metropolis–Hastings

$$M_{ji} = q(\boldsymbol{\xi}_j|\boldsymbol{\xi}_i)\,\alpha(\boldsymbol{\xi}_j|\boldsymbol{\xi}_i) + \delta_{ij}\left(1 - \sum_k q(\boldsymbol{\xi}_k|\boldsymbol{\xi}_i)\,\alpha(\boldsymbol{\xi}_k|\boldsymbol{\xi}_i)\right). \tag{30.17}$$

So far the result is valid for arbitrary acceptance probability. Next we will tailor the acceptance probability such that it yields the desired PDF ρ^0. For $i \neq j$ the detailed balance condition shall read

$$q(\xi_j|\xi_i)\,\alpha(\xi_j|\xi_i)\,\rho_i^0 = q(\xi_i|\xi_j)\,\alpha(\xi_i|\xi_j)\,\rho_j^0,$$

or rather

$$\frac{\alpha(\xi_j|\xi_i)}{\alpha(\xi_i|\xi_j)} = \frac{q(\xi_i|\xi_j)\,\rho_j^0}{q(\xi_j|\xi_i)\,\rho_i^0}.$$

Apparently, one choice for the acceptance probability that yields the correct PDF is given by equation (30.8). The detailed balance condition can easily be confused with the product rule of probability theory. In a more thorough notation, the detailed balance condition should rather be written

$$P(x_{\tau+1}{=}\xi_j|x_\tau{=}\xi_i,\mathcal{I})P(x_\tau{=}\xi_i|\mathcal{I}') = P(x_{\tau+1}{=}\xi_i|x_\tau{=}\xi_j,\mathcal{I})P(x_\tau{=}\xi_j|\mathcal{I}').$$

The subtle but crucial difference from the product rule is the difference in the background information $\mathcal{I}$ and $\mathcal{I}'$. The former contains the details of the Markov process and the latter defines which scientific problem we are interested in, or rather which PMF we are aiming at. $\mathcal{I}$ and $\mathcal{I}'$ are completely independent. This stresses again the utter importance of the background information.

Eigenvalue problem of the Markov matrix

The eigenvalue problem of the Markov matrix is of crucial importance for the detailed understanding of the performance of MCMC. Here we present a slightly different approach from that usually found in the literature, e.g. [74], which is more appropriate for our purposes and examines the MCMC properties from a different perspective. The Markov matrix is not symmetric, and right and left eigenvectors will be different. We start out from the right eigenvalue problem

$$\mathbf{M}\,\boldsymbol{x}_\nu = \lambda_\nu\,\boldsymbol{x}_\nu. \tag{30.18}$$

The normalization of the Markov matrix (30.11) has important consequences. Firstly, there is always one eigenvalue $\lambda_1 = 1$. The corresponding right eigenvector is, according to the definition in equation (30.12), given by the PMF ρ, i.e. $X_{i1} = \rho(\xi_i)$. A second consequence of the normalization ensues from the summation of the eigenvalue equation

$$\sum_{i=1}^{\mathcal{N}} \underbrace{\sum_{j=1}^{\mathcal{N}} M_{ji}}_{=1}\; X_{i\nu} = \lambda_\nu \sum_{j=1}^{\mathcal{N}} X_{j\nu},$$

$$\left(\sum_{i=1}^{\mathcal{N}} X_{i\nu}\right) = \lambda_\nu \left(\sum_{i=1}^{\mathcal{N}} X_{i\nu}\right).$$

This implies

$$\sum_{i=1}^{\mathcal{N}} X_{i\nu} \neq 0 \Rightarrow \lambda_\nu = 1. \tag{30.19}$$

Next we switch to the more compact matrix notation and introduce the diagonal matrix

$$\mathbf{\Delta} = \text{diag}\left\{\rho_1^{1/2}, \rho_2^{1/2}, \ldots, \rho_{\mathcal{N}}^{1/2}\right\}.$$

Based on the detailed balance equation in matrix form $\mathbf{\Delta}^2\mathbf{M}^T = \mathbf{M}\mathbf{\Delta}^2$, we modify $\mathbf{M}$ to obtain a real symmetric matrix:

$$\mathbf{A} := \mathbf{\Delta}^{-1}\mathbf{M}\mathbf{\Delta} = \mathbf{\Delta}\mathbf{M}^T\mathbf{\Delta}^{-1} = \mathbf{A}^T. \tag{30.20}$$

Consequently, the Markov matrix $\mathbf{M}$ has real eigenvalues, since it has the same eigenvalues as the real symmetric matrix $\mathbf{A}$. Starting from the eigenvalue equation $\mathbf{AU} = \mathbf{UD}$, with $\mathbf{D}$ being the diagonal matrix of eigenvalues, and using equation (30.20) we get

$$\begin{aligned} \mathbf{AU} &= \mathbf{\Delta}^{-1}\mathbf{M}\mathbf{\Delta}\mathbf{U} = \mathbf{UD}, \\ \mathbf{M}(\mathbf{\Delta U}) &= \underbrace{(\mathbf{\Delta U})}_{:=\mathbf{X}}\mathbf{D}, \\ \mathbf{M} &= \mathbf{XDX}^{-1}. \end{aligned} \tag{30.21}$$

Obviously, the eigenvalues are the same and the corresponding matrix $\mathbf{X}$ of right eigenvectors of $\mathbf{M}$ is related to the orthonormal matrix $\mathbf{U}$ of eigenvectors of $\mathbf{A}$ through

$$\mathbf{X} = \mathbf{\Delta U}. \tag{30.22}$$

As mentioned before, an eigenvector to $\lambda = 1$ is $X_{i1} = \rho_i$. Hence the corresponding vector U_{i1} is given by

$$U_{i1} = \rho_i^{1/2},$$

which has the normalization $\left(\sum_i U_{i1}^2 = 1\right)$ in agreement with the fact that it is a column of an orthonormal matrix. Since $\mathbf{A}$ is real symmetric, the eigenvalues are real. Now we can exploit the third consequence of the normalization of the Markov matrix. To this end we recall Gershgorin's theorem [126]. For a square matrix with a real-valued spectrum, along with the fact that M_{ji} is a probability, i.e. $M_{ji} \geq 0$ and $\sum_j M_{ji} = 1$, it yields that all eigenvalues lie in the following union of intervals:

$$\lambda_\nu \in \bigcup_i \left[M_{ii} - \sum_{j\neq i} M_{ji}\ ,\ M_{ii} + \sum_{j\neq i} M_{ji}\right] = \bigcup_i \left[2M_{ii} - 1,\ 1\right].$$

Hence, all eigenvalues are restricted to the interval $[-1, 1]$. The eigenvectors $\mathbf{X}$ of $\mathbf{M}$ do not form an orthogonal set of vectors. Instead we have, according to equation (30.22), the relation

$$\mathbb{1} = \mathbf{U}^\dagger \mathbf{U} = \mathbf{X}^\dagger \boldsymbol{\Delta}^{-2} X, \tag{30.23a}$$

$$\delta_{\nu,\nu'} = \sum_i \frac{1}{\rho_i} X^*_{i\nu} X_{i\nu'}. \tag{30.23b}$$

Along with $X_{i1} = \rho_i$ we find for $\nu' = 1$:

$$\sum_i X_{i1} = 1, \tag{30.23c}$$

$$\sum_i X_{i\nu} = 0; \quad \forall \nu > 1. \tag{30.23d}$$

Here $\mathbf{X}^\dagger$ stands for the adjoint, or rather the conjugate transpose of matrix $\mathbf{X}$. If $\lambda = 1$ is degenerate, the other degenerate eigenvectors also have the property in equation (30.23d). From equation (30.23a) we readily obtain the inverse of $\mathbf{X}$:

$$\mathbf{X}^{-1} = \mathbf{X}^\dagger \boldsymbol{\Delta}^{-2},$$

and we modify the spectral representation in equation (30.21):

$$\mathbf{M} = \mathbf{X}\mathbf{D}\mathbf{X}^\dagger \boldsymbol{\Delta}^{-2}. \tag{30.24}$$

From the orthonormality of $\mathbf{U}$ we can derive a further property of $\mathbf{X}$, which is related to the completeness relation of the eigenvectors:

$$\mathbb{1} = \mathbf{U}\mathbf{U}^\dagger = \boldsymbol{\Delta}^{-1}\mathbf{X}\mathbf{X}^\dagger \boldsymbol{\Delta}^{-1} \Rightarrow \tag{30.25}$$

$$\mathbf{X}\mathbf{X}^\dagger = \boldsymbol{\Delta}^2,$$

$$\sum_l X_{il} X^*_{jl} = \delta_{ij}\, \rho_i. \tag{30.26}$$

We introduce a modified representation of the spectral representation in equation (30.21) that will be useful in the sequel:

$$(\mathbf{M}^\tau)_{ij} = \sum_{l=1}^{\mathcal{N}} X_{il}\, \lambda_l^\tau\, \frac{X^*_{jl}}{\rho_j} = \rho_i + \left(\mathbf{M}'^\tau\right)_{ij} \tag{30.27}$$

$$\text{with} \quad \left(\mathbf{M}'^\tau\right)_{ij} := \sum_{l=2}^{\mathcal{N}} X_{il}\, \lambda_l^\tau\, \frac{X^*_{jl}}{\rho_j}. \tag{30.28}$$

30.3.3 Convergence towards the invariant distribution

We can use these results to determine the probability of finding the walker at time τ in state $\boldsymbol{\xi}_j$. Along with the marginalization over the initial state $\boldsymbol{x}_0$, we derive

$$\rho_j^{(\tau)} := P\left(\boldsymbol{x}_\tau{=}\boldsymbol{\xi}_j|p^0,\mathcal{I}\right) = \sum_i^{\mathcal{N}} P(\boldsymbol{x}_\tau{=}\boldsymbol{\xi}_j|\boldsymbol{x}_0{=}\boldsymbol{\xi}_i,\mathcal{I})\ \underbrace{P\left(\boldsymbol{x}_0{=}\boldsymbol{\xi}_i|p^0,\mathcal{I}\right)}_{p_i^0},$$

$$\rho_j^{(\tau)} = \sum_i^{\mathcal{N}} \left(\mathbf{M}^\tau\right)_{ji} p_i^0 \quad = \rho_j + \sum_{l=2}^{\mathcal{N}} X_{jl}\lambda_l^\tau\ \tilde{p}_l^0. \tag{30.29}$$

We have introduced a new quantity

$$\tilde{p}_l^0 := \sum_{i=1}^{\mathcal{N}} \left(\mathbf{X}^{-1}\right)_{li} p_i^0 = \sum_{i=1}^{\mathcal{N}} X_{il}^* \frac{p_i^0}{\rho_i}, \tag{30.30}$$

which is the overlap of the initial PMF p^0 with the left eigenvectors corresponding to the eigenvalue λ_l. For $l = 1$ it has the important property

$$\tilde{p}_1^0 = \sum_{j=1}^{\mathcal{N}} \frac{p_j^0}{\rho_j}\rho_j = 1. \tag{30.31}$$

Just as a consistency test, we consider $p_i^0 = \rho_i$ in equation (30.30):

$$\forall l > 1: \qquad \tilde{p}_l^0 = \sum_{i=1}^{\mathcal{N}} X_{il}^* \overset{(30.23d)}{=} 0$$

and $P(\boldsymbol{x}_\tau{=}\boldsymbol{\xi}_j|p^0,\mathcal{I}) \equiv \rho_i$, irrespective of τ. Equation (30.29) is nothing but the power method, which is known to converge towards the dominant eigenvector, provided it is not degenerate. Without degeneracy, the contribution of the various eigenvectors dies out exponentially, governed by λ_l. Therefore, the convergence is dominated by the second largest eigenvalue λ_2. We can give a rough estimate of the order of magnitude of the sample size, or rather the number of Monte Carlo steps τ, required to reduce the dominant contribution below a threshold ε^* by

$$\tau^* = \frac{\ln(\varepsilon^*)}{\ln(\lambda_2)}. \tag{30.32}$$

It guarantees that $\lambda_l^\tau < \varepsilon^*$, $\forall l > 1$ and for all $\tau > \tau^*$. Figure 30.1 illustrates the convergence of the Markov process towards the invariant distribution in the case of a Gaussian target PMF. Within each of the panels (a$_1$)–(a$_3$) snapshots $\rho^{(\tau)}$ for various ratios $\frac{\tau}{\tau^*}$ are represented. Here τ^* is determined such that an accuracy $\varepsilon^* = 10^{-3}$ is reached. The panels display results for different initial distribution p^0, which are (a$_1$) a delta peak at the centre of the target distribution, (a$_2$) a uniform distribution and (a$_3$) another delta peak, this time located outside the target distribution at the far right of the figure. At time $\tau/\tau^* = 1$ all

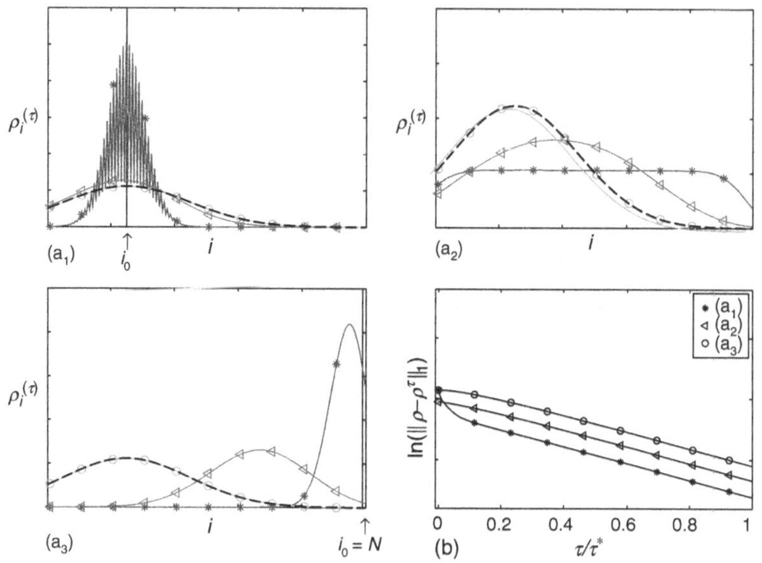

Figure 30.1 (a$_1$)−(a$_3$) Convergence of the distributions $\rho^{(\tau)}$ towards the Gaussian target PMF ρ (dashed line) for different initial distributions: (a$_1$) $p_i^0 = \delta_{i,i_0}$, (a$_2$) $p_i^0 = 1/\mathcal{N}$, (a$_3$) $p_i^0 = \delta_{i,N}$. The results for three different times $\frac{\tau}{\tau^*} = \{0.01, 0.1, 1.0\}$ are marked by $\{*, \triangleleft, \circ\}$. (b) The L_1 distance between the target PMF and $\rho^{(\tau)}$ is plotted versus $\frac{\tau}{\tau^*}$. In all cases $\mathcal{N} = 200$ and $\tau^* = 9811$.

distributions are indistinguishable from the target PMF in the plots. In Figure 30.1(b) the distance $\|\rho^{(\tau)} - \rho\|_1$ to the target distribution is depicted. Initially, the uniform PMF has the smallest distance to the target distribution, while the two delta peaks have almost the same distance, irrespective of the initial location. For long times, only the second dominant eigenvalue survives and determines the convergence rate (slope of the logarithm of the distance versus time), which is therefore the same in all three cases. Initially, however, the convergence rate is fastest for the Markov process that starts at the centre of the target PMF and this chain shows the best convergence. Next comes the flat PMF and the poorest is the Markov process started far away from the centre of the Gaussian. The behaviour can be understood by the simple random walk necessary to bring the walker to the important region of the target PMF.

30.3.4 On the degeneracy of the largest eigenvalue

We have pointed out that the Markov process generates distributions corresponding to the power method for the Markov matrix. The latter has eigenvalues in the interval $[-1, 1]$. The power method generally projects into the subspace spanned by the eigenvectors corresponding to the dominant eigenvalues, which are $+1$ and maybe -1. The goal of MCMC is, however, to generate samples according to a target distribution that corresponds to one particular eigenvector of the Markov matrix with eigenvalue 1. This goal can only be achieved if we can ensure that the largest eigenvalue ($\lambda_1 = 1$) is not degenerate on the one hand, and that the other dominant eigenvalue ($\lambda = -1$) can be eliminated on the other hand. Firstly, we will show that the degeneracy of $\lambda = 1$ implies that the Markov process is not ergodic. In other words, there are two or more subspaces of the state-space and the Markov chain always stays in the subspace it started in.

Ergodicity implies the eigenvalue $\lambda = 1$ is not degenerate

In order to prove this statement, we assume that the largest eigenvalue is degenerate. In this case and in view of the above there exists an eigenvector $\boldsymbol{x}$ obeying the following equations:

$$\mathbf{M}\boldsymbol{x} = \boldsymbol{x}, \tag{30.33a}$$

$$\sum_i x_i = 0. \tag{30.33b}$$

The latter equation implies that we can decompose the entire index set $\mathcal{J} := \{1, \ldots, \mathcal{N}\}$ into disjoint subsets $\mathcal{J} = \mathcal{J}_+ \cup \mathcal{J}_- \cup \mathcal{J}_0$, defined as

$$\begin{aligned}
\mathcal{J}_+ &:= \{i | x_i > 0\}, \\
\mathcal{J}_- &:= \{i | x_i < 0\}, \\
\mathcal{J}_0 &:= \{i | x_i = 0\}.
\end{aligned}$$

In order to fulfil equation (30.33b) the set $\mathcal{J}_0$ may be empty, but not so the other two sets, because otherwise the eigenvector would be the null vector. We split equation (30.33a) into

$$\sum_j M_{ij} x_j = \sum_{j \in \mathcal{J}_+} M_{ij} \underbrace{x_j}_{>0} + \sum_{j \in \mathcal{J}_-} M_{ij} \underbrace{x_j}_{<0} = \underbrace{\sum_{j \in \mathcal{J}_+} M_{ij} |x_j|}_{:= m_i^+} - \underbrace{\sum_{j \in \mathcal{J}_-} M_{ij} |x_j|}_{:= m_i^-}. \tag{30.34}$$

By construction, $m_i^{\pm} \geq 0$. If there is an index i for which $(m_i^+ > 0 \wedge m_i^- > 0)$ holds, that implies

$$\left| \sum_j M_{ij} x_j \right| < |m_i^+| + |m_i^-| = \sum_{j \in \mathcal{J}_+} M_{ij} |x_j| + \sum_{j \in \mathcal{J}_-} M_{ij} |x_j| = \sum_j M_{ij} |x_j|.$$

Otherwise, if $(m_i^+ = 0 \vee m_i^- = 0)$ then

$$\left| \sum_j M_{ij} x_j \right| = |m_i^+| + |m_i^-| = \sum_{j \in \mathcal{J}_+} M_{ij} |x_j| + \sum_{j \in \mathcal{J}_-} M_{ij} |x_j| = \sum_j M_{ij} |x_j|.$$

Hence, if we sum over all indices i and if among these there is at least one index with $(m_i^+ > 0 \wedge m_i^- > 0)$, then we obtain

$$\sum_i \left| \sum_j M_{ij} x_j \right| < \sum_j |x_j| .$$

This stands in contradiction to equation (30.33a), which claims

$$\sum_i \left| \sum_j M_{ij} x_j \right| = \sum_i |x_i| .$$

Hence, assuming degeneracy for $\lambda = 1$, there cannot be an index i for which both contributions $m_i^{\pm}$ are nonzero. This implies for equation (30.33a), which can be modified according to equation (30.34):

$$x_i = \sum_j M_{ij} x_j = m_i^+ - m_i^- ,$$

that

$$\text{if } m_i^+ > 0 \Rightarrow \qquad i \in \mathcal{J}_+ \qquad \wedge \qquad x_i = m_i^+ = \sum_{j \in \mathcal{J}_+} M_{ij} x_j \, , \tag{30.35a}$$

$$\text{if } m_i^- > 0 \Rightarrow \qquad i \in \mathcal{J}_- \qquad \wedge \qquad x_i = -m_i^- = \sum_{j \in \mathcal{J}_-} M_{ij} x_j . \tag{30.35b}$$

We conclude that the two index sets $\mathcal{J}_{\pm}$ are decoupled if the stochastic matrix has a degenerate eigenvalue $\lambda = 1$. Hence the indices of the three subsets never mix. In other words, if we start with a configuration belonging to an index in $\mathcal{J}_+$, we can never reach a configuration belonging to indices in $\mathcal{J}_-$. Degeneracy of the largest eigenvalue can therefore be ruled out if the Markov process is ergodic, that is if each configuration can be reached from any other configuration by the Markov process in a finite number of time steps. For non-ergodic processes the set of states divides into subsets. The elements of a Markov chain all belong to only one of these subsets. Accordingly, there are as many eigenvalues $\lambda = 1$ as there are disjoint subsets.

A second dominant eigenvalue $\lambda = -1$

We still need to get rid of a possible second dominant eigenvalue, namely $\lambda = -1$. If $\lambda = -1$ can be ruled out, which is the case in so-called *aperiodic* Markov chains, one can prove as in the power method that

$$\lim_{\tau \to \infty} (\mathbf{M}^{\tau})_{ij} = \rho_i . \tag{30.36}$$

We will first elucidate the features associated with $\lambda = -1$, guided by a very simple example. Let the PMF of interest be uniform, $\rho_i = 1/\mathcal{N}$ on the index set $i \in \mathcal{J}$. For the proposal distribution we use a symmetric nearest-neighbour random walk with periodic boundary condition (pbc), that is

$$P(j|i) = \frac{1}{2}\left(\delta_{j,[i-1]_\mathcal{N}} + \delta_{j,[i+1]_\mathcal{N}}\right),$$
$$[i]_\mathcal{N} := \mathrm{mod}(i-1,\mathcal{N}) + 1.$$

The square bracket function ensures pbc. Put differently, the random walk describes nearest-neighbour moves on a ring with $\mathcal{N}$ sites. The Markov matrix, resulting from the Metropolis–Hastings algorithm, is simply

$$\mathbf{M} = \frac{1}{2}\overbrace{\begin{pmatrix} 0 & 1 & & & & 1 \\ 1 & 0 & 1 & & & \\ & 1 & 0 & \ddots & & \\ & & \ddots & \ddots & 1 & \\ & & & 1 & 0 & 1 \\ 1 & & & & 1 & 0 \end{pmatrix}}^{\mathcal{N}},$$

which is ergodic but *periodic*. The eigenvalues are given by

$$\lambda_l = \cos(k_l), \qquad k_l := \frac{2\pi(l-1)}{\mathcal{N}}, \qquad \text{for } l \in \mathcal{J}.$$

The eigenvalue $\lambda = 1$ corresponds to $n = 0$. The eigenvalue $\lambda = -1$ exists if $\mathcal{N}$ is even and corresponds to $l = \mathcal{N}/2 + 1$. In the periodic case, $\boldsymbol{\rho}^{(\tau)}$ switches back and forth between two distributions in the limit $\tau \to \infty$. In terms of the random walk the reason behind the alternation is the following. A walker starting at a site with an even index i will always be at an even site in an even number of steps, provided $\mathcal{N}$ is even. This is not the case for odd $\mathcal{N}$, due to hopping processes across the boundary. In general, the existence of an eigenvalue $\lambda = -1$ can easily be checked based on the matrix $\mathbf{M}^2$, which describes a Markov process that proceeds in double steps. Both matrices $\mathbf{M}$ and $\mathbf{M}^2$ have the same set of eigenvectors and the eigenvalues of $\mathbf{M}^2$ are given by λ_i^2. Hence, the spectrum of $\mathbf{M}^2$ lies in the interval [0, 1]. In order to check the existence of $\lambda = -1$ for $\mathbf{M}$, we merely have to test for degeneracy of $\lambda = 1$ for $\mathbf{M}^2$. From what we have worked out so far, $\lambda = -1$ can be ruled out if $\mathbf{M}^2$ is ergodic. In general, the ergodicity of the single and double-step Markov process can be tested easily.

In order to illustrate this aspect we consider again the *random walk* process with periodic boundary conditions for an even number of sites. The probability $(M^2)_{ji}$ is only nonzero if $i = j$ or if $|j - i| = 2$. In other words, if the walker starts at a site with an even index, he can never reach a site with an odd index and vice versa. Hence, $\mathbf{M}^2$ is not ergodic, corroborating the existence of an eigenvalue $\lambda = -1$ and of a periodic behaviour.

In general, the eigenvalue $\lambda = -1$ can be removed easily by a spectral shift combined with a rescaling[1] of the Markov matrix:

$$M_{ji} \to \tilde{M}_{ji} := (M_{ji} + a\delta_{ji})/(1+a).$$

[1] In order to retain the correct normalization.

The corresponding eigenvalues in the present example are

$$\lambda_l := \lambda(k_l) := \big(\cos(k_l) + a\big)/(1+a). \tag{30.37}$$

Now the even/odd sites are no longer correlated to an even/odd number of steps. This Markov matrix is obtained in the Metropolis–Hastings scheme if the proposed distribution is

$$q(j|i) = \big(\tfrac{1}{2}[\delta_{j,[i+1]_{\mathcal{N}}} + \delta_{j,[i-1]_{\mathcal{N}}}] + a\delta_{ij}\big)/(1+a). \tag{30.38}$$

Now the Markov process is ergodic and aperiodic, that is the dominant eigenvalue $\lambda = 1$ is not degenerate and the Markov process converges in the long run towards the desired (uniform) distribution.

However, the gap between the dominant eigenvalue and the next one is very small if $\mathcal{N} \gg 1$, namely

$$\Delta\lambda := 1 - \lambda_2 = 1 - \left(\cos\left(\frac{2\pi}{\mathcal{N}}\right) + a\right)/(1+a) = \mathcal{O}\left(\frac{1}{\mathcal{N}^2}\right).$$

As we will see in the next sections, a small gap causes severe slowing down of the MCMC process. A more interesting and realistic Markov process for test purposes is given by the Ehrenfest model [114], also known as the dog-flea problem, which was invented to study the fundamental properties of thermodynamics. The model has been analysed in great detail in the context of Markov processes in [4, 18]. For a wider-ranging discussion on the general behaviour of Markov chains, we refer the reader to [67, 74, 218].

30.4 Expectation value of the sample mean

The generation of a sample is in itself seldom of central interest. Rather, it is the expectation value

$$\langle f \rangle = \sum_{i=1}^{\mathcal{N}} \underbrace{f(\boldsymbol{\xi}_i)}_{:=f_i} \underbrace{\rho(\boldsymbol{\xi}_i)}_{:=\rho_i} = \sum_{i=1}^{\mathcal{N}} f_i \, \rho_i , \tag{30.39}$$

of some function $f(\boldsymbol{x})$ for a given PMF $\rho(\boldsymbol{x})$. As the exact evaluation is impossible for most real-world problems, the expectation value is estimated by the sample mean

$$\hat{S} = \frac{1}{N} \sum_{\tau=1}^{N} f(\boldsymbol{x}_\tau) . \tag{30.40}$$

When the sample is generated by the Metropolis–Hastings algorithm it is highly correlated, with a disadvantageous impact on the expectation value and variance of the sample mean of f. To begin with, we will analyse the expectation value of the sample mean:

$$\langle \hat{S} \rangle = \frac{1}{N} \sum_{\tau=1}^{N} \langle f(\boldsymbol{x}_\tau) \rangle = \frac{1}{N} \sum_{\tau=1}^{N} \sum_{j=1}^{\mathcal{N}} f_j \; P(\boldsymbol{x}_\tau = \boldsymbol{\xi}_j | p^0, \mathcal{I}) . \tag{30.41}$$

The Markov process starts with an initial state $\boldsymbol{x}_0$, which is drawn from some PMF p^0, with $p_i^0 = p^0(\boldsymbol{\xi}_i)$. As a special case, the chain could also always start from the same state, $\boldsymbol{x}_0 = \boldsymbol{\xi}_{i_0}$ say. The argument p^0 in the conditional part of the probability specifies which initial distribution is being used. All other information about the MCMC details is packed up into the background information $\mathcal{I}$. We employ the marginalization rule in order to specify the initial state $\boldsymbol{x}_0$:

$$\begin{aligned}\left\langle \hat{S}\right\rangle &= \frac{1}{N}\sum_{\tau=1}^{N}\sum_{j=1}^{\mathcal{N}}\sum_{i=1}^{\mathcal{N}} f_j \; P(\boldsymbol{x}_\tau{=}\boldsymbol{\xi}_j|\boldsymbol{x}_0{=}\boldsymbol{\xi}_i,\mathcal{I}) \; \underbrace{P(\boldsymbol{x}_0{=}\boldsymbol{\xi}_i|p^0,\mathcal{I})}_{=p_i^0} \\ &= \frac{1}{N}\sum_{\tau=1}^{N}\sum_{j=1}^{\mathcal{N}}\sum_{i=1}^{\mathcal{N}} f_j \; (\mathbf{M}^\tau)_{ji} \; p_i^0 .\end{aligned} \tag{30.42}$$

In the unlikely situation that the initial distribution p^0 is equivalent to the distribution of the underlying physical problem ρ, we don't actually need MCMC at all, equation (30.12) yields

$$\sum_i M_{ji}^\tau \; p_i^0 = \sum_i M_{ji}^\tau \; \rho_i = \rho_j \, ,$$

independent of the 'Markov time' τ. The expectation value of the sample mean is identical to the true mean:

$$\langle \hat{S}\rangle = \sum_{j=1}^{\mathcal{N}} f_j \; \rho_j = \langle f\rangle \, .$$

We proceed with equation (30.42) by inserting the spectral representation of equation (30.24) [p. 549] for $\mathbf{M}$:

$$M_{ji} = \sum_{l=1}^{\mathcal{N}} X_{jl} \; \lambda_l \; \frac{X_{il}^*}{\rho_i}, \tag{30.43}$$

$$\langle \hat{S}\rangle = \frac{1}{N}\sum_{l=1}^{\mathcal{N}} \underbrace{\Big(\sum_{i=1}^{\mathcal{N}} f_j \; X_{jl}\Big)}_{:=\tilde{f}_l} \Big(\sum_{\tau=1}^{N} \lambda_l^\tau\Big) \underbrace{\Big(\sum_{i=1}^{\mathcal{N}} \frac{p_i^0}{\rho_i} \; X_{il}^*\Big)}_{:=\tilde{p}_l^0},$$

$$\langle \hat{S}\rangle = \tilde{f}_1 \; \tilde{p}_1^0 + \frac{1}{N}\sum_{l=2}^{\mathcal{N}} \tilde{f}_l \left(\lambda_l \frac{1-\lambda_l^N}{1-\lambda_l}\right) \tilde{p}_l^0 . \tag{30.44}$$

The contributions of the eigenvalue $\lambda = 1$ have been split off and, similarly to the definition of $\tilde{p}_l^0$ given in equation (30.30) [p. 550], we introduce

$$\tilde{f}_l = \sum_j f_j X_{jl} = \sum_j f_j \sqrt{\rho_j} \; U_{jl},$$

the projection of the function f on the right eigenvectors X. These functions have some useful properties:

$$\sum_{l=1}^{\mathcal{N}} \tilde{f}_l^2 = \sum_{i,j} f_i f_j \sum_{l=1}^{\mathcal{N}} X_{il} X_{jl} \overset{(30.26)}{=} \sum_i f_i^2 \rho_i = \langle f^2 \rangle,$$

$$\tilde{f}_1 = \sum_i f_i X_{i1} = \sum_i f_i \rho_i = \langle f \rangle.$$

Hence, the restricted sum yields

$$\sum_{l=2}^{\mathcal{N}} \tilde{f}_l^2 = \langle (\Delta f)^2 \rangle. \tag{30.45}$$

Finally we obtain the sought-for result:

EXPECTATION VALUE OF THE SAMPLE MEAN

$$\langle \hat{S} \rangle = \langle f \rangle + B, \tag{30.46a}$$

$$B := \frac{1}{N} \sum_{l=2}^{\mathcal{N}} \tilde{f}_l \underbrace{\left(\lambda_l \frac{1 - \lambda_l^N}{1 - \lambda_l} \right)}_{:=G_N(\lambda_l)} \tilde{p}_l^0 . \tag{30.46b}$$

There is a bias B due to the Markov process and the choice of the initial PMF that vanishes as $1/N$. The function $G_N(\lambda)$ has two important features: firstly, $G_N(\lambda \to 1) = N$ and secondly, for large sample size N, the function $G_N(\lambda)$ is sharply peaked at $\lambda = 1$ such that the eigenvalues λ_l in the proximity of $\lambda = 1$ predominate the sum. This shows at the same time that MCMC is less efficient for Markov matrices with eigenvalues close to the dominant one. As pointed out earlier, the bias vanishes identically if p^0 coincides with the target PMF ρ. The bias depends not only on the details of the Markov matrix and the initial distribution p^0, but also on the observable f. It is conceivable that some contributions to the bias drop out due to symmetry ($\tilde{f}_l = 0$). A particular case is $f_i = 1$, with $\tilde{f}_l = \sum_i x_{i,l} = 0 \; \forall l > 1$. In this case the bias vanishes. This example is, however, not really important since it corresponds to the normalization of ρ: $\langle f \rangle = \sum_i \rho_i$. Nonetheless, it reveals that the Markov process conserves the normalization exactly.

30.4.1 An illustrative example

It is very revealing to study the properties of the Markov process for a uniform target PMF with nearest-neighbour moves and periodic boundary conditions, as outlined in Section 30.3.4. The walkers shall start their random journey always at position $i = 1$. For the sake of simplicity and transparency we use a simple observable f, namely the density $\langle n_I \rangle$ at site $i = I$. That is

$$p_i^0 = \delta_{i,1} \quad \text{and} \quad f_i = \delta_{i,I}.$$

Obviously, if the number of MC steps is smaller then I, the walker can never reach site $i = I$ and the MCMC estimator for $\langle n_I \rangle$ yields zero instead of $1/\mathcal{N}$, which would be the exact value. Moreover the variance, and hence the standard error, is also zero, resulting in the misleading conclusion

$$\langle n_I \rangle = 0 \pm 0.$$

The wrong error estimate will be the topic of the next section. Here we will discuss the wrong expectation value in more detail. In random walks, the mean number of steps it takes the walker to cover the distance I is proportional to I^2, as discussed in Section 4.2 [p. 57]. But even for $N > I^2$ the bias is not necessarily small. Consider the case $I = 1$, that is all walkers start at the very site where the measurement takes place. One might be inclined to expect zero bias. Evaluating equation (30.46b) reveals that this is not true. On second thoughts, the behaviour is reasonable since the distribution produced by the random walk under consideration is approximately Gaussian, centred around the initial site, and the width increases with $\sqrt{N}$ (after all, it's a diffusion process). Hence the probability of finding the walker at the initial site is overrated. These aspects are accounted for in equation (30.44). To illustrate this point we will evaluate equation (30.44) for $\mathcal{N} \gg 1$, where the slowing down is particularly severe. Moreover, we use the parameter $a = 1$ in equation (30.38), which simplifies the evaluation further. The eigenvalues are then, according to equation (30.37),

$$\lambda_l := \lambda(k_l) = (\cos(k_l) + 1)/2 = \cos^2\left(\frac{k_l}{2}\right).$$

In the present example the Markov matrix is real symmetric and the eigenvectors are:

$$U_{jl} = \frac{1}{\sqrt{\mathcal{N}}} e^{ik_l j}, \qquad k_l := \frac{2\pi}{\mathcal{N}}(l-1), \quad l \in \mathcal{J}.$$

Furthermore we have

$$\tilde{f}_l = \sqrt{\rho_I}\, U_{Il} = \frac{1}{\mathcal{N}} e^{ik_l I},$$

$$\tilde{p}_l^0 = \frac{1}{\sqrt{\rho_{i=0}}} U_{0l}^* = 1.$$

Equation (30.44) for $\langle \hat{S} \rangle$ then reads

$$\langle \hat{S} \rangle = \frac{1}{N\mathcal{N}} \sum_{l=1}^{\mathcal{N}} e^{ik_l I} \lambda_l \frac{1 - \lambda_l^N}{1 - \lambda_l}.$$

Since $\mathcal{N} \gg 1$ we approximate the sum by an integral:

$$\langle \hat{S} \rangle = \frac{1}{N}\frac{1}{2\pi}\int_{-\pi}^{\pi} e^{ikI}\, d(k)\frac{1-d(k)^N}{1-d(k)}\, dk$$

$$= \frac{1}{N}\frac{1}{\pi}\int_{-\pi/2}^{\pi/2} e^{i2Ix}\, \cos^2(x)\frac{1-\cos^{2N}(x)}{1-\cos^2(x)}\, dx \tag{30.47}$$

$$= \frac{1}{\pi}\int_{-\pi/2}^{\pi/2} e^{i2Ix}\, \underbrace{\left(\frac{1}{N}\cos^2(x)\frac{1-\cos^{2N}(x)}{1-\cos^2(x)}\right)}_{:=G_{N(x)}} dx. \tag{30.48}$$

If, in addition, the number of MCMC steps is large ($N \gg 1$) the function $G_N(x)$ can be approximated quite accurately by the Gaussian $G_N(x) \approx e^{-\frac{1}{2}x^2 N}$. Moreover, the integrand is negligible for $x > \pi/2$ and we can safely extend the integration limits to infinity. The remaining integral is well known and yields

$$\langle \hat{S} \rangle = \sqrt{\frac{2}{\pi N}} e^{-\frac{2I^2}{N}}.$$

Particularly interesting is the relative uncertainty

$$\varepsilon_{rel} := \frac{\langle \hat{S} \rangle - \langle f \rangle}{\langle f \rangle} = \mathcal{N}\sqrt{\frac{2}{\pi N}} e^{-\frac{2I^2}{N}} - 1,$$

which is worst for $I = 0$, corroborating the above qualitative discussion.

This example is nice from the theoretical point of view as it allows us to compute all quantities analytically and it sheds some light on several aspects of MCMC. In particular, it illustrates the random walk behaviour present in more realistic target distributions when the walk starts outside the important regions. On the whole, a flat or almost flat PMF is quite unrealistic and by far the worst application for MCMC. MCMC is tailored for the opposite situation, where the variability of the observable f is much smaller than that of the PMF.

30.4.2 A Gaussian prior PMF

A more realistic example would therefore be a bell-shaped PMF of small width. We retain the nearest-neighbour proposed distribution, however, with open boundary condition. The latter implies that a walker at site 1 either stays at site 1 or moves to site 2, both with probability 1/2, and similarly for $i = \mathcal{N}$. The Markov matrix is tridiagonal and the matrix elements along the diagonal and the lower and upper off-diagonal are depicted in Figure 30.2. We see that moves in the direction of the peak are always accepted, corresponding to $M_{ji} = \frac{1}{2}$. Numerical simulations corroborate that generally $\tau^* = \frac{\ln(\varepsilon^*)}{\ln(\lambda_2)}$, defined in equation (30.32) [p. 550], represents the typical number of MCMC steps required to

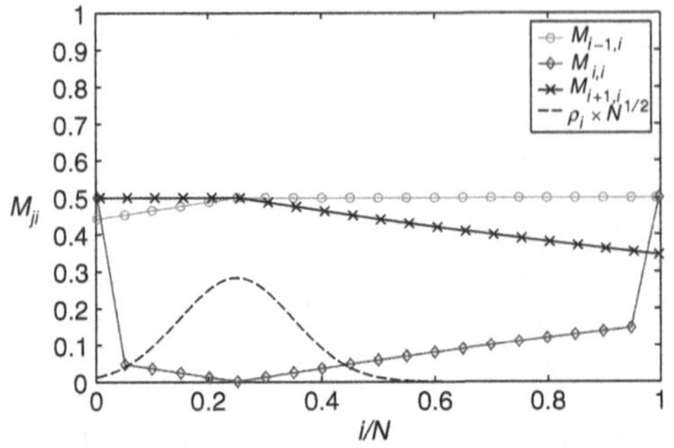

Figure 30.2 Elements of the Markov matrix (see key) for the main diagonal as well as the first lower and upper off-diagonal based on a nearest-neighbour probability distribution for a Gaussian PMF (ρ_i) (dashed line). The result is obtained for $\mathcal{N} = 400$.

suppress the bias. In the case of a bare random walk, $\lambda_2 = \cos\left(\frac{2\pi}{\mathcal{N}}\right) \approx 1 - \frac{2\pi^2}{\mathcal{N}^2}$ and hence $\tau^* \propto \mathcal{N}^2$. The same behaviour is observed in the case of a Gaussian PMF, if p_0 is centred far away from the important region of ρ.

30.5 Equilibration

Equilibration, as far as MCMC is concerned, is understood as ignoring the first K steps of the Markov chain in the sample mean. Skipping K steps is equivalent to starting the random walk with $\tilde{\boldsymbol{p}}^0 = \mathbf{M}^K \boldsymbol{p}^0$, resulting in a modification of equation (30.42):

$$\left\langle \hat{S} \right\rangle = \frac{1}{N} \sum_{\tau=1}^{N} \sum_{j=1}^{\mathcal{N}} \sum_{i=1}^{\mathcal{N}} f_j \, (\mathbf{M}^{\tau+K})_{ji} \, p_i^0 \, .$$

Apparently, the bias declines exponentially in K. If the total number N_0 of Markov steps is predefined, it is therefore expedient to use a certain fraction ($\kappa := K/N_0$) of them for equilibration. This lead directly to

$$B(\kappa) = \frac{1}{(1-\kappa)N_0} \sum_{l=2}^{\mathcal{N}} \tilde{f}_l \, \tilde{p}_l^0 \, \lambda_l^{1+\kappa N_0} \, \frac{1 - \lambda_l^{(1-\kappa)N_0}}{1 - \lambda_l} \, .$$

$B(\kappa)$ is a monotonically decreasing function in κ. In other words, as far as the bias is concerned, it is best to use only the last element of the Markov chain for the estimate. In contrast, as we will see below, the variance of the sample mean increases with decreasing sample size. Clearly, there is a trade-off between bias and variance reduction.

In closing this section, we conclude that the purpose of equilibration is not just to bring the walker closer to the important region of the target PMF ρ, but to bring the entire initial PMF p_0 close to the target PMF.

30.6 Variance of the sample mean

Apart from the fact that the sought-for expectation value $\langle f \rangle$ can be estimated by the sample mean, it is important to assess the efficiency or rather to assign confidence limits. To this end we analyse the variance of the sample mean:

$$\langle (\Delta \hat{S})^2 \rangle = \langle (\hat{S})^2 \rangle - \langle \hat{S} \rangle^2.$$

We assume that the Markov chain is properly equilibrated and the bias is negligible, i.e. $\langle \hat{S} \rangle = \langle f \rangle$, otherwise there is no point in computing the variance. In the formalism it means that we start the measurement after the equilibration phase and then $p_0 = \rho$. The variance is derived as follows:

$$\begin{aligned}
\left\langle (\Delta \hat{S})^2 \right\rangle &= \frac{1}{N^2} \sum_{\tau,\tau'} \left(\langle f(\boldsymbol{x}_{\tau'}) f(\boldsymbol{x}_\tau) \rangle - \langle f \rangle^2 \right) \\
&= \frac{2}{N^2} \sum_{\tau=1}^{N-1} \sum_{\tau'=\tau+1}^{N} \left(\langle f(\boldsymbol{x}_{\tau'}) f(\boldsymbol{x}_\tau) \rangle - \langle f \rangle^2 \right) \\
&\quad + \frac{1}{N^2} \sum_{\tau=1}^{N} \left(\langle f(\boldsymbol{x}_\tau)^2 \rangle - \langle f \rangle^2 \right) \\
&= \frac{2}{N^2} \sum_{\tau=1}^{N-1} \sum_{\Delta\tau=1}^{N-\tau} \Big(\underbrace{\langle \Delta f(\boldsymbol{x}_{\tau+\Delta\tau})\, \Delta f(\boldsymbol{x}_\tau) \rangle}_{:=a(\tau+\Delta\tau,\tau)} \Big) + \frac{1}{N^2} \sum_{\tau=1}^{N} \langle (\Delta f(\boldsymbol{x}_\tau))^2 \rangle.
\end{aligned}$$

Once the Markov chain is equilibrated, the expectation values are homogeneous in time τ, in other words $\langle f(\boldsymbol{x}_\tau) \rangle = \langle f \rangle$ and the autocorrelation depends only on the lag $\Delta\tau$, i.e. $a(\tau + \Delta\tau, \tau) = a(\Delta\tau)$, and $\langle (\Delta f(\boldsymbol{x}_\tau))^2 \rangle = \langle (\Delta f(\boldsymbol{x}))^2 \rangle := \sigma_f^2$ leading to

$$\begin{aligned}
\left\langle (\Delta \hat{S})^2 \right\rangle &= \frac{\sigma_f^2}{N} + \frac{2}{N^2} \sum_{\tau=1}^{N-1} \sum_{\Delta\tau=1}^{N-1} a(\Delta\tau) \theta\big(\Delta\tau \le N - \tau\big) \\
&= \frac{\sigma_f^2}{N} \left(1 + \frac{2}{N} \sum_{\Delta\tau=1}^{N-1} (N - \Delta\tau) \tilde{a}(\Delta\tau) \right).
\end{aligned}$$

The normalized autocorrelation $\tilde{a}(\Delta\tau) := a(\Delta\tau)/\sigma_f^2$ is a function that starts with one and decreases monotonically. The correction factor represents a kind of integrated autocorrelation. In the extreme case where the autocorrelation does not decrease within the Markov chain, i.e. $\tilde{a}(\Delta\tau) = 1$, resulting in

$$\left\langle (\Delta \hat{S})^2 \right\rangle = \sigma_f^2,$$

then the result is the same as if there was only one element in the Markov chain. The normalized autocorrelation as well as the variance of the sample mean is computed in Appendix B.4 [p. 616], and the result reads

Autocorrelation and variance of sample mean for an equilibrated Markov chain

$$\tilde{a}(\Delta\tau) = \sum_{l=2}^{\mathcal{N}} \frac{|\tilde{f}_l|^2}{\sigma_f^2}\, \lambda_l^{\Delta\tau}, \tag{30.49}$$

$$\left\langle (\Delta\hat{S})^2 \right\rangle = \frac{\sigma_f^2}{N}\Bigg[1 + \underbrace{\sum_{l=2}^{\mathcal{N}} \frac{|\tilde{f}_l|^2}{\sigma_f^2}\, \kappa_l}_{:=\kappa}\Bigg], \tag{30.50}$$

$$\kappa_l := \frac{2\lambda_l}{1-\lambda_l}\left(1 - \frac{1}{N}\frac{1-\lambda_l^N}{1-\lambda_l}\right).$$

The first term of the variance is the result predicted by the central limit theorem for independent random variables, identically distributed according to ρ. MCMC, however, generates random numbers which are highly correlated, even after the equilibration phase, which is described by the correction term κ. It can be detrimental to many realistic applications for two reasons. Firstly, the actual value of the correction is not known and secondly, it can be very large. The ubiquitous supposed cure to get rid of correlations is to thin the sample by retaining only every kth element in the Markov chain. This amounts to replacing the sample mean $\hat{S}$ by

$$S' := \frac{1}{N'} \sum_{\tau=1}^{N'} f(\boldsymbol{x}_{k\cdot\tau}),$$

with $N' = \frac{N}{k}$. A brief inspection of the derivation of the variance reveals that we merely have to replace the eigenvalues λ_l by λ_l^k. The change leads to

$$\left\langle (\Delta\hat{S})^2 \right\rangle = \frac{\sigma_f^2}{N'}\Bigg[1 + \underbrace{\sum_{l=2}^{\mathcal{N}} \frac{|\tilde{f}_l|^2}{\sigma_f^2}\, \kappa_l'}_{:=\kappa'}\Bigg], \tag{30.51}$$

$$\kappa_l' := \frac{2\lambda_l^k}{(1-\lambda_l^k)}\left(1 - \frac{1}{N'}\frac{1-\lambda_l^N}{1-\lambda_l^k}\right). \tag{30.52}$$

Finally, for sufficiently large N' the last term can be ignored and we get a simplified form

$$\kappa_l' = \frac{2\lambda_l^k}{\left(1-\lambda_l^k\right)}.$$

The value for k is determined such that the resulting nearest-neighbour autocorrelation is less than a sensible threshold. In order to illustrate this aspect, we consider a simple case with only two eigenvalues:

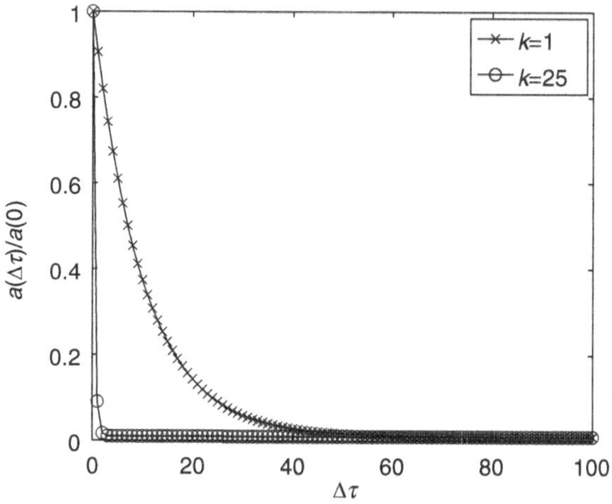

Figure 30.3 Autocorrelation function composed of two eigenvalues with $\lambda_2 = 1 - 10^{-6}$, $\lambda_3 = 0.9048$ and $\tilde{c}_3 = 100$ for $k = 1$ and $k = 25$.

$$\tilde{a}(\Delta\tau) = \frac{\lambda_2^{\Delta\tau} + \tilde{c}_3 \lambda_3^{\Delta\tau}}{1 + \tilde{c}_3}.$$

The corresponding autocorrelation is depicted in Figure 30.3. In this case, the correction factor for the correlated chain is $\kappa = 4 \times 10^4$. If we choose k such that $\tilde{a}(k) \approx 0.1$, we obtain in the present example $k = 25$. The figure also contains the autocorrelation for a Markov chain (MC*) thinned by a factor $k = 25$. If this were a realistic situation, we would not have the exact autocorrelation but only an MCMC estimate of it. The latter would pretend that the curve corresponding to MC* (marked by circles) has dropped to zero at $\Delta\tau \geq 1$ and we would be inclined to believe that we had cured the correlated Markov chain. But, if we compute the correction factor for MC* exactly, we still get $\kappa = 8 \times 10^3$. The reason is that the second dominant eigenvalue is very close to one, resulting in an awkwardly slow decrease of its contribution to the autocorrelation. On top of that, it is buried under the rapidly decaying contribution of the third eigenvalue which has a much larger relative weight $\tilde{c}_3$. Unfortunately, this behaviour is not unusual in realistic applications and the autocorrelation requires a thorough analysis in actual MCMC application.

30.6.1 Influence of the second dominant eigenvalues

We have seen that the second dominant eigenvalue ($\lambda_2 = 1 - \varepsilon_2$, with $\varepsilon_2 \ll 1$) causes the problems. We will analyse its contribution to κ. Starting from equation (30.50) and using the Taylor expansion for $\lambda_2^N = e^{N \ln(1-\varepsilon_2)}$, the contribution reads to leading order in ε_2

$$\kappa_2 := \frac{2\lambda_2}{(1-\lambda_2)}\left(1 - \frac{1-\lambda_2^N}{N(1-\lambda_2)}\right)$$
$$= \frac{2}{\varepsilon_2}\left(1 - \frac{1-e^{-N\varepsilon_2}}{N\varepsilon_2}\right).$$

Apparently, $1/\varepsilon_2$ defines the relevant scale for the required number of Markov steps. For $N \gg 1$ but still $N < 1/\varepsilon_2$, the behaviour of the leading-order term is

$$\kappa_2 \approx N.$$

Consequently, $\langle(\Delta S)^2\rangle \approx \sigma_f^2$, which does not decrease with increasing sample size. As a matter of fact, this case corresponds to a constant autocorrelation. If, however, the sample size N is large on the scale set by $1/\varepsilon_2$, i.e. $N\varepsilon_2 \gg 1$, then

$$\kappa_2 \approx \frac{2}{\varepsilon_2} \quad \Rightarrow \langle(\Delta S)^2\rangle \approx \frac{\sigma_f^2}{N}\,\frac{2}{\varepsilon_2}.$$

This is the result we would have obtained by an effective sample size $\tilde{N} := N\varepsilon_2/2$. In other words, elements of the Markov chain that are further apart than $1/\varepsilon_2$ are uncorrelated, i.e. $1/\varepsilon_2$ defines the autocorrelation length.

In summary, in equation (30.50) the contributions κ_l of eigenvalue λ_l are weighted by $|\tilde{f}_l|^2$, which is a measure of the overlap of the observable with the corresponding eigenvector. The autocorrelation time is determined by the maximum of all $1/\varepsilon_l$ that have a significant weight.

30.7 Taming rugged PDFs by tempering

We have seen repeatedly that the Bayesian posterior probability for parameters, say $\boldsymbol{x}$, is composed of prior and likelihood

$$p(\boldsymbol{x}|D,\mathcal{I}) = \frac{1}{Z}\,L(\boldsymbol{x})\;p_0(\boldsymbol{x}) \tag{30.53}$$

and the posterior expectation value of an observable $O(\boldsymbol{x})$ reads

$$\langle O\rangle := \int O(\boldsymbol{x})\;\frac{L(\boldsymbol{x})\;p_0(\boldsymbol{x})}{Z}\;d^E x,$$

where $d^E x$ stands for the integration measure. MCMC provides the means to estimate such integrals by the sample mean of the elements of a Markov chain $\{\boldsymbol{x}_1,\ldots,\boldsymbol{x}_N\}$:

$$\langle O\rangle^* = \frac{1}{N}\sum_i^N O(\boldsymbol{x}_i).$$

Unfortunately, MCMC does not always work satisfactorily. It has severe problems with rugged and multimodal posterior functions. In this case, correct sampling of the different peaks in the PDF can only be guaranteed by an infinite sample size. In particular, samples may not decouple from the initial configuration for a long time and autocorrelation times may be hard to determine. Here the concept of tempering in its two implementations, simulated tempering [137, 138] and parallel tempering [100, 155], comes in handy. The key idea behind this approach is borrowed from statistical physics, where the likelihood corresponds to the Boltzmann factor which contains an inverse temperature β that controls the smoothness of the posterior. Inspired by the Boltzmann factor, we define a temperature-dependent likelihood

$$p(\boldsymbol{x}|\beta, D, \mathcal{I}) := \frac{1}{Z(\beta)} L(\boldsymbol{x})^{\beta} \, p_0(\boldsymbol{x}) \tag{30.54}$$

with a β-dependent partition function or evidence $Z(\beta)$, defined as

$$Z(\beta) = \int L(\boldsymbol{x})^{\beta} \, p_0(\boldsymbol{x}) \, d^E x. \tag{30.55}$$

For $\beta = 1$ the original likelihood is recovered while in the opposite case, $\beta = 0$, the prior is retained. Now we extend the space of the random variables $\boldsymbol{x}$ by the auxiliary inverse temperature β and define the joint PDF

$$p(\boldsymbol{x}, \beta|D, \mathcal{I}) := \frac{1}{Z} L(\boldsymbol{x})^{\beta} \, p_0(\boldsymbol{x}) \, p_{\beta}(\beta),$$

with an as yet arbitrary PDF $p_{\beta}(\beta)$. The marginal probability for β becomes

$$p(\beta|D, \mathcal{I}) := \frac{Z(\beta)}{Z} \, p_{\beta}(\beta). \tag{30.56}$$

We can also determine the conditional PDF for β:

$$p(\beta|\boldsymbol{x}, D, \mathcal{I}) := \frac{L(\boldsymbol{x})^{\beta} \, p_{\beta}(\beta)}{Z_x(\boldsymbol{x})}. \tag{30.57}$$

Here $Z_x(\boldsymbol{x})$ is the appropriate normalization, which needs no further specification for our purposes. Now standard MCMC can be used to generate a sample in the extended parameter space $(\boldsymbol{x}, \beta)$ by doing alternating moves in $\boldsymbol{x}$ according to equation (30.54) and in β according to equation (30.57). As a matter of fact, the normalization $Z(\beta)$, like $Z_x(\boldsymbol{x})$, is not required in this context.

The advantage of introducing the temperature degree of freedom is that the joint PDF is smooth for small β values and has the ruggedness of $L(\boldsymbol{x})$ for $\beta = 1$. The walkers can therefore move freely between the various modes of $L(\boldsymbol{x})$ by a detour via small β values. Instead of using continuous values for β, it is necessary to allow only a set of discrete values $\beta \in \{\beta_1 < \beta_2 < \ldots < \beta_M\}$, with $\beta_1 = 0$ and $\beta_M = 1$. For each value β_i of the inverse temperature we generate a separate Markov chain. Whenever the walker has an inverse temperature β_l, the parameters $\boldsymbol{x}$ are added to the Markov chain for that temperature. The entries represent a valid sample for the β-dependent posterior defined in

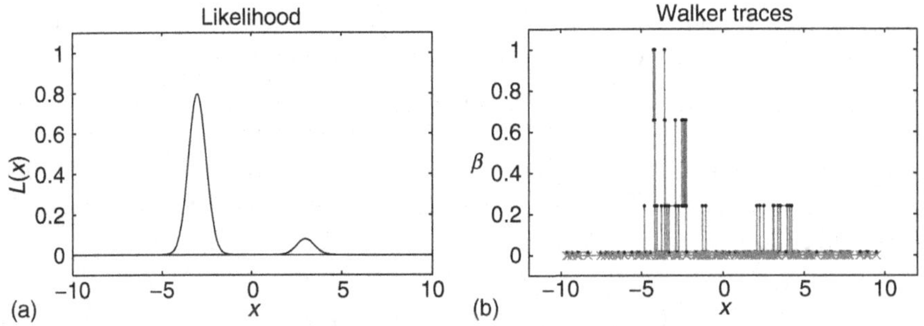

Figure 30.4 (a) Test likelihood with parameters $\sigma_{1/2} = 0.5, x_{1/2} = \pm 3, c_1 = 1, c_2 = 0.1$. The traces of the walkers in extended $(\beta, \boldsymbol{x})$-space are depicted in (b).

equation (30.54), which could be exploited to estimate thermodynamic expectation values of the form

$$\langle O \rangle_{\beta_l} := \int O(\boldsymbol{x})\; p(\boldsymbol{x}|\beta_l, D, \mathcal{I})\; d^E x.$$

This thermodynamic generalization of equation (30.53) occurs at the heart of statistical physics. In Bayesian inference we would not really be interested in values of β other than $\beta = 1$.

Some further details may be worth mentioning. The proposal distribution for $\boldsymbol{x}$ moves should be adjusted for each inverse temperature such that the acceptance rates are approximately 60%. Moreover, it is expedient to perform $\boldsymbol{x}$ moves more frequently than β moves, in other words between β moves a certain number $N_x \gg 1$ of $\boldsymbol{x}$ moves is performed. Finally, the β values have to be chosen such that neighbouring β-dependent likelihood functions have significant overlap, so that the acceptance probability for the various β moves is again in the 60% range. All these arrangements are necessary to guarantee reasonable mixing.

30.7.1 Bimodal test case

We will illustrate the features of simulated tempering guided by a simple yet challenging example. For the sake of clarity, we consider merely a one-dimensional parameter space and use a uniform prior over the interval $I = [-10, 10]$. The likelihood function shall consist of two well-separated Gaussian peaks:

$$L(x) = c_1\, \mathcal{N}(x|x_1, \sigma_1) + c_2\, \mathcal{N}(x| - x_1, \sigma). \tag{30.58}$$

The likelihood is depicted in Figure 30.4(a). For this test case we chose five β values $\{0, 0.02, 0.24, 0.66, 1\}$. The values have been chosen such that the acceptance probability for β moves is about 60%. For each β value a Gaussian proposal distribution for the $\boldsymbol{x}$ moves with individual variances is adjusted such that the acceptance rates are roughly 60%. For a short random walk, the path of the walker through $(\beta, \boldsymbol{x})$-space is depicted in

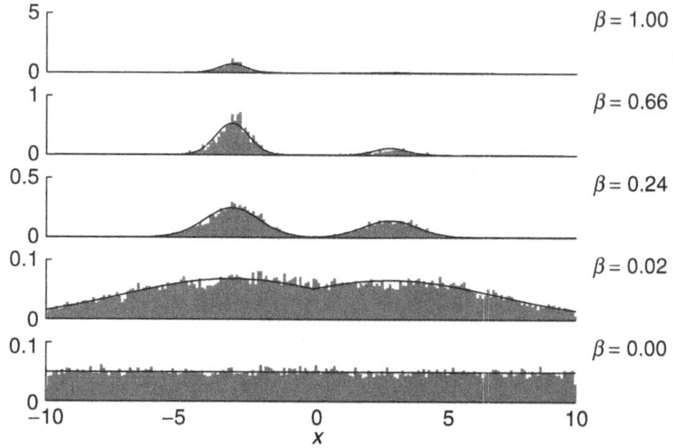

Figure 30.5 Normalized β-dependent histograms of the Markov points for the various inverse temperatures for $N_x = 100$ and $N_\beta = 10^5$ compared with the conditional PDF $p(\boldsymbol{x}|\beta, D, \mathcal{I})$ (solid line).

Figure 30.4(b) by solid lines and the $\boldsymbol{x}$ values entering the Markov chains are marked by dots. Clearly, at low temperatures ($\beta \approx 1$) the walker never tunnels directly between the two maxima of the PDF. He would rather take a detour through small β values.

In Figure 30.5 the normalized histogram of the $\boldsymbol{x}$ values entering the Markov chains for the various inverse temperatures are depicted and compared with the conditional probabilities $p(\boldsymbol{x}|\beta, D, \mathcal{I})$. The agreement illustrates that the walkers behave in the desired way. In this example, between the β moves we performed $N_x = 100$ $\boldsymbol{x}$ moves. The number of β steps is $N_\beta = 10^5$.

30.7.2 Parallel tempering

In passing we want to explain 'parallel tempering', also called replica exchange Monte Carlo, in a nutshell. It is based essentially on the same idea as simulated tempering. The difference being that one generates and keeps many (replicas) of walkers at different temperatures in parallel. For a suitably chosen number of time steps, the walkers within each β-slice move in $\boldsymbol{x}$-space driven by standard MCMC. Periodically, walkers on neighbouring β-slices are proposed to be swapped. Let the inverse temperatures and parameters of the two walkers be $(\beta_1, \boldsymbol{x}_1)$ and $(\beta_2, \boldsymbol{x}_2)$, respectively. The proposed move would result in a new configuration $(\beta_1, \boldsymbol{x}_2)$ and $(\beta_2, \boldsymbol{x}_1)$. Accordingly, the acceptance probability is given by

$$\alpha = \min\left(1, \frac{L(\boldsymbol{x}_1)^{\beta_1} L(\boldsymbol{x}_2)^{\beta_2}}{L(\boldsymbol{x}_2)^{\beta_1} L(\boldsymbol{x}_2)^{\beta_1}}\right).$$

For more details we refer the reader to [100, 137, 155].

30.7.3 Perfect tempering

A slight modification of simulated or parallel tempering, which we call perfect tempering [38], allows us to draw exact and independent samples from the posterior distribution $p(\boldsymbol{x}|D,\mathcal{I})$. All we need is a way to draw an exact and uncorrelated sample from the prior. If this is not possible, we can always introduce a different β-dependent posterior

$$p(\boldsymbol{x}|\beta, D, \mathcal{I}) := \frac{1}{Z}\left(L(\boldsymbol{x})\; p_0(\boldsymbol{x})\right)^{\beta}\left(\tilde{p}_0(\boldsymbol{x})\right)^{(1-\beta)},$$

which for $\beta = 0$ reduces to an arbitrary new prior, from which we know how to draw independent samples. For $\beta = 1$ we still retain the original posterior. Now, whenever the walker returns to $\beta = 0$, a new exact sample from the prior is drawn independent from the current state of the Markov chain. In doing so, the walker loses his memory (correlation) and the next time he reaches $\beta = 1$ a new uncorrelated and exact element in the Markov chain has been created. As a matter of fact, we can exploit not only the first arrival at $\beta = 1$ but all others as well, if we perform a careful analysis of the variance as outlined in [38].

30.8 Evidence integral and partition function

A task that is equally important in Bayesian inference as in thermodynamics is the evaluation of high-dimensional integrals of the form

$$Z = \int L(\boldsymbol{x})\, p(\boldsymbol{x})\, d^E x. \tag{30.59}$$

As outlined along with equation (30.55), such integrals occur in statistical physics as partition functions, where $L(\boldsymbol{x})$ is the Boltzmann factor and $p(\boldsymbol{x})$ the prior probability of microstate $\boldsymbol{x}$. In the Bayesian context such integrals are, for example, required in model selection and are called evidence integrals, where $p(\boldsymbol{x})$ and $L(\boldsymbol{x})$ now represent prior and likelihood, respectively. In both cases, the PDF $p(\boldsymbol{x})$ is slowly varying compared with $L(\boldsymbol{x})$, even on a logarithmic scale. It is therefore not possible to evaluate such integrals by standard MCMC means, since the likelihood/Boltzmann factor is much more restrictive than the prior. Consequently, sampling from the prior would result in a huge variance. In special cases one may have accumulated sufficient knowledge about the structure of the posterior such that reweighting can solve the problem. In general, however, a more sophisticated approach is needed. There have been various proposals on how such integrals could be tackled numerically. One of them is simulated tempering, an approach we have just been discussing.

30.8.1 Tempering

According to equation (30.56), the ratio of the marginal probabilities $p(\beta|D,\mathcal{I})$ for different inverse temperatures is

$$\frac{p(\beta_2)}{p(\beta_1)} = \frac{p_\beta(\beta_2|D,\mathcal{I})\; Z(\beta_2)}{p_\beta(\beta_1|D,\mathcal{I})\; Z(\beta_1)}.$$

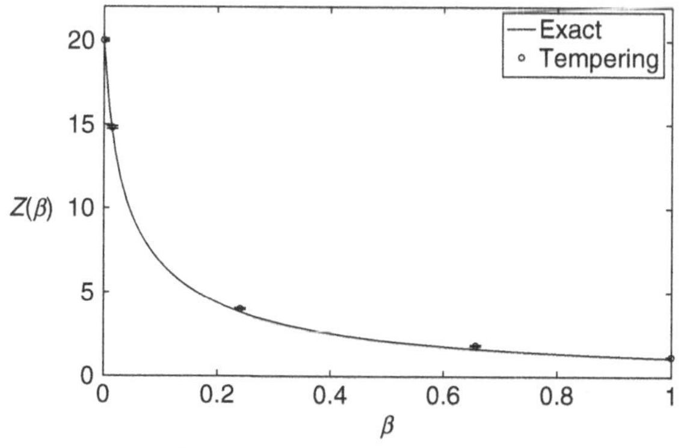

Figure 30.6 Evidence corresponding to the likelihood in equation (30.58) as a function of the inverse temperatures β. The exact result (solid line) is compared with the estimation based on simulated tempering (circles with error bars).

Clearly, the number of times N_β the walker reaches a specific β_l is proportional to $p(\beta_l|D,\mathcal{I})$. As a side note it should be mentioned that it is worthwhile adjusting $p_\beta(\beta)$ dynamically such that $p(\beta_{n+1}|D,\mathcal{I}) \approx p(\beta_n|D,\mathcal{I})$ in order to have a balanced variance for all β-slices. Owing to the normalization of the prior $p_0(\boldsymbol{x})$ we have $Z(\beta=0)=1$ and we can determine the partition function for any inverse temperature β through

$$Z(\beta) = \frac{N_\beta}{N_{\beta=0}} \frac{p_\beta(\beta=0)}{p_\beta(\beta)}.$$

The rest is mere MCMC routine. In Figure 30.6 the results corresponding to the data in Figure 30.5 are depicted and compared with the exact values of the partition function. Again the results demonstrate convincingly that simulated tempering indeed works.

30.8.2 Thermodynamic integration

Another approach, also based on the temperature-dependent posterior of equation (30.54), is thermodynamic integration, which plays an integral role in statistical physics. Upon differentiation of $\ln(Z(\beta))$ with respect to β, we find

$$\frac{d\ln Z(\beta)}{d\beta} = \int \ln(L(\boldsymbol{x}))\, p(\boldsymbol{x}|\beta, D, \mathcal{I})\, d^E x = \langle \ln L\rangle_\beta. \tag{30.60}$$

Now the task has changed significantly, in that we merely have to compute an expectation value based on the conditional posterior defined in equation (30.54) [p. 565]. This can be done by MCMC without knowing the normalization $Z(\beta)$. The new observable, $\ln(L(\boldsymbol{x}))$, is of course much more diffuse than the likelihood itself and simultaneously the PDF $p(\boldsymbol{x}|\beta, D, \mathcal{I})$ becomes more localized. The most unfavourable situation is $\beta=0$. In this case the PDF corresponds to the bare prior, but the likelihood is still replaced by the log-likelihood. The situation becomes progressively more effective as we move away from

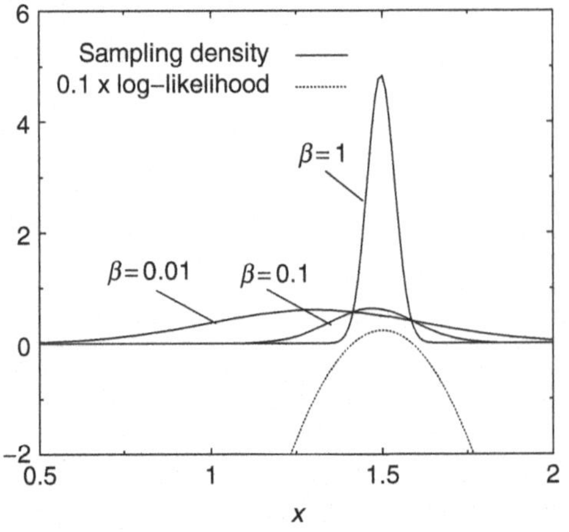

Figure 30.7 Behaviour of the factors in the integrand of equation (30.60). Solid lines stand for the sampling density $p(x\,|\,\beta, D, \mathcal{I})$ for different β values and the dotted line represents the log-likelihood, scaled by 0.1.

$\beta = 0$. The sought-for partition function can then be determined by integration over β, resulting in

$$\ln(Z(1)) = \int_0^1 d\beta \frac{d\ln(Z(\beta))}{d\beta} = \int_0^1 \Big\langle \ln(L) \Big\rangle_\beta \, d\beta. \tag{30.61}$$

The integrand turns out to be a fairly smooth function and the integral can be determined by standard numerical quadrature methods expressed as a weighted sum over a few properly chosen β values:

$$\ln(Z(1)) = \sum_{j=1}^{M} \omega_j \Big\langle \ln(L) \Big\rangle_{\beta_j}. \tag{30.62}$$

The expectation values of the log-likelihood, $\langle \ln(L) \rangle_{\beta_j}$, can be determined by standard MCMC.

As an illustrative one-dimensional example, a Gaussian centred at $x = 1.5$ with $\sigma = 0.04$ was chosen as the likelihood. The prior is 12.5 times more diffuse ($\sigma_p = 0.5$) and centred at $x = 1.0$. Figure 30.7 shows the sampling density $p(\boldsymbol{x}\,|\,\beta, D, \mathcal{I})$ for three different values of β. Note the shift of the sampling density towards $x = 1.0$ as the prior becomes more important. The log-likelihood scaled by a factor of 0.1 is shown as the dotted line. For $\beta = 1$ the sampling density covers essentially the positive part of the log-likelihood. For ever smaller values of β the sampling density is shifted towards the position of the peak of the prior ($x = 1$) and therefore the rapidly increasing negative part of the likelihood is sampled. The mean log-likelihood (equation (30.60)) therefore starts at huge negative values for $\beta = 0$ and increases rapidly as a function of β until it becomes positive for $\beta = 1$. In order to improve the evaluation of equation (30.61) it is therefore advantageous

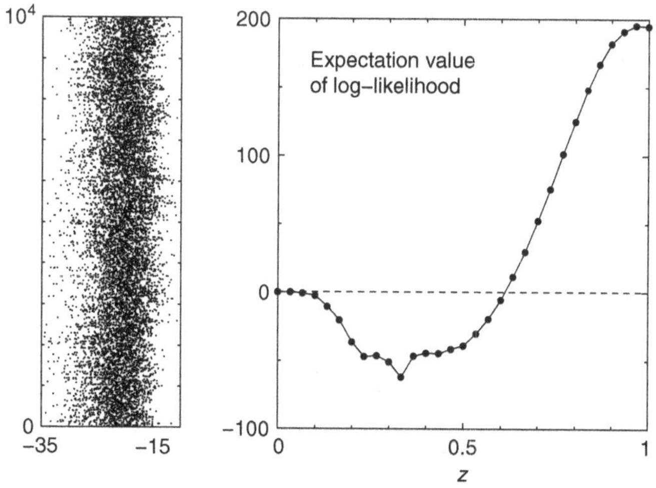

Figure 30.8 Sampling density after equilibration (left) and integrand in equation (30.64) after variable transformation equation (30.64) (right).

to apply a nonlinear transformation

$$\beta(z) = \frac{2\, z^n}{1 + z^n}, \qquad z \in [0, 1]. \tag{30.63}$$

Then the β-integral turns into

$$\int_0^1 \langle \ln(L) \rangle_\beta \, d\beta = \int_0^1 \langle \ln(L) \rangle_{\beta(z)} \, \frac{d\beta}{dz} \, dz. \tag{30.64}$$

This representation worked well, for example, for calculating the evidence in the analysis of the mass spectrometry data as discussed in Section 27.5 [p. 484]. For this problem Figure 30.8 shows in the right panel the integrand of equation (30.64) as a function of the new variable z. The exponent was chosen to be $n = 6$ in this calculation. The integrand is rather well behaved in the new variable and the integral in equation (30.61) can be computed by employing the trapezoidal rule preferentially with an adaptive technique. The left panel shows a section of the sampling density after equilibration.

31
Nested sampling

31.1 Motivation

An interesting and promising complementary approach for the stochastic estimation of high-dimensional integrals of the form

$$Z := \int d^N x \; L(\boldsymbol{x}) \; p(\boldsymbol{x}),$$

which may be evidence integrals or partition functions, has been proposed by John Skilling [65, 103, 188, 191–193]. It is widely applicable and allows us also to treat problems where the likelihood is extremely peaked in tiny regions in $\boldsymbol{x}$-space on an otherwise almost flat background. In such cases all MCMC techniques discussed so far are inefficient. Lebesgue once compared the brute-force evaluation of integrals via Riemann summation over $\boldsymbol{x}$ with an unsystematic merchant, who counts his money in the random order he receives it. A prudent merchant however, he claimed, would count his money by first grouping the coins and banknotes according to their respective values and counting how often the individual values occur [61]. This is precisely what Skilling suggested [188, 191]. He expresses the integral as

$$Z = \int d\lambda \; X(\lambda), \tag{31.1a}$$

$$X(\lambda) := \int d^N x \; p(\boldsymbol{x}) \; \boldsymbol{1}(L(\boldsymbol{x}) > \lambda)\,, \tag{31.1b}$$

where $X(\lambda)$ is the *prior mass* of regions where the likelihood exceeds a threshold λ. One verifies equation (31.1a) immediately:

$$\int d\lambda \; X(\lambda) = \int d^N x \; p(\boldsymbol{x}) \underbrace{\int d\lambda \; \boldsymbol{1}(\lambda < L(\boldsymbol{x}))}_{=L(\boldsymbol{x})} = Z.$$

For a general discussion on the mapping of high-dimensional integrals to one-dimensional integrals, see Section 7.3 [p. 96]. Clearly, $X(\lambda)$ decreases monotonically with increasing λ. For the following considerations we need strict monotonicity, which can always be achieved by adding a suitable infinitesimal function $X(\lambda) \to X(\lambda) + \eta(\lambda)$ with $\eta(\lambda) \to 0$ in the final results. Now we can define the inverse function $L(X)$ by $L(X(\lambda)) = \lambda$ and vice

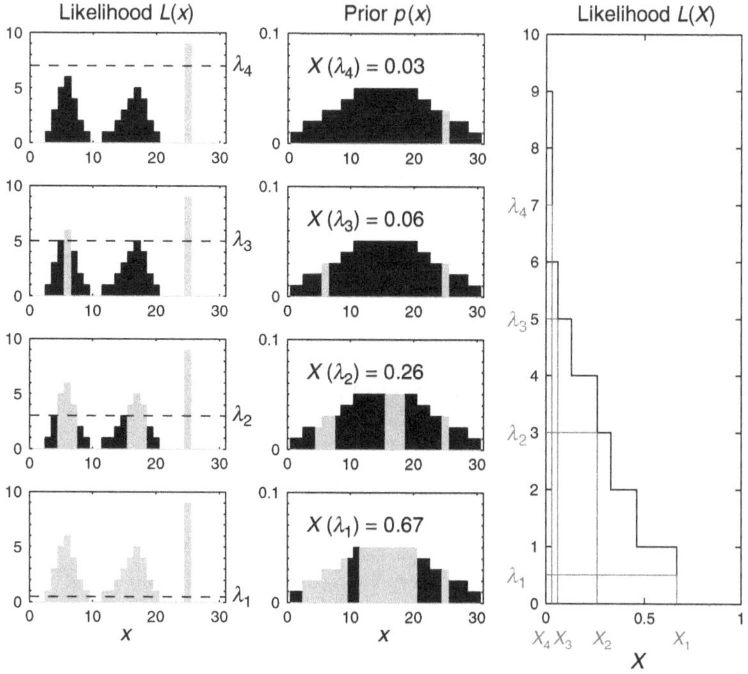

Figure 31.1 Sketch of the nested sampling concept. Meaning of columns from left to right: $L(x)$, $p(x)$ and $L(X)$. For four different threshold values of the likelihood $\lambda_4 = 7, \lambda_3 = 5, \lambda_2 = 3$ and $\lambda_1 = 0.5$ the corresponding prior masses $X_1, \ldots, X_4$ are indicated. For further discussion see text.

versa, $X(L(\xi)) = \xi$. Note that we here overload the symbol L for likelihood value, allowing it to be not just a function of position $\boldsymbol{x}$ but also (and equivalently) a function of prior mass X, according to the type of its argument. Then integration by parts of equation (31.1a) yields

$$Z = \int_0^1 dX \, L(X). \tag{31.2}$$

These ideas can be illustrated by a simple one-dimensional problem with only discrete likelihood values. $L(x)$ is depicted in the left panel of Figure 31.1. The prior PDF $p(x)$ is shown in the middle panel. The rows of the left panel correspond to different threshold values λ_ν, indicated by dashed horizontal lines. The constraint intervals of x values for which $L(x) > \lambda$ are shown in a lighter shade, both in the likelihood and prior graphs. The integral of the prior over these marked areas yields the prior mass $X(\lambda)$ for the underlying λ value. The result is printed in the corresponding subfigure. Finally, the rightmost figure depicts $L(X)$ vs X. In this figure we can read off the relation between λ_ν and the corresponding $X_\nu = X(\lambda_\nu)$, as indicated by the connecting lines. The inverse relation $X(\lambda)$ vs λ is obtained by merely swapping the axes.

In order to evaluate the integral numerically, we convert it into a Riemann sum. To this end we introduce a sequence X_n of X values

$$X_n := \zeta^n, \qquad \zeta := \frac{K}{K+1}, \tag{31.3}$$

which will be more fully explained shortly. The values X_n become dense in the limit $K \to \infty$ and we express the integral by

RIEMANN SUM

$$Z_K := \sum_{n=1}^{\infty} L(X_n) \left(X_n - X_{n+1}\right), \tag{31.4a}$$

$$X_n - X_{n+1} = \zeta^n \frac{1}{K+1}. \tag{31.4b}$$

Since $L(X)$ decreases monotonically over X, Z_K is a lower bound for Z. Likewise, an upper bound is obtain by

$$Z_K^u := \sum_{n=1}^{\infty} L(X_n)\,(X_{n-1} - X_n) = \zeta^{-1}\, Z_K.$$

If L has a nonzero minimum value ($L(X_0) = L(1) \neq 0$), this lower limit can be improved by summing from $n = 0$. Hence, the quadrature error is well controlled through lower and upper bounds:

UPPER AND LOWER BOUNDS FOR Z

$$Z_K \le Z \le \left(1 + \tfrac{1}{K}\right) Z_K. \tag{31.5}$$

The quadrature error can easily be improved to order $O\left(1/K^2\right)$ by the trapezoidal rule, but that is unnecessary as the statistical uncertainties will dominate the numerical error. So far, this is very nice but not really useful, since $L(X_n)$ is not at our disposal yet.

31.1.1 Nested sampling

Skilling proposed a stochastic approach to sample $L(X)$ based on an order statistic, which he called *nested sampling*. It is based on the (conditional) prior

$$p(\boldsymbol{x}|\lambda) := \frac{p(\boldsymbol{x})}{X(\lambda)}\, \boldsymbol{I}(L(\boldsymbol{x}) > \lambda), \tag{31.6}$$

which is proportional to the original one, however, restricted to regions in parameter space $\boldsymbol{x}$ for which $L(\boldsymbol{x})$ is above the given threshold λ. The nested sampling algorithm can be summarized in the following few lines:

NESTED SAMPLING ALGORITHM

- Initiate $\lambda_0^* = 0$, $n = 0$.
- Until convergence, iterate the following lines:
  1. Increment the iteration count $n \to n + 1$.
  2. Draw a sample $\{\boldsymbol{x}_i\}$ of size K from $p(\boldsymbol{x}|\lambda_{n-1}^*)$.
  3. Compute the corresponding L values $\lambda_i = L(\boldsymbol{x}_i)$.
  4. Determine the sample minimum $\lambda_n^* = \min_i \lambda_i$.
- Set $n_{\max} = n$.

Then the stochastic estimator is

SKILLING ESTIMATE OF Z

$$\hat{Z}_K := \sum_{n=1}^{n_{\max}} \lambda_n^* \left(X_n^* - X_{n+1}^*\right). \tag{31.7}$$

Nested sampling only yields the likelihood values λ_n^*. These values uniquely define the corresponding prior mass values X_n^* through the relation

$$\lambda_n^* = L(X_n^*). \tag{31.8}$$

However, since the functional form of $L(X)$ is not known, we replace X_n^* by X_n, defined in equation (31.4b):

$$\tilde{Z}_K := \sum_{n=1}^{n_{\max}} \lambda_n^* \left(X_n - X_{n+1}\right). \tag{31.9}$$

The reason will become clear later on. The sample size K, or rather the number of walkers as we will call them in this context, defines the quantity ζ that in turn determines the quadrature points in equation (31.4a). Needless to say, $\tilde{Z}_K$ is a random variable just like the sample minima λ_n^*. It is noteworthy that step 2 of the nested sampling algorithm can be simplified significantly. Instead of generating an entirely new sample, for $n > 1$ the elements of the previous sample can be recycled and merely the sample minimum λ_{n-1}^* of the previous iteration needs to be replaced. The other elements constitute already a valid random sample drawn from the conditional prior. Since λ_n^* is a monotonically increasing series, we can estimate the residue based on equation (31.7):

$$R_{n_{\max}} := \sum_{n=n_{\max}+1}^{\infty} \lambda_n^* \left(X_n - X_{n+1}\right).$$

Once the sample minima λ_n^* have converged, i.e. λ_n^* no longer changes over the course of the iterations, we can estimate the residue of the sum:

$$R_{n_{\max}} \approx \lambda_{n_{\max}}^* \sum_{n=n_{\max}+1}^{\infty} \Delta X_n = \lambda_{n_{\max}}^* X_{n_{\max}}.$$

Then, the error introduced by stopping the run after a finite number of steps reads

TRUNCATION ERROR

$$R_{n_{\max}} \approx \lambda_{n_{\max}}^* \zeta^{n_{\max}}. \tag{31.10}$$

$R_{n_{\max}}$ is a possible stopping criterion for the nested sampling run, provided λ_n^* levels off in an acceptable number of steps. This can, however, be dangerous if the likelihood consists of narrow peaks on an otherwise smooth background. Then λ_n^* can be rather flat for several iterations until the contributions from the peaks set in. In some cases the maximum of the likelihood $L_{\max}$ is known and can be used to decide on convergence. This would be a safe stopping criterion, but as we will see below, it might be overcautious. The computational cost required for a desired accuracy, which is proportional to $n_{\max}$, depends strongly on the problem. A different convergence criterion will be given in Section 31.3 [p. 591].

By now we know the quadrature and truncation error. But we still have to prove that equation (31.7) defines a sensible estimator for Z_K and on top of that we need to assess the stochastic uncertainty. Before elaborating on these aspects, we will illustrate the features of nested sampling guided by a simple and comprehensible example.

31.1.2 Illustrative example

A suitable model to study the behaviour of nested sampling is given by the 1-D example introduced for the assessment of simulated tempering, given by equation (30.58) [p. 566]. The likelihood possesses two well-separated peaks, that might be associated with two equivalent minima in the free energy, mimicking broken symmetry states but only in one (effective) dimension. High-dimensional problems will be discussed later on. In Figure 31.2 we sketch the L-dependence of the prior masses X_i, represented by the shaded areas above the lines λ_i. Clearly, for $L > L_{\max}$ ($L_{\max} \approx 0.8$ in this example), X vanishes. Lowering L below $L_{\max}$ yields a slow increase in X corresponding to the increasing interval size (prior mass) defined by the width of the highest peak. The decrease

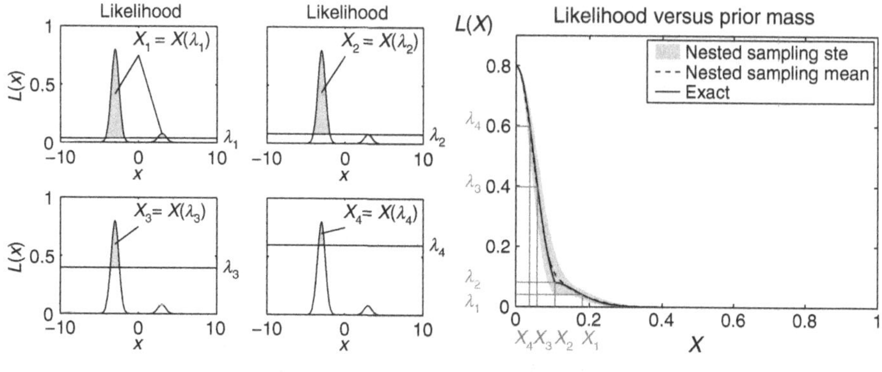

Figure 31.2 Illustration of the L-dependence of the prior mass. Left panels contain the likelihood $L(x)$ with different thresholds in L. Right panel contains exact results for $L(X)$ (solid line) and those obtained by nested sampling (dashed line) (see text).

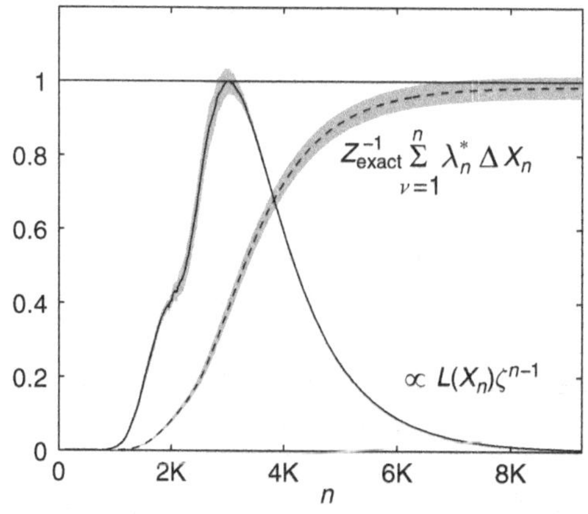

Figure 31.3 Nested sampling for the likelihood depicted in Figure 30.4 [p. 566] for $K = 100$ walkers. Mean values (solid/dashed lines) and error bands are computed from 40 independent nested sampling runs. The scaled summands (solid line) and the scaled partial sum (dashed curve) are plotted against the iteration count n.

in L below the top of the second peak at about $L \approx 0.1$ is indicated by a kink in $L(X)$ (see (λ_2, X_2) in the right panel).

In Figure 31.3 some diagnostic results are presented for $K = 100$ walkers. From 40 independent nested sampling runs, the mean and standard error are determined. The figure shows the dependence of the partial sum $\hat{Z}_K^{(n)}$ and the summands $\lambda_n^* \zeta^{n-1}$ on the iteration count n. The partial sum has been divided by the exact value of Z and the curve for the summands has been rescaled appropriately to fit into the same graph. We observe that the summands have a pronounced maximum at $\frac{n}{K} \approx 3$ and then they drop rapidly. From the partial sum we can tell that nested sampling yields the correct result within the error band. The likelihood dependence on the prior mass, $L(X)$, can also be determined from the same nested sampling data, namely by simply plotting $\hat{\lambda}_n^*$ vs X_n. The results corresponding to the data in Figure 31.3 are shown in the right panel of Figure 31.2. We see that nested sampling is indeed capable of providing $L(X)$ fairly accurately. This model, though only one dimensional, contains already the typical difficulties. It is multimodal, the modes are well separated and the volumes of the modes are different. Standard MCMC, based on short-range proposed distributions, has severe problems in such cases. First of all, walkers will get trapped in one of the modes, depending on the initial position of the walkers. The result will only describe the properties of this single mode correctly. A common cure is independently repeated MCMC runs, which eventually sample all modes but not necessarily with the correct relative weight. As discussed in Chapter 30 [p. 537], the latter will be biased by the way the initial positions of the walkers are chosen in combination with the relative apertures of the modes when communication ceases. In the present low-dimensional example these problems in MCMC can be overcome by using random numbers from a Cauchy distribution. The PDF in equation (29.26) [p. 519] has half of its area between $-\gamma$ and $+\gamma$ and the other half in its wings. Adjustment of the deviates to a finite support $[a, b]$ can easily be made. This cure is, however, inefficient in higher dimensions.

31.1.3 Sampling from a constrained prior

A crucial part of Skilling's approach is the sampling from a constrained prior. In the present test case, the allowed regions satisfying the likelihood constraint $L(\boldsymbol{x}) > \lambda^*$ can be determined easily and it is straightforward to sample uniformly within these (disconnected) intervals. In the left panel of Figure 31.2 we see that λ_1 defines two disjoint regions in parameter space. In general high-dimensional problems, the disconnected subregions may have a much more complex topology and the sampling from the constrained prior poses a real challenge. Various approaches have been suggested (see e.g. [181]). Most of them pick out at random one of the living walkers, as they are definitely in allowed regions, and modify the position randomly such that the walker stays inside the union of subregions (peaks) under consideration. If allowed to leave the peaks, it would in general be hard for the walker to ever get back into an allowed region.

One might expect that the result of nested sampling also depends on the initial distribution of walkers into the various subregions, as is the case in MCMC. Initially, the subregions will be populated according to their enclosed prior masses, as illustrated in [188]. This seed population drawn from the prior is unlikely to represent the distribution of posterior masses that we actually want, so we require some mechanism that does not simply allow a fixed population of walkers to wander around. Nested sampling is that mechanism.

As in the *Great Deluge* algorithm [58], used to search for the maxima of high-dimensional functions, we start out with $\lambda^* = 0$ and then gradually increase it. Already in the second iteration step the two modes in our test case are disjoint and living walkers will not be allowed to swap regions. However, unlike MCMC, walkers at small peaks will gradually die out and get reborn in regions with higher likelihood. The position of these regions is known from the walkers moving around in them. Eventually, all walkers gather

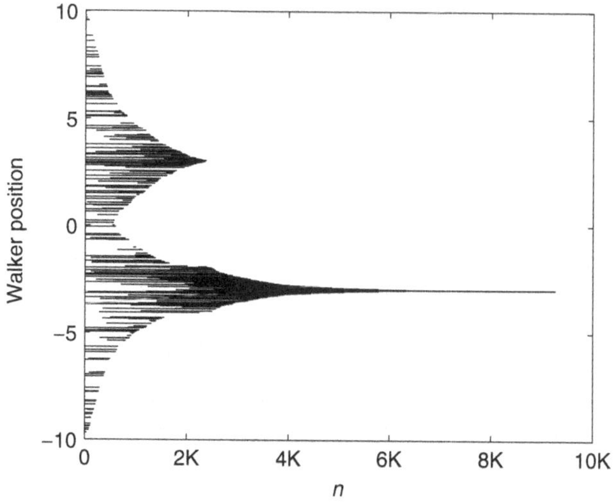

Figure 31.4 'Time evolution' of the walker positions for an ensemble of $K = 100$ walkers in a single nested sampling run for the test likelihood depicted in Figure 30.4 [p. 566].

in the highest-likelihood region. Obviously, it is crucial in multimodal problems to have enough walkers to populate each mode.

In Figure 31.4 the positions of the walkers versus the iteration count are depicted. By construction, the number of walkers is conserved in each iteration step. The structure resembles the shape of the likelihood, and we see clearly how the walkers eventually concentrate in the maximum of the likelihood. This dynamics ensures that the peaks of different heights are properly sampled.

As a matter of fact, the results in Figures 31.2 and 31.3 were generated by such a type of sampling, in which a living walker is selected at random and replaced by a new walker at random somewhere in the union of allowed subregions. So we see that nested sampling indeed works very well for this mock problem, and yields the correct answer within the error bars.

31.2 The theory behind nested sampling

Now we want to outline the theory behind nested sampling. At each iteration step L values increase, which is equivalent to decreasing X values. A remarkable feature of nested sampling is that – irrespective of the likelihood function and the prior PDF – the algorithm generates a sequence of successive X_n values that are uniform within the restricted interval $(0, X_{n-1})$. Most of the statistical analysis of nested sampling is based purely on these random variables. To begin with, we rearrange the sum in equation (31.7):

$$\hat{Z}_K = \sum_{n=1}^{n_{\max}} \lambda_n^* \left(X_n^* - X_{n+1}^*\right) = \sum_{n=1}^{n_{\max}} X_n^* \underbrace{\left(\lambda_n^* - \lambda_{n-1}^*\right)}_{\Delta\lambda_n^*},$$

where we used $L(X_0^*) = 0$. So, an alternative form of the estimator reads

ESTIMATOR FOR Z

$$\hat{Z}_K = \sum_{n=1}^{n_{\max}} X_n^* \, \Delta\lambda_n^* . \tag{31.11}$$

31.2.1 *Statistic of the prior masses X_n^**

For the proof that equation (31.11) converges towards the correct result and for the analysis of the statistical uncertainties, we need the mean and covariance of the prior masses X_n^* corresponding to the sample minima of the likelihood values λ_n^*. As already shown in equation (7.9) [p. 98], the PDF of the prior masses of the K walkers in the first iteration step is uniform in the unit interval:

$$p(X) = \mathbf{1}(X \in [0, 1]) . \tag{31.12}$$

Owing to monotonicity, the sample minimum likelihood λ_n^* in iteration n corresponds to the sample maximum X_n^* of the corresponding prior masses. Therefore, the values X of the prior masses at iteration step $(n + 1)$ are constrained by $X < X_n^*$. Then

$$p(X|X_n^*) = \frac{\mathbf{1}\big(X < X_n^*\big)}{X_n^*}, \tag{31.13}$$

and X_{n+1}^* is determined by the maximum of a sample of size K drawn from a uniform distribution in $(0, X_n^*)$. Therefore, each ratio $\frac{X_{n+1}^*}{X_n^*}$ is independently distributed in the same way as X_1^*.

SUCCESSIVE SAMPLE MAXIMA X_n^*

The prior mass range is compressed at each iteration identically and independently by X_1^*, which is the maximum of a sample of size K drawn from the uniform unit interval PDF.

According to equation (7.45) [p. 120], the maximum X_1^* of the sample of size K obeys a β distribution (equation (7.12) [p. 100]). For notational ease we define $\theta := X_1^*$, then

$$p(\theta) = p_\beta(\theta|\alpha = K, \rho = 1) = K\theta^{K-1} \quad \text{with} \quad \langle \theta^\nu \rangle = \frac{K}{K+\nu}. \tag{31.14}$$

Therefore, the mean *compression rate* per step is $K/(K + 1)$. For instance $X_2^* = X_1^* \, X_1^*$, but the two factors are independently drawn, that is the maximum of the first sample times the maximum of the second sample. Generally,

$$X_n^* = \prod_{\nu=1}^{n} \theta_\nu, \tag{31.15}$$

where the index enumerates the independent samples. Now it is an easy task to find the mean

$$\left\langle X_n^* \right\rangle = \prod_{\nu=1}^{n} \left\langle \theta_\nu \right\rangle = \left(\frac{K}{K+1}\right)^n = \zeta^n = X_n. \tag{31.16}$$

The covariance can be determined similarly (see Appendix B.5 [p. 617]).

MEAN AND COVARIANCE MATRIX OF X_n^*

$$\langle X_n^* \rangle = \zeta^n = X_n, \tag{31.17}$$

$$\langle \Delta X_{n+m}^* \Delta X_n^* \rangle = X_{n+m} X_n \left(X_n^{-\kappa} - 1\right), \tag{31.18}$$
$$\kappa := \frac{\ln(1 + 1/K(K+2))}{\ln(1+1/K)} = \frac{1}{K} + O\left(\frac{1}{K^2}\right).$$

Interestingly, the properties of X_n^* are independent of the likelihood and the prior PDF, respectively. We see that the mean of the prior mass values X_n^* associated with the sample minima of likelihood values λ_n^* is equal to the quadrature points X_n defined in equation (31.11).

31.2.2 Approaching Z directly, if admissible

Let us suppose that the problem at hand is challenging, but reliable estimates for Z are required and doable. The analysis will reveal the limitations of this approach.

Based on the likelihood values $\boldsymbol{\lambda} = \{\lambda_n^*\}$ of a single nested sampling run, we want to infer Z. In the Bayesian approach we compute the probability $p(Z|\boldsymbol{\lambda}, K, \mathcal{I})$. To this end we introduce the unknown prior mass values $\boldsymbol{X} = \{X_n^*\}$ corresponding to the likelihood values via the marginalization rule

$$p(Z|\boldsymbol{\lambda}, K, \mathcal{I}) = \int d\boldsymbol{X}\; p(Z|\boldsymbol{X}, \boldsymbol{\lambda}, K, \mathcal{I})\; p(\boldsymbol{X}|\boldsymbol{\lambda}, K, \mathcal{I}). \tag{31.19}$$

The first factor is a Dirac delta function, as the background information contains knowledge about equation (31.11):

$$p(Z|\boldsymbol{X}, \boldsymbol{\lambda}, K, \mathcal{I}) = \delta\left(Z - \sum_{n=1}^{n_{\max}} X_n^*\, \Delta\lambda_n^*\right).$$

The second factor in equation (31.19) in principle contains information about $\boldsymbol{\lambda}$, but that does not help us since we do not know the functional form of $L(X)$ or its inverse. This is also part of the background information. Otherwise the Bayesian inference would be unnecessary, as we could compute Z directly from $\boldsymbol{\lambda}$ alone. We therefore have

$$p(Z|\boldsymbol{\lambda}, K, \mathcal{I}) = \int dX\, \delta\Big(Z - \sum_{n=1}^{n_{\max}} X_n^*\, \Delta\lambda_n^*\Big)\, p(X|K, \mathcal{I}),$$

where $p(X|K, \mathcal{I})$ is the beta PDF for the prior masses. Now we easily determine the mean and variance of Z. For the νth moment we get

$$\big\langle Z^\nu \big\rangle = \Big\langle \Big(\sum_{n=1}^{n_{\max}} X_n^*\, \Delta\lambda_n^* \Big)^\nu \Big\rangle.$$

The mean is

$$\big\langle Z \big\rangle = \sum_{n=1}^{n_{\max}} \big\langle X_n^* \big\rangle\, \Delta\lambda_n^* = \sum_{n=1}^{n_{\max}} X_n\, \Delta\lambda_n^*.$$

This explains the replacement of X_n^* by X_n in equation (31.9) [p. 575]. For the variance we obtain

$$\big\langle (\Delta Z)^2 \big\rangle = \sum_{n,n'=1}^{n_{\max}} \Delta\lambda_n^*\, \Delta\lambda_{n'}^* \big\langle \Delta X_{n'}^*\, \Delta X_{n'}^* \big\rangle.$$

Along with the covariance given in equation (31.18), we obtain

$$\big\langle (\Delta Z)^2 \big\rangle = \sum_{n}^{n_{\max}} \Delta\lambda_n^*\, X_n \left(X_n^{-\kappa} - 1\right) \sum_{n' \geq n} \Delta\lambda_{n'}^* \left(2 - \delta_{nn'}\right) X_{n'}. \tag{31.20}$$

This expression can easily be computed based on the data $\boldsymbol{\lambda}$ of the nested sampling run. Owing to equation (31.18), for large K the variance is proportional to $1/K$.

The variance can be determined from a single nested sampling run and, like in MCMC, it is of order $1/K$.

For a quick-and-dirty estimate we can replace $\Delta\lambda_n^*$ by $\Delta L(mvXn)$ and that, provided $L(X)$ is continuous, by $L(X_n)(X_n - X_{n+1})$. Then the sample variance can be transformed back into an integral:

$$\big\langle (\Delta Z)^2 \big\rangle \approx \frac{2}{K} \int_0^1 dX\, L'(X) X \big(X^{-\kappa} - 1\big) \int_0^X dY\, L'(Y) Y.$$

The function $L(X)$ is sharply peaked close to $X = 0$. For a rough estimate we approximate $L(X)$ by an exponential $L(X) \approx L_{\max} e^{-X/W}$. Since the width $W := e^{-\xi}$ is significantly smaller than 1, the integrals over X can be extended to infinity and we obtain $Z = L_{\max} W$ and

$$\left\langle (\Delta Z)^2 \right\rangle_{\text{qd}} = Z^2 \left(e^{\frac{\xi}{K}} \left(1 - \frac{0.78}{K} \right) - 1 \right). \tag{31.21}$$

We see that $\frac{\xi}{K}$ has to be small compared with 1, otherwise the relative uncertainty is not small. For $\frac{\xi}{K} \leq 1$ we can expand the exponential and obtain

$$\frac{\text{std}(Z)_{\text{qd}}}{Z} \propto \sqrt{\frac{\xi}{K}}.$$

We will encounter later on a similar result for the uncertainty in $\ln(\hat{Z})$, which will not be restricted to $\frac{\xi}{K} < 1$.

Numerical results

In Table 31.1 the numerical results for Z and various estimates for the variance are depicted. The results correspond to the nested sampling runs presented in Figures 31.2 and 31.3 [p. 577]. We have used $K = 100$ walkers and the runs are stopped according to the criterion in equation (31.10) [p. 576] when the residue corresponds to a relative accuracy of less then 10^{-4}.

The first row contains the average and standard error of 40 independent nested sampling runs. This result shows that nested sampling provides an accurate estimate. Here, the standard error is proportional to $1/\sqrt{N_{\text{rep}}}$ and the smallness is due to $N_{\text{rep}} = 40$ repeated runs, which should not be confused with the number of walkers K. The other results in the table are the expected single-run estimates, averaged over 40 independent runs. The standard deviation $\text{std}(\hat{Z}_K)^{(a)}$ is determined from the variance of the estimates of $\hat{Z}_K$ of

Table 31.1 Numerical results for nested sampling

$\frac{Z}{Z_{\text{ex}}}$	=	1.00 ± 0.02
$\frac{n_{\max}}{K}$	=	9.30
$L_{\max}$	=	0.80
$\frac{\text{std}(Z)^{(a)}}{Z}$	=	0.18
$\frac{\text{std}(Z)^{(b)}}{Z}$	=	0.14
$\frac{\text{std}(Z)^{(c)}}{Z}$	=	0.32
$\frac{\text{std}(Z)^{(d)}}{Z}$	=	0.3

the independent runs, while $\text{std}(\hat{Z}_K)^{(b)}$ is the result of equation (31.20) averaged over the independent runs. Both results are in fairly good agreement. Even the rough estimate of the variance given in equation (31.21) yields, at least in the present example, a reasonable result, which is denoted by $\text{std}(\hat{Z}_K)^{(c)}$. The last row contains the result for the upper bound that will be discussed later in equations (31.34) [p. 593] and (31.30) [p. 591].

Most importantly, in the present example and for $K = 100$ the statistical error is already in the 10% regime. For a statistical accuracy of 1% we would therefore need $K = 10^4$ walkers.

This example was for a (maybe effective) one-dimensional problem that consisted of two well-separated peaks in the likelihood. Otherwise it was not really challenging. As pointed out in [193], most real-world problems involve high-dimensional situations resulting in spiky structures in $L(X)$ of extremely small width $W = e^{-\xi}$, or rather very large ξ. So there is no way to get ξ/K small, and the presented approach will break down. That does not, however, imply that nested sampling is bound to fail. Let us consider such a challenging problem next.

31.3 Application to the classical ideal gas

For the classical ideal gas, the partition function is proportional to

$$Z := \int d^N p \; e^{-\mathbf{p}^2/(2\sigma^2)}$$

where $\mathbf{p}$ is an N-dimensional vector containing the components of the momenta of all particles and $\sigma^2 = mk_BT$ corresponds to the product of mass, Boltzmann constant and temperature. Trivial prefactors stemming from the real-space volume and quantum corrections, respectively, have been omitted for the sake of simplicity. The integral can be computed exactly:

$$Z = \left(2\pi\sigma^2\right)^{N/2}. \tag{31.22}$$

In a strict sense, equilibrium thermodynamics requires $N \to \infty$, which is prohibitive for numerical approaches. In any event, the particle number has to be large enough to reach convergence for the physical quantities of interest. This is not really a problem for systems with non-interacting particles, but it turns into a severe restriction as soon as particle interactions come into play. Then, we need to be able to solve the integral for $N \ggg 1$.

For the application of nested sampling, or any other numerical integration scheme, we restrict the integral to an N-dimensional hypersphere $S(R)$ of radius R, such that the integral converges within $S(R)$ to the desired accuracy. This is the case for

$$R = 2\sqrt{N}\sigma. \tag{31.23}$$

Along with the volume of the hypersphere $S(R)$,

$$V_N(R) = \frac{R^N \pi^{\frac{N}{2}}}{\Gamma\left(\frac{N}{2}+1\right)}, \tag{31.24}$$

the partition function assumes the form

$$Z = V_N(R) \underbrace{\int_{S(R)} d^N x \; e^{-x^2/2\sigma^2} \cdot p_0(\boldsymbol{x})}_{:=\tilde{Z}} .$$

Here $p_0(\boldsymbol{x})$ is a uniform prior in $S(R)$, with $e^{-x^2/2\sigma^2}$ as the corresponding likelihood. The remaining numeral task is the evaluation of $\tilde{Z}$, for which the exact result is

$$\tilde{Z} = \frac{\left(2\pi\sigma^2\right)^{N/2}}{V_N(R)}.$$

Using the cutoff radius R of equation (31.23) and Stirling's formula for large N, we obtain

$$\ln\left[\tilde{Z}\right] = -N \ln[2] - \frac{N}{2}. \tag{31.25}$$

Obviously, $\tilde{Z}$ is independent of σ. The σ-dependence of the original partition function Z is hidden in $V_N(R)$, as the cutoff radius R is related to σ through equation (31.23).

Owing to the rotational symmetry, the nested contours are high-dimensional onion shells. The prior mass corresponding to radius r is $X(r) = \left(\frac{r}{R}\right)^N$ and in turn $r(X) = RX^{1/N}$. Then

$$L(X) = L(r(X)) = \exp\left(-\frac{X^{2/N}}{2(\sigma/R)^2}\right).$$

The Riemann sum in equation (31.4a) [p. 574] is

$$Z_K = \left(1-\zeta\right) \sum_{n=1}^{n_{\max}} L(X_n)\, X_n,$$

with $X_n = \zeta^n$. In equation (31.3) [p. 574] we have defined $\zeta = K/(K+1)$. Alternatively, we can also use $\zeta = e^{-1/K}$. Both choices yield correct results, as we will see, but the latter is more advantageous in the present context. In Figure 31.3 [p. 577] we have seen that the summands $S_n := L(X_n)\, X_n$ plotted versus n represent a bell-shaped curve.

In Figure 31.5 the summands are depicted for two independent nested sampling runs for the ideal-gas case with $N = 1000$ and $K = 10$. The nested sampling summands are defined as

$$\hat{S}_n := \lambda_n^* X_n .$$

We observe that the positions of the relevant contributions agree roughly, but the heights differ by several orders of magnitude. Analytically, the maximum can be derived from the root of the derivative of $L(X_n)X_n$, resulting in

$$X_{n^*} = \left(N\frac{\sigma^2}{R^2}\right)^{N/2}.$$

Since the cutoff radius R is chosen according to equation (31.23), we obtain

$$X_{n^*} = 2^{-N} \qquad \text{or rather} \qquad n^* = KN\ln(2). \tag{31.26}$$

The width of the peak, which can be defined as the position where the summands drop below a threshold ε, is $\Delta n = K\sqrt{|\ln(\varepsilon)N|}$. In Figure 31.5 we see that in actual nested sampling runs the peaks can be shifted about the true maximum by $O(\Delta n)$. In the numerical simulation discussed below, we will employ 10^8 particles in three spatial dimensions, i.e. $N = 3 \times 10^8$. Then $X_{n^*} = 2^{-N} \approx 10^{-10^8}$! It is obviously advantageous to introduce the *logarithmic compression* $\boldsymbol{\xi} := |\ln[X_{n^*}]|$.

The numerical simulation has to find this incredibly tiny target region. The number of steps it will take is roughly $n_{\max} = 3K \times 10^8 + K \times 10^4$. The last term ensures that the summands have dropped below 10^{-5}. This can be used in the present example as another stopping criterion. The peak of the summands corresponds to a likelihood value

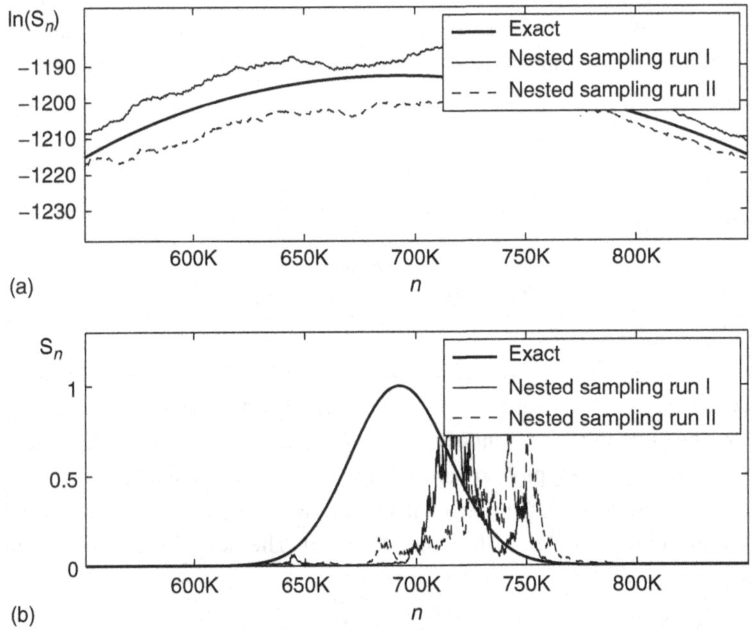

Figure 31.5 Sequence of summands occurring in the Riemann sum. The results of two independent nested sampling runs (jagged lines) are compared with the exact result (thick solid line). The underlying parameters are $N = 1000$ and $K = 10$. The upper panel contains $\ln(S_n)$. The lower panel shows S_n scaled such that all curves have the same height.

$$\ln\left[L(X_{n^*})\right] = -\frac{X_{n^*}^{2/N}}{2(\sigma/R)^2} = -\frac{1}{8(\sigma/R)^2} = -\frac{N}{2},$$
$$L(X_{n^*}) = e^{-N/2},$$

which is still far away from the maximum value $\mathcal{L}(X = 0) = 1$. Hence it is not always necessary to reach the upper limit of $L(X)$, if it is known at all. Moreover, there is not necessarily an upper bound on $L(X)$.

Computation of H

In general problems, where the functional form of $L(X)$ is not known and hence n^* is not accessible, Skilling introduced the *logarithmic prior–posterior compression* H:

$$H := \int \tilde{p}(\boldsymbol{x}) \ln\left[\frac{\tilde{p}(\boldsymbol{x})}{p(\boldsymbol{x})}\right] d^N x, \tag{31.27}$$

$$\tilde{p}(\boldsymbol{x}) := \frac{1}{Z} L(\boldsymbol{x})\, p(\boldsymbol{x}), \tag{31.28}$$

known in statistics as the Kullback–Leibler distance. The bulk of the integral for Z is usually found around the compression rate e^{-H} [193]. That is, the width of the peak of $L(X)$ is $W = e^{-H}$. H can also be expressed as

$$\begin{aligned} H &= \frac{1}{Z} \int L(\boldsymbol{x}) \ln\left[\frac{L(\boldsymbol{x})}{Z}\right] p(\boldsymbol{x})\, d^N x \\ &= \frac{1}{Z} \int L(X) \ln\left[L(X)\right] dX - \ln\left[Z\right]. \end{aligned}$$

From the statistical physics point of view it is very reasonable to choose H as a measure for the logarithmic compression. In statistical physics problems, L corresponds to the Boltzmann factor $e^{-\beta E(\boldsymbol{x})}$ and $\ln[Z] = -\beta F$ to the Helmholtz free energy F times the inverse temperature β. Then H is

$$\begin{aligned} H &= \frac{1}{Z} \int e^{-\beta E(\boldsymbol{x})}\Big(- \beta E(\boldsymbol{x})\Big) p(\boldsymbol{x})\, d^N x - \ln\left[Z\right] \\ &= -\beta U + \beta F = \beta\Big(F - U\Big) = -\beta T S = -\frac{S_B}{k_B}, \end{aligned}$$

with U being the inner energy and S_B the Boltzmann entropy. The latter is the logarithm of the relevant phase-space volume, which in turn is identical to the logarithm of the relevant prior mass. This strongly supports the choice $\ln(W) = H$.

In the ideal-gas case, the H value corresponding to $\tilde{Z}$ is

$$H = -\ln(\tilde{Z}) + \frac{\int_0^\infty e^{-\frac{r^2}{2\sigma^2}}\left(-\frac{r^2}{2\sigma^2}\right) r^N \frac{dr}{r}}{\int_0^\infty e^{-\frac{r^2}{2\sigma^2}} r^N \frac{dr}{r}}.$$

Along with equation (7.27) [p. 108], we obtain

$$H = -\ln(\tilde{Z}) - \frac{N}{2},$$

and equation (31.25) eventually leads to

$$H = N \ln 2.$$

So we see that e^{-H} in this example is precisely the position X_{n^*} of the maximum of the summands in equation (31.26).

Nested sampling simulation

Now let us compute $\tilde{Z}$ by nested sampling, based on equation (31.9) [p. 575]. For the same parameters as in Figure 31.5, we obtain from 100 independent runs

$$\tilde{Z} = \left(4.68 \pm 5.11\right) \times 10^{-508},$$

whereas the exact result is 5.85×10^{-517}. The nested sampling outcome for $\tilde{Z}$ is wrong by orders of magnitude and it gets even worse in higher dimensions N. The failure to compute the bare $\tilde{Z}$ by nested sampling has been pointed out by Skilling [193]. He argues that, in any case, the natural quantity to head for is $\ln(\tilde{Z})$. This is indeed the case for the statistical problem under consideration. As a matter of fact, it is the free energy F rather than the partition function that defines the thermodynamic potential:

$$-\frac{F}{k_B T} = \ln(Z).$$

This allows us to compute other thermodynamic variables, for example the Boltzmann entropy, via partial derivatives with respect to T. We take the same output of 100 independent nested sampling runs, compute $\ln(Z)$ for each run and determine the mean and standard deviation. The outcome is

$$\ln[Z] = -1187.9 \pm 7.9,$$

in strikingly good agreement with the exact value $Z_{\text{ex}} = -1188.6$. The reason for the inconsistent behaviour when estimating Z or $\ln[Z]$ is due to

$$\left\langle \ln\left[\hat{Z}\right]\right\rangle \neq \ln\left[\left\langle \hat{Z}\right\rangle\right].$$

In this case we can even compute the Riemann sum exactly. The outcome $Z_R = -1188.7$ corroborates that the bias due to quadrature and truncation is much smaller than the statistical uncertainty.

Table 31.2 contains results for various particle numbers $N_p = N/3$ (first column) in three spatial dimensions. The corresponding logarithmic width of the relevant prior mass $\xi = |\log(W)|$ and the exact result are depicted in columns two and three, respectively. The next two columns show the mean and standard deviation of $\ln(\tilde{Z})$ obtained from 100 independent runs. It is impressive to see how accurately nested sampling reproduces the

Table 31.2 Nested sampling results for the classical ideal gas for $K = 10$ walkers for various particle numbers N_p. For further details see text

N_p	ξ	$\ln(\tilde{Z}_{\text{ex}})$	$\overline{\ln(\tilde{Z}_{\text{ns}})}$	$\overline{\text{std}(\ln(\tilde{Z}_{\text{ns}}))}$	std*
10^2	462	–354.01	–353.56	4.32	5.63
10^3	6119	–3574.37	–3574.06	14.43	15.57
10^4	76188	–35788.19	–35785.41	46.62	46.79

Table 31.3 As Table 31.2 but for $\ln(\tilde{Z})/N$

$N/3$	ξ	$\ln(\tilde{Z}_{\text{ex}})$	$\ln(\tilde{Z}_{\text{ns}})$	$\text{std}(\ln(\tilde{Z}_{\text{ns}}))$	std*
10^2	462	–1.18006	–1.17852	0.01440	0.02655
10^3	6119	–1.19146	–1.19135	0.00481	0.00734
10^4	76188	–1.19294	–1.19285	0.00155	0.00156
10^6	10618772	–1.19314	–1.19310	0.00013	0.00015
10^8	1361877187	–1.19315	–1.19315	0.00002	0.00002

increasingly larger results. Although the relevant prior mass region shrinks rapidly with the particle number, the relative accuracy gets increasingly better. It is well known in statistical physics that the relative uncertainty of observables declines as $1/\sqrt{N}$. Moreover, the free energy and hence the log-partition function is extensive, i.e. proportional to N. It is therefore expedient to compute $\ln(\tilde{Z})/N$ (Table 31.3). It should be pointed out, however, that this quantity is not constant since the factor $V_N(R)$ is missing for the real partition function $\ln(Z)$. One can easily check that the relative uncertainty drops indeed as $1/\sqrt{N}$. As a matter of fact, with increasing dimension we don't even need more than a single walker in this unimodal case. The unimportance of the number of walkers K is surprising as the starting point was the Riemann sum and the quadrature error should depend on K. How come? Skilling proposed a generic model to study the origin of the statistical uncertainty in nested sampling, which we will outline next.

A generic example for challenging problems

In mega-dimensional problems, $L(X)$ will often have the form of a narrow spike at $X = 0$ rising out of a shallow background. Unfortunately, the spike yields the dominant contribution to the integral. In order to elucidate the behaviour of nested sampling under these circumstances, Skilling proposed [193] the following model: $L(X) = \boldsymbol{1}(0 \leq X \leq W)$ with exponentially small width $W = e^{-\xi}$. Correspondingly, $Z = e^{-\xi}$. A single nested sampling run generates a sequence X_n^* of prior mass values. The contribution to the integral is zero until $X_{\hat{n}}^* \leq W$ at step $\hat{n}$. In all remaining steps $L(X_{n'}^*) = 1$, so the estimated value is

$$\hat{Z}_K = (1-\zeta)\sum_{n=\hat{n}}^{\infty} X_n = \zeta^{\hat{n}}.$$

Here it is again expedient to use $\zeta = e^{-1/K}$. Then, from a single run,

$$\ln(\hat{Z}_K) = \hat{n}\,\ln(\zeta) = -\frac{\hat{n}}{K}. \tag{31.29}$$

The variability enters through $\hat{n}$. We will show in the next section that $\hat{n}$ has a Poisson PDF $p(\hat{n}|\mu)$ with mean $\langle n\rangle = K\xi := \mu$ and as such it turns into a Gaussian for large μ. Consequently, $\ln(\hat{Z})$ is Gaussian as well. The mean value is

$$\left\langle \ln(\hat{Z})\right\rangle = -\frac{\langle\hat{n}\rangle}{K} = -\xi = \ln(Z).$$

The result is unbiased and for the variance we obtain

$$\left\langle\left(\Delta\ln(\hat{Z})\right)^2\right\rangle = \xi K\left(\ln(\zeta)\right)^2 = \frac{\xi}{K}.$$

Hence,

$$\ln\left[Z\right] = -\xi \pm \sqrt{\frac{\xi}{K}}.$$

As indicated above, the result is unbiased for all K and the relative uncertainty of the ξ-dependent term is $1/\sqrt{K\xi}$, again in agreement with the observation in the ideal-gas example. For moderate ξ the uncertainty can be improved by increasing K, which is not necessary if ξ itself is already very large. Let us see what happens if we analyse the bare $\hat{Z}$, given by

$$\hat{Z} = e^{-\frac{\hat{n}}{K}}.$$

The moments follow from equation (31.29):

$$\left\langle\hat{Z}^\nu\right\rangle = \sum_{n=0}^{\infty}\zeta^{\nu n}\,p(n|\mu) = e^{-\mu}\sum_{n=1}^{\infty}e^{-\frac{\nu n}{K}}\frac{\mu^n}{n!} = e^{-\mu(1-e^{-\nu/K})}.$$

Expanding the exponent to leading order in $1/K$, we obtain

$$\left\langle\hat{Z}\right\rangle = e^{-\xi(1-\frac{1}{2K})}, \qquad \left\langle\hat{Z}^2\right\rangle = e^{-2\xi(1-\frac{1}{K})}.$$

The mean is off by a factor $e^{\xi/2K}$, which is gigantic, since in real-world applications $\xi \gg 1$ and K can never be chosen big enough to make up for it. For a concrete example, let us suppose that $\xi = 1000$ and $K = 10$, then the logarithmic estimate yields

$$\ln\left[Z\right] = -1000 \pm 10,$$

which is a fairly satisfactory and unbiased result.

The estimate for the bare Z, however, yields

$$Z = e^{-951.6} \pm e^{-906.3}.$$

The mean is orders of magnitude smaller than the standard deviation. Using the prior knowledge ($Z > 0$), we have

$$Z \in [0, e^{-906.3}]$$

or rather

$$\ln(Z) \in [-\infty, -906.3].$$

This useless result shows that $\ln(Z)$ is preferred to Z.

The impact of $\ln(Z)$ being Gaussian on the PDF for Z deserves a closer look. We consider the Gaussian random variable $x = \ln(Z)$ with

$$p(x) \propto e^{-(x-x_0)^2/2\sigma_x^2}.$$

Along with $p(Z) = p(x)\frac{dx}{dZ}$ we obtain for Z:

$$\langle Z \rangle = e^{x_0+\sigma_x^2/2},$$

$$\sigma_Z = \sqrt{\langle (\Delta Z)^2 \rangle} = e^{x_0+\sigma_x^2}\left(1 - e^{-\sigma_x^2}\right)^{1/2}.$$

Then the relative uncertainty is

$$\frac{\sigma_Z}{\langle Z \rangle} = e^{\sigma_x^2/2}\left(1 - e^{-\sigma_x^2}\right)^{1/2}.$$

Obviously, the relative uncertainty in Z is only small if $\sigma_x < 1$. Then we have

$$\frac{\sigma_Z}{\langle Z \rangle} = \sigma_x. \tag{31.30}$$

How long does it take to reach convergence?

An interesting question concerns the number $\hat{n}$ of steps it takes to enter the spike region, defined by $X^*_{\hat{n}} \leq W = e^{-\xi}$. Based on equation (31.17) [p. 581], we can at least conclude

$$\hat{n} \approx \frac{|\ln(W)|}{|\ln(\zeta)|} = K\xi.$$

For a detailed analysis and for later discussion it is expedient to introduce $\Phi := -\ln(X^*_1)$. Along with equation (31.14) [p. 580], the corresponding PDF is

$$p(\Phi) = p(\theta)\frac{d\theta}{d\Phi} = Ke^{-K\Phi}. \tag{31.31}$$

Then the sample maximum X^*_n of the prior masses in iteration step n, which is according to equation (31.15) [p. 581] a product of n independent factors θ_ν, translates into a sum

$t_n^* = \sum_{\nu=1}^{n} \Phi_\nu$ of independent and exponentially distributed random variables. This qualifies the values t_n^* as Poisson points (see Section 7.8.5 [p. 129]).

In a nested sampling run, let the first step to enter the spike be the $\hat{n}$th. Then $\hat{n} - 1$ Poisson points have to precede that, with their values t_ν not yet having reached the spike edge at ξ. On average, that interval will cover $\mu = K\xi$ steps. Hence, $\hat{n} - 1$ is Poisson distributed with mean μ:

$$p(\hat{n}|\mu) = e^{-\mu} \frac{\mu^{\hat{n}-1}}{(\hat{n}-1)!}.$$

Consequently, the number of steps it takes the walker to enter the relevant prior mass region can be summarized by

$$\hat{n} - 1 = K\xi \pm \sqrt{K\xi}.$$

This is what we have observed in the ideal-gas case.

31.4 Statistical uncertainty

Now for a realistic problem where we do not know the precise form of $L(X^*)$ and hence we do not know ξ, how can we estimate the error? The following approach is based on the ideas outlined in [193]. We start out from equation (31.11) [p. 580]:

$$\hat{Z}_K = \sum_{n=1}^{n_{\max}} X_n^* \, \Delta\lambda_n^* = \sum_{n=1}^{n_{\max}} e^{-t_n^*} \, \Delta\lambda_n^*. \tag{31.32}$$

We would obtain the correct result within the quadrature error if we knew the correct position t_n^* corresponding to the measured λ_n^*. We do not know these values, but we can derive their distribution. According to equation (9.6) [p. 152] t_n^*, the sum of independent and exponentially distributed random numbers, is Gamma distributed $p(t_n^*) = p_\Gamma(t_n^*|n, K)$, with mean and variance given in equations (7.14c) and (7.14e) [p. 102] $\langle t_n^* \rangle = \frac{n}{K}$ and $\text{var}(t_n^*) = \frac{n}{K^2}$. Then a Bayesian analysis, analogous to equation (31.19) [p. 581], yields

$$\left\langle \ln\left[Z_K\right]^\nu \right\rangle = \int DT \left(\ln\left[\sum_{n=1}^{n_{\max}} e^{-t_n^*} \, \Delta\lambda_n^* \right] \right)^\nu p(T|K, \mathcal{I}),$$

with $T := \{t_n^*\}$. We can easily generate Gamma samples according to $p(T|K, \mathcal{I})$ and evaluate first and second moments of $\ln(Z)$ numerically. In order to gain insight into the uncertainty of nested sampling, we will also present an analytic approach based on ideas outlined in [193].

We will compute the uncertainty that is introduced if we replace the true t_n^* in equation (31.32) by the mean values $t_n = n/K$, resulting in $\overline{Z}_K$. The mean squared difference between the true and the estimated log-Z value is

$$\left\langle \left(\ln(\hat{Z}_K) - \ln(\overline{Z}_K) \right)^2 \right\rangle = \left\langle \left(\ln\left[\frac{\sum_{n=1}^{n_{\max}} e^{-t_n^*} \, \Delta\lambda_n^*}{\sum_{n=1}^{n_{\max}} e^{-\frac{n}{K}} \, \Delta\lambda_n^*} \right] \right)^2 \right\rangle.$$

Since $\Delta\lambda_n^*$ is by construction positive, we can define normalized weights

$$\rho_n := \frac{e^{-\frac{n}{K}}\,\Delta\lambda_n^*}{\sum_{n=1}^{n_{\max}} e^{-\frac{n}{K}}\,\Delta\lambda_n^*}, \qquad \rho_n > 0, \qquad \sum_n \rho_n = 1.$$

Then

$$\left\langle \left(\ln(\hat{Z}_K) - \ln(\overline{Z}_K)\right)^2 \right\rangle = \left\langle \left(\ln\left[\sum_{n=1}^{n_{\max}} \rho_n\, e^{-\Delta t_n^*}\right]\right)^2 \right\rangle, \tag{31.33}$$

with $\Delta t_n^* := t_n^* - \langle t_n^* \rangle$. Now, let $e^{-\Delta t_{n'}^*}$ be the greatest contribution to the inner sum of equation (31.33). Then we obtain an upper bound if we replace all other exponentials by the largest one. By virtue of the normalization of the weights, we obtain an upper bound for the variance

$$\left\langle \left(\ln(\hat{Z}_K) - \ln(\overline{Z}_K)\right)^2 \right\rangle \le \left\langle \left(\ln\left[e^{\Delta t_{n'}^*}\right]\right)^2 \right\rangle = \left\langle (\Delta t_{n'}^*)^2 \right\rangle = \frac{n'}{K^2}.$$

Obviously, the upper bound comes from the last term $n' = n_{\max}$, which is defined by the *logarithmic compression* factor $n_{\max} = K\xi$.

Upper bound for the variance of $\ln[Z]$

$$\sqrt{\left\langle \left(\ln(\hat{Z}_K) - \ln(\overline{Z}_K)\right)^2 \right\rangle} \le \frac{\sqrt{n_{\max}}}{K} = \sqrt{\frac{\xi}{K}}. \tag{31.34}$$

Since ξ is determined by the prior-to-posterior compression, this upper limit demonstrates nested sampling is convergence to $\ln(Z)$ as $K \to \infty$. The last column in Table 31.2 and equation (31.33) contains this upper bound for the ideal-gas example. We see that it is actually very close to the true uncertainty, because firstly we can estimate $n_{\max}$ very well and secondly the inner sum in equation (31.33) seems to be dominated by one term. We also observe that the upper bound correctly describes the size dependence and that a more accurate determination of the bound is immaterial.

It should be pointed out, however, that equation (31.34) is the distance from the true truncated Riemann sum. It is assumed that the truncation $n_{\max}$ is under control. Needless to say, this is not always possible. From Figure 31.5 [p. 586] we see that it might be tough to tell from the summands of an individual run whether convergence has been reached, or whether we can use convergence in the likelihood as we have noticed that the run is converged long before the maximum likelihood is reached. For these high-dimensional problems the H function seems to be the best choice for the scale ξ. More details on the stopping criterion and its influence on the accuracy can be found in [193].

31.5 Concluding remarks

We have seen that nested sampling works very well for high-dimensional applications. It has no problem zooming into isolated peaks of the likelihood function and it provides the relative weights of the peaks correctly. To this end, however, it is decisive that either the walkers can tunnel freely between the peaks or, preferably, there are enough walkers for all separate modes. In the two examples it was simple to generate exact independent samples for the restricted prior. If this is not possible and MCMC has to be invoked for that purpose, then the bias and variance of MCMC have to be taken into account as well. But most of all it will slow down the simulation significantly. We refer the reader to the original publications for the various proposed algorithms [31, 181].

The presented didactic examples did not exploit the full potential of nested sampling. It is straightforward to use nested sampling in order to determine posterior expectation values as well. But more importantly, nested sampling really does open up an important class of multiphase problems, related to phase transitions and mixed phases, that have been inaccessible to standard tempering techniques.

Appendix A

Mathematical compendium

A.1 Schur complement

In connection with the marginalization rule, the 'Schur complement' can be very useful in dealing with matrices in general or covariance matrices in particular, as they occur for example in connection with Gaussian integrals. We start out from a 2×2 block representation of the matrix under consideration:

$$\mathbf{M} = \begin{pmatrix} \mathbf{M}_{11} & \mathbf{M}_{12} \\ \mathbf{M}_{21} & \mathbf{M}_{22} \end{pmatrix}.$$

The key idea is to transform the matrix into a diagonal block form by upper and lower triangular matrices. In a first step we demand

$$\begin{pmatrix} \mathbf{M}_{11} & \mathbf{M}_{12} \\ \mathbf{M}_{21} & \mathbf{M}_{22} \end{pmatrix} \begin{pmatrix} \mathbb{1} & -\mathbf{F}_{12} \\ 0 & \mathbb{1} \end{pmatrix} \stackrel{!}{=} \begin{pmatrix} \mathbf{M}_{11} & 0 \\ \mathbf{M}_{21} & \tilde{\mathbf{M}}_{22} \end{pmatrix},$$
$$\mathbf{F}_{12} := \mathbf{M}_{11}^{-1}\mathbf{M}_{12},$$
$$\tilde{\mathbf{M}}_{22} := \mathbf{M}_{22} - \mathbf{M}_{21}\mathbf{M}_{11}^{-1}\mathbf{M}_{12}.$$

The quantity $\tilde{\mathbf{M}}_{22}$ is the so-called Schur complement. In a second step we reach the diagonal form by

$$\begin{pmatrix} \mathbb{1} & 0 \\ -\mathbf{F}_{21} & \mathbb{1} \end{pmatrix} \begin{pmatrix} \mathbf{M}_{11} & 0 \\ \mathbf{M}_{21} & \tilde{\mathbf{M}}_{22} \end{pmatrix} \stackrel{!}{=} \begin{pmatrix} \mathbf{M}_{11} & 0 \\ 0 & \tilde{\mathbf{M}}_{22} \end{pmatrix},$$
$$\mathbf{F}_{21} := \mathbf{M}_{21}\mathbf{M}_{11}^{-1}.$$

Hence, in total we have

$$\begin{pmatrix} \mathbb{1} & 0 \\ -\mathbf{F}_{21} & \mathbb{1} \end{pmatrix} \begin{pmatrix} \mathbf{M}_{11} & \mathbf{M}_{12} \\ \mathbf{M}_{21} & \mathbf{M}_{22} \end{pmatrix} \begin{pmatrix} \mathbb{1} & -\mathbf{F}_{12} \\ 0 & \mathbb{1} \end{pmatrix} = \begin{pmatrix} \mathbf{M}_{11} & 0 \\ 0 & \tilde{\mathbf{M}}_{22} \end{pmatrix},$$

or along with

$$\begin{pmatrix} \mathbb{1} & \mathbf{F}_{12} \\ 0 & \mathbb{1} \end{pmatrix}^{-1} = \begin{pmatrix} \mathbb{1} & -\mathbf{F}_{12} \\ 0 & \mathbb{1} \end{pmatrix},$$
$$\begin{pmatrix} \mathbb{1} & 0 \\ \mathbf{F}_{21} & \mathbb{1} \end{pmatrix}^{-1} = \begin{pmatrix} \mathbb{1} & 0 \\ -\mathbf{F}_{21} & \mathbb{1} \end{pmatrix}$$

we obtain

Schur complement block diagonalization of 2 × 2 block matrices

$$\begin{pmatrix} \mathbf{M}_{11} & \mathbf{M}_{12} \\ \mathbf{M}_{21} & \mathbf{M}_{22} \end{pmatrix} = \begin{pmatrix} \mathbb{1} & 0 \\ \mathbf{F}_{21} & \mathbb{1} \end{pmatrix} \begin{pmatrix} \mathbf{M}_{11} & 0 \\ 0 & \tilde{\mathbf{M}}_{22} \end{pmatrix} \begin{pmatrix} \mathbb{1} & \mathbf{F}_{12} \\ 0 & \mathbb{1} \end{pmatrix}, \tag{A.1a}$$

$$\mathbf{F}_{12} = \mathbf{M}_{11}^{-1}\mathbf{M}_{12}, \tag{A.1b}$$

$$\mathbf{F}_{21} = \mathbf{M}_{21}\mathbf{M}_{11}^{-1}, \tag{A.1c}$$

$$\tilde{\mathbf{M}}_{22} = \mathbf{M}_{22} - \mathbf{M}_{21}\mathbf{M}_{11}^{-1}\mathbf{M}_{12}. \tag{A.1d}$$

So we immediately get

$$\det(\mathbf{M}) = \det(\mathbf{M}_{11})\det(\tilde{\mathbf{M}}_{22}) \tag{A.2}$$

and

$$\mathbf{M}^{-1} = \begin{pmatrix} \mathbb{1} & -\mathbf{F}_{12} \\ 0 & \mathbb{1} \end{pmatrix} \begin{pmatrix} \mathbf{M}_{11}^{-1} & 0 \\ 0 & \tilde{\mathbf{M}}_{22}^{-1} \end{pmatrix} \begin{pmatrix} \mathbb{1} & 0 \\ -\mathbf{F}_{21} & \mathbb{1} \end{pmatrix}. \tag{A.3}$$

For some applications it is useful to determine the inverse matrix explicitly:

$$\begin{aligned} \mathbf{M}^{-1} &= \begin{pmatrix} \mathbb{1} & -\mathbf{F}_{12} \\ 0 & \mathbb{1} \end{pmatrix} \begin{pmatrix} \mathbf{M}_{11}^{-1} & 0 \\ -\tilde{\mathbf{M}}_{22}^{-1}\mathbf{F}_{21} & \tilde{\mathbf{M}}_{22}^{-1} \end{pmatrix} \\ &= \begin{pmatrix} \mathbf{M}_{11}^{-1} + \mathbf{F}_{12}\tilde{\mathbf{M}}_{22}^{-1}\mathbf{F}_{21} & -\mathbf{F}_{12}\tilde{\mathbf{M}}_{22}^{-1} \\ -\tilde{\mathbf{M}}_{22}^{-1}\mathbf{F}_{21} & \tilde{\mathbf{M}}_{22}^{-1} \end{pmatrix}. \end{aligned}$$

Inverse of a 2 × 2 block matrix

$$\begin{pmatrix} \mathbf{M}_{11} & \mathbf{M}_{12} \\ \mathbf{M}_{21} & \mathbf{M}_{22} \end{pmatrix}^{-1} = \begin{pmatrix} B_{11} & B_{12} \\ B_{12}^T & B_{22} \end{pmatrix}, \tag{A.4a}$$

$$B_{11} := \left[\mathbf{M}_{11}^{-1} + \mathbf{M}_{11}^{-1}\mathbf{M}_{12}\tilde{\mathbf{M}}_{22}^{-1}\mathbf{M}_{21}\mathbf{M}_{11}^{-1}\right], \tag{A.4b}$$

$$B_{12} := -\left[\mathbf{M}_{11}^{-1}\mathbf{M}_{12}\tilde{\mathbf{M}}_{22}^{-1}\right], \tag{A.4c}$$

$$B_{22} := \left[\mathbf{M}_{22} - \mathbf{M}_{21}\mathbf{M}_{11}^{-1}\mathbf{M}_{12}\right]^{-1}. \tag{A.4d}$$

A.1.1 Application of the Schur complement

Let us assume we are dealing with a multivariate Gaussian likelihood

$$p(\boldsymbol{x}|\boldsymbol{x}_0, \mathbf{C}, \mathcal{I}) = \frac{1}{(2\pi)^{\frac{N}{2}}} |\mathbf{C}|^{-\frac{1}{2}} \exp\left\{-\frac{1}{2}\Delta\boldsymbol{x}^T\mathbf{C}^{-1}\Delta\boldsymbol{x}\right\},$$
$$\Delta\boldsymbol{x} := \boldsymbol{x} - \boldsymbol{x}_0$$

and we are interested in the marginal distributions of some elements x_i of $\boldsymbol{x}$. Without loss of generality, we consider the marginal distribution of the first l elements of $\boldsymbol{x}$. To this end, we introduce the following abbreviations:

$$\boldsymbol{x} = \begin{pmatrix}\boldsymbol{x}_1\\ \boldsymbol{x}_2\end{pmatrix}, \quad \boldsymbol{x}_0 := \begin{pmatrix}\boldsymbol{x}_{1,0}\\ \boldsymbol{x}_{2,0}\end{pmatrix} \text{ and } \Delta\boldsymbol{x} := \begin{pmatrix}\Delta\boldsymbol{x}_1\\ \Delta\boldsymbol{x}_2\end{pmatrix}, \quad \text{with } \Delta\boldsymbol{x}_\alpha = \boldsymbol{x}_\alpha - \boldsymbol{x}_{\alpha,0}.$$

The dimensions of the subvectors $\boldsymbol{x}_1$ and $\boldsymbol{x}_2$ are l and m, respectively. Then the sought-for marginal PDF is

$$p(\boldsymbol{x}_2|\boldsymbol{x}_0, \mathbf{C}, \mathcal{I}) \propto \int dx_2^m \exp\left\{-\frac{1}{2}\Delta\boldsymbol{x}^T\mathbf{C}^{-1}\Delta\boldsymbol{x}\right\}. \tag{A.5}$$

In order to avoid unnecessary computations, it is expedient to suppress all prefactors and determine the normalization at the end. We proceed by first splitting the real symmetric matrix $\mathbf{C}$ into a 2×2 block structure and, invoking equation (A.2), we obtain

$$\begin{aligned}
\Delta\boldsymbol{x}^T\mathbf{C}^{-1}\Delta\boldsymbol{x} &= \begin{pmatrix}\Delta\boldsymbol{x}_1\\ \Delta\boldsymbol{x}_2\end{pmatrix}^T \begin{pmatrix}\mathbf{C}_{11} & \mathbf{C}_{12}\\ \mathbf{C}_{21} & \mathbf{C}_{22}\end{pmatrix}^{-1} \begin{pmatrix}\Delta\boldsymbol{x}_1\\ \Delta\boldsymbol{x}_2\end{pmatrix}\\
&= \begin{pmatrix}\Delta\boldsymbol{x}_1\\ \Delta\boldsymbol{x}_2\end{pmatrix}^T \begin{pmatrix}\mathbf{1} & -\mathbf{F}_{12}\\ & \mathbf{1}\end{pmatrix} \begin{pmatrix}\mathbf{C}_{11}^{-1} & 0\\ & \tilde{\mathbf{C}}_{22}^{-1}\end{pmatrix} \begin{pmatrix}\mathbf{1} & 0\\ -\mathbf{F}_{21} & \mathbf{1}\end{pmatrix} \begin{pmatrix}\Delta\boldsymbol{x}_1\\ \Delta\boldsymbol{x}_2\end{pmatrix}\\
&= \begin{pmatrix}\Delta\boldsymbol{x}_1\\ \Delta\boldsymbol{x}_2 - \mathbf{F}_{21}\Delta\boldsymbol{x}_1\end{pmatrix}^T \begin{pmatrix}\mathbf{M}_{11} & 0\\ & \tilde{\mathbf{M}}_{22}\end{pmatrix} \begin{pmatrix}\Delta\boldsymbol{x}_1\\ \Delta\boldsymbol{x}_2 - \mathbf{F}_{21}\Delta\boldsymbol{x}_1\end{pmatrix}.
\end{aligned}$$

We have used the fact that $\mathbf{F}_{21} = \mathbf{F}_{12}^T$ for real symmetric $\mathbf{C}$. For the following calculation we don't need the definition of $\mathbf{F}_{12}$ and $\tilde{\mathbf{C}}_{22}$. Next we substitute $\boldsymbol{x}_2 \to \boldsymbol{x}_2' := \Delta\boldsymbol{x}_2 - \mathbf{F}_{21}\Delta\boldsymbol{x}_1$ to obtain

$$\begin{aligned}
\Delta\boldsymbol{x}^T\mathbf{C}^{-1}\Delta\boldsymbol{x} &= \begin{pmatrix}\Delta\boldsymbol{x}_1\\ \boldsymbol{x}_2'\end{pmatrix}^T \begin{pmatrix}\mathbf{C}_{11}^{-1} & 0\\ & \tilde{\mathbf{C}}_{22}^{-1}\end{pmatrix} \begin{pmatrix}\Delta\boldsymbol{x}_1\\ \boldsymbol{x}_2'\end{pmatrix}\\
&= (\Delta\boldsymbol{x}_1)^T\mathbf{C}_{11}^{-1}\Delta\boldsymbol{x}_1 + (\boldsymbol{x}_2')^T\tilde{\mathbf{C}}_{22}^{-1}\boldsymbol{x}_2'.
\end{aligned}$$

Now the integral in equation (A.5) over $\boldsymbol{x}_2'$ is independent of $\Delta\boldsymbol{x}_1$ and is therefore merely part of the normalization. The result reads

$$p(\boldsymbol{x}_2|\boldsymbol{x}_0, \mathbf{C}, \mathcal{I}) \propto \exp\left\{-\frac{1}{2}(\Delta\boldsymbol{x}_1)^T\mathbf{C}_{11}^{-1}\Delta\boldsymbol{x}_1\right\}. \tag{A.6}$$

MARGINAL GAUSSIAN PDF

$$p(\boldsymbol{x}_1|\boldsymbol{x}_0, \mathbf{C}, \mathcal{I}) = \frac{|\mathbf{C}_{11}|^{-\frac{1}{2}}}{(2\pi)^{\frac{l}{2}}} \exp\left\{-\frac{1}{2}(\Delta\boldsymbol{x}_1)^T \mathbf{C}_{11}^{-1} \Delta\boldsymbol{x}_1\right\}. \tag{A.7}$$

It is noteworthy that $\mathbf{C}_{11}^{-1}$ is the inverse of the first block $\mathbf{C}_{11}$ of $\mathbf{C}$ and not the first block of the inverse of $\mathbf{C}^{-1}$. In particular, if $\boldsymbol{x}_1$ has size $l = 1$ and we want the marginal PDF for a single variable, x_i say, then the result is

$$p(x_i|\boldsymbol{x}_0, \mathbf{C}, \mathcal{I}) = \frac{1}{\sqrt{2\pi C_{ii}}} e^{-\frac{(\Delta x_i)^2}{2C_{ii}}}. \tag{A.8}$$

A.2 Saddle-point approximation

One of the ubiquitous mathematical tasks in Bayesian probability theory is the evaluation of multidimensional integrals of the form

$$I_f := \int_V f(\boldsymbol{x})\ p(\boldsymbol{x})\ d^N x,$$

where $p(\boldsymbol{x})$ stands for a PDF and is innately positive, and as such can be expressed as $p(\boldsymbol{x}) = \exp[\Phi(\boldsymbol{x})]$. Usually $f(\boldsymbol{x})$ is slowly varying compared with $p(\boldsymbol{x})$. If $f(x)$ is also positive, it can be incorporated into $\Phi(\boldsymbol{x})$. An approximation often used in statistical physics for such integrals is the 'saddle-point approximation', which consists of two steps. The first is the Taylor expansion of $\Phi(\boldsymbol{x})$ about its maximum up to second order. The position $\boldsymbol{x}^*$ of the maximum is, as usual, determined via

$$\nabla_x \Phi(\boldsymbol{x})\big|_{x=x^*} = 0$$

and $\Phi(\boldsymbol{x})$ is then replaced by

$$\Phi(\boldsymbol{x}) \approx \Phi(\boldsymbol{x}^*) + \frac{1}{2}(\boldsymbol{x} - \boldsymbol{x}^*)^T \mathbf{H}(\boldsymbol{x} - \boldsymbol{x}^*),$$
$$\mathbf{H} := \nabla\nabla^T \Phi(\boldsymbol{x})\big|_{x=x^*}.$$

The matrix $\mathbf{H}$ is the Hesse matrix. The second step of the saddle-point approximation is to replace the volume of integration V by $\mathbb{R}^N$. Then the remaining integral is a standard Gaussian integral. We will exemplify the saddle-point approximation guided by Stirling's formula for the Gamma function.

A.2.1 Saddle-point approximation for $\Gamma(x)$

We start out from the definition of the Gamma function

$$\Gamma(z) = \int_0^\infty x^{z-1}\, e^{-x}\, dx = \int_0^\infty e^{-x+(z-1)\ln(x)} dx.$$

Next we determine the maximum of the argument

$$\Phi(x) = -x + (z-1)\ln(x)$$

of the exponential

$$\begin{aligned}\frac{d\Phi(x)}{dx} &= -1 + \frac{z-1}{x} = 0\\ \Rightarrow \qquad x^* &= z-1\\ \Phi(x^*) &= (z-1)\ln(z-1) - (z-1).\end{aligned}$$

The second derivative yields

$$H = \left.\frac{d^2\Phi(x)}{dx^2}\right|_{x^*} = -\frac{z-1}{(x^*)^2} = -\frac{1}{z-1},$$

and according to the first step of the saddle-point approximation the exponential is replaced by

$$e^{\Phi} \approx e^{\Phi(x^*)} e^{-\frac{1}{2(z-1)}(x-x^*)^2}$$

and the integral along with the second step of the saddle-point approximation yields

$$\begin{aligned}\Gamma(z) &\approx e^{\Phi(x^*)} \cdot \int_{-\infty}^{\infty} e^{-\frac{1}{2(z-1)}(x-x^*)^2} dx\\ &\approx (z-1)^{(z-1)} e^{-(z-1)} \cdot \sqrt{2\pi(z-1)}\\ &\approx (z-1)^{(z-1/2)} e^{-(z-1)} \sqrt{2\pi}.\end{aligned}$$

For integer values $z = N+1$ we obtain Stirling's formula for the factorial

$$N! = \Gamma(N+1) \approx N^{(N+1/2)} e^{-N} \sqrt{2\pi}.$$

A.3 Useful expressions for det(A)

A.3.1 ln(det(A)) = tr(ln(A))

We first prove the relation

DETERMINANT OF SQUARE MATRICES

$$\ln(\det(\mathbf{A})) = \operatorname{tr}(\ln(\mathbf{A})) \qquad \text{(A.9)}$$

for a $N \times N$ square matrix **A**. This representation can be very useful for the numerical evaluation of determinants of huge matrices or for the derivative of determinants w.r.t. some parameters.

For the proof we invoke the Jordan normal form of the square matrix $\mathbf{A}$:

$$\mathbf{A} = \mathbf{Q}\mathbf{J}\mathbf{Q}^{-1}, \tag{A.10}$$

where the Jordan normal form $\mathbf{J}$ is block diagonal and consists of so-called Jordan matrices, i.e. $\mathbf{J} = \operatorname{diag}(\{\mathbf{J}_\nu\})$, for $\nu = 1, \ldots, L$. The Jordan blocks $\mathbf{J}_\nu$ have the special form

$$\mathbf{J}_\nu = \lambda_\nu \mathbb{1} + \Delta_\nu, \tag{A.11}$$

$$\Delta_\nu := \begin{pmatrix} 0 & 1 & & \\ & 0 & 1 & \\ & & \ddots & \ddots \\ & & & \end{pmatrix}.$$

The size of the Jordan block $\mathbf{J}_\nu$ is equal to the degeneracy n_ν of the eigenvalue λ_ν. Consequently, $\mathbf{J}$ can be decomposed as

$$\mathbf{J} = \mathbf{D} + \boldsymbol{\Delta} \tag{A.12}$$

into the diagonal part $\mathbf{D} = \operatorname{diag}(\{\lambda_i\})$ containing the eigenvalues λ_i and an off-diagonal part $\boldsymbol{\Delta}$ which has only nonzero entries on the superdiagonal. The entries on the superdiagonal are either zero or one. Hence the derivative w.r.t. some parameter affects only the diagonal and hence the eigenvalues

$$\frac{d}{dt}\mathbf{J}(t) = \frac{d}{dt}\mathbf{D}(t). \tag{A.13}$$

Based on the property $\operatorname{tr}(\boldsymbol{\Delta}^m) = 0$, for $m \in \mathbb{N}$, straightforward Taylor expansion yields ($\lambda_\nu \neq 0$ since $\det(\mathbf{A}) \neq 0$)

$$\operatorname{tr}\big[\ln(\mathbf{J}_\nu)\big] = \operatorname{tr}\big[\ln(\lambda_\nu \mathbb{1})\big] = n_\nu \ln(\lambda_\nu).$$

Consequently,

$$\begin{aligned}\operatorname{tr}\big[\ln(\mathbf{A})\big] &= \operatorname{tr}\big[\ln(\mathbf{Q}\mathbf{J}\mathbf{Q}^{-1})\big] = \operatorname{tr}\big[\mathbf{Q}\ln(\mathbf{J})\mathbf{Q}^{-1}\big] = \operatorname{tr}\big[\ln(\mathbf{J})\big] \\ &= \sum_{\nu=1}^{L} n_\nu \ln(\lambda_\nu) = \sum_{i=1}^{N} \ln(\lambda_i).\end{aligned}$$

In the last sum degenerate eigenvalues occur multiply according to their degeneracy. Moreover,

$$\begin{aligned}\sum_{i=1}^{N} \ln(\lambda_i) &= \ln\Big[\prod(\lambda_i)\Big] = \ln\Big[\det(\mathbf{J})\Big] = \ln\Big[\det(\mathbf{Q}^{-1}\mathbf{A}\mathbf{Q})\Big] \\ &= \ln\Big[\det(\mathbf{Q}^{-1})\det(\mathbf{A})\det(\mathbf{Q})\Big] = \ln\Big[\det(\mathbf{A})\Big].\end{aligned}$$

This finishes the proof for equation (A.9).

A.3.2 Derivative of determinants w.r.t. parameters

In many cases we need the derivative of some function that contains the determinant of a matrix $\mathbf{A}(t)$. We concentrate therefore on the evaluation of

$$\frac{d}{dt}\det\big[\mathbf{A}(t)\big].$$

Based on an alternative form of equation (A.9),

$$\det(\mathbf{A}(t)) = \exp\Big\{\mathrm{tr}\big[\ln(\mathbf{A}(t))\big]\Big\},$$

we obtain

$$\begin{aligned}\frac{d}{dt}\det(\mathbf{A}(t)) &= \exp\Big\{\mathrm{tr}\big[\ln(\mathbf{A}(t))\big]\Big\}\ \frac{d}{dt}\ \mathrm{tr}\big[\ln(\mathbf{A}(t))\big]\\ &= \det(\mathbf{A}(t))\ \mathrm{tr}\left[\frac{d}{dt}\ \ln(\mathbf{A}(t))\right]\\ &= \det(\mathbf{A}(t))\ \mathrm{tr}\left[\big[\mathbf{A}(t)\big]^{-1}\frac{d}{dt}\mathbf{A}(t)\right].\end{aligned} \tag{A.14}$$

Next, we will prove that

$$\mathrm{tr}\left[\big[\mathbf{A}(t)\big]^{-1}\frac{d}{dt}\mathbf{A}(t)\right] = \mathrm{tr}\left[\big[\mathbf{D}(t)\big]^{-1}\frac{d}{dt}\mathbf{D}(t)\right].$$

Based on the similarity transformation equation (A.10), we get

$$\mathrm{tr}\left[\big[\mathbf{A}(t)\big]^{-1}\frac{d}{dt}\mathbf{A}(t)\right],$$

for which the two factors can be transformed according to equation (A.10):

$$\big[\mathbf{A}(t)\big]^{-1} = \mathbf{Q}\mathbf{J}^{-1}\mathbf{Q}^{-1}$$

and

$$\frac{d}{dt}\big[\mathbf{A}(t)\big] = \left(\frac{d}{dt}\mathbf{Q}\right)\mathbf{J}\mathbf{Q}^{-1} + \mathbf{Q}\left(\frac{d}{dt}\mathbf{J}\right)\mathbf{Q}^{-1} + \mathbf{Q}\mathbf{J}\left(\frac{d}{dt}\mathbf{Q}^{-1}\right).$$

Multiplying the two factors and using the cyclic invariance of the trace yields

$$\mathrm{tr}\left[\big[\mathbf{A}(t)\big]^{-1}\frac{d}{dt}\mathbf{A}(t)\right] = \underbrace{\mathrm{tr}\left[\mathbf{Q}^{-1}\left(\frac{d}{dt}\mathbf{Q}\right)\right] + \mathrm{tr}\left[\left(\frac{d}{dt}\mathbf{Q}^{-1}\right)\mathbf{Q}\right]}_{=\frac{d}{dt}\ \mathrm{tr}[\mathbf{Q}^{-1}\mathbf{Q}]=0} + \mathrm{tr}\left[\mathbf{J}^{-1}\left(\frac{d}{dt}\mathbf{J}\right)\right].$$

Now, according to equations (A.13) and (A.12), we have

$$\mathrm{tr}\left[\mathbf{J}^{-1}\frac{d}{dt}\mathbf{J}\right] = \mathrm{tr}\left[\mathbf{J}^{-1}\frac{d}{dt}\mathbf{D}\right] = \sum_{i=1}^{N}(\mathbf{J}^{-1})_{ii}\frac{d}{dt}\lambda_i.$$

Since $\mathbf{\Delta}^m$ has no diagonal entries for $m > 0$, only $\mathbf{D}$ contributes to the diagonal of $\mathbf{J}^{-1}$ and we obtain

$$\mathrm{tr}\left[\big[\mathbf{A}(t)\big]^{-1}\frac{d}{dt}\mathbf{A}(t)\right] = \mathrm{tr}\left[\mathbf{D}^{-1}\frac{d}{dt}\mathbf{D}\right].$$

Eventually, we continue with equation (A.14) and find

$$\frac{d}{dt}\det(\mathbf{A}(t)) = \det(\mathbf{A}(t))\ \mathrm{tr}\Big[\big[\mathbf{D}(t)\big]^{-1}\frac{d}{dt}\mathbf{D}(t)\Big] \tag{A.15}$$

$$= \det(\mathbf{A}(t))\ \sum_{i=1}^{N}\frac{d}{dt}\ln(\lambda_i(t)), \tag{A.16}$$

which is a generalization of the Hellmann–Feynman theorem that can be useful both for numerical and analytical calculations.

A.4 Multivariate Gauss integrals

We will often encounter multivariate Gauss integrals, either because the PDF under consideration is Gaussian or because the saddle-point approximation is invoked. The most general form of a Gaussian integral is

$$I = \int \exp\Big\{-\frac{1}{2}\boldsymbol{x}^T\mathbf{A}\boldsymbol{x} + \boldsymbol{b}^T\boldsymbol{x}\Big\}\, d^N x. \tag{A.17}$$

Without loss of generality we assume that $\mathbf{A}$ is real symmetric, otherwise we use the fact that $(\boldsymbol{x}^T\mathbf{A}\boldsymbol{x})$ is real and therefore $(\boldsymbol{x}^T\mathbf{A}\boldsymbol{x})^* = (\boldsymbol{x}^T\mathbf{A}^T\boldsymbol{x})$ holds. Hence

$$(\boldsymbol{x}^T\mathbf{A}\boldsymbol{x}) = \frac{1}{2}\Big((\boldsymbol{x}^T\mathbf{A}\boldsymbol{x}) + (\boldsymbol{x}^T\mathbf{A}^T\boldsymbol{x})\Big) = \boldsymbol{x}^T\underbrace{\frac{\mathbf{A}+\mathbf{A}^T}{2}}_{\to\mathbf{A}}\boldsymbol{x}.$$

The argument of the Gaussian

$$\Phi(\boldsymbol{x}) = -\frac{1}{2}\boldsymbol{x}^T\mathbf{A}\boldsymbol{x} + \boldsymbol{b}^T\boldsymbol{x}$$

has its maximum at

$$\boldsymbol{x}^* = \mathbf{A}^{-1}\boldsymbol{b},$$

with

$$\Phi(\boldsymbol{x}^*) = \frac{1}{2}\boldsymbol{b}^T\mathbf{A}^{-1}\boldsymbol{b}.$$

Hence

$$\Phi(\boldsymbol{x}) = \Phi(\boldsymbol{x}^*) - \frac{1}{2}\big(\boldsymbol{x}-\boldsymbol{x}^*\big)^T\mathbf{A}\big(\boldsymbol{x}-\boldsymbol{x}^*\big).$$

For the integration over $\boldsymbol{x}$ we introduce the substitution $\boldsymbol{x} \to \boldsymbol{z} = \boldsymbol{x} - \boldsymbol{x}^*$ and get

$$I = \exp\Big\{\frac{1}{2}\boldsymbol{b}^T\mathbf{A}^{-1}\boldsymbol{b}\Big\}\int \exp\Big\{-\frac{1}{2}z^T\mathbf{A}z\Big\}\, d^N z.$$

Since $\mathbf{A}$ is real symmetric, it can be cast into the form

$$\mathbf{A} = \mathbf{U}\mathbf{D}\mathbf{U}^T,$$

with $\mathbf{U}$ being a unitary matrix, which contains the eigenvectors of $\mathbf{A}$ in its columns, and $\mathbf{D} = \text{diag}\{\lambda_i\}$ the diagonal matrix of the eigenvalues of $\mathbf{A}$. The next step is a further substitution $\mathbf{U}^T \boldsymbol{z} := \boldsymbol{y}$. Since $\mathbf{U}$ is unitary we find

$$I = e^{\frac{1}{2}\boldsymbol{b}^T \mathbf{A}^{-1}\boldsymbol{b}} \prod_{i=1}^{N} \left\{ \int e^{-\frac{1}{2}\lambda_i y_i^2} \, dy_i \right\}.$$

Obviously, all eigenvalues have to be greater than zero, otherwise the corresponding integral diverges. In other words, $\mathbf{A}$ has to be strictly positive. If this is the case, we obtain

$$\begin{aligned} I &= \exp\left\{\frac{1}{2}\boldsymbol{b}^T \mathbf{A}^{-1}\boldsymbol{b}\right\} (2\pi)^{N/2} \prod_{i=1}^{N} \lambda_i^{-1/2} \\ &= (2\pi)^{N/2} \, |\mathbf{A}|^{-1/2} \, \exp\left\{\frac{1}{2}\boldsymbol{b}^T \mathbf{A}^{-1}\boldsymbol{b}\right\}. \end{aligned}$$

The final result reads

GAUSS INTEGRALS

$$\int \exp\left\{-\frac{1}{2}\boldsymbol{x}^T \mathbf{A}\boldsymbol{x} + \boldsymbol{b}^T \boldsymbol{x}\right\} d^N x = (2\pi)^{\frac{N}{2}} \, |\mathbf{A}|^{-\frac{1}{2}} \, \exp\left\{\frac{1}{2}\boldsymbol{b}^T \mathbf{A}^{-1}\boldsymbol{b}\right\}. \tag{A.18}$$

If $\mathbf{A} \neq \mathbf{A}^T$ *replace it by* $(\mathbf{A} + \mathbf{A}^T)/2$.

A.5 Sums containing binomial coefficients

Here we list some useful relations containing (negative) binomial coefficients, that will be used in this book, along with the proofs. If not stated otherwise, we use the convention $n, k, \nu, \mu \in \mathbb{N}_0$ and $a, b, r \in \mathbb{R}$. To begin with we repeat the definition of the binomial coefficient:

$$\binom{r}{n} := \begin{cases} \frac{r_{(n)}}{n!} & \text{for } n \geq 0 \\ 0 & \text{otherwise.} \end{cases} \tag{A.19}$$

We begin with the relation

$$\binom{r}{n-1} + \binom{r}{n} = \binom{r+1}{n}. \tag{A.20}$$

Proof:

$$\binom{r}{n-1} + \binom{r}{n} = \frac{r_{(n-1)}}{(n-1)!} + \frac{r_{(n)}}{n!}$$
$$= \frac{n\, r_{(n-1)} + (r-(n-1))r_{(n-1)}}{n!}$$
$$= \frac{r_{(n-1)}(r+1)}{n!}$$
$$= \frac{(r+1)_{(n)}}{n!}.$$

□

$$\sum_{\nu=0}^{k} \binom{n}{\nu} \binom{n-\nu}{k-\nu} x^{\nu} = \binom{n}{k} (1+x)^k. \tag{A.21}$$

Proof:

$$\sum_{\nu=0}^{k} \binom{n}{\nu} \binom{n-\nu}{k-\nu} x^{\nu} = \sum_{\nu=0}^{k} \frac{n!}{\cancel{(n-\nu)!}\nu!} \frac{\cancel{(n-\nu)!}}{(n-k)!(k-\nu)!} x^{\nu}$$
$$= \binom{n}{k} \sum_{\nu=0}^{k} \binom{k}{\nu} x^{\nu}$$
$$= \binom{n}{k} (1+x)^k.$$

□

$$\binom{-r}{n} = (-1)^n \binom{n+r-1}{n}. \tag{A.22}$$

Proof:

$$\binom{-r}{n} = \frac{(-r)(-r-1)\cdots(-r-(n-1))}{n!}$$
$$= (-1)^n \frac{(r+n-1)(r+n-2)\cdots r}{n!}$$
$$= (-1)^n \binom{n+r-1}{n}.$$

□

$$\binom{-\frac{1}{2}}{n} = \left(-\frac{1}{4}\right)^n \binom{2n}{n}. \tag{A.23}$$

Proof:

$$\begin{aligned}
\binom{-\frac{1}{2}}{n} &= \frac{\left(-\frac{1}{2}\right)\left(-\frac{1}{2}-1\right)\left(-\frac{1}{2}-2\right)\cdots\left(-\frac{1}{2}-(n-1)\right)}{n!} \\
&= \left(-\frac{1}{2}\right)^n \frac{(1)(1+2)(1+4)\cdots(1+2(n-1))}{n!} \\
&= \left(-\frac{1}{2}\right)^n \frac{(2n-1)!!}{n!} \\
&= \left(-\frac{1}{2}\right)^n \frac{(2n)!}{2\cdot 4\cdots(2n)\; n!} \\
&= \left(-\frac{1}{2}\right)^n \frac{(2n)!}{2^n\; n!\; n!} \\
&= \left(-\frac{1}{4}\right)^n \binom{2n}{n}.
\end{aligned}$$

□

$$\binom{\frac{1}{2}}{n} = -\frac{1}{2n-1}\left(-\frac{1}{4}\right)^n \binom{2n}{n}, \tag{A.24a}$$

$$\binom{\frac{1}{2}}{n} = -\frac{1}{2n-1}\binom{-\frac{1}{2}}{n}. \tag{A.24b}$$

Proof: For $n = 1$ we find directly that both equations are correct. For $n > 1$ we obtain

$$\begin{aligned}
\binom{\frac{1}{2}}{n} &= \frac{\left(\frac{1}{2}\right)\left(\frac{1}{2}-1\right)\left(\frac{1}{2}-2\right)\cdots\left(\frac{1}{2}-(n-1)\right)}{n!} \\
&= \frac{\left(\frac{1}{2}\right)\left(-\frac{1}{2}\right)\left(-\frac{1}{2}-1\right)\cdots\left(-\frac{1}{2}-(n-2)\right)}{n!} \\
&= -\left(-\frac{1}{2}\right)^n \frac{(1)(1)(1+2)(1+4)\cdots(1+2(n-2))}{n!} \\
&= -\left(-\frac{1}{2}\right)^n \frac{(2n-3)!!}{n!} \\
&= -\left(-\frac{1}{2}\right)^n \frac{(2n-2)!}{2\cdot 4\cdots(2(n-1))\; n!}
\end{aligned}$$

$$= -\left(-\frac{1}{2}\right)^n \frac{(2n-2)!(2n-1)(2n)}{2^{n-1}\,(n-1)!\,n!\,(2n-1)(2n)}$$
$$= -\left(-\frac{1}{4}\right)^n \frac{(2n)!}{n!\,n!\,(2n-1)}$$
$$= -\frac{1}{2n-1}\left(-\frac{1}{4}\right)^n \binom{2n}{n}.$$

Equation (A.24b) follows from equations (A.24a) and (A.23). □

$$\sum_{\nu=0}^{n} \binom{r}{\nu} (-1)^\nu = (-1)^n \binom{r-1}{n}. \tag{A.25}$$

Proof: We prove the relation by induction on n. The relation is fulfilled trivially for $n = 0$. We assume it is correct for all n with $n \leq m$. Then

$$\sum_{\nu=0}^{m+1} \binom{r}{\nu} (-1)^\nu = \sum_{\nu=0}^{m} \binom{r}{\nu} (-1)^\nu + \binom{r}{m+1} (-1)^{m+1}$$
$$= (-1)^m \binom{r-1}{m} - (-1)^m \binom{r}{m+1}.$$

According to equation (A.20) this yields

$$= (-1)^m \left[\binom{r-1}{m} - \binom{r-1}{m} - \binom{r-1}{m+1}\right]$$
$$= (-1)^{m+1} \binom{r-1}{m+1}.$$

Hence, the relation is also true for $m + 1$. □

$$\sum_{\nu=0}^{n} \binom{\nu+k}{k} = \binom{n+k+1}{k+1}. \tag{A.26}$$

Proof: We use again an inductive proof on n. The relation is obviously correct for $n = 0$. We assume it is correct for all n with $n \leq m$. Then

$$\sum_{\nu=0}^{m+1} \binom{\nu+k}{k} = \sum_{\nu=0}^{m} \binom{\nu+k}{k} + \binom{\nu+m+1}{m+1}$$
$$= \binom{m+k+1}{k+1} + \binom{m+1+k}{k}.$$

Based on equation (A.20), we finally find

$$\sum_{\nu=0}^{m+1} \binom{\nu+k}{k} = \binom{m+1+k+1}{k+1}$$

and therefore the relation holds for $m+1$ as well. □

Relation (A.26) is equivalent to

$$\sum_{\nu=0}^{n} \binom{\nu}{k} = \binom{n+1}{k+1}. \tag{A.27}$$

Proof: According to equation (A.19), the binomial coefficients $\binom{\nu}{n}$ are zero for $\nu < n$. Hence the sum on the l.h.s. vanishes for $n < k$, as does the r.h.s. For $n = k$, the only remaining term on the l.h.s. is $\binom{n=k}{k} = 1$, which again agrees with the r.h.s., $\binom{k+1}{k+1} = 1$. Finally, we consider the case $n > k$:

$$\sum_{\nu=0}^{n} \binom{\nu}{k} = \sum_{\nu=k}^{n} \binom{\nu}{k} = \sum_{\mu=0}^{n-k} \binom{k+\mu}{k} \overset{\text{(A.26)}}{=} \binom{k+(n-k)+1}{k+1},$$

which finishes the proof. □

Another relation of binomial coefficients that we want to discuss is

$$S_n := -\sum_{\nu=1}^{n} \binom{n}{\nu} \frac{(-1)^\nu}{\nu} = \sum_{\nu=1}^{n} \frac{1}{\nu}. \tag{A.28}$$

Proof: The value for $n = 1$ is simply $S_1 = 1$ in both expressions. In order to determine the other values, we consider the differences

$$\begin{aligned}
S_{n+1} - S_n &= -\sum_{\nu=1}^{n+1} \binom{n+1}{\nu} \frac{(-1)^\nu}{\nu} + \sum_{\nu=1}^{n} \binom{n}{\nu} \frac{(-1)^\nu}{\nu} \\
&= -\binom{n+1}{n+1} \frac{(-1)^{n+1}}{n+1} - \sum_{\nu=1}^{n} \left[\binom{n+1}{\nu} - \binom{n}{\nu} \right] \frac{(-1)^\nu}{\nu} \\
&\overset{\text{(A.20)}}{=} \frac{(-1)^n}{n+1} - \sum_{\nu=1}^{n} \binom{n}{\nu-1} \frac{(-1)^\nu}{\nu}
\end{aligned}$$

$$
\begin{aligned}
&= \frac{(-1)^n}{n+1} - \sum_{\nu=1}^{n} \frac{n!}{(n-\nu+1)!\nu!}(-1)^\nu \\
&= \frac{(-1)^n}{n+1} - \frac{1}{n+1}\sum_{\nu=1}^{n} \binom{n+1}{\nu}(-1)^\nu \\
&= \frac{(-1)^n}{n+1} - \frac{1}{n+1}\Bigg[\sum_{\nu=0}^{n+1} \binom{n+1}{\nu}(-1)^\nu \\
&\qquad \cdots - \binom{n+1}{n+1}(-1)^{n+1} - \binom{n+1}{0}(-1)^0\Bigg] \\
&= \frac{(-1)^n}{n+1} - \frac{1}{n+1}\Big((1-1)^{n+1} - (-1)^{n+1} - 1\Big) \\
&= \frac{1}{n+1}.
\end{aligned}
$$

This completes the proof, as the differences of $\sum_{\nu=1}^{n} \frac{1}{\nu}$ yield the same result and both have the same initial value for $n = 1$. □

The last relations involving binomial coefficients that shall be discussed are related to the hypergeometric distribution, given by

$$
P(k_I|k, n_I, n_{II}) = \frac{\binom{n_I}{k_I}\binom{n_{II}}{k_{II}}}{\binom{n}{k}}, \tag{A.29}
$$

with $k = k_I + k_{II}$ and $n = n_I + n_{II}$. All quantities are integers in this context.

The normalization reads

$$
\sum_{k_I=0}^{k} \binom{n_I}{k_I}\binom{n-n_I}{k-k_I} = \binom{n}{k}.
$$

We will generalize this relation to real numbers a and b:

$$
\sum_{\nu=0}^{n} \binom{a}{\nu}\binom{b}{n-\nu} = \binom{a+b}{n}. \tag{A.30}
$$

Proof: To begin with we note that the sum can be extended all the way up to ∞, since by definition the second binomial is zero for $\nu > n$. The sum is then a convolution and the generating function is therefore the product of the generating functions of the individual factors, which are given by

$$\phi_a(z) := \sum_{n=0}^{\infty} \binom{a}{n} x^n = (1+z)^a.$$

Hence, the generating function of the convolution is

$$\phi(z) = (1+z)^a\ (1+z)^b = (1+z)^{a+b},$$

which agrees with the generating function of $\binom{a+b}{n}$. □

A special case is $a = b = n$, for which we find

$$\sum_{n=0}^{n} \binom{n}{\nu}^2 = \binom{2n}{n}. \tag{A.31}$$

Next we want to show that the mean of the hypergeometric distribution (equation (A.29)) is given by

$$\langle k_I \rangle = n_I \frac{k}{n}. \tag{A.32}$$

Proof: Based on

$$\binom{n}{k} = \frac{n}{k}\binom{n-1}{k-1}, \tag{A.33}$$

which holds for $n \geq 1$ and $k \geq 1$, we find

$$\binom{n}{k}\langle k_I \rangle = \sum_{k_I=0}^{k} k_I \binom{n_I}{k_I}\binom{n_{II}}{k-k_I} = n_I \sum_{k_I=1}^{k} \binom{n_I - 1}{k_I - 1}\binom{n_{II}}{k-k_I}.$$

Next we shift the summation index $k_I \to k_I + 1$:

$$= n_I \sum_{k_I=0}^{k-1} \binom{n_I - 1}{k_I}\binom{n_{II}}{(k-1)-k_I} \overset{(A.30)}{=} n_I \binom{n-1}{k-1}.$$

So finally we obtain, with equation (A.33),

$$\langle k_I \rangle = n_I \frac{k}{n}.$$

□

Next comes the proof that the variance of the hypergeometric distribution (equation (A.29)) is given by

$$\mathrm{var}(k_I) = \frac{n_I\, k\,(n-k)\, n_{II}}{n^2\,(n-1)}. \tag{A.34}$$

Proof: The first step is identical to that in the derivation of the mean:

$$\begin{aligned}\binom{n}{k}\left\langle k_I^2\right\rangle &= \sum_{k_I=0}^{k} k_I^2 \binom{n_I}{k_I}\binom{n_{II}}{k-k_I}\\ &= n_I \sum_{k_I=0}^{k-1} (k_I+1)\binom{n_I-1}{k_I}\binom{n_{II}}{k-1-k_I}.\end{aligned}$$

Along with equation (A.30), we get

$$= n_I \binom{n-1}{k-1} + n_I \sum_{k_I=0}^{k-1} k_I \binom{n_I-1}{k_I}\binom{n_{II}}{k-1-k_I}.$$

Repeating these steps yields

$$\binom{n}{k}\left\langle k_I^2\right\rangle = n_I \binom{n-1}{k-1} + n_I(n_I-1)\binom{n-2}{k-2},$$

and finally

$$\left\langle k_I^2\right\rangle = n_I \frac{k}{n} + n_I(n_I-1)\frac{k(k-1)}{n(n-1)}.$$

Then the variance of the hypergeometric distribution follows from

$$\begin{aligned}\mathrm{var}(k_I) &= \left\langle k_I^2\right\rangle - \langle k_I\rangle^2\\ &= \frac{n_I\,k}{n} + \frac{n_I\,k\,(k-1)\,(n_I-1)}{n\,(n-1)} - \frac{n_I^2\,k^2}{n^2}\\ &= \frac{n_I\,k\,(n-k)\,n_{II}}{n^2\,(n-1)}.\end{aligned}$$

□

Appendix B

Selected proofs and derivations

B.1 Proof of the Neyman–Pearson lemma

Here we want to give a rigorous proof of the Neyman–Pearson (NP) lemma for a simple null tested against a simple alternative. Let the data sample be X. For a predefined significance level α (probability for type I error),

$$P(X \in R|H_0) = \alpha, \tag{B.1}$$

an optimal rejection region R is obtained by minimizing the probability for type II error, or equivalently by maximizing the power

$$\max_A P(X \in A|H_1). \tag{B.2}$$

The NP lemma claims that the optimal rejection region of size α is given by

$$\mathcal{R} = \left\{ D \middle| \frac{p(D|H_a, \mathcal{I})}{p(D|H_0, \mathcal{I})} \geq K_\alpha \right\}, \tag{B.3}$$

where K_α is a constant that has to be adjusted such that equation (B.1) is fulfilled. For the proof we introduce the abbreviations $P_\alpha(A) := P(X \in A|H_\alpha)$ for $\alpha \in \{0, a\}$. Equation (B.3) implies

$$P_a(X) \geq K_\alpha P_0(X) \qquad \forall X \in \mathcal{R}, \tag{B.4a}$$

$$P_a(X) \leq K_\alpha P_0(X) \qquad \forall X \in \overline{\mathcal{R}}. \tag{B.4b}$$

First we consider $X \in \overline{A} \cap \mathcal{R} \subseteq \mathcal{R}$. According to equation (B.4a), we have

$$P_a(\overline{A} \cap \mathcal{R}) \geq K_\alpha P_0(\overline{A} \cap \mathcal{R}).$$

Next we consider $X \in A \cap \overline{\mathcal{R}} \subseteq \overline{\mathcal{R}}$, for which equation (B.4b) holds, resulting in

$$P_a(A \cap \overline{\mathcal{R}}) \leq K_\alpha P_0(A \cap \overline{\mathcal{R}}).$$

Let $\mathcal{R}$ be the optimal rejection region, defined by equation (B.3). For any subset A of the data space of size α, based on the elementary rules of probability theory we perform the following transformations:

$$P_a(A) = P_a(A \cup \mathcal{R}) + P_a(A \cup \overline{\mathcal{R}}) = P_a(\mathcal{R}) - P_a(\overline{A} \cup \mathcal{R}) + P_a(A \cup \overline{\mathcal{R}}).$$

Invoking equations (B.4a) and (B.4b), we get

$$P_a(A) \leq P_a(\mathcal{R}) - K_\alpha \left[P_0(\overline{A} \cup \mathcal{R}) - P_0(A \cup \overline{\mathcal{R}})\right]$$
$$\leq P_a(\mathcal{R}) - K_\alpha \left[P_0(\mathcal{R}) \underbrace{- P_0(A \cup \mathcal{R}) - P_0(A \cup \overline{\mathcal{R}})}_{=-P_0(A)}\right].$$

Finally we use the fact that both $\mathcal{R}$ and A have the same size. Therefore, the expression in the bracket vanishes, which finishes the proof:

$$P_a(A) \leq P_a(\mathcal{R}) \qquad \forall A \text{ of size } \alpha.$$

B.2 Proof of chi-squared for distributions

Here we outline the proof that the sampling distribution for the random variable

$$x = \sum_{i=1}^{L} \frac{(n_i - \langle n_i \rangle)^2}{\langle n_i \rangle}, \qquad \langle n_i \rangle = N P_i$$

approaches the chi-squared distribution for large N. In order to make the points very clear we consider a specific example, namely the distribution of N balls in L boxes. Let the probability for a ball to end in box i be P_i. We introduce a random variable for a single throw of the ball $b_i \in \{0, 1\}$, which is 1 if the ball lands in box i and 0 otherwise. Now we repeat the experiment N times and obtain for the occupation number of box i

$$n_i = \sum_{\nu=1}^{N} b_i^{(\nu)},$$

where $b_i^{(\nu)}$ is the random variable for box i in the νth throw. There is a simple relation between the covariances:

$$c_{ij}^{(n)} := \left\langle \Delta n_i \Delta n_j \right\rangle = \sum_{\nu} \left\langle \Delta b_i^{(\nu)} \Delta b_j^{(\nu)} \right\rangle + \sum_{\nu \neq \nu'} \left\langle \Delta b_i^{(\nu)} \Delta b_j^{(\nu')} \right\rangle.$$

Since the variables for different throws are uncorrelated, we find

$$C_{ij}^{(n)} = \underbrace{\sum_{\nu=1}^{N}}_{N} \underbrace{\left\langle \Delta b_i^{(\nu)} \Delta b_j^{(\nu)} \right\rangle}_{:=C_{ij}^{(b)}} + \sum_{\nu \neq \nu'} \underbrace{\left\langle \Delta b_i^{(\nu)} \right\rangle}_{=0} \underbrace{\left\langle \Delta b_j^{(\nu')} \right\rangle}_{=0},$$

$$C^{(b)} = \frac{1}{N} C^{(n)} \overset{(4.17\text{d})}{=} \delta_{ij} P_i - P_i P_j. \tag{B.5}$$

Now it is expedient to introduce the relative occupation of box i, i.e.

$$\frac{n_i}{N} = \frac{1}{N} \sum_{\nu=1}^{N} b_i^{(\nu)} = \overline{b_i}.$$

The occupation numbers n_i are related to the sample means b_i, which is the point where the central limit theorem comes into play. In terms of $\overline{b_i}$, the test statistic reads

$$x = N \sum_{i=1}^{L} \frac{(\overline{b_i} - P_i)^2}{P_i} = N \sum_{i=1}^{L} y_i^2,$$
$$y_i := \left(\overline{b_i} - P_i\right) P_i^{-1/2}.$$

Now we switch to vector notation $\boldsymbol{y} = \left(y_1, \ldots, y_L\right)^T$, and the test statistic simply reads

$$x = N \boldsymbol{y}^T \boldsymbol{y}. \tag{B.6}$$

If it can be shown that the random variables y_i are asymptotically i.i.n.d., then the sampling distribution of x is chi-squared. Let us check this assumption. The random variable y_i is actually the sample mean of $\{y_i^{(\nu)}\}$, where the upper index enumerates the members of the sample:

$$y_i = \frac{1}{N} \sum_{\nu=1}^{N} y_i^{(\nu)},$$
$$y_i^{(\nu)} := \left(b_i^{(\nu)} - P_i\right) P_i^{-1/2}.$$

The mean of these random variables $y_i^{(\nu)}$ is zero:

$$\left\langle y_i^{(\nu)} \right\rangle = \Big(\underbrace{\left\langle b_i^{(\nu)} \right\rangle}_{P_i} - P_i \Big) P_i^{-1/2} = 0,$$

and their covariance vanishes if the upper indices are different. The covariance matrix for diagonal upper indices reads

$$C_{ij}^{(y)} := \left\langle y_i^{(\nu)} y_j^{(\nu)} \right\rangle = P_i^{-1/2} \underbrace{\left\langle \left(b_i^{(\nu)} - P_i\right)\left(b_j^{(\nu)} - P_j\right) \right\rangle}_{:= C_{ij}^{(b)}} P_j^{-1/2}.$$

The covariance matrix is independent of the upper index and has been given in equation (B.5). We obtain

$$C_{ij}^{(y)} = P_i^{-1/2} \left(P_i \delta_{ij} - P_i P_j \right) P_j^{-1/2} = \delta_{ij} - \sqrt{P_i P_j},$$
$$\mathbf{C}^{(y)} = \mathbb{1} - \sqrt{\boldsymbol{P}} \sqrt{\boldsymbol{P}}^T, \tag{B.7}$$

where $\sqrt{\boldsymbol{P}}^T$ is the row vector $(\sqrt{P_1}, \ldots, \sqrt{P_L})$. The covariance matrix $\mathbf{C}^{(y)}$ is real symmetric and idempotent $[\mathbf{C}^{(y)}]^2 = \mathbf{C}^{(y)}$, as one can easily see. So it only has eigenvalues $\lambda_l = \{0, 1\}$. Moreover, the trace is

$$\operatorname{tr}\left(\mathbf{C}^{(y)}\right) = \underbrace{\operatorname{tr}\left(\mathbb{1}\right)}_{=L} - \underbrace{\sum_{i=1}^{L} P_i}_{=1} = L - 1 \stackrel{!}{=} \sum_{l=1}^{L} \lambda_l,$$

and it follows that the eigenvalue $\lambda = 1$ occurs $L-1$ times and the value $\lambda = 0$ occurs only once. We enumerate the eigenvalues and eigenvectors such that the last eigenvalue is zero, i.e. $\lambda_L = 0$. Now we introduce the $\mathbf{U}$ that contains column-wise the eigenvectors of $\mathbf{C}^{(y)}$. Hence

$$\mathbf{U}^T\mathbf{C}^{(y)}\mathbf{U} = \text{diag}\big(\{1, \ldots, 1, 0\}\big).$$

Since the covariance matrix is real symmetric, $\mathbf{U}$ is orthogonal: $\mathbf{U}^T\mathbf{U} = \mathbf{1}$. Finally, we introduce new random variables $z_l^{(\nu)}$ by the following orthogonal transformation:

$$z_l^{(\nu)} = \sqrt{N} \sum_i U_{li}^T y_i^{(\nu)}. \tag{B.8}$$

These variables also have zero mean $\langle z_l^{(\nu)} \rangle = 0$, and the covariance matrix is

$$\begin{aligned}\left\langle z_l^{(\nu)} z_{l'}^{(\nu)} \right\rangle &= N \sum_{ij} U_{li}^T \left\langle y_i^{(\nu)} y_j^{(\nu)} \right\rangle U_{jl} = N \left(\mathbf{U}^T\mathbf{C}^{(y)}\mathbf{U}\right)_{ll'} \\ &= \text{diag}(\{N, \ldots, N, 0\}).\end{aligned}$$

Since the mean and variance of $z_L^{(\nu)}$ are zero, this quantity is no longer a random variable, instead it is identical to zero. All other random variables $z_l^{(\nu)}$ are now i.i.d., with zero mean and $\sigma^2 = N$. So the sample mean

$$z_l := \frac{1}{N} \sum_\nu z_l^{(\nu)} \tag{B.9}$$

fulfils all requirements for the central limit theorem and converges for large N towards the centred unit variance normal $\mathcal{N}(0, 1)$.

Finally, we merely need to express the chi-squared statistic in terms of z_l.

Owing to equations (B.8) and (B.9), we get

$$z_l = \sqrt{N} \sum_{i=1}^{L} U_{li}^T y_i,$$

$$\sum_{l=1}^{L} z_l^2 = N \sum_{i,j} \underbrace{\sum_l U_{il} U_{lj}^T}_{=\delta_{ij}} y_i y_j = N \sum_i y_i^2 = x.$$

If we take into account that z_L is zero, then the result is

$$x = \sum_{l=1}^{L-1} z_l^2.$$

Hence we are summing up the squares of $L-1$ central unit variance normal random variables. This proves that the sampling distribution of x converges for large L towards the chi-squared distribution with $L-1$ degrees of freedom.

B.3 Derivation of Student's t-distribution for the difference of means

The topic of this section is the derivation of the sampling distribution of the t-statistic for difference in means, as defined in equation (19.32) [p. 301]. We can streamline the proof since it contains the same elements as the proof in Section 19.4.1 [p. 299]. Basically the steps are introduction of missing parameters $(\overline{x}_\alpha, v_\alpha, \sigma, \mu_0)$, substitution $s_\alpha := (\overline{x}_\alpha - \mu_0)/\sigma$, $w_\alpha := v_\alpha/\sigma^2$, elimination of the nuisance parameters σ and μ_0 and evaluation of the remaining integral. For the sake of a lean notation we utilize $\overline{\boldsymbol{x}}$ for the set $\{\overline{x}_1, \overline{x}_2\}$ and likewise for the other sample-dependent quantities $\boldsymbol{N}, \boldsymbol{v}, \boldsymbol{s}, \boldsymbol{w}$. In addition, we will use

$$\mathrm{SE}^* := \sqrt{\tfrac{N}{(N-2)N_1N_2}(w_1+w_2)}.$$

The proof proceeds as follows:

$$\begin{aligned}
p(t|\boldsymbol{N},\mathcal{I}) &= \int \cdots \int d\mu_0\, d\sigma\, d\overline{\boldsymbol{x}}\, d\boldsymbol{v}\; p(t|\overline{\boldsymbol{x}}, \boldsymbol{v}, \mu_0, \sigma, \boldsymbol{N}, \mathcal{I}) \cdots \\
&\qquad \cdots p(\overline{\boldsymbol{x}}|\not{\boldsymbol{v}}, \mu_0, \sigma, \boldsymbol{N}, \mathcal{I})\, p(\boldsymbol{v}|\not\mu_0\not\sigma, \sigma, \boldsymbol{N}, \mathcal{I})\; p(\mu_0, \sigma|\boldsymbol{N}, \mathcal{I}) \\
&= \int \cdots \int d\boldsymbol{s}\, d\boldsymbol{w}\; \delta\left(t - \tfrac{s_2 - s_1}{\mathrm{SE}^*}\right) \prod_\alpha \mathcal{N}\left(s_\alpha|0, \frac{1}{\sqrt{N_\alpha}}\right) \cdots \\
&\qquad \prod_\alpha p_\Gamma\left(w_\alpha|\tfrac{N_\alpha - 1}{2}, \frac{1}{2}\right) \underbrace{\iint d\mu_0\, d\sigma\; p(\mu_0, \sigma|\boldsymbol{N}, \mathcal{I})}_{=1} \\
&\propto \int d\boldsymbol{w}\; \underbrace{\int ds_1\, \mathcal{N}\left(s_1|0, \frac{1}{\sqrt{N_1}}\right) \mathcal{N}\left(s_1 + t\,\mathrm{SE}^*|0, \frac{1}{\sqrt{N_2}}\right)}_{\propto \exp\left\{-\frac{t^2(w_1+w_2)}{2(N-2)}\right\}} \cdots \\
&\qquad \cdots \sqrt{w_1 + w_2} \prod_\alpha w_\alpha^{\frac{N_\alpha - 3}{2}}\, e^{-\frac{w_\alpha}{2}} \\
&\propto \int \frac{dw_1}{w_1} \frac{dw_2}{w_2} \sqrt{w_1 + w_2} \prod_\alpha w_\alpha^{\frac{N_\alpha - 1}{2}}\, e^{-\frac{w_\alpha}{2}\left(1 + \frac{t^2}{N-2}\right)}.
\end{aligned}$$

Finally, the substitution $w_\alpha(1 + \frac{t^2}{N-2}) = x_\alpha$ yields

$$\begin{aligned}
p(t|\boldsymbol{N},\mathcal{I}) &\propto \left(1 + \frac{t^2}{N-2}\right)^{-\frac{N-2+1}{2}} \underbrace{\int \frac{dx_1}{x_1} \frac{dx_2}{x_2} \sqrt{x_1 + x_2} \prod_\alpha x_\alpha^{\frac{N_\alpha - 1}{2}}\, e^{-\frac{x_\alpha}{2}}}_{=\text{const.}} \\
&\propto \left(1 + \frac{t^2}{\nu}\right)^{-\frac{\nu+1}{2}}.
\end{aligned}$$

This finishes the proof that t is distributed as $p_\Gamma(t|\nu)$, with $\nu = N - 2$.

B.4 Variance of MCMC sample mean

Here we outline the computation of the variance of the sample mean depicted in equation (30.50) [p. 562]:

$$
\begin{aligned}
a(\Delta\tau) &= \left\langle \Delta f(\boldsymbol{x}_{\Delta\tau})\Delta f(\boldsymbol{x}_0)\right\rangle = \left\langle f(\boldsymbol{x}_{\Delta\tau}) f(\boldsymbol{x}_0)\right\rangle - \langle f\rangle^2 \\
&= \sum_{i,j} f_i f_j \; P(\boldsymbol{x}_{\Delta\tau} = \boldsymbol{\xi}_j, \boldsymbol{x}_0 = \boldsymbol{\xi}_i | p^0, \mathcal{I}) \; - \; \langle f\rangle^2 \\
&= \sum_{i,j} f_i f_j \; P(\boldsymbol{x}_{\Delta\tau} = \boldsymbol{\xi}_j | \boldsymbol{x}_0 = \boldsymbol{\xi}_i, \mathcal{I}) \; \underbrace{P(\boldsymbol{x}_0 = \boldsymbol{\xi}_i | p^0, \mathcal{I})}_{\rho_i} - \langle f\rangle^2 \\
&= \sum_{i,j} f_j \left((M'^{\Delta\tau})_{ji} + \rho_j \right) \rho_i f_i - \langle f\rangle^2 \\
&= \sum_{i,j} (\rho_j f_j)\,(M'^{\Delta\tau})_{ji}\,(\rho_i f_i) + \cancel{\langle f\rangle^2} - \cancel{\langle f\rangle^2} \\
&= \sum_{l=2}^{\mathcal{N}} \underbrace{\left(\sum_j f_j \, x_{jl} \right)}_{=\tilde{f}_l} \lambda_l^{\Delta\tau} \underbrace{\left(\sum_i \rho_i \, f_i \, \frac{X_{il}^*}{\rho_i} \right)}_{=\tilde{f}_l^*} \\
&= \sum_{l=2}^{\mathcal{N}} \left| \tilde{f}_l \right|^2 \lambda_l^{\Delta\tau}.
\end{aligned}
$$

Hence the normalized autocorrelation reads

> **NORMALIZED AUTOCORRELATION**
>
> $$\tilde{a}(\Delta\tau) = \sum_{l=2}^{\mathcal{N}} \frac{|\tilde{f}_l|^2}{\sigma_f^2} \lambda_l^{\Delta\tau}.$$

Then we find for the variance of the sample mean

$$
\left\langle (\Delta S)^2 \right\rangle = \frac{\sigma_f^2}{N} \left(1 + \sum_{l=2}^{\mathcal{N}} \frac{|\tilde{f}_l|^2}{\sigma_f^2} \underbrace{\frac{2}{N} \sum_{\Delta\tau=1}^{N-1} (N - \Delta\tau)\, \lambda_l^{\Delta\tau}}_{:=\kappa(\lambda_l)} \right).
$$

Now, we merely need to compute $\kappa(\lambda)$:

$$
\frac{N}{2} h(\lambda) = N \sum_{\mu=1}^{N-1} \lambda^{\mu} - \sum_{\mu=1}^{N-1} \mu \lambda^{\mu} = N\lambda \frac{1 - \lambda^{N-1}}{1 - \lambda} - \sum_{\mu=1}^{N-1} \mu \lambda^{\mu},
$$

$$\sum_{\mu=1}^{N-1} \mu\,\lambda^{\mu} = \lambda \frac{\partial}{\partial \lambda} \sum_{\mu=1}^{N-1} \lambda^{\mu} = -\frac{N\lambda}{1-\lambda}\left(\lambda^{N-1} - \frac{1-\lambda^N}{N(1-\lambda)}\right),$$

$$\kappa(\lambda) = \frac{2\lambda}{(1-\lambda)}\left(1 - \frac{1}{N}\frac{1-\lambda^N}{1-\lambda}\right).$$

Collecting all terms, we have the desired result:

Variance of sample mean for equilibrated Markov chains

$$\left\langle (\Delta S)^2 \right\rangle = \sigma_f^2 \left[\frac{1}{N} + \sum_{l=2}^{\mathcal{N}} \frac{|\tilde{f}_l|^2}{\sigma_f^2}\,\kappa_l\right],$$

$$\kappa_l := \frac{2\lambda_l}{(1-\lambda_l)}\left(1 - \frac{1}{N}\frac{1-\lambda_l^N}{1-\lambda_l}\right).$$

B.5 Covariance for nested sampling

Here we outline the derivation of the covariance required for nested sampling. The covariance matrix $\left\langle \Delta X^*_{n'} \Delta X^*_n \right\rangle$ is symmetric and we can restrict the discussion to $n' = n + m$, with $m \geq 0$. In the following derivation we use equation (31.14) [p. 580]:

$$\begin{aligned}
\left\langle X^*_{n+m} X^*_n \right\rangle &= \left\langle \prod_{\nu=1}^{n+m} \theta_\nu \prod_{\nu'=1}^{n} \theta_{\nu'} \right\rangle \\
&= \prod_{\nu=n+1}^{n+m} \left\langle \theta_\nu \right\rangle \prod_{\nu'=1}^{n} \left\langle \theta_{\nu'}^2 \right\rangle \\
&= \left(\frac{K}{K+1}\right)^m \left(\frac{K}{K+2}\right)^n \\
\left\langle \Delta X^*_{n+m} \Delta X^*_n \right\rangle &= \left(\frac{K}{K+1}\right)^m \left(\frac{K}{K+2}\right)^n - \left(\frac{K}{K+1}\right)^{m+n} \left(\frac{K}{K+1}\right)^n \\
&= X_{n+m} X_n \left(\left[1 + \frac{1}{K(K+2)}\right]^n - 1\right) \\
&= X_{n+m} X_n \left(e^{n \ln(1+1/K(K+2))} - 1\right).
\end{aligned}$$

Now

$$X_n = \left(\frac{K}{K+1}\right)^n \qquad \Rightarrow \ln(X_n) = -n \ln(1 + 1/K),$$

and hence

$$\begin{aligned}
\langle \Delta X^*_{n+m} \Delta X^*_n \rangle &= X_{n+m} X_n \left(e^{-\ln(X_n)\kappa} - 1 \right), \\
\kappa &:= \frac{\ln(1 + 1/K(K+2))}{\ln(1 + 1/K)} \\
&= \frac{1}{K}\left(1 - \frac{3}{2K} + O\left(\frac{1}{K^2}\right)\right).
\end{aligned}$$

So the final result for $n' \geq n$ reads

$$\langle \Delta X^*_{n'} \Delta X^*_n \rangle = X_{n'} X_n \left(X_n^{-\kappa} - 1 \right).$$

Appendix C

Symbols and notation

Table C.1 Most important symbols used in this book

$\boldsymbol{I}(.)$	indicator function, Heaviside
$\mathbf{1}$	unit matrix
$\vee$	logical or
$\wedge$	logical and
$O(X)$	of order X
$\exists \ldots$	there exists ...
$\delta(.)$	Dirac delta function
$\delta_{i,j}$	Kronecker delta function
$x \in X$	x is an element of X
$X \ni x$	X contains x
$\overline{y}$	sample mean of $\{y_i\}$
$\overline{(\Delta y)^2}$	sample variance of $\{y_i\}$
$\langle x \rangle$	mean value of random variable x
$\text{var}(x)$	variance of random variable x
$\stackrel{\curlywedge}{x}$	median
$\breve{x}$	sample median
$\boldsymbol{v}$	bold lowercase symbols represent column vectors
$\mathbf{M}$	bold uppercase symbols represent matrices
$\mathbf{M}^T$	transposed matrix
$\lvert\mathbf{M}\rvert$	determinant of matrix $\mathbf{M}$
$\arg\max_x f(x)$	value of x for which $f(x)$ attains its largest value
$\arg\min_x f(x)$	value of x for which $f(x)$ attains its smallest value
CDF	cumulative distribution function
$D(p:q)$	Kullback–Leibler distance
i.i.d.	independent and identically distributed
i.n.d.	independent and normally distributed, with individual normal distribution
i.i.n.d.	independent and identically normally distributed, all with the same normal distribution
$\max_x f(x)$	largest value of $f(x)$ under variation of x
$\min_x f(x)$	smallest value of $f(x)$ under variation of x
PDF	probability distribution function
PMF	probability mass function
SE	standard error
$\text{KL}(p\Vert q)$	Kullback–Leibler distance
$p(.)$	lowercase p for PDF
$P(.)$	uppercase P for probability
$P(A\vert B, \mathcal{I})$	probability for A given B and background information $\mathcal{I}$

References

[1] Abramowitz, M., and Stegun, I. A. 1965. *Handbook of Mathematical Functions*. Washington DC: National Bureau of Standards.

[2] Akaike, H. 1974. A new look at statistical model identification. *IEEE Transactions on Automatic Control*, **19**, 716–723.

[3] Alimov, V. Kh., Mayer, M., and Roth, J. 2005. Differential cross-section of the D(^{3}He, p)^{4}He nuclear reaction and depth profiling of deuterium up to large depths. *Nuclear Instruments and Methods in Physics B*, **234**, 169–175.

[4] Ambegaokar, V., and Troyer, M. 2010. Estimating errors reliably in Monte Carlo simulations of the Ehrenfest model. *American Journal of Physics*, **78**(2), 150.

[5] Amsel, G., and Lanford, W. A. 1984. Nuclear reaction techniques in materials analysis. *Annual Review Nuclear and Particle Science*, **34**, 435–460.

[6] Anton, M., Weisen, H., Dutch, M. J., and von der Linden, W. 1996. X-ray tomography on the TCV tokamak. *Plasma Physics and Controlled Fusion*, **38**, 1849–1878.

[7] Atkinson, A. C., Donev, A., and Tobias, R. 2007. *Optimum Experimental Designs, with SAS*. Oxford: Oxford University Press.

[8] Bailey, N. T. J. 1990. *The Elements of Stochastic Processes*. New York: John Wiley & Sons.

[9] Ballentine, L. E. 2003. *Quantum Mechanics: A Modern Development*. Singapore: World Scientific Publishing.

[10] Bayes, Rev. T. 1763. Essay toward solving a problem in the doctrine of chances. *Philosophical Transactions of the Royal Society London*, **53**, 370–418.

[11] Bellman, R. E. 1957. *Dynamic Programming*. Princeton, NJ: Princeton University Press.

[12] Berger, J. O. 1985. *Statistical Decision Theory and Bayesian Analysis*. New York: Springer.

[13] Berger, J. O., and Wolpert, R. L. 1988. *The Likelihood Principle*, 2nd edn. Hayward, CA: IMS.

[14] Bernardo, J. M. 1979. Expected information as expected utility. *Annals of Statistics*, **7**(3), 686–690.

[15] Bernardo, J. M., and Smith, A. F. M. 2000. *Bayesian Theory*. Chichester, UK: John Wiley & Sons.

[16] Binder, K. 1986. *Monte Carlo Methods in Statistical Physics*, 2nd edn. Topics in Current Physics, vol. 1. Berlin: Springer-Verlag.

[17] Borth, D. M. 1975. A total entropy criterion for the dual problem of model discrimination and parameter estimation. *Journal of the Royal Statistical Society, Series B*, **37**(1), 77–87.

[18] Bremaud, P. 1999. *Markov Chains, Gibbs Fields, Monte Carlo Simulation, and Queues*. New York: Springer.

[19] Bretthorst, G. L. 1988. *Bayesian Spectrum Analysis and Parameter Estimation*. New York: Springer.

[20] Bretthorst, G. L. 1990. Bayesian analysis I. Parameter estimation using quadrature NMR models. *Journal of Magnetic Resonance*, **88**, 533–551.

[21] Bretthorst, G. L. 2001. Generalizing the Lomb–Scargle periodogram. In Mohammad-Djafari, A. (ed.), *Bayesian Inference and Maximum Entropy Methods in Science and Engineering*, vol. 568. Melville, NY: AIP; pp 241–245.

[22] Bretthorst, G. L. (ed.). 1988. *Bayesian Spectrum Analysis and Parameter Estimation*. Berlin: Springer-Verlag.

[23] Bretthorst, G. L. 1988. Excerpts from Bayesian spectrum analysis and parameter estimation. In Erickson, G. J., and Smith, C. R. (eds), *Maximum Entropy and Bayesian Methods in Science and Engineering*, vol. 1. Dordrecht: Kluwer Academic Publishers; p. 75.

[24] Bretthorst, G. L. 1993. On the difference in means. In Grandy, W. T̃. and Milonni, P. W̃. (eds), *Physics and Probability Essays in Honor of Edwin T. Jaynes*, vol. 1. Cambridge: Cambridge University Press; pp 177–194.

[25] Brockwell, A. E., and Kadane, J. B. 2003. A gridding method for Bayesian sequential decision problems. *Journal of Computational and Graphical Statistics*, **12**(3), 566–584.

[26] Bryson, B. 2010. *At Home: A Short History of Private Life*. London: Transworld Publishers.

[27] Buck, B., and Macaulay, V. A. (eds). 1991. *Maximum Entropy in Action*. Oxford: Clarendon Press.

[28] Chaloner, K., and Verdinelli, I. 1995. Bayesian experimental design: A review. *Statistical Science*, **10**, 273–304.

[29] Chernoff, H. 1972. *Sequential Analysis and Optimal Design*. Philadelphia, PA: SIAM.

[30] Chick, S. E., and Ng, S. H. 2002. Joint criterion for factor identification and parameter estimation. In Yücesan, E., Chen, C.-H., Snowdon, J. L., and Charnes, J. M. (eds), *Proceedings of the 2002 Winter Simulation Conference*; pp 400–406.

[31] Chopin, N., and Robert, C. P. 2010. Properties of nested sampling. *Biometrika*, **97**(3), 741–755.

[32] Clyde, M., Müller, P., and Parmigiani, G. 1993. Optimal design for heart defibrillators. In *Case Studies in Bayesian Statistics, II*. Berlin: Springer-Verlag; pp 278–292.

[33] Cohen, E. R., and Taylor, B. N. 1987. The 1986 adjustment of the fundamental physical constants. *Reviews of Modern Physics*, **59**(4), 1121–1148.

[34] Cornu, A., and Massot, R. 1979. *Compilation of Mass Spectral Data*. London: Heyden.

[35] Cox, R. T. 1946. Probability, frequency and reasonable expectation. *American Journal of Physics*, **14**, 1–13.

[36] Cox, R. T. 1961. *The Algebra of Probable Inference*. Baltimore, MD: Johns Hopkins Press.

[37] Currin, C., Mitchell, T., Moris, M., and Ylvisaker, D. 1991. Bayesian predictions of deterministic functions, with applications to the design and analysis of computer experiments. *Journal of the American Statistical Association*, **86**(416), 953–963.

[38] Daghofer, M., and von der Linden, W. 2004. *Perfect Tempering*. AIP Conference Proceedings, vol. 735, p. 355.

[39] DasGupta, A. 1996. Review of optimal Bayes design. In Gosh, S., and Rao, C. R. (eds), *Handbook of Statistics 13: Design and Analysis of Experiments*. Amsterdam: Elsevier.

[40] DeGroot, M. H. 1962. Uncertainty, information and sequential experiments. *Annals of Mathematical Statistics*, **33**(2), 404–419.

[41] DeGroot, M. H. 1986. Concepts of information based utility. In Daboni, L., Montesano, A., and Lines, M. (eds), *Recent Developments in the Foundations of Utility and Risk Theory*. Dordrecht: Reidel; pp 265–275.

[42] Devroye, L. 1986. *Non-uniform Random Variate Generation*. New York: Springer.

[43] Dittrich, W., and Reuter, M. 2001. *Classical and Quantum Dynamics*. New York: Springer.

[44] Dlugosz, S., and Müller-Funk, U. 2009. The value of the last digit: Statistical fraud detection with digit analysis. *In Advances in Data Analysis and Classification*, vol. 3. New York: Springer.

[45] Dobb, J. L. 1953. *Stochastic Processes*. New York: John Wiley & Sons.

[46] Doetsch, G. (ed.). 1976. *Einführung und Anwendung der Laplace-Transformation*. Basel: Birkhäuser-Verlag.

[47] Dose, V. 2007. Bayesian estimate of the Newtonian constant of gravitation. *Measurement Science and Technology*, **18**, 176–182.

[48] Dose, V. 2009. Analysis of rare-event time series with application to Caribbean hurricane data. *Europhysics Letters*, **85**, 59001.

[49] Dose, V., and Menzel, A. 2004. Bayesian analysis of climate change impacts in phenology. *Global Change Biology*, **10**, 259–272.

[50] Dose, V., Fauster, Th., and Gossmann, H. J. 1981. The inversion of autoconvolution integrals. *Journal of Computational Physics*, **41**(1), 34–50.

[51] Dose, V., von der Linden, W., and Garrett, A. 1996. A Bayesian approach to the global confinement time scaling in W7AS. *Nuclear Fusion*, **36**, 735–744.

[52] Dose, V., Preuss, R., and Roth, J. 2001. Evaluation of chemical erosion data for carbon materials at high ion fluxes using Bayesian probability theory. *Journal of Nuclear Materials*, **288**, 153–162.

[53] Dose, V., Neuhauser, J., Kurzan, B., Murmann, H., Salzmann, H., and ASDEX Upgrade Team. 2001. Tokamak edge profile analysis employing Bayesian statistics. *Nuclear Fusion*, **41**(11), 1671–1685.

[54] Dreier, H., Dinklage, A., Fischer, R., Hirsch, M., Kornejew, P., and Pasch, E. 2006. Bayesian design of diagnostics: Case studies for Wendelstein 7-X. *Fusion Science and Technology*, **50**, 262–267.

[55] Dreier, H., Dinklage, A., Fischer, R., Hirsch, M., and Kornejew, P. 2006. Bayesian design of plasma diagnostics. *Review of Scientific Instruments*, **77**, 10F323.

[56] Dreier, H., Dinklage, A., Fischer, R., Hirsch, M., and Kornejew, P. 2008. Bayesian experimental design of a multi-channel interferometer for Wendelstein 7-X. *Review of Scientific Instruments*, **79**, 10E712.

[57] Dreier, H., Dinklage, A., Fischer, R., Hirsch, M., and Kornejew, P. 2008. Comparative studies to the design of the interferometer at W7-X with respect to technical boundary conditions. In Hartfuss, H. J., Dudeck, M., Musielok, J., and Sadowski, M. J. (eds), *PLASMA 2007*, vol. 993. Melville, NY: AIP; pp 183–186.

[58] Dueck, G. 1993. New optimization heuristics: The Great Deluge algorithm and the record-to-record travel. *Journal of Computational Physics*, **104**(1), 86–92.

[59] Economou, E. N. 1979. *Green's Functions in Quantum Physics*. Springer Series in Solid State Physics, vol. 1. Berlin: Springer-Verlag.

[60] Edwards, W., Lindman, H., and Savage, L. J. 1963. Bayesian statistical inference for psychological research. *Psychological Review*, **70**, 193–242.

[61] Elstrodt, J. 1999. *Maß- und Integrationstheorie (Das Lebesgue-Integral, S. 118-160.)*. 2. edn. Heidelberg: Springer.

[62] Emsh, G. G., Eggerfeldt, G. C., and Streit, L. (eds). 1994. *On Klauder's Path: A Field Trip*. Singapore: World Scientific.

[63] Enzensberger, H. M. 2011. *Fatal Numbers: Why Count on Chance*. New York: Upper Westside Philosophers.

[64] Ertl, K., von der Linden, W., Dose, V., and Weller, A. 1996. Maximum entropy based reconstruction of soft X-ray emissivity profiles in W7-AS. *Nuclear Fusion*, **36**, 1477–1488.

[65] Evans, M. 2007. Discussion of nested sampling for Bayesian computations by John Skilling. In Bernardo, J. M., Bayarri, M. J., Berger, J. O., Dawid, A. P., Heckermann, D., Smith, A. F. M., and West, M. (eds), *Bayesian Statistics*, vol. 8. Oxford: Oxford University Press; pp 491–524.

[66] Fedorov, V. V. 1972. *Theory of Optimal Experiments*. New York: Academic Press.

[67] Feller, W. 1968. *An Introduction to Probability Theory and Applications I and II*. New York: John Wiley & Sons.

[68] Fischer, R. 2004. Bayesian experimental design – Studies for fusion diagnostics. In Fischer, R., Preuss, R., and von Toussaint, U. (eds), *Bayesian Inference and Maximum Entropy Methods in Science and Engineering*, vol. 735. Melville, NY: AIP; pp 76–83.

[69] Fischer, R., von der Linden, W., and Dose, V. 1996. On the importance of α marginalization in maximum entropy. In Silver, R. N., and Hanson, K. M. (eds), *Maximum Entropy and Bayesian Methods*. Dordrecht: Kluwer Academic Publishers; p. 229.

[70] Fischer, R., Hanson, K. M., Dose, V., and von der Linden, W. 2000. Background estimation in experimental spectra. *Physical Review E*, **61**, 1152.

[71] Fischer, R., Dreier, H., Dinklage, A., Kurzan, B., and Pasch, E. 2005. Integrated Bayesian experimental design. In Knuth, K. H., Abbas, A. E., Morris, R. D., and Castle, J. P. (eds), *Bayesian Inference and Maximum Entropy Methods in Science and Engineering*, vol. 803. Melville, NY: AIP; pp 440–447.

[72] Fisher, R. A. 1925. *Statistical Methods for Research Workers*. Edinburgh: Oliver and Boyd.

[73] Fisher, R. A. 1990. *Statistical Methods and Scientific Inference*. Oxford: Oxford University Press.

[74] Fishman, G. S. 1996. *Monte Carlo: Concepts, algorithms and applications*. Berlin: Springer-Verlag.

[75] Freedman, D., Pisani, R., and Purves, R. 2007. *Statistics*, 4th edn. New York: W. W. Norton & Co.

[76] Frieden, B. R. 1991. *Probability, Statistical Optics, and Data Testing*. Springer Series in Information Sciences, no. 10. Berlin: Springer-Verlag.

[77] Frieden, R., and Gatenby, R. A. 2007. *Exploratory Data Analysis Using Fisher Information*. New York: Springer.

[78] Fröhner, F. H. 2000. *Evaluation and Analysis of Nuclear Resonance Data, JEFF Report 18*. Paris: OECD Nuclear Energy Agency.

[79] Fukuda, Y. 2010. Appearance potential spectroscopy (APS): Old method, but applicable to study nano-structures. *Analytical Sciences*, **26**, 187–197.

[80] Garrett, A. J. M. 1991. Ockham's razor. *Physics World*, **May**, 39.

[81] Gauss, C. F. 1873. Göttingische gelehrte Anzeigen 1823. In *C. F. Gauss, Werke IV*. Göttingen: Königliche Gesellschaft der Wissenschaften zu Göttingen; pp 100–104.

[82] Gautier, R., and Pronzato, L. 1998. Sequential design and active control. In *New Developments and Applications in Experimental Design*, vol. 34. Hayward, CA: Institute of Mathematical Statistics; pp 138–151.

[83] Geyer, C. L., and Williamson, P. P. 2004. Detecting fraud in data sets using Benford's law. *Communications in Statistics – Simulation and Computation*, **33**(1), 229–246.

[84] Gilks, W. R., Richardson, S., and Spiegelhalter, D. J. 1997. *Markov Chain Monte Carlo in Practice*. London: Chapman Hall.

[85] Goggans, P. M., and Chi, Y. 2007. Electromagnetic induction landmine detection using Bayesian model comparison. In Djafari, A. M. (ed.), *Bayesian Inference and Maximum Entropy Methods in Science and Engineering*, vol. 872. Melville, NY: AIP; pp 533–540.

[86] Goldstein, M. 1992. Comment on 'Advances in Bayesian experimental design by I. Verdinelli'. In Bernardo, J. M., Berger, J. O., Dawid, A. P., and Smith, A. F. M. (eds), *Bayesian Statistics*, vol. 4. Oxford: Oxford University Press; p. 477.

[87] Goodman, S. N. 1999. Toward evidence-based medical statistics. 1: The *P* value fallacy. *Annals of Internal Medicine*, **130**(12), 995–1004.

[88] Gralle, A. 1924. Über die geodätischen Arbeiten von Gauss. In *C. F. Gauss, Werke XI Abt. 2*. Göttingen: Gesellschaft der Wissenschaften zu Göttingen; p. 14.

[89] Grandy, W. T. 1987. *Foundations of Statistical Mechanics I, II*. Amsterdam: Kluwer Academic Publishers.

[90] Gregory, P. 2005. *Bayesian Logical Data Analysis for the Physical Sciences*. Cambridge: Cambridge University Press.

[91] Grinstead, C. M., and Snell, J. L. 1998. *Introduction to Probability*. Providence, RI: American Mathematical Society.

[92] Gull, S. F. 1989. Bayesian data analysis: straight-line fitting. In Skilling, J. (ed.), *Maximum Entropy and Bayesian Methods (1988)*. Dordrecht: Kluwer Academic Publishers; p. 511.

[93] Gull, S. F. 1989. Developments in maximum entropy data analysis. In Skilling, J. (ed.), *Maximum Entropy and Bayesian Methods*. Dordrecht: Kluwer Academic Publishers; p. 53.

[94] Gull, S. F., and Daniell, G. J. 1978. Image reconstruction from incomplete and noisy data. *Nature*, **272**, 686.

[95] Gull, S. F., and Skilling, J. 1984. Maximum entropy method in image processing. *IEEE Proceedings*, **131 F**, 646.

[96] Hammersley, H. 1965. *Monte Carlo Methods*. London: Methiens Monographs.

[97] Harney, H. L. 2003. *Bayesian Inference: Parameter Estimation and Decisions*. Berlin: Springer-Verlag.

[98] Hjalmarsson, H. 2005. From experiment design to closed-loop control. *Automatica*, **41**, 393–438.

[99] Horiguchi, T., and Morita, T. 1975. Note on the lattice Green's function for the simple cubic lattice. *Journal of Physics C*, **8**, L232.

[100] Hukushima, K., and Nemoto, K. 1996. Exchange Monte Carlo method and application to spin glass simulations. *Journal of the Physical Society of Japan*, **65**, 1604–1608.

[101] Ito, K. 1993. *Encyclopedic Dictionary of Mathematics 2*, 2nd edn. Cambridge, MA: MIT Press.

[102] Jacob, W. 1998. Surface reactions during growth and erosion of hydrocarbon films. *Thin Solid Films*, **326**(1–2), 1–42.

[103] Jasa, T., and Xiang, N. 2005. Using nested sampling in the analysis of multi-rate sound energy decay in acoustically coupled rooms. *AIP Conference Proceedings*, **803**(1), 189–196.

[104] Jaynes, E. T., and Bretthorst, L. 2003. *Probability Theory, The Logic of Science*. Oxford: Oxford University Press.

[105] Jaynes, E. T. 1973. The well-posed problem. *Foundations of Physics*, **3**, 477–493.

[106] Jaynes, E. T. 1968. Prior probabilities. *IEEE Transactions on Systems Science and Cybernetics*, **SSC4**, 227–241.

[107] Jeffreys, H. 1961. *Theory of Probability*. Oxford: Oxford University Press.

[108] Jones, P. D., New, M., Parker, D. E., Martin, S., and Riger, I. G. 1999. Surface air temperature and its variations over the last 150 years. *Reviews of Geophysics*, **37**, 173–199.

[109] Kaelbling, L. P., Littman, M. L., and Moore, A. W. 1996. Reinforcement learning: A survey. *Journal of Artificial Intelligence Research*, **4**, 237–285.

[110] Kaelbling, L. P., Littman, M. L., and Cassandra, A. R. 1998. Planning and acting in partially observable stochastic domains. *Artificial Intelligence*, **101**, 99–134.

[111] Kapur, J. N., and Kesavan, H. K. (eds). 1989. *Maximum-Entropy Models in Science and Engineering*. New York: John Wiley & Sons.

[112] Kapur, J. N., and Kesavan, H. K. 1992. *Entropy Optimization Principles with Applications*. Boston, MA: Academic Press.

[113] Kass, R. E., and Raftery, A. E. 1995. Bayes factors. *Journal of the American Statistical Association*, **90**(430), 773–795.

[114] Kelly, F. P. 1979. *Reversibility and Stochastic Networks*. Chichester, UK: John Wiley & Sons.

[115] Kendall, M. G., and Moran, P. A. P. 1963. *Geometrical Probability*. London: Griffin.

[116] Kennedy, M. C., and O'Hagan, A. 2001. Bayesian calibration of computer models (with discussion). *Journal of the Royal Statistical Society, Series B*, **63**(3), 425–464.

[117] Knuth, K. H., Erner, P. M., and Frasso, S. 2007. Designing intelligent instruments. In Knuth, K. H., Caticha, A., Center, J. L., Giffin, A., and Rodrigues, C. C. (eds), *Bayesian Inference and Maximum Entropy Methods in Science and Engineering*, vol. 954. Melville, NY: AIP; pp 203–211.

[118] Knuth, K. H., and Skilling, J. 2012. Foundations of inference. *Axioms*, **1**, 38–73.

[119] Koch, D. G. *et al.* 2010. Kepler mission design, realized photometric performance, and early science. *Astrophysical Journal Letters*, **713**(2), L79–L86.

[120] Kolmogorov, A. N. 1956. *Foundations of the Theory of Probability*. New York: Chelsea Publishing Company.

[121] Kornejew, P., Hirsch, M., Bindemann, T., Dinklage, A., Dreier, H., and Hartfuss, H. J. 2006. Design of multichannel laser interferometry for W7-X. *Review of Scientific Instruments*, **77**, 10F128.

[122] Landau, L. 1944. On the energy loss of fast particles by ionization. *Journal of Physics-USSR*, **8**, 201.

[123] Laplace, P. S. 1951. *A Philosophical Essay on Probabilities* (translated into English by F. W. Truscott, and F. L. Emory). New York: Dover Publications; p. 4.

[124] Lehmann, E. L. 1993. The Fisher, Neyman–Pearson theories of testing hypotheses: One theory or two? *Journal of the American Statistical Association*, **88**(424), 1242–1249.

[125] Leonard, T., and Hsu, J. S. J. 1999. *Bayesian Methods: An Analysis for Statisticians and Interdisciplinary Researchers*. Cambridge: Cambridge University Press.
[126] Lewis, D. W. 1991. *Matrix Theory*, 3rd edn. Singapore: World Scientific.
[127] Lindley, D. V. 1956. On a measure of the information provided by an experiment. *Annals of Mathematical Statistics*, **27**(4), 986–1005.
[128] Lindley, D. V. 1972. *Bayesian statistics – a review*. Philadelphia, PA: SIAM.
[129] Linsmeier, C. 1994. Auger electron spectroscopy. *Vacuum*, **45**, 673–690.
[130] Loredo, T. J. 1999. Computational technology for Bayesian inference. In Mehringer, D. M., Plante, R. L., and Roberts, D. A. (eds), *ASP Conference Series 172: Astronomical Data Analysis Software and Systems VIII*, vol. 8. San Francisco, CA: ASP.
[131] Loredo, T. J. 2004. *Bayesian Adaptive Exploration*.
[132] Loredo, T. J., and Chernoff, D. F. 2003. Bayesian adaptive exploration. In Feigelson, E. D., and Babu, G. J. (eds), *Statistical Challenges in Astronomy*. Berlin: Springer-Verlag, pp 57–70.
[133] Loredo, T. J., and Chernoff, D. F. 2003. Bayesian adaptive exploration. In Erickson, G., and Zhai, Y. (eds), *Bayesian Inference and Maximum Entropy Methods in Science and Engineering*, vol. 707. Melville, NY: AIP; pp 330–346.
[134] Luttrell, S. P. 1985. The use of transinformation in the design of data sampling schemes for inverse problems. *Inverse Problems*, **1**, 199–218.
[135] MacKay, D. J. C. 1992. Information-based objective functions for active data selection. *Neural Computation*, **4**(4), 590–604.
[136] MacKay, D. J. C. 2003. *Information Theory, Inference and Learning Algorithms*. Cambridge: Cambridge University Press.
[137] Marinari, E. 1996. *Optimized Monte Carlo Methods*. Lectures at the 1996 Budapest Summer School on Monte Carlo Methods.
[138] Marinari, E., and Parisi, G. 1992. Simulated tempering: A new Monte Carlo scheme. *Europhysics Letters*, **19**, 451–458.
[139] Matthews, R. 1998. Data sleuths go to war. *New Scientist*, **2135**.
[140] Mattuck, R. D. 1976. *A Guide to Feynman Diagrams in the Many-Body Problem*. New York: McGraw-Hill Book Co.
[141] Meroli, S., Passeri, D., and Servoli, L. 2011. Energy loss measurement for charged particles in very thin silicon layers. *Journal of Instrumentation*, **6**, P06013.
[142] Metropolis, N., Rosenbluth, A. W., Rosenbluth, M. N., Teller, A. H., and Teller, E. 1953. Equations of state calculations by fast computing machines. *Journal of Chemical Physics*, **21**, 1087–1091.
[143] Mohr, J. P., and Taylor, B. N. 2005. CODATA recommended values of the fundamental physical constants: 2002. *Reviews of Modern Physics*, **77**, 42–47.
[144] Möller, W., and Besenbacher, F. 1980. A note on the ^{3}He–^{2}D nuclear reaction cross section. *Nuclear Instruments and Methods*, **168**(1), 111–114.
[145] Montgomery, D. C., and Peck, E. A. (eds). 1982. *Introduction to Linear Regression Analysis*. New York: John Wiley & Sons.
[146] Müller, P. 1999. Simulation based optimal design (with discussion). In Bernardo, J. M., Berger, J. O., Dawid, A. P., and Smith, A. F. M. (eds), *Bayesian Statistics*, vol. 6. Oxford: Oxford University Press; pp 459–474.
[147] Müller, P., and Parmigiani, G. 1995. Numerical evaluation of information theoretic measures. In Berry, D. A., Chaloner, K. M., and Geweke, J. F. (eds), *Bayesian Statistics and Econometrics: Essays in Honor of A. Zellner*. New York: John Wiley & Sons; pp 397–406.

[148] Müller, P., Sansó, B., and Delorio, M. 2004. Optimal Bayesian design by inhomogeneous Markov chain simulation. *Journal of the American Statistical Association*, **99**, 788–798.

[149] Müller, P., Berry, D., Grieve, A., Smith, M., and Krams, M. 2007. Simulation-based sequential Bayesian design. *Journal of Statistical Planning and Inference*, **137**(10), 3140–3150.

[150] Murray, C. D., and Dermott, S. F. 1999. *Solar System Dynamics*. Cambridge: Cambridge University Press.

[151] Myers, R. H., Khuri, A. I., and Carter, W. H. 1989. Response surface methodology: 1966–1988. *Technometrics*, **31**(2), 137–157.

[152] Neal, R. 1999. Regression and classification using Gaussian process priors. In Bernardo, J. M., Berger, J. O., Dawid, A. P., and Smith, A. F. M. (eds), *Bayesian Statistics*, vol. 6. Oxford: Oxford University Press; pp 69–95.

[153] Neyman, J., and Pearson, E. S. 1933. On the problem of the most efficient tests of statistical hypotheses. *Philosophical Transactions of the Royal Society of London*, **231A**, 289–337.

[154] Ng, A. Y., and Jordan, M. I. 2000. PEGASUS: A policy search method for large MDPs and POMDPs. *In Uncertainty in Artificial Intelligence: Proceedings of the Sixteenth Conference*. Stanford, CA: Morgan Kaufmann; pp 406–415.

[155] Nussbaumer, A., Bittner, E., and Janke, W. 2008. Make life simple: Unleash the full power of the parallel tempering algorithm. *Physical Review Letters*, **101**, 130603.

[156] Oakley, J. E., and O'Hagan, A. 2004. Probabilistic sensitivity analysis of complex models: A Bayesian approach. *Journal of the Royal Statistical Society, Series B*, **66**(3), 751–769.

[157] O'Hagan, A. 1994. *Distribution Theory*. Kendall's Advanced Theory of Statistics, vol. 1. London: Edward Arnold.

[158] Padayachee, J., Prozesky, V. M., von der Linden, W., Nkwinika, M. S., and Dose, V. 1999. Bayesian PIXE background subtraction. *Nuclear Instrument of Methods B*, **150**, 129–135.

[159] Papoulis, A., and Pillai, S. U. 2002. *Probability, Random Variables and Stochastic Processes*. London: McGraw-Hill Europe.

[160] Petz, F. and Hiai, D. 2000. *The Semicircle Law, Free Random Variables and Entropy*. Mathematical Surveys and Monographs, vol. 77. Providence, RI: American Mathematical Society.

[161] Press, W. H. 1996. Understanding data better with Bayesian and global statistical methods. *astro-ph/9604126*.

[162] Press, W. H., Teukolsky, S. A., Vetterling, W. T., and Flannery, B. P. (eds). 1992. *Numerical Recipes*, 2nd edn. Cambridge: Cambridge University Press.

[163] Press, W. H., Teukolsky, S. A., Vetterling, W. T., and Flannery, B. P. (eds). 2007. *Numerical Recipes*, 3rd edn. Cambridge: Cambridge University Press.

[164] Press, W. H., Vetterling, W. T., Teukolsky, S. A., and Flannery, B. P. 1992. *Numerical Recipes in C*, 2nd edn, vol. 1. Cambridge: Cambridge University Press.

[165] Preuss, R., Muramatsu, A., von der Linden, W., Assaad, F. F., Dieterich, P., and Hanke, W. 1994. Spectral properties of the one-dimensional Hubbard model. *Physical Review Letters*, **73**, 732.

[166] Preuss, R., Hanke, W., and von der Linden, W. 1995. Quasiparticle dispersion of the 2D Hubbard model: From an insulator to a metal. *Physical Review Letters*, **75**, 1344.

[167] Preuss, R., Dose, V., and von der Linden, W. 1999. Dimensionally exact form-free energy confinement scaling in W7-AS. *Nuclear Fusion*, **39**, 849–862.

[168] Preuss, R., Dreier, H., Dinklage, A., and Dose, V. 2008. Data adaptive control parameter estimation for scaling laws for magnetic fusion devices. *EPL*, **81**(5), 55001.

[169] Pronzato, L. 2008. Optimal experimental design and some related control problems. *Automatica*, **44**, 303–325.

[170] Protter, P. E. 2000. *Stochastic Integration and Differential Equations*. 2.1 edn. New York: Springer.

[171] Prozesky, V. M., Padayachee, J., Fischer, R., von der Linden, W., Dose, V. and Ryan, C. G. 1997. The use of maximum entropy and Bayesian techniques in nuclear microprobe applications. *Nuclear Instrument and Methods B*, **130**, 113–117.

[172] Pukelsheim, F. 1993. *Optimal Design of Experiments*. New York: John Wiley & Sons.

[173] Rodriguez, C. C. 1999. Are we really cruising a hypothesis space. In von der Linden, W., Dose, V., Fischer, R., and Preuss, R. (eds), *Maximum Entropy and Bayesian Methods*. Dordrecht: Kluwer Academic Publishers; pp 131–140.

[174] Rosenkrantz, R. D. (ed.). 1983. *E. T. Jaynes: Papers on Probability, Statistics and Statistical Physics*. Dordrecht: Reidel.

[175] Ryan, K. J. 2003. Estimating expected information gains for experimental designs with application to the random fatigue-limit model. *Journal of Computational and Graphical Statistics*, **12**(3), 585–603.

[176] Sacks, J., Welch, W. J., Mitchell, T. J., and Wynn, H. P. 1989. Design and analysis of computer experiments. *Statistical Science*, **4**(4), 409–435.

[177] Salsburg, D. 2002. *The Lady Tasting Tea*. New York: Henry Hold and Co.

[178] Saltelli, A., Tarantola, S., and Campolongo, F. 2000. Sensitivity analysis as an ingredient of modeling. *Statistical Science*, **15**, 377–395.

[179] Schwarz-Selinger, T., Preuss, R., Dose, V., and von der Linden, W. 2001. Analysis of multicomponent mass spectra applying Bayesian probability theory. *Journal of Mass Spectroscopy*, **36**, 866–874.

[180] Sebastiani, P., and Wynn, H. P. 2000. Maximum entropy sampling and optimal Bayesian experimental design. *Journal of the Royal Statistical Society, series B*, **62**(1), 145–157.

[181] Shaw, J. R., Bridges, M., and Hobson, M. P. 2007. Efficient Bayesian inference for multimodal problems in cosmology. *Monthly Notices of the Royal Astronomical Society*, **378**, 1365–1370.

[182] Shore, J. E., and Johnson, R. W. 1980. Axiomatic derivation of the principle of maximum entropy and the principle and the minimum of crossentropy. *IEEE Transactions on Information Theory*, **26**(26).

[183] Silver, R. N., and Martz, H. F. 1994. Applications of quantum entropy to statistics. In Robinson H. C. (ed.), *Presented at the American Statistical Association, Toronto, Ontario, 14–18 August 1994*; pp 14–18.

[184] Silver, R. N., Sivia, D. S., and Gubernatis, J. E. 1990. Application of maxent and Bayesian methods to many-body physics. *Physical Review B*, **41**, 2380.

[185] Silverman, B. W. 1986. *Density Estimation for Statistics and Data Analysis*. London: Chapman and Hall.

[186] Singh, V. P. 2013. *Entropy Theory and its Application in Environmental and Water Engineering*. Oxford: Wiley-Blackwell.

[187] Sivia, D. S., and David, W. I. F. 1994. A Bayesian approach to extracting structure-factor amplitudes from powder diffraction data. *Acta Crystallographical A*, **50**, 703–714.

[188] Sivia, D. S. and Skilling, J. 2006. *Data Analysis: A Bayesian Tutorial*. Oxford: Oxford University Press.

[189] Skilling, J. 1990. Quantified maximum entropy. In Fougère, P. F. (ed.), *Maximum Entropy and Bayesian Methods*. Dordrecht: Kluwer Academic Publishers; p. 341.

[190] Skilling, J. 1991. Fundamentals of MaxEnt in data analysis. In Buck, B., and Macaulay, V. A. (eds), *Maximum Entropy in Action*. Oxford: Clarendon Press; p. 19.

[191] Skilling, J. 2004. Nested sampling. In Erickson, G., Rychert, J. T., and Smith, C. R. (eds), *Bayesian Inference and Maximum Entropy Methods in Science and Engineering*, vol. 735. Melville, NY: AIP; pp 395–405.

[192] Skilling, J. 2006. Nested sampling for general Bayesian computation. *Bayesian Analysis*, **1**(4), 833–859.

[193] Skilling, J. 2009. Nested sampling's convergence. In *Bayesian Inference and Maximum Entropy Methods in Science and Engineering: The 29th International Workshop on Bayesian Inference and Maximum Entropy Methods in Science and Engineering. AIP Conference Proceedings*, vol. 1193; pp 277–291.

[194] Skilling, J., and Gull, S. F. 1991. *Bayesian Maximum Entropy Image Reconstruction*. Spatial Statistics and Imaging, vol. 20. Institute of Mathematical Statistics.

[195] Skinner, C. H., Haasz, A. A., Alimov, V. Kh., Bekris, N., Causey, R. A., Clark, R. E. H., Coad, J. P., Davis, J. W., Doerner, R. P., Mayer, M., Pisarev, A., Roth, J., and Tanabe, T. 2008. Recent advances on hydrogen retention in ITER's plasma facing materials: Beryllium, carbon, and tungsten. *Fusion Science and Technology*, **54**(4), 891–945.

[196] Stanley, R. P. 2011. *Enumerative Combinatorics*, vol. I, 2nd edn. Cambridge: Cambridge University Press.

[197] Steinerg, D. M., and Hunter, W. G. 1984. Experimental design: Review and comment. *Technometrics*, **26**(2), 71–97.

[198] Stone, M. 1959. Application of a measure of information to the design and comparison of regression experiment. *Annuals of Mathematical Statistics*, **30**, 55–70.

[199] Strauss, C. E. M., Wolpert, D. H., and Wolf, D. R. 1993. Alpha, evidence, and the entropic prior. In Heidbreder, G. (ed.), *Maximum Entropy and Bayesian Methods*. Dordrecht: Kluwer Academic Publishers.

[200] Stroth, U., Murakami, M., Dory, R. A., Yamada, H., Okamura, S., Sano, F., and Obiki, T. 1996. Energy confinement scaling from the international stellarator database. *Nuclear Fusion*, **36**(8), 1063–1078.

[201] Tesmer, J., and Nastasi, M. (eds). 1995. *Handbook of Modern Ion Beam Analysis*. Pittsburgh, PA: Material Research Society.

[202] Thrun, S., and Fox, D. 2005. *Probabilistic Robotics*. Boston, MA: MIT Press.

[203] Titterington, D. M. 1985. General structure of regularization procedures in image reconstruction. *Astronomy and Astrophysics*, **144**, 381–387.

[204] Toman, B. 1996. Bayesian experimental design for multiple hypothesis testing. *Journal of the American Statistical Association*, **91**(433), 185–190.

[205] Toman, B. 1999. Bayesian experimental design. In Kotz, S., and Johnson, N. L. (eds), *Encyclopedia of statistical sciences, Update vol. 3*. New York: John Wiley & Sons; pp 35–39.

[206] Tribus, M. 1969. *Rational Descriptions, Decisions and Design*. New York: Pergamon.

[207] von der Linden, W. 1995. Maximum-entropy data analysis. *Applied Physics A*, **60**, 155.

[208] von der Linden, W., Donath, M., and Dose, V. 1993. Unbiased access to exchange splitting of magnetic bands using the maximum entropy method. *Physical Review Letters*, **71**, 899.

[209] von der Linden, W., Dose, V., Matzdorf, R., Pantförder, A., Meister, G., and Goldmann, A. 1997. Improved resolution in HREELS using maximum-entropy deconvolution: CO on $Pt_xNi_{1-x}(111)$. *Journal of Electron Spectroscopy and Related Phenomena*, **83**, 1–7.

[210] von der Linden, W., Dose, V., Memmel, N., and Fischer, R. 1998. Probabilistic evaluation of growth models based on time dependent Auger signals. *Surface Science*, **409**, 290–301.

[211] von der Linden, W., Dose, V., Padayachee, J., and Prozesky, V. 1999. Signal and background separation. *Physical Review E*, **59**, 6527–6534.

[212] von der Linden, W. 1992. A QMC approach to many-body physics. *Physics Reports*, **220**, 55.

[213] von Neumann, J. 1951. Various techniques used in connection with random digits. Monte Carlo methods. *National Bureau of Standards*, **12**, 36–38.

[214] von Toussaint, U. 2011. Bayesian inference in physics. *Review of Modern Physics*, **11**(3), 943–999.

[215] von Toussaint, U., Gori, S., and Dose, V. 2004. Bayesian neural-networks-based evaluation of binary speckle data. *Applied Optics*, **43**, 5356–5363.

[216] von Toussaint, U., Schwarz-Selinger, T., and Gori, S. 2008. Optimizing nuclear reaction analysis (NRA) using Bayesian experimental design. In Lauretto, M. S., Pereira, C. A. B., and Stern, J. M. (eds), *Bayesian Inference and Maximum Entropy Methods in Science and Engineering*, vol. 1073; Melville, NY: AIP. pp 348–358.

[217] Werman, M., and Keren, D. 2001. A Bayesian method for fitting parametric and nonparametric models to noisy data. *IEEE Transactions on Pattern Analysis and Machine Intelligence*, **23**, 528–534.

[218] Woess, W. (ed.). 2009. *Denumerable Markov Chains*, vol. 1. European Mathematical Society Publishing House.

[219] Wright, J. T., and Howard, A. W. 2009. Efficient fitting of multiplanet Keplerian models to radial velocity and astrometry data. *Astrophysical Journal Supplement*, **182**(1), 205–215.

Index